Undergraduate Lecture Notes in Physics

Undergraduate Lecture Notes in Physics (ULNP) publishes authoritative texts covering topics throughout pure and applied physics. Each title in the series is suitable as a basis for undergraduate instruction, typically containing practice problems, worked examples, chapter summaries, and suggestions for further reading.

ULNP titles must provide at least one of the following:

- An exceptionally clear and concise treatment of a standard undergraduate subject.
- A solid undergraduate-level introduction to a graduate, advanced, or non-standard subject.
- A novel perspective or an unusual approach to teaching a subject.

ULNP especially encourages new, original, and idiosyncratic approaches to physics teaching at the undergraduate level.

The purpose of ULNP is to provide intriguing, absorbing books that will continue to be the reader's preferred reference throughout their academic career.

Series editors

Neil Ashby
University of Colorado, Boulder, CO, USA

William Brantley
Department of Physics, Furman University, Greenville, SC, USA

Matthew Deady
Physics Program, Bard College, Annandale-on-Hudson, NY, USA

Michael Fowler
Department of Physics, University of Virginia, Charlottesville, VA, USA

Morten Hjorth-Jensen
Department of Physics, University of Oslo, Oslo, Norway

Michael Inglis
SUNY Suffolk County Community College, Long Island, NY, USA

Heinz Klose
Humboldt University, Oldenburg, Niedersachsen, Germany

Helmy Sherif
Department of Physics, University of Alberta, Edmonton, AB, Canada

More information about this series at http://www.springer.com/series/8917

Fabrizio Cleri

The Physics of Living Systems

 Springer

Fabrizio Cleri
IEMN, CNRS UMR8520
Université de Lille I Sciences et
    Technologies
Villeneuve d'Ascq
France

ISSN 2192-4791   ISSN 2192-4805   (electronic)
Undergraduate Lecture Notes in Physics
ISBN 978-3-319-80858-1   ISBN 978-3-319-30647-6   (eBook)
DOI 10.1007/978-3-319-30647-6

Printed on acid-free paper

This Springer imprint is published by Springer Nature
The registered company is Springer International Publishing AG Switzerland

A mio padre e mia madre,
che mi hanno lasciato in eredità
la dolcezza di un mondo antico.

You'll get mixed up
with many strange birds as you go.
So be sure when you step.
Step with care and great tact
and remember that
Life's a Great Balancing Act.
Just never forget to be dexterous and deft,
and never mix up your right foot with your left.
Dr. Seuss, Oh, the places you'll go! Random House, *1990*

Preface

Before diving head to feet into the subject matter, a few words are in order to provide the moral excuse for writing this book and to give me the opportunity to tell a story that goes back to the year 2006.

When I arrived as a full professor in the Department of Physics of the University of Lille, the Director of that time, Michel Foulon, wanted the newly hired professor (it would be great if I could add "young" here, but I was already 45 at that time) to break some walls and open some original routes in research and teaching. I will always express my warmest thanks and deepest gratitude to Michel, for the wide freedom he allowed me in the choice and organisation of the new enterprise.

A quick look around Lille, a medium-sized, very lively city in the upper north of France, right next to the border with Belgium, gave me the obvious answer. With its about seventy institutes and laboratories revolving around biology, genetics, biotechnology, medical and clinical R&D, especially (but not exclusively) focussed on cancer research and therapy, *biophysics* was the way to go. It would have been a somewhat new field for me, but not far from my scientific interests at that time.

Thus, I started assembling an undergraduate school ('master', according to European nomenclature) with some colleagues from the Institut Pasteur, the Institut de Biologie and the Interdisciplinary Research Institute from CNRS, of course the Department of Physics, and my own CNRS Institut de Microélectronique et Nanotechnologie. The key idea was to shake up a cocktail of fundamental research in biophysics and applications in medical physics. The latter had its stronghold in the Oscar Lambret Cancer Therapy Center, just at the opposite end of the subway from the university. It was there that I met the brilliant Thierry Sarrazin, soon to become my best partner in crime, with whom I could put together a master programme in medical physics to be coupled with the programme in biophysics. Since its opening in 2010, this has been a successful story. Our new master in Biological and Medical Physics has attained a stable number of students, improving both the teaching throughput of the Department and the research potential of the laboratories, which now have a privileged route to attract some of our best students to the enormously exciting area of biophysical research.

Then, what is the place of this book? Already before the formal creation of the master course, I had started teaching biophysics units at various levels. As a theorist, I had chosen this way to learn about the basics, while increasingly redirecting my own research interests towards two main subjects: molecular mechanics of cell constituents, and microscopic radiation effects on the nucleic acids DNA and RNA. However, since the very beginning I realised that it was the very youngest students, second- or third-year undergraduate, who had to be exposed to introductory subjects of biophysics as early as possible. Without such an early exposure, there would be no 'feeding' for the students towards the more advanced subjects, and the master courses of the 4th and 5th years would have fallen from the sky, onto the shoulders of completely unprepared students. For that reason, I created from scratch a course of Introduction to the Physics of Living Systems for the physics sophomore. The idea was not to introduce much new physics for these students, but rather use their already acquired, albeit still elementary, knowledge about thermodynamics, mechanics, fluid physics, and electricity, to start *seeing* the physics behind the biology. Second-year students are still enough close to the high school to have some basic biology in their backpack, and that's all that was needed. The course was original in its layout, trying not to follow the much abused path of 'unveiling hidden physical principles underlying biological facts'. Rather, in the footsteps of D'Arcy Thompson, J.S.B. Haldane, Archibald Hill, and the more modern Knut Schmidt-Nielsen, Steven Vogel, J.C. Pennycuick and few others, I wanted to *start from physics* and show how living organisms *must conform* to the inevitable bounds imposed by gravity, light, temperature, atmosphere, oceans of the Earth, and by the more general constraints deriving from such life-setting variables as the water phase diagram, oxygen diffusivity, molecular elasticity, to name just a few.

This book evidently stems from the lecture notes for that course. Clearly, this is not a book for the research scientist in biophysics: the level is too elementary, the maths goes little beyond high school basics (at least for the *franco-français* students), and the subjects are well assessed and could not truly represent the last cry in biophysics research. It is primarily intended as a biophysics primer for young students, and, by just skipping a few pages too dense in formal math developments, it should be a pleasurable reading also for educated professionals working in the area of life sciences.

The physics inherent to living systems is immense and challenging. Where a physicist seeks mathematical rigour and experimental repeatability under extremely well-controlled conditions, the biologist rather seeks inductive proof, statistical correlations and performs hugely complex experiments with a whole bunch of competing (and often ill-known) free parameters. Biophysics is sometimes considered with a bit of a raised eyebrow by 'purist' colleagues: it may be felt that it requires sometimes too simple experiments, and too little theory, to keep the pace with 'big-time' physics such as superstring theory, tokamak magnetohydrodynamics, or the quest for the Higgs boson. However, the more I delve into the subject, the more interesting questions and puzzling connections I discover. To me, the fact that a simple experiment assembled in the backyard of the laboratory, or a crystal clear piece of non-quantum, non-relativistic theory, or even a back-of-the-envelope calculation of a dimensionless number, could reveal a crucial information about the living, represents instead a great advantage and a fascinating opportunity.

Experienced readers will notice that the subject matter treated in this book is partly covered, and often with much deeper scope, in several other texts, such as (to cite just a few prominent ones) *Physics of life* by Clas Blomberg, *Biological physics* by Philip Nelson, *How animals work* by Knut Schmidt-Nielsen, *Comparative biomechanics, or Life's physical world* by Steven Vogel, *Physical biology of the cell* by Phillips, Kondev & Theriot, *Newton rules biology* by C.J. Pennycuick.

However, both the breadth of subjects touched upon and the pedagogical approach followed here should be unlike any of the above, highly commendable and respected works. The keywords behind the present effort can be summarised in the following three concepts:

1. use the least possible amount of mathematics and molecular chemistry, and provide a minimum necessary knowledge of cell and structural biology;
2. propose a wide subject coverage, with a macro → micro → macro logical path: start from the macroscopic world, namely the thermodynamics of the Universe and the Solar system, and, via such subjects as the greenhouse effect and energetics of metabolism, step down to the microscopic world (physics of bacteria and unicellular life, cells and tissues, biomolecules); hence, move again upwards in length- and timescales to the physics of organs and whole organisms, and end up with subjects in zoology (e.g. simplified aerodynamics of insect flight, energy budget for the survival and reproduction of a flock of animals), and planetary ecology (species competition in the Biosphere, limits of ecosystems);
3. exploit as much as possible the physics knowledge base of second-year undergraduate students (elementary thermodynamics, classical mechanics and electrokinetics, elementary fluid mechanics), without need to introduce more complex notions, unless strictly necessary.

The book can be approached at least at two different levels, by different groups of readers, namely: as an undergraduate textbook in introductory biophysics and as a "case of curiosities" for professionals working in the vast life sciences and biomedical domains. For the first approach, each chapter contains the necessary background and tools, including exercises and Appendices, to form a progressive course. In this case, the chapters can be used in the order proposed by the index, eventually split over two semesters (Chaps. 5–9 covering somewhat more advanced subjects, susceptible of further developments). For the second approach, the curious but less physics-oriented reader might skip the first chapter (if school memories of thermodynamics are still haunting his/her dreams), as well as all the grey boxes containing the more formal developments, and create his/her own menu of chapters *á la carte* (with the only author's suggestions of reading Chap. 8 before Chap. 9, and Chap. 10 before Chap. 11). Also, note that the bibliographical references at the end of each chapter are not intended to provide a fully detailed support for all the subjects treated, as it would be the case for a scientific paper, but rather to merely propose some possible directions of development.

No book will ever be complete and definitive, and this one can be no exception. In particular this one, I should say. The material has been expanding over the years, some subjects leaving the place to newer ones according to my own curiosity, or to the discovery of interesting scientific papers amenable to an easily accessible level. Had I kept going with including any new items that came to my attention as a teacher, I would have never written this book. However, one has to stop somewhere, to give account of the state of the house at a given point. Hopefully, others will continue this effort and provide it with more motivation, better writing, deeper substance, nicer examples and smarter problems. Hopefully, among these there could be one of my former students or one of the readers of this book. In any case, it is my hope that in reading this book, be it for an introductory course as a student, or for a curiosity refresher as a practicing life scientist, your interest and attention towards biophysics could only increase.

Todi, Italy   Fabrizio Cleri
April 2016

This work has been possible also thanks to the many colleagues and students with whom I have been interacting, during all these years. Whereas discussing with colleagues is (almost) always a pleasure and a good occasion for funny jokes, the daily exchange with students is definitely the most refreshing and challenging moment. I am grateful to all those people who had the patience of listening to me, advising and correcting my mistakes, and helping me to find better and better ways to transmit the message. And I know it is not over yet.

Special thanks are in order for those colleagues and friends who took the burden of reading early versions of the various chapters and could bring their precious comments, criticisms, enlightenments to my ongoing, often immature work. In alphabetical order: Angela Bartolo, Bruno Bastide, Ralf Blossey and Jean Cosleou (Lille), Enrico Carlon (Leuven), Dominique Collard (Tokyo), Luciano Colombo (Cagliari), Antonio Di Carlo (Rome), Bahram Djafari-Rouhani, Alessandro Faccinetto, Stefano Giordano and Frank Lafont (Lille), Rob Phillips (Pasadena), Felix Ritort (Barcelona), Paola Salvetti (Dubai). However, for any error, imprecision or misprint still lurking in the text, the responsibility must be fully charged to the author.

I am grateful to the people at Springer who invested their energies and resources to propose this book to the public, especially to Maria Bellantone, who constantly encouraged and guided me in the pursuit of this project, and to the editorial assistants Annelies Kersbergen and Mieke van der Fluit.

Contents

List of Appendices

Since this book is intended as an introduction to biophysics for a public of undergraduate students, as well as for the physics-curious professional in life and health sciences, the continuity of the main text was privileged. Therefore, background notions are relegated in ample and illustrated Appendices, including the minimum mathematical toolbox (Appendix A, and F–H), and some essential notions in molecular and cell biology (Appendix B–E).

In particular the latter could appear excessively superficial and stripped-down versions of the immense amount of knowledge embodied by modern biology. I am including here such a very synthetic account for the sake of completeness, with the aim of providing the uneducated readers with just some basic vocabulary. Those who have a more substantial knowledge of biology will hopefully spare a benevolent look at my humble effort.

List of Greyboxes

Greyboxes are special sections of a chapter, in which mathematical treatments and/or more advanced concepts are developed. They are marked with a grey background and may be skipped by the reader less interested in the formal developments, without compromising the readability of the main text.

Editing Notes

The spelling and hyphenation in this book follows the UK English conventions. Uppercase is used only for specific names and in chapter titles, as well as in section and subsection titles. Moreover, in the text body the words 'Chapter', 'Section', 'Figure', 'Table' and 'Equation' are usually capitalised, to allow easy identification. In both the main text and figure captions, boldface characters are used when a technical term is encountered for the first time (for example, a muscle **fibre** is a multinuclear cell), and the corresponding entry is found in the subject index. Italic characters are used to highlight a word in the context (for example: a muscle performing an *isometric* effort), but the corresponding entry may or may not appear in the index. Single quotes are used to highlight a word or adjective attributed in a context different from its current usage (for example: a muscle can be 'rapid' or 'slow'). Double quotes are used to highlight a word or adjective used with a meaning different from its usual one (for example: a wing muscle "snaps" from its position). I tried to stick to these conventions as much as possible; however, there can be exceptions.

In general, measurement units are indicated according to the international system (S.I.) nomenclature, with the usual prefixes indicating multiples (kilo = 10^{-3}, mega = 10^{-6}, giga = 10^{-9}) and submultiples (milli = 10^{-3}, micro = 10^{-6}, nano = 10^{-9}). However, whenever useful the more specific units, such as electron-Volts (eV) or the temperature-energy equivalent ($k_B T$), are used. According to the standard convention, names of units are written in lowercase, while the symbol can be uppercase (for example 'coulomb' and 'C'). A mole of substance is indicated by 'mol', not to be confused with molecule. When needed, the latter is written in full, such as in 'kcal/molecule'.

Physical dimensions are written between square brackets, e.g. [M], [L^2], [T^{-1}] and so on. Grouped dimensions, such as [Energy] or [Pressure], are sometimes used, by writing them in full and uppercase.

Standard mathematical notation is used, with $\cdot$ indicating the scalar product, $\times$ the vector product, $\otimes$ the tensor product. Vectors are indicated in boldface and without any arrows (**v**); tensors and matrices are indicated in underscored boldface (**M**).

The symbol $\times$ is also used to write real numbers in scientific notation, such as 5.6×10^{-3}.

The website physicsoflife.net contains many additional resources for both the students and the instructors:

- high-resolution version of all non-copyrighted images in various formats, including OpenOffice slides;
- more complete solutions to all the end-of-chapter problems, as well as additional "questions-and-answers" students' self-examination checklists;
- a set of C++ computer programs for some of the mathematical models described in the book.

Part of the resources of the website is restricted by a password distributed to Springer customers.

Chapter 1
Introduction

To confront the study of living systems for a physicist or an engineer is, at the same time, an exciting challenge, as well as a complex and sometimes frustrating endeavour. The differences between the living systems as studied by the biologists, and the inorganic matter that is traditionally the domain of the so-called hard sciences, are discouraging, already starting from the catalog of the basic materials. Inorganic materials, such as rocks and soil, as well as artificial man-made objects, span at large Mendeleev's periodic table of the elements, from the lightest to the heaviest: iron, nickel, chromium, aluminium, silicon, lead, and tens of less common elements such as beryllium, germanium, gallium, arsenic, lanthanum, uranium... Biological materials, on the other hand, are based on a handful of light elements, just carbon, oxygen, nitrogen and hydrogen, plus an allowance in minor concentrations of a few heavier elements, such as calcium, sodium, potassium, phosphorus, usually in the form of salts and ions.

Also the 'fabrication' methods are vastly different. Materials of engineering, as well as those taken from the Earth's crust, are (at least, up to now) invariably produced in a top-down way by highly energetic processes: melting, casting, stamping, high-pressure and high-temperature sintering, moulding, laser cutting and drilling, chemical extraction and separation, and so on, all designed to produce substances, objects and parts of machines with predefined functions and well-defined shapes. The objects issued from an assembly line all have rigorously the same identical design and character, all parts fit in exactly the same positions, and move in the only way they were designed for. Once a part is broken it must be replaced by external actions, or the whole object will stop functioning.

Biological organisms, on the other hand, are capable of growing from a small amount of matter, by using ridiculously small amounts of energy per unit mass; they can self-assemble in a bottom-up way, by placing atoms and molecules one next to

© Springer International Publishing Switzerland 2016
F. Cleri, *The Physics of Living Systems*, Undergraduate Lecture
Notes in Physics, DOI 10.1007/978-3-319-30647-6_1

each other, to such a degree of precision that a single amino acid missing from a protein may signify the death of the entire organism; they adjust and change their size, shape and capabilities, during their development and all along their lifespan; they can regenerate and self-heal broken parts, to a good extent. All this thanks to an internally defined plan, the genetic code, which works more as a blueprint than as a rigid design. All sunflowers in a field look the same, yet not two flowers are identical.

The mobile phones that we carry in our pockets seem capable of wonders, e.g. when placing our position on Google maps, yet they consume about 1 W per cm^3 of computing material (the silicon chip), to perform a rather simple trigonometric calculation (that's easy when you have available hundreds of satellite data). By comparison a human brain, whose material is mostly water and salts, uses less than 1/100th of that power to perform infinitely more complex 'computations', such as tracking and hitting in less than 100 ms a tennis ball approaching at 100 km/h, while keeping full body balance and preparing to fool the opponent with another uncatchable shot.

The internal structures of engineering objects are based on a usually simple and repeated microstructure of the assembled materials, whose homogeneity and stability over long times are necessary and demanded requisites. The engineering design aims at predicting the conditions under which the object will be utilised, and tries to select or concoct the most appropriate materials, for the object to sustain mechanical stresses, temperatures, pressures, electromagnetic perturbations, during its lifetime. The shape and size are decided according to the functions the object will fulfil, with reassuring extra thickness, length, width, against premature ageing or failure. By contrast, living organisms display a complex and interconnected hierarchical structure, with substructures organised over all length scales, from the molecule, to the cell, to the tissues, to the whole organs. Their cells can multiply and differentiate into hundreds of different types, each appropriate to one or more specific functions, starting from just a few basic types. They can respond dynamically to variations of temperature and pressure, to reduction or increase in oxygen levels, they can remodel and reshape parts of their body during development or accidental damage, and—most importantly—they are capable of reproducing themselves endlessly.

Nevertheless, despite such vast differences, many of the important advances in biological and medical sciences have been made possible by technologies developed from the different domains of physics. It is possible today to scan thousands of genes in one single experiment, by squeezing an entire laboratory within a microfluidic chip of a few cm^2, designed on a silicon wafer with standard methods

Same engineering design   **Same genetic code**

taken from semiconductor industry; the enormous increase in data flow after the success of genome decoding has pushed biologists to take interest in the methods of high-energy physicists and astronomers, well-used to treat immense amounts of experimental data; sources of synchrotron radiation, confocal microscopy and spectroscopy, nuclear magnetic resonance, are more and more in demand by structural biologists, to deduce the structure and function of proteins, enzymes, and entire cell organelles.

Therefore, the contact between physics and life sciences should be restricted to such "utilitaristic" activities, in which physics is offering her precious services in the form of advanced technologies for diagnostic and imaging?

Enter Biophysics. Physicists and their methods have met biology a long time ago, already in the early XIX century. Probably, the first biophysicist in the modern sense could have been Luigi Galvani, from the University of Bologna, who around 1780 discovered animal electricity with his famous experiments on frogs (from which the term 'galvanic current' for the electric currents generated by acid-salts solutions). Many prominent physicists of the past had found interest in biological phenomena, notably in what has become known as the Berlin School of the mid-XIX century, including such people as Hermann von Helmholtz, Emil DuBois-Reymond, Ernst von Brücke, and Carl Ludwig. Their scientific approach to *organic physics* still sounds as an ideal description of the scope and aims of modern biophysics: "a vital phenomenon can only be regarded as explained, when it has been proven that it appears as a result of the material components of living organisms, interacting according to the laws which those same components would follow in their interaction outside of living systems".[1]

From a fundamental point of view, biophysics aims at explaining biological phenomena exactly with the same laws that apply to the rest of the Universe. The progressive discovery of similitudes between physical phenomena, notably in mechanics, energetics, electricity, and corresponding phenomena occurring inside living cells, has been the important motor of the recent, increasing interest in biophysics. As physicists, we always try to explain the essential features underlying an ensemble of observations by proposing synthetic, unifying theories. At the beginning of the XX century, the chemical view of living organisms was that they looked rather like

---

[1] Adolf Fick, *Gesammelte Schriften*, vol.3, pp. 492 and 767, Würzburg, 1904.

a bowl of soup, although capable of performing complex and sometimes amusing tasks, usually forbidden to soup bowls. At that time, scientists had but a vague idea of the way living organisms are capable of creating ordered structures starting from food and energy. Around the mid-XX century, it started to become clear that the answers to such questions should be found in giant molecules, found in the nucleus of each cell, capable of self replicating and of producing other giant molecules for all the cell functions; therefore, scientists invested themselves in the job of learning the largest possible number of details on all such molecules and their functions. Today, we find ourselves in a sort of opposite situation: we have way too much information about such molecules, but we are lacking conceptual schemes and analytical tools to organise all this messy information.

Some biologists still tend to reject us physicists as having a too reductionist approach, with our naïve tendency to eliminate all the details that make the difference between a bag of molecules and a frog. Unfortunately, this is a notorious attitude (or shouldn't I say a *defect*) that can make a physicist a quite intolerable presence among a group of friends discussing politics or football, and obnoxious to colleagues scientists who spend years of research in their field just to be baffled by a presumptuous guy with his simplistic ideas, as the cartoon by Randall Munroe nicely puts it. [Courtesy of www.xkcd.com.]

Nevertheless, our overarching scope as physicists should be that of finding fundamental principles underlying the organisation of all those molecules into a frog, since we physicists are seriously convinced that those molecules obey exactly the same Schrödinger's equation that atoms obey in a dead crystal. It would be easy at this point to piggyback on the change of paradigm that the discovery of the molecular structure of DNA brought about, and the undeniable success of molecular biology that followed. However, the problem is not to reduce all biology to physics—a scientific program which would fail even before starting, since not even all of physics could be reduced into one single scheme. But at least, a good start could be to attach more rigorous tests and quantitative measurements, to some cartoons that are presented in biology textbooks as explanation of molecular and cell phenomena, while being at best qualitative descriptions of the known facts.

Since the second half of the XX century, physics has been increasingly rejecting a strictly reductionist approach, and the idea that phenomena may *emerge* at each level of observation, instead of being bottom-up dictated by "laws", has been gaining ground. We started seeing stochastic processes operating everywhere in the universe, at every level, from subatomic particles to weather systems, to ocean currents, to galaxies. Deterministic physical laws on the macroscopic scale leave the room to random behaviour of molecules on the smaller scale. Diffusion of a dye in water looks like an ordered process, when the colour get evenly distributed inside the water jar, but it is indeed caused by a desperately random movement of the molecules. In this apparent destruction of order into disorder, the so-called Second Law of Thermodynamics dictates the direction of evolution: once the process of diffusion is scrutinised, we realise that the overall disorder of the dye-plus-water system has increased, and in fact the system has spontaneously evolved into a state of higher disorder. Living systems seem to evade this fundamental principle of physics, in that order is truly created from disorder: cells synthesise and organise their proteins into new structures, and split into two, four, eight new identical cells, and arrange into bones and muscles and brains, up to creating the now-famous frog. This—only apparent—contradiction of one of the most revered "laws" of physics by biological entities has been at the origin of a long-standing controversy between the respective scientific communities, which could be briskly summarised in the statement *Boltzmann and Darwin cannot be right at the same time*. In the course of this book I hope to provide the reader a large body of evidence that this is not the case, even if physical models of biological processes may seem at times exceedingly simplistic.

Since Boltzmann is more often than not called into cause, it seems to me that a correct understanding of such concepts as free energy and entropy, exothermic and endothermic, what is meant by open systems, and so on, should lay a good foundation to start the marriage between physics and biology. For this reason the opening Chapter of this book, somewhat unusually, starts with a synthetic but hopefully useful review of the basic concepts of thermodynamics and of their application in the context of living systems, ending with a discussion of the Earth's biosphere as a thermal engine; an appendix provides a small set of mathematical tools that will be used throughout the book. Chapter 3 elaborates on the concepts of entropy and probability, leading to a discussion on the current theories trying to explain the origins of life on Earth. The fourth Chapter starts again from the thermodynamics concepts of energy and heat of a transformation, to introduce some basic element of chemistry of the metabolism; even without going into much details, the main cell cycles of glycolysis and respiration are introduced, as key to the ATP-ADP cycling that represents the fuel of all active processes in all cells, from unicellular organisms to plants and animals. This will also give us the opportunity of discussing the problem of how heat is generated and evacuated from the human body.

With Chap. 5 we take a leap into the world of moving fluids, to understand how water, ions, nutrients, proteins and foreign substances, are trafficked in and out of the cells, by diffusion and transport across the many membranes that make a cell to resemble a multi-compartment chemical reactor. At this point, the appendices B to E should have already provided the reader with some elementary knowledge of the cell components, biomolecules, cytoskeleton and membranes, nucleic acids and the genetic code. In Chap. 6, some of these biomolecules are seen to operate like molecular motors since, by consuming ATP, they can develop at the molecular scale all the functions of a mechanical engine: motor proteins can transport a cargo, apply a force to deform a membrane, generate the powerful strokes that make unicellular animals to swim in a sea of water that, at the micron-length scale, appears as viscous as molasses. Chapter 7 introduces the theme of bioelectricity, by looking at specialised cells capable of producing, transmitting and detecting electrical signals; while such electric symphony occurs to a variable extent in almost all animal cells, the neurons of the brain and the cardiac cells in the heart will be the tenors of the show, without forgetting some interesting quirks that are starting to appear in the still little explored domain of plant electricity.

Chapters 8 and 9 give a broad view of the mechanics of cells and tissues. We start from the peculiar elasticity of one-dimensional filaments and two-dimensional membranes, which at the molecular length- and energy scales is profoundly influenced by thermal fluctuations and therefore reveals its entropic origin; the mechanics of cell division represents a key topic around which many of these concepts are put at work. In Chap. 9, the elasticity of biological materials that make up animal and plant tissues is investigated, showing that the exceptional properties of such materials originate from their tightly designed hierarchical structure, which couples the molecular scale to the micron-scale to the macroscopic; in this way, it will be clear that no artificial polymer could replace the fabulous response of cartilage as the best shock absorber ever invented, or why trees fare much better against winds with their cellulose-based structure, rather than with a metal-reinforced one.

Chapter 10 presents the mechanics and dynamics of muscles, with their hierarchical molecular structure remarkably preserved and transmitted nearly identical through the entire evolutionary tree of all animals, from insects to the elephants; muscles actuate forces in the animal body, and we use the example of insect flight as a playground. With the second part of Chap. 9 we leave the microscopic to step up again into the macroscopic. Scaling laws and dimensional analysis provide a theoretical tool to deduce interesting features and correlations of animal and plant metabolism, from the scale of the single organs to the whole body, and scaling up to the size of flocks and populations, in Chaps. 11 and 12. The differences between walking, flying and swimming animals are seen to arise from the physical bounds imposed to them by their respective environments, the same environmental constraints that are shown to drive the choice of the better shapes for living organism. The shape of the animal and plant body, of course, but also the size of the flock, the areal density of a forest, and the time-scale for the reproduction of the offspring, as a function of the availability of food, presence of predators, average temperature, precipitations, exposure to the sunlight, and so on. Then, what else than a final touch at the mathematical modelling of ecosystems could provide the ideal wrap-up of the entire subject of the physics of living systems?

Now you may just sit back, start reading and, hopefully, enjoy the show. But to preemptively comfort the most skeptical reader, I will leave the stage to a quote from one of our most acute colleagues, Freeman J. Dyson:

> Since I am not an expert in this subject, I can say things that the experts might find outrageous. Don't be surprised if some of the things I say will turn out to be wrong. In science it is better to be wrong than to be vague. Often we find the right way only after we tried all the wrong ways first. That is why it is fun to be a scientist: you don't need to be afraid of being wrong.

Chapter 2
Thermodynamics for Living Systems

Abstract We start with an introductory chapter containing a concise and selected summary of the formalism of thermodynamics. Some basic concepts and the "principles" of macroscopic thermodynamics are firstly recalled, and put in an original perspective against the microscopic definitions of the same physical quantities, such as energy, temperature, pressure. The notions necessary for making the link between physics and biology are then introduced, and the prominent role of the entropy in describing the physics of living systems is highlighted. Should physics and biology be considered as two profoundly distinct camps, or rather two faces of one and a same reality? While such a hard question could hardly find an answer here, we show that even some of the most abstract conceptualisations of physics, such as the ideal gas, can be of great value in biology.

2.1 Macroscopic and Microscopic

Thermodynamics is not, strictly speaking, a physical theory on the same level as electromagnetism, gravity, quantum mechanics and so on. Besides the capability of giving a synthetic and formalised description of an ensemble of phenomena, the distinctive character of a theory is that of being able to formulate predictions. Moreover, a theory must include the possibility of confirming or refuting its own predictions. In this respect, thermodynamics is rather a powerful formalism, capable of describing diverse phenomena, allowing to unify under a rich conceptual structure experimental data from widely different domains, as well as proposing analogies between apparently disconnected phenomena.

Thermodynamics establishes relationships among physical quantities pertaining to a system, allowing to describe its behaviour under different conditions, however it needs to know nothing about the microscopic behaviour of the constituents of the system. As Steven Weinberg put it, *Thermodynamics is more like a mode of reasoning than a body of physical laws* [1]. The system of interest may be very often a machine, such as a refrigerator, a heater, an engine, a windmill. This is indeed how thermodynamics was invented, in the XIX century. However, a 'system' can be also any kind of experimental device, from the simplest one like a box filled of soap

© Springer International Publishing Switzerland 2016
F. Cleri, *The Physics of Living Systems*, Undergraduate Lecture
Notes in Physics, DOI 10.1007/978-3-319-30647-6_2

bubbles, to the most complicated plasma chamber of a nuclear fusion reactor. And it could as well be a completely theoretical, idealised situation, such an ensemble of perfect mathematical points, each infinitely small but endowed with a finite mass, or an assembly of hard spheres riding on a frictionless plane, bumping into each other in straight trajectories.

Anyway, to give a thermodynamical description of our system of interest it is supposed that the system is in some way isolated from the external world, so that its behaviour can be studied in almost ideal conditions; or at least, it is arranged so as to be in contact with a reservoir (of energy, heat, matter) so big that such almost ideal conditions can be imposed from the exterior. An example of a well isolated system could be the interior of a glass balloon filled with gas; an example of a system in contact with a reservoir could be the interior of a refrigerator, kept at a constant temperature considerably lower than the surroundings.

In comparison to such ideally, or nearly-ideally controllable systems, any living organism looks like an extremely complex (and messy) system, inside which many transformations of all kind occur all the time, with a continuous exchange of energy, heat, materials, information with the surrounding environment. In this respect life would be regarded, from the point of view of thermodynamics, as not belonging to any of the above categories. Rather, almost going against the typical behaviour of thermal machines. We will do our best to cancel such a wrong impression from the mind of our reader.

Most people have (or think they have) an intuitive understanding of basic quantities such as temperature, pressure, volume. On the other hand, there are some other physical quantities in thermodynamics which require a bit more of reasoning to be understood, for example the energy, or the work. Eventually there are yet some other quantities, such as the entropy or the chemical potential, which seem rather obscure from the point of view of our daily experience, and require a deeper analysis in order to grasp their meaning and usefulness. In this Chapter we will follow a kind of hierarchical approach, introducing these various quantities starting from the most basic ones, and moving progressively up to the less intuitive ones. We will avoid as much as possible mathematical formalism if not strictly necessary, but will try to be as rigorous as possible whenever mathematics is needed.

A **macroscopic state** of a system is specified by global thermodynamic variables, or functions of state, which describe a property of the system as a whole.

The various functions of state are categorised into **extensive** and **intensive**. Extensive variables are a function of the system volume V: in other words, if we double V, the value of an extensive variable is also doubled; if we halve V, the value of an extensive variable is also divided by two. Moreover, the value of an extensive variable for a system is equal to the sum of that extensive variable for all of its subsystems: for example, the number of objects in a drawer ranged into compartments is the sum of the objects in all the compartments. Examples of extensive variables are:

- the volume itself, V
- the energy, E
- the entropy, S

- the number, N
- the deformation $\varepsilon = \Delta L / L$, of a length L

On the other hand, intensive variables are not a function of the system volume: they do not change upon a variation of V. Moreover, the value of an intensive variable for a system subdivided into subsystems is not the sum of the corresponding variables: the overall temperature of a flat divided in two rooms is just the common value of temperature, not the sum of the two temperatures in each room. Examples of intensive variables are:

- the temperature, T
- the pressure, P
- the chemical potential, μ
- the stress tensor, σ

Although not necessary for the development of thermodynamic relationships between the different macroscopic functions, the thermodynamic system can also be thought of being made up of some microscopic components (e.g., cells, atoms, molecules). In order to be given a proper mathematical treatment, the basic properties of such microscopic components are that they must be all identical and interchangeable.

Any particular arrangement (i.e., the ensemble of values specifying the details) of each and every microscopic components, corresponding to given values of the macroscopic thermodynamics variables, $N, P, V, T, ...$, represents a **microscopic state** of the system, or microstate. A microstate of the system is specified by a set of variables describing the arrangement of its microscopic components: for example, the exact positions and velocities of each molecule in a balloon filled with gas.

Despite such an overly detailed microscopic accounting may seem an impossible task to carry out in a real experiment, the corresponding conceptual formulation is at the basis of all the developments of *statistical mechanics*, a powerful physical theory whose objective is to give an interpretation of macroscopic physical quantities in terms of their microscopic constituents. Since a macroscopic piece of matter contains a number of atoms or molecules of the order of the Avogadro's number, $N_{Av} = 6.02 \times 10^{23}$, and even one cubic centimetre of living tissue contains millions of cells, this connection cannot be made on a one-by-one basis, but must rather exploit some statistical properties of the ensemble. Hence the denomination of 'statistical' mechanics, as opposed to the deterministic character of Newtonian mechanics. On the other hand, the modern developments of computer simulations (see "Further reading" at the end of the chapter) allow to verify the predictions of statistical mechanics, by following the detailed dynamics of a simulated system of microscopic objects, and to prove the correspondence between microscopic and macroscopic quantities to an exceptional extent.

In classical thermodynamics there is no necessary relationship between the macroscopic state of a system and its microscopic states. In fact, thermodynamics *does not need* microscopic variables in order to work. It will be immediately apparent that, for any macroscopic state, there can be a large number of equivalent microscopic states. Think of a large ensemble of N molecules of a gas enclosed in a balloon of volume V, for each of which you know exactly the positions and velocities at a given instant t. This is a microstate. Then, think of exchanging the velocities between two molecules, while leaving each of the two molecules at its place. This second configuration represents a different microstate, however the macroscopic condition of the system has hardly changed. Such a fundamental difference between the microscopic and macroscopic states will be the basis for introducing later on the physical variable entropy.

Notably, the fact that a seemingly very large number of microstates can correspond to one same macrostate, suggests that there should be some kind of statistics, such as a probability distribution, of these microstates. Think of the thermal agitation of the molecules in our gas, at any instant of time. For any given macroscopic condition of the gas (a set of values of temperature, pressure and volume) the many microstates composing it could indeed have different probabilities of occurring.

2.1.1 Isolated System

In thermodynamics, an isolated system is any ensemble of N objects, enclosed in a well defined volume of space V, which does not perform any exchanges of matter or energy with the space external to the volume V. It should be stressed that, up to this point, it is not at all necessary to specify the nature, nor the internal structure of the objects making up the system, in order to use the formalism of thermodynamics to describe it. The only basic requirements is that the objects are distinguishable, and that may be grouped if necessary into subsets according to some criteria. Our system can be composed by any number of independent subsystems (for example, the moving parts of an engine), provided the appropriate thermodynamical quantities can be defined for the different subsystems.

It is important to observe that, both in an experiment and in theory, we can never measure the absolute value of either extensive or intensive variables, but only differences of such variables between a current macroscopic state and a reference condition. This statement may appear apodictic, but it can be intuitively grasped by observing that to know the absolute system's energy we should be able to measure it in every possible microstate corresponding to a macrostate, and it is practically impossible to enumerate all the microscopic configurations available even for a system of a few objects enclosed in a rather small volume. However, by measuring relative differences between such quantities a relative scale can be established, and variations of quantities such as energy, temperature, pressure between different systems can be measured and compared.

The **Statistical Postulate of Equilibrium** states that: any isolated system spontaneously evolves towards a state, defined as the thermodynamic equilibrium, characterised by the fact that any differences of the functions of state between its subsystems go to zero. In practice this means that, once arrived at the equilibrium condition, the macroscopic functions of the system remain constant.

2.1.2 Energy

In very general terms, the energy of a system E describes its ability to operate transformations. For example, for a purely mechanical system the energy describes its capability to realise a mechanical work, W, such as lifting a mass against the gravity:

$$E = W = mgh \tag{2.1}$$

mg being the weight of the mass m in the Earth's gravitational field with constant acceleration $g = 9.807 \, \mathrm{m\,s^{-2}}$, and h is the relative height with respect to the point of start. As said above, we can only define differences of functions of state with respect to a reference value. If we take as zero the energy of the mass when it is lying on the floor, any position of the mass above the ground will correspond to a positive energy difference, and thus to the possibility of performing a non-zero work. The old pendulum clocks worked exactly with this principle. In the age when there was no electricity available in the household, at any recharge of the mechanism one or more weights were raised inside the clock; during the operation of the clock, the weights, attached to the wheels of the clock by a thin metal chain, would slowly descend toward the ground, setting in motion the spheres indicating the hours and minutes; once the weights reached the bottom, the clock would have exhausted all the available potential energy and would stop, until the next recharge. This was a fine and precise mechanism, capable of converting gravitational energy into mechanical energy of motion of the clock wheels.

If a system is composed of subsystems 1, 2, 3, ...k, the total energy of the system is given by the sum of the energies of each subsystem:

$$E_{tot} = E_1 + E_2 + E_3 + ... + E_k \tag{2.2}$$

Since the energy is additive, it must necessarily be an extensive thermodynamical variable. To specify its difference with respect to other energy-like quantities, to be introduced later on, this form of energy is usually called the **internal energy**.

2.1.3 Heat

Believed in ancient times to be some sort of fluid pervading the bodies, heat is just another form of energy. On a microscopic level, it is associated with the thermal agitation of the atoms and molecules: the faster the thermal agitation, the hotter will appear the body.

While remaining conserved, energy can however change of form: for example the mechanical energy of a system can be turned partly or totally into heat, Q. Every change of the form of energy bringing the system to a different macrostate with final energy E_{fin}, must be reported to the initial energy, E_{in}, which is the reference state. Energy can be transformed into mechanical work, or any other kind of work (electrical, magnetic...), always indicated by W, while the system can transfer part of its energy to the surroundings as heat, Q, and end up in the final value E_{fin}, such that the initial energy E_{in} is conserved. Such an equivalence between different forms of energy is primarily grounded in countless experimental observations on the most various kind of systems, which are summarised by the classical equation:

$$\Delta E = E_{fin} - E_{in} = -\Delta W + \Delta Q \tag{2.3}$$

This is the so-called **First Principle** of thermodynamics, nothing but another way of expressing the universal conservation of energy, by stating that mechanical work and heat are equivalent forms of the energy.[1]

It is customary to separate the energy turned into heat from the other forms of energy, in the balance of energy transformations, for the reason that, differently from other forms of energy, heat is a sort of "dead end" for the energy: once a part of energy is turned into heat, that energy is lost forever and cannot be recovered to perform any more work. This point will become more clear a bit later, once the concept of temperature is introduced.

The very important experimental observation here, for which not a single violation has ever been observed yet in the history of physics and chemistry, is that any transformation of energy is always accompanied by some loss in the form of heat.

A transformation of energy without any heat loss would be an ideal transformation, sometimes called **adiabatic**: this is defined as a transition between two states via a virtually infinite sequence of infinitesimally small and slow transformations, to make a system go from an initial to a final state without any heat loss.

2.2 Perfect Gas

A perfect gas is a very simple, idealised thermodynamical system, therefore very practical to study as a bookkeeping example of more complex situations. We consider a volume V, filled by a number N of microscopic particles, i.e. each with a negligible size compared to the volume V. Each one of these point-like particles has a mass m. We choose N and V such that the particle density:

$$\rho = N/V \tag{2.4}$$

---

[1] It may be interesting to note that the first definite statement about this, eminently physical, principle came in fact from medicine. Around 1840, Julius Robert Mayer, then a physician in Java, deduced the energy equivalent of heat by observing differences in venous blood colour, which he attributed to different oxygen concentrations, and hence to different amounts of heat produced by the body. His empirical calculations led him to a value of 3.58 J/cal, not too far from the more accurate value measured by James Joule just a few years later, 4.16 J/cal, by means of calorimetry experiments [2, 3].

is so low that collisions between particles are practically highly unlikely. Each particle of our perfect gas is characterised by a position and a velocity $\mathbf{v}_i$ in the volume V. Particles make perfectly elastic collisions with the walls of V, meaning that their energy is the same before and after the collision, only their direction changes because of momentum conservation. Additionally, we consider that there are no interactions whatsoever between the particles (chemical, electrical, magnetic, gravitational), therefore their position in space is not relevant to their energy. As a consequence, the only pertinent form of energy to each particle $i = 1, ..., N$ is the kinetic energy:

$$E_i = \frac{1}{2}mv_i^2 \tag{2.5}$$

The total energy of the gas, E, is given by the sum of the kinetic energies of all the N particles:

$$E = \sum_{i=1}^{N} E_i = \sum_{i=1}^{N} \frac{p_i^2}{2m} \tag{2.6}$$

where we have introduced the particle momentum vector $\mathbf{p} = (p_x, p_y, p_z) = m\mathbf{v}$.

Every microstate of the perfect gas corresponds to a set of values of positions and momenta $\{\mathbf{r}_i, \mathbf{p}_i\}_{i \in N}$, different at every instant, provided that E remains constant since the system is isolated. In principle, the number of microstates for a given combination of $\{NVE\}$ is infinite: we may change the momentum of a particle in infinitesimal increments (taking care of changing two $\mathbf{p}$ vectors at a time by equal and opposite amounts, so that the energy remains constant), and obtain infinitely many different microstates of the system. Instead of the absolute number of microstates for an energy E, which is clearly impossible to count, we can look at the function expressing the *density of microstates* in an energy interval dE around E. This latter can be rather easily estimated by considering the very special situation of the perfect gas. (This function can be calculated just for a few other simple systems, but it becomes a very complicate task for any realistic system, with arbitrary interactions among the particles.)

2.2.1 Counting Microstates

The peculiarity of the perfect gas is that particles do not have any mutual interaction. Therefore, their energy is given by a simple sum of squares of the particle momenta, $E = \frac{1}{2m}\Sigma_i p_i^2$. We can also detail this expression according to the Cartesian components of each particle's momentum vector:

$$E = \frac{1}{2m} \sum_{i=1}^{N} p_{ix}^2 + p_{iy}^2 + p_{iz}^2 \tag{2.7}$$

the sum containing $3N$ terms.

This formula for the total energy looks like a sort of Pythagoras' theorem for an ideal "triangle" with $3N$ sides. In fact, one could read the above formula as: $\sqrt{2mE}$ *is the distance from the origin of the point* $\mathbf{p}$, *with coordinates* $(\mathbf{p}_1,...\mathbf{p}_N)$, *in a space with* $3N$ *dimensions*.

This is a very interesting interpretation of the energy of a perfect gas. If we imagine $\mathbf{p}$ as a vector with $3N$ components centred at the origin, and we imagine to change arbitrarily any of its components in every possible way that give the same final value of E, the vector $\mathbf{p}$ will describe a $3N$-dimensional sphere centred at the origin with radius $R = \sqrt{2mE}$.

Then, it may be thought that the number of available microstates (i.e., different combinations of the $\mathbf{p}_i$) for a given energy, let us call it $\Omega(E)$, should be proportional to the surface of this sphere: a larger valuer of the energy E corresponds to a larger sphere surface, and therefore to a larger Ω. A system with assigned values of $\{NVE\}$ (the macrostate) must conserve its energy, so it will ideally "move" on this constant-energy surface, exploring all the Ω microscopic configurations whose E is compatible with the macrostate.

What is the surface of a sphere in $3N$ dimensions? We can reason by analogy:

- $2\pi R$ is the perimeter of a circle (sphere in 2 dimensions)
- $4\pi R^2$ is the surface of a sphere (3 dimensions)

The surface of dimension n appears to be proportional to the volume divided by R, $S(n) \approx V(n)/R$, therefore we may write:

$$S(3N) = cR^{3N-1} \tag{2.8}$$

for the surface of a hypersphere in $3N$ dimensions. With a little algebra, it can be shown that the exact proportionality coefficient is $c = (2\pi)^{3N/2}/\Gamma(3N/2)$ (see the Appendix A for the properties of the special Γ function). For very large values of N, one can approximate $N - 1 \simeq N$. Therefore, replace $3N - 1$ by $3N$, and $\Omega \approx R^{3N}$. Since $R = \sqrt{2mE}$, it is finally $\Omega(N, E) = (2\pi)^{3N/2}(2mE)^{3N/2}/\Gamma(3N/2)$.

Now, we note that each microstate is specified by assigning also the values of positions $\mathbf{r}_i$ for each particle, besides their momenta or velocities: two microstates could have the same distribution of velocities, but differ in the positions of some particles. Since each particle of the perfect gas can be found anywhere in the volume V, independently of the others, the number Ω should be multiplied by a factor V contributed from each particle. The complete expression of Ω is therefore:

$$\Omega(N, V, E) = V^N \frac{(2\pi)^{3N/2}(2mE)^{3N/2}}{\Gamma(3N/2)} \tag{2.9}$$

2.3 Entropy and Disorder

Disorder in real life is conceived as opposed to order. A sequence, a pattern, a bookshelf can be quickly seen to be ordered or disordered. Both order and disorder seem easy to recognise, when looking at an array of objects, or at your children's room. However, the view of the contrast between order and disorder as just an attribute of the spatial organisation is somewhat limited. A spatially well-ordered crystalline lattice of magnetic atoms can turn out to be completely disordered, if we look at the values of magnetisation in that same crystal above some critical temperature. In physics, a macroscopic state of disorder of a system consists in the inability to formulate predictions about the behaviour of the system or, in other words, a very limited knowledge about its actual state. For the microscopic states of this same system, the condition of disorder would correspond to the microscopic variables (positions, velocities of the particles) assuming a wide spectrum of completely random values.

Strange as it may seem, this very concept of disorder will allow us to establish a conceptual link between the macroscopic system and its microscopic constituents. Moreover, it should not be thought that the condition of equilibrium is synonymous to some kind of order. Indeed, the macroscopic state of equilibrium of a system is not represented by one particular microstate, for example one having a very peculiar distribution of velocities, or one with a very regular arrangement of the positions of the particles. Microscopic equilibrium is associated to an equal probability distribution of all the admissible micro states (i.e., those corresponding to a same macrostate), *provided the disorder of the system is maximised.* Where such a quite surprising observation comes from? How come that equilibrium should correspond to a maximum disorder, in parallel to the disappearing of any differences between all the state variables of the system?

We are now ready to propose a definition of the order and disorder condition for a system, by constructing a mathematical function explicitly dedicated to this purpose. The characters of such a function should be at least: (1) that it be linked to the constant energy of the system, and (2) that it be proportional to the number Ω of microscopic realisations of the same value of macroscopic energy. The latter condition implies that such a quantity must be extensive.

Since V and E appear as multiplicative factors in the expression for Ω, Eq. (2.9), such a function must be necessarily be constructed from the logarithm of Ω, so that if we break the system into subsystems with energies $E_1, ... E_k$, the value of such a quantity can be obtained as the sum of the corresponding quantities for each subsystem k. For historical reasons, this new physical quantity is called the **entropy** of the system, and is indicated by a letter S:

$$S = k_B \ln \frac{\Omega}{N!} \tag{2.10}$$

The factor k_B is the Boltzmann constant, equal to 1.38×10^{-23} J/K, or 8.5×10^{-5} eV/K, allowing to express the entropy in energy-like units (we note that entropy may be also used in different contexts, such as information theory, in which no energy-like units are necessary). The $1/N!$ factor was introduced by J. W. Gibbs, one of the 'founding fathers' of statistical physics, to discount the fact that the particles in the perfect gas are all identical. This implies that microstates with identical positions and velocities of all particles, but in which the "label" of any two particles is exchanged, are identical, therefore the number Ω must be divided by the number $N!$ of equivalent permutations of the N identical particles. On the other hand, this multiplicative factor has no physical basis for classical particles, and was introduced by Gibbs merely on empirical grounds (because without it the entropy would not strictly be additive [4, 5]). The presence of the $1/N!$ can be fully justified only in the framework of quantum mechanics, which admits the non-distinguishability to be a fundamental property of quantum particles.

Therefore, from the above semi-analytical derivation, the entropy of the perfect gas is obtained as:

$$S(N, V, E) = k_B \ln \left(\frac{(2\pi)^{3N/2} V^N (2mE)^{3N/2}}{\Gamma(3N/2)N!} \right) =$$

$$= k_B[N \ln V + N \ln(2mE)^{3/2} - \ln \Gamma(3N/2) - \ln N! + c] \simeq \qquad (2.11)$$

$$\simeq Nk_B \left\{ \ln \left[\left(\frac{V}{N} \right) \left(\frac{2mE}{N} \right)^{3/2} \right] + c \right\}$$

This is the Sackur-Tetrode equation for the *absolute entropy* of the perfect gas,[2] independently derived in 1912 by the physicists Otto Sackur in Germany, and Hugo Tetrode in the Netherlands [6, 7]. Its utility lies in the fact that it is one of the rare examples for which we can calculate explicitly the absolute value of the entropy, which is otherwise a rather elusive quantity to grasp in real life.

It is an experimental fact that the entropy of an isolated system can only increase with any spontaneous transformation. Thermodynamics was born in the early XIX century, with the purpose of knowing better how thermal machines worked [8, 9]. In the old experiments designed to understand the relationships between energy and heat, the French physicist Sadi Carnot defined around the year 1824 the concept of a cyclic thermal engine, for which he established that it was necessary to have at least two parts of the engine at different temperatures in order to extract useful work. About twenty years later, the German physicist Rudolf Clausius realised that any transformation in which only heat was exchanged, between a "hot" and a "warm" body, would be *irreversible*. Please notice the "..." in the previous sentence. Indeed, this is a bit of a circular definition, since we do not know exactly what is a hot or a warm

---

[2]By using arguments from quantum mechanics, it is shown that the constant appearing in the equation must be $c = \frac{5}{2} + \frac{3}{2} \ln \frac{2\pi}{3h^2}$, with h the Planck's constant.

body, unless we measure a difference between two bodies in contact.[3] The definition of a quantity to be used as temperature, in fact, is intimately linked to the definition of the entropy (although the practical notion of temperature, and tools to measure it, were already known since at least two centuries earlier). What the experiments actually measured was the fact that the flow of heat is always unidirectional, and this direction of the flow allows to establish which body is "hotter" than the other, thus making for a temperature scale.

Sometime during the first half of the XIX century, all the experimental observations about the loss of energy into heat, expansion, and irreversible phase changes, were summarised in the so-called **Second Principle** of thermodynamics:

$$\Delta S \geq 0 \tag{2.12}$$

which states that in any spontaneous transformation the entropy of the system must increase. The function entropy was introduced just to give this principle a formal mathematical statement. Besides, we note that the second principle was in fact the first to be established, in 1824 by Carnot. The 'disorder' interpretation of entropy were to be laid out only about 60 years later, by Ludwig Boltzmann. According to our construction of the entropy function of state, in terms of the size Ω of the available space of microstates, an increase in S corresponds to an ever larger number of microstates, all becoming equally probable at equilibrium. Therefore, as far as the entropy increases, the information about the microscopic state of the system spreads out onto a larger and larger ensemble. We can interpret this as a net loss of information, or an increase of the disorder of the system.

2.3.1 Irreversibility and Probability

Let us consider a perfect gas initially confined in a volume V. If we let the gas expand so as to double its volume in such a way that the gas does not perform any work, the only term that changes in the Eq. (2.11) above is V^N. The corresponding entropy change is:

$$\Delta S = S_{fin} - S_{in} = k_B[\ln(V)^N - \ln(V/2)^N] = Nk_B \ln 2 \tag{2.13}$$

always positive. One may ask why we implicitly assumed that the approach to equilibrium implies the *spontaneous* expansion of the gas from $V/2$ to V. The idea of a spontaneous transformation is linked to the concept of **irreversibility**. In a classical definition, irreversible means that the system could never go back spontaneously to its initial state. However, a more subtle interpretation can be posed in terms of the

---

[3] Already in the middle of XVIII century the English natural philosopher John Locke, in his *Essay on the human intellect*, had defined the temperature as a relative concept, stating that a body could be hot or warm only in relation to another body.

probability of occurrence of the accessible system states. This is not just an academic disquisition, since the macroscopic irreversibility contrasts with the perfect reversibility (called *time invariance* in physics) of the microscopic events.

It may be considered that the N microscopic particles, in their continuous thermal agitation, should explore any possible microscopic configuration. As a result, a microstate with $N/2$ particles in the half-volume $V/2$ and $N/2$ particles in the other half-volume $V/2$, should have the same probability as another microstate in which all the N particles are in the half-volume $V/2$, and zero in the other half. That may be true, provided the entropy of the system is increased in the transformation, according to the Second Principle. It is easy to see that the entropy of the gas in a doubled volume increases. On the other hand, the probability that starting from a volume V, one mole of gas ($N \sim 10^{23}$ particles) goes back spontaneously by random fluctuations inside a volume $V/2$ is:

$$prob = \left[\frac{(V/2)}{V} \right]^{10^{23}} \approx 10^{-8,240,000,000,000,000,000,000} \qquad (2.14)$$

Evidently, for a system with a very small number of particles, at the limit of one single atom or molecule, the theoretical reversibility (originating from the fact that both the classical and quantum mechanics equations are unchanged if the time t is exchanged with $-t$) should be possible also in practice. In the recent years, it has indeed become feasible to perform experiments in which one single molecule is tracked in time and, after accurate measurements, local violations of the Second Principle have been found, when the number of degrees of freedom is reduced to a minimum [10].

Such an observation is common to any systems with a reduced number of degrees of freedom. If we flip ten coins, the probability of obtaining the same face on all of them is small but non negligible: this probability is simply $(1/2)^{10} = 0.001$. However, if we flip an Avogadro's number of coins, the probability of having the same face on all coins is $(1/2)^{10^{23}}$: in this sense, there are no strictly irreversible transformations, only very much improbable ones!

We will come back on this relationship between entropy, reversibility and disorder in Chap. 3, when discussing the probability of assembling complex proteins and nucleic acids starting from the simpler molecular building blocks.

2.4 Closed Systems

A system which can exchange energy but not matter with its environment, is called a **closed system**. The ideal example of a closed system is given by considering a box divided into two parts by a wall taken to be infinitely rigid, except for a small ideally elastic portion of negligible size compared to the whole wall surface. When two particles coming from the opposite parts of the box collide against this flexible wall, which is said to be ideally elastic, they can exchange their respective kinetic energies as if undergoing a perfectly elastic collision, however without the possibility of jumping in the opposing part of the box. As required by the above definition, the two parts of the box can exchange some energy but not matter.

Such a highly idealised system will now allow us to introduce the concept of **temperature**. Let us imagine that in the two parts of the box, which we call A and B, there are N_A and N_B particles, respectively, with energies E_A and E_B. The energy of the combined system is:

$$E_{tot} = E_A + E_B \qquad (2.15)$$

The total energy must be constant. However, the subdivision into E_A and E_B can take any combinations resulting in the same value of E_{tot}.

The total entropy of the combined system is:

$$S_{tot} = S_A(E_A) + S_B(E_B) \qquad (2.16)$$

But once E_{tot} is fixed, the two values E_A and E_B are no longer independent. Therefore:

$$S_{tot} = S_A(E_A) + S_B(E_{tot} - E_A) = S_{tot}(E_A) \qquad (2.17)$$

a writing which underscores the dependence of S on a single energy variable. The entropy equation gives for the two gases:

$$S_{tot}(E_A) = k_B \left[N_A \left(\frac{3}{2} \ln E_A + \ln V_A \right) + N_B \left(\frac{3}{2} \ln(E_{tot} - E_A) + \ln V_B \right) \right] + const \qquad (2.18)$$

Now we ask: which is the most probable value for E_A? The statistical postulate seems to suggest that any value should be equally probable at equilibrium. On the other hand, since the number of microstates increases with energy, we should have the maximum possible energy on each side, in order to simultaneously maximise the entropy. To find the maximum of entropy from the above equation, it is sufficient to take its derivative with respect to the unique variable E_A and set it equal to zero:

$$\frac{dS_{tot}}{dE_A} = \frac{3}{2} k_B \left[\frac{N_A}{E_A} - \frac{N_B}{(E_{tot} - E_A)} \right] = 0 \qquad (2.19)$$

This clearly shows that the above conditions can be satisfied only if the energy is shared between A and B in a way proportional to the number of particles in each subsystem (actually, proportionally to the number of degrees of freedom of the particles in each subsystem, which is $3 \times$ the number of particles, one for each cartesian coordinate). In particular, if the number of particles in the two subsystems is equal, the energy is also equally shared, $E_A = E_B$. If this is not the case, the maximum of the entropy corresponds to:

$$\frac{E_A}{N_A} = \frac{E_B}{N_B} \qquad (2.20)$$

This result was experimentally known well before the introduction of the microscopic ideas about the constitution of matter. It is the **Equipartition law**, which was based on the experimental observations performed by the French chemists A.-T. Dulong and P.-L. Petit (1819) on the constancy of the specific heat of solids at high temperature.

2.4.1   Temperature

Indeed, it is just on the basis of this result that we can introduce the concept of temperature, a quantity for which most of us have an intuitive perception, but which is just a bit more complicate to put on formal grounds.

A formal definition, eventually valid for any system and not just for a perfect gas, is the following. Starting from the entropy equation, we want to find the variation of entropy as a function of an infinitesimal change of energy from E to $E + dE$, all the other thermodynamic functions remaining unchanged. We have:

$$dS(E) = Nk_B d(\frac{3}{2} \ln E + \ln V) = \frac{3}{2} Nk_B (\ln(E + dE) - \ln E) = $$

$$\tag{2.21}$$

$$= \frac{3}{2} Nk_B \ln(1 + dE/E) \simeq \frac{3}{2} Nk_B \frac{dE}{E}$$

The temperature T is identified as the numerical coefficient in the right-hand side of the equation:

$$\frac{dS}{dE} = \frac{1}{T} \tag{2.22}$$

This is the fundamental definition of the temperature, which we postponed up to this point (although the practical notion of temperature would come first, historically) in order to firstly have well clear in mind the concepts of energy, heat and entropy. From this definition, we see that a small variation of the energy makes a large variation of entropy when the system is at low temperature; on the other hand, as the temperature is increased, the increase in entropy as a function of energy gets smaller and smaller. (For the perfect gas, this should be already evident from the relationship $S \propto \ln E$, hence $T \propto E$.)

By the last definition, the law of equipartition reads:

$$\frac{E}{N} = \frac{3}{2} k_B T \tag{2.23}$$

This equation says that the average energy per particle is equal to $\frac{3}{2} k_B T$. By considering that each point particle is defined by three degrees of freedom (v_x, v_y, v_z), it also says that each degree of freedom in the system contributes $\frac{1}{2} k_B T$, which was exactly the experimental deduction of Dulong and Petit in their study of solids.

For a perfect gas the temperature coincides with its kinetic energy, apart from a numerical coefficient with appropriate dimensions:

$$\frac{E}{N} = \frac{3}{2}k_B T = \frac{1}{N}\sum_{i=1}^{N}\frac{1}{2}mv_i^2 \tag{2.24}$$

This is an important consideration, in that the above equation puts into correspondence the macroscopic vision of thermodynamics, represented by the temperature of the perfect gas, with the statistical properties of its microscopic constituents, that is the kinetic energy of an ensemble of particles.

Let us consider two systems A and B at two different temperatures, $T_A \neq T_B$, both at their respective equilibrium. By the moment we put them in contact, the particles from the two systems will start colliding, and redistributing their energy until no further exchanges are capable of changing the distribution of velocities. The new equilibrium state is therefore characterised by the condition $T_A = T_B$. This is sometimes called the **Zeroth Principle** of thermodynamics. In simple words: *temperature is the quantity that becomes equal when two macroscopic closed systems, freely exchanging energy, attain a common equilibrium.*

2.4.2 Caloric Definition of the Entropy

Up to now we have been constantly looking for microscopic definitions of known macroscopic thermodynamic quantities. However, as we noticed above, thermodynamics does not need such microscopic definitions in order to work. Therefore, it may be surprising that at this point we have only a microscopic definition of the entropy. Entropy has been defined in terms of the number Ω of microscopic realisations of a given macroscopic state. Is there a macroscopic function which corresponds to such a microscopic definition?

In fact, the most ancient definitions of entropy given by Clausius (1854) and others (Maxwell 1867, Gibbs 1902, etc.), which largely preceded the microscopic formulation above, were suggested by experiments of transformation of energy into work. As already discussed, it was (and still is) experimentally observed that in every transformation some part of the energy was irreversibly transformed into heat, dispersed in the environment. For an energy transformation at constant temperature, Clausius defined the quantity:

$$\Delta S = \frac{\Delta Q}{T} \tag{2.25}$$

as the fraction of initial energy lost into heat, and introduced the term entropy for the quantity labelled S (with dimensions of [Energy]/[Temperature]). Therefore, by going back to Eq. (2.3), we can write:

$$\Delta E = E_{fin} - E_{in} = -\Delta W + \Delta Q = -\Delta W + T\Delta S \tag{2.26}$$

Mechanical equilibrium of a perfect gas under pressure

A wall of the box of volume V is free to move against a fixed spring, with constant k, which holds it in place at equilibrium (see Fig. 2.1a). The lateral section of the box is S, and the equilibrium length of the box is L, in the direction parallel to the force $F = -kL$ applied by the spring. Therefore, $V = SL$.

The total energy of the perfect gas is the sum of the kinetic energy of the molecules, plus the elastic potential energy $U = kL^2/2 = FL$ stored in the elastic spring:

$$E_{tot} = E_{kin}^{GP} + FL \tag{2.27}$$

If the wall is moved from L to L', with a change of volume from $V = SL$ to $V' = SL'$, the entropy of the gas changes by a quantity:

$$\Delta S = Nk_B \, \Delta(\ln(E_{kin})^{3/2} + \ln V) = Nk_B \left(\frac{3}{2} \frac{\Delta E_{kin}}{E_{kin}} + \frac{\Delta V}{V} \right) =$$

$$\tag{2.28}$$

$$= \frac{1}{T} \left(\Delta E_{kin} + \frac{Nk_B T}{L} \right) (L - L')$$

since $\Delta(\ln x) \simeq \Delta x/x$, and $E_{kin}/N = 3/2(k_B T)$. Energy conservation implies:

$$E_{kin}^{GP} + FL = E_{kin}^{'GP} + FL' \tag{2.29}$$

or:

$$\Delta E_{kin} = E_{kin}' - E_{kin} = -F(L' - L) \tag{2.30}$$

Therefore:

$$\Delta S = \frac{1}{T} \left(-F + \frac{Nk_B T}{L} \right) \Delta L \tag{2.31}$$

At equilibrium the variation of entropy stops, $\Delta S = 0$. The new equilibrium position of the wall is:

$$L' = \frac{Nk_B T}{F} \tag{2.32}$$

The force of the spring divided by the surface S is the pressure, $P = F/S$ (a force per unit surface). If we multiply Eq. (2.26) by S, with $V = SL$, we find:

$$V = \frac{S}{F} Nk_B T \tag{2.33}$$

or:

$$PV = Nk_B T \tag{2.34}$$

This is just the *equation of state* of the perfect gas. It should be noted that, in a isolated system, every volume variation implies a variation of the temperature (since the latter represents in a perfect gas a variation of the kinetic energy, $\Delta E_{kin} = -P\Delta V$).

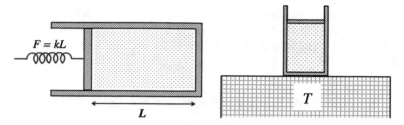

Fig. 2.1 *Left* Schematic of a gas-containing box of cross section S and variable length L. The total volume is $V = SL$. The walls are assumed to be infinitely rigid so that the gas molecules perform only perfectly elastic collisions. The *left-side* wall can move without friction, to adjust against the force $F = kx$ imposed by the mobile spring. The corresponding internal gas pressure is $P = F/S$. *Right* The gas box in contact with a thermostat at temperature T. By definition, the volume of the thermostat is so much bigger than the volume V of the gas, to be unperturbed by whatever happens inside the volume V

This 'caloric' definition of the entropy was the one chemists and physicists had been familiar with for the whole XIX-th century, and still is the most useful definition for engineers and whoever works with macroscopic thermal machines. It should be absolutely remarkable, then, that we can reobtain all the well known properties of a perfect gas, such as the equipartition, or the equation of state (see next greybox on p. 24), starting from Boltzmann's statistical formulation of the entropy. Boltzmann had already used a similar equation in his early work on the mechanical interpretation of the Second Principle, yet without thinking of it as a probability. The identification of the phase space density Ω with a microscopic probability came a few years later, in his work of 1871 on the kinetic theory of gases [11]. Either way, the theoretical definition of entropy written as Eq. (2.10) was a purely conceptual endeavour, compared to the fully experimentally-grounded caloric definition above. In his time, Boltzmann had long disputes with his colleagues scientists, and even with journal editors, who refused to take his assumptions about the microscopic behaviour of atoms as nothing more than practical speculations. His pioneering views about the microscopic connection with the macroscopic world would be confirmed only many years later, by J. Perrin's experiments on colloidal suspension that allowed to measure with high precision the values of both the Avogadro's number and of Boltzmann's own k_B constant [12], and were finally vindicated by the revolution of quantum physics.

2.5 Free Energy

Now we move on, to considering our perfect gas enclosed in a volume V and in contact with a surface at constant temperature T. It may be imagined that the volume V is placed over a very large block of material, as in Fig. 2.1b, so large to not be affected by the transformations eventually occurring inside the much smaller volume V. The perfect gas can exchange kinetic energy across the contact surface with the block, however according to our definition the temperature of the latter will remain unchanged. The sum of the subsystem represented by the volume V plus the whole block is again an isolated system. The block is called a thermal reservoir, or **thermostat**.

2.5.1 Exchanges of Energy at Constant Volume

The greybox on p. 24 shows that the temperature of the perfect gas is linked to the changes of volume for an isolated system. Since $\Delta E = T\Delta S - P\Delta V$, we should have $T\Delta S = P\Delta V$ for a transformation occurring at constant internal energy, $\Delta E = 0$. This is a well known result of thermodynamics: the spontaneous expansion of a gas ($\Delta V > 0$) implies an increase in the entropy ($\Delta S > 0$), which also implies that it should be impossible to observe a spontaneous contraction of the volume ($\Delta V < 0$), since this would violate the Second Principle. A compression of the gas must be accompanied by a parallel variation in the energy, $\Delta E < 0$, the negative sign of the energy meaning that this is supplied to the system from an external source, in an amount such that $\Delta S = (P\Delta V - \Delta E)/T > 0$.

Let us now imagine that the volume V is a closed (sub)system, exchanging energy with the block at constant temperature T (a *thermostat*, Fig. 2.1b). In this way, we can imagine a transformation, for example an expansion or a compression of the volume, during which the temperature of the subsystem V also remains constant, the difference in energy being compensated by the (positive or negative) exchange with the block. It would seem that in this way the Second Principle could be violated: a volume compression could lead to a diminution of the entropy. In fact, we must not fool ourselves and look at the overall entropy variation. During the transformation, the block transferred a quantity of energy $-\Delta E$ to the volume V, to keep its temperature equal to T. Therefore, the entropy of the block changed by $\Delta S_{block} = -\Delta E/T$, so that the total entropy variation is:

$$T\Delta S_{tot} = T\Delta S_V + T\Delta S_{block} = T\Delta S_V - \Delta E \qquad (2.35)$$

For a closed system, this is the quantity that has to be maximised during a transformation. In the place of the internal energy, we can introduce a more complete function of state, F, which is called (Helmoltz) **free energy**:

$$F = E - TS \qquad (2.36)$$

which is minimised during the approach to equilibrium at constant V and T.

The concept of a "free" energy is connected with the idea that in thermodynamics we always think of the capability of our system to develop some work (mechanical, or other). The difference between E and TS means that when equilibrium is reached at constant temperature, E is not yet at its minimum, and therefore we could think of using this residual energy to do some more work. Instead, in such conditions, the quota of work actually available is the difference $E - TS$.

2.5.2 Exchanges of Energy at Constant Pressure

Still in the same conditions of contact with the surface of the large block, we could think that our perfect gas could use the energy supplied to perform an expansion at constant pressure P. In this case we would let the volume V change so that the pressure can remain at a constant value. The energy supplied by the block in this case would be $\Delta E + \Delta W = \Delta E + P \Delta V$, namely a part necessary to keep $T = const$ as before, plus a part to keep $P = const$. The corresponding change in entropy is:

$$T \Delta S_{tot} = T \Delta S - \Delta E - P \Delta V \qquad (2.37)$$

This leads to the introduction of yet another function of state, which is also called (Gibbs) free energy:

$$G = E + PV - TS \qquad (2.38)$$

to be minimised in the approach to equilibrium at constant P and T (actually, this is just another way of saying that entropy has always to be maximised, no matter what you do to your system).

The quantity $H = E + PV$ is called the **enthalpy** of the system. It is a quantity often measured in experiments, since it takes into account all of the energy traded in the transformation of a sample (for example, the melting of an ice cube into water, accompanied by a reduction of volume at constant pressure).

At equilibrium, when $\Delta G = 0$, the system pressure can be formally defined as:

$$P = T \left(\frac{dS}{dV} \right)_E \qquad (2.39)$$

(the subscript E indicating to calculate the derivative at constant total energy) in good analogy with the formal definition of temperature above.

2.6 Open Systems

A thermodynamic system capable of exchanging both energy and matter with its surroundings is called an **open system**. This is evidently the most general situation, and it is the typical context of living systems, as opposed to either isolated or closed systems, which are typical representations of a sealed laboratory experiment or a well isolated thermal machine, such as an oven or a refrigerator. An open system is characterised by the possibility of varying the number of constituents N (atoms, molecules, cells...), or their concentration $c = N/V$, according to a given value of **chemical potential** μ.

The Gibbs-Duhem equation

The derivation of the equivalence between free energy and chemical potential offers a nice example about how the thermodynamical relationships are derived. Sometimes this may seem a blind mathematical procedure which must just be followed strictly. Even if the mathematical steps may have not much of a meaning in itself, the final result may often show unsuspected correlations between different physical variables, or experimental measurements.

Let us start with the equation expressing the First Principle for an open system in the most general way, i.e. by including also the possible variation of the number of particles. For even more generality, let us also include the possibility that there are k different families of microscopic particles in the system, each with its own chemical potential μ_k and concentration $c_k = N_k/N$:

$$E(S, V, N) = TS - PV + \sum_{i=1}^{k} \mu_i N_i \tag{2.40}$$

The total differential of the energy with respect to all of its independent variables is formally written as:

$$dE(S, V, N) = TdS - PdV + \sum_{i=1}^{k} \mu_i dN_i \tag{2.41}$$

The total differential of each term composed by two variables is, in fact, $d(TS) = TdS + SdT$, and $d(PV) = PdV + VdP$, so that the previous equation becomes:

$$dE(S, V, N) = d(TS) - SdT - d(PV) + VdP + \sum_{i=1}^{k} \mu_i dN_i \tag{2.42}$$

Now, let us regroup all the total differentials at the left-hand side:

$$d(E - TS + PV) = -SdT + VdP + \sum_{i=1}^{k} \mu_i dN_i \tag{2.43}$$

However, by looking at the definition of Gibbs free energy, Eq. (2.38), it is also:

$$dG = -SdT + VdP + \sum_{i=1}^{k} \mu_i dN_i \tag{2.44}$$

This is the famous *Gibbs-Duhem equation*, which is at the basis of all the chemical thermodynamics. Of particular interest for our purposes is the case of constant temperature and pressure, which is typically realised in living systems: for that condition, the dT and dP terms are equal to zero, and the following relationship holds:

$$(dG)_{T,P} = \sum_{i=1}^{k} \mu_i dN_i \tag{2.45}$$

The above equation is also very useful to study equilibrium concentrations of multi-component systems. For example, the relative concentration of a binary mixture of A and B at equilibrium ($dG = 0$) is: $N_A/N_B = -(\mu_B/\mu_A)$.

The chemical potential can be thought of representing a sort of external field, like a gravitational potential, but concerning concentration adjustments. In the case of gravity, a mass would be attracted towards the center of gravity of another body. By analogy, chemical entities like atoms or molecules are attracted from a region of high concentration to a region of low concentration, until a condition of equal concentration is reached. At that point, the chemical potential has the same value everywhere, and the attraction stops. In microscopic terms, the role of chemical potential in defining the particle flow direction will become very clear when describing the process of diffusion across a membrane (Chap. 5).

From the point of view of the formal theory of thermodynamics, it is worth noting that, for constant pressure and temperature, the chemical potential at equilibrium equals the partial molar Gibbs free energy (see the greybox on p. 28).[4]

2.6.1 Entropy of a Mixture

Let us consider a system with a mixture of different elements, for example particles of two different types, A and B. The volume V is divided in two parts, V_A and V_B, such that $V_A + V_B = V$. The two groups of particles, N_A and N_B, are initially confined in the two subsystems V_A and V_B. The total number of particles is $N = N_A + N_B$, correspondingly we can define the concentrations $c_A = N_A/N$ and $c_B = N_B/N$, so that $c_A + c_B = 1$. We imagine that the two volumes can be connected, for example by a mobile wall. Removing the wall corresponds to an expansion at constant temperature, each of the two gases of particles now occupying simultaneously the entire volume V. Once the equilibrium is reached, the change in free energy of the mixture is:

$$\Delta G = \Delta E + P \Delta V - T \Delta S = 0 \tag{2.46}$$

On the other hand, the two non interacting gases are described only by their kinetic energy. At constant T the kinetic energy does not change ($\Delta E = 0$), therefore:

$$\Delta S = \left(\frac{P}{T}\right) \Delta V = N k_B \frac{\Delta V}{V} = N k_B \ln \left(\frac{V + \Delta V}{V}\right) \tag{2.47}$$

Now we write the entropy for each component of the gas separately:

$$\Delta S_A = N_A k_B \ln \left(\frac{V_A + V_B}{V_A}\right)$$

$$\Delta S_B = N_B k_B \ln \left(\frac{V_A + V_B}{V_B}\right)$$

---

[4]Nearly all the concepts and mathematical functions for the study of systems at constant pressure or temperature, including the enthalpy, free energy, chemical potential, were developed and formalised in the monumental paper by J.W. Gibbs [13], which by common consensus marks the beginning of the modern vision of thermodynamics and its connections with chemical and electrical phenomena.

and sum the two to obtain the total entropy:

$$\Delta S_{tot} = \Delta S_A + \Delta S_B = N_A k_B \ln\left(\frac{V_A + V_B}{V_A}\right) + N_B k_B \ln\left(\frac{V_A + V_B}{V_B}\right) =$$

$$= -k_B \left[N_A \ln\left(\frac{V_A}{V_A + V_B}\right) + N_B \ln\left(\frac{V_B}{V_A + V_B}\right) \right] = \qquad (2.48)$$

$$= -k_B \left[N_A \ln\left(\frac{N_A}{N_A + N_B}\right) + N_B \ln\left(\frac{N_B}{N_A + N_B}\right) \right]$$

since for either component of the gas it is $V = N(k_B T/P)$. Remembering the definition of concentration c:

$$\Delta S_{tot} = -k_B (N_A \ln c_A + N_B \ln c_B) = -N k_B (c_A \ln c_A + c_B \ln c_B) =$$

$$(2.49)$$

$$-N k_B [c_A \ln c_A + (1 - c_A) \ln(1 - c_A)]$$

In the limit of small concentrations, either c_A or $c_B \ll 1$, the limit of the previous equation for the minority component is:

$$\Delta S \simeq -N k_B c \ln c \qquad (2.50)$$

This equation will be very useful to study the osmotic behaviour of ions and molecules in small concentrations inside the cellular fluid.

2.7 The Biosphere as a Thermal Engine

The **biosphere** is defined as the ensemble of the Earth's matter that is actively employed in the fabrication of living organisms. A part of this matter is found directly in the organisms (animals, plants, bacteria...) and the remaining part is found in the recycled material. Liquid and solid water from oceans, as well as the air of the atmosphere, are part of the biosphere, in that they are necessary components of living systems. The environment comprises all the remaining matter of the Earth (the solid lithosphere), and the Universe, including the radiation. In thermodynamic terms, the biosphere is an open system, receiving a constant flux of energy and entropy from the outside environment (it also receives a small flux of matter, from the cosmic radiation). If we exclude the incoming flux, the biosphere can be considered a closed system. We will now delve a bit deeper in the thermodynamics of the energy and entropy flowing from the environment into the Earth's biosphere. To this purpose, we must firstly introduce the concept of **thermal engine**.

Transformation of free energy into mechanical work

Let us consider again a perfect gas inside a box, closed by a mobile upper surface on which some weights (for example w_1, w_2) can exert their gravitational force. The lower wall of the box is in contact with a thermostat at temperature T (see the schematic in Fig. 2.2a, b). The pressure is:

$$P_{in} = \frac{(w_1 + w_2)}{S} \tag{2.51}$$

To keep things simple, we take that there is no air surrounding and therefore absence of friction, and moreover that the weight of the moving wall is negligible compared to the w_i's. At the initial equilibrium, the mobile wall is found at a position h_{in} (Fig. 2.2a).

Now the weight w_2, for example by sliding horizontally without friction. Under the reduced weight, the internal pressure will raise the wall to the new position h_{fin} (Fig. 2.2b), with the pressure decreasing to:

$$P_{fin} = \frac{w_1}{S} \tag{2.52}$$

At constant temperature, and with no change of internal energy (perfectly reversible transformation), the free energy of the perfect gas equals the entropy. From the Sackur-Tetrode equation:

$$\Delta G = S_{fin} - S_{in} = -N k_B T \ln \left(\frac{V_{fin}}{V_{in}} \right) = -N k_B T \ln \left(\frac{h_{fin}}{h_{in}} \right) \tag{2.53}$$

On the other hand, the perfect gas equation of state gives for the final state pressure:

$$P_{fin} S h_{fin} = w_1 h_{fin} = N k_B T \tag{2.54}$$

By setting $x = (h_{fin} - h_{in})/h_{fin}$, $0 < x < 1$, we can compare the mechanical work performed by the perfect gas, to the corresponding free energy variation. The work is just:

$$P_{fin} \Delta V = \left(\frac{w_1}{S} \right) S(h_{fin} - h_{fin}) = w_1(h_{fin} - h_{fin}) \tag{2.55}$$

Therefore we have:

$$w_1(h_{fin} - h_{fin}) = -N k_B T \ln(1 - x) \tag{2.56}$$

for the work, and

$$\Delta G = x N k_B T \tag{2.57}$$

for the free energy. Since for $0 < x < 1$ it is always $x < -\ln(1 - x)$: as predicted, the quantity of work performed by the perfect gas is always less than the available free energy.

The ratio between the two:

$$r = \frac{-x}{\ln(1 - x)} \tag{2.58}$$

is bigger the smaller is x, i.e. for small variations $(h_{fin} - h_{in})$ or, again, for small values of w_2 compared to w_1. We can imagine to split w_2 into many smaller weights $w_i << w_2$, to increase the efficiency (quantity of work extracted for a given available energy). This is an interesting conclusion: the transformation of energy into work is more effective if performed in many small steps, rather than in one big step. At the limit of a virtually infinite number of infinitesimal steps, this approaches the definition of a quasi-static, or **adiabatic** process, in which no heat is exchanged at all, and all the energy is turned into work.

By pursuing the reasoning from the situation displayed in the greybox on p. 31, consider the right half of Fig. 2.2. Let us imagine to slide the box V above another thermostat at a temperature $T' < T$ (always horizontally, without friction and without doing work against the gravity). The perfect gas will cool down, and therefore its volume will decrease until getting back to the height h_{in} (Fig. 2.2c). At this point, we put back the weight w_2 into place (Fig. 2.2d), and slide back the box above the first thermostat at $T > T'$ (Fig. 2.2a). The cycle can start over in this weird machine, which is in fact an idealised representation of a cyclic thermal engine. At every cycle, a fraction of thermal energy is turned into mechanical work (lifting the weight). Every cycle increases a fraction of the entropy of the surrounding environment (the thermostats), with a transfer of heat from the reservoir at high temperature T, to the reservoir at lower temperature T'. In the real world, without an external intervention, such a process cannot continue indefinitely: the entropy of the two thermostats will at some point become identical, $\Delta S = 0$, and so will their temperature.

Such a process is completely reversible, meaning that the energy is the same at the beginning as at the end of the cycle, $\Delta E = 0$. Let us restart from the 'caloric' definition of the entropy, Eq. (2.25), from which we can compute the net variation of entropy between the two thermostats:

$$\Delta S_{tot} = \Delta Q \left(\frac{1}{T'} - \frac{1}{T} \right) = \frac{\Delta W}{T} \qquad (2.59)$$

Here $\Delta W = P(S\Delta h)$ is the total mechanical work resulting from the displacement of the mobile wall, and $\pm \Delta Q$ is the quantity of heat taken from the reservoir at temperature T and transferred to the reservoir at temperature T', with the appropriate algebraic sign. Since the variation of entropy must always be positive, in order to have

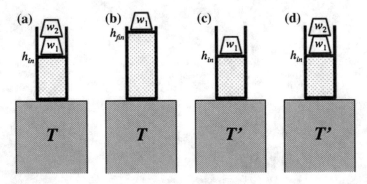

Fig. 2.2 Schematic of an ideal cyclic thermal engine, constituted by perfect gas enclosed in a box of cross section S and variable height h, with volume $V = Sh$. The top-side wall, of negligible mass compared to the weights w_i, can move without friction to adjust against the changing internal pressure of the gas. The volume V can be cyclically moved atop thermostats at different temperatures $T > T'$, steps (a) through (d)

a net work $\Delta W > 0$ it must clearly be $T' < T$. In other words, *it is the difference of temperature which keeps the thermal engine working*. This was just the principle established by Carnot in his experiments.

An ideal engine could convert all the energy into work, by an ideal (adiabatic) transformation with $\Delta Q = 0$. In all practical cases, this is forbidden by the Second Principle. However, the equation suggests that the efficiency of the engine can be increased by increasing the temperature difference between the two thermostats.

We will see in the next Chapter that the most important molecules making up the living systems often have a positive enthalpy of formation, i.e. the enthalpy of the products is less negative than that of the precursors, as well as a negative entropy of formation. As a result, also the ΔG of complex molecules such as proteins is positive, compared to the simpler building blocks (the amino acids) from which they are assembled. With a positive ΔG, such a chemical reaction would be forbidden. Therefore, to not violate the Second Principle, the formation and maintenance of living systems demands a continuous flux of free energy, as well as a supply of 'negative entropy', from the external environment, just like a refrigerator needs a continuous supply of energy to keep the temperature lower than the outside. In other words, the biosphere is not an isolated system, and is constantly maintained in a condition of non-equilibrium.

As suggested by the Eq. (2.59) above, the large temperature difference between the Earth surface (280–300 K) and the Sun (5,800 K) is at the basis of any process of energy transformation on our planet (Fig. 2.3). The radiant energy is transported by electromagnetic radiation, with its combined oscillating electric and magnetic fields. The amount of energy contained in the electromagneticwave is directly proportional

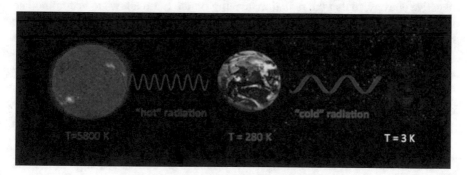

Fig. 2.3 Exchanges of radiation energy. The Sun irradiates 'hot' radiation, at a temperature of 5,800 K and short wavelength (visible and ultraviolet). A very small fraction of this energy is captured by the Earth, and reemitted at the temperature of 280–300 K and longer wavelength (infrared). This heat of radiation is dispersed towards the dark sky, which is at the temperature of 3 K. (The temperature of the sky being a remnant of the Big Bang, and constantly cooling with the expansion of the universe.)

to its frequency, v, or inversely proportional to its wavelength λ (another result which, besides having a classical interpretation, must wait quantum mechanics for a thorough explanation).

The energy of the solar radiation is not delivered just at one frequency, but extends over a continuous distribution of intensity covering a wide range of frequencies. The maximum of the solar energy distribution (see Fig. 2.7 in Problem 2.9 below) is centred approximately around the green band of the visible portion of the electromagnetic spectrum ($\lambda \sim 550$ nm), with also an important fraction of energy delivered in the ultraviolet (UV) spectral region (λ's of a few tens nm). On the other hand, the distribution of the radiation energy emitted by the Earth, being associated to a much lower temperature, is rather centred on the infrared (IR) band, at wavelengths around a few micrometers (see again Fig. 2.7). From the point of view of a biological system, this difference is extremely important. The visible and UV light is useful for initiating the photochemical process of chemical synthesis (and it is moreover dangerous for photo-labile proteins). On the other hand, the energy of IR waves is not useful for biological processes, other than keeping the organisms at a reasonably warm temperature. The reason is in the quantum mechanical energy-wavelength correspondence. The energy of a radiation corresponding to a temperature of 5,800 K is $k_B T = 8.6 \times 10^{-5} \times 5,800 \sim 0.5$ eV, a figure comparable with the difference between the discrete (stationary) energy levels of atoms and molecules, therefore capable of inducing chemical reactions. On the other hand, the energy of the IR waves at $T = 280$ K is about $k_B T = 8.6 \times 10^{-5} \times 280 \sim 0.03$ eV, rather comparable with the vibrational energy levels of the molecules, thus capable of producing just some waste heat.

Due to the substantial temperature difference, the Sun appears as a source of negative entropy, in that the radiation arriving on the Earth allows to accommodate the apparently negative difference in the entropy balance ("apparently" only if the biosphere is considered as a closed system, remember the discussion on p. 20 about the introduction of the concept of free energy). The foremost case is that of photosynthetic reactions, for which it can be $\Delta S_{ph} < 0$ provided this is accompanied by enough Sun-produced entropy, such that:

$$\Delta S = \Delta Q \left(\frac{1}{300} - \frac{1}{5,800} \right) > -\Delta S_{ph} \qquad (2.60)$$

After the works of E. Schrödinger and others (see "Further reading" at the end of the Chapter), it has become familiar the idea that in biology it would be common to observe the transfer from a high-entropy source (the Sun) to a low-entropy drain (the Earth and its biosphere), without any violation of the Second Principle [14].

2.7.1 A Synthesis of Photosynthesis

Photosynthesis is an extremely complex mechanism, including very many chemical reactions in chain, which can be conveniently subdivided into two phases: the *light* phase, during which the photochemical reactions powered by the sunlight take place; and the *dark* phase (also called *Calvin's cycle*), whose main characteristic is the fixation of carbon. A highly simplified, synthetic writing of this chain of reactions, looking only at the start and endpoints, would be:

$$6CO_2 + 6H_2O + nh\nu \rightarrow C_6H_{12}O_6 + 6O_2$$

carbon dioxyde + water + radiation → glucose + oxygen

Once the energy of the solar radiation, $E = nh\nu$, is captured by the plant leaves, a fraction η is utilised in the photosynthetic cycles and will appear as $\Delta G_{ph} > 0$. On the other hand, the fraction $(1 - \eta)$ is reemitted at room temperature, in the form of heating of the glucose and oxygen molecules. According to Eq. (2.60), the entropy balance must accommodate the negative quantity ΔS_{ph}, by a fraction η of the solar entropy at least equal or larger:

$$\eta(n \cdot 1.986 \cdot 5,800) \left(\frac{1}{300} - \frac{1}{5,800} \right) + (-\Delta S_{ph}) > 0 \qquad (2.61)$$

with $E = h\nu = RT$, and $R = N_{Av}k_B = 1.986$ cal $K^{-1}mol^{-1}$, or 8.31 J $K^{-1}mol^{-1}$, the universal gas constant. The factor n comes from the number of *photons* typically needed for the light phase (in quantum physics, the energy of the electromagnetic radiation is defined in terms of finite packets of energy, called photons, each carrying an energy $E = h\nu$, with $h = 6.62 \times 10^{-34}$ J-s the Planck constant). This number is found to be somewhere between $n = 5$ and 10. An estimate of ΔS_{ph} can be given by calculating the difference between the free entropies of the (products) – (reactants), all taken in the gaseous state but water, thus obtaining $\Delta S_{ph} = (209 + 6 \cdot 205) - (6 \cdot 213.6 + 6 \cdot 69.9) = -262$ J $K^{-1}mol^{-1}$. Therefore, for the photosynthesis to be compatible with the Second Principle, the yield of photochemical reactions must be:

$$\eta > 0.35 \qquad (2.62)$$

or even less if $n > 5$. In fact, in dedicated laboratory experiments it is found that the photosynthesis has a global yield of about $\eta = 0.50$. On the other hand, naturally occurring photosynthesis has a much lower efficiency because of several factors, such as the many different molecules participating in the ensemble of reactions (more complex than the simple picture above), or the losses from light-harvesting molecules, which may decay rapidly by electron-transfer reactions and other competing mechanisms, in addition to fluorescence and stimulated emission. However, the core of our discussion is not changed: the total system entropy is globally increased, and the inflow of entropy from the Sun is indeed compensating the apparent decrease in entropy during the chemical synthesis process.

It is equally important to underscore that the simple availability of energy (and entropy) is not enough to support the living systems, in the same way that it is not enough to sit on a tank of gasoline to make a car run. It is necessary to have a converter, which transforms the available energy in useful work (mechanical, chemical, electromagnetic), the equivalent of a combustion engine for a car, which turns the chemical energy of gasoline into rotating motion of the wheels of the car. For the case of biological systems, the appropriate use of the "negative" entropy flux is done through encoding and decoding the information in the genetic code, stored in the DNA of every living being.

Terrestrial organisms have learned, during the evolution, to extract with a good efficiency the energy and entropy from the solar blackbody radiation. From the point of view of the thermodynamics balance, life on Earth is organised in the form of a pyramid. At the bottom of the pyramid we find the species capable of synthesising the base organic compounds, such as the carbohydrates, by directly using the solar radiation via the photosynthesis. Without such organisms (plants), life would not exist as we know it. Notably, the energy stored in the biomass by the photosynthesis is a ridiculous 0.023 % of the total energy received on the Earth's surface. Nevertheless, this small amount is enough to sustain the growth and development of all the living organisms in the upper levels of the pyramid (including the energy consumption of all the human-made machines, thanks to the energy stored in fossil fuels throughout the ages).

Animals of the upper levels of the pyramid cannot directly use the solar energy and entropy, as they depend on the photosynthesising species. During the assimilation and digestion of food, energy and entropy stored by the plants in the base organic compounds are extracted and used by the herbivores, and from these they pass on to the other animal species (carnivores and omnivores). During the geological ages, natural selection has been operating in such a way to favour the species which are most effective in the process of thermodynamic extraction of the pristine stored energy and entropy. In the next Chap. 3 we will see how entropy can be stored in DNA, and used to make the building blocks of cells and tissues. Notably, sexual recombination of the genetic material has proven to be a very effective way of protecting the 'information' entropy stored in the DNA, against the unavoidable information degradation due to the steady increase of entropy of the surrounding environment.

The energy stored in a tree

Let us take a tree of 20 years of age, which has been growing up to a mass of 500 kg, of which 400 kg in wood and the rest in leaves and circulating water. We know that burning dry wood gives about 4.05 kWh/kg, or 14,580 kJ/kg. Can we estimate from such a simple data the energy stored in the Earth's biomass?

Starting from the value of the solar constant, $C_S = 1366 \, \mathrm{W \, m^{-2}}$, we can calculate that a surface of 1 m^2 of leaves receives:

$$E = 1{,}366 \cdot (20 \cdot 365 \cdot 12) = 119.7 \times 10^6 \ \mathrm{Wh} = 430.8 \times 10^6 \ \mathrm{kJ} \tag{2.63}$$

of energy over the 20 years of its life (we multiplied the number of hours in a day only by 12, to consider that on average half of the day is actually night).

Let us take that the photosynthesis yield is about 2 %, therefore the energy store would be 8.616×10^6 kJ per m^2 of leaves over the 20 years. A typical tree should have rather 200–300 m^2 of leaves (a gross estimate, not taking into account the tree shape, living latitude, etc.), which gives a total available energy of:

$$E = 300 \cdot 8.616 \times 10^6 \approx 2.585 \times 10^9 \ \mathrm{kJ} \tag{2.64}$$

This energy is necessary for al the vital functions of the tree including its growth, therefore our calculation will be an underestimate, when considering that all the energy is instead used only for growth. The energy accumulated in the wood is $E_{acc} = 14{,}580 \cdot 400 = 5.83 \times 10^6$ kJ. This represents a fraction:

$$f = E_{acc}/E = (5.83 \times 10^6)/2.585 \times 10^9 = 0.0022 \tag{2.65}$$

i.e., 0.22 % of the total energy stored. By accounting that about 30 % of the Earth's surface is covered by forests, we get about 0.066 %, i.e., only 2–3 times bigger than the accepted value of 0.023 %. This seems a rather decent estimate, given our very rough approximations.

All this energy stored in the trees that lived on the early Earth would be found later deep in the geological layers, under the form of coal. In fact, all hydrocarbons of coal are derived from the cellulose of the biomass, fossilised during the millions of years of life on Earth, mostly during the Carboniferous age (about 300 millions of years ago). Likely, the first stage of the conversion was initiated by bacteria, which digested the organic matter by producing methane, CO_2 and oxygen; this may have been followed by stages of anaerobic decomposition, in which acids should be the waste, thus increasing the pH up to levels at which all bacteria would die. Once this proto-carbon material got buried under hundreds of meters of soil, the temperature would rise to $\simeq 100 \,^\circ$C and the pressure to tens of atmospheres. In such conditions, carbonification could take place by turning into less and less volatile compounds, firstly forming lignin, then coal, and finally anthracite. The composition of such materials and the chemistry of their processes are still largely unknown (Fig. 2.4).

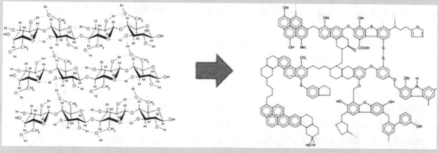

Fig. 2.4 *Left* From cellulose (*left*) to coal (*right*). Bonds between sugar molecules in cellulose long chains are protected against chemical attack, because cellulose tangles itself into tight microfibrils. Harsh conditions such as high temperatures and pressures are required to complete the carbonification

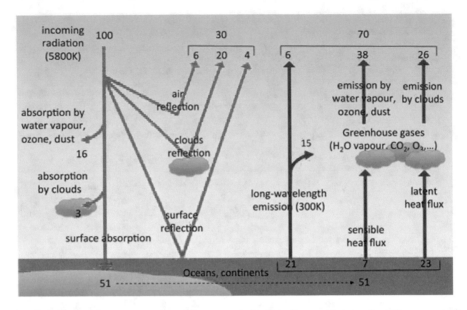

Fig. 2.5 Schematic representation of the energy balance in the greenhouse effect. An amount of solar radiation energy equal to 100 (*on the left*) is assumed to reach the Earth. In *yellow arrows*, the ways this incoming energy is reflected or stored in the atmosphere. The 51 % absorbed by the Earth's surface is re-irradiated at lower radiation temperature (indicated with *red arrows*, on the *right of the figure*). Some of this outgoing radiation is again captured and stored in the atmosphere. The sum of the *yellow* and *red* fractions stored is at the origin of the greenhouse effect

2.8 Energy from the Sun

As shown in the Fig. 2.5, the biosphere receives the solar energy in the form of a 'gas of photons' with high energy and low entropy, and emits residual thermal energy in the form of a similar gas of radiation, with lower energy and higher entropy. The transformation from short- to long-wavelength photons increases the entropy of the radiation (this could be calculated from the equations of the quantum radiation theory), and this entropy compensates the decrease of entropy in building organic molecules.

A theoretical calculation of the total energy arriving on the Earth is affected by a considerable uncertainty, since it requires a number of approximations. This total energy is termed **solar constant**. Such a quantity can be measured by a bolometer[5] mounted on a satellite, and integrating all the radiation (IR, visible, UV, X, gamma,...) which arrives on the Earth in the direction perpendicular to the external surface of the atmosphere. Although such a quantity may have quite large fluctuations as a function

---

[5]A bolometer is a detector of electromagnetic radiation. Its principle is simple: it converts the absorbed light radiation into heat. By choosing an absorbing material whose electrical resistance changes with the temperature, the incident energy can be estimated by measuring the impedance variation in the detector.

of the latitude, of the time of the day, and of the period of the year, an average value for the whole Earth surface has been assessed, $C_S = 1366 \, \text{W m}^{-2} = 1.366 \, \text{GW km}^{-2}$. By taking the average Earth's radius $R_T = 6366$ km, the total fluence (energy flux times the total surface, i.e., $[\text{E}][\text{T}^{-1}]$) of solar energy captured by the Earth is:

$$P_T = C_S \pi R_T^2 = 1.7 \times 10^8 GW \tag{2.66}$$

It will be noted that, on average, the flux on the surface of our planet is only 1/4 of this value, since the flux across the circle of surface πR_T^2 must be distributed over the sphere of surface $4\pi R_T^2$. Overall, the Earth captures only a fraction equal to 2×10^{-9} of the total energy emitted from the Sun, and this is already not bad!

Moreover, we note that the energy balance is zero on average: as much energy is received from the Sun, as it is sent back from the Earth in the form of low-frequency radiation.[6] Since the energy of emitted photons is lower, their density must be higher, however the integral of flux times energy is constant.

2.8.1 The "Greenhouse" Effect

By looking at the schematic diagram in Fig. 2.5, which represents the balance of the incoming and outgoing radiant energy, we see that the energy arriving on the Earth's surface from the Sun is reflected by about 30 % by the molecules making up the air, the water vapour of the clouds, and the white parts of the surface (snow, glaciers). This fraction represents the **albedo**, A, of the Earth's surface.

Of the remaining 70, 19 % is absorbed by the molecules of the atmosphere and clouds. Therefore, only 51 % of the incident radiation energy is delivered to the surface. This is the same quantity of energy which is reemitted, however with an energy spectrum (density of energy as a function of radiation wavelength) very different from that of the incident radiation.

Of the 51 % reemitted, 21 % is directly radiant energy. Of this, 6 % goes directly into the outer space, and 15 % is again recaptured by the molecules in the atmosphere. Another 7 % of the reemitted energy is used to heat the lower layers of the atmosphere, say the troposphere (500 m), by convection. The remaining 23 % (21 + 7 + 23 = 51 %) is used as latent heat in the phase transformations of the water cycle, namely evaporation and condensation of clouds.

By summing up the various contributions, we find that molecules in the atmosphere and clouds capture 64 % of the total energy (19 % direct + 45 %). This fraction is reemitted in the form of infrared (IR) radiation. The energy captured is irradiated in all directions, notably both towards the upper space and the lower Earth's surface.

---

[6]A very small quantity of energy is contributed by the internal heat of the Earth itself, due to the primordial heat, the decay of radioactive elements in the Earth's interior, and the heat of crystallisation of the core materials. This contribution is about 1/10,000 of the amount of external energy received by the Sun.

This descendent flux of energy adds to the incident solar energy, and allows the surface temperature to attain an average value of about $+15\,°C$, against the about $-18\,°C$ predicted in the absence of a partially absorbing atmosphere. Moreover, it should be considered that with such lower average temperature, the extent of ice caps on the surface would be increased, with a corresponding increase in the albedo. More reasonable calculations predict an average surface temperature in the absence of the atmosphere around $-100\,°C$.

This effect of heating of the Earth's surface, originating form the reflection of part of the energy by the atmosphere, is called **greenhouse effect**, since it makes the Earth surface to resemble to a covered greenhouse. This is obviously a very beneficial and desirable effect by all the living organisms. It is due to the "greenhouse gases" contained in the terrestrial atmosphere: mainly water vapour (the principal contributor to the heating and cooling of Earth's surface), carbon dioxide CO_2, and methane CH_4, plus a number of minor constituents.

The name of "greenhouse" comes from the analogy of the Earth's atmosphere with a real greenhouse. For such a kind of construction, once exposed to the sunlight, the inside air temperature is higher than the outside even in the absence of internal heat generation (which in a real greenhouse could be added). This is due to the different transparency of the glass of the walls to the radiation, that is quite good for the high-frequency light coming from the Sun, and pretty bad for the infrared radiation emitted by the Earth's surface at lower frequency. In practice, the glass behaves as a sort of energy valve, letting easily the energy to get in while being less good at letting energy out. In the analogy, the entire solid and liquid mass of the Earth's crust is the greenhouse, and the atmosphere is the equivalent of the glass. The possible problems in the equilibrium originate from the fact that the infrared transparency is even more decreased when the concentrations of greenhouse gases increase, thereby increasing to higher values the surface temperature.

Natural or artificial perturbations of the atmospheric concentrations of the greenhouse gases can alter the equilibrium of the radiation exchanges, with the effect of changing the amount of energy stored in the atmosphere. Such a disequilibrium could entail a long-term change in the atmospheric temperature, and therefore in the surface temperature. While glacial periods and warmer periods have naturally alternated on the Earth surface for millions of years, there has been in recent years a concern about the possible long-term effects of man-made alterations of the greenhouse gases concentrations, especially due to the atmospheric increase of CO_2 and CH_4 levels following the burning of large quantities of fossil fuels. This effect has been termed the **global heating** problem, since most of the indications point towards an increase of the average surface temperature, although there are also predictions based on computer models which would rather indicate a decreasing trend. This is a very complicate problem, involving the contribution of widely ranging scientific knowledge, from physics, mathematics, geography, oceanography, geology, and so on. In the following we will develop a simple model to describe some of the basic effects linked to the variation of greenhouse gases concentration.

Before going into the details of the greenhouse effect, however, a more basic question should be posed, on the basis of the thermodynamics. It was amply demonstrated in the beginning of this chapter that two parts of a system at different temperatures put in contact will evolve in the direction of attaining a common equilibrium temperature, pushed by the maximisation of the total entropy. Why the system composed by the Sun and Earth is not at the same temperature? How it is possible that the two systems, which can exchange energy without limits, should have not yet attained the equilibrium after about 10^{10} years of the age of the universe? The answer is "no way". The radiation continuously emitted from the Sun is an expanding gas, which changes its energy density as it expands, somewhat like a perfect gas that cools down while continuously expanding. As a consequence, the system is *never* at thermodynamic equilibrium. However, it must be noted that the 'cooling' of the radiation cannot be explained by the concepts of classical physics, but can be understood only by making recourse to quantum mechanics.[7]

2.8.2 The Temperature of the Earth's Surface

We consider the Sun as an ideal black body at a temperature $T_S = 5,800$ K. This is a good approximation, despite the fact that the Sun is not black at all! (That is a joke, in fact by the term *black body* in physics it is meant an object ideally capable of absorbing all the radiation it receives, therefore appearing black at low temperatures. On the other hand, a black body also radiates energy at the same temperature, since it is in equilibrium.)

For an ideal black body with spherical shape, the Stefan-Boltzmann law[8] gives the following expression for the power emitted, as a function of the temperature T and the radius R of the emitting sphere:

$$P = 4\pi R^2 \sigma T^4 \tag{2.67}$$

This equation states that the power emitted (energy/unit time) is the product between the emitting surface (a sphere, in this case) and the fourth power of the temperature, times the Stefan-Boltzmann constant, $\sigma = 5.67 \times 10^{-8}$ W m^{-2} K^{-4}.

Now let us try to estimate the Earth's surface temperature by considering also the Earth as a black body, but at a lower temperature than the Sun. Consider that the Sun

---

[7]It is interesting to note that, even after a very complex mathematical treatment based on quantum mechanics, the result for the difference in entropy in cooling from a temperature T_1 to a temperature $T_2 < T_1$ is very close to the classic result for the entropy variation of a perfect gas [15]. The important difference between the two treatments is the concept of temperature, which is completely different for a classical gas and a quantum "gas of radiation".

[8]The Stefan-Boltzmann law is physically justified only in the quantum mechanical treatment of the radiation; however, Boltzmann derived it by a fully classical argument, see Problem 2.6.

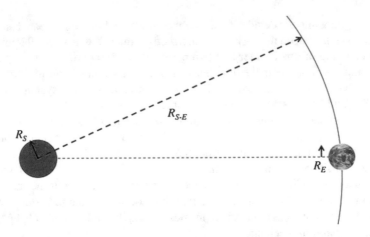

Fig. 2.6 The Sun emits all of its radiant power across its surface, which measures $(4\pi R_S^2)$ m^2. This same energy, traveling across the empty space, is distributed over the concentric sphere with radius R_{S-E}. The Earth intercepts a fraction of this energy, equal to the ratio of the projected circle with the same radius of the Earth, R_E, to the surface irradiated at the Sun-Earth distance. Then, this energy is distributed over the entire Earth's surface, again a sphere with radius R_E

distributes all of the power P_S irradiated across its surface, with radius $R_S = 6.9 \times 10^9$ m, in all the radial directions:

$$P_S = 4\pi R_S^2 \sigma T_S^4 \tag{2.68}$$

The Earth is located at the Sun-Earth distance, $R_{S-E} = 1.5 \times 10^{11}$ m. When the radiation reaches such a distance, it is distributed over a sphere with this same radius, concentric with the Sun (see the schematic in Fig. 2.6).

Moreover, consider that the Earth intercepts only a fraction of the Sun power, on a circle of radius $R_E = 6.366 \times 10^6$ m projected on this large sphere, and multiplied by the fraction $(1 - A)$ since the effect of the albedo is to reduce the amount of radiation P_E absorbed by the Earth:

$$P_E^{abs} = P_S \left(\frac{\pi R_E^2}{4\pi R_{S-E}^2} \right)(1 - A)\sigma T^4 \tag{2.69}$$

Finally, consider that Earth reemits the power absorbed, over all of its surface:

$$P_E^{emit} = 4\pi R_E^2 \sigma T_E^4 \tag{2.70}$$

We impose that at equilibrium, $P_E^{abs} = P_E^{emit}$. Therefore, the Earth's temperature is easily found:

$$T_E = \left(\frac{1}{4}\right)^{1/4} \left(\frac{R_S}{R_{S-E}}\right)^{1/2} (1 - A)^{1/4} T_S = 255K \tag{2.71}$$

This shows that, to a first approximation, the observed surface temperature of the Earth is only determined by the distance and structure (temperature, size) of the Sun. As anticipated, without any account for the presence of the atmosphere we end up with an estimate of about 20° below the zero Celsius. By our standards, this would be a very inhospitable planet!

To improve the predictions of this simple model, which gives a quite cold surface temperature, we must add the effect of the atmospheric layers. For the sake of simplicity, let us take just one layer of atmosphere, at the temperature T_A, at approximately the same distance from the Sun as the Earth, R_{S-E}. This layer captures a fraction $f < 1$ of the power P_E^{emit} emitted by the Earth surface. This fraction of power sequestered by the atmosphere will be reemitted toward the space, one half back in the direction of the Earth (thus contributing to the surface heating), and the other half toward the upper sky. The balance of the power between the Earth and the atmosphere is then:

$$f(\sigma T_E^4) = 2f(\sigma T_A^4) \tag{2.72}$$

On the other hand, the balance equation between absorbed and emitted power from the Earth, $P_E^{abs} = P_E^{emit}$, is modified as:

$$P_E^{abs} = P_S \left(\frac{\pi R_E^2}{4\pi R_{S-E}^2} \right)(1 - A) = (1 - f)\left(4\pi R_E^2\right)\sigma T_E^4 + f\left(4\pi R_E^2\right)\sigma T_A^4 \tag{2.73}$$

which can be simplified, by using the Eq. (2.72), as:

$$\left(\frac{R_S}{R_{S-E}} \right)^2 \frac{1 - A}{4} T_S^4 = (1 - f)T_E^4 + fT_A^4 = \left(1 - \frac{f}{2} \right) T_E^4 \tag{2.74}$$

The corrected expression for the Earth's surface temperature is:

$$T_E = \left\{ \left(\frac{R_S}{R_{S-E}} \right)^2 \frac{(1 - A)}{4(1 - \frac{f}{2})} \right\}^{1/4} T_S \tag{2.75}$$

This equation contains f as an unknown parameter. We can use it in reverse, to obtain the average temperature at the Earth surface, $T_E = 288$ K. For this, we must set $f = 0.78$ for the fraction of power absorbed by the atmosphere. As a quality check, by imposing this condition we can derive from Eq. (2.72) the average temperature of the atmosphere layer as $T_A = 242$ K, which is actually a very good estimate for the temperature of the troposphere at the height of about 7 km above the surface.

Besides, the fitted value of $f = 0.78$ does not coincide with the fraction of solar power sequestered and reemitted by the greenhouse gases, which is rather 0.38 (see Fig. 2.5 above). However, apart from the more or lessaccurate numerical values, it is interesting to note that already with such a very simple model, we can obtain a correct qualitative correlation in the temperature response: by increasing the gas concentration of the atmospheric layer, for example in methane or CO_2, we would increase the absorption fraction f, which would entail a parallel increase of the temperature at the Earth's surface from Eq. (2.75). On the other hand, such simple correlations should not be pushed too far. The modelling of Earth's climate is a dauntingly more complex task, comprising a wealth of physical-chemical, atmospheric, oceanographic, and geophysical phenomena, which we have not even hinted at here, and commands the use of the largest computers in the world.

In particular, our very simplified model takes the Earth as a sphere with a homogeneous surface and homogeneous atmosphere layers, which is far from truth. The large-scale differences in the surface distributions of land and biomass, water, ice, clouds, are the very motors of Earth's climate, and cannot be neglected. The water vapour, which represents the major barrier to cooling by radiation emission by the T^4 law Eq. (2.67), generally is maximum at the surface near the tropics, and sharply decreases with both altitude and latitude. Because of this layer mostly opaque to infrared radiation, heat is firstly carried away from the surface by fluid convection, starting from the cloud towers of the tropics, which then carry most of the heat upward and to the poles, whence it is possible for thermal radiation emitted from these levels to escape into space.

The qualitative variations of the Earth surface also have great implications for the definition of the average surface temperature, a concept that has been recently popularised also by the media. Climate cannot be associated to a single temperature, rather the differences of temperatures drive the climate processes and create the storms, winds, sea currents, and everything that makes up the climate. The Earth surface has a large number of interacting components, which one cannot just add up and average: it would be as meaningless as calculating the "average phone number" in the phone book. If temperature decreases at one point and increases at another, the "average temperature" would be unchanged, but the thermodynamic forces would be totally different, and so would be the climate. If, for example, we measure $20\,°C$ at one point on the surface and $30\,°C$ at another point $40\,km$ away, we would be tempted to attribute an average temperature of $25\,°C$ to that area; but if we measured $25\,°C$ at both places, the average would be still $25\,°C$. However, these two situations would give rise to two entirely different climate reactions, because in the former case one would have an air pressure difference and strong winds, while in the latter case there would be a calm and pleasant day.

Appendix A: Some Useful Mathematical Tools

The Gamma Function

The mathematical Gamma function $\Gamma(x)$ is an extension of the factorial function, valid for both real and complex numbers. Its analytical definition, due to the French mathematician Adrien-Marie Legendre (1752–1833), is:

$$\Gamma(t) = \int_0^\infty x^{t-1}e^{-x}dx \tag{2.76}$$

with t a real or complex number. This (only apparently) difficult integral can be integrated by parts:

$$\int_0^\infty x^{t-1}e^{-x}dx = \left[\frac{x^t}{t}e^{-x}\right]_0^\infty + \frac{1}{t}\int_0^\infty x^t e^{-x}dx \tag{2.77}$$

(the term within [...] being equal to 0), to obtain the recurrence formula of the Gamma function:

$$\Gamma(t+1) = t\Gamma(t) \tag{2.78}$$

From this result, it is immediately obtained $\Gamma(n) = (n-1)!$ when t is an integer n, which justifies the definition of *generalised factorial*. From the same integral, it is also easily seen that $\Gamma(1) = 1$.

In thermodynamics, it is often necessary to calculate the Gamma function for half-integer argument, $\Gamma(n/2)$. The recurrence formula (2.78) above can still be used, but ending up with the task of calculating the last term, $n = 1/2$. In the next Section, this will be shown to be:

$$\Gamma(1/2) = \sqrt{\pi} \tag{2.79}$$

Dirac's Delta Function

The Dirac delta function, indicated as $\delta(x)$, is a real function that is zero everywhere except at $x = 0$, and with an integral equal to 1:

$$\int_{-\infty}^{+\infty} \delta(x)dx = 1 \tag{2.80}$$

It was introduced by the English physicist Paul A. Dirac, and it may be physically interpreted as the density of an idealised point mass or point charge situated at the origin. Although it makes little sense mathematically, the delta "function" becomes meaningful only when inside an integral, a the limit of a distribution becoming infinitely narrow about $x = 0$ while preserving its unitary integral.

The delta function admits any n-th order derivatives $\delta^{(n)}(x)$, a Fourier transform, and several other analytical manipulations typical of a true function. Some interesting properties of the delta are:

$$\int_{-\infty}^{+\infty} \delta(ax)dx = \int_{-\infty}^{+\infty} \delta(u)\frac{du}{|a|} = \frac{1}{|a|} \tag{2.81}$$

$$\int_{-\infty}^{+\infty} \delta'(x)f(x)dx = -\int_{-\infty}^{+\infty} \delta(x)f'(x)dx \tag{2.82}$$

$$\int_{-\infty}^{+\infty} f(x)\delta(x - x_0)dx = f(x_0) \tag{2.83}$$

$$\int_{-\infty}^{+\infty} f(x)\delta^{(n)}(x - x_0)dx = (-1)^n f^{(n)}(x_0) \tag{2.84}$$

The last two properties present the delta function as a kind of filter, by which a particular value of another function $f(x)$ can be extracted.

The Dirac delta may be thought as a continuous-x analog of the discrete Kronecker delta, $\delta_{ij} = 1$ for $i = j$ and 0 if $i \neq j$, which selects a discrete value out of a series $\{a_i\}, i = 1, ..., n$:

$$\sum_{i=-\infty}^{\infty} a_i \delta_{ik} = a_k \tag{2.85}$$

Among the useful applications of the Dirac function, for a discrete distribution consisting of a set of points $x = \{x_1, ..., x_n\}$, with corresponding probabilities $\{p_1, ..., p_n\}$, a continuous probability density function $f(x)$ can be written as:

$$f(x) = \sum_{i=1}^{n} p_i \delta(x - x_i) \tag{2.86}$$

Gauss and Euler Integrals

Throughout this book, we will encounter several times integrals of the type:

$$I(n) = \int_{u}^{v} x^n e^{-\alpha x^2} dx \tag{2.87}$$

with $[u, v] = [-\infty, +\infty]$ or $[0, +\infty]$.

The simplest of these integrals, $I(0)$, can be solved on the infinite real axis by an ingenious trick, due to Siméon-Denis Poisson (1781–1840), if we start from its square:

$$I^2(0) = \left[\int_{-\infty}^{+\infty} e^{-\alpha x^2} dx\right]^2 = \int_{-\infty}^{+\infty} \int_{-\infty}^{+\infty} e^{-\alpha(x^2+y^2)} dx dy =$$

$$= \int_0^{2\pi} d\theta \int_0^{+\infty} r e^{-\alpha r^2} dr \qquad (2.88)$$

The last identity follows from the change of Cartesian to polar coordinates, $x = r\cos\theta$, $y = r\sin\theta$, $dxdy = rdrd\theta$. The integral is now easily calculated:

$$I^2(0) = \int_0^{2\pi} d\theta \int_0^{+\infty} r e^{-\alpha r^2} dr = \int_0^{2\pi} d\theta \left[-\frac{1}{2\alpha} e^{-\alpha r^2}\right]_0^{+\infty} = 2\pi \cdot \frac{1}{2\alpha} = \frac{\pi}{\alpha}$$

$$(2.89)$$

from which we get the basic result $I(0) = \sqrt{\pi}$, for $\alpha = 1$.
If we take the integral on the interval $[0, +\infty]$ and make the substitution $x = \sqrt{t}$, we get the first Euler integral, $E(0)$:

$$\sqrt{\pi} = 2 \int_0^{+\infty} e^{-x^2} dx = 2 \cdot \frac{1}{2} \int_0^{+\infty} t^{-\frac{1}{2}} e^{-t} dt = \Gamma(1/2) \qquad (2.90)$$

which proves the Eq. (2.79) of the previous Section.
The integrals (2.87) for odd n are equal to 0 on the interval $[-\infty, +\infty]$, since being the product between an even and an odd function.
The integrals for even n are obtained by differentiation with respect to the parameter α. By taking the first derivative of both sides of Eq. (2.87) we get:

$$\int_0^{+\infty} x^2 e^{-\alpha x^2} dx = \frac{\pi^{1/2}}{2\alpha^{3/2}} \qquad (2.91)$$

By sequentially taking higher order derivatives, the following general result is obtained:

$$\int_0^{+\infty} x^n e^{-\alpha x^2} dx = \frac{1 \cdot 3 \cdot 5 \cdots (n+1)\pi^{1/2}}{2^{n/2}\alpha^{(n+1)/2}} = \frac{\Gamma[(n+1)/2]}{2\alpha^{(n+1)/2}} \qquad (2.92)$$

The Stirling Approximation

When deriving Eq. (2.11) above, and several other times, in Chap. 3 and in other places in this book, we need to compute the factorial $n!$ of very large values of n, as well as its logarithm. Especially in the latter case, a very useful formula is the Stirling's approximation:

$$\ln n! = n \ln n - n + \frac{1}{2} \ln(2\pi n) + \mathcal{O}(1/n) \qquad (2.93)$$

This formula is very accurate already for small n. For example, the relative error for $n = 20$ is less than 10^{-4}.

Stirling's formula can be applied also to the Gamma function, provided its argument t is real:

$$\ln \Gamma(t) \simeq (t - \tfrac{1}{2}) \ln t - t + \tfrac{1}{2} \ln 2\pi \tag{2.94}$$

We also note the important property, with α a real constant:

$$\ln(\alpha n)! = \ln(\alpha^n n!) = n \ln \alpha + \ln n! \tag{2.95}$$

Vector Calculus and Analysis

A scalar is a quantity a characterised only by its magnitude (a number). Specification of a vector $\mathbf{v}$, instead, requires stating its direction as well as its magnitude $|\mathbf{v}| = v$. Unit vectors are vectors of unit length, while the zero vector has zero length and arbitrary direction. The unit vectors of the Cartesian coordinate system are written as $\mathbf{i}, \mathbf{j}$ and $\mathbf{k}$, respectively along the axes $\{x, y, z\}$. Any other vector in the 3D space can then be expressed by giving its scalar components, $\{v_x, v_y, v_z\}$, as $\mathbf{v} = v_x\mathbf{i} + v_y\mathbf{j} + v_z\mathbf{k}$. The magnitude (or modulus) of the vector is: $v = \sqrt{v_x^2 + v_y^2 + v_z^2}$.

Vectors $\mathbf{u}$ and $\mathbf{v}$ can be added, as $\mathbf{w} = \mathbf{u} + \mathbf{v}$, by adding their components: $\mathbf{w} = (u_x + v_x)\mathbf{i} + (u_y + v_y)\mathbf{j} + (u_z + v_z)\mathbf{k}$. A vector can also be multiplied by a scalar s, by multiplying its components: $s\mathbf{v} = sv_x\mathbf{i} + sv_y\mathbf{j} + sv_z\mathbf{k}$.

The *scalar product* (or dot product) of two vectors is defined as:

$$\mathbf{u} \cdot \mathbf{v} = u_x v_x + u_y v_y + u_z v_z = uv \cos \theta \tag{2.96}$$

with u, v the modulus of the vectors, and θ the angle comprised between the directions of two vectors, joined at a common origin. The result of the scalar product of two vectors is a scalar (a number).

The *vector product* (or cross product) of two vectors is defined as:

$$\mathbf{u} \times \mathbf{v} = (u_y v_z - u_z v_y)\mathbf{i} + (u_z v_x - u_x v_z)\mathbf{j} + (u_x v_y - u_y v_x)\mathbf{k} = \mathbf{w} \tag{2.97}$$

and its result is another vector $\mathbf{w}$, perpendicular to the plane containing $\mathbf{u}$ and $\mathbf{v}$.

Consider a scalar function of the coordinates, $\psi(x, y, z)$, such as a temperature or a pressure distributed in the volume of a body. The **gradient** of ψ is a vector defined as:

$$grad \ \psi = \nabla \psi = \frac{\partial \psi}{\partial x}\mathbf{i} + \frac{\partial \psi}{\partial y}\mathbf{j} + \frac{\partial \psi}{\partial z}\mathbf{k} \tag{2.98}$$

It should be evident that the gradient of the scalar function (or *field*) is pointing to the direction where ψ changes more rapidly.

The operation producing the gradient can be thought of coming from the application of an operator ∇ on the scalar ψ:

$$\nabla = \mathbf{i}\frac{\partial}{\partial x} + \mathbf{j}\frac{\partial}{\partial y} + \mathbf{k}\frac{\partial}{\partial z} \tag{2.99}$$

If we now apply the same operator ∇ but to a vector $\mathbf{v}$, we obtain the **divergence** of that vector:

$$div \ \mathbf{v} = \nabla \cdot \mathbf{v} = \frac{\partial \mathbf{v}}{\partial x} \cdot \mathbf{i} + \frac{\partial \mathbf{v}}{\partial y} \cdot \mathbf{j} + \frac{\partial \mathbf{v}}{\partial z} \cdot \mathbf{k} \tag{2.100}$$

The physical meaning of the divergence is to calculate the net amount of the vector $\mathbf{v}(x, y, z)$ (actually a vector field) flowing in or out a closed surface. Imagine a vector field running parallel to the x-axis, such as water flowing at constant speed: the divergence calculates the integral of the flux across a surface perpendicular to x.

Finally, the vector product of ∇ times a vector $\mathbf{v}$, gives the **rotor**, or "curl" of the vector:

$$curl \ \mathbf{v} = \nabla \times \mathbf{v} = (\frac{\partial v_y}{\partial z} - \frac{\partial v_z}{\partial y})\mathbf{i} + (\frac{\partial v_z}{\partial x} - \frac{\partial v_x}{\partial z})\mathbf{j} + (\frac{\partial v_x}{\partial y} - \frac{\partial v_y}{\partial x})\mathbf{k} \tag{2.101}$$

The rotor of a vector field is non-zero only if the field turns around some point, like in a vortex. For a vector field forming a vortex spinning circularly around a vertical line, the rotor calculates the value of $\mathbf{v}(x, y, z)$ along the perimeter of any circle drawn about the central line.

The square of the operator ∇^2 is called the **Laplacian**. Applied to a scalar, it gives another scalar:

$$\nabla^2 \psi = \frac{\partial^2 \psi}{\partial^2 x} + \frac{\partial^2 \psi}{\partial^2 y} + \frac{\partial^2 \psi}{\partial^2 z} \tag{2.102}$$

The Laplacian of a vector is a vector:

$$\nabla^2 \mathbf{v} = (\frac{\partial^2}{\partial^2 x} + \frac{\partial^2}{\partial^2 y} + \frac{\partial^2}{\partial^2 z})(v_x\mathbf{i} + v_y\mathbf{j} + v_z\mathbf{k}) =$$
$$= (\frac{\partial^2 v_x}{\partial^2 x} + \frac{\partial^2 v_x}{\partial^2 y} + \frac{\partial^2 v_x}{\partial^2 z})\mathbf{i} + (\frac{\partial^2 v_y}{\partial^2 y} + \frac{\partial^2 v_y}{\partial^2 y} + \frac{\partial^2 v_y}{\partial^2 z})\mathbf{j} + (\frac{\partial^2 v_z}{\partial^2 x} + \frac{\partial^2 v_z}{\partial^2 y} + \frac{\partial^2 v_z}{\partial^2 z})\mathbf{k} \tag{2.103}$$

Some useful formulae of vector analysis are:

$$\nabla \cdot (\mathbf{u} \times \mathbf{v}) = \mathbf{v} \cdot \nabla \times \mathbf{u} - \mathbf{u} \cdot \nabla \times \mathbf{v} \tag{2.104}$$
$$\nabla(\mathbf{u} \cdot \mathbf{v}) = \mathbf{u} \cdot \nabla\mathbf{v} + \mathbf{v} \cdot \nabla\mathbf{u} + \mathbf{u} \times (\nabla \times \mathbf{v}) + \mathbf{v} \times (\nabla \times \mathbf{u}) \tag{2.105}$$
$$\nabla \times (\nabla \psi) = 0 = curl \ (grad \ \psi) \tag{2.106}$$

$$\nabla \cdot (\nabla \times \mathbf{u}) = 0 = div \ (curl \ \mathbf{u}) \tag{2.107}$$

$$\nabla \cdot (\nabla \psi_1 \times \nabla \psi_2) = 0 \tag{2.108}$$

$$\nabla \times (\nabla \times \mathbf{u}) = curl \ (curl \ \mathbf{u}) = grad \ (div \ \mathbf{u}) - \nabla^2 \mathbf{u} \tag{2.109}$$

Note that a vector field with zero divergence is said to be *solenoidal* (a field with no point source, such as a magnetic field). A vector field with zero curl is said to be *irrotational* (such as a tube of water flowing in laminar flux without any turbulence). A scalar field with zero gradient is said to be *constant* (such as a temperature uniform everywhere in a body).

Simple Tensor Algebra

If we multiply a vector by a scalar, $\mathbf{u}' = a\mathbf{u}$, the vector changes in magnitude (each of the components are multiplied by a) but not in direction. On the other hand, as shown in the previous Section, by multiplying two vectors we get either a scalar (dot product) or another vector perpendicular to the first two (cross product). But how do we get to change both the direction and magnitude of a vector into an arbitrary direction and different magnitude? This is obtained by introducing a more complex entity, the **tensor**, indicated by an underline:

$$\mathbf{u} = \underline{\mathbf{M}} \otimes \mathbf{v} \tag{2.110}$$

By writing the vector components explicitly, with $\mathbf{u} = u_x \mathbf{i} + u_y \mathbf{j} + u_z \mathbf{k}$ and $\mathbf{v} = v_x \mathbf{i} + v_y \mathbf{j} + v_z \mathbf{k}$, we see that the new mathematical object $\underline{\mathbf{M}}$ can be obtained by multiplying two original vectors component by component:

$$\underline{\mathbf{M}} = (u_x \mathbf{i} + u_y \mathbf{j} + u_z \mathbf{k}) \otimes (v_x \mathbf{i} + v_y \mathbf{j} + v_z \mathbf{k}) =$$
$$= u_x v_x \mathbf{ii} + u_x v_y \mathbf{ij} + u_x v_z \mathbf{ik} + u_y v_x \mathbf{ji} + ... \tag{2.111}$$

The new symbol $\otimes$ represents this idea of distributing the product among all the vector components. By rewriting the 3×3 scalar quantities as M_{ij}, with $i, j = 1, 2, 3$:

$$\underline{\mathbf{M}} = M_{11}\mathbf{ii} + M_{12}\mathbf{ij} + M_{13}\mathbf{ik} + M_{21}\mathbf{ji} + ... \tag{2.112}$$

this new mathematical entity appears as a square matrix, also called a tensor of rank 2, which can be multiplied by a vector or by another tensor, by using the usual rules of matrix algebra:

$$\underline{\mathbf{M}} = \begin{pmatrix} M_{11} & M_{12} & M_{13} \\ M_{21} & M_{22} & M_{23} \\ M_{31} & M_{32} & M_{33} \end{pmatrix} \tag{2.113}$$

$$\mathbf{u} = \underline{\mathbf{M}} \otimes \mathbf{v} = \begin{pmatrix} M_{11} & M_{12} & M_{13} \\ M_{21} & M_{22} & M_{23} \\ M_{31} & M_{32} & M_{33} \end{pmatrix} \begin{pmatrix} v_x \\ v_y \\ v_z \end{pmatrix} = \begin{pmatrix} u_x \\ u_y \\ u_z \end{pmatrix} \tag{2.114}$$

$$\underline{\mathbf{M}} \otimes \underline{\mathbf{N}} = \begin{pmatrix} M_{11} & M_{12} & M_{13} \\ M_{21} & M_{22} & M_{23} \\ M_{31} & M_{32} & M_{33} \end{pmatrix} \begin{pmatrix} N_{11} & N_{12} & N_{13} \\ N_{21} & N_{22} & N_{23} \\ N_{31} & N_{32} & N_{33} \end{pmatrix} = \sum_{k=1}^{3} M_{ik} N_{kj} \tag{2.115}$$

In the following (see e.g. Chaps. 8–10) we will meet with tensors of rank 2, such as the stress and the strain of a deformed material element, and tensors of rank 4, such as the matrices of the elastic constants and elastic compliances of a material. However, in this book we will be exclusively concerned with quantities defined in orthogonal reference frames (i.e., the angles between the x, y, z axes are always at $90°$), which simplifies much the tensor analysis, getting rid of (usually important) details such as covariant and contravariant components.

By extension, it might be tempting to deduce that a vector is nothing else but a tensor of rank 1, a mathematical entity with just one subscript index. However, care must be exercised. Although a seemingly mathematical trickery, tensor calculus is instead at the core of the concepts of invariance in physics. It turns out that all rank-1 tensors are also vectors, however *not* all vectors are rank-1 tensors: to be a tensor of any rank, a mathematical entity must be *invariant* with respect to the coordinate system. For example, if we consider the position vector $\overrightarrow{OP}$ joining a point P with the origin O of a $\{x, y, z\}$ reference frame, and the position vector $\overrightarrow{O'P}$ joining the same point with the origin O' of a different reference frame $\{x', y', z'\}$, it is immediately seen that this vector depends on the reference frame, therefore it is not a rank-1 tensor. However, the *difference* between two position vectors $\overrightarrow{PQ} = \overrightarrow{OP} - \overrightarrow{OQ}$ does not depend on the reference frame, therefore the vector of the distance between two points is a rank-1 tensor.

Similarly, a tensor can always be written as a matrix, but a matrix is not necessarily a tensor. By multiplying a column vector (u_1, u_2, u_3) by a matrix $\mathbf{M}$, a new column of coefficients (q_1, q_2, q_3) is obtained: if these numbers are the components of another vector, then the matrix is a tensor, $\mathbf{M} = \underline{\mathbf{M}}$. That the resulting (q_1, q_2, q_3) is a vector can be easily checked: just change the basis (reference frame) to trasform the first vector (u_1, u_2, u_3) into another vector (u'_1, u'_2, u'_3); then apply the same change of basis to (q'_1, q'_2, q'_3); if $\mathbf{q'} = \mathbf{Mu'}$, then $\mathbf{q}$ is a vector, and $\mathbf{M}$ is a tensor.

As it was already said, the tensor notation allows to expose underlying symmetries and invariances of the corresponding physical quantities. The *tensor invariants* are quantities derived from the tensor that do not changeupon changing or rotating the

reference frame. They are defined as the coefficients of the characteristic polynomial
of the tensor $\underline{\mathbf{M}}$:

$$p(\lambda) = \text{Det}[\underline{\mathbf{M}} - \lambda\underline{\mathbf{I}}] \tag{2.116}$$

where $\underline{\mathbf{I}}$ is the identity tensor (with 1 on the diagonal and 0 everywhere) and λ is an
indeterminate quantity. For a 3×3 rank-2 tensor, the most commonly encountered
in this book, there are only three invariants:

$$M_I = \text{Tr}\{\underline{\mathbf{M}}\} = M_{11} + M_{22} + M_{33} \tag{2.117}$$

$$M_{II} = \frac{1}{2}\left[(\text{Tr}\{\underline{\mathbf{M}}\})^2 - \text{Tr}\{\underline{\mathbf{M}}^2\}\right] =$$

$$= M_{11}M_{22} + M_{22}M_{33} + M_{33}M_{11} - M_{12}M_{21} - M_{23}M_{32} - M_{13}M_{31} \tag{2.118}$$

$$M_{III} = \text{Det}\{\underline{\mathbf{M}}\} \tag{2.119}$$

The first one is called also the *trace* of the tensor (sum of the diagonal components
of the corresponding matrix); the third one coincides with the determinant of the
matrix; the second one has no obvious interpretation. Note that, since a tensor is
written as a matrix, its eigenvalues can also be calculated; they would be M_1, M_2, M_3
for the 3×3 tensors. However, the number of invariants for a $n \times n$ tensor is just n,
therefore the three invariants previously defined must be expressed in terms of the
three eigenvectors, and vice versa:

$$M_I = M_1 + M_2 + M_3 \tag{2.120}$$

$$M_{II} = M_1 M_2 + M_2 M_3 + M_1 M_3 \tag{2.121}$$

$$M_{III} = M_1 M_2 M_3 \tag{2.122}$$

The importance of invariants becomes evident when considering *objective functions* (i.e., not depending on the change of coordinates) of the tensor. Such objective
functions depend only on the n invariants of the tensor, instead of its components.
For example, when calculating the elastic potential energy as function of the strain
tensor (see Appendix H to Chap. 9), this reduces to a function of three parameters
rather than six (the strain tensor $\underline{\varepsilon}$ has $3 \times 3 = 9$ components, however by symmetry
$\varepsilon_{ij} = \varepsilon_{ji}$, therefore its independent components are just six). Moreover, within the
framework of linear elasticity, the energy must be a quadratic function of the strain,
which eliminates an additional scalar. This is the reason why, for an isotropic material, only two independent parameters are needed to describe the elastic properties,
known as *Lamé coefficients*.

Simple Fourier Analysis

The Fourier transform decomposes a function of a variable x into the normal components of a conjugate variable y. Physically useful pairs of conjugate variables are, e.g., time and frequency, or position and momentum. For example, a time signal can be decomposed into the frequencies that make it up; or, a movement in space can be decomposed into the wavevectors that correspond to elementary oscillation modes.

The basic rule to obtain the **Fourier transform** of a function $f(x)$ in the y-space is:

$$g(y) = \frac{1}{(2\pi)^{1/2}} \int_{-\infty}^{+\infty} f(x)e^{-ixy}dx \tag{2.123}$$

The function $g(y)$ can be anti-transformed, to obtain back the $f(y)$:

$$f(x) = \frac{1}{(2\pi)^{1/2}} \int_{-\infty}^{+\infty} g(y)e^{+iyx}dy \tag{2.124}$$

The Fourier transform is closely connected with the **Fourier series**, both named after the French mathematician J. J. Fourier (1768-1830). Any function $f(x)$ can be expressed by a series of Fourier components, as:

$$f(x) = \frac{a_0}{(2\pi)^{1/2}} + \sum_{n=1}^{\infty} a_n \cos(nx) + \sum_{n=1}^{\infty} b_n \sin(nx) \tag{2.125}$$

with the coefficients:

$$a_n = \frac{1}{L} \int_{-L}^{+L} f(x) \cos\left(\frac{n\pi x}{L}\right) dx \quad ; \quad b_n = \frac{1}{L} \int_{-L}^{+L} f(x) \sin\left(\frac{n\pi x}{L}\right) dx \tag{2.126}$$

With a little algebra, it can be shown that upon substituting these coefficients in the series development for $f(x)$, one obtains exactly the definition (2.123) of the Fourier transform.

Some useful FT of elementary functions and operators (pairs (x, y) = generic; (t, ω) = time-frequency, or pulsation; $(\mathbf{x}, \mathbf{k})$ = position – momentum, or wavevector):

$$f(\omega) = \frac{1}{(2\pi)^{1/2}} \int_0^{+\infty} e^{-at}e^{-i\omega t}dt = \frac{2a}{(a^2 + 4\pi^2\omega^2)} \quad (a > 0) \tag{2.127}$$

$$f(\omega) = \frac{1}{(2\pi)^{1/2}} \int_0^{+\infty} e^{-at^2}e^{-i\omega t}dt = e^{-a\omega^2} \tag{2.128}$$

$$f(\omega) = \frac{1}{(2\pi)^{1/2}} \int_0^{+\infty} e^{-at}u(t)e^{-i\omega t}dt = \frac{1}{(a + i\omega)} \quad (a > 0) \tag{2.129}$$

$$f(\omega) = \frac{1}{(2\pi)^{1/2}} \int_0^{+\infty} e^{i\omega_0 t}e^{-i\omega t}dt = 2\pi\delta(\omega - \omega_0) \tag{2.130}$$

$$f(y) = \frac{1}{(2\pi)^{1/2}} \int_0^{+\infty} \sin^2(x) e^{-ixy} dx = \tfrac{1}{4}[2\delta(y) - \delta(y - \tfrac{1}{\pi}) - \delta(y + \tfrac{1}{\pi})]$$

(2.131)

$$f(y) = \frac{1}{(2\pi)^{1/2}} \int_0^{+\infty} \cos^2(x) e^{-ixy} dx = \tfrac{1}{4}[2\delta(y) + \delta(y - \tfrac{1}{\pi}) + \delta(y + \tfrac{1}{\pi})]$$

(2.132)

$$f(k) = \frac{1}{(2\pi)^{1/2}} \int_{-\infty}^{+\infty} [\nabla g(x)] e^{-i\mathbf{k}\cdot\mathbf{x}} d\mathbf{x} = ikg(k) \qquad (2.133)$$

$$f(k) = \frac{1}{(2\pi)^{1/2}} \int_{-\infty}^{+\infty} [\nabla^2 g(x)] e^{-i\mathbf{k}\cdot\mathbf{x}} d\mathbf{x} = -k^2 g(k) \qquad (2.134)$$

Problems

2.1 Basic nomenclature
Identify which of the following systems are either isolated, closed or open systems.
(a) a car tyre; (b) the Milky Way; (c) a brain cell; (d) a refrigerator; (e) a hammer;
(f) a frog.

2.2 Formal identities
(a) Show that, for a perfect gas, the pressure is $P = \frac{2}{3}\frac{E}{V}$.
(b) Show that the Helmoltz free energy, $F = U - TS$, is also equal to $F = -pV + \mu N$.
(c) If the Gibbs free energy is $G = F + PV$, state the thermodynamic independent
variables on which G depends explicitly.

2.3 Thermal engine
An air conditioner is made of a sealed piping circuit, in which a coolant fluid flows.
The piping goes through electrically-powered cooling stages: a condenser, which
turns the hot liquid into a cold gas by expansion, and a compressor, which takes
the room-temperature fluid and turns it to a high pressure, high temperature gas. A
simplified scheme is represented in the figure below.
(a) Is this system closed or open? Can part of the system be identified as a closed or
open subsystem?
(b) Describe the exchanges of energy, heat and entropy in the system, according to
the thermal engine model.
(c) Could the system work without the condenser, by just disposing of the air from
room temperature, to the exterior?

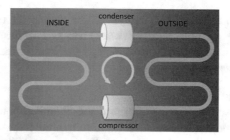

2.4 Exchanges of entropy

You add 50 L of hot water at 55 °C, with 25 L of cold water at 10 °C. What is the equilibrium temperature? How much entropy is produced by the time equilibrium occurs? Can you attribute part of the entropy to the hot water and part to the cold water?

2.5 Boiling, temperature and pressure

Boiling is the process by which a heated fluid turns into vapour. The *boiling point* is the temperature at which the pressure exerted by the evaporating liquid is equal to that of the surroundings. An open pot of water at the sea level, where the ambient pressure is $P_0 = 101.32$ kPa, boils by definition at the normal temperature of $T_0 = 100\,°C$. The Clausius-Clapeyron equation:

$$T_B = \left(\frac{1}{T_0} - \frac{R \ln \left(\frac{P}{P_0} \right)}{\Delta H_{vap}} \right)^{-1}$$

relates the boiling point at a different pressure P, with ΔH_{vap} the heat of vaporisation of the liquid, equal to 40.65 kJ/mol for water (note that this is more than five times the energy required to heat the same quantity of water from 0 to 100 °C). Let us take a pressure cooker of volume $V = 6$ l, and fill it by half with water at $T = 23\,°C$. At what temperature will the water boil?

2.6 Stefan-Boltzmann T^4 law

After Josef Stefan presented in 1879 his experimental T^4 law for the radiative heat transfer from a surface, Ludwig Boltzmann set out to give an explanation based on classical thermodynamics. He considered radiation "particles" to behave as a classical fluid, with energy density e. The idea that radiation could exert a pressure was quite new at that time, but it was a logical outcome of Maxwell's equations. The pressure of this radiation fluid inside the familiar piston-cylinder ideal experiment (see Fig. 2.1a) would have been be $p = e/3$. By writing the internal energy as $E = eV$, use the fundamental thermodynamic relation $dU = T dS - p dV$ to reobtain the T^4 law, as Boltzmann found out.

2.7 A negative temperature

Consider an ideal gas of N particles, each of which can exist in a "ground state" with energy $\varepsilon_i = 0$, or in an "excited state" with energy $\varepsilon_i = +e$. By taking that there are no interaction in the ideal gas, can you show that in the thermodynamic limit ($N \rightarrow \infty$ and $V \rightarrow \infty$, with $N/V = const$) this system has a region of negative temperatures, as a function of the population of particles in the excited state? Does such a result make sense?

2.8 Greenhouse gases 1

Assuming the atmosphere is at equilibrium, the chemical potential of each gas is constant at any altitude h. This observation allows to predict the concentration of each greenhouse gas as a function of the altitude. The chemical potential of a molecular species is defined:

$$\mu = -T \left(\frac{\partial S}{\partial N} \right)_{E,V}$$

that is the derivative of the entropy with respect to N, at constant E and V. You know already the expression of entropy as a function of (N, V, E), it is the Eq. (2.11), or Sackur-Tetrode formula. Use it to derive an expression for $\mu(h)$. Then by imposing that μ is constant at any h, obtain the number of molecules N as a function of h and of the mass of the molecule.

2.9 Greenhouse gases 2

The Earth's atmosphere is made up by 78 % of N_2, 21 % of O_2 and 1 % of Ar_2, not counting the fraction of water vapour which, according to the height and tempera- ture, can vary from 1 to 4 %. Diatomic homonuclear molecules, N_2, O_2 or Ar_2, do not contribute to the greenhouse effect since their radiation absorbing power is zero. The "greenhouse effect" comes from the other gaseous species found in the atmosphere, in smaller concentration. To a first approximation, the efficacy of each gas in the greenhouse effect depends on: (1) its concentration, and (2) its capability of absorb- ing the electromagnetic radiation. In turn, the *global-warming potential* (GWP) of each gas depends on its efficacy and of its lifetime in the atmosphere. For the five main greenhouse gases (CO_2; methane; N_2O; ozone; water vapour), the table below gives typical concentration and lifetime; their radiation absorption capability can be deduced from the following figure.
(a) Which of the five gases is mostly effective in absorption? Explain why.
(b) Rank in order of importance the five greenhouse gases, by taking into account the combination of the three parameters (concentration, lifetime, absorption).
(c) Why the gas apparently the most important in this ranking seems to be, conversely, the least important in global warming? (Fig. 2.7)

Gas	Concentration (ppm)	Lifetime (years)
CO_2 (carbon dioxyde)	400	30–90
CH_4 (methane)	2	12
N_2O (nitrogen protoxyde)	0.3	115
O_3 (ozone)	0.03	0.05
H_2O (water vapour)	($^\circ$)	0.02–0.06

($^\circ$) water vapour makes about 1–4 % of the atmosphere

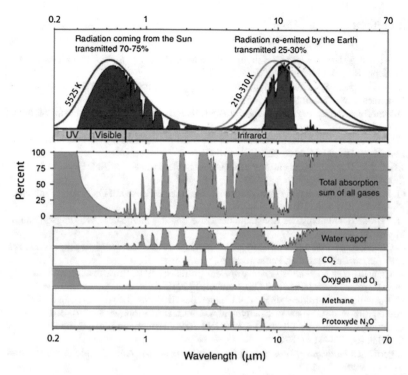

Fig. 2.7 *Top row* spectra of the electromagnetic radiation emitted by the Sun (*red*) and by the Earth (*blue*). *Middle row*: total absorption fraction by wavelength, for all the greenhouse gases. *Bottom row* absorption fraction by wavelength for the five main greenhouse gases. [Image © by R. A. Rohde, Global Warming Art project, repr. under CC-BY-SA 3.0 licence, see (*) for terms.]

References

1. S. Weinberg, *Dreams of a final theory: the scientist's search for the ultimate laws of nature* (Vintage, Random House, New York, 1992)
2. R. Newburgh, H.S. Leff, The Mayer-Joule principle: the foundation of the first law of thermodynamics. Phys. Teach. **49**, 484–487 (2011)
3. H.J. Steffens, *James Prescott Joule and The Concept of Energy* (Science History Publications, New York, 1979)
4. R.H. Swendsen, Statistical mechanics of classical systems with distinguishable particles. J. Stat. Phys. **107**, 1143–1166 (2002)
5. J.F. Nagle, Regarding the entropy of distinguishable particles. J. Stat. Phys. **117**, 1047–1062 (2004)
6. O. Sackur, Die Anwendung der kinetischen Theorie der Gase auf chemische Probleme. Ann. Physik **36**, 958–980 (1911)
7. H. Tetrode, Die chemische Konstante der Gase und das elementare Wirkungsquantum. Ann. Physik **38**, 434–442 (1912)
8. S. Carnot, *Réflexions sur la puissance motrice du feu et sur les machines propres à développer cette puissance* (Bachelier, Paris, 1824)
9. R. Clausius, *The Mechanical Theory of Heat* (transl. by W.R. Browne of nine Clausius' papers from German) (MacMillan and Co., London, 1879)
10. C. Bustamante, Unfolding single RNA molecules: bridging the gap between equilibrium and non-equilibrium statistical thermodynamics. Quart. Rev. Biophys. **38**, 291–301 (2006)
11. L. Boltzmann, Über das Wärmegleichgewicht zwischen mehratomigen Gasmolekülen. Wiener Berichte **63**, 397–418 (1871)
12. J. Perrin, *Le mouvement Brownien et la réalité moléculaire*. Ann. Chimie Phys. **18** (8me ser.), 5–114 (1909)
13. J.W. Gibbs, On the equilibrium of heterogeneous substances. Trans. Connecticut Acad. Arts Sci. III, 198-248 and 343–520 (1874-1878) [The voluminous (300 pages) Gibbs' paper was published in several parts in the Connecticut Academy bulletin, and remained largely unknown until it was translated in German and French, between 1891 and 1899.]
14. A. Kleidon, A basic introduction to the thermodynamics of the Earth system far from equilibrium and maximum entropy production. Philos. Trans. Roy. Soc. (London) B: Biol. Sci. **365**, 1303–1315 (2010)
15. S.G. Brittin, G.A. Gamow, Negative entropy and photosynthesis. Proc. Nat. Acad. Sci. USA **47**, 724–728 (1961)

Further Reading

16. S.B. Braun, J.P. Ronzheimer, M. Schreiber, S.S. Hodgman, T. Rom, I. Bloch, U. Schneider, Negative absolute temperature for motional degrees of freedom. Science **339**, 52–55 (2013)
17. E. Schrodinger, *What is life?* (Cambridge University Press, Cambridge, 1944)
18. W. Yourgrau, A. Van der Merwe, Entropy balance in photosynthesis. Proc. Nat. Acad. Sci. USA **59**, 734–737 (1958)
19. E. Fermi, *Thermodynamics* (Dover Books on Physics, Reprint of the 1937 edn.) (Dover, New York, 1965)
20. D. Kondepudi, I. Prigogine, *Modern Thermodynamics: From Heat Engines to Dissipative Structures* (John Wiley, New York, 1998)
21. D. Schroeder, *An Introduction to Thermal Physics* (Pearson, Boston, 1999)
22. D. Frenkel, B. Smit, *Understanding Molecular Simulation. From Algorithms to Applications*, 2nd edn. (Academic Press, New York, 2002)

23. S.E. Jorgensen, YuM Svirezhev, *Towards a Thermodynamic Theory for Ecological Systems* (Pergamon Press, New York, 2004)
24. J.D. Neelin, *Climate Change and Climate Modeling* (Cambridge University Press, 2011)

Chapter 3
Energy, Information, and The Origins of Life

Abstract This chapter deals mainly with probability and statistics. How life with all its complicated mechanisms could have emerged on Earth is a long standing question, full of implications going beyond the strictly scientific attempts at providing an answer. Physics poses limits as to how things may happen, beginning with the fundamental principles of thermodynamics. Some of these limits can be appreciated by looking at the emergence of life as a sequence of probabilistic events, trying to understand how an astonishingly improbable order could emerge from the statistically overwhelming chaos. Modern theories about the origins of life are still rather speculative. However many important bricks are falling into place as far as our knowledge deepens, showing how biological things in the Universe can indeed evolve from the simple to the complicated.

3.1 Thermodynamics, Statistics and the Microscopic

In the preceding Chapter we introduced the entropy, a thermodynamic function related to the dissipated heat and the irreversibility of spontaneous transformations, on one side; and a statistical quantity expressing the multiplicity of microscopic states of a system with a fixed amount of energy, number of particles, and volume, on the other. These two points of view, complementary to each other, allow to establish a conceptual link between **irreversibility** and **probability**. Namely, irreversible macroscopic phenomena associated to a variety of microscopic phenomena, each perfectly reversible until they are considered one by one, but with such an uneven distribution of probability to become practically irreversible when considered as a collective ensemble. Therefore, we turn over to the *microscopic* interpretation of the entropy: a measure of the dispersion of the energy of the macroscopic state (specified by E, N, V, etc.) among all the available microstates. The lesser the number of microstates available for a given energy content (at the lower limit just one, such as in a perfectly ordered crystal at zero temperature), the smaller the entropy of the system.

© Springer International Publishing Switzerland 2016
F. Cleri, *The Physics of Living Systems*, Undergraduate Lecture
Notes in Physics, DOI 10.1007/978-3-319-30647-6_3

Fig. 3.1 The gravestone of
Ludwig Boltzmann
(1844–1906) in the Vienna
main cemetery
(Zentralfriedhof)

The Boltzmann equation for the entropy (2.10), $S = k_B \ln \Omega$, so famous that
it was even written on his grave (Fig. 3.1), expresses just this equivalence. It is,
strictly speaking, valid only when every microstate is equally accessible or, in other
words, each microstate has the same probability of being visited by the system,
during its thermal fluctuation about the average macroscopic values of the state
variables. Recall that this was exactly the definition of microscopic equilibrium (see
p. 17), parallel to the definition of macroscopic equilibrium, which states that all
the state variables attain a constant average value. Macroscopic equilibrium seems a
rather obvious condition, once we have appropriate devices to measure temperature,
pressure, quantity of matter. But, besides the above qualitative statement, is there a
way of quantifying the microscopic probabilities?

The biological systems that we are going to study, usually operate in conditions
of constant temperature, a constraint that is relatively easy to reproduce also in laboratory experiments. In a system at constant temperature, the energy is a variable and
can fluctuate among any possible value. In this case, it is found that the equilibrium
probability of a microstate with energy E, is proportional to the so-called Boltzmann
factor, $\exp(-E/k_BT)$. Let us see how we can show this property by means of simple, intuitive arguments. We start from a closed system S in contact with a reservoir
(thermostat) R. While the energy $E^{(s)}$ of the closed system S can fluctuate, the sum
of S and R makes an isolated system, for which at any instant $E_{tot} = E^{(r)} + E^{(s)}$
is constant, therefore at equilibrium all the microstates in it must have the same
probability of occurrence.

Consider now a microstate 1 of S, with energy $E_1^{(s)}$, reservoir energy $E_1^{(r)}$ and a
number of allowed microstates in the reservoir $\Omega_1^{(r)}$; and a different microstate 2,
with $E_2^{(s)}$, $E_2^{(r)}$ and $\Omega_2^{(r)}$. The relative probability of finding the system S in the state
1 or 2 is, by definition, given by the ratio:

$$\frac{P_1^{(s)}}{P_2^{(s)}} = \frac{\Omega_1^{(s)}\Omega_1^{(r)}}{\Omega_2^{(s)}\Omega_2^{(r)}} = \frac{\Omega_1^{(r)}}{\Omega_2^{(r)}} \tag{3.1}$$

The second equality says that, since when we are looking at two particular microscopic states of S, it is $\Omega_1^{(s)} = \Omega_2^{(s)} = 1$. Therefore, what happens to the system S depends on the distribution of microstates in the reservoir R. In terms of the reservoir entropy, $S^{(r)} = k_B \ln \Omega^{(r)}$:

$$\frac{P_1^{(s)}}{P_2^{(s)}} = \frac{\exp(S_1^{(r)}/k_B)}{\exp(S_2^{(r)}/k_B)} = \exp[(S_1^{(r)} - S_2^{(r)})/k_B] \tag{3.2}$$

By recalling the thermodynamic relation $TdS = dE + PdV - \mu dN$, in these conditions for which $dV = dN = 0$, and by using the $E_{tot} = E^{(r)} + E^{(s)}$, we obtain:

$$S_1^{(r)} - S_2^{(r)} = \frac{1}{T}[E_1^{(r)} - E_2^{(r)}] = -\frac{1}{T}[E_1^{(s)} - E_2^{(s)}] \tag{3.3}$$

which merely stipulates the obvious fact that the energy $\Delta E = E_1^{(r)} - E_2^{(r)}$ passed from R to S, is the negative of the energy received in S from R. Eventually, the ratio of probabilities in Eq. (3.2) is obtained as:

$$\frac{P_1^{(s)}}{P_2^{(s)}} = e^{-\Delta E/k_B T} \tag{3.4}$$

The whole point of this derivation is to show that the relative probability of occurrence of two microscopic configurations of the closed system S is a function of only the S-energy difference between these two configurations: these probabilities do not care about what is happening in the reservoir R.

Without losing any generality, we can always choose one particular state of the system as the zero of our energy scale, for example let us set $E_2 = 0$. Then, the above reasoning can be repeated for any microstate i different from 2. By introducing the normalisation factor:

$$Z = \sum_i e^{-E_i/k_B T} \tag{3.5}$$

representing the sum of the probabilities of all the microstates of the system, the absolute probability of any microstate, correctly normalised $0 < p_i < 1$, is written as:

$$p_i = \frac{e^{-E_i/k_B T}}{Z} \tag{3.6}$$

The factor Z is called the **partition function** of the system. Note that we do not know exactly how many states appear in the sum (except for some special cases, like the perfect gas); we only assume that they can be counted someway, to give the correct normalisation. However, we are sure that the form of the function is that of Eq. (3.5). In practice, the knowledge of the absolute value of Z is never required, since all we want to know and use are ratios of microscopic probabilities, i.e. quantities related to differences in energy or entropy. In the macroscopic description, this translates in the fact that all measurable quantities are always represented by derivatives of functions of state.

3.1.1   A Probability Interlude

Let us consider a system which can take any of a discrete set of N states (whose number can be arbitrarily small or large, provided it remains finite). If each state i has a probability $0 < p_i < 1$ of being visited by the system, by analogy with the thermodynamic meaning of the entropy, we can define an equivalent entropy function for this system as:

$$S = -\sum_{i=1}^{N} p_i \ln p_i \tag{3.7}$$

to measure the spreading of the system among the N states (the constant k_B served only for giving the entropy the right energy-like units, but can be omitted here). Note that since the probabilities are all $p_i < 1$, we have to put a minus sign for the entropy to be positive. If we ask what is the distribution of values p_i for which the entropy S is a maximum, it is found that this corresponds to the condition of equal probabilities, $p_i = 1/N$ (see greybox on p. 65).

Furthermore, if the all-equal p_i are taken to be the equilibrium probabilities of the microstates of a thermodynamical system, $p_i = 1/\Omega$, it is seen that the definition (3.7) is perfectly equivalent to the (2.10):

$$S = -\sum_{i=1}^{N} p_i \ln p_i = \sum_{i=1}^{N} \Omega^{-1} \ln \Omega \tag{3.8}$$

and by considering a large density of microstates in the domain C, normalised so that $\int_C dC = \Omega$, the sum can be turned into an integral:

$$S = \int_C dC \, \Omega^{-1} \ln \Omega = \ln \Omega \tag{3.9}$$

that is the (2.10), after multiplying by k_B.

The method of Lagrange multipliers

To determine the set of p_i corresponding to the maximum of the entropy, amounts to finding the p_i's for which the total derivative of S with respect to all p_i is zero, subject to the constraint of unitary normalisation $C = \sum_{i=1}^{N} p_i = 1$. In mathematics, the method of Lagrange multipliers is an effective way of solving such problems. In this case, we would put together the entropy definition, Eq. (3.7), with the normalisation condition, multiplied by a free parameter λ, like $S - \lambda C$. Taking the derivative of this with respect to all p_i independently, requires solving the set of equations:

$$\left(\frac{dS}{dp_i}\right)_{c \neq i} - \lambda \left(\frac{dC}{dp_i}\right)_{c \neq i} = (-1 - \ln p_i - \lambda) = 0 \tag{3.10}$$

for each i. The writing "$c \neq i$" means that we take each derivative with respect to one parameter p_i while all the others are kept constant. The coefficient λ is called a Lagrange multiplier. Now, it is easy to see that the solution is $p_i = e^{-1-\lambda}$ for every i. Putting this back in the normalisation condition $C = 1$, one gets $\lambda = \ln N - 1$, and $p_i = 1/N$ for all i, which proves that the maximum of entropy is obtained when all the microstates are equiprobable.

We may want to slightly complicate our job, and ask what would be the distribution of values p_i, upon imposing that the average is equal to a number ξ. Well, average of what? Let us assume that each state is associated with the value of an extensive quantity ε_i (for simplicity, you may think of ε as the energy). According to the unknown probabilities p_i, the average value of this quantity 'energy' would be:

$$E = \sum_{i=1}^{N} p_i \varepsilon_i \tag{3.11}$$

We want to know the distribution of p_i for which 'energy' has a given value E. We must then add a second constraint, i.e. a second multiplier in the equation:

$$\left(\frac{dS}{dp_i}\right)_{c \neq i} - \lambda \left(\frac{dC}{dp_i}\right)_{c \neq i} - \xi \left(\frac{dE}{dp_i}\right)_{c \neq i} = (-1 - \ln p_i - \lambda - \xi \varepsilon_i) = 0 \tag{3.12}$$

with solution $p_i = e^{-1-\lambda-\xi\varepsilon_i}$. At least the constant λ can be eliminated, by using the normalisation $C = 1$:

$$p_i = \frac{p_i}{\sum_{i=1}^{N} p_i} = \frac{e^{(-1-\lambda)} e^{-\xi\varepsilon_i}}{e^{(-1-\lambda)} \sum_{i=1}^{N} e^{-\xi\varepsilon_i}} = \frac{e^{-\xi\varepsilon_i}}{\sum_{i=1}^{N} e^{-\xi\varepsilon_i}} \tag{3.13}$$

But this is just the same expression we found above, Eq. (3.6), with the partition function Z appearing as the normalisation factor, and $\xi = 1/k_B T$. With such a solution, the average energy would be:

$$E = \sum_{i=1}^{N} p_i \varepsilon_i = \frac{\sum_{i=1}^{N} \varepsilon_i e^{-\xi\varepsilon_i}}{\sum_{i=1}^{N} e^{-\xi\varepsilon_i}} \tag{3.14}$$

This shows that imposing a given value for the energy at constant V and T, necessarily results in a Maxwell-Boltzmann distribution of the probabilities.

Having a set of states for which we know the probabilities, allows us to estimate quantities relative to these states whose outcome is known only in probabilistic terms. If we imagine tossing a coin, we might think of calculating exactly the outcome (head or cross), if only we could know with absolute precision the initial position and velocities of all the atoms in the coin and in the surrounding air. This is of course a meaningless task. We know for sure, however, that the outcome can be only either head or cross, and if we assume that the coin is reasonably symmetric about its faces, we know that the probability p_i of either outcome is 0.5. Intuitively, if we think of tossing the coin an enormous amount of times, we may predict with reasonable accuracy that we will likely get head half of the times, and cross the other half. But, based on the values of the p_i's (in this case 0.5 and 0.5), could we estimate the probability of getting so many head or so many cross, after *just a few* tosses of the coin?

For a system undergoing a sequence of jumps between discrete states, each one independent on the others, the probability of being found in one given state is given by the **binomial distribution**:

$$P(n, m) = \binom{n}{m} p^m (1 - p)^{n-m} = \frac{n!}{m!(n - m)!} p^m (1 - p)^{n-m} \qquad (3.15)$$

If we imagine taking random snapshots of our system, the coefficients $P(n, m)$ give the probability of observing m times a 'success' out of n 'attempts', like finding the system for m times in the desired microstate, out of n snapshots. A wide variety of situations can be described as a sequence of such 'yes/no' experiments, in which each occurrence is independent on the others, and has a probability p. When some of the probabilities of each occurrence are different, we speak of Poisson's discrete distribution; if all the p are the same, we speak instead of a binomial distribution. The simplest example of binomial distribution is, again, the flipping of a coin for n times, and asking what is the probability of getting, e.g., head m times and cross $n - m$ times. For example, if we make $n = 4$ tosses, the probabilities from Eq. (3.15) of getting head for $m = 0, 1, 2, 3$ or 4 times, are: $P(4, 0) = 0.0625$, $P(4, 1) = 0.25$, $P(4, 2) = 0.375$, $P(4, 3) = 0.25$, $P(4, 4) = 0.0625$.

The sum of the probabilities is 1 and the maximum of the probability, $P = 0.375$, is obtained for an even split of occurrences, 2 heads and 2 crosses. No surprise. But these numbers also tell us what is the probability, for instance, of getting head in all the 4 tosses, that is 0.0625, a not negligible value (symmetrically equal to the probability of getting all crosses). It can be seen that if we increase the number of tosses n, the probability of this event is quickly decreasing as $(0.5)^n$, i.e. about 1 in 1,000 for $n = 10$, and 1 in a million already for $n = 20$. This is an important message: for an increasing number of attempts, the probability distribution becomes increasingly narrow around the average value, and the tails of the distribution get rapidly smaller and smaller. Our system is found most of the time around one state, the one of evenly split occurrences, despite the fact that all the states are equivalent. In other words, this tells us that the sheer number of states around the average is much larger, therefore we see the 'macroscopic' probability of the average result to be much higher. This is exactly the same situation we encountered in defining the microscopic equilibrium: a condition in which all microstates have the same individual probability, however their number is narrowly clustered about an exceedingly frequent average value.

Given the two probabilities p_1 and p_2, the binomial coefficients $\binom{n}{m}$ represent the coefficients of the terms of the binomial $(p_1 + p_2)$ raised to the power m:

$$(p_1 + p_2)^m = p_1^m + m p_1^{m-1} p_2 + \frac{m(m-1)}{2} p_1^{m-2} p_2^2 + \cdots + m p_1 p_2^{m-1} + p_2^m$$

$$(3.16)$$

If the two probabilities are equal, $p_1 = p_2 = p$, all the terms of the binomial expansion take the same value p^m, which is easily shown to coincide also with the sum $\sum_m \binom{n}{m}$, and becomes therefore the normalisation factor turning each of the coefficients $\binom{n}{m}$ into the probabilities $P(n, m)$.

The operation of raising to power n a polynomial $(p_1 + p_2 + p_3 + \cdots + p_m)$ is given by a general algebraic expression:

$$(p_1 + p_2 + \cdots + p_m)^n = \sum_{k_1 + k_2 + \cdots + k_m = n} \binom{n}{k_1, k_2, \ldots k_m} \prod_{i=1}^m p_i^{k_i} \qquad (3.17)$$

The object $\binom{n}{k_1 \cdots k_m}$ appearing in the previous expression is a 'multinomial' coefficient, whose explicit value is:

$$\binom{n}{k_1, k_2, \ldots k_m} = \frac{n!}{k_1! k_2! \ldots k_m!} \qquad (3.18)$$

Multinomial coefficients are another very useful tool of probability theory. For example, they can be related to the number of ways of arranging a number n of objects into m bins, with k_1 objects in the first bin, k_2 in the second, and so on, up to $k_1 + \cdots + k_m = n$. In fact, when putting the first k_1 objects we have n bins available, so the number of combinations is just $\binom{n}{k_1}$; for putting the second group of k_2 objects, we now have only $n - k_1$ places left, so the number of combinations is $\binom{n-k_1}{k_2}$; for the third group the places left are $n - k_1 - k_2$, and the number of combinations is then $\binom{n-k_1-k_2}{k_3}$, and so on. The total number is the product of all the possible combinations, that is:

$$\binom{n}{k_1}\binom{n-k_1}{k_2}\binom{n-k_1-k_2}{k_3}\cdots$$

$$= \frac{n!}{(n-k_1)! k_1!} \frac{(n-k_1)!}{(n-k_1-k_2)! k_2!} \frac{(n-k_1-k_2)!}{(n-k_1-k_2-k_3-k_4)! k_4!} \cdots \qquad (3.19)$$

$$= \frac{n!}{k_1! k_2! \ldots k_m!} = \binom{n}{k_1, k_2, \ldots k_m}$$

Note that we are not prescribing *one* specific arrangement of the n objects, but just counting the number of *all* possible arrangements. This difference can be easily understood when thinking in terms of words and letters. Assume we have an alphabet of 26 letters, and three bins. If we ask how many three-letter words we can write with this alphabet, this is just the binomial coefficient $\binom{26}{3} = 2600$ (of which only a small fraction will have a true meaning, when this writing is used as a language).

But, what if we ask the number of different sentences, e.g. of length 270 characters, that can be made with the same alphabet (plus 1 blank)? We need a frequency distribution for each of the 26+1 letters, in other words we need the p_i's. For simplicity, let us assume that all letters including the blank have the same frequency, i.e. the same probability of appearing in a word, $p_i = 1/27$ (in real human languages this would be highly unlikely). Therefore, we find $k_1 = 10$ times the letter a, $k_2 = 10$ the b, $k_3 = 10$ the c, and so on. Now, this number of different sentences obtained by just throwing the 270 characters at random, is given by the multinomial

$$\binom{270}{10, 10, \ldots 10} = \frac{270!}{(10!)^{27}} \simeq 10^{386}$$

a rather astronomical number. Note that very few among all these sentences will actually carry a meaning, in the chosen language; very likely, just one will be the 'right' sentence. On the other hand, there will be very many sentences which differ from the right one just by a small number of mistaken characters here and there. The right sentence carries evidently the maximum of information, while the random sentences do not carry information at all. The mistaken sentences, however, can be often interpreted correctly if the number of misplaced letters is not too large: they are the equivalent of the states fluctuating about the mean value, with a much higher probability (of being understood) than the mass of states lying in the tails of the distribution. Moreover, it can happen that a sentence could be ambiguous if the context is not properly specified. For example, the Latin sentence *I Vitelli Dei Romani Sono Belli* translates into English as "Go, Vitellius, to the Roman Gods at the music of war". However, the same sentence if read in Italian means something like "The young roman cows are very beautiful": the same combination of letters carries an entirely different meaning in two different cultural contexts. Therefore, partly wrong sentences are like a signal affected by some noise. Such considerations will be quite useful in the following, when discussing the information contained in DNA and proteins.

3.2   Life and the Second Principle

The second principle of thermodynamics describes the flow of energy, as it is every day observed in natural, irreversible processes. The physical interpretation of the second principle is that energy flows always in the direction of creating a uniform energy distribution in the system. The physical quantity entropy S (from the ancient

Greek words ἐν + τροπή, for "transformation") was introduced to describe this ubiquitous trend in a more quantitative way. From the point of view of thermodynamics, the Universe is a isolated system whose energy—as far as we know—is constant. However, energy is being continuously redistributed and homogenised in every spontaneous transformation, thus increasing the entropy at constant energy. Taken to its extreme consequences, this process will lead to the so-called "thermal death" of the Universe: nothing could ever happen anymore spontaneously in the Universe, once the temperature would be the same everywhere and the entropy would be at its maximum.

On the other hand, calorimetric measurements performed on many living molecular systems, such as DNA, proteins, even some macromolecular aggregates like a virus,[1] and elementary, however already very complex, microorganism like a bacterium, demonstrate that their energy contents is larger than that of their elementary precursors, or "molecular building blocks". We have already seen an example of this in the Chap. 2, when briefly discussing the ensemble of chemical reactions underlying the complex mechanisms of the photosynthesis. In other words, the ensemble of chemical reactions by which simple, small molecules are spontaneously assembled into the more complex molecular structures typical of the living systems, up to arriving at the assembly of a whole living organisms, seem to be quite generally **endoergic**, in that they require $\Delta G > 0$ to be assembled: more free energy to make the products than actually available from the reactants. From the thermodynamics point of view such reactions should be forbidden, while from the point of view of statistical mechanics they should be extremely unlikely, having a vanishingly small probability. But then, how to explain the ubiquitous expansion and evolution of all life forms without any apparent external intervention? It has even been said that *Clausius and Darwin cannot be right at the same time* [1].

And an even stronger statement can be obtained by replacing Herr Clausius by Sir Isaac Newton. Eventually, classical Newtonian mechanics is the rigorous outcome of the existence of gravity in the universe. It predicts cyclic orbits and recurring motions for all celestial bodies, and its laws are perfectly time-reversible, i.e., if we were able to invert someway the course of time we would see exactly the same universe. How can an universe ruled by gravitational forces give rise to the astonishing complexity of biological structures? In this respect, thermodynamics breaks the time invariance by introducing the idea of irreversibility and evolution. A strict Darwinist, however, cannot be happy with this kind of 'evolution' since, as we saw in Chap. 2, entropy only increases in spontaneous transformations, meaning that things in the universe invariably evolve from complicated to simpler, from order to chaos, from thermal non-equilibrium to thermal equilibrium (and ultimately, to the "thermal death" for the universe as a whole). By looking at the living, instead, we have the net impression that order is constantly generated out of chaos.

---

[1] The position of viruses in the scale of living systems is still under debate. They are considered at the limit between living and non-living systems, since the carry a DNA, and thus genetic information, but are unable to reproduce themselves without the transcription machinery, which they find in infected cells.

Our observation in Sect. 2.9, that the Earth is not a closed system but rather it constantly receives a flux of energy and "negative" entropy (i.e., a flow of $\Delta G > 0$) was eminently qualitative. However, it allows to conceptually move across the complexity of the organisation of the living matter. Formally, we can write the combination of the first and second principle in one single equation as:

$$\Delta S \geq \frac{1}{T}(\Delta E + p\Delta V) \tag{3.20}$$

or $\Delta G \leq 0$. This is the defining criterion for a transformation, such as a chemical reaction or a phase transformation, to be *spontaneous*. If the transformation takes place in a time Δt, it must also be:

$$\frac{\Delta G}{\Delta t} \leq 0 \tag{3.21}$$

The notation with 'Δ' underscores that the differences between thermodynamic quantities are taken between the initial and final states of the transformation, which includes the possibility that during the transformation there could be intermediate steps for which the variation is positive (if we were in the presence of a monotonic variation we would use the normal derivative symbol). The approach to equilibrium is signalled by the condition:

$$\frac{\Delta G}{\Delta t} \to 0 \tag{3.22}$$

The more proper interpretation of the Eq. (3.21) is found by reversing the terms:

$$\frac{\Delta S}{\Delta t} - \frac{1}{T}\left(\frac{\Delta E}{\Delta t} + p\frac{\Delta V}{\Delta t}\right) \geq 0 \tag{3.23}$$

or, by using the function enthalpy:

$$\frac{\Delta S}{\Delta t} - \frac{1}{T}\frac{\Delta H}{\Delta t} \geq 0 \tag{3.24}$$

The first term in this equation represents the variation of entropy for processes internal to the system (like a chemical reaction or a phase change), and the second term the entropy variation due to mechanical or thermal exchanges with the external world. For example, if we look at the transformation of water into ice, calorimetry experiments measure a variation of enthalpy $\Delta H = -80$ cal/g (exothermal reaction, with release of heat toward the exterior), while the entropy variation is $\Delta S = -0.293$ cal/g/K, over a large interval of temperature. At a temperature $T \leq 273$ K the transformation becomes spontaneous, that is the temperature at which ΔG goes negative:

$$-0.293 - \frac{(-80)}{T} > 0 \tag{3.25}$$

What's in a bacterium

Escherichia coli, also called colibacillus or simply *E. coli*, is a bacterium living in the intestine of mammals, very common among the humans. Discovered in 1885 by the French biologist Théodore Escherich, in the faeces of baby humans, it is a coliform usually commensal (meaning that it lives out of the human dejection, without interfering with the functions of our superior organs). However, some strains of *E. coli* can be occasionally pathogenic.

   E. coli is frequently used in biology studies as a model organism for procaryotes (living cells without a nucleus). A biological model organism is a species which is fully characterised and studied in the most complete way to understand specific biological phenomena, in the hope that the results of such a study could be transferable, at least in part, also to other living systems. Such a concept is founded on the observation that many of the fundamental biological principles, such as metabolic pathways, regulatory and developmental cycles, as well as the genes carrying the corresponding codes, are largely conserved (i.e. transmitted between different species) across the evolution. From the photo in Fig. 3.2, obtained from a scanning electron microscope, we can estimate a typical **diameter** of $D = 0.6\,\mu$m and **length** $L = 2\,\mu$m, from which the **volume** of the pseudo cylindrical bacterium is $V = \pi(D/2)^2 L = 0.56\,\mu$m^3. From biological data, we know that one of these bacteria contains mostly carbon, oxygen and hydrogen atoms. For simplicity, we may take glucose, $C_6H_{12}O_6$ as a typical chemical structure likely to be found in the bacterium, thus finding an average atomic mass $A = 7.5$. By considering that living matter has a **density** very close to that of water, $\rho \approx 1$ g/cm^3, or $n = \rho(N_{Av}/A) = 8 \times 10^{22}$ atoms/cm^3, we get that the **number of atoms** in the bacterium is about $N = (Vn)^{-1} = 2.23 \times 10^{10}$.

   The measured **protein concentration** in the cytoplasm of most cells is of about 180 g/l. Then, in a volume of $0.56\,\mu$m^3 or 0.56×10^{-12} ml, the mass taken up by proteins is $M_P = 180 \times 0.56 \times 10^{-12} \approx 10^{-10}$ mg. By taking the typical **mass of a protein** equal to 60,000 Da (1 dalton is a mass unit equal to $1/N_{Av}$ grams), we have room for about 1 million proteins in the bacterium body. Since there are typically **2,000 different proteins** in a prokaryotic cell (they would be about 10,000 in a eukaryotic cell), we estimate that each different protein is present in about **500 copies**. A 60 kDa protein could be made up of 60,000/7.5 = 8,000 atoms. The total number of atoms involved in building proteins in the bacterium would thus be $N_P = 8,000 \cdot 2,000 \cdot 500 = 8 \times 10^9$. The rest of the atoms, $2.23 \times 10^{10} - 8 \times 10^9 = 1.33 \times 10^{10}$, is involved in fabricating the nucleic acids DNA and RNA, and the cytoplasm, mostly water, sugar, nutrients and so on.

   The **genetic material** of the bacterium is represented by a single loop of a long DNA molecule, formed by 4.6 millions pairs of bases, each base plus phosphate chain (see DNA structure on p. 100) made by about 30 atoms with a mass of about 335 Da. Then, the **number of atoms in the DNA** is about 2.76×10^8, for a mass of 3.1×10^9 Da. In a bacterium like *E. coli*, there is also an amount of **mitochondrial RNA**, which totals 1.31×10^9 atoms or 1.46×10^{10} Da. By summing up all the known components we have about 9.6×10^9 atoms, supporting the general idea that most of the volume in the cell (about 57 % in this case) is occupied by **water**. However, in practical terms, the interior of a cell, even a simple one like that of a *E. coli*, is a very crammed environment, with thousands of bulky proteins swimming among long swirling branches of nucleic acids, with an average distance of not more than 5–10 nm from each other.

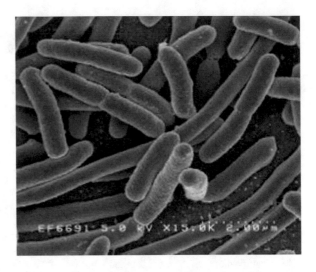

This highlights the fact that by properly setting the external conditions, a transformation can become spontaneous, while it was forbidden under different constraints. But then, could we set up a parallelism between the crystallisation of water into ice upon lowering the temperature, and the aggregation of simple molecules into a living organism? In other words, is there a "transition temperature", below or above which the appearance of life on Earth becomes spontaneous?

3.3 Impossibility of Spontaneous Aggregation

A very idealised example taken from biology is the attempt at calculating the probability of spontaneously forming a bacterium like *Escherichia coli*, starting from its molecular building blocks. This also offers us an occasion for quantifying some of the molecular components of a cell (see grey box on p. 71), for example we learn there that our little bacterium is made up of something like 22 billion atoms.

By measuring the heat of vaporisation of a known mass of bacteria in a calorimeter, an energy of $+9.5 \times 10^{-10}$ J/bact is obtained. Since 1 eV $= 1.602 \times 10^{-19}$ J, this means a positive supply of energy of 5.93×10^{9} eV, or $+0.31$ eV/atom on average.

Therefore, binding together the atoms into the molecular and protein structures that make up a bacterial cell, requires a positive ΔH contribution. By looking at the Eq. (3.24), this ensemble of reactions could not be spontaneous at whichever value of temperature, since $\Delta G > 0$ for any T: while in the case of water, the chemical binding may force water molecules into a more energetically favorable crystalline order, below that temperature for which the term $\Delta H/T$ becomes sufficiently negative to overcome the entropy term, a random bag of biomolecules seems to resist the organisation at any temperature. Therefore, the answer to the last question of the previous Section is a round "no".

But then, given the possibility of an equilibrium fluctuation of at least ΔG for a very small time Δt, may we at least hope that the necessary biomolecules could be formed by such instantaneous fluctuations? Even if the question does not make much sense, since the Earth is a system constantly out of thermal equilibrium, we can nevertheless calculate the probability of having such a spontaneous fluctuation of the free energy, with the only scope of showing that such an idea cannot have nothing to do with the appearing of life. On the basis of Eq. (2.14) of Chap. 2, this probability would be of the order of:

$$prob \propto [\exp(\Delta E/k_B T)]^{N_{atoms}}$$
$$= [\exp(0.31/0.025)]^{-2.23 \times 10^{10}} \approx 10^{-120,000,000,000} \qquad (3.26)$$

at $T = 300$ K, an evidently meaningless value. Therefore, as it could have been easily guessed without much calculating, life does not assemble itself spontaneously by chance, at any temperature.

Since the free energy of living systems is so large and positive, the formation and survival of such systems demands a constant supply of energy, accompanied by a reduction of the entropy, from the outside world. On the other hand, maintaining a living system would not be possible just by providing the necessary energy, in the absence of a mechanism capable of converting this energy into useful work. A car is an ensemble of an engine, transmission and driving chain, necessary to convert the chemical energy of gasoline into the mechanical work of locomotion: as we already noticed, just placing a tank of gasoline on the front seat of a car does not turn into locomotion. In the same way, without a well organised structure to transform food into work (mechanical, chemical, electrical, etc.), no living system could ever realise any of its vital functions. In other words, the simple principle of following the entropy reduction (which in the case of spontaneous water-to-ice transformation is just enough to generate order) is unable to arrange the very complex structures required by the living systems, randomly chosen among the practically endless number of possibilities.

The comparison between the living and the non-living cannot be merely based on order versus disorder. Statistical sampling is not enough: what is needed is **information**. This is the function of the genetic code in any living system.

3.4   Complexity and Information

The acknowledgement that the distinctive character of living systems is their **complexity** (in quite a mathematical sense), rather than their degree of order, is a relatively recent cultural acquisition. The distinction between the two concepts is based on the observation that both the two necessary ingredients for self-replicating systems, i.e. enzymes and nucleic acids (DNA and RNA), are molecules that carry encoded information. On the other hand, a crystal or an inorganic polymer, such as nylon, are systems displaying an exceptional degree of order, however with a minuscule amount of information. It could be objected that self-replicating polymers have been discovered, however this makes the simple self-replication ability a necessary, but clearly not sufficient characteristic to speak of a living system. Nucleic acids are *aperiodic* structures: at first sight they may appear quite disordered, but it is just their *lack* of periodicity which makes them the ideal carriers of information. In this view, a periodic structure carries order, an aperiodic structure carries complexity.

In terms of information, a crystal or an organic polymer resembles a book with all pages filled with one single word: the sequence of the few letters composing the word is highly ordered, and the position of every word in every page can be predicted with perfect accuracy, however the little information contained in this book is extremely redundant. On the other hand, an aperiodic structure can carry information—not necessarily a useful or readable one: only some specific sequences of letters make up useful phrases in such a book. Only some sequences of bases in the DNA, or some sequences of amino acids in a protein, correspond to useful messages (see Appendix B). Therefore, another fundamental characteristic of living systems is their **specificity**: it is not enough to be complex, such a complexity must be accurately specified.

If, in an ideal laboratory, we wish to build a crystal atom by atom all that is needed would be a very short sequence of information. Namely, to specify the positions in space of the basic unit of the crystal (three coordinates (x, y, z), multiplied by the number of atoms in the basic unit, that is rarely more than five or six), followed by the single instruction "to be repeated in the three directions of space, for an infinite number of times". To build a random polypeptide, all that is needed is a bunch of letters to specify the proportion of the different amino acids, followed by the instruction "to mix at random, in unlimited amounts". But if we want to produce the bacterium *E. coli*, it is necessary to specify with the highest accuracy, the type and spatial position of each and every one of its 2.23×10^{10} atoms. As it is well known, even one single missing atom in a protein may dramatically alter its function, and potentially compromise the life of the organism.

Is there a way to measure such a complexity? Can we establish in non-ambiguous terms a scale, on which it would be numerically clear that a disordered polymer carries nearly zero information, a crystal a little information, and a piece of DNA a lot? And can we measure the difference between the amount of information between different parts of DNA, in terms of the ability to produce a working protein? It turns out that such a measure can be someway established, by using a particular variation of the concept of entropy.

Let us think of a random polymer formed by a long chain of amino acids linked by peptide bonds (that is, a polypeptide), and imagine to turn it into a useful (active) protein by just switching positions between different amino acids along the chain, as shown in the scheme of Fig. 3.3. Each different combination of the sequence can be counted as a microscopic state of the polypeptide, such that the number of combinations in the random macroscopic state is Ω^R and the number of combinations

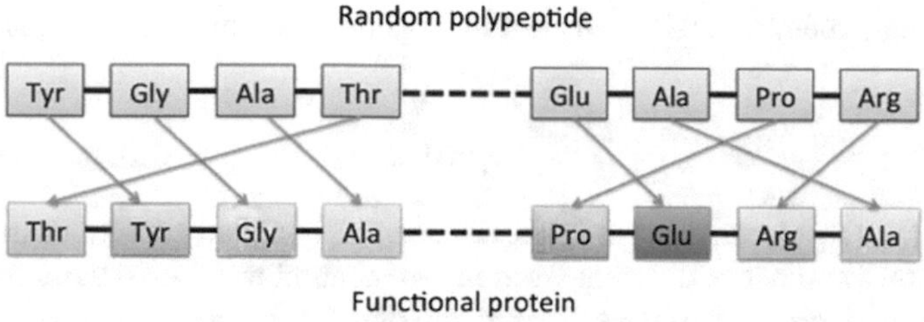

Random polypeptide

Functional protein

Fig. 3.3 Schematic of a long polypeptide made by a sequence of amino acids arranged at random (*above*), and turned into a functional protein by switching the order of the various components (*below*)

of the same amino acids giving a working protein is Ω^A. Then, we can calculate a kind of *configurational* entropy, S_C, and take the difference between the two states (labelled as 'A' for active, and 'R' for random) as:

$$\Delta S_I = \left(S_C^A - S_C^R\right) = k_B T \left(\ln \Omega^A - \ln \Omega^R\right) \tag{3.27}$$

Such an entropy difference, indicated as ΔS_I, is the **information entropy**, a concept introduced in 1948 by Claude Shannon. The amount of information is maximised for the case of a unique useful configuration, $\Omega^A = 1$, which implies $S_C^A = 0$. Then, we obtain a negative sign for the information entropy, meaning that this is to be supplied by the external environment, notably by the genetic code of the cell. Equation (3.27) quantifies the notion that only macromolecules formed by non-periodic and specific sequences (of nucleotides, or amino acids) can carry enough information, as required for the living systems. In particular, this classical definition specifies that the information content of the biopolymer is the amount of information required to specify a *unique* sequence or structure. Note that the constant k_B, previously necessary in the context of thermodynamics to get the right physical dimensions of the entropy, could be omitted here since we are mostly interested in measuring bits of information; in this case, the logarithm to use would be base-2, to measure the amount of 0/1 bits of information. However, it can as well be useful to maintain physical units to S_I, if we want to give a meaning to the information on a per-volume or per-mass basis.

According to the rules of combinatorial algebra discussed in Sect. 3.1 above, the number of different sequences that can be realised starting from an assembly of N objects is:

$$\Omega = N! \tag{3.28}$$

However, if some of the N objects belong to the same type (for example, the 4 DNA bases, or the 20 amino acids making up the proteins), this number is reduced by the consideration that by exchanging two or more objects of the same type the sequence is unchanged, just like in the word *physics* the two 's' can be exchanged always giving the same word. If the N objects are subdivided into k groups, each with $n_1, \ldots n_k$ members, the number of independent random sequences is:

$$\Omega^R = \frac{N!}{n_1! n_2! \ldots n_k!} \qquad (3.29)$$

This is nothing else than the multinomial coefficient already described in Sect. 3.1. Note that this number is different from the *total* number of sequences that can be realised by assigning any of the k types to any of the N objects, which would instead be equal to:

$$\Omega^T = k^N \qquad (3.30)$$

In the former case we must know from the outset the numbers n_i, while in the latter the n_i can take any value between 0 and N. Note also that, since in practical cases it is $N \gg k$, it is as well $\Omega > \Omega^T > \Omega^R$.

For a protein formed by amino acids we have an alphabet of $k = 20$ "letters", while the size N of the "words" can range from a few 100s to many 1,000 s. Conversely, for DNA and RNA it is $k = 4$ (the four A,C,T,G bases), while N is extremely large, of the order of 5×10^6 pairs for a simple bacterium, and up to 6×10^9 for the human genome. Let us take as an example a protein with $N = 600$ amino acids, and for the sake of simplicity let us assume that the type of amino acids composing the protein is evenly distributed among all the 20 possible species, i.e. all the n_k are identical to $600/20 = 30$. The number of independent random sequences is:

$$\Omega^R = \frac{600!}{(30!)^{20}} \qquad (3.31)$$

The configurational entropy for the random sequence is therefore:

$$S^R = k_B \ln \Omega^R = k_B \left(\ln 600! - 20 \ln 30! \right) \qquad (3.32)$$

By using the Stirling approximation (see Appendix A), the logarithm of the factorials can be easily calculated, giving:

$$S^R \simeq k_B \left(600 \ln 600 - 600 - 20(30 \ln 30 - 30) \right)$$

$$\qquad (3.33)$$

$$= 600 k_B (\ln 600 - \ln 30) = 1797 k_B$$

For the sake of comparison, let us take that this protein is to be synthesised by the DNA of our old friend, the bacterium *E. coli*, with its 4,600,000 base pairs. Now we calculate the information entropy of the entire DNA. Again supposing that all the nucleotides are evenly distributed among the 4 possible base types, we have:

$$\Omega^R = \frac{4.6 \times 10^6!}{(1.15 \times 10^6!)^4} \qquad (3.34)$$

and:

$$S^R = k_B \left(\ln(4.6 \times 10^6!) - 4\ln(1.15 \times 10^6!) \right)$$

$$\simeq 4.6 \times 10^6 k_B \left(\ln(4.6 \times 10^6) - \ln(1.15 \times 10^6) \right) = 6.38 \times 10^6 k_B \tag{3.35}$$

For both the protein and the DNA, we can take that the active state corresponds to a unique conformation and sequence, such that for both cases the entropy of the active state is $S^A = k_B \ln 1 = 0$. Therefore, the information entropy at $T = 300$ K is obtained as $T\Delta S_I = k_B T (0 - S^R) = (-1.38 \times 10^{-23}) \cdot 300 \cdot 1797 = -7.44 \times 10^{-18}$ J/molecule, and $(-1.38 \times 10^{-23}) \cdot 300 \cdot (6.38 \times 10^6) = -2.64 \times 10^{-14}$ J/molecule, respectively for the protein and the coding DNA. As we see, the amount of information available in the DNA is much bigger than needed for the protein. In fact, if we take the size of this protein as average for the bacterium, we see that the information contained in the DNA is more than enough to describe all of the about 2,000 different proteins, in fact more than twice the amount needed. Actually, if we carry out a similar calculation for the human DNA, with its 6×10^9 base pairs, it would be found that there is enough information to code for more than 4 million different proteins of average size. However, it has been established that the actual DNA expressing genes is but a small fraction of the total, and only 2 % is actually dedicated to the coding of proteins, while the remaining of the genes is used for the genetic regulatory mechanisms. Even reduced to 2 %, the human DNA seems enough to code for the about 50,000 different proteins and variants. The largest amount of DNA, however, is not currently identified. This is a mystery of modern genetics, seemingly common to all living organisms. Indeed, next to the fragments of the genes (called **introns**) that code all the necessary proteins of the cell, the DNA of any organism contains also an overwhelmingly large fraction of non-coding sequences (**exons**), whose genetic function is yet unknown. All this information is apparently wasted, to the point that until recently it was common to define such portions as "junk DNA". More recent studies are starting to revise the definition of what actually is a gene, and it could well be that the regulatory portions of a gene are indeed much bigger than the coding portions themselves, with large regions of overlap between such different "extended genes". According to the experimental results of the ENCODE project [2], at least 80 % of the DNA should display signs of activity correlated with genetic expression.

It is maybe worth noting that if, by following the convention of the chemists, we turn the $T\Delta S$ values from Joules per molecule into calories per gram, we would find a quite different, and somewhat deceptive answer. By taking an average mass $A = 100$ amu for the amino acids, the mass of the protein would be 60,000 amu. On the other hand, the average mass of a nucleotide is 339 amu, giving $A = 1.56 \times 10^9$ amu for the mass of DNA. By multiplying by N_{Av}/A, and dividing by 4.168 J/cal, we find $T\Delta S_I = -17.6$ cal/g, and -0.68 cal/g, for the protein and the DNA respectively: the fact that the DNA is a much heavier molecule than the average protein completely hinders the fact that its length allows much more information to be stored.

Most importantly, however, the **redundancy** of the information should be taken into account. Let us think back to our previous observation (see end of p. 69), that next to the perfect sequence there could be many other sequences, which differ very little from the right one, but whose "message" can be correctly guessed even if being somewhat wrong. In the biological context, this corresponds to the fact that many related protein sequences are structurally and functionally equivalent. The recently developed genome sequencing capabilities are providing thousands of examples of related but different sequences encoding essentially identical structures and functions, as well as examples of both RNA and proteins with entirely different structures, but similar biochemical functions (such as the many structurally distinct protease enzymes). On the other hand, structurally similar sequences can arise because of random errors in the copying process. For DNA transcription the error rate is estimated about 1 in 10^9, while for RNA translation the error rate is somewhat larger, about 1 in 10^4.

Let us calculate the amount of information in bits, by taking the base-2 logarithms. The above defined classical information, Eq. (3.30), gives the maximum amount of bits needed to specify a unique sequence of DNA of length n from a random distribution of the 4 nucleotides, as $(\ln_2 4^n) = 2n$. By analogy, the number of bits of the "message" information would be equal to $-\ln_2$ of the probability that one or more of the random sequences, close to the exact one, will encode a protein with a given function. In the very unlikely case that all the possible sequences correspond to an active protein, this probability would be equal to 1, therefore the corresponding "message" information content may vary between $(-\ln_2 1) = 0$ and $2n$ bits. As an example, the probability that a random RNA sequence of 70 nucleotides will bind ATP at micromolar affinity, has been experimentally determined to be about 10^{-11}. This corresponds to a message-information content of $(-11 \ln_2 10) \simeq 37$ bits, compared with $(70 \ln_2 4) = 140$ bits to specify the unique 70-mer sequence. If there are multiple sequences with a given activity, then the corresponding "message" information will always be less than the amount of information required to specify strictly one particular sequence.

Contrary to the famous statement by Marshall McLuhan, in the case of DNA the medium *is not* the message.

3.4.1 Free energy for the Synthesis of Biomolecules

If we want to estimate the free energy necessary for the synthesis of biomolecules, we must look at the actual free energy, $\Delta G = \Delta H - T \Delta S$ for polypeptide assembly. Even if the chemistry of the peptide bond is well understood, not much experimental data are available. By looking at compilations of experimental data (see for example [3]), it is found that in order to form a dipeptide by starting from different pairs of amino acids, one must supply an enthalpy of the order of $\Delta H = 5 - 8$ kcal/mol, or 0.2–0.35 eV/bond. This may be taken as a measure of the binding energy of an average peptide bond. We can also estimate the enthalpy by unit mass, based on an average mass A of the amino acids, such that the total formation enthalpy of a protein formed by q amino acids, once normalised by N_{Av}/qA, becomes roughly independent on the particular protein chosen. The entropy term is correspondingly more difficult to assess. It is suggested that the polymerisation reaction, because of the increase of mass, tends to progressively suppress the translational degrees of freedom, while the increasing structural complexity should favour the increase in the number of vibrational and rotational degrees of freedom. The overall result should be a reduction in the number of ways the total energy of the molecule can be redistributed, or a lowering of the entropy. Both a positive enthalpy and a negative entropy variation indicate that the ΔG of formation of biomolecules starting from simple amino acids is always positive. In other words, these chemical reactions are endothermic, or thermodynamically unfavourable. Moreover, it has been estimated [4] that the critical step requiring the largest amount of free energy is the assembly of a pair of amino acids, $\Delta G = 3.6$ kcal/mol, while the addition of a third, fourth etc. costs progressively less, ΔG decreasing from 2.5 to $\sim$1.4 kcal/mol.

On the other hand, it is found that the enthalpy of formation of the amino acids themselves, starting from even simpler molecules under reducing atmosphere (methane, ammonia, water) is generally negative, with values ranging from $\Delta H = -50$ to -250 kcal/mol, indicating exothermic reactions. This is the reason why different amino acids are formed with relative ease in the experiments of "prebiotic soups", to be discussed in the next Section. However, this is clearly not enough to speak of the spontaneous self-assembly of living systems, firstly because the subsequent step, i.e. the super-assembly of amino acids into proteins, is endothermic; and secondly because, when the same "prebiotic" experiments are carried out in a less reducing atmosphere (for example CO_2, nitrogen, water), the enthalpy becomes positive also for the formation of single amino acids.

An estimate of the binding energy for our typical protein of 600 amino acids would give $\Delta G = 1200$ kcal/mol. Even discounting that a large amount of stabilisation energy is gained by the folding of the protein and the interaction with the solvent, let us say the value is reduced by a factor of 10, the free energy remains exceedingly large compared to the isolated amino acids in solution. An equivalent chemical constant for the ideal reaction of self-assembly of the protein can be calculated (see greybox on p. 128) as $K = \exp(-\Delta G/RT) = \exp(-120 \times 10^3/1.986 \cdot 300) \simeq 10^{-87}$; if the amino acids are present in the solution at the concentration of 1 M (a rather high value by biological standards), this same figure would be the equilibrium concentration of the protein, i.e. zero by any measure.

That the building blocks of living organisms cannot possibly assemble just by chance appears obvious, just from such simple estimates. However, it is worth noting that such a view was deemed plausible until about the 1950s, the vastly improbable being considered achieved on this planet only because the immensity of geological time, which could have converted the nearly impossible into the virtually certain. The best evidence for this view was the outstanding lack of fossils representing the whole first half of the history of Earth: this allowed for a time span of about 2 billion years, to try all the huge number of variants before finding the "right one", which could have led to the start of everything from the assembly of a small chunk of primal living matter. Such a myth was debunked when the paleontologists Tyler and

Barghoorn in 1954 found cellular remains of bacteria in the (now famous) Gunflint beds between Canada and Minnesota, allowing to push back the early origin of life on Earth at least to 3.5–3.6 billion years ago. (Before that it seems impossible to go, since these are the oldest surviving rock strata that could still have kept intact the fossil records.)

3.5 Against All Odds

But then, if the self-assembly is clearly out of question, what other possible mechanisms could have done the unfeasible? Clearly, an external source of free energy is needed, or some mechanism that lowers the free-energy barrier, or both. At this stage, speculations and partial answers (which inevitably raise further questions) abound. Multiple sources of energy were indeed available for chemical reactions on the early Earth, heat from geothermal processes, sunlight, and atmospheric electrical discharges, among others. Could such energy contribution have been enough to promote the self-assembly of primordial molecules? The Miller-Urey experiment and its many variants, to be described in the next Section, is one step forward in this direction, despite some important criticism.

On the other hand, lowering or reversing the free-energy barrier could result from **autocatalytic** chemical reactions. A case in point is the *selection and amplification*, a spontaneous mechanism by which a small fragment which is 'successful' according to some criterion gets automatically sorted out of the chaotic mass and is replicated in great many copies, until some 'more successful', bigger fragment comes up and is in turn replicated, and so on.

There is a now classical experiment, performed for the first time in 1993 by Bartel and Szostak (Fig. 3.4), whose goal was to see if a completely random system of molecules could undergo selection in such a way that some defined species of molecules would emerge, with specific properties. The experiment began by synthesising some 10^{15} different RNA molecules, each about 300 nucleotides long, but all with random nucleotide sequences. The hope was that buried in those trillions there could be a few catalytic RNA sequences (called *ribozymes*) that happened to catalyse a **ligation** reaction, in which one strand of RNA is linked to a second strand. This is a key result, since it proves the ability of RNA to attach nucleotides (i.e., catalyse its own self-polymerisation), whereas in the cell nucleus this reaction is operated by the RNA-ligase enzyme. In the experiment, these longer RNA fragments are identified, then manually amplified (by a technique called PCR, protease chain reaction) and put back in the mixture. After repeating many times such a procedure a strong growth of the RNA auto-catalytic activity was observed: after only 10 rounds of amplification, the frequency of the ligation reaction went from 5×10^{-7} to 3.8 per h. What this experiment tells us is that, once some small population of catalytic molecules has grown in the mixture, it is capable of auto-selecting and reproducing itself going

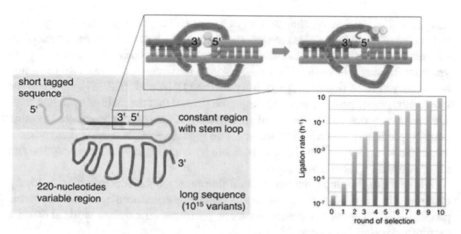

Fig. 3.4 Schematic of the RNA-ligation amplification experiment. *Left* A large pool of RNA molecules is synthesised (*blue-green*), all have in common the stem-loop region (*blue*), while the 220-long part (*green*) is randomly assembled. Such RNA are put in solution with a short tagged RNA, which attach the complementary fragment to the long one (*orange-blue*). *Middle* Rarely, some of the green sequences might be able to catalyse the 5'–3' ligation reaction (*red*), liberating the diphosphate blocks (*pale-blue spheres*) that provide the necessary activation enthalpy. These 'successful' sequences are separated, amplified, and reinjected in the next step, for 10 subsequent cycles. *Right* Selection and amplification of the successful sequences: after only 10 steps the concentration of ligand-catalysing molecules increased by 7 orders of magnitude

enormously beyond the manual amplification.[2] Moreover, it means that such a property is passed down to the "progeny", i.e., self-catalytic molecules have the ability of transmitting information, pretty much like genes in DNA can do. The conclusion is that transmissible information could emerge from a random population, provided there is an initial, even tiny fraction of catalytic members.

Information in DNA is processed and replicated by mechanisms taking place at the molecular level (see Appendix B at the end of the chapter), and it is important to note that such mechanisms always operate in the presence of thermal fluctuations. These fluctuations are due to the random motion of the atoms and molecules composing the DNA, the proteins of transcription or replication machinery, and their environment (water and ions). Biological information processing is thus subject to the statistical laws of thermal motion. If such processes were to occur in thermodynamic equilibrium, no information could be spontaneously processed or generated, because each random move would be statistically balanced by the corresponding reverse move (see also the greybox on Brownian motion, p. 224).

From the point of view of a physicist or a chemist, thermal fluctuations are a big deal: they represent noise for the self-assembly of systems at the molecular scale. A central problem with all theories on the origin of life has been the difficulty in

---

[2]Amplification was only meant to speed up the clock, with respect to the Earth's geological scale. It contributes negligibly to the efficiency increase, since it is a linear process—each amplification adds roughly the same amount of new copies—whereas the auto-catalysis process is exponential.

demonstrating efficient reaction pathways, for producing high yields of the primary molecules of life. High yields are important, since the half-lives of these molecules are relatively short at high temperature: a few hours for the ribose, a few days to years for the single DNA and RNA bases, with a maximum of 12 years for the uracil found only in RNA (the lifetimes of fully-formed DNA and RNA are longer). In the late 1960s, Nobel-prize winner Ilya Prigogine proved that the yield of a product from a chemical reaction can be increased enormously, if the reaction is coupled to other entropy-producing irreversible processes. The typical example is the photosynthesis: a reaction with $\Delta G > 0$ that is made possible by coupling it with the absorption and dissipation of high-energy photons.

Along these footsteps, the "fluctuation theorem" of statistical mechanics [5, 6], elaborated in the mid-late 1990s, states that the entropy produced by a thermodynamic process $A \rightarrow B$ in which the environment performs an external work ΔW on a system, corresponds to a simple ratio:

$$\frac{P(A \rightarrow B)}{P(A \leftarrow B)} = e^{(\Delta W - \Delta F)/k_B T} \tag{3.36}$$

that is, the probability that the atoms will undergo that process divided by their probability of undergoing the reverse process. By remembering that the Helmholtz free energy is $F = U - TS$, and that for a closed system it is $\Delta U = -\Delta W + \Delta Q$, the exponent in the previous equation is also equal to $(\Delta W - \Delta F)/k_B T = -\Delta Q/k_B T + \Delta S/T$ (note that to reverse the process, the sign of ΔW changes, since the work is done *by* the system). Put in this form, the ratio of the probabilities resembles the thermodynamic description of free-energy transduction by biochemical reactions developed in the 1970s by Hill. [7] Such thermodynamic relations state that as entropy production increases, so does the ratio of the probability for the direct/reverse process, and the system behaviour becomes more and more irreversible.

Starting from these results, it can be shown that a system fluctuating in conditions far from equilibrium may present a *time asymmetry*, in which the random path followed the time evolution in one direction is more probable than the time-reversed one. Remarkably, such a "temporal ordering" of non-equilibrium fluctuations is just another consequence of the Second Law. This suggests that dynamical order could be naturally generated in molecular motion under non-equilibrium conditions.

Recently, [8] the MIT physicist J. England proposed that for a system driven by an external source of energy *and* in contact with a thermal bath, the more likely evolutionary paths are those that absorb and dissipate more energy from (i.e., produce more entropy into) the environment. And self-replication is one such mechanism by which a system can dissipate increasing amounts of energy: certainly, a good way of dissipating more, is to make more copies of yourself.

For a simple chemical system that self-replicates at an exponential rate g, and is destroyed (by thermal fluctuations) at a rate δ, England derived a simple entropy equation from the fluctuation theorem, as:

$$\Delta S_{tot} = \frac{\Delta Q}{k_B T} + \Delta S_i \geq \ln \frac{g}{\delta} \tag{3.37}$$

where ΔS_i is the internal entropy increase by each replication event, and ΔQ is the heat dissipated. By rearranging the terms of this equation, the maximum replication rate for the chemical system at $\Delta S_{tot} = 0$ can be expressed as:

$$g_{max} = \delta \; e^{(\Delta Q/k_B T + \Delta S_i)} \tag{3.38}$$

In other words, the maximum net growth rate of a self-replicating system is fixed by its internal entropy, its durability $(1/\delta)$, and the heat thrown into the environment during the process of replication.

As a quite convincing example, one can estimate the heat dissipation for a self-catalysed replication experiment of RNA of the same type described in Fig. 3.4. The RNA half-life measured in solution is on the order of 4 years ($\delta = 1/4y^{-1}$), and the doubling time is estimated at $\sim$1 h. Thus, the lower bound for the dissipated heat, for the situation (realistic in this case) of a negligible entropy production, is:

$$\Delta Q \geq k_B T \ln \frac{4y}{1h} = 6.5 \text{ kcal mol}^{-1} \tag{3.39}$$

Since experimental data indicate an enthalpy for the ligation reaction of $\sim$10 kcal mol^{-1}, RNA appears to be close to the limit of the thermodynamic efficiency. For comparison, we may consider what the bound would be if this same reaction were achieved using DNA, which is much more kinetically stable, with a half-life estimated at 3×10^7 years. The same calculation would give 16 kcal mol^{-1}, much larger than the ligation enthalpy. This simple example illustrates a significant difference between DNA and RNA, regarding the respective ability to participate in self-catalysed replication reactions, powered by external energy (triphosphate building blocks): the far greater durability of DNA demands a much higher cost per base, for the growth rate to match that of RNA. As a relatively cheap and highly efficient building material, once RNA could arise in the primordial Earth environment, its quick takeover might not be too surprising.

3.6 Modern Theories About the Origins of Life on Earth

Firstly, why *modern*? The antiquity abounds with ideas and myths about the origin of the universe and life, and this is a subject which, even after the advent of the scientific method by Galilei & Co. in the 17th century, has been open to the most interesting (and wildest) speculations. In this—necessarily concise—Section, I will try to give account of some theories and views developed in the past half-century or so, which have some kind of experimental support, enough to attribute them the temporary status of credibility. Which does not mean that they would be free of contradictions, or possible loopholes, or even flaws, and that could be proved wrong and replaced by something better in the future, or even in a few weeks from now. But, *helás*, this is how science works, while waiting for the mystery to be solved.

In fact, we have no 'standard' model to describe the origins of life, and every strict definition of *life* meets with unexpected difficulties. Starting from what we believe to know to qualify as living organisms, and going backwards, the huge ensemble of collected data and most theoretical models seem to converge on the following, short list of necessary occurrences:

1. All living organisms are composed by one or more cells, capable of duplicating. Cells contain DNA as the support of the transmissible (genetic) information, which is copied by RNA and transferred to ribosomes. Ribosomes, composed in turn by a mix of RNA and proteins, are the machines which perform the protein assembly.
2. Ribosomes originated from ribozymes, a more primitive form of molecule capable of enzymatic activity, made only from RNA. Such special RNA fragments should have been produced by chance, under some peculiar environmental conditions, but once appeared they could take over by self-catalytic action.
3. Such self-catalytic and reproductive ability should have occurred in isolated compartments, created by the spontaneous assembly of phospholipids into closed-shape bilayers, which constitute the basic structure of all modern cell membranes.
4. Plausible prebiotic conditions should have initially created simple organic molecules, which would subsequently constitute the basic building blocks of all life-supporting molecules.

Each of these steps could have taken several hundreds of millions of years to develop, and their sum should cover the time approximately between 4.2 and 3.4 billions years ago, although step 4 could have also started well before, in the extraterrestrial space. We do not know where to exactly put the barrier of 'living' at any of the steps below 1, which is the actual evidence for any organism that is both capable of self-replication and transmission of its character to a progeny (even if viruses fall somewhere across this category, and cannot be entirely considered as living things). And we do not know if some other intermediate characters could have taken the stage. For example between steps 3 and 4, some scientists proposed the appearance of peptide nucleic acids, or PNA, as intermediates, with the nucleotide bases attached to a backbone made of peptide bonds, instead of sugar and phosphate. Or, it could be that even earlier precursors of RNA were molecules with no resemblance to nucleic acids, for example some organic polymers for which self-replicating capability has been demonstrated in laboratory experiments.

In parallel with the experimental studies on the Darwinian theory of evolution, the second half of the XX century witnessed the birth of several models of the molecular evolution of life. Such theories try to find evolutionary path leading to the observed structure of life-supporting molecules (nucleic acids, proteins, enzymes), starting from simpler and simpler molecules, up to arriving at the very elementary chemical constituents, such as water, methane, ammonia.

In 1953, Stanley Miller and Harold Urey tried to reproduce the conditions of the primordial Earth's atmosphere in a famous experiment (Fig. 3.5). They sealed

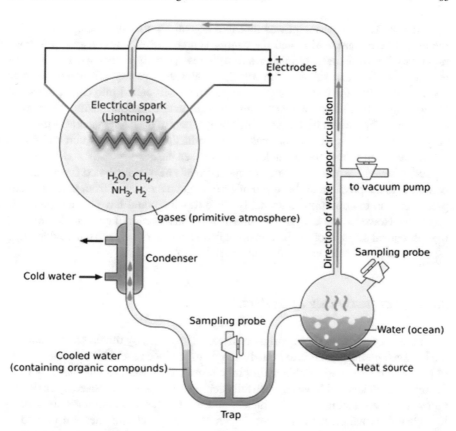

Fig. 3.5 The Urey and Miller experiment. A mixture of water, methane, ammonia and hydrogen is sealed in a flow circuit. The bottle half-full of water represents the ocean. A larger bottle including two electrodes represents the atmosphere. Water in the first bottle is heated to induce evaporation, and sparkles are fired between the electrodes in the second bottle, to simulate lightning in the primordial atmosphere. The atmospheric vapour is then cooled, water can condense and go back in the first bottle to start over the cycle. Water contents can be sampled through the trap. [Image ©by Y. Mrabet, repr. under CC-BY-SA-3.0, see (*) for terms.]

a mixture of methane, ammonia, gaseous hydrogen, and water in a closed flowing circuit. Water was evaporated on one side, and the vapour containing a gaseous mixture of all the four species was let to circulate in a volume, in which electric sparks simulated the lightning in the primordial atmosphere. After which, the vapour was cooled and condensed, to continue the cycle. After seven days of such treatment, the water was found to contain some molecules, such as urea (CON_2H_4), formaldehyde (H_2CO), cyanhydric acid (HCN), together with some bases, and proto-amino acids. Some of the compounds were present in more than 2 % concentration.

Miller and Urey used a reducing atmosphere (CH_4, NH_3, H_2, H_2O) instead of an oxidising atmosphere, which is the condition supposedly existing on Earth in the times in which life should have started to appear, and this was one of the main sources

of criticism. Their experiment was repeated a great number of times, by changing
the atmosphere composition and the energy source (for example using ultraviolet
radiation). Unfortunately, oxidising atmospheres typically composed by CO_2, N_2
and water, as it could originate from volcanism, always gave inconclusive or negative
results. Moreover, the reliability of the experiment was called into cause, since the
observed organic species could have come from an external contamination. However,
the reactions implicated in the sequence always require elevated concentrations and
very narrow pH intervals, which makes contamination less likely. Such conditions
could have been found, for example, in shallow sea basins.

Today, several experiments and chemical models have contributed further pieces
of answer to the problem of the origin of early organic molecules from basic gaseous
precursors, up to conditions reasonably close to what could have been a prebiotic
condition. However, the Urey-Miller line of experiments cannot give any hints about
the subsequent steps, such as the transition from a bunch of organic molecules to the
assembly of proto-cells having a simple metabolism.

3.6.1   Not just a Bag of Molecules

One accessory problem for the origin of life is represented by the first appearance of
the lipid cell membrane. It is extremely simple to produce in laboratory an aggregate
of long-chain fatty acids with a hydrophilic termination, which spontaneously form
flat or spherical layers. However, such fatty acids can only be synthesised by enzymes,
therefore we are facing a classic chicken-and-egg question: membranes are needed
to isolate enzymatic reactions from the surroundings, but enzymes are needed to
make the membranes. And, as usual, the answer to chicken-and-egg questions in
evolution is that there was neither chicken nor an egg to start with, but something
simpler than both. Moreover, an isolated compartment bound by a closed membrane
is not enough for a proto-cell. According to Szathmáry and Maynard-Smith [9],
two minimal conditions are required to form a proto-cell: (1) molecules capable of
replicating must be linked tighter in a kind of 'chromosome', thus making a structural
unit which will be conserved after the replication; (2) the membrane must possess
some mechanisms of exchanging with the external medium (much more primitive
than the complex ion channels of modern cells).

The permeability of phospholipid membranes to ionic nutrients such as amino
acids, nucleotides, and phosphate is in the range 10^{-11}–10^{-12} cm/s, a value so low
that only a few solute ions can cross the lipid bilayer of a given vesicle per minute.
However, bacteria take up millions of nutrient solute molecules per second during
active growth, using specialised membrane proteins to facilitate nutrient transport
across the lipid bilayer barrier. In the absence of such highly evolved transport proteins, how might an early form of cellular life gain access to nutrient solutes? One
possibility is that primitive membranes composed of simple amphiphiles were significantly more permeable to ionic solutes. For example, if the chain lengths of
phospholipids composing a lipid bilayer are reduced from the 18 carbons of modern
cell membranes to just 14 carbons, thereby thinning the membrane, the permeability

to ionic solutes increases by three orders of magnitude [10]. This would be sufficient to allow molecules as large as ATP to cross the barrier at a useful rate, while still maintaining macromolecules in the encapsulated environment.

Thus, it seems reasonable to suggest that the earliest membranes could have been composed of amphiphiles with relatively short chain lengths. Such membranes could capture and concentrate macromolecules, yet still provide access to ionic nutrient solutes in the external aqueous phase. Later, at some point early in evolution, a primitive transport system would have evolved, perhaps in the form of a polymeric compound able to penetrate the bilayer structure and provide a channel. It is interesting to note that selected RNA species have been demonstrated to interact with lipid bilayers and produce structures resembling ion-conducting channels [11].

3.6.2 The RNA World

The concept of a living organism based only on ribonucleic acid stems from the idea that RNA was the principal, most likely the only one form of life preceding the first DNA-based cell. We owe to Walter Gilbert the first use of the expression "RNA world", in 1986. The hypothesis of an archaic world based on RNA today meets a large favour in the scientific community, and it is grounded on various supporting elements. Firstly, the fact that RNA is in theory capable of assuring metabolic functions, as well as being a carrier of genetic information. In fact, RNA has the capability of storing and transmitting information by a genetic code similar to that of DNA. Secondly, as demonstrated by the Bartel-Szostak experiment (Fig. 3.4), RNA can also work as a **ribozyme** (contraction of the words ribose and enzyme) to catalyse chemical reactions, being in this case more similar to a protein. From the point of view of genetics, therefore, this molecule holds together the two primordial functions: the storage of information, and the catalysis necessary for self-replication.

DNA can also copy itself, but only with the help of some specific proteins. On the other hand, proteins are very good at catalysing reactions, but they are incapable of storing the information necessary to create themselves. RNA instead, is capable of both catalysis and self-replication. The ribosome itself, the very protein-stamping machine of the cell, in part composed of RNA, is a ribozyme. Ribozymes can fold in space, being composed of a highly flexible single strand of nucleic acids, and can in this way build up an active site for catalysis, pretty much with the same structural mechanism found in proteins. DNA, with its much stiffer double helix, cannot fold over lengths much shorter than about 50 nm, therefore precluding the possibility of making up a catalytic site along its structure.

For the concept of the RNA-world to be viable, it must be hypothesised the early appearance of 'viroids', structures similar to self-catalytic RNA, kept inside some form of isolated compartment. This could have been a primitive membrane, or even a inorganic crystalline pocket. From such structures, proto-cells capable of some archaic form of metabolism should have evolved (in the Darwinian sense), until a primitive, however functional RNA-cell was born, with its varied and complex metabolic activity ready.

Even though most indications go in support of the idea that RNA should have originated DNA in the building process of cell metabolism, such a primordial transformation from RNA to DNA is very difficult to occur. Indeed, it cannot happen in most living organisms without the help of specialised proteins, the ribonucleotide-reductase. Moreover, such an enzymatic reaction consumes a lot of energy to chemically reduce ribose, and releases highly reactive free radicals. Since RNA is by itself a very fragile molecules, it seems unlikely that a transformation involving liberation of lots of free radicals could be supported without other proteins coming to help. Therefore, DNA could have evolved from RNA only after the appearance of some proteins, necessary to the various steps of the synthesis from a RNA precursor.

Proteins are excellent catalysers, much better than ribozymes. Thinking that all proteins are made from twenty amino acids, while RNA is made from only four nucleotides, it should be obvious that proteins exhibit a vastly superior diversification compared to RNA. From an evolutionary point of view, it is therefore unlikely that a protein-enzyme should have been replaced by a RNA-enzyme. On the contrary, if RNA appeared well before proteins, it is plausible that it would have been later replaced by some proteins, more effective as far as the catalytic functions. Such an argument is also supported by the observation that RNA plays indeed a role in protein synthesis, via the ribosome. Someway, RNA should be as well at the origin of proteins. Proteins should have appeared later in the panorama of molecular evolution, initially with the function of ameliorating the operation of ribozymes, before eventually replacing them.

Intermediate between a mere aggregate of proteins, and a properly functioning cell, we find the **virus**. A virus is a self-replicating structure, minimally constituted by a nucleic acid and some proteins, enclosed in a hard and highly symmetric protein shell structure (the 'capsid'). Some viruses are found to contain RNA instead of DNA ('ribovirus'), usually as single strand and more rarely in double-strand form. Carriers of some of the most important and lethal human diseases, such as Ebola, polio, hepatitis-C, and many others, such riboviruses have been supposed to be relics of the ancient RNA-world. The RNA-to-DNA transformation could have occurred in a virus, instead than inside a cell. On this basis, DNA-based viruses should be preceding modern DNA cells: the first DNA cell could have 'stolen' the new form of nucleic acid (the DNA, in fact), from one such viruses. The French geneticists Didier Raoult and Jean-Michel Claverie discovered the *mimivirus*: a giant DNA-virus (its genome being more than twice longer than that of the smallest known bacterial genome), whose peculiarity lies in the fact that it can produce directly some proteins implicated in the DNA-RNA-protein translation process (for example, some enzyme which loads amino-acids on the structure of a tRNA). This mimivirus could have as an ancestor, a virus older that the first DNA-cell. The enzymes that replicate a DNA virus are very much different from each other, as well as from the cell enzymes playing their same role. Such indications allow to suppose that enzymes bound to DNA could have first appeared during an ancient era of the modern DNA world, in which there was simultaneous existence of RNA-cells with both RNA- and DNA-viruses. According to this theory, only much later the full-fledged DNA cells, which today make up all the superior organisms, could have appeared on the Earth.

3.6.3   Abiotic Hypotheses

Some research evidence suggests that, even prior to the appearance of organic enzymes (RNA, proteins), the lowering of the free energy barrier for the synthesis of protomolecules could have occurred also with the intervention of inorganic components [12]. Reaction sequences that resemble essential reaction cascades of metabolism, such as the glycolysis (see Chap. 4), could have occurred spontaneously in the Earth's ancient oceans. In particular the presence of ferrous iron, which was abundant in the early oceans, accelerates many of the chemical reactions leading to precursors of RNA and proteins. The thermodynamic conditions for such metal-catalysed reactions to occur is the absence of free oxygen, and the presence of relatively high temperatures, between 50 and 100 °C, such as existing in hydrothermal vents on the ocean floor. Moreover, a strongly alkaline environment could have provided an ideal ecological niche [13], with micro-caverns concentrating the newly synthesised molecules, steep temperature and concentration gradients, isolation of the reacting species even in the absence of a cell proto-membrane.

In the 1980s, G. Wächtershäuser, had proposed the "iron-sulphur" theory, in which the energy released from redox reactions of the surface of metallic sulphides (such as pyrite) is available for the synthesis of organic molecules. Such systems could therefore be able to evolve into self-replicating, metabolically active proto-molecules. Although such hypothesis is too remote from the synthesis of life building blocks, the emergence of a primitive metabolism which could provide a safe environment for the later emergence of RNA is contemplated also in other models. The basic molecules needed to initiate the CO_2 reduction, key to the Krebs' cycle (see Chap. 4) producing energy in all known aerobic organisms, could have formed much earlier than the enzymatic complex using them. In this way, the development of an independent metabolism would precede the development of genetics.

Inorganic surfaces have been called into cause as suitable environments for the early biosynthesis also by other models, notably the "clay hypothesis", which postulates that complex organic molecules arose gradually on a pre-existing, non-organic replication surfaces of silicate crystals in solution. Some studies confirmed that clay minerals of montmorillonite catalyse RNA synthesis in aqueous solution, by joining nucleotides to form chains up to about 50 units. Inorganic surfaces are also relevant to the problem of the chirality of biological matter (see Appendix B), namely the fact that between specularly-symmetric amino acids and nucleic acids, only one of the variants is always observed. In recent years many studies have addressed the phenomenon of chiral molecular adsorption on mineral surfaces such as inorganic silicates, a process that might have jumpstarted the homochirality of life molecules.

3.6.4   Between Quiet and Thunder

By all the present evidence, the first steps of the appearance of life on Earth took at least 3 billions of years. All the hypotheses reported above focus on the crucial step of the earliest passage from the 'non-living' to the 'living'. But even solving this complicate puzzle will not represent the final word on the evolution of life. An even bigger enigma is the evidence that, after such a long and slow incubation state, an exceptional acceleration known as the **Cambrian explosion**, spread all the modern life forms in the primitive oceans. It is known from fossil records that the process leading from the microscopic single cells, to the multicellular organisms, to the first plants and animals, up to the giant vertebrates, occurred with a singular rapidity.

Prior to the Cambrian explosion, about 543 millions years ago, organisms were simple, composed of individual cells occasionally organised into colonies. Over the following 70–80 million years, the rate of formation of new species accelerated by an order of magnitude, and the diversity of life began to resemble that of today.

The appearance of the modern **body plan** is the story of a sudden divergence. Earth's first community of animals, which held nearly exclusive sway from the time of their appearance up to the early Cambrian period, consisted of strange species with no clear relation to modern forms. These organisms, whose fossils were first found in the Australian region of Ediacara, contained neither complex internal organs nor even any recognisable body openings of mouth, anus, and so on. Ediacaran creatures were flattened forms, in a variety of shapes and sizes, built of numerous tubelike sections fitted together into a single structure. Such animals shared the relative simplicity of all primitive life forms. However, the big difference is in their body design: these animals were **diploblastic**, their bodies being made up of one outer layer and one inner layer. Such a construction does not allow to develop internal cavities and organs, just one single cavity which acts as stomach and performs all the basic functions. A similar structure is observed in very few modern life forms, such as jellyfish, corals, sea pens, sea anemones and the like. All modern animals are instead **triploblastic**, meaning that their bodies are made by three layers of tissues (exoderm, mesoderm and endoderm), the middle layer allowing the development of all the internal organs for the various functions. Although it may be imagined that triploblastic body plans may have evolved from diploblastic, it is nevertheless amazing to observe that Ediacaran fauna had its own "explosion", by rapidly invading the oceans about 35 million years before the Cambrian (an event dubbed the *Avalon explosion*), to disappear quite suddenly. It seems that the advent of entirely new life forms is made possible each time by the sudden (on geological time scales) development of a new layer of tissues: before the Ediacarans the only complex forms existing were sponges, amorphous colonies of identical cells living together in a single multicellular structure, but yet unable to turn into a whole organised individual. These had just one type of tissue, that of the individual cells fused together. Like the differentiation of ectoderm and endoderm, which lead to diploblasts, the differentiation of the mesoderm leading to the triploblastic shapes was a rapid and selective event: it came up suddenly, and took over all the new life forms, leaving behind but faint traces of the preceding, 30-million years long archaic age.[3]

---

[3]We will come back to the interesting issue of diplo- versus triploblastic body plans in Chap. 11.

Radiochemical dating of the ancient earth

Each chemical element consists of atoms with a specific number of protons in their nuclei, as identified by the atomic number Z from Mendeleev's periodic table, However, atoms of a same element can have different atomic weights owing to variations in the number of neutrons that make up the nucleus. Atoms of the same element with differing atomic weights are called **isotopes**.

For each chemical element, there may be several different isotopes occurring in nature, some of which are stable and part of its natural composition. For example, carbon has $Z = 6$, so it has 6 protons in its nucleus, which can be accompanied by either 6 or 7 neutrons, thereby giving ^{12}C or ^{13}C as stable isotopes (the number to the upper-left indicates the sum of protons and neutrons). The two variants of natural carbon have masses, respectively, of 12 and 13.003355 atomic units (identical to daltons, or Da, when the same unit is used to weigh molecules), and are present in the Earth's crust in the proportions of about 99 and 1 %, hence the mass of atomic carbon of 12.0107 units reported in the periodic table. However, neutrons in the carbon atom can be present in various numbers, from 2 to 16, giving ^{8}C–^{22}C: all these isotopes except the 12 and the 13 are unstable, and disintegrate into some other nuclei and elements, by various processes of **radioactive decay**. For example, the isotope ^{14}C transforms into ^{14}N, a stable isotope of nitrogen, by a nuclear decay in which one of the neutrons of the carbon nucleus turns into one proton, plus one electron and another subatomic particle, the anti-neutrino. This is the "beta-decay", one of the most common radioactive transformations. The total number of protons plus neutrons is conserved, however since the new nucleus has one more proton, its atomic number has changed to $Z = 7$, i.e. the carbon atom has turned into a nitrogen, in a truly alchemical transformation.

The rate at which each isotope of any element decays is expressed in terms of an isotope's **half-life**, $T_{1/2}$, or the time it takes for one-half of a particular radioactive isotope sample to decay. Most radioactive isotopes have rapid rates of decay (that is, short half-lives) and lose their radioactivity within a few seconds; some take days or years. To continue our previous example, ^{9}C gives off ^{9}B with a high probability, with a half-life of 127 ms; ^{11}C turns into ^{11}B with a half-life of 20.33 min; ^{14}C is the longest-lived of carbon isotopes, performing its beta-decay with a half-life of 5,730 years.

The empirical equation governing the decay of an isotope at a time t starting from an initial sample size N_0 is an exponential:

$$N(t) = N_0 \exp(-\lambda t) \tag{3.40}$$

The parameter λ is the **decay constant** of the isotope, related to the half-life as $\lambda = \ln 2/T_{1/2}$. Interestingly, some of the product isotopes during a radioactive decay can be unstable as well, and will decay by some other mode, with their own half-life or λ. This gives rise to what is called a "radioactive decay chain", which stops only when the final isotope of some element down the chain is stable. For example, starting from the ^{9}C, a first decay is:

$$^{9}C \rightarrow {}^{9}B + e^+ + \nu \tag{3.41}$$

(the e^+ is an electron with positive charge, a *positron*, and ν is a neutrino). The daughter isotope ^{9}B almost immediately decays by spewing off a proton:

$$^{9}B \rightarrow {}^{8}Be + p \tag{3.42}$$

and, in turn, the ^{8}Be decays into two stable Helium atoms:

$$^{8}Be \rightarrow {}^{4}He + {}^{4}He \tag{3.43}$$

Some isotopes, however, decay very slowly with half-lives of millions, or billions of years, and several of these can be used as true **geologic clocks**. For such isotopes, only the one in the chain with the longest half-life is taken as reference, while all the intermediate ones have much faster decay rates. For the geologic clock to be measured, it is necessary to perform a chemical separation of the elements in the rock, and identify the relative fractions of the long-lived parent, and of the stable isotope. The underlying assumption is that, at the moment the rock was formed by geological events, like a volcanic eruption, the stable daughter isotope was not yet present, $N_D = 0$, and the parent was at its maximum concentration, say $N_P = N_0$. Therefore, the number of nuclei of the parent isotope still present in the sample after a time t is:

$$N_P(t) = N_0 \exp(-\lambda t) \tag{3.44}$$

and, by conservation, the number of stable daughter isotopes is equal to the difference between the number of initial parents and the current ones:

$$N_D(t) = N_0[1 - \exp(-\lambda t)] \tag{3.45}$$

Then, by taking the ratio of the two expressions, the age t_0 of the rock can be determined by the following equation:

$$t_0 = \frac{1}{\lambda} \ln\left(1 + \frac{N_D(t_0)}{N_P(t_0)}\right) \tag{3.46}$$

where the t_0 values in the ratio are the concentrations of parent and daughter isotope currently detected in the rock sample.

The parent isotopes and corresponding daughter products most commonly used to determine the ages of ancient rocks are listed below:

Parent Isotope	Stable Daughter Isotope	$T_{1/2}$ ($\times 10^9$ years)
^{238}U	^{206}Pb	4.5
^{235}U	^{207}Pb	0.704
^{232}Th	^{208}Pb	14
^{87}Rb	^{87}Sr	48.8
^{40}K	^{40}Ar	1.25
^{147}Sm	^{143}Nd	106

Dating rocks by these radioactive timekeepers is simple in theory, but the laboratory procedures are complex. The numbers of parent and daughter isotopes in each specimen are determined by various kinds of analytical methods. The principal difficulty lies in measuring precisely very small amounts of isotopes. Moreover, not all rocks can be dated by radiometric methods. For a radiometric date to be useful, all minerals in the rock must have formed at about the same time. Sedimentary rocks can rarely be dated directly by radiometry, since they may contain particles including radioactive isotopes, which may be not the same age as the rock in which they occur (e.g., sediments cemented together into a sedimentary rock, but weathered from older rocks). Radiometric dating of metamorphic rocks may also be difficult, because in this case the age of a particular mineral does not necessarily represent the time when the rock first formed, but the time when the rock was metamorphosed.

The Potassium-Argon method can be used on rocks as young as a few thousand years as well as on the oldest rocks known. Potassium is found in most rock-forming minerals, the half-life of its radioactive isotope ^{40}K is such that measurable quantities of Ar (daughter) have accumulated in K-bearing minerals of nearly all ages, and the amounts of K and Ar isotopes can be measured accurately, even in very small quantities. Where feasible, two or more methods of analysis are used on the same specimen of rock to confirm the results (Fig. 3.6).

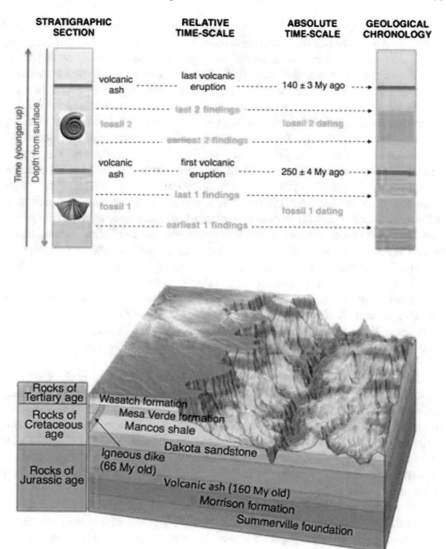

Fig. 3.6 *Above* How relative dating of events and radiometric dating are combined to produce a calibrated geological chronology. In this idealised example, the ensemble of data demonstrate that "fossil 1 time" was somewhere between 250 and 140 My ago, and that "fossil 1 time" is older than 250 My ago. Actual radiometric dating would be carried out only on the volcanic ashes strata. *Below* Example of geological dating in a region of the Grand Canyon National Park (USA). The age of the volcanic ash bed and the igneous dike are determined directly by radiochemical methods. The layers of sedimentary rocks *below* the ash bed are obviously older than the ash, and all the layers above the ash are younger. The igneous dike is younger than the Mancos shale and Mesa Verde formation, but older than the Wasatch formation because the dike does not intrude the Tertiary rocks. From this kind of evidence, geologists estimate that the end of the Cretaceous period and the beginning of the Tertiary period took place between 63 and 66 million years ago. Similarly, a part of the Morrison formation of Jurassic age was deposited about 160 million years ago as indicated by the age of the ash bed

About 150 million years after the Cambrian explosion, life made another great jump, when it moved from the ocean onto land. To survive on land, oxygen-breathing lungs, and weight-carrying bones needed to be invented, and a skin that could prevent loss of water and protect from the sunlight. The first animals that moved on land did not have impermeable skin, they were amphibians that live only part of their life outside water. It took another 50 million years for the descendants of the amphibians to become reptiles fully adapted to living on land. In the poetic words of Freeman Dyson: *The liberation of life from the oceans made possible all the later inventions that make the land beautiful, fur and feathers, and forests and flowers.* The reptiles with their impermeable skins spread all over the Earth and made it their home, until a further 100 million years would bring about the largest reptiles ever seen. Dinosaurs ruled the Earth for 150 millions years, before going suddenly extinct. In the meantime, mammals, birds, flowering plants, bees had started to invade the planet. The great apes appeared about 40 millions years after the last Tyrannosaurus roared in the Cretaceous forests, and in the 23 millions years that followed one very special branch of apes evolved into *Homo sapiens*. Between *H. habilis*, who lived about 2.8 My ago, and *H. erectus*, about 1 My later, a rapid process of encephalisation took place, with the cranial capacity nearly doubled from 500–650 to 850–1100 cm^3 (we modern *Sapiens* have 1100–1300 cm^3, however the *Neanderthal* had an even larger 1900 cm^3). If taken as a progressive distribution of cerebral matter over $\sim 10^5$ generations, such a rapid increase would correspond to each generation having $\sim 500,000$ more neurons than their parents.

Many hypotheses and theories have been set forth, to try to explain such an explosive development. Palaeontology was dominated for the three quarters of the XX century by the picture of **phyletic gradualism**, according to which the evolution from one species to another (*speciation*) arise by slow transformative events involving entire populations. This theory required an unbroken fossil record as a verification, to prove that each form was a minor variant of a preceding form. The existence of many gaps in the fossil lineages was seen in this context simply as a missing evidence, to be adjusted by the progress of the research (Fig. 3.7).

About 1954 the zoologist Ernst Mayr proposed a theory of 'allopatric' speciation in the context of ecological competition. [14] The underlying idea was that at some point in its history a population may be split because of various events, and could separately evolve in distinct ecological niches, subject to different evolutionary pressure. In this way, the species could differentiate and evolve in isolation, and an entirely new species could result from the original one.

The most definitive leap in this direction came in 1972, with the Eldredge-Gould theory of **punctuated equilibria** [15]. The two palaeontologists relied on the well known observation that the fossil record of an evolutionary progression shows species that suddenly appear, and ultimately disappear, even millions years later, without any change in external appearance. During their existence, new competing species appear at random intervals, also lasting very long before disappearing. Extending the allopatric speciation theory into the domain of palaeontology could explain such evidence, by the fact that mutations are diluted in a large population, thereby making the transformation of an entire species a much rare event, and that only when smaller

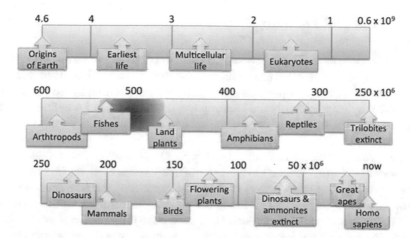

Fig. 3.7 A storyline of the evolution of life on Earth, indicating some key events relative to the appearance or extinction of major life forms. The time line (approx. log-scale) runs *left-right* and *top-down* (note that the *upper* bar is in billions of years, and the two *lower* ones in millions of years. The *orange* cloud marks the time extent of the Cambrian explosion

populations remain isolated, the evolution can accelerate. The punctuated equilibrium brings this idea to its extreme consequences, by proposing that a species show no evolution for long periods (*stasis*), and the isolation of a subpopulation brings about a rapid sequence of mutations under the changed evolutionary pressure. Once the condition of isolation were removed, the new species mixes back and competes with the old one, without possibility of inter breeding.

Whether the fossil records truly display a predominant pattern of stasis, the gaps being thus a real feature and not just missing information, continues to be an active area of research. Such evolutionary patterns have been indeed observed in the fossil records of many organisms. For example, some fossil records of foraminiferans (unicellular organisms with shells) are consistent with a punctuated pattern. On the other hand, also examples of gradual, non-punctuated, evolution are found.

In conclusion, it may be noted that the history of life on Earth has more than once gone through sharp turns, with sudden mass extinctions. Although the Cretaceous event of 65 million years ago is the most famous, because of the drastic disappearance of dinosaurs, other mass extinction events, even more devastating, have become evident from the sharp breaks in the fossil record: short bursts of time in which abnormally large numbers of species die out simultaneously. The most severe event known up to now occurred at the end of the Permian period, about 250 million years ago, when 96 % of all the living species perished; other dramatic events of this kind have been identified, at about 443 (Ordovician-Silurian), 359 (late Devonian) and 200 million years ago (Triassic-Jurassic). Changes in sea level, asteroid impacts, climate change, new kinds of plants altering the soil, asteroid impact, flood basalt eruptions, catastrophic methane release, drop in oxygen levels, have all been blamed for these extinctions.

3.6.5   And Still Thinking

The last Sections gave a drastically succinct account of theories and experiments that have been puzzling scientists and philosophers for centuries. Unsolved dilemmas plague just about every proposed feature of life. Why is defining life so frustratingly difficult? Why have scientists and philosophers failed for centuries to find a specific physical property, or a set of properties that could clearly separate the living from the inanimate?

The ancient Greek medical school of Hippocrates, Diocles, Praxagoras, defined the *pneuma* (πνεῦμα, "breath") as the material that sustains consciousness and maintains the functioning of all organs in the body, connecting the heart and the brain. Then, Aristotle theorised that all living things have one or more of three kinds of soul: vegetative, animal and rational. Humans are the only ones endowed with the rational. A few centuries later another Greek, the physician and philosopher Galenus merged the theories of pneuma and triple soul into a similar, organ-based system of tripartite souls. Thanks to his knowledge of anatomy and medicine, he assigned specific components of the soul to locations in the body, giving birth to the idea of "localisation of functions".

The idea of pneuma pervaded Middle-Age and Renaissance science in Europe, where alchemists began to formulate the concepts of *aether* (αιϑήρ, "upper air") as the invisible and intangible medium that transmits forces in the empty space, and *phlogiston* (φλογιστόν, "burning"), to explain the phenomena of combustion. The concept of an invisible medium supporting life processes was common and independently born in many ancient cultures, such as India with the *prana*, also interpreted as the vital, life-sustaining force of both the individual body and the universe; or ancient China, Korea and Japan, where the *qi* or *ki* was believed to be the active principle forming part of any living thing. In the Europe of XVII century, Georg Ernst Stahl and other alchemists/physicians began to describe a doctrine that would eventually become known as *vitalism*. Vitalists maintained that the living and the inanimate are governed by different principles, and that organic matter (i.e., materials containing carbon and hydrogen, produced by living things) could not arise from inorganic matter (materials lacking carbon, primarily resulting from geological processes). The experiments of XIX century chemists proved that inorganic matter can indeed be converted into organic, and discredited such a vision.

However, it is worth remembering that outstanding scientists sometimes persisted in supporting such theories, because of the many counterexamples that could still leave the door open to doubts. For example Lavoisier, despite being the one to prove wrong the phlogiston theory, took the caloric theory seriously for long time, and he certainly was not alone in this; the idea of aether resisted until the late XIX century, in the works of Maxwell and Kelvin; and Louis Pasteur maintained (and believed to have proved) that only living organisms could catalyse fermentation. Scientific theories rarely are absolutely correct or absolutely wrong. The history of science plentifully demonstrates how it is easy to be wrong, how brilliant minds could have set on the wrong track. With the benefit of time passed by, we can see why some scientists thought these were reasonable ideas. Such ideas and theories did not become failures overnight, they were legitimate proposals in their time and supported by intelligent people, not the fantasy of some weird minds.

Maybe defining what is life is so difficult, because such a property does not exist. "Life" might as well be a concept that we invented out of our brains. As physicists we are led to believe that, at the most fundamental level, all matter is an arrangement of atoms and their constituent particles, with increasing degrees of complexity. When trying to define life, we must draw a line at some arbitrary level of this complexity. Does such a division exist anywhere else besides our mind?

Going back to the ideas proposed in the first half of this chapter, the great difference between the various degrees of assembly of matter is *complexity*, rather than *life*. It may seem an easy way out to say that a bacterium can reproduce itself while a rock cannot. However, we have built chemical machines in the laboratory which are capable of reproducing themselves, and yet we do not dare to call such stuff "alive". By reasoning with examples and counterexamples we would always get stuck somewhere. It is my personal idea that we will probably never find this sharp dividing line, because *life* is more a concept than a reality. It is something embedded in our human way of perceiving the world around us. Maybe someplace out there, other minds different from ours could perceive this ensemble of organised atoms and molecules on entirely different scales of length and time, and come up with a very different dividing line. Maybe it is a matter of human perspective, just like looking at a pond. In that greenish pool of water, bursting with thousands of unseizable life forms, many of which will be born or die by the time we walk away, we see only the narcissistic reflection of our human image.

Appendix B: From DNA to Proteins (and Back)

Our current molecular-scale knowledge of biology comes from several disciplines: genetics, cytology, chemistry, molecular physics, biochemistry. Molecular genetics was born when Avery, McLeod and McCarthy suggested in 1943, and Hershey and Chase proved definitively in 1952, that deoxyribonucleic acid (DNA) is the molecular support of genetic information. At the time, Avery et al. discovery was regarded with skepticism, because most scientists still believed that genetic information should have a protein-based nature. Such experiments were brought to their logical completion with the elucidation of the DNA structure, by James Watson and Francis Crick in 1953 (Fig. 3.8), based on the x-ray DNA diffraction patterns of crystallographers Maurice Wilkins and Rosalind Franklin. This discovery earned the three men the Nobel Prize for Medicine in 1962. (It is generally agreed that Rosalind Franklin would have deserved the prize as well, but she had already died in 1958—the Nobel prize cannot be awarded to a dead scientist—from an ovary cancer probably induced by the x-rays with which she had worked for her whole life.) The existence of a genetic code for translating the information contained in DNA in a sequence of amino acids was firstly described by Khorana and Nirenberg in 1961. With the advent of genetic

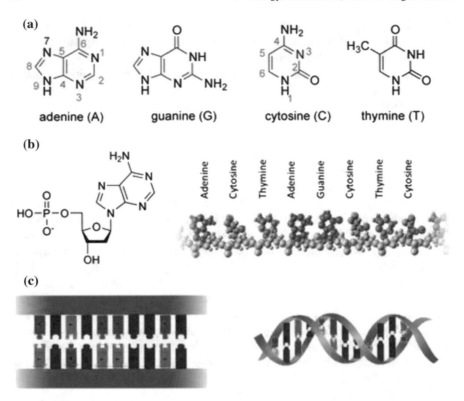

Fig. 3.8 Assembly of the DNA structure. **a** Molecular structure of the four DNA bases: Adenine (A), Thymine (T), Cytosine (C) and guanine (G). **b** Nucleotides (*left*) are formed when a base (in this case A) binds to a ribose sugar (pentagon), linked to a phosphate group. Since the ribose has lost one Oxygen in the binding, the nucleoside is called a deoxyribo-nucleoside, and becomes a deoxyribo-nucleotide when the phosphate (PO_4H) is attached. A polymer holding many bases (example on the right) is formed when a nucleotide phosphate loses the OH group, and can bind to the -OH hanging group of a ribose from another nucleotide; this ribose loses its H, which forms a H_2O molecule with the other OH, and the phosphate-sugar-phosphate-sugar- backbone of a single-strand DNA is thus formed. **c** Since A and T can form each two hydrogen bonds, while G and C can form three, two sequences of single-strand DNA can come together, if their respective sequences are complementary, pairing at every site two bases A-T or G-C on either side of the double backbone (*left*). Structural energy minimization, together with water and ion interactions in the nucleoplasm, force the paired double-polymer to assume the characteristic twisted double-helix shape (*right*), which won the Nobel prize to Wilkins, Watson and Crick, with the crucial help of Rosalind Franklin. Despite repeated attempts, we have been unable to obtain a response from the copyright holder. If notified the publisher will be pleased to rectify any errors or omissions

engineering in the 1970s, the genes of the most complex organisms could be decoded and directly analysed, until in 2003 the entire human genome was decoded.

The DNA macromolecule is a polymer formed by two oligonucleotide antiparallel strands, helically wound around each other in the famous 'double helix', as shown in Fig. 3.8. A strand of DNA consists of a sequence of four basic units, the **nucleotides**. Each of these bears one of the four different nucleobases: adenine (A), guanine (G), cytosine (C) and thymine (T). Each base is attached to a 2-deoxyribose (pentagonal sugar ring) by a N-glycoside bond, to form a **nucleoside**. When a nucleoside is linked to at least one phosphate, it becomes a nucleotide. A and G are called purine bases, while C and T pyrimidine bases. Subsequently, many phosphate-sugar groups can attach to each other, by losing one OH and one H each, respectively, which is liberated as a H_2O molecule. In this way, a long polymer of many nucleotides can be formed, a **single-strand** DNA (ssDNA). Since the OH terminus of the sugar is indicated by the symbol 3', and the CH_2 terminus (where the PO_4 sits) is indicated as 5', the order of bonding in the single-strand polymer is said to proceed in the 3'-to-5' progression. However, when DNA is being read in the transcription stage, the enzyme RNA-polymerase always proceeds in the 5'-to-3' direction.

As shown in the Figure, two single strands of DNA can come together and form a **double-strand** polymer (dsDNA). Because of their hydrogen-bond forming ability, A can pair with T (two H-bonds each), but not with C or G; on the other hand, C can pair with G (three H-bonds each), but not with A or T. If the two single polymer strands have complementary sequences on either side, for example:

$$5' - AGTCCAGCATG - 3'$$

$$3' - TCAGGTCGTAC - 5'$$

hydrogen bonds can be formed, and the antiparallel (i.e., with ends and heads reversed) double strand becomes a real DNA:

$$5' - AGTCCAGCATG - 3'$$
$$|| \ ||| \ || \ ||| \ ||| \ || \ ||| \ ||| \ || \ || \ |||$$
$$3' - TCAGGTCGTAC - 5'$$

Antiparallel here means that one strand is attached in the 3'-to-5' sense and the other in the 5'-to-3' sense. The spacing between each nucleotide is 0.34 nm, and the average twist at each base pair is about 35°, so that a complete turn of the helix requires about 10.5 base pairs, and a helix pitch of 3.6 nm; the bases also have a slight tilt by about -1 deg towards the 3' direction. A human DNA can contain about 3 billion base pairs, making for about 1 m of length (2 m if considered each strand separately), all packaged in the cell nucleus of about 1 µm diameter. These are the canonical Watson-Crick base pairings. In principle, other pairs may form (and are indeed observed in DNA, and more often in RNA) with "non standard" couplings, for example $C \equiv C$, since both cytosines can form three hydrogen bonds. However, such non standard pairs are comparatively quite rare, since their formation energy is higher than for standard pairs: they represent defects in the coding structure, and are called *mismatches*.

The characteristic double-helical DNA shape comes about from the molecular interactions besides the hydrogen bonding, notably: (1) a stacking interaction, of Van der Waals type, between the nearly parallel, twisted bases lying on top of each other along each side of the helix; (2) an electrostatic screening of the negative PO_4^- charges (counterions) along the backbone, mostly from the Na^+ ions in the physiological solution; (3) interaction with water molecules, surrounding and stacking along and

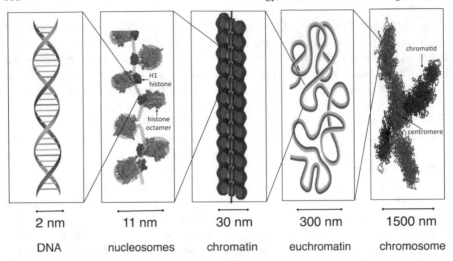

| 2 nm | 11 nm | 30 nm | 300 nm | 1500 nm |
| DNA | nucleosomes | chromatin | euchromatin | chromosome |

Fig. 3.9 The DNA molecule carries a large negative charge, due to the phosphates in the backbone. Histones are positively charged proteins that wrap up DNA. Double-stranded DNA loops around packs of 8 histones twice, forming the nucleosome, the building block of chromatin packaging. Further densification occurs by forming coils of nucleosomes, the chromatin fibre. These fibres are even more condensed into chromosomes during mitosis, the process of cell division. For most of the time in the cell cycle, however, DNA is loosely packaged into chromatin fibre

inside the helical grooves. (4) Elastic torsion and twisting energy of the backbone bonds. The sum of all these interactions makes the helical structure to be preferred with respect to the straight parallel strands. Upon forming the helix, it can also be noted (see again Fig. 1.8 above) that the grooves are not equally spaced, but alternate in a **major** and a **minor groove**, of width respectively 0.22 and 0.12 nm. Structural water molecules (micro-hydration) tend to nest preferentially along the grooves, while counterions may also bind close to the center of the helix. Divalent cations (Ca^{2+}, Mg^{2+}) have a higher affinity for the DNA grooves than monovalent ones (Na^+, K^+).

Whenever a cell is duplicated, its DNA must be identically copied to the daughter cell. Moreover, when some protein needs to be fabricated in the cell, some portion of the DNA must be read. The so-called *Central Dogma* of molecular biology states that the flow of information goes from DNA to RNA, and from RNA to proteins, and never can go in the reverse direction.

DNA Compaction, Chromatin and Chromosomes

For most of the time during the cell life, DNA is packed in a dense fibre of about 30 nm thickness called **chromatin**, nearly filling the entire cell nucleus (nucleoplasm) (Fig. 3.9). The 3-billion base pairs of the DNA double strand are regularly wrapped around blocks of 8 similar proteins, the **histones**, by making nearly two turns (147 base pairs) around each block. The ensemble of the 8 histones plus the 1.75 turns of DNA forms the **nucleosome**, the building block of chromatin. Nucleosomes are separated by a free DNA stretch (*linker*) of up to about 80 base pairs (27 nm). Nucleosomes can be densely or loosely packed, giving rise to different forms of chromatin: the dense 30-nm fibre called **heterochromatin**, and a lighter fibre of about 11-nm thickness, called **euchromatin**.

One big question is: since each cell has exactly the same DNA arranged into exactly the same genes, how come that cells, e.g., of the liver are so different from cells, e.g., in the brain? In more technical terms, how different portions of the DNA (*genome*) are used (*expressed*) in each different cell?

When the cell nucleus of a multicellular organism is considered, its most striking feature is probably the coexistence of denser chromatin regions next to less compact regions (Fig. 3.10). Such density differences are persistent, not just the result of random fluctuations in chromatin density. DNA density is strongly correlated with the transcription activity. Actively copied portions of DNA (*expressed genes*) tend to be found toward the center of the nucleus, in a region where chromatin is less dense and more accessible (the euchromatin). Inactive genes are found instead in the more compact regions of heterochromatin, most often situated at the nuclear periphery. In agreement with this description, it is observed that the activation of a genome portion results in an evident change of its topology. Gene transcription by RNA polymerase enzyme has been shown to occur at well-defined sites, called *transcription factories*. These factories are located within euchromatin, and each factory usually deals with genes that are expressed together (*coregulated*) by the cell. Therefore, it can be deduced that the differences (*phenotypes*) between the various cell types can be related to the specific way the genome is folded, in the nucleus of each different cell type.

The transcription machinery requires access to the genetic information throughout the cell cycle, to pick up protein-building instructions. On the other hand, the replication machinery needs to copy the DNA during the mitotic (cell duplication) phase. At the start of the cell duplication phase, the already dense chromatin fibre is further compacted into a double set of 23 + 23 **chromosomes**, which are subsequently distributed to the daughter cells. After the relatively rapid duplication phase, the chromatin returns to its disordered state inside the cell nucleus. During the phase of cell division, duplicated chromosomes form the well-known X shape, with each DNA copy forming one of the two rods (the sister *chromatids*, bound together at the *centromere*). The rest of the time (*interphase*) chromosomes are less condensed and fill the whole nucleus, more or less homogeneously [17]. From a physical point of view, chromosomes are more than just a sequence of DNA codes. They are giant

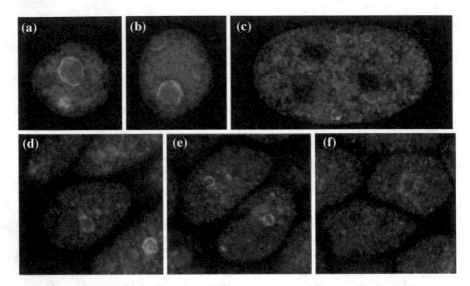

Fig. 3.10 Distribution of chromatin inside the cell nuclei, visualised with the FISH technique (Fluorescence In-Situ Hybridization), which attaches fluorescent probes to specific parts of the chromosomes (actually, to histone modifications corresponding to active or repressed genes); here, green is euchromatin, *red* and *blue* is two kinds of heterochromatin. **a, b** Retinal ganglion cell and **c** fibroblast nuclei from mouse. **d–f** Retinal rod cell nuclei from chipmunk, pig and macaque. [From Ref. [16], adapted w. permission.]

polymers formed by a long sequence of monomers, the nucleosomes. In analogy with the multi-level structural description adopted for proteins (see Appendix D), the structural conformation of the chromosome at different length scales can be described in terms of: a *primary* structure, the mere string of nucleosomes; a *secondary* structure, that is the conformation adopted by an array of 50–100 successive nucleosomes; and a *tertiary* structure, constituted by the 3-dimensional arrangement of several secondary arrays of compact nucleosomes.

DNA Transcription and Translation

One special variant of RNA, called messenger-RNA or mRNA, which takes the form of a long filament analogous to the copied DNA, is the molecule that is in charge of making a copy (*transcription*) of the chosen DNA sequence, and to carry it outside the cell nucleus, where the conversion of the original DNA sequence (*translation*) into a useful protein will be performed, in the special cell units called **ribosomes**. In the copying, RNA makes an exact duplicate of the DNA, with the difference that each T base is replaced by a U (uracyl). Therefore, for the above sequence, AGTCCAGCATG, the mRNA produces the sequence AGUCCAGCAUG (and similarly for the complementary sequence, on the opposite half-helix). Other variants of RNA are the transfer-RNA or tRNA, which is a small, trefoil-shaped

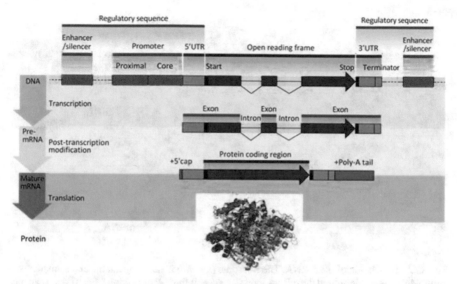

Fig. 3.11 The structure of a eukaryotic protein-coding gene. The DNA sequence is schematised by the (*horizontal*) chain of coloured blocks, and is divided into a 'regulatory' and a 'reading' part. The regulatory sequence controls when and where expression occurs for the protein coding region (*red*). Promoter and enhancer regions (*yellow*) regulate the transcription of the gene into a pre-mRNA, which is subsequently modified by adding a 5′ cap and poly-A tail (*grey*), and by removing introns. The mRNA-5′ and -3′ untranslated regions (*blue*) will regulate the translation into the final protein product. [Courtesy of T. Shafee (http://en.wikipedia.org/wiki/Regulatory_sequence) under CC-BY-SA 4.0 licence, see (*) for terms.]

fragment of about 80 bases, whose role is to transfer the necessary amino acids to the ribosome; and the ribosomal-RNA or rRNA, which does not participate in the duplication process but makes up the essential of the ribosome enzyme. The three types of RNA are found in the cells of any organism, and the rRNA is by large the most abundant.

The mechanism of copying DNA is extremely complex, both at the molecular and at the genetic scale (Fig. 3.11). Within the chromosomes, the portions of DNA coding for specific proteins are grouped into **genes**. About 25,000 genes have been identified in the human DNA. For example, chromosome-22, one of the smallest human chromosomes, contains about 50 million DNA base pairs; a subsection of 1 % of the chromosome contains 4 genes on average, of variable length (typically, 30- to 60,000 base pairs), separated by large portions of unknown function. The gene itself is made of small portions with the useful coding sequences (**exons**), and of large portions which do not participate in the coding (**introns**). The mechanism by which a gene is activated/enhanced, or repressed/silenced, goes under the name of **gene expression**, and is regulated by intricate chains of chemical signals. As shown in the figure below, a large part of the gene is also dedicated to such regulatory functions.

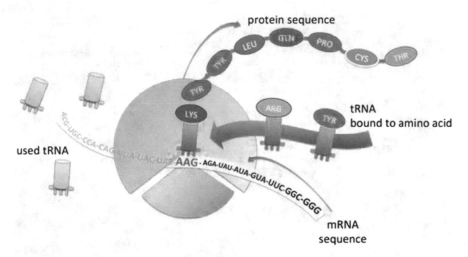

Fig. 3.12 Translation of the mRNA. The ribosome (*middle*) is made by a large and a smaller sub-units, which assemble around the mRNA long sequence of three-letter codons. Each codon can bind one specific copy of tRNA, carrying the right amino acid that fits in the sequence. Once in the active site of the ribosome, the amino acid is fed into the nascent protein sequence, and the used tRNA is expelled. The mRNA advances by one codon, and the whole process is repeated for the next tRNA to bring up its amino acid. The already read sequence is shown in *light grey*, and the sequence to be read in *black*; the *blue* codon is the one under current processing. The whole translation machinery implicates a number of auxiliary proteins, to deliver the *right* tRNA into the ribosome. The protein sequence coming out of the ribosome is eventually injected in the membrane of the endoplasmic reticulum, where it can fold into the proper native structure

The mRNA actually **transcribes** only the exon portions of DNA necessary to encode the given protein, and skips the non-coding (intron) portions, by cutting and rejoining the non-contiguous fragments in the process called **splicing**. The nascent mRNA is further processed by different chemical pathways, with the help of specific enzymes, such that the final (mature) mRNA is made only of a precise copy of the adjoining DNA portions expressing the protein sequence. The mRNA can thus exit the nucleus through pores in the nuclear membrane, and reach out in the cell cytoplasm for the ribosome, where its message will be **translated** from the 4-letter alphabet of the nucleic acids, into the 20-letter alphabet of amino acids.

The **ribosome** is a multi-unit structure, made by a special ribosomal RNA (rRNA) and proteins. It is the factory where amino acids are assembled into proteins, carried into the ribosome by transport-RNA (tRNAs), small noncoding RNA chains (74–93 nucleotides) that transport the "right" amino acid to the ribosome. Once the mRNA is outside the cell nucleus, the sub-units of the ribosome directly assemble around it, identifying the starting point for translation. Then, the mRNA is "read" and the chain of amino acids is linked together (Fig. 3.12). Translation and transcription rates vary between organisms. In prokaryotes, the two processes occur in a tight sequence, so the rates must be very close in order to avoid bottlenecks. In fact, in the bacterium *E. coli*, transcription is carried out by about 10,000 polymerase molecules, at a rate of 40–80 nucleotides/s; translation is performed by as many as 50–100,000 ribosomes, at a rate of about 20 amino-acids/second. Since each amino acid corresponds to three nucleotides (see below), the two rates are actually comparable. In eukaryotes the two processes are disjointed, since transcription occurs inside the nucleus and translation in the cytoplasm next to the endoplasmic reticulum; the corresponding rates vary according to the organism and cell type, for human cells about 6–70 nucleotides/second.

The four nucleic acids (U, C, A, G) in mRNA can be ordered in triplets, termed **codons**. These triplets make up the genetic code for transcription and translation into amino acids. For example, in the above sequence AGUCCAGCAUG, the codons would be AGU-CCA-GCA-UG. However, by starting at the next position, also A-GUC-CAG-CAU-G is a valid translation; as well as the AG-UCC-AGC-AUG. Therefore, each sequence has 3 possible *open reading frames*, which become 6 when considering the complementary half of the double helix. Since the biochemical reading process is statistical, it is not known a priori which sequence corresponds to a "good" protein, therefore the ribosomes attempt at reading the fragments wherever they find a starting point, and keep going until the stop signal is found. In principle, there are $4^3 = 64$ possibilities of forming unique triplet sets of the RNA or DNA bases. In this way, there are numerous redundancies in the correspondence such that more than one codon corresponds to each of the 20 naturally occurring standard amino acids (plus three combinations corresponding to a "stop" codon, and one combination corresponding to a "start" codon, which also codes for the amino acid methionine). In total, these give rise to 22 elementary building blocks (distinct pieces of information) for the construction of proteins.

The translation table of codons into amino acids is the equivalent of a sort of "Rosetta stone" of life, shown in Fig. 3.13. The table must be read starting from the left side, then on the top, and finally on the right: each pick of three bases identifies one line/row in the table, and a corresponding amino acid. Notably, of the more than 500 natural amino acids discovered up to now, only 20 are found in the proteins assembled by the machinery of the living cells, for any known living organism (three more variants are observed in special cases, which however are not directly coded by DNA). In the past 40 years, however, some twenty variations about this general code have been found, all special cases clearly derived from this original "frozen accident" (according to Francis Crick [18]), the most common being the replacement of the stop codon with tryptophan. (See "Further reading" at the end of the chapter.)

The chemical structure of the 20 amino acids is shown in Fig. 3.14. Each amino acid has a $-NH_2$ (amine) and a $-COOH$ (acidic) terminal; what changes from one to another is the **side chain**. Some side chains are hydrophilic while others are hydrophobic. Since these side chains stick out from the backbone of the molecule, they help determine the properties of the protein made from them. The side chains exhibit a wide chemical variety and can be grouped into three categories: non-polar, uncharged polar, and charged polar. The sequence of amino acids in each polypeptide or protein is unique, giving each molecule its characteristic three-dimensional shape, or *native conformation*.

Second Position

		U		C		A		G		
		UUU	phe	UCU	ser	UAU	tyr	UGU	cys	U	
	U	UUC		UCC		UAC		UGC		C	
		UUA	leu	UCA		UAA	**STOP**	UGA	**STOP**	A	
		UUG		UCG		UAG	**STOP**	UGG	trp	G	
		CUU	leu	CCU	pro	CAU	his	CGU	arg	U	
First Position	C	CUC		CCC		CAC		CGC		C	Third Position
		CUA		CCA		CAA	gln	CGA		A	
		CUG		CCG		CAG		CGG		G	
		AUU	ile	ACU	thr	AAU	asn	AGU	ser	U	
	A	AUC		ACC		AAC		AGC		C	
		AUA		ACA		AAA	lys	AGA	arg	A	
		AUG	met	ACG		AAG		AGG		G	
		GUU	val	GCU	ala	GAU	asp	GGU	gly	U	
	G	GUC		GCC		GAC		GGC		C	
		GUA		GCA		GAA	glu	GGA		A	
		GUG		GCG		GAG		GGG		G	

Fig. 3.13 Table of correspondence between DNA codons and amino acids. To be noted the sequences AUG, or methionine, signaling the starting point of a new protein, and UAG/UAA/UGA, signaling the point of stop of the sequence reading

A few Words on Epigenetics

Epigenetics is the reason why this Appendix was titled *From DNA to proteins (and back)*. In fact, such a statement seems to go against the so-called "central dogma", which states that information flows from DNA to RNA to proteins, and never the other way around. In the late 1990s, experiments on rats showed that the diet of pregnant mothers could alter the behaviour of genes in the offspring, that these changes could last the whole lifetime and—most importantly—be passed on to the children. Those rats' genes had been switched on or off by something that was happening in the environment, and not by mechanisms inherent to the gene itself. Such modifications were therefore called *epigenetic* (from the ancient Greek word επὶ, "above"). In the following years, similar findings in human cells started to create a true revolution in genetic thinking.

Epigenetic changes modify the activation/inactivation of certain genes, but not the DNA code sequence. The molecular structure of DNA, or the associated chromatin proteins, can be modified, and induce gene activation or silencing. Some epigenetic changes in the DNA script are implicit, a sort of "reprogramming" of the genetic message. Other changes are external, such as by mechanical or radiation damage of the DNA, or by some food components that alter the chemical reactivity at certain sites. For example, the formation of double-strand breaks in nearby sites of the double helix by UV radiation has been shown to leave epigenetic marks.

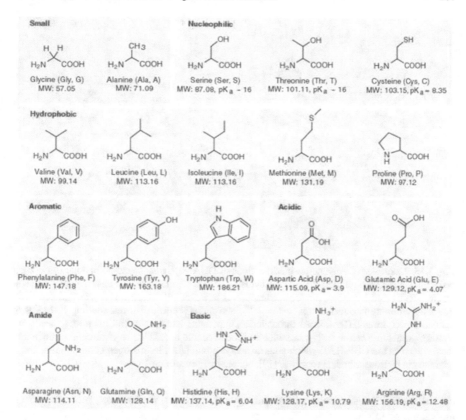

Fig. 3.14 The twenty amino acids constituting the building blocks for all known proteins of the living cells. The name is indicated next to the 3-letter or 1-letter shorthand for each molecule. Amino acids can be grouped according to their chemical behaviour, such as nucleophilic, acidic, hydrophobic etc. The acidic a.a. have a negative charge, while the basic a.a. have a positive charge, the others are electrically neutral. However, amide, aromatic and nucleophilic have a non-zero dipole moment. Both the charged and polar a.a. are hydrophilic, whereas all the aromatic a.a. (including proline) are hydrophobic. Note that each amino acid has two terminations, the amine NH_2 and the carboxyl COOH, through which the peptide bond can be formed, by removing one H atom from the amine and an OH from the carboxyl in the form of a water molecule. Since this binding process is endothermic, it is also reversible, namely by supplying water molecules the bond can be broken (hydrolysis of proteins)

Covalent modifications of DNA or of the histone proteins are among the most important markers of epigenetic inheritance. At the molecular scale, methylation, hydroximethylation, acetylation and phosphorylation occurring at particular sites are known to produce epigenetic gene regulation. These highly reactive, negatively-charged functional groups, $-CH_3$, $-CH_3OH$, $-COCH_3$, $-PO_3$ respectively, can react with well defined sites of DNA and histones, assisted by enzymes that catalyse the typically uphill reaction (Fig. 3.15). Similar epigenetic modifications in histones occur along the "tails", terminal amino acids sequences floating outside of the nucleosome main body, which can also interact with DNA and with the host of proteins that

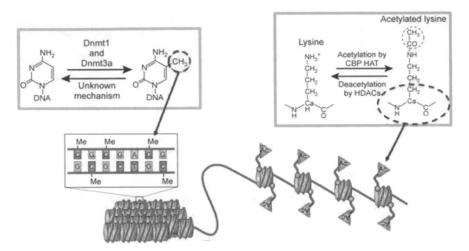

Fig. 3.15 Examples of epigenetics modifications of DNA (*left, blue box*) and histone (*pink, right box*). On the *left*, methylation (Me) of the cytosine base is promoted by the enzymes methyltransferase (Dnmt1 and Dnmt3a); the reverse reaction is however still unclear. Hypermethylated DNA promotes chromatin condensation (*below left*), and recruits chromatin remodelling factors with histone deacetylases (HDACs). At the opposite, hypomethylated DNA unfolds into a "beads-on-a-string" structure (*below right*), in which histones become accessible to chromatin remodelling factors such as the CBP-HAT, which adds an acetyl group (Ac). In this open configuration, genes are ready to be transcribed. [From Ref. [19], adapted w. permission.]

crowd around the chromatin fibre during transcription, replication, recombination, and repair of DNA.

The consequences of such epigenetic modifications of the genome (or "post-translational", since they occur after mRNA has been translated into proteins) are vast and largely unknown, potentially enormous, so as to produce a large debate among the scientific community. The molecular mechanisms by which such chemical mutations can be inherited are not yet clear, most notably for histone modifications. For the particular case of DNA methylation, at least, a more complete picture is just starting to emerge. The field is quickly growing, and with it the understanding that both the environment and individual lifestyle could also directly interact with the genome, to influence epigenetic change. These changes may be reflected at various stages throughout an individual's life, and even in later generations. Human epidemiological studies have provided evidence that prenatal and early-postnatal environmental factors may influence the adult risk of developing some chronic diseases and behavioural disorders. In one often-cited example, studies have shown that children born during the period of the Dutch famine from 1944–1945 had increased rates of coronary heart disease and obesity after their mothers' exposure to famine during early pregnancy, compared to children of mothers not exposed to famine.

Problems

3.1 Thermodynamic and probabilistic entropy are the same
Prove that $S = k_B \ln \Omega$ is fully equivalent to $S = -N k_B \sum_i p_i \ln p_i$.

3.2 Information entropy
Explain why both a disordered polymer and a nicely ordered crystal have a small value of information entropy, while a protein has a much larger value of either one.

3.3 Entropy of erasure
What is the value of entropy associated with destroying the entire information of a human DNA sequence? How this compares with the thermal entropy from the chemical breaking of bonds between the bases in the same sequence?

3.4 Genetic mistakes
The enzyme DNA-polymerase, responsible for the replication (transcription) of the DNA, puts a wrong base about every 10^9 nucleotides synthesised. By contrast, the RNA-polymerase, which translates DNA into RNA, makes a mistake about every 10^4 nucleotides. This may suggest that a translation error is less dangerous for the cell than a replication error. Why?

3.5 Peptide bonds in proteins
Which arrows in the following figure indicate a peptide bond? Which kind of bonds are the other ones? Which are the α-carbons?

3.6 The Solar system has a negative heat capacity
The heat capacity is defined as the temperature derivative of the energy of a thermo-dynamic system, $c_p = (\partial E / \partial T)$, and by its definition it is a positive quantity: the amount of energy necessary to *raise* the temperature by $1°$. The Sun and the planets in the Solar system are bound together by their own gravity. Since gravity force is purely attractive, the only thing that prevents the planets from collapsing onto the Sun is their motion, which can be related to their "temperature": the faster the planets move, the higher their temperature, and the more they can resist the pull of gravity. Can you show that the heat capacity of the solar system is negative?

References

1. R. Caillois, *Cohérences aventureuses* (Gallimard, Paris, 1976)
2. ENCODE Consortium, The: Identification and analysis of functional elements in 1% of the human genome by the ENCODE pilot project. Nature **447**, 799–816 (2007)
3. R.L. Lundblad, F.M. MacDonald, *Handbook of Biochemistry and Molecular Biology* (4th ed.). (CRC Press, 2010)
4. R.B. Martin, Free energies and equilibria of peptide bond hydrolysis and formation. Biopolymers **45**, 351 (1998)
5. D.J. Evans, E.G. Cohen, G.P. Morriss, Probability of second law violations in shearing steady states. Phys. Rev. Lett. **71**, 2401 (1993)
6. G. Crooks, Entropy production fluctuation theorem and the nonequilibrium work relation for free energy differences. Phys. Rev. E **60**, 2721 (1999)
7. T.L. Hill, *Free-Energy Transduction and Biochemical Cycle Kinetics* (Reprinted by Dover, New York, 2004)
8. J. England, Statistical physics of self-replication. J. Chem. Phys. **139**, 121923 (2013)
9. E. Szathmáry, J. Maynard-Smith, From replicators to reproducers: the first major transitions leading to life. J. Theor. Biol. **187**, 555–571 (1997)
10. S. Paula et al., Permeation of protons, potassium ions, and small polar molecules through phospholipid bilayers as a function of membrane thickness. Biophys. J. **70**, 339 (1996)
11. A. Vlassov, A. Khvorova, M. Yarus, Binding and disruption of phospholipid bilayers by supramolecular RNA complexes. Proc. Natl. Acad. Sci. **98**, 7706 (2001)
12. M. Keller, A.V. Turchyn, M. Ralser, Non-enzymatic glycolysis and pentose phosphate pathway-like reactions in the plausible Archean ocean. Mol. Syst. Biol. **10**, 725 (2014)
13. W. Martin, M.J. Russell (2005) On the origins of cells: a hypothesis for the evolutionary transitions from abiotic geochemistry to chemoautotrophic prokaryotes, and from prokaryotes to nucleated cells. Phil. Trans. Roy. Soc. London, B **358**, 59–85 (2005)
14. E. Mayr, Change of genetic environment and evolution, p. 157, in *Evolution as a Process*, ed. by J. Huxley, A.C. Hardy, E.B. Ford (Allen & Unwin, London, 1952)
15. N. Eldredge, S.J. Gould, Punctuated equilibria: an alternative to phyletic gradualism, p. 82, in *Models in Paleobiology*, ed. by T.J.M. Schopf (Freeman, San Francisco, 1972)
16. I. Solovei, M. Kreysing, C. Lanctôt, S. Kösem, L. Peichl, T. Cremer, J. Guck, Boris Joffe, B.: Nuclear architecture of rod photoreceptor cells adapts to vision in mammalian evolution. Cell **137**, 356–368 (2009)
17. T. Misteli, Beyond the sequence: cellular organization of genome function. Cell **128**, 787–800 (2007)
18. F.H. Crick, The origin of the genetic code. J. Mol. Biol. **38**, 367–379 (1968)

19. E. Korzus, Manipulating the brain with epigenetics. Nature Neurosci. **13**, 405 (2010)
20. J. Avery, *Information Theory and Evolution* (World Scientific, Singapore, 1993)
21. D.P. Bartel, J.W. Szostak, Isolation of new ribozymes from a large pool of random sequences. Science **261**, 1411–1418 (1993)
22. T. Cavalier-Smith, Obcells as proto-organisms: membrane heredity, lithophosphorylation, and the origins of the genetic code, the first cells, and photosynthesis. J. Mol. Evol. **53**, 555–595 (2001)
23. G.F. Joyce, The antiquity of RNA-based evolution. Nature **418**, 214–221 (2002)
24. R. Egel, D.-H. Hankenau, G. Labahn (eds.), *Origins of Life: The Primal Self-organization* (Springer, New York, 2002)
25. E.V. Koonin, A.S. Novozhilov, Origin and evolution of the genetic code: the universal enigma. IUBMB Life **61**, 99–111 (2009)
26. S.J. Gould, *Wonderful Life*, ed. by W.W. Norton ed. by (New York, London, 2002)
27. D. Stuart, *The Mechanisms of DNA Replication*. Intech Open, Rijeka, Croatia (2013) [This book is open-access under CC-BY-3.0 license, at the website www.intechopen.com/books/the-mechanisms-of-dna-replication, see (**) for terms.]
28. U. Meierhenrich, *Amino Acids and the Asymmetry of Life* (Springer, New York, 2008)
29. H.J. Morowitz, *Entropy for Biologists: An Introduction to Ihermodynamics* (Elsevier, Amsterdam, 2012)

Chapter 4
Energy Production and Storage for Life

Abstract This will definitely be the most "chemical" chapter of the entire book. While we are interested in describing the physics of living organisms, one cannot escape the fact that, at the most microscopic level, a variety of molecules and vastly complex chemical reactions constitute the basis of all life processes. Understanding some basic principles of how energy is obtained and stored by the cells in what constitutes the vast book of metabolism is very helpful, to understand how this energy is then transported and used, turned into work and heat, for all the functions of the body. It is just amazing to realise how deeply rooted are all such chemical mechanisms: the fact that we can observe the same chemical synthesis pathways in such distant organisms as a bacterium, an oak tree, and a giraffe, tells that these fundamentals were already well established in the early days of the evolutionary path of life on Earth.

4.1 From Food to ATP

Differently from plants, which can directly convert solar light into energy, animals need chemical intermediates to extract energy from food. Every living organism, including plants, employs the adenosine triphosphate (or ATP) as the universal currency to transport and exchange the energy necessary for survival and reproduction. Animals, however, cannot obtain energy directly in the form of ATP: they start from the lipids (fats), carbohydrates and, to a lesser extent, the proteins contained in the food, to extract the energy to be stored in the ATP molecules. The food fuel reserve is progressively decomposed by the living organism, to synthesise the ATP molecules according to both its instantaneous and overall needs.

ATP is not a long-lived molecule: it is obtained from the phosphorylation (i.e., addition of one phosphate PO_4^{3-} group) to its sister molecule ADP (see Appendix C), and it goes back to ADP when it is consumed in cell work. Therefore the sum of ATP plus ADP molecules is about constant, ATP being produced from ADP by an excess of available energy stored, and ADP being produced in turn by consuming the energy stored in ATP. Nevertheless, only a quite small amount of ATP, around 250 g in a whole human body, are available at steady state for the cells. Therefore,

© Springer International Publishing Switzerland 2016

F. Cleri, *The Physics of Living Systems*, Undergraduate Lecture
Notes in Physics, DOI 10.1007/978-3-319-30647-6_4

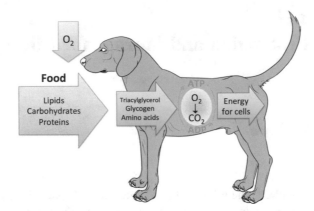

Fig. 4.1 Food enters the body as a mixture of lipids, carbohydrates and proteins. These compounds are digested and assimilated (i.e., degraded and broken into simpler components), and stored in the form of triacylglycerol and glycogen with the help of oxygen, mostly in the adipose tissue, the liver and (especially proteins) in the muscles. More than 90 % of the total energy stock is in the form of triacylglycerol for the ordinary, long-term metabolism. The glycogen stock is employed chiefly when there is a local need for a quick production of ATP, or when there is not enough oxygen immediately available to the cell (for example, during a very intense muscular effort). When the organism is not eating, the energy contained in the triacylglycerol and glycogen is progressively stored in the form of ATP molecules, by a reaction ATP $\leftrightarrows$ ADP + Pi. The energy contained in the ATP is extracted by breaking one phosphate bond, liberating the inorganic phosphate group (symbolised by Pi), and turning back the ATP into ADP. Examples of cellular work include: the contraction of muscles; actioning of molecular motors to move cilia or flagella; pumping of ions across the neuron cell membrane, to transmit the nerve information; synthesis of proteins in the ribosome; synthesis of macromolecules to generate new tissues during the organism growth (morphogenesis), and so on

several metabolic pathways exist, specifically arranged to maintain the correct ATP concentration wherever needed, with the aim of keeping such a limited resource at values constantly adjusted to the required level (Fig. 4.1).

Compared to the about 250 g present at equilibrium, a human body of 70 kg cycles between ATP + H_2O $\leftrightarrows$ ADP + Pi + H^+ the equivalent of its weight each day: whenever it is necessary to do some work (mechanical, electrical, chemical) somewhere in the cell, the energy necessary is rapidly obtained by breaking one of the phosphate bonds in ATP, which goes then from tri- onto diphosphate (ADP), thus liberating the equivalent of 30.5 kJ/mol. The spent diphosphate (ADP) plus one inorganic phosphate ion (usually indicated as Pi[1]) are subsequently recycled into ATP, at the level of mitochondria, thus storing in reverse the same equivalent of 30.5 kJ/mol, obtained from the degradation of the food.

How is it that the energy from metabolic food is stored, transported, and eventually used, to sustain all the living processes?

---

[1]At physiologic (neutral) pH $\simeq$ 7, Pi is a mixture of HPO_4^{2-} and $H_2PO_4^-$ ions, and the nucleotides are fully deprotonated. In such conditions, the hydrolysis reaction is properly written as ATP^{4-} + H_2O $\leftrightarrows$ ADP^{3-} + Pi + H^+, see e.g. Ref. [1] for details.

4.2   Storage of Energy in the Cell

Food enters the body intermittently, before being digested and assimilated. All the energy not immediately needed is stored in the strategic reserves of the body, which can be gradually exploited at later times, such as between feeding or during prolonged fasting. For an animal in normal condition and feeding regularly, more than 90 % of its fuel reserve is stored in the body in the form of lipids, mainly as triacylglycerol (or triglyceride, TAG) in the fat tissue, liver and muscles.

Lipids are favoured for the long-term synthesis of ATP, since they can be stored without water, in *anhydrous* form. They represent the lightest and most concentrated form of biochemical energy. Even if it is possible to store also carbohydrates and proteins in anhydrous form, lipids provide nearly twice as much ATP per gram of fuel (see next greybox on p. 120). Such an advantage is particularly exploited by migrating birds and hibernating mammals, which build their lipid reserves in excess of up to half their normal weight, to perform long distance flight, or to make it through the winter without eating.

Besides lipids, organisms maintain some less important energy reserves thanks to **carbohydrates**, such as the glycogen, a polymer of glucose mainly stored in the liver and muscles. Carbohydrates have the capability to support the highest rates of ATP production (although by smaller specific amounts) compared to lipids or proteins, and differently from other fuels they can also be employed in absence of oxygen (*anaerobic* conditions). Such unique characteristics make carbohydrates the essential compounds for short-term work and energy-demanding tasks, such as escaping a predator, catching a prey, or running the olympic races. Moreover, some important parts of the body, such as the nervous system or red blood cells, require glucose as their exclusive fuel.

Cell **proteins** play several important roles in the structural integrity (for example, the contractile proteins actin and myosin in the muscles, see Chap. 6), and in the regulatory mechanisms (such as enzymes and ion pumps, see Chap. 7). They are broken down into subunits, to be used in tissue regeneration and growth, and in making up the amino acid reserve. Comparatively, they play a smaller role in energy production and would not be employed to synthesise ATP under normal conditions. However, particular circumstances such as extremes of prolonged starvation in dry areas, or hugely energy-consuming functions such as the upstream swimming of some salmon species, can exhaust any other energy sources and lead to the utilisation of proteins, this being often the last resort of an organism before its death.

Figure 4.2 provides a sketch of the main steps and protagonists in the conversion of food into energy, starting from lipids, carbohydrates and proteins, down to the final main product, that is ATP molecules. The following greybox on *Beta oxidation* provides some information about the process of degradation of lipids, while the two important metabolic phases called the Krebs' cycle and the respiratory chain will be discussed in more details in the next sections. A very basic description of some of the most important molecules involved in these chemical reactions is given in the Appendix C to this chapter.

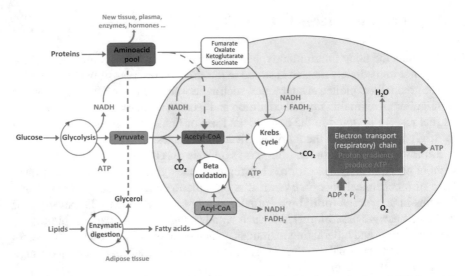

Fig. 4.2 The principal metabolic aerobic pathways implicated in the production of ATP from primary food sources. Cyclic chains of chemical reactions are indicated by *circles* with *arrows*. *Dashed arrows* represent minor pathways. Glucose (*on the left*) is firstly broken down into pyruvate molecules (two per each glucose) in the **glycolysis**; pyruvate enters the mitochondria (*light blue ellipse*) through its double membrane, and is turned into Acetyl-CoA. The same coenzyme can be formed in large amounts by the decomposition of lipids (*below*), via the **beta-oxidation** cycle; this latter cycle works the many molecules of Acyl-CoA obtained by the decomposition of fatty acids (long chains of hydrocarbons, $-(CH_2)_n-$). Proteins (*above*) have a minor part in the production of such coenzymes, however their decomposition (transamination) prepares, among many other compounds, a number of intermediates (fumarate, oxalate, etc.) which will enter the crucial Krebs' cycle. Acetyl-CoA is the main actor of the **Krebs' cycle**, whose main result is to produce a large amount of NADH and FADH$_2$. These will be completely oxidised in the subsequent cell respiratory cycle (or **electron transport chain**), which consumes oxygen and produces a high number of ATP molecules, from the available ADP+Pi. Note that some smaller amounts of ATP and NADH are also produced along the various connected cycles. Moreover, entry of the pyruvate inside the mitochondria consumes some ATP. At the end of the various cycles, CO$_2$ and water are the waste products. To feed the cycles, ATP and ADP molecules must be exported/imported in the mitochondria by specialised transport proteins

All organisms must be able to mobilise their different energy reserves at the right time, and with the appropriate rates, according to their life rhythms. The elementary bricks of the reserves, **triacylglycerol** and **glycogen**, must be further broken, before they can be used locally or transported to other tissues. The shuffling of fuels all around the body is commanded by neural and hormonal mechanisms, which activate the enzymes catalysing the hydrolysis[2] of triacylglycerol and glycogen. Such specific enzymes are the lipases, which decompose the triacylglycerol into simple glycerol and fatty acids, and the glycogen phosphorilase, which cuts glucose subunits one by one from a long glycogen molecule.

---

[2]Any chemical reaction in which bonds in a species are cleaved by adding water molecules.

Glycogen molecules and the long-chain fatty acids can therefore be further decomposed, either in the interior of the cells where they are produced, or be transported elsewhere by the blood circulation. For example, mammals under intensive effort supply energy to the muscles either by the local sources (TAG and glycogen reserves in the muscle), as well as from distant sources (TAG from the adipose tissue, and glycogen from the liver). Glucose can be easily transported in the blood plasma (the liquid part of the blood) and by the cytoplasm (the liquid part of each cell), since it is hydrosoluble in such aqueous fluids. On the other hand, fatty acids are only liposolubles (i.e., they are soluble in lipids or organic solvents), therefore they must be bound to hydrosoluble proteins, such as the plasma-albumin or other cytosolic proteins, to be transported into the tissues and cells.

The ensemble of food degradation into its basic components, and their further processing into smaller and smaller units until producing energy, building new tissues, or storing the excess in the form of reserves (most notably, fat or adipose tissue) is described in modern biology by a complex sequence of intertwined chemical cycles, making up the **metabolism**. Such a vast ensemble of finely coordinated chemical reactions can be read as being organised into *metabolic pathways*, eventually supported by enzymes, since many of the chemical steps are endothermic (see the greybox on p. 128).

4.3 Energy-Converting Membranes

The large majority of ATP in animal cells, and to a lesser extent in plant cells, is produced by membrane-bound proteins (enzymes), which are limited to a very particular class of **double-layer membranes**. Such *energy-converting* membranes are: (i) the plasma membrane of simple prokaryotes such as bacteria or blue-green algae; (ii) the innermost of the two membranes enveloping the mitochondria of eukaryote cells; (iii) the membrane of the thylakoids (small bags contained in the chloroplasts of plant cells). All these double membranes share a common evolutionary origin: in fact, both the chloroplasts and the mitochondria are generally considered as having evolved from a symbiotic relationship between a primitive eukaryotic cell, which was unable to breathe, and an invading prokaryote, an oxygen-breathing bacterium. Normally, the eukaryote cell would have "eaten" the bacterium. But at least once (likely more than once) it happened that the cell found it more useful to incorporate the bacterium, and started exploiting its oxygen-breathing capability. In this way, a proto-organism with a double layer of membrane was born, and was replicated in the mitochondria and chloroplast. Therefore, all such membrane-bound mechanisms of ATP synthesis, and the ion transport associated with these different membranes, are connected despite the different nature of their primary energy sources.

Beta-oxidation and the degradation of lipids

Beta-oxidation is the principal metabolic way leading to the degradation of fatty acids, to produce (i) the enzyme acetyl-CoA, which is subsequently oxidised during the Krebs cycle, and (ii) molecules of NADH and $FADH_2$ (see Appendix C in this chapter), whose high-potential electrons stoke up the respiratory chain. In eukaryote cells, beta-oxidation takes place in aerobic conditions, inside the mitochondrial double membrane.

Saturated fatty acids—Inside the mitochondria, the degradation of saturated fatty acids by the beta-oxidation cycle involves four chain reactions, which all take place within the mitochondria matrix. The complete degradation of the fatty acid is pursued, until the carbon chain is completely cut into single units of acetyl-CoA: this sequential process is called the *Lynen's helix*. Each turn of the "helix" shortens the fatty acid chain by two carbon atoms, and produces one molecule of acetyl-CoA, while regenerating one molecule of $FADH_2$ from FAD^{2+} and one molecule of NADH from NAD^+. Such a sequence of reactions occurs by oxidising the beta carbons of the acid, hence the name of **beta-oxidation**.

For the case of saturated fatty acids with an even number of carbon atoms, the last turn of Lynen's helix makes two acetyl-CoA, besides the $FADH_2$ and NADH. Each of the acetyl-CoA may be used as the key enzyme in the Krebs cycle, or serve in some biosynthesis process. For the case of saturated fatty acids with an odd number of carbons, the last turn also produces one molecule of the enzyme propionyl-CoA.

Unsaturated fatty acids—The beta-oxidation in this case has a peculiar difficulty, in that the presence of double carbon bonds along the beta-carbon chain hampers the proper functioning of the Lynen's helix enzymatic cycle. Other enzymatic substrates must intervene, to firstly convert the double bonds into single bonds, which overall slows down the degradation cycle compared to the case of saturated fatty acids.

Energy yield—The conversion of each fatty acid into acetyl-CoA by the enzyme acetyl-CoA-synthetase (often called an "activation") consumes the equivalent of two ATP molecules, since one ATP is hydrolysed into AMP (mono-phosphate) instead of ADP (di-phosphate) via this chemical reaction:

$$\text{Fatty acid} + \text{ATP} + \text{CoA-SH} \rightarrow \text{acetyl-CoA} + \text{AMP} + 2\text{Pi} \tag{4.1}$$

On the other hand, the NADH and $FADH_2$ produced accordingly can yield a maximum of 3 and 2 ATP molecules, respectively, in the cycle called the "respiratory chain" (in practice, about 2.5 and 1.5). The acetyl-CoA produced in the beta-oxidation is principally decomposed by the Krebs cycle:

$$\text{Acetyl-CoA} + 3\,NAD^+ + \text{CoQ10} + \text{GDP} + \text{Pi} + 2\,H_2O \rightarrow$$
$$\text{CoA-SH} + 3\,(NADH + H^+) + \text{CoQ10}H_2 + \text{GTP} + 2\,CO_2 \tag{4.2}$$

The $\text{CoQ10}H_2$ produces two more ATPs, thanks to the respiratory chain, such that the complete oxidation of the acetyl residue of the acetyl-CoA can make a maximum of 11 ATPs, which brings to $5 + 11 - 2 = 14$ the number of ATP molecules produced by each turn of the Lynen's helix.

Moreover, in the case of a fatty acid with an even number of carbons, also the last acetyl-CoA fragment can be oxidised, thus liberating 9 more ATPs. So, the beta-oxidation followed by the Krebs' cycle degradation of a saturated fatty acid with $n = 2p$ carbon atoms makes, for example, 112 ATPs for one molecule of palmitic acid, $CH_3(-CH_2)_{14}-COOH$, which contains 16 carbon atoms ($n = 16$, thus $p = 8$); or 42 ATPs for a molecule of caproic acid $CH_3(-CH_2)_4-COOH$, which contains 6 carbon atoms ($n = 6$, thus $p = 3$). When turned into energy contents (one mole of ATP giving off 30.5 kJ), each gram of palmitic or caproic acid give 13.3 and 11 kJ, respectively. For comparison, the complete oxidation of one glucose molecule, which contains 6 carbon atoms, corresponds to a maximum of 36 ATPs, that is 6.1 kJ per gram. Therefore, compared to glucose, caproic acid liberates 17 % more energy (on a per-molecule basis, that is more than 80 % on a per-gram basis) for a same number of carbon atoms in the pristine molecule. On a general basis, all fatty acids are much more energetic than carbohydrates, although sensibly slower in their rate of energy release.

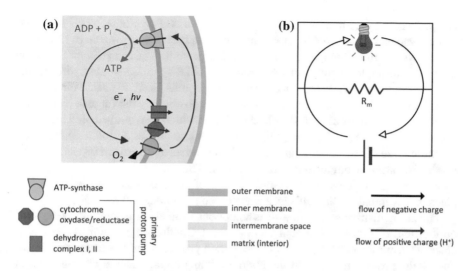

Fig. 4.3 A simple electrical circuit lighting a bulb is equivalent to the proton circuit in across the double membrane (e.g., in mitochondria, gram-negative bacteria, thylakoid membrane of chloroplasts) with its primary and secondary proton pumps. The primary pump is activated by electrons issued from the respiratory cycle (in animal cells), or by light energy (in plant cells), and pushes several protons into the intermembrane space. The electrical gradient so created, powers the secondary pump (ATP-synthase) to push back H^+ inside the matrix, while catalysing the ADP+Pi→ATP reaction (phosphorylation)

Both animal and plant cells contain mitochondria, although these organelles are much more numerous in animal cells. Animal cells get most of their ATP from mitochondria, whereas plant cells get most of their ATP from chloroplasts, and ATP generated from the mitochondria is only used when the plant cannot generate ATP directly from the light-dependent reaction.

Compared to other cell membranes, these energy-converting double-membranes share a number of distinctive characteristics. Each such membrane has two different types of specialised proteins working as **proton pumps** (Fig. 4.3). The function of a proton pump is to pull H^+ ions from one side to the other of the membrane, going against the chemical equilibrium, which would impose equal concentrations on the two sides, and the electrostatic equilibrium, which would impose zero charge difference. The result of the operation of such pumps is to create a **gradient**, both in H concentration and electric charge, between the two sides. Of course, the total charge must be conserved, and if positive charges go somewhere, there must be negative charge that comes back. Therefore, the electrons issued by the respiratory cycle (see below) provide the balance to the flow of protons.

The detailed nature of the proteins composing the *primary proton pump* depends on the nature of the energy source (light in plants, food in animals). By contrast, all the various energy-converting membranes share the same kind of *secondary proton pump*, formed by a highly conserved protein called ATP-synthase.[3] This big enzymatic complex is composed by a fixed and a turning part (see Fig. 4.8 on p. 128). In its normal way of functioning (observed when this protein works as an isolated system inserted in a simple fragment of membrane), the ATP-synthase would hydrolyse the ATP molecules into ADP+Pi, thereby pushing protons along the same direction as the primary pump. However, according to the proposition by the English biochemist Peter Mitchell and his **chemiosmotic theory** (for which he was attributed the Nobel prize in chemistry in 1978), the primary proton pump creates a proton gradient sufficient to force the secondary pump to work in reverse, and perform instead the ATP synthesis starting from available ADP and Pi. This concept was controversial for more than twenty years, because of a number of small details that did not fit into place, but eventually Mitchell's idea was proven right under all respects.

The mechanism of chemiosmosis is analogous to an electrical circuit, as depicted on the right of Fig. 4.3. The equivalent of the DC voltage generator is the flux of electrons from the cell respiratory cycle, which produces the electric potential that pushes the protons to move in the intermembrane region. The equivalent load of the circuit (the lightbulb) is represented by the secondary pump, the ATP-synthase, which uses this potential energy to move back the protons in the matrix, and their kinetic energy to activate the catalysis of new ATP molecules. Note that to avoid short-circuits, the naked cell membrane has a very high value of resistance (equivalent of the R_m) towards the ions.

It may be interesting to note that, from an evolutionary point of view, the disparity between primary and secondary proton pumps could imply that the first organisms to evolve should have relied on *external* proton gradients. Likely, they just shared some primitive versions of the ATPase, which was used to flow protons from some environmental source, into the primitive cell membrane. Later, organisms differentiated and evolved the double-membrane mechanism and the specialised primary proton pumps, which enabled them to produce mitochondrial proton gradients of their own.

Figure 4.4 shows the average structure of a mitochondrion from a eukaryote cell. The shape of mitochondria is not fixed but changes continuously inside the cytoplasm, and the appearance of *cristae* (the repeated folds of the inner membrane) can be very different, either for mitochondria isolated from different tissues, or when the same mitochondria are suspended in different media. For example, cardiac cell mitochondria, for which periods of high respiratory activity are necessary, tend to have a more important density of folds compared, e.g., to liver cell mitochondria.

Mitochondria are often considered the cell's power plants. Indeed they produce almost all the ATP in animal cells, starting from elementary molecules obtained from the decomposition of carbohydrates and lipids. To summarise the main steps, which will be fully described later in this chapter:

(a) inside the mitochondrial matrix, pairs of carbon atoms are sequentially cut from the long-chain fatty acids, in the cyclic beta-oxidation pathway, each pair producing one molecule of acetyl-coenzyme A (acetyl-CoA);

---

[3]In the evolutionary sense, a conserved element (a protein, a gene) is maintained across different species, even very distant from each other.

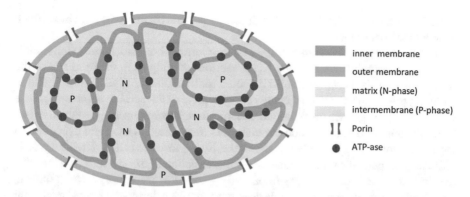

Fig. 4.4 Schematic representation of a mitochondrion. The two membranes (*inner and outer*) are shown. The 'matrix' is the region inside the inner membrane, and is considered negatively polarised (N), while the space between the two membranes is positively polarised (P). The *cristae* are the deep folds of the inner membrane, necessary to increase the surface/volume ratio

(b) in this same place, the pyruvate produced by the glycolysis of glucose is also oxidised, producing as well acetyl-CoA molecules;

(c) all the acetyl-CoA produced from the various pathways are then metabolised, by the ensemble of chemical reactions constituting the Krebs cycle, producing some ATP directly, plus larger amounts of NADH and $FADH_2$;

(d) during the subsequent cell respiratory chain, which takes place as well inside the mitochondrial membrane, the NADH is converted into NAD^+ and the $FADH_2$ into FAD^{2+}, upon the availability of oxygen, thus producing large amounts of ATP molecules; in parallel, electrons made available in the process reduce O_2 molecules (hence the name of "respiration").

Eventually, a good part of the energy contents from the carbohydrates and lipids is, through all these combined cycles, recovered and stored in the ATP. Taken together, all these reactions allow to utilise the energy contained in the metabolic fuel by reforming new phosphate bonds (ADP→ATP). This energy, transported anywhere by the diffusion of ATP, will be released when the ATP molecule is hydrolysed, to produce useful cellular work.

4.4 Krebs' Cycle and the Production of ATP

In the energy production process, the choice of one particular metabolic path (a *substrate*) depends: (i) on the availability of oxygen, (ii) on the rate at which ATP is demanded, and (iii) on the availability of fuel (food). Whenever very high rates of ATP synthesis are required, or under reduced oxygen availability, the anaerobic pathways are the only ones possible. On the other hand, for the resting or normal conditions, in which energy is consumed at regular and lower rates, and when oxygen is readily available, ATP is mostly produced by aerobic (oxidative) pathways, which have a much better yield.

The **Krebs cycle**, discovered in 1937 by the German-born biochemist Hans Adolf Krebs,[4] is a complex chain of chemical reactions starting from the products of the aerobic glycolysis. However it must be noted that, despite the glycolysis is found in all living organisms, the Krebs' cycle is only typical of the aerobic organisms, those which breath oxygen via the pulmonary respiration. Macroscopically, oxygen enters from the lungs and via the blood circulation is transported to each and every cell in the body; from there the CO_2 returns to the lungs, to be expelled. Kreb's cycle is ultimately responsible for the molecular transformation of oxygen into CO_2, for this reason it is sometimes—but very improperly—called a "respiratory" cycle. Indeed, the actual respiration step of the cells does not take place during the Krebs' cycle, but at the subsequent stage, appropriately dubbed the 'cell respiratory chain', in which electrons and protons are transported across the inner mitochondrial membrane. The other name of the Krebs' cycle, 'cycle of tricarboxylic acids', originates from the fact that two of the earliest substrates of the chemical reaction chain, the citrate and the isocitrate, are in fact acids, carrying three –COOH groups each.

At the outset of Krebs' cycle we find the **pyruvate** molecule, CH_3-CO-COOH, produced in the glycolysis pathway (two units for each molecule of glucose consumed). The global formula of the cycle is:

$$CH_3\text{-CO-COOH} + 3\ H_2O \rightarrow 3\ CO_2 + 10\ H^+ + 10\ e^- \qquad (4.3)$$

The energy liberated by the processing of the various molecular species is stored temporarily in the reduced species NADH and $FADH_2$ (plus one ATP directly produced during the cycle). Subsequently, the cell respiratory chain will extract this energy, to regenerate a much larger number of ATP molecules.

The Krebs' cycle is coupled uphill to the glycolysis, and downhill to the electron transport (or respiratory) chain (see again Fig. 4.2). The glycolysis cycle occurs in the cytoplasm, and its main product is the pyruvate, through this reaction[5]:

---

[4]H.A. Krebs was born and educated in Germany, completing his studies of medicine in the universities of Göttingen, Hamburg, and Berlin where he studied chemistry. Coming from a jewish family, he was forced to leave Germany in 1933 and emigrated to England, where he remained for the rest of his life. Therefore he and his work are often considered British, including the Nobel prize in medicine which was awarded to him in 1953 for his studies on metabolism. His manuscript on the citric acid cycle, still known under the name of "Krebs' cycle", was refused by the journal *Nature* in 1937, under the excuse of lack of space for publication. It is just one more example of big-name journals missing fundamental works, due to poor judgement.

[5]The graphic symbols in the glucose molecule indicate that two groups (full thick bonds) lie above, and three (dashed bonds) lie below the hexagonal ring; in the pyruvate, the dashes indicate that the electron is delocalised between the two oxygens.

The pyruvate can be consumed either in a fermentation (anaerobic) cycle, or be transformed into lactic acid, or—most importantly—enter the mitochondria, where the aerobic Krebs' + respiratory cycles will take place.

The complex chain of chemical reactions in the **glycolysis** can be formally grouped into two phases (each one occurring through several intermediate steps):

(1) the *oxidation* of glucose into pyruvate:

$$C_6H_{12}O_6 + 2\ NAD^+ \rightarrow 2(CH_3\text{-}CO\text{-}COO^-) + 2\ NADH + 4H^+ \qquad (4.4)$$

(2) followed by the *phosphorylation* of the ADP into ATP:

$$(2\times)\quad ADP^{3-} + Pi^{2-} + H^+ \rightarrow ATP^{4-} + H_2O \qquad (4.5)$$

If we take the combined sum of the glycolysis and the subsequent Krebs cycle, we get the synthetic formula:

$$C_6H_{12}O_6 + 6\ H_2O \rightarrow 6\ CO_2 + 24\ H \qquad (4.6)$$

from which the coenzymes and the ADP/ATP have been excluded, since they are recycled between the beginning and the end of the combined cycles. This can be further simplified to:

$$C_6H_{12}O_6 + 3\ O_2 \rightarrow 6\ CO_2 + 12\ H \qquad (4.7)$$

Now, if this cellular reaction is compared to the complete oxidation of glucose as it can be done in the laboratory:

$$C_6H_{12}O_6 + 6\ O_2 \rightarrow 6\ CO_2 + 6\ H_2O \qquad (4.8)$$

it is seen that the oxidation carried out in the cell with the glycolysis plus the Krebs' cycle is only partial: the 12 hydrogens are still in the reduced state. The last step will take place in the subsequent phase, the respiratory chain, when the hydrogens will be oxidised into water, and ATP will be produced.

On a side, it may be noted that the reaction (4.8) is exactly the complementary of the photosynthesis (Eq. 2.48 of Chap. 2). However, one is not simply the inverse of the other. In fact, the free energy balance is overall favourable for the glycolysis (its $\Delta G < 0$, in fact there is more energy contained in the glucose than it is effectively extracted), whereas photosynthesis has $\Delta G > 0$, therefore it requires a supplement of energy from the exterior (sunlight photons), in order to occur.

4.4.1   The Role of the Enzymes

The Krebs' cycle, or cycle of tricarboxylic acids, is a typical **enzymatic cycle**. Use of the word *cycle* emphasises the fact that the first molecule (a "substrate", in the language of biologists) coincides with the last one. Moreover, all the components of the cycle are recycled. This is the role of the enzymes, molecular species that are not consumed in the reaction but have the ability to ease the development of a reaction, by lowering its free-energy barrier (see the greybox on p. 128).

In prokaryotic organisms the whole cycle takes place in the cell cytoplasm. In eukaryotes, instead, it takes place within the matrix of mitochondria (i.e., deep inside the double membrane), which also implies that breathing would be impossible without mitochondria. The pyruvate produced by the glycolysis outside the mitochondria must pass the two membranes by a **symporter**, a membrane protein making a channel that allows the passage of two species at the same time: in this case, one pyruvate can enter while letting out one OH^-. Once the pyruvate is inside the matrix, Krebs' cycle takes places in the same way as for the prokaryotes.

The first substrate of the Krebs cycle is not just the pyruvate, but its byproduct the acetyl-coenzyme-A (or acetyl-CoA). This species is produced both from lipids (via beta-oxidation) and from carbohydrates (via glycolysis). The first step therefore consist in transforming the pyruvate into acetyl-CoA. This occurs by the intermediation of the enzyme pyruvate-dehydrogenase (all enzyme names end with "ase"), a giant protein which can include up to about 60 subunits, in a microorganism like *E. coli*.

During this first reaction, one CO_2 is liberated and one NAD^+ is reduced to NADH:

$$CH_3\text{-}CO\text{-}COO^- + NAD^+ + CoA \xrightarrow{PDH} Ac\text{-}CoA + NADH + CO_2 \qquad (4.9)$$

The product is an acetyl group ($CH_3C{=}O$) bonded to the coenzyme-A (see in the Appendix C what it looks like) by means of a highly energetic thiol bond (a sulphur atom bridging two carbons). It is this molecule that will actually start the Krebs cycle.

Even without paying too much attention to the complicated names and detailed reactions of the chemical species that make up the sequence of transformations along the cycle, we can just keep track of the number of carbon atoms in each molecule to see what happens (Fig. 4.5). The Krebs cycle can be seen as a sequence of decarboxylation reactions (i.e., loss of one C=O group, with release of a CO_2). Already the initial transformation of pyruvate into acetyl-CoA is the conversion of a 3-carbon species into one with only 2 carbon atoms.

The acetyl-CoA with its 2 carbons can start its cycle, by forming a bond with a 4-carbon molecule, the oxaloacetate: the CoA is liberated, and the new species containing now 6 carbon atoms is the citrate. After one first decarboxylation, one CO_2 is released and one NAD^+ is reduced to NADH, giving isocitrate with 5 carbon atoms. This will undergo another decarboxylation giving alpha-ketoglutarate, which now contains 4 carbon atoms, after releasing a second CO_2 and production of a

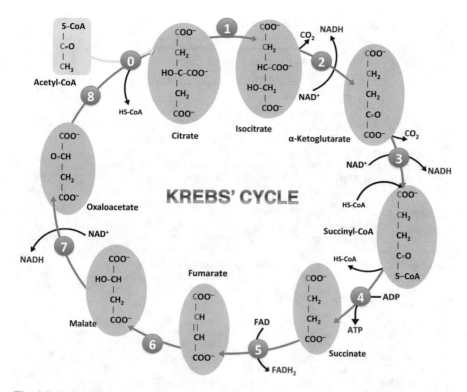

Fig. 4.5 Simplified scheme of Krebs cycle. Acetyl-CoA enters the cycle at step 0, it reacts with the oxaloacetate and turns into citrate, which is the step 1 of the cycle. In each turn of the cycle, the two "*red*" carbon atoms will be oxidised into two CO_2 molecules. The two "*green*" carbon atoms provided by the Acetyl-CoA will turn into "*red*" (oxidisable) carbons before the oxaloacetate is reformed from malate, in the last step of the cycle. Each reaction is catalysed by a different enzyme. Participating water molecules are not indicated. All the eight reactions of the cycle occur in the mitochondrial matrix. The overall yield of each cycle is 1 ATP molecule, plus 1 $FADH_2$ and 3 NADH molecules that will be fed into the electron transport chain. Note that the cycle is repeated twice for one initial glucose molecule, since glycolysis produces two pyruvates that give off two acetyls

second NADH. No carbons are lost in the subsequent reaction, in which succinyl-CoA binds back the CoA with its thiol bond. Subsequent release of the CoA leads to one ADP→ATP regeneration, and gives succinate. These 4 carbon atoms will survive until the end, however the molecule is going to be triply oxidised, subsequently going into fumarate, then malate, and finally oxaloacetate. Its three protons H^+ come back during the reduction of a third NAD^+ and one FAD, thus giving the third NADH plus one $FADH_2$. The end-product is still a 4-carbon molecule, an oxaloacetate identical to the starting one, which can now recombine with another acetyl-CoA and start over a new Krebs cycle.

The thermodynamic function of enzymes and the ΔG of chemical reactions

A generic chemical reaction, $A + B \rightarrow C$, among reactants A and B giving a product C, can practically take place (even if the free energy difference between reactants and products is advantageous, i.e. $\Delta G < 0$) only as a function of the activation energy E_a necessary to break the bonds in the reactants, and bring the chemical species in the reactive state. In terms of the reaction coordinate shown in the figure below, this corresponds to taking the system free energy at the top of the barrier in the $\Delta G + E_a$ curve. The activation energy represents a barrier to be passed, translating into an effective energy value the sum of all the kinetic constraints in the breaking and rebuilding of chemical bonds in the molecules, in order to observe the chemical reaction. In the case of $E_a = 0$ the reaction has no barrier and occurs spontaneously, at the corresponding values of temperature and pressure.

The ΔG of the reaction is intrinsically fixed by the difference between the free energies of the products and the reactants:

$$\Delta G = G(products) - G(reactants) \tag{4.10}$$

therefore it cannot change, whereas the barrier E_a can actually be modified as a function of the thermodynamic constraints (the values of chemical potential, or T, or P, or the local pH, and so on), and/or the kinetic path taken by the reaction, just as to go from your home to the supermarket you can take different paths, each with advantages and disadvantages and a different energetic cost.

A completely different way of lowering (to facilitate the chemical reaction), or in some cases increasing E_a, is to add an **enzyme** or catalyser, namely a chemical species which will take part in the reaction only as a 'bystander', and will be found intact at the end of the reaction:

$$A + B + (\text{cata}) \rightarrow C + (\text{cata})$$

The presence of the catalyser allows to take reaction paths that are more favourable from the kinetics point of view, while the ΔG obviously remains unchanged. With the values of the free energies of formation found in the literature (the following Table gives a sample, see e.g. [2] for the ATP/ADP series) we can calculate the free energy difference for any desired chemical reaction. However, such values do not tell us anything about the kinetic feasibility of that reaction, namely about the value of the eventual E_a existing between the initial and final states. To obtain that we must know the details of the reaction kinetics, and eventually the microscopic mechanism by which the enzyme would operate.

One must pay attention to the fact that the values of the formation energies allow to establish the ΔG under standard conditions, indicated by ΔG_0, corresponding to $T = 300$ K, $P = 1$ bar, and 1M concentrations of the reactants [2]. For example, for the oxidation of glucose, $C_6H_{12}O_6 + 6O_2 \rightarrow 6CO_2 + 6H_2O$, we would calculate:

$$\Delta G_0 = [6(-394.9) + 6(-237.57)] - [-917.61 + 0.] = -2877.21\text{kJ/mol} \tag{4.11}$$

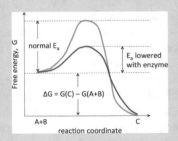

However, for a generic stoichiometry of the reaction, $aA + bB + ... \rightarrow cC + dD + ...$, we should calculate:

$$\Delta G = \Delta G_0 + RT \ln \left(\frac{[C]^c [D]^d \dots}{[A]^a [B]^b \dots} \right)$$ (4.12)

with $R = 0.00829$ kJ/mol/K, and T = temperature in Kelvin degrees. The term within parenthesis in the equation is defined as the **equilibrium constant** of the reaction, equal to the ratio of the concentrations of the products and reactants:

$$K_{eq} = \frac{[C]^c [D]^d \dots}{[A]^a [B]^b \dots}$$ (4.13)

It is thus seen that, even for a negative value of ΔG_0, the actual value of ΔG can be smaller or larger. The ΔG of the reaction approaches the value 0 at equilibrium, as soon as the concentrations of products and reactants spontaneously adjust to their equilibrium values. On the other hand, as we have already seen, for an endothermic reaction with its $\Delta G > 0$, the only way to make it possible would be to contribute free energy (enthalpy + entropy) from the exterior.

Table 4.1 Free energy of formation ΔG_0 (kJ/mol) for some biologically relevant substances

C-based			Inorganics	
CO	-137.34	Mannitol	-942.61	H_2	0.
CO_2	-394.4	Methanol	-175.39	H^+ pH7	0.
H_2CO_3	-623.16	Oxalate	-674.04	O_2	0.
HCO_3	-586.85	Phenol	-47.60	OH^- pH7	-198.76
CO_3^{2-}	-527.90	n-Propanol	-175.81	H_2O	-237.17
Acetate	-369.41	Propianate	-361.08	H_2O_2	-134.1
Alanine	-371.54	Pyruvate	-474.63	PO_4^{3-}	-1026.55
Aspartate	-700.4	Ribose	-757.30	Se	0.
Benzoic acid	+245.60	Succinate	-690.23	H_2Se	-77.09
Butyrate	-352.63	Sucrose	-370.90	SeO_4^{2-}	-439.95
Caproate	-335.96	Urea	-203.76	S	0.
Citrate	-1168.34	Valerate	-344.34	SO_3^{2-}	-486.6
Crotonate	-277.40			SO_4^{2-}	-744.6
Cysteine	-339.80	**Metals**		$S_2O_3^{2-}$	-513.4
Ethanol	-181.75	Cu^+	+50.28	H_2S	-27.87
Formaldehyde	-130.54	Cu^{2+}	+64.94	HS^-	+12.05
Formate	-351.04	Fe^{2+}	-78.87	S^{2-}	+85.8
Fructose	-915.38	Fe^{3+}	-4.6		
Fumarate	-604.21	$FeCO_3$	-673.23	**N-based**	
Gluconate	-1128.3	FeS_2	+150.84	N_2	0.
Glucose	-917.22	$FeSO_4$	-828.62	NO_2	+51.95
Glutamate	-699.6	PbS	-92.59	NO_2^-	-37.20
Glycerate	-658.10	Mn^{2+}	-227.93	NO_3^-	-111.34
Glycerol	-488.52	Mn^{3+}	-82.12	NH_3	-26.57
Glycine	-314.96	MnO_4^{2-}	-506.57	NH_4^+	-79.37
Glycolate	-530.95	MnO_2	-456.71	N_2O	+104.18
Guanine	+46.99	MnSO4	-955.32		
Lactate	-517.81	HgS	-49.02		
Lactose	-151.24	MoS2	-225.42		
Malate	-845.08	ZnS	-198.60		

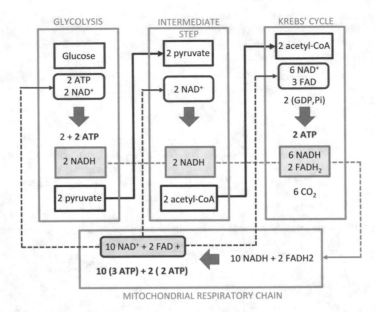

Fig. 4.6 Accounting of the ATP molecules produced during the coupled cycles of glycolysis, Krebs and respiration. The input is glucose, the output is carbon dioxide and water molecules (not shown). The intermediate redox couples $NAD^+/NADH$ and $FAD/FADH_2$ are completely recycled, but in the coupled cycling each mole of either produce respectively 3 and 2 moles of ATP

The global accounting of the cycle is shown in Fig. 4.6 Starting from the acetyl-CoA initially obtained from the pyruvate (Eq. 4.9):

$$Ac\text{-}CoA + 3\ NAD^+ + FAD + ADP + Pi + 2\ H_2O \rightarrow$$
$$\rightarrow 2\ CO_2 + 3\ NADH + FADH_2 + 2\ H^+ + ATP \qquad (4.14)$$

For each initial glucose molecule two pyruvates were produced, so we must multiply by 2 the above values. Each (double) Krebs' cycle produces therefore 2 ATPs, which are summed to the 2 ATPs directly produced in the cytoplasm by the glycolysis (reaction (4.4)): at this stage, 4 ATPs, 2 $FADH_2$ and 10 NADHs (including the two from reaction (4.9), and the two from glycolysis) are already produced, the waste being the 6 CO_2 molecules to be expelled by the pulmonary respiration, plus water molecules (which are not properly speaking a waste). Note that the two NADH produced outside the mitochondria will be treated in a different way from the eight produced inside.

4.5   Electrons and Protons Flowing

The outer mitochondrial membrane hosts a quantity of special proteins, the **porins** (see again Fig. 4.4), which act as non-specific pores, allowing the passage of any

Fig. 4.7 Three-dimensional structure of cytochrome-c (*green*), with the four α-helices evidenced in the cartoon representation. The central heme molecule (composed by the four pentagonal rings joined together, the *blue* vertex indicating a nitrogen atom), coordinates a Fe atom (*orange sphere in the middle*)

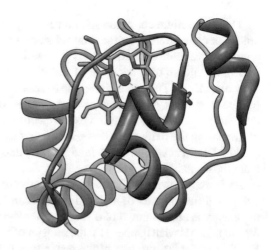

solute with a molecular weight below 10 kDa. Therefore, this membrane is freely permeable to ions and most metabolites.

As explained in the previous Sect. 4.4, the inner membrane is the true energy transducer. It hosts five different integral protein complexes: the NADH-dehydrogenase (complex I); the succinate dehydrogenase (complex II); the cyto-chrome-c reductase (complex III); the cytochrome-c oxidase (complex IV); and the ATP-synthase (complex V). Such different protein complexes with the function of proton pumps, implanted in the inner membrane, allow the electron removal from (i.e., oxidation) and transfer to (i.e., reduction) the different species involved in the respiratory cycle:

NADH $\rightarrow$ NAD$^+$	Oxidation
FADH$_2$ $\rightarrow$ FAD	Oxidation
Cytochrome-c ox $\rightarrow$ cytochrome-c red	Reduction
Cytochrome-c red $\rightarrow$ cytochrome-c ox	Oxidation
ADP $\rightarrow$ ATP	Oxidation

The critical step controlling the equilibrium of such **redox reactions** (see the greybox on p. 133) is the cycle of the cytochrome-c, a small metalloprotein including a heme group and a complexed Fe^{3+} ion (Fig. 4.7a):

$$\tfrac{1}{2}\text{NADH} + \text{cyt-c}_{ox} + \text{ADP} + \text{Pi} \leftrightarrow \tfrac{1}{2}\text{NAD}^+ + \text{cyt-c}_{red} + \text{ATP} \qquad (4.15)$$

Indeed, it is the availability of cytochrome-c in the reduced state in the inner space of the mitochondrial matrix, which forces the oxidation of the NADH and FADH$_2$ species. Equation (4.15) can be rewritten as a function of the ratio of the relative concentrations, thus giving:

$$\frac{[\text{cyt-c}_{red}]}{[\text{cyt-c}_{ox}]} = \left(\frac{[\text{NADH}]}{[\text{NAD}^+]}\right)^{1/2} \left(\frac{[\text{ADP}][\text{Pi}]}{[\text{ATP}]}\right) K_{eq} \qquad (4.16)$$

K_{eq} being the equilibrium constant of the above redox pair.

This equilibrium condition should be interpreted in the following way: the availability of cytochrome-c in the reduced state may increase either in the presence of a large concentration of NADH (as it would happen after many turns of the Krebs cycle), or in case of a decrease of the ATP concentration (for example, during a muscular effort). Therefore, either the excess of production of NADH, or the excess of consumption of ATP, unbalance the systems towards the increase of reduction of cytochrome-c, which in turn forces the electron transport chain to push in the sense of increasing the oxidation. The result is that of transforming the excess of NADH (and $FADH_2$) back into NAD^+ (and FAD^{2+}) and, in parallel, synthesising new ATP from the available ADP+Pi.

As it was shown in Chap. 2 (see the greybox on p. 31), the transfer of energy between different parts of a closed system can be made much more efficient when the energy is broken down into small amounts, instead of transferring it in just one big step. This is what happens for the transfer of energetic electrons from NADH (and $FADH_2$) to the last product of the respiratory cycle, i.e. water. The species NADH liberates electrons in the oxidative reaction:

$$NADH \leftrightarrows NAD^+ + H^+ + 2e^-$$

From Table 4.2 in the greybox *REDOX reactions*, we see that this corresponds to a $\Delta E_0 = -0.32$ V. By directly using the reaction products to couple to hydrogen oxidation into water:

$$2H^+ + 2e^- + \tfrac{1}{2}O_2 \leftrightarrows H_2O$$

we see that the $\Delta E_0 = +0.82$ V. The direct transfer of electrons from NADH to oxygen therefore corresponds to a free-energy difference of $\Delta G_0 = -2F(0.82 - (-0.32)) = 216$ kJ/mol, or about 53 kcal/mol. A similar redox-pair reaction occurs for $FADH_2$ going into FAD, giving off 43 kcal/mol. These are indeed huge amounts of energy (they correspond to 70–90 $k_B T$) that, if transferred directly, would be likely wasted into heat. Instead, a more gentle "electron transport chain" is being set up inside the mitochondria, by which this large energy jump is broken down into several steps, with the result of increasing the yield, and of displacing more protons.

This is the function of the various protein complexes, schematically depicted in Fig. 4.8a, which distribute the free energy step by step. The ensemble of such proteins constitutes the primary proton pump of most eukaryotic cell mitochondria. In the figure, the green arrows indicate the electron transfer: each of the complexes involved receives the energetic electrons, and transfers one or two H^+ from the matrix into the inter-membrane space. Electrons progressively lose their energy along the path and, for each initial electron, several protons (up to 10 for one NADH, equal to 2 electrons) are eventually transferred. It must be noted the key role in this process of the coenzyme-Q10, or ubiquinone (central red box): CoQ is the enzyme that transfers electrons from complexes I-II to complex III, no other molecule (except Vitamin A, in some conditions) is able to perform this function.

Electron transfer and REDOX equilibrium

A fundamental concept for chemical reactions occurring with transfer of electrons between the species (called oxidation-reduction reactions, or *REDOX*) is the difference ΔE_0 between the so-called reduction potentials of the donor and acceptor species. Measured in volt units, the relationship between the reduction potential and the free energy (in joules) is simply:

$$\Delta G_0 = -nF\Delta E_0 \tag{4.17}$$

for n moles of electrons transferred, with F the Faraday constant, equal to 96.63 J/V-mol (or 23.06 kcal/V-mol). A redox reaction must be equilibrated on the donor and acceptor side. Let us consider the 'reduction' of oxygen into water:

$$O_2 \rightarrow H_2O$$

Firstly, we must equilibrate the number of moles of the element to be oxidised (or reduced) by the right amount of water molecules, to have the same amount of oxygen on either side of the reaction:

$$O_2 \rightarrow 2H_2O$$

Secondly, we adjust the number of protons H^+ on either side:

$$O_2 + 4H^+ \rightarrow 2H_2O$$

Finally, we have to adjust the electric charge on either side, by adding the right number of electrons:

$$O_2 + 4H^+ + 4e^- \rightarrow 2H_2O$$

In chemical terms, we may think of the above as a "half-reaction", in which the two oxygen atoms in the O_2 molecule are ready to accept 4 electrons to be *reduced*. In electrical terms, free electrons do not exist, so any electron accepted must be donated by some other species. Where such electrons should come from? This reducing half-reaction must be coupled to another oxidising half-reaction, in which some other species will be oxidised, giving of the required 4 electrons. We speak of **redox couples** exactly for such a reason.

Let us take then for example glucose oxidation, and repeat the same steps. Firstly, adjust oxygen contents by means of water molecules:

$$C_6H_{12}O_6 + 6H_2O \rightarrow 6CO_2$$

Then the protons:

$$C_6H_{12}O_6 + 6H_2O \rightarrow 6CO_2 + 24H^+$$

And finally the charge:

$$C_6H_{12}O_6 + 6H_2O \rightarrow 6CO_2 + 24H^+ + 24e^-$$

Now, we can combine (=sum) the two half-reactions (reduction of oxygen + oxidation of glucose), under conditions of same number of transferred electrons from the donor to the acceptor species, therefore by counting 6 times the oxygen half-reaction and eliminating the excess water on the left/right sides:

$$C_6H_{12}O_6 + 6O_2 \leftrightarrow 6CO_2 + 6H_2O$$

It should be noted that the complete reaction could proceed in either direction, as a function of the relative concentrations (see the greybox on p. 128).

As an exercise, we may calculate the relative probability of a redox reaction under different 'breathing' conditions: aerobic (i.e., in oxygen, with water as a waste product), nitrous (with NO as a waste), sulphuric (with SO_2 as a waste), or methanogenic (with methane as a waste). The four acceptor species, to be reduced, are therefore: O_2, NO_3^-, SO_4^{2-} and CO_2.

We will test three different donors, to be oxidised: hydrogen (H_2), acetate (CH_3COO^-, or AcO^-), and methane (CH_4).

The following Table 4.2 allows to obtain the $\Delta E_0 = E_0(acceptor) - E_0(donor)$ (values in Volts). Therefore, for hydrogen we find:

$\Delta E_0(O_2–H_2) = 1.23$, $\Delta E_0(NO_3^-–H_2) = 0.84$, $\Delta E_0(SO_4^{2-}–H_2) = 0.19$, $\Delta E_0(CO_2–H_2) = 0.17$.

For the acetate:

$\Delta E_0(O_2–Ac) = 1.11$, $\Delta E_0(NO_3^-–Ac) = 0.72$, $\Delta E_0(SO_4^{2-}–Ac) = 0.07$, $\Delta E_0(CO_2–Ac) = 0.05$

And for the methane:

$\Delta E_0(O_2–CH_4) = 1.06$, $\Delta E_0(NO_3^-–CH_4) = 0.67$, $\Delta E_0(SO_4^{2-}–CH_4) = 0.02$, $\Delta E_0(CO_2–CH_4) = 0$.

By multiplying by the Faraday constant, $F = 96485.309$ C/mol, such values translate into free-energy differences, ΔG_0, with $n = 2$ for hydrogen (since there are two electrons in the donor molecule, $H_2 \rightarrow 2H^+ + 2e^-$), and $n = 8$ for both the acetate and the methane, thus obtaining the following values (in kJ/mol):

Donor	Acceptor			
	O_2	NO_3^-	SO_4^{2-}	CO_2
Hydrogen	−237.7	−162.3	−37.3	−32.8
Acetate	−858.1	−556.6	−56.4	−38.7
Methane	−819.4	−517.9	−17.8	0

The values show that the biological oxidation of one species (the *substrate*) is easier, from a thermodynamical point of view, in the presence of strong acceptors (i.e., larger values of ΔE_0), oxygen being evidently the most effective in all cases. The available free energy ΔG_0 decreases accordingly, from the stronger to the weaker acceptors, which also shows that aerobic pathways are usually preferred, since they permit to develop more energy.

Table 4.2 Reduction potential E_0 for some biologically relevant couples, in Volts

Redox couple					
SO_4^{2-}/HSO_3^-	−0.52	Pyruvate$^-$/lactate$^-$	−0.19	DMSO/DMS	+0.16
CO_2/formate	−0.42	FMN/FMNH	−0.19	Fe(OH)$_3$+HCO$_3^-$/FeCO$_3$	+0.20
$2H^+/H_2$	−0.41	$HSO_3^-/S_3O_6^{2-}$	−0.17	$S_3O_6^{2-}/S_2O_3^{2-}$+HSO$_3^-$	+0.225
$S_2O_3^-/HS^-$ + HSO_3^-	−0.40	HSO_3^-/HS^-	−0.116	cytochrome-c1 ox/red	+0.23
Ferrodoxine ox/red	−0.39	menaquinone ox/red	−0.075	NO^{2-}/NO	+0.36
Flavodoxine ox/red	−0.37	APS/AMP+HSO$_3^-$	−0.075	cytochrome a3 ox/red	+0.385
$NAD^+/NADH$	−0.32	mubredoxine ox/red	−0.057	NO_3^-/NO_2^-	+0.43
Cytochrome-c3 ox/red	−0.29	acyl-CoA/propionyl-CoA	−0.015	SeO_4^{2-}/SeO_3^{2-}	+0.16
CO_2/acetate	−0.29	glycine/acetate$^-$ +NH$_4^+$	−0.010	Fe^{3+}/Fe^{2+}	+0.16
S^0/HS^-	−0.27	$S_4O_6^{2-}/S_2O_3^{2-}$	+0.024	Mn^{4+}/Mn^{2+}	+0.16
CO_2/CH_4	−0.24	fumarate^{2-}/succinate^{2-}	+0.033	O_2/H_2O	+0.16
$FAD^{2+}/FADH$	−0.22	cytochrome-b ox/red	+0.035	ClO^{3-}/Cl^-	+0.16
SO_4^{2-}/HS^-	−0.217	ubiquinone ox/red	+0.113	NO/N_2O	+0.16
Acetaldehyde/ethanol	−0.197	AsO_4^{3-}/AsO_3^{3-}	+0.139	N_2O/N_2	+0.16

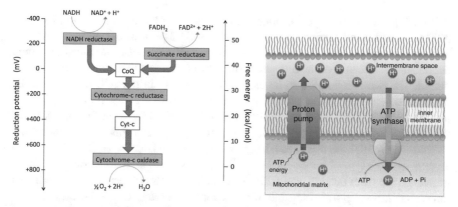

Fig. 4.8 **a** The electron transport chain taking place among the protein complexes integral to the inner mitochondrial membrane, indicated by the *light blue boxes*. The *red-lined boxes* indicate the coenzyme-Q10 and the cytochrome-c species, respectively. *Green arrows* indicate the progressive transfer of the two electrons liberated in the oxidation of NADH or FADH$_2$. Note that the three steps shown in *blue rectangles* actually represent several multiple steps, one for each species oxidised. Reduction-potential (*left*) and free-energy (*right*) scales approximately indicate the relative values at each different reaction step. **b** Schematic of the oxidative phosphorylation across the inner mitochondrial membrane. The primary proton pumps (*on the left*) increase the H$^+$ concentration in the intermembrane space, thus creating a large electrical and chemical gradient across the inner membrane. Subsequently, the protons flow back into the matrix (innermost region of the mitochondria) across the secondary pump ATP-synthase. The lower part (spheroid) of this protein rotates very fast (up to 500 Hz) providing the free energy to catalyse the ATP synthesis from ADP+Pi. The rotating ATP-synthase can catalyse up to about 100 ATP per second

As described in this complex chain of biochemical steps (which would be even more complex, if all the details were included), the flow of electrons from the matrix into the inter-membrane space during the respiratory chain establishes the corresponding gradient of protons across the inner mitochondrial membrane. At this stage there are strong electrical and pH gradients set across the inner membrane, because of the excess of protons accumulated in the inter-membrane space, while the interior of the matrix is largely negative and high-pH.[6] Therefore the protons, which exited the matrix through the primary proton pump, are now forced to cross the membrane in the opposite direction, going back into the matrix through the protein ATP-synthase (see Figs. 4.3 and 4.8b), thus stimulating this enzymatic complex to synthesise the new ATP molecules. As shown in the figure, this protein complex is a kind of helicoidal rotatory engine, which can turn under the positive charge flow. The name of **oxidative phosphorylation** given to this process originates from the fact that upon the turning of the ATP-synthase, a phosphate (Pi) group is attached to the ADP, to make a new ATP molecule. The ATP-synthase is divided into 10 identical subunits, each of which liberates one H$^+$ in a partial turn, therefore to make a complete 360°

---

[6]While this is a general statement, the situation can be different in particular cases. Due to the simultaneous influx of other charged species, as for example in the thylakoid membrane of chloroplasts, the relative contribution of the electric gradient and pH gradient can be largely different. A large pH gradient requires the membrane to be little permeable to anything but protons.

turn each proton crossing the channel pushes the rotor by about $36°$. On average, 3 ATP are synthesised at each turn, with a ratio of 3.33 protons per ATP.

It will be noted that, at this stage, the new ATP molecules produced are found deep inside the matrix of the mitochondrion, and someway they must be transported outside the two membranes, to be used anywhere else in the cell. In fact, also the opposite is necessary, namely the ADP and Pi must find their way from the cytoplasm into the mitochondria and across the two membranes, to arrive in the matrix where this whole "battery recharging" process can take place. For this scope, other trans-membrane proteins are used: the adenine nucleotide translocase (ADP/ATP carrier), an *antiporter* (a type of ion channel that allows passage in both directions) which can transfer ADP from outside, and ATP from the inside; the Pi instead uses a different ion channel, a *symporter* (capable of allowing the passage of two species simultaneously) which carries the Pi together with the H^+ flow. All such mechanisms are still under active research, and only for a few of them we have a rather complete microscopic explanation.

4.6   Energy Yield in the Cycle

As shown in the preceding Fig. 4.2, the Krebs' cycle is situated at the intersection of several metabolic pathways. At any given time during the daily life of an organism, the various reactants could be present in right amounts, or else be depleted towards some other pathway or cycle, because of a number of physiological reasons. Under normal conditions, the various substrates are maintained at the right concentrations by the **transamination** of proteins occurring in the liver, some of the amino acids being able to turn into fumarate, ketoglutarate, succinate, oxaloacetate, some other being able to produce more pyruvate and acetyl-CoA.

The very crucial Krebs' cycle can function only if some basic constraints are respected. Firstly, the right amount of oxaloacetate, which is to react with the acetyl-CoA at the start, must be correctly regenerated at the end. If for whatever reason the oxaloacetate is diverted to other metabolic paths, it must be quickly produced by complementary ways; for example, by breaking down phosphoenol-pyruvate, the intermediate molecule which precedes the pyruvate during the glycolysis. As well as the oxaloacetate produced from proteins, this externally-produced CoA must then cross the mitochondrial membrane, to restart the Krebs' cycle.

Another important condition is that the right amount of NAD^+ is available. Under conditions of lack of oxygen, enough NAD^+ must be regenerated from NADH upon oxidation of pyruvate into lactate (subsequently resulting in the annoying excess of lactic acid, giving spikes of pain in the muscles for example after a prolonged running effort): in this case, the direct (oxygen-less) conversion of glucose into lactate is called **anaerobic glycolysis**. When oxygen is scarce the electron transport chain is unusable. Therefore energy is converted only in the first step of the glycolysis, which produces 2 NADHs and 2 ATPs in the cytoplasm, after which the two NADH molecules must be recycled back to NAD^+. The enzyme lactate-dehydrogenase (LDH) provides the necessary help in this case, by the reaction:

$$CH_3\text{-}CO\text{-}COO^- + NADH + H^+ \xrightarrow{LDH} CH_3\text{-}CH\text{-}OH\text{-}COO^- + NAD^+ \quad (4.18)$$

in which pyruvate is transformed into lactic acid. This reaction belongs to the category of **fermentation**, with the peculiarity of being one of the rare cases of fermentation in which the waste product is not a gas.

It is worth noting that Krebs' cycle itself does not produce much ATP directly: only 1 ATP per cycle is produced, after the reduction of coenzyme CoA. Most importantly, the key role of this cycle is to reduce the various participating species, with NAD^+ turning into NADH, and FAD turning into $FADH_2$.

The 4 NADH produced during the cycle contain lots of energy, since each one of them can give back 3 ATPs in the next step, the electron transport chain, i.e. $2 \times 12 = 24$ ATPs for one glucose molecule (which powers two cycles, with the two pyruvate produced by the glycolysis). Similarly, $FADH_2$ is capable of producing 2 ATPs, i.e. 4 total. The grand total is therefore a maximum of $24 + 4 = 28$ ATP molecules, produced inside the mitochondria during the cellular respiration.

We can now sum up the glycolysis, the Krebs cycle, and the respiratory cycle (electron transport chain), to find the maximum theoretical amount of ATP produced by one molecule of glucose: the two NADH produced in the cytosol in the anaerobic glycolysis step (see reaction (4.4)), and which are not consumed here (contrary to what happens in the anaerobic fermentation) can give 6 ATPs. The subsequent entry of each of these NADH in the mitochondria consumes 1 ATP, for a total balance of $(2 + 6 - 2)$ ATPs in the glycolysis, 2 in the Krebs, and 28 in the respiration = 36 ATPs (see again Fig. 4.6). Each mole of ATP corresponds to a stored energy of 30.5 kJ/mol, to be compared to the pristine energetic contents of one mole of glucose molecules (as measured for example in an experiment of calorimetry), equal to 2871 kJ/mol. Therefore, 36×30.5 kJ represents about 38 % of stored energy for each mole of glucose consumed, that is largely superior to the yield from the fermentation (anaerobic) pathway, equal to only about 4 %. Also, note that since each molecule of glucose yields 10 NADH and 2 $FADH_2$, the energy initially stored in these coenzymes is $10 \times 52.6 + 2 \times 43.4 = 612.8$ kcal/mol, or 2565.2 kJ/mol, that is 89 % of the available glucose enthalpy. The greatest loss of efficiency of the energy conversion, which brings the 89 down to 38 %, is to be attributed to the complex series of reactions inside the mitochondrial matrix, necessary to get ATP from the coenzymes.

In the simpler prokaryotes, all the reactions are carried out in the cytoplasm, the entry fee across the mitochondrial membrane does not exists, and the maximum theoretical production would therefore be of 38 ATPs, with an even slightly better yield of about 40 %. Such values are close to the best yields of machines using chemical energy, such as internal combustion engines. However, it should be underscored that the utilisation of NADH and $FADH_2$ to produce ATP does not depend on the Krebs' cycle, but on the respiratory chain whose actuation enzymes (in the eukaryotes) are situated within the intermembrane cleft of the double mitochondrial membrane. Therefore, such theoretical estimates are never fully obeyed, for various reasons the ATP yield is not optimal, and the total number of molecules is rather close to about 30 ATP/glucose.

Each mole of NADH or FADH$_2$ consumes half a mole of oxygen and makes one mole of water (see Fig. 4.8a). In biochemistry experiments, it is a standard procedure to correlate the measured consumption of oxygen to the measured production of ATP (see also Problem 4.6). With the above calculations, this amounts to about 5–6.3 moles of ATP produced from one mole of glucose, per mole of oxygen consumed. Notably, the fact that this number is not fixed but variable, points to the fact that the whole conversion process is *non-stoichiometric*. This was the original motivation that pushed the biochemists to search for alternate mechanisms, until Mitchell proposed the proton-gradient theory. The flow of H$^+$ through the proton pumps powers the synthesis of ATP, in much the same way that the flow of water through turbines generates electricity. This explains why respiration is not stoichiometric: a gradient, by its very nature, is a continuous variation of a quantity.

4.7 Temperature and Heat in the Animal Body

Thermoregulation is the complex mechanisms by which animals maintain a constant body temperature under changing external conditions. Given the variety of Earth's ecosystems, ranging from the polar ice packs to the equatorial savannah, each species has a preferred body temperature at which its metabolic functioning is optimal. Cold-blooded animals regulate their body temperature via the body surface, which exchanges heat with the external environment. On top of this, warm-blooded animals also employ physiological mechanisms which can autonomously produce and dissipate heat in the body (i.e., cells) volume. Under steady conditions, the power balance requires that the sum of the body's basal metabolism (for the normal functions like heartbeat, respiration, digestion etc.) plus the work done by muscles, $M + W$, equals the sum of all heat gains or losses, H (including convection, conduction and radiation), plus the latent heat E from water evaporated through the body surface, and the amount of energy S eventually stored in the body, by chemical conversion into fat and other tissues:

$$M + W = H + E + S \quad \text{(Watts)} \tag{4.19}$$

Note that when the external temperature is higher than that of the body, $T_{ext} > T_B$, conduction, convection and radiation actually transfer heat from the outside environment to the body, while the opposite occurs when the external temperature falls below T_B.

Conduction occurs via the contact of the skin to the air, according to the simple balance equation for the power, i.e. amount of heat flux released in a unit time through a surface A:

$$\frac{\Delta Q}{\Delta t} = \frac{\kappa A(T_{ext} - T_B)}{d} \tag{4.20}$$

where $\kappa = 2.4 \times 10^{-2}$ W/(m K) is the air thermal conductivity, and $d \simeq 5$ cm is a typical distance over which the temperature of the body T_B goes over the (higher or lower) air temperature T_{ext}. A minor contribution of convection comes from the fact that the surrounding air is moving, and it can be accounted by slightly adjusting the distance d. Taken together, the two mechanisms account for about 10 W (input or output) for a temperature difference of $\pm 10°$. (So small only thanks to the low conductivity of air. Heat conduction is a totally different story in water, since $\kappa = 0.6$ W/(m K) for water, and heat flow occurs much faster. This is why divers always must wear a technical suit, even in relatively warm waters around 20°.)

Radiation is a major source of heat flux, to and from the body surface. According to the Stefan-Boltzmann equation (which we already saw in Chap. 2):

$$\frac{\Delta Q}{\Delta t} = \varepsilon \sigma A (T_{ext}^4 - T_B^4) \tag{4.21}$$

with ε the emissivity coefficient of the skin, and σ the Stefan-Boltzmann constant. For infrared radiation, which is the largest component around temperatures of 300 K, the human skin behaves nearly as a perfect blackbody, i.e., with $\varepsilon \simeq 1$ and equal emission/absorption efficiency. Radiation may account for about 120 W power, for temperature differences of about $\pm 10°$. However, due to the T^4 dependence (note that temperature in Eq. (4.21) must be given in K, not in °C), this contribution is rapidly varying as a function of temperature, from only 20 W for 2°–3° difference, to more than 400 W if emitted by a human body put at $T_{ext} = 273$ K (or 0 °C).

Evaporation of water from the skin pores is a very important means of heat loss, through the equation:

$$\frac{\Delta Q}{\Delta t} = r_{wH} \lambda_H \tag{4.22}$$

with r_{wH} the water evaporation rate (mass or volume per unit time). The *latent heat of evaporation* of water at 37 °C at the skin surface is $\lambda_H = 0.58$ kcal/g, i.e. the evaporation of 1 g of water from the body surface removes 580 calories (2430 J) of heat. Besides a smaller fraction emitted as vapour, breathed out or diffused through the skin, water loss occurs mostly in liquid form through sweating or panting. This liquid deposited at the skin surface then has to evaporate. The rate of evaporation depends most importantly on the relative humidity (RH) of the air, and can only occur when RH < 100 %. The minimum rate of water loss from the human body (the "insensitive" loss cited in physiology books, meaning that this is not directly related to sensing of external temperature differences) is about $r_{wH} = 600$ g/day, or 0.025 l/h, to which water lost by sweat or panting must be eventually added. Overall, this amounts to about 17 W. However, note that under medium-rate exercise the perspiration rate is about 1.5 L/h, and it can attain 3.5 L/h in a tropical climate, corresponding to a heat loss rate of more than 2 kW. (At such rate of sweating, the unfortunate human must drink water at about the same rate, to avoid dehydration.)

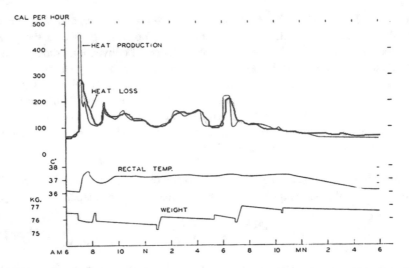

Fig. 4.9 An estimation of the variations which occur in the daily life of a human, obtained partly by direct measurements on several days and partly by estimation from calorimeter experiments on other days. The subject runs for jogging at 7:00 a.m. daily; other peaks in the curve of heat production occur when he travels from his home to the job in the morning, engages in the usual work activities, and walks home in the evening. Heat loss highlighted in *red*. The variations in weight are caused by meals, voiding, insensible perspiration. [Adapted from E. DuBois, 1938 Harvey Society lecture, with kind permission of the New York Academy of Medicine.[7]]

Evidently, water evaporation is by far the most efficient way of dissipating excess heat, either produced from inside the body or from the environment (Fig. 4.9).

4.7.1 Temperature Monitoring

But, how much heat is produced by an animal body?[7] The Basal Metabolic Rate (BMR) is defined as the heat production by a human at external temperature of 33 °C (*thermoneutral* environment), and is measured under conditions of steady state (typically 12 h after the last meal). The standard BMR for a 70 kg man is approximately 1.2 W/kg, or $M = 84$ W in Eq. (4.19). This value can be altered by changes in active body mass, diet, endocrine levels but, contrary to intuition, it is much less affected by environmental changes. Considering a body surface of about 2 m², this corresponds to a lower limit of the heat flux of about 40 W/m², which can increase by up to a factor of 20 under intense effort. Despite substantial differences in body shape, composition, functions, the BMR is outstandingly conserved through

---

[7]The study *Heat loss from the human body* by the American pathologists Eugene DuBois and Graham Lusk, albeit somewhat dated in terms of modern physiology, still represents a magnificent account of these phenomena. It was reported in DuBois' lecture to the Harvey Society of Dec. 15, 1938, and published in the 1939 Bulletin of the New York Academy of Medicine [3].

the evolution of superior animals, and has nearly the same value for all species, from birds to mammals.

In warm-blooded animals, a variety of physiological systems provide automatic feedback to maintain the reference temperature. Temperature sensors throughout the body respond to the central nervous controller, situated in the medial preoptic/anterior hypothalamic region of the brainstem, which then adjusts heat production and loss accordingly. In humans, sweating starts when the body temperature T_B is above 37 °C, and stops immediately when T_B is lowered below this value. The human forehead skin can detect temperature differences as small as +0.003 °C, with a response time of about 3s, and just a slightly worse sensitivity for negative temperature changes.

Thermal receptors are distributed all over the skin, in the form of free nerve terminations penetrating the skin from below, and ending just under the outer keratin and lipid envelope (less than 100 μm thick). Notably, warm-sensing receptors (the C-fibers) have a much slower response than cold-sensing ones (Aδ-fibers): C-fibers lack the myelin sheath (see details in Chap. 7), and the sensory stimulus travels at 2 m/s, compared to 20–30 m/s for the Aδ-fibers. Such sensors compose the normal sensory system, but their exact reaction mechanisms are still unclear. On the other hand, specialised receptors are devoted to signalling sudden changes of temperature that are harmful to the body: TRPV1 hot receptors have a pain-signalling threshold at 42 °C, while TRPM8 cold receptors activate a progressive response at temperatures of 20 °C and below. These sensors are actually special proteins (again ion channels) located at the nerve cell membrane, which can quickly change shape in response to temperature, and activate a sudden flux of ions (Na^+ and Ca^{2+}), thus firing a rapid electrical pulse, propagating along the neuron to the peripheral nervous system.

Additional thermoregulatory mechanisms may include: changing the diameter of blood vessels (*vasomotor control*) to increase/decrease the flow of heat to the skin; *shivering*, to increase heat production in the muscles; *secretion* of molecules such as norepinephrine, epinephrine, and thyroxine, to increase heat production; erection of the *hairs and fur*, to increase insulation (indeed most effective in animals different from humans).

Skin represents about 15 % of the body mass in humans (the "largest organ" of the body), while muscles take up about 40 %. For external temperatures far from 37 °C, the skin temperature is usually 2°–4° below that of the body. Under conditions of warming, as much as 30 % of the blood flow can go to the skin, to increase the cooling rate. A human has about 5 L of blood, which take about 1 min to completely circulate in the body, therefore about 1.5 L/min flow to the skin under normal conditions. The blood circulation rate is called the **cardiac index**: it originates from the heart pumping rate, which in humans is about 65 mL/beat at 1.15 Hz (i.e., about 70 beats/min), resulting in a cardiac index of about 80 mL/(min kg). By reducing the size of the animal, the amount of blood decreases, the heartbeat frequency increases and so does the cardiac index. For example, in a hamster of mass 100 g the cardiac index is 200 mL/(min kg). By scaling blood volume with the human/hamster body mass ratio, the pumping rate of hamster's heart should be about 0.08 mL per beat, from which a heartbeat frequency of 4–5 Hz (>250 beats/min) can be guessed. This is not far from the real values, which are about 300 beats/min.

Tissue conductance refers to the combined effect of conduction through layers of muscle and fat, and convective heat transfer by the blood. In a hot environment, peripheral vessels are expanded (*vasodilation*) since the body temperature is higher than the surroundings. The **heat capacity** of a material is the amount of energy necessary to increase by 1° the temperature of one kg of material. With a heat capacity of 3.6 kJ/(kg K) and density 1.05 g/cm^3, each litre of blood at 37 °C that flows to the skin and returns 1 °C cooler, releases about 3.4 kJ (0.8 kcal). If we consider 30 % flowing to the skin, that is 1.5 L/min, this corresponds to a heat power of 3400 × 1.5/60 = 85 W removed per °C difference, or a heat flow through the skin surface of about 50 W/m^2 per °C. During vigorous exercise, or when running away from an ominous predator, peripheral blood flow can increase up to 6–8 L/min, coupled to increased heartbeat rate, to eliminate the metabolic heat produced by stressed muscles. (This is why the skin on thighs and arms looks much more red after gym.)

In cold environments, conversely, the surface lower temperature makes heat loss even more important. To reduce radiation loss and blood cooling, the skin temperature must be brought closer to outer temperature and blood flow must be reduced at the surface. Both these effects are obtained by *vasoconstriction*: the shrinking of vases, to limit heat loss from the body interior to the skin through peripheral vasculature, such as in the hands and feet. This is commonly experienced in the fact that hands and feet are the first to feel cold on a cold day. Residual heat flow from the skin is reduced to 5–9 W/m^2 per °C difference between the inner body and skin, in the peripheral areas, which may lead to frosting under extreme conditions. Shivering of the muscles is also a response to cold, in the attempt to generate additional heat.

4.8 Heat from the Cells

Despite the very small temperature fluctuation and the sub-micrometer spatial scale, it has become possible in recent years to image the temperature inside living cells, by a number of techniques. Because of the very small length scales and of very small temperature variations involved, this is indeed a challenging problem. But we are not sure whether it may be also a truly meaningful one.

Since the cell is a chemical engine that burns combustible, it may be interesting to study the balance between energy accumulation and dissipation to the surrounding medium. At the macroscopic scale, we just described how the plasmic component of the blood circulates as a thermoregulating fluid in the body, and how water evaporation is responsible for the most part of excess heat evacuation. But at the small scale of the cell, blood red cells rather carry oxygen molecules to power the respiratory chain, and it is the rate of oxygen availability that regulates the rate of burning fuel, i.e., the rate of heat production from inside the cell.

Heat equation for macroscopic bodies

Let us consider a volume V with arbitrary shape, bounded by the closed surface A. For an amount of heat $S(t)$ released in V, the temperature field inside V is described by a distribution $T(\mathbf{r}, t)$, with $\mathbf{r}$ any point in V. From the definition of specific heat of the material:

$$S(t) = \int_V \rho c_V T(\mathbf{r}, t) d\mathbf{r} \tag{4.23}$$

hence the time-variation of S (indicated by the superscripted dot) is:

$$\dot{S}(t) = \int_V \rho c_V \dot{T}(\mathbf{r}, t) d\mathbf{r} \tag{4.24}$$

Fourier's law (4.30) says that heat flows from hot regions to cold regions, at a rate proportional to the temperature gradient. The only way heat can flow from V is by crossing the surface A. By integrating over the surface, Fourier's law is written:

$$\dot{S}(t) = \int_A \kappa \nabla T(\mathbf{r}, t) \cdot \mathbf{n} dA \tag{4.25}$$

In this integral, the unit vector $\mathbf{n}$ describes the local normal to the surface element dA, necessary to project, by means of the scalar (dot) product, the temperature gradient ∇T (another vector) along the direction of the heat flow. Since the shape of the body is arbitrary, so is the direction of $\mathbf{n}$ at every dA, provided its orientation is always pointing outside the body V. We can now equate the two expressions for the heat rate:

$$\int_V \rho c_V \dot{T}(\mathbf{r}, t) d\mathbf{r} = \int_A \kappa \nabla T(\mathbf{r}, t) \cdot \mathbf{n} dA \tag{4.26}$$

Now, consider the right-side integral: it is a scalar quantity (dot-product of two vectors) exiting a closed surface A. Let us imagine this as a plastic balloon filled with gas. If we increase the gas by some amount, this will expand the balloon flowing across A to a larger surface. If we look at the molecules in the gas, only those with a velocity vector $\mathbf{v}$ pointing toward the surface contribute to the expansion. These can be identified by calculating the *divergence* of the vector field in the whole volume V, written as $\nabla \cdot \mathbf{v}$ (to distinguish this from the gradient operation, we write a dot between the two symbols). In our case the vector field is $\kappa \nabla T$, and the surface integral is equal to the volume integral of its divergence, as:

$$\int_V \nabla \cdot (\kappa \nabla T(\mathbf{r}, t)) \, d\mathbf{r} = \int_A \kappa \nabla T(\mathbf{r}, t) \cdot \mathbf{n} dA \tag{4.27}$$

This is an example of the *divergence theorem*, by which many physical laws can be written either in differential form (one quantity is the divergence of another) or in integral form (the flux of one quantity through a closed surface is equal to some other quantity).

By comparing the two last equations, we see two integrals of some argument over the same volume V, both equal to the same surface integral. Therefore, the two arguments of the volume integrals must be equal, giving the partial differential equation:

$$\rho c_V \dot{T}(\mathbf{r}, t) = \nabla \cdot (\kappa \nabla T(\mathbf{r}, t)) \tag{4.28}$$

If ρ, c_V and κ are constants, the *heat equation* it thus obtained:

$$\dot{T}(\mathbf{r}, t) = \frac{\kappa}{\rho c_V} \nabla^2 T(\mathbf{r}, t) \tag{4.29}$$

The Laplacian operator in Cartesian coordinates reads $\nabla^2 = \frac{\partial^2}{\partial x^2} + \frac{\partial^2}{\partial y^2} + \frac{\partial^2}{\partial z^2}$.

Firstly, how much heat is generated in a cell? Most animal cells are happy with glucose concentrations between 5 and 6 mM. Given the molecular weight of glucose, $A = 180$ Da, and a typical cell volume of 4 pL (4×10^{-12} L, for a spheroidal diameter of 20 μm), this corresponds to about 23 fM of glucose in each cell, or about 14×10^9 glucose molecules. As we saw in the previous Sect. 3.5, the ATP yield from glucose is between 30 and 40 % under aerobic conditions. The rest of glucose available energy is wasted into heat at the level of mitochondria. Some more heat is released when ATP is turned back into ADP, in the muscle fibers (see Chap. 7), with a thermodynamic efficiency of about 60–70 %. However, to give an upper bound let us imagine that *all* the glucose energy goes into heat, so that the maximum available heat source from the cell is about 50 nJ (50×10^{-9} J). The reported glucose consumption rate for typical eukaryote cells is in the range of 0.2 pM/h, meaning that the 23 fM are consumed in about 400 s. The power release is thus 50 nJ/400 s = 125 pW for a cell of this size, a value broadly confirmed by several experimental measurements. Since the human body is estimated to contain about 10^{13}–10^{14} cells, the total body power consumption of the order of 10^2–10^3 W is also retrieved.

The second issue would be, then, how efficiently this heat is transferred to the surrounding medium to attain a steady temperature? Even down to the scale of a cell, heat flows in a medium of given thermal conductivity κ according to the Fourier equation:

$$J = \kappa \nabla T \tag{4.30}$$

The equation stipulates that the heat flux $J = Q/A$, amount of heat flowing across a surface A in a unit time, is equal to the spatial gradient of the temperature times the conductivity. This is a sort of Ohm's law for the heat, if we interpret the power Q/t as the electrical current, the temperature difference as the voltage difference, and the quantity $\kappa A/\Delta x$ as a "heat resistance".

As shown in the greybox on p. 143, starting from the Fourier equation we can derive a heat equation to describe how the temperature distribution $T(\mathbf{r}, t)$ evolves inside a dense body:

$$\frac{\partial T(\mathbf{r}, t)}{\partial t} = \alpha \nabla^2 T(\mathbf{r}, t) \tag{4.31}$$

with α the thermal diffusivity (in m^2/s), defined as $\alpha = \kappa/(\rho c_V)$, the ratio between the thermal conductivity and the specific heat c_v of the material, in J/(kg K). In other words, α is a measure of the relative ability of the material to *conduct* heat compared to its ability to *store* heat. This is a quite complicate equation to solve, especially when there is an internal source of heat as it is the case for a cell.

A mathematically much simpler description of heat flow can however be given by Newton's law:

$$\frac{dQ(\mathbf{r}, t)}{dt} = -h(\mathbf{r}) A \left[T(t) - T_{ext} \right] \tag{4.32}$$

in which the space $\mathbf{r}$ has disappeared from the temperature. This equation is applicable for a body all kept at the same temperature, thereby quantifying just how quickly (or slowly) its overall temperature $T(t)$ goes to the outside value T_{ext}, starting from some initial value T_0 at the instant $t = 0$. The parameter h is now the heat-transfer coefficient, in W/(m^2 K), which depends on a number of things: the thermal properties of the body, the heat flow geometry, as well as the relative importance of convection vs. conduction. Note also that this equation looks formally similar to Eq. (4.20), which makes sense since Newton's model is perfectly adapted to the case in which conduction is the dominating process (although the parameters as well as their $\mathbf{r}$-dependence are different). By recalling that the definition of the specific heat of a substance is the amount of heat to raise its temperature by a unit temperature, i.e. $\rho V c_v = dQ/dT$ (ρV being the total mass of substance), the left-side member of Newton's equation can be transformed as $dQ/dt = (dQ/dT)(dT/dt)$, or $dQ/dt = \rho V c_v(dT/dt)$. The transformed equation has now the simple solution:

$$T(t) = (T_0 - T_{ext})\exp(-t/\tau) \tag{4.33}$$

that is, the temperature attains exponentially the external temperature T_{ext}, with a characteristic relaxation time $\tau = \rho V c_v / hA$.

Judging whether the simpler Newton law (4.32) can be used, in place of the more complicated heat equation (4.31), can be in many cases decided by looking at the nondimensional **Biot number**, $Bi = hL/\kappa$, with L a characteristic length (typically the volume/surface ratio, V/A, already appearing in the solution above). This parameter defines how important is the resistance to thermal transfer across the body surface compared to heat flow inside the body. Values of Bi less than 1 mean that heat flows easily inside the body, thus quickly making the inside temperature uniform, while the limit to flow is represented by heat transfer across the surface. In such cases, Newton's law is readily applicable. Despite the difficulty in estimating the heat-transfer coefficient h, it is known that its value in water is of the order of 10^3–10^4 W/(m^2 K), therefore the Biot number for a cell should be largely below 1.

We can thus estimate that the heat $Q_0 = 50$ nJ from inside the cell is transferred to the outside extracellular fluid (and to neighbouring cells) over a typical time $\tau \simeq 10^{-5}$ s (10 ms), for a cell of radius $R = 10\,\mu$m, and by taking the values $\rho = 1$ g/cm^3, $c_v = 4.186$ J/(g K) for water.

The third and last question we must answer is, therefore, what is the value of T_0, or by how much the cell temperature is increased because of the internally generated heat? In principle, to know this value one should solve the non-homogeneous heat equation (4.31), by including a heat source term $S(\mathbf{r}, t)$. For example, we could fix a spherical volume V for the cell, and indicate a series of coordinates $(\mathbf{r}_1, ..., \mathbf{r}_q)$ where mitochondria are located inside V, with an integrated heat rate equal to 125 pW, and attempt a numerical solution of such a (very complicate) problem on a computer.

In the light of the previous discussion, we may in this case approximate the power generation to occur homogeneously in V, and to be steady in time. Then, Fourier's law (4.30) above, allows us to write $\Delta T = (\Delta Q/\Delta t)/(\kappa L)$, with L as above. Therefore, the expected temperature increase should be calculated as:

$$\Delta T = \frac{125 \quad (\text{pW})}{10^{-5} \quad (\text{m}) \times 0.6 \quad (\text{W/(m K)})} \simeq 2 \times 10^{-5} \quad \text{K}$$

This is indeed a very modest temperature increase. Together with the characteristic $\tau \simeq 10^{-5}$ s, which is much faster than the fuel burning rate, this tells us that heat from the cell is transferred very efficiently from the interior to the surrounding environment,[8] i.e. cells maintain thermal equilibrium.

4.8.1   Fever and Hyperthermia

But, if cells cannot heat themselves very much, how can animal body temperature increase above ambient, then?

As we hinted above, the central nervous system reacts to external stimuli by adjusting the temperature setting, pretty much as a thermostat would do in our homes. The hypothalamus does so by reacting to some hormones, cytokines or other chemicals, generically named **pyrogens**, which are liberated for example after a bacterial attack to the organism. In response, the nervous system can issue "orders" directed to increase the body temperature, thus starting what we call a fever. Other situations in which the body requires to increase the temperature are related to cold shock (already a few degrees below 23° elicit hormonal response), or hyperthermia. This latter is distinct from fever or cold-shock response, in that the hypothalamus in this case does not react by changing the temperature set point. In other words, hyperthermia is caused by external sources, such as heat shock, or unpredicted reaction to drugs (among which some anaesthetics). In all cases, the ways temperature can be increased are generally: (i) by restricting, or by insufficient, blood flow to the outer parts of the body, (ii) by increasing the rate of fuel burning, and (iii) by "wasting" some fuel into heat.

Concerning the mechanism (i), it is worth noting that blood carries at the same time glucose and oxygen to the cells, i.e. both the fuel and the oxidiser to burn it. Therefore, reducing blood supply to certain regions of the body amounts to reducing the microscopic fuel supply to cells, as well as to restricting the amount that can be burned. At the more macroscopic level, reduced blood flow also reduces the convective heat removal towards the surface, thus increasing even more the internal temperature. This is also the case for insufficient heat dissipation, when normal and even surface-dilated vessel blood flow is unable to cool the inner body. Under such conditions, it may be considered that inner body cells become thermally insulated from the external surface, shielded under layers of colder skin. As a consequence, heat is poorly dissipated. We can make the extreme hypothesis that *all* the heat produced inside the cells will remain there, and deduce an upper extreme value of temperature increase from the specific heat:

---

[8]The cell membrane, made of a double layer of lipid molecules, cholesterol and some other proteins (see Appendix D), represents a negligible interface resistance to thermal flow.

$$\Delta T = \frac{Q}{\rho V c_v} \tag{4.34}$$

i.e., a good few degrees C or K, for $Q = 50$ nJ, $V \sim 10^{-9}$ cm^3, and ρ, c_v as above. This would be already enough to generate a fever.

4.8.2 Metabolic Rate and Thermogenesis

Further temperature increase can come from the two other mechanisms. Concerning (ii), the fuel-burning rate, let us note again that a normal level of blood glucose is about 5 mM. This is the available concentration that is being carried around for every cell to pick up. Under normal conditions, a human body consumes 3–4 moles of glucose daily, of which about 60 % goes to the brain. (Note that 3 moles correspond to about 540 g of glucose.) This makes for a raw estimate of 0.2 moles/h, which should be compared to the previously quoted value of 0.2 pM/h for an average cell. By accounting about 10^{13} cells in our body, not all of which are capable of storing directly energy (such as brain cells, which have to be continuously supplied with glucose and oxygen to avoid death within a few minutes), the two figures are quite consistent. Glucose is captured in the cell by specialised proteins, **glucose transporters** called GLUTn, with $n = 1, 2, 3...$, which are necessary to move the glucose against the concentration gradient (see next Chap. 5), i.e., from inside the blood vessel, where its concentration is higher, to the cells where it is consumed.

There are different ways to measure the rate of energy consumption by an organism, or its **metabolic rate**. One important parameter is the volume of oxygen consumed, usually given in mL/(min kg), which is a direct consequence of ADP-to-ATP conversion, and therefore should be directly linked to ATP consumption. In Fig. 4.10, left, the oxygen consumption measured in experiments on a group of laboratory mice [4] is plotted as a function of the external temperature. The normal value is labelled RMR for *resting metabolic rate*. At normal to high temperatures the RMR is constant; however, if the animal is put at a temperature below its lower critical temperature (vertical dashed line), the metabolic rate increases linearly, with a slope that indicates the degree of extra metabolic activity for self-protection (animals living in cold climates would have a smaller slope). Note that the experimental data are plotted as oxygen consumption divided by body mass to power 3/4. What this scaling law says here,[9] is that larger animals can better defend themselves from cold, compared to smaller animals. In fact, for some still unclear reason, it is very generally observed that the metabolic rate always grows with the body mass to power 3/4 (which means that the ratio (metabolic rate)/M$^{3/4}$ should be a sort of "universal value", valid for most animals). Since the body surface across which heat is dissipated increases as the power 2/3 of the mass (actually it should be $S \propto V^{2/3}$, since $S \propto L^2$ and $V \propto L^3$, but

---

[9]We will discuss in detail this "metabolic 3/4-scaling law" in Chap. 12, when dealing with the scaling of energy and power consumption, as a function of body mass.

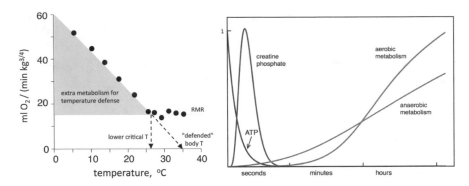

Fig. 4.10 *Left* Thermoregulatory metabolic response to environmental temperature. Points are experimental mouse data from Ref. [4]. RMR is the resting metabolic rate. The slope of the thermoregulatory line below the lower critical T is a measure of the insulation (more insulation, smaller slope). The *grey area* denotes the extra metabolism needed for body temperature defence; if this amount of heat is not produced, the body temperature cannot be defended, the animal will get hypothermic and eventually die. *Right* Sources of ATP in muscle cells during prolonged effort. In the first few seconds of exercise, energy is provided by ATP and creatine phosphate. After those initial sources of ATP are consumed, ATP must be regenerated from metabolic pathways, firstly the anaerobic, then the "normal" aerobic ones

we can also write $S \propto M^{2/3}$, if the body density is taken as constant), it turns out that the ratio between heat produced and dissipated decreases as $M^{(3/4-2/3)} = M^{-1/12}$, quite a slow power but definitely decreasing, for increasing animal mass.

Shivering, i.e. fast muscular contraction is a normal way of increasing the rate of ATP burning in muscle cells. The major fuels for muscle cells are glucose, fatty acids, and ketone bodies. Muscle differs from the brain in having a large storage of glycogen (about 1–2 % of the muscle mass, providing about 1200 kcal, or 5000 kJ). In fact, about three-fourths of all the glycogen in the body is stored in muscles, ready to be converted into glucose-6-phosphate (the first brick of glycolysis) for use within muscle cells. In resting muscle, fatty acids are the major fuel, meeting 85 % of the energy needs. Since the muscle cell produces in this state more ATP than needed, this is temporarily stored as creatine phosphate (PCr). In actively contracting skeletal muscle, the normally available ATP is consumed in a very few seconds (Fig. 4.10, right); at subsequent times, PCr is converted back to ATP (by transferring the phosphate to one ADP, thus giving back a simple creatine + 1 ATP). However, also PCr is rapidly consumed, and if the effort continues for a few minutes or more, the rate of glycolysis far exceeds that of the citric acid cycle, and much of the pyruvate formed is more quickly reduced to lactate in the anaerobic pathway (see Eq. (4.18) above). Only at later times, for a continued effort, the correct supply of oxygen, triacylglycerol and glycogen are restored.

Fat, cholesterol and other similar molecules are contained in small capsules bounded by a single-layer lipid membrane, called **lipid droplets**. For reasons which will become clear in the following (see Appendix D in Chap. 5), such a single-layer lipid membrane can store and transport fat molecules inside an aqueous medium, and make the molecules available when needed. Lipid droplets are present almost in any cell type, and notably in muscle cells. However, body fat is mostly stored in **adipocytes**, specialised cells in which one unique, giant lipid droplet occupies most of the cell volume, which can attain 10^6 μm^3. These cells make up the so-called *white fat*, and the energy stored there can be made available to the other cells (e.g., the muscles) by progressively releasing the broken lipids into the bloodstream. This is a rather slow process, not very effective unless the organism is subject to consistent phases of reduced fuel (dieting?) and oxygen supply.

4.8.3   Of Brown Fat, Alternative Respiration, and Thermogenic Plants

Finally, it is also possible that temperature in the cell is increased by redirecting some of the fuel energy directly into waste heat, as (iii), instead of producing ATP [5]. When sudden heat is needed, rapid burn-out of calories into heat is assured by the **brown fat**. In the brown adipocytes, fat is stored in many small lipid droplets (**adiposomes**) scattered in the whole cell volume. The brown color comes from the high presence of mitochondria, which can burn locally the lipids. Stimulated by the production of noradrenaline hormone from the central nervous system, lipids in brown adipocytes are broken down into triglycerides and long-chain fatty acids, to be turned into a large number of NADH and $FADH_2$ molecules by the beta-oxidation cycle (see greybox on p. 199). Under normal metabolism, these would be converted to ATP, by the protons flowing across the ATP-synthase channels (remember, flow of protons that is established by the opposite flow of electrons in the respiratory chain). However, in brown adipocytes the protons can leak, and return into the mitochondrial matrix via specialised uncoupling proteins, such as UNCn, $n = 1, 2, 3...$ These proteins "uncouple" the ATP-producing reactions: the protons flowing across the inner membrane can bypass the ATP-synthase and dissipate all their energy locally, into heat. It may be noted that, when animals are exposed for long times to low temperatures they tend to develop a larger mass of brown fat, therefore the shivering is progressively replaced, or accompanied, by such endogenous thermogenesis.

   Brown fat mitochondria demonstrate the use of normal respiratory chains but with a high proton permeability of the membrane, thanks to the uncoupling proteins, which leads to "efficient" energy dissipation. Note that in the normal operation of the organism, the production of heat is a waste going under the chapter of efficiency reduction, since the target is to extract as many ATP units as possible from a single glucose or fatty acid molecule. In the case of endogenous heat production, or thermogenesis, the efficiency is measured by how much heat can be extracted from the conversion process, of which the Eq. (4.34) above represents the extreme hypothesis of 100 % efficiency. However, we must always remember that the system free energy, ΔG, and not its enthalpy, is the available supply, and on the other hand, the animal body can never be considered as a truly isolated thermodynamic system.

However, **alternative respiration pathways** also exist. We already saw the possibility of supplementing ATP via anaerobic pathways, which can nevertheless be used for a quite short time by the animal, since it will need to breathe oxygen again, sooner or later. A high respiratory rate in a cell without concomitant ATP synthesis may also come from a normal proton conductance of the mitochondrial membrane, but concomitant lack of coupling between electron flow and proton translocation. The standard description given in Sect. 4.8, with complexes I-IV coupling electrons liberated in NADH and $FADH_2$ oxidation, to protons flowing in the intermembrane space, thus creating the electrochemical gradient that in turn powers complex V to flow protons back into the mitochondrial matrix, is indeed mostly adapted to mammals. To break off this chain, they have selected (in evolutionary sense) the brown fat proton-leaking proteins.

This description is however incomplete for the mitochondria of many species of oysters, mussels, bivalves, marine worms, some algae, and most species of terrestrial plants, as well as for prokaryotic respiratory systems. All these organisms rely also on additional electron transport components that add more points of entry and/or exit for electrons, thus creating a *branched* structure for the pathways of the electron-transfer process. Such alternative respiration pathways are often resistant to cyanides, nitrides, or sulphides, all of which are potent inhibitors of the functioning of cytochrome-c oxidase (hydrogen cyanide, HCN, was employed as an asphyxiant gas by many armies during WWI), therefore their presence in marine species could be linked to adaptation to particular underwater environments rich in such harmful chemicals.

Notably, the same alternative respiration pathways may be also linked to **thermogenesis**, especially is some plants. The most thoroughly investigated example of a modified respiratory chain is the *Arum maculatum* (Fig. 4.11), which can increase

Fig. 4.11 Examples of thermogenic plants. *Left Arum maculatum* can increase the temperature of its spadix by 15 °C above ambient, to distill an insect attractant as a pollination aid. *Middle Symplocarpus foetidus*, found in Nova Scotia, Quebec and other cold regions of North America, can raise the temperature of its inflorescence up to 35 °C above air temperature, to break its way through snow and ice. *Right Amorphophallus titanum*, the largest known flower on Earth with its 3 m of height, diffuses its carrion-like foul smell by increasing the temperature of the tip of the central spadix to about 35 °C. [Photos © by **a** Sannse Carter Cushway, **b** Susan Sweeney, repr. under CC-BY-SA 3.0 licence, see (*) for terms; **c** U.S. Botanic Garden archive (public domain)]

the temperature of its spadix by 15° above ambient. The mitochondria of the spadix possess a highly active cyanide-insensitive alternative respiratory chain, that does not translocate protons. Ubiquitous in all plants, the alternative oxidase (or AOX) is a mitochondrial inner membrane protein which functions as a component of electron transport. It catalyses the reduction of O_2 to H_2O, thus representing a branching point in the respiratory chain. Significantly, the alternative electron flow from ubiquinone (Q10 in Fig. 4.8) to O_2 via AOX is not coupled to proton translocation. Hence, AOX represents a non-energy-conserving branch of electron transport, bypassing the last two sites of proton translocation (complexes III and IV).

Although AOX as far as we know is present in all plants, it is not always linked to heat generation. Several other examples of **thermogenic plants** exist (see Bibliography at the end of this chapter), such as the *Symplocarpus foetidus*, for which the explanation of heat production is most likely the protection of the frost-sensitive spathe and spadix from freezing in below $-0\,°C$ temperatures. During sustained cold spells of $-10\,°C$ and colder, the inflorescence temperature may be kept just above freezing for a number of days. Longer cold spells may cause the plants to sacrifice inflorescences in order to conserve plant energy.

Other explanations for such endogenous heating phenomenon might include heat aiding in the release of odours to attract insect pollinators, and heat aiding in the growth of pollen tubes, or of the inflorescence itself. These scents, typically released from the spadix, often mimic carrion or dung, such as in the giant inflorescence of *Amorphophallus titanum*, or other essences such as garlic, apple, or turnip. These odours may help attract insects, and the heat itself may prove enticing for invertebrate pollinators. Whatever the purposes of self-heating are, these plants display surprisingly high respiratory rates, equivalent to that of similarly sized small mammals.

Appendix C: The Molecules of Life

Water, ions, and a quantity small organic molecules, such as sugars, vitamins, fatty acids, account for about 80 % of living matter by weight. Of these small molecules, water is by far the most abundant. The remaining 20 % consists of macromolecules: proteins, polysaccharides, and nucleic acids. In Chap. 3 the attention was focused on these latter. In Chap. 4, ATP and ADP have taken the stage, together with a number of important enzymes among which $FADH_2$ and NADH. Notably, the nucleotides making up the nucleic acids DNA and RNA (see Appendix B) share the same basic chemical structure of ATP and ADP, and the coenzymes have chemical structures strictly derived from these same ones. Phospholipids and the multi-layered membranes they can form will be an important part of Chap. 5. And Chaps. 6–10 will be dominated by a quantity of highly specialised proteins.

The chemical structure of the **nucleoside** is shown in yellow, in Fig. 4.12: it is composed by a pentagonal sugar ring, the **ribose**, to which one of the five possible nucleosidic bases (in blue: A, G, T, C for DNA, and U replacing T for RNA) are attached by a glycosidic bond. The chemical difference between DNA and RNA is visible in the carbon atom labelled 2': DNA has a H atom, while RNA has an hydroxyl (OH) group. The carbon 3', where another OH is attached in both nucleic acids, is the site where a link with an adjacent base along the chain can be formed. Both DNA and RNA are composed by joining together **nucleotides** (a nucleoside plus its lateral phosphate chain) in the monophosphate form, i.e. carrying only one PO_3^- side group (red in the Figure). In the monophosphate, one of the oxygen atoms is doubly bound to the phosphorus, as P = O, a second one is saturated by a hydrogen, as OH, and the last one is unsaturated and therefore negatively charged. When the next base is attached to either a DNA or RNA chain at the 5' end, the OH from one base reacts with the H at the 3' position of the other, and the two form a H_2O molecule while the two bases are covalently bonded together. This is called a **phosphodiester** bond, the newly formed PO_4^{3-} group being overall negatively charged. Such a bond can be broken by adding back the water molecule, in a process of **hydrolysis**. Because of the presence of the OH in the 2' position, hydrolysis of the -O-P-O- phosphodiester bond is much easier (energetically less costly) in RNA than in DNA. This is one reason for the higher catalytic activity of RNA.

As shown in Fig. 4.12, nucleotides can also occur in the diphosphate and triphosphate form, with two or three PO_4^{3-} groups consecutively attached to the 5' carbon atom. This structure is the same found in the ATP and ADP. Figure 4.13 (top row) shows the structure of these molecules, which have one adenine base attached to the ribose and, respectively, three or two phosphate groups (their names, adenosin-*tri*-phosphate, and adenosin-*di*-phosphate, specify exactly this feature). It should be noted that in their 'naked' form ATP and ADP have a large negative charge of -4 and -3 respectively. Despite the coordination with water molecules, such large charges are difficult to stabilise, therefore in biologically-relevant conditions these species

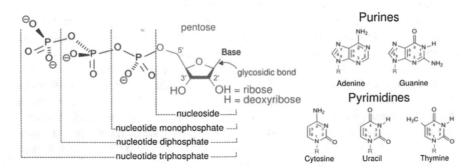

Fig. 4.12 Chemical structure of nucleic acids. A nucleoside plus one, two or three phosphates makes a nucleotide. DNA and RNA are distinguished by the different bases attached to the central ribose by the glycosidic bond

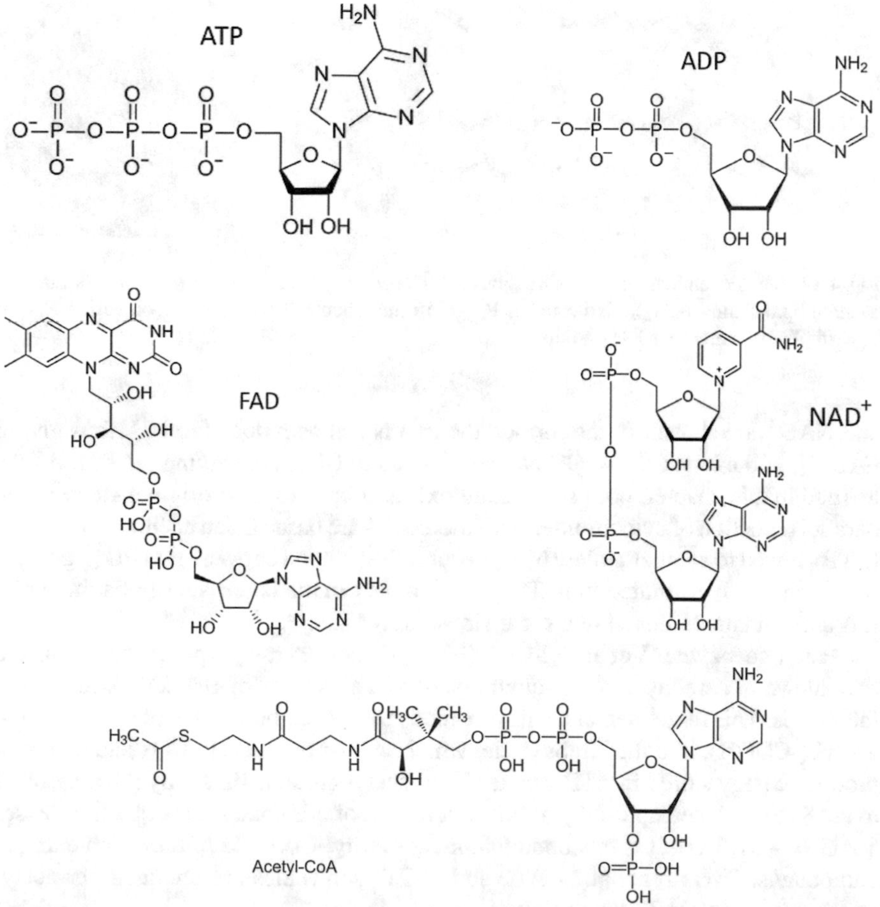

Fig. 4.13 The ATP (*above, left*) and ADP (*right*) nucleotides. Each contains adenine as the base, and respectively three or two phosphate groups, which are charged in the former, and neutralised by H in the latter. The coenzymes FAD (*middle, left*), NAD$^+$ (*right*), and acetyl-CoA (*below*). Note the similarity in the chemical structure, built from an adenine base and two phosphate groups (like in ATP and ADP), and a second moiety (riboflavine in FAD, nicotinamide in NAD, mercapto-ethylamine in the acetyl-CoA) linked by a phosphodiester bond to the phosphates

are always complexed, typically with Mg^{2+} ions, to [ATP-Mg]$^{2-}$ and [ADP-Mg]$^-$. Also the other bases (G, T, C) can form di- or tri-phosphates, however these species will be rarely encountered in the subjects discussed here, with the possible exception of GTP, guanosine triphosphate.

A subset of important players is represented by the **coenzymes**. Flavin adenine dinucleotide, or FADH$_2$, is a **redox cofactor** that is created during the Krebs cycle and utilised during the last part of respiration, the electron transport chain. Nicotinamide adenine dinucleotide, or NADH, is a similar compound actively used as well in the electron transport chain. In Fig. 4.13 (middle row) the chemical structures of FAD

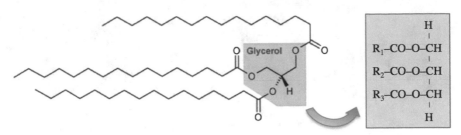

Fig. 4.14 Stereochemical formula of a triglycerid. The central glycerol (*in the green box*) is attached to three lipid chains, $(CH_2)_n$, indicated as $R_{1,2,3}$. In the schematic on the *left*, each vertex $\diagup\diagdown$ or $\diagdown\diagup$ implicitly indicates a CH_2 group

and NAD^+ are shown. In the former, the two N can be reduced to NH, thus giving $FADH_2$; in the latter, the $4'$ CH can be reduced to CH_2, thus giving NADH. As we learned in this chapter, both species are oxidised back to their original state, in the parallel reduction of cytochrome. Both molecules are based on an adenine nucleotide (ATP), linked to another moiety by a phosphodiester bond between two PO_4^{3-} groups, which are partly saturated in $FADH_2$. The moiety in the latter is a riboflavin, while it is a nicotinamide linked to a sugar ribose in NADH.

Acetyl-coenzyme-A or acetyl-CoA (Fig. 4.13, bottom row) is produced during the breakdown of carbohydrates through glycolysis, as well as by the beta-oxidation of fatty acids. This fundamental coenzyme feeds the two carbon atoms of its end-group acetyl (-COOH), into the Krebs cycle, which will be oxidised to CO_2 and water, to produce energy stored in ATP. The terminal acetyl group is linked by a strong bond to the S atom of mercapto-ethylamine; hydrolysis of this bond is exoergic (it releases a $\Delta G = -31.5$ kJ). It is this bond that makes acetyl-CoA one of the "high energy" compounds. Overall, about 11 ATP and 1 GTP molecules are obtained per acetyl group that enters the Krebs cycle.

Note that all these coenzymes have a similar structure shared with ATP and ADP, in that one ADP moiety is common to all of them.

The **triglycerides** (also called triacylglycerols, TG, or triacylglycerides, TAG) are compound molecules in which the three hydroxyls (OH) of a glycerol are linked to three fatty acids. Such molecules are the main constituents of the vegetable oils and of animal fat. In Fig. 4.14, the chemical structure and general formula of triglycerides is given. Here R_1, R_2 and R_3 are three, generally non-identical fatty acids, with general formula $(-CH_2)_n$-COOH, and length ranging from $n = 4$ to 22 carbon atoms. However, a length between 16 and 18 is the most commonly observed. Shorter carbon chains are observed in the butyric acid, the principal component of home butter. The glycerol is a polyol, familiarly know as glycerine, usually produced as side-product in the glycolysis. Liver and adipose tissue can supply glycerol when glycolysis is scarce, by using an alternate metabolic path involving amino acids.

Practically all naturally occurring fatty acids have an even number of carbon atoms, because they are all bio-synthesised starting from acetic acid, the smallest carboxylic acid with chemical formula CH_3COOH. The carbon chains in triglyc-

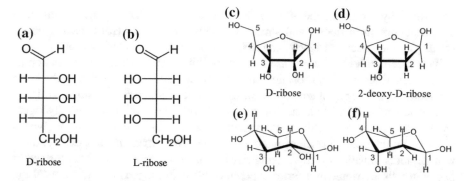

Fig. 4.15 Different representations of the five-carbon sugar ribose. **a**, **b** Fischer projection of D-ribose and L-ribose. In this representation it is evident the enantiomerism, the two molecules being mirror images of each other. **c**, **d** Furanose and **e**, **f** pyranose molecular structure of D-ribose and 2-deoxy-D-ribose. These two sugars link in the furanose form to phosphate groups (via C3 and C5) to build up the backbone of the RNA or DNA chain, and provide the linkage (via C1) between the backbone and the nucleobases

erides can be *saturated* or *unsaturated*, indexsaturated/unsaturated, fatty chain i.e., they can contain one or more double carbon bonds C=C, in each chain (see also Appendix D, about the chemical nature of phospholipids).

Most natural fats, such as butter, lard, tallow, are made from a complex mixture of triglycerides. Due to this, they melt progressively over a wide interval of temperatures. Cocoa butter is atypical, since it is made of only one type of triglyceride, in which the three chains are a palmitic, an oleic and a stearic acid, and has therefore a well-defined melting point. This is likely the reason why chocolate melts in the mouth without giving off a too "fatty" feeling.

Carbohydrates are a widely diverse group of compounds that are ubiquitous in nature. More than 75% of the dry weight of the plant world is carbohydrate in nature, particularly cellulose, hemicellulose and lignin. Among carbohydrates, **sugars** occupy a special position, due to their variety of structures and bonding, allowing a chemical flexibility vastly superior even to proteins, with combinations ranging from simple monosaccharides to polymers made of millions of units (Fig. 4.15).

Monosaccharides are linear or ring-shaped molecules with four to seven carbon atoms. Because these molecules have multiple asymmetric carbons, they can exist as isomers that are not mirror images of each other (enantiomers), indicated by the symbols D and L. Among the most important sugars to be found in the cell environment, we find the five-carbon D-ribose that is at the heart of RNA, and of DNA in the deoxyribose form (one O is lost from the OH group in the 1′ carbon); and the six-carbon D-glucose, produced in the photosynthesis and at the centre of the glycolysis cycle (note that L-ribose does not exist in nature, and also L-glucose is rarely found). Such sugars in solution are nearly always in the closed ring form, with only ~0.1–0.5% of the molecules in the open-chain structure. When the chain closes into a ring, a pentagonal (*furanose*) or hexagonal (*pyranose*) structure can be formed, by excluding one of the carbons from the ring. The hexagonal structure is more common in ribose, and almost exclusive in glucose. However, the ribose making up the structure of DNA and RNA is always in the pentagonal form.

Polysaccharides or *glycans* are formed by joining together any combination of monosaccharides, via a glycosidic bond (the same name is given also to the bond between a sugar and any other molecule). They range in structure from linear to highly branched. Examples include storage polysaccharides such as starch and glycogen, and structural polysaccharides such as cellulose and chitin. A dense layer of glycans is found on the outer surface of many cells, the *glycocalyx*. Glycans can combine with proteins in various forms, giving rise to **peptidoglycans**. The outer surface of bacterial cells is covered by a cortex of peptidoglycans arranged in a nearly crystalline form, providing a kind of exoskeleton that gives the bacterium structural strength and resistance against osmotic pressure.

Proteins are the other majority component of cells. The assembly of proteins from a sequence of amino acids translated by the mRNA was briefly described in the Appendix B. Once linked by peptide bonds in the *primary structure*, the long sequence of amino acids must however get folded into a three-dimensional structure, for the protein to become fully functional.

Starting from the primary structure, a variety of local interactions (van der Waals and electrostatic interactions between charges and dipoles, π-stacking, hydrogen bonds, hydrophobic effect) make up contacts between different portions of the peptide chain, and make it fold and bend into the *secondary structure*. The properties of the amino acids are listed in Fig. 3.14 of Appendix B. According to their degree of hydrophobicity, parts of the structure of the protein can adjust to minimise contact with water. Although it is impossible to provide an exhaustive list of the thousands different proteins necessary for the functioning of living organisms, there are some structural motifs in the secondary structure, which are recurrently found in the architecture of diverse proteins. Such "universal" motifs allow to recognise and classify common functional substructures, even in proteins performing completely different functions. The most important such motifs are the **alpha helix** and the **beta sheet**, shown in Fig. 4.16(a, b). The alpha helix gets its name from the central C atom of each amino acid, called the "alpha" carbon. In some regions of the protein, neighbouring N-H and C=O groups can form hydrogen bonds, which make for a more solid bonding than provided by the longer-range Van der Waals and electrostatic forces. A locally helical structure can arise from such bonds, every fourth α-C being found on top of each other, with a typical helix period of 0.54 nm (compare to 0.34 nm in DNA). The β-sheet is also formed by the same type of hydrogen bonds between N-H and C=O, but in this case the repeated structure is a multiply-folded flat pattern of roughly aligned strips of amino acids. The alignment of the strips can be parallel or antiparallel, according to the arrangement of the residues, either facing or opposing each other on each side of the strip. Such easily recognisable structures as the α-helix and the β-sheet are ubiquitous in all proteins, and are usually represented by a helical ribbon, and by a flat arrow, respectively, in the *ternary structure*, the actual 3-D form of the protein (Fig. 4.16c).

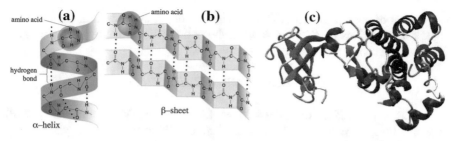

Fig. 4.16 Schematic of **a** α-helix and **b** β-sheet subunits of protein secondary structure. **c** Tertiary 3-D structure of the protein Src-kinase, drawn in the "cartoon" style, to highlight the presence of α-helices (*purple*) and β-sheets (*yellow*). The strings drawn in *grey* correspond to subunits with undefined (random) structure

Any protein can be composed by several different *domains*, or subunits that can fold independently into the 3-D structure. Some proteins can be composed by several repeats of one same domain. For example titin, the largest protein found in the human body, whose sequence of about 27,000 amino acids is made for about 90 % of two modules, the Ig (immunoglobulin) and the FN3. Note that these same domains are observed, with some variants, also in several other proteins.

When the protein functions as enzyme, its structure hosts one (rarely more than one) *active site*. These may appear as "pockets" or "holes" in the tertiary structure, in which a small organic molecule (the *ligand*) can fit and be temporarily bound (for example, myosin binding ATP to perform the power stroke in muscle contraction, see Chap. 9). The portion of the active site where the ligand binds is the *binding site*. The "lock-and-key" model of the ligand-enzyme interaction predicts that there is a perfect geometrical fitting between the two, such that the binding does not induce any further structural change in the couple. In the "induced-fit" model, the active site is modified by the entry of the ligand, and returns to its unperturbed shape when the ligand is released.

Problems

4.1 The ΔG of metabolic reactions

Consider a typical metabolic reaction in the form A$\rightarrow$B. Its standard free energy change is 7.5 k Jmol^{-1}.

(a) Calculate the equilibrium constant for the reaction at 25 °C.
(b) Calculate the ΔG at 37 °C, when the concentration of A is 0.5 mM and the concentration of B is 0.1 mM. Is the reaction spontaneous?
(c) Under which conditions might the reaction proceed in the cell?

4.2 Switching from ATP to ADP

Adenylate kinase (ADK) is a phosphotransferase enzyme that catalyses the interconversion of adenine nucleotides, and plays an important role in maintaining the right concentrations of ATP and ADP in the cell ("cellular energy homeostasis"). The reaction can be schematised as:

$$ATP + AMP \leftrightarrows 2\,ADP$$

Given the concentrations of [ATP] = 5 mM, [ADP] = 0.5 mM, calculate the [AMP] concentration at pH = 7 and 25 °C, under the condition that the adenylate kinase reaction is at equilibrium.

4.3 Energy harvesting

Calculate the absolute yield of ATP per mole, when a substrate is completely oxidised to CO_2, in the case of:
  (a) pyruvate (CH_3-CO-COO$^-$),
  (b) lactate (CH_3-CH-OH-CO$_2^-$),
  (c) glucose ($C_6H_{12}O_6$),
  (d) fructose 1,6-diphosphate ($C_6H_{14}O_6(PO_3^{2-})_2$).

4.4 Human blood

The kidneys help control the amount of phosphate in the blood. Extra phosphate is filtered by the kidneys and passes out of the body in the urine. A high level of phosphate in the blood is usually caused by a kidney problem. Normal levels of potassium in human blood should be in the range 3.5–5 mM. However, the phosphate ion can be found in any of its protonation states, H_3PO_4 (neutral), $H_2PO_4^-$, HPO_3^{2-} and PO_4^{3-} (this is the mix called inorganic phosphate, indicated with Pi, see also footnote to p. 114). Given the equilibrium constants for the three reactions:

$$H_3PO_4 + H_2O \leftrightarrow H_3O^+ + H_2PO_4^-$$
$$H_2PO_4^- + H_2O \leftrightarrow H_3O^+ + HPO_4^{2-}$$
$$HPO_4^{2-} + H_2O \leftrightarrow H_3O^+ + PO_4^{3-}$$

respectively equal to $K_1 = 7.5 \times 10^{-3}$, $K_2 = 6.2 \times 10^{-8}$, $K_3 = 2.2 \times 10^{-13}$, calculate the relative concentrations of the different phosphate ions in human blood at physiologic pH = 7.4.

4.5 Gym doesn't slim

Fats are usually metabolised into acetyl-CoA and then further processed through the citric acid (Krebs') cycle. However, glucose also could be synthesised from oxaloacetate, one of the intermediates during the citric acid cycle. Why, then, after some hours of exercise depleted our carbohydrate reserve, do we need to replenish those stores by eating again carbohydrates? Why do we not simply replace them, by converting some stored fats into carbohydrates?
(*Hint: look at the number of carbon atoms entering and exiting the Krebs' cycle*)

4.6 Pigeon muscles love citrate

The activity of the citric acid cycle can be monitored by measuring the amount of O_2 consumed. The greater the rate of O_2 consumption, the faster the rate of the cycle, the faster the rate of ATP production. Hans Krebs in 1937 used this type of experiments, working with fragments of pigeon breast muscle, very rich in mitochondria. In one set of experiments, he measured O_2 consumption in the presence of carbohydrate only, and in the presence of carbohydrate plus 3 μmol of citrate ($C_6H_8O_7$). After 2h30 he measured a consumption of 49 μM with glucose only, and 85 μM when citrate was added. Complete oxidation of citrate follows this chemical equation:

$$C_6H_8O_7 + x \ O_2 \rightarrow y \ CO_2 + z \ H_2O$$

(a) What is x, y and z, and how many moles of oxygen are consumed in the experiment after adding the citrate?
(b) Given the experimental result, what implications does this have for metabolism?

4.7 Antibiotics

Oligomycin-A is a natural antibiotic, isolated from the *Streptomyces* bacteria, which works by inhibiting the action of the ATP-synthase pump. It is sometimes used in laboratory research about ion channels, but it is never adopted in any pharmaceutical prescription drugs, because it is highly toxic. What is the main reason for it being so dangerous for animals?

4.8 Transmembrane proteins

Transmembrane proteins are quite big molecules that cross the cell membrane, exposing part of the structure both to the inside and the outside of the cell. By looking at the amino acid sequence of one such proteins, it is seen that it includes four regions characterised by strongly hydrophilic amino acids, separated by regions containing mostly hydrophobic amino acids. Draw a sketch of the tertiary structure arrangement across the membrane.

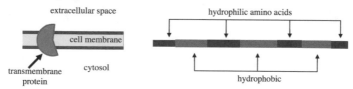

References

1. J.M. Berg, J.L. Tymoczko, L. Stryer, *Biochemistry*, 5th edn. (Freeman, New York, 2002)
2. R.A. Alberty, R.N. Goldberg, Standard thermodynamic formation properties for the adenosine 5'-triphosphate series. Biochemistry **31**, 10610–10615 (1992)
3. E. DuBois, Heat loss from the human body: Harvey Lecture. Biochemistry **15**, 143 (1939)
4. B. Cannon, F. Niedergaard, Brown adipose tissue: function and physiological significance. Physiological Reviews **84**, 278–337 (2004)
5. D. Rolfe, G.C. Brown, Cellular energy utilization and molecular origin of standard metabolic rate in mammals. Physiological Reviews **77**, 732–753 (1997)

Further Reading

6. R.S. Seymour, Plants that warm themselves. Scientific American **279**, 104–109 (1997)
7. R.M. Knutson, Plants in heat. Scientific American **88**, 42–47 (1979)
8. B.B. Lowell, B.M. Spiegelman, Towards a molecular understanding of adaptive thermogenesis. Nature **404**, 652–660 (2000)
9. N.R. Pace, The universal nature of biochemistry. Nature **30**, 805–808 (2001)
10. D. Metzeler, *Biochemistry: The Chemical Reactions of Living Cells* (Academic Press, New York, 2003)

Chapter 5
Entropic Forces in the Cell

Abstract In physics we are accustomed to four fundamental forces governing every phenomenon in the Universe. However, when dealing with heterogeneous, multi-phase systems, showing aggregation and self-organisation at length scales between nanometers and micrometers, other interactions seem to appear mysteriously, inducing strange effects such as osmosis, diffusion, depletion, hydrophobicity, settling, viscous drag, and so on. Certainly, also these effects must ultimately find their origins in the four fundamental forces. But in order to master them we need to introduce statistical thermodynamics concepts, conveniently embodied in the notion of "entropic" forces. The internal dynamics of a cell, a dense fluid crowded by hundreds of different proteins, molecules, charged ions, multiple lipid membranes, appears as an ideal laboratory to study such exotic physical phenomena.

5.1 Thermodynamic Forces

In the realm of classical mechanics, a force is expressed as the variation of a potential energy function, with respect to a position variable:

$$f = -\frac{\Delta U}{\Delta s} \tag{5.1}$$

The meaning of such a mathematical procedure is that, for a mechanical system in a state of equilibrium, characterised by a potential energy profile $U(s)$ (for the sake of simplicity we take the system behaviour to depend on one single variable, s, for example the gravitational field attracting Newton's apple to the ground), any displacement Δs requires the application of a force f working against the variation of potential energy ΔU. Therefore, the operational method to define a force is to perform a controlled displacement of the system from its equilibrium state, and to measure the resulting variation in energy. The ratio between the energy variation and the imposed displacement is a measure of the force exerted by the potential field or, when taken with a minus sign, of the equal and opposite force needed to perform the displacement. Note that the variation ΔU is necessarily positive, since we start from a local equilibrium point, i.e. one whose value of potential U is lower than any other

© Springer International Publishing Switzerland 2016
F. Cleri, *The Physics of Living Systems*, Undergraduate Lecture
Notes in Physics, DOI 10.1007/978-3-319-30647-6_5

point in the immediate neighbourhood. When in Eq. (5.1) the Δs is taken to the limit of an infinitesimal displacement ds, the finite variation turns into the infinitesimal derivative of U with respect to s, written as dU/ds.

It is worth noting that in classical mechanics there is no use for the concept of temperature. Mechanical forces, being usually much larger than thermal fluctuations, work in an ideal condition of zero temperature. For example the kinetic energy of the rotational movement of the Moon in the gravitational field of the Earth, Sun and all the other bodies of the solar system, is given only by the square of its translational plus rotational speed, while the thermal motions of the atoms making up these bodies are entirely negligible. All the mechanical quantities are defined and conceived at $T = 0$ K.

On the other hand, in our study of thermodynamics in Chap. 2 we introduced a whole host of additional potentials, such as TS, H, F, G, all with dimensions of energy, $[E]=[M][L^2][T^{-2}]$. The important character of these additional quantities is their typical dependence on the temperature. Therefore, we could ask what would be the equivalent of a "force", if we were to perform a constrained variation of any of the above potentials. In that case, we would need to identify some control variable, λ (the equivalent of the displacement s for the mechanical force), whose variation would impose a change in the corresponding potential function Λ. This would be the equivalent definition of a **generalised force**:

$$f_\lambda = -\frac{\Delta\Lambda}{\Delta\lambda} \tag{5.2}$$

Such a generalised force is often called an *entropic force*, since in the most interesting cases it is the variation of the entropy (multiplied by T, to obtain the dimensions of an energy) to be implicated. Indeed, in thermodynamics it is always interesting to look at the change of the free energy, F or G, rather than the internal energy U (the latter being nearly equal to the free energy at the very lowest temperatures). The total variation would be written as:

$$f_\lambda = -\frac{\Delta F}{\Delta\lambda} = -\frac{\Delta U}{\Delta\lambda} + T\frac{\Delta S}{\Delta\lambda} \tag{5.3}$$

If the variable λ affects only, or mostly, the entropy S, and has a null, or negligible, effect on the internal energy U, the variation of free energy translates into a variation of the sole entropy, hence the denomination of entropic force for f_λ.

For example, in the greybox on p. 24, the free energy of the system of volume V, held by the mobile wall at the position λ, is:

$$F = E - TS = const - Nk_BT \ln V = const - Nk_BT \ln(L^2\lambda) \tag{5.4}$$

Equilibrium in this case is the result of the balance between an internal pressure, coming from the hits of the gas molecules against the wall, and an external pressure coming from the spring holding the wall at the position λ, corresponding to a fixed value of the total energy (and by consequence of the temperature T). Since the only variable here is λ, the corresponding generalised force can be defined as:

$$f_\lambda = -\frac{\Delta F}{\Delta \lambda} = \frac{Nk_BT}{\lambda} \qquad (5.5)$$

Because the free energy comprises both the kinetic energy and the entropy of the gas, we can speak of entropic force, as the generalisation of the mechanical concept of force as originated by the displacement of a test body. Note that the parameter λ (in this case the position of the mobile wall) can be any control variable allowing to change a thermodynamic potential, such as a chemical concentration, the pH of a solution, the magnetization of a component, and so on.

5.2 The Strange Case of Osmosis

Let us consider a thermodynamic open system (i.e., one capable of exchanging both energy and matter with the surroundings) divided into two parts like in Fig. 5.1a. A large container A holds a cylinder B inside. Both A and B are filled with water. The cylinder B is sealed at its bottom by a semi-permeable membrane, such that the level of water inside is equal to that in the contained A. Here, by semi-permeable we mean a porous membrane that lets water molecules to pass freely, while stopping other molecules with a bigger size than that of H_2O (about 1 nm). Such a membrane may be thought of as a sort of sponge, with a network of connected pores of average size about 1 nm.

Now, let us drop some glucose in the cylinder, with concentration c_B. Since the glucose molecule ($C_6H_{12}O_6$) is about 10 times bigger than the water molecule, it cannot pass through the membrane pores, therefore the concentration in the water

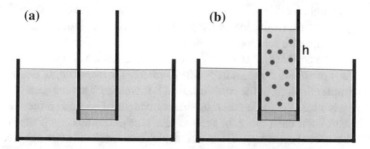

Fig. 5.1 Schematic of the osmotic pressure experiment. **a** The initial condition: the fluid filling the cylinder and the container is pure water, the level of the fluid in the two parts of the systems is the same. **b** The system after adding glucose (*blue particles*) in the cylinder: the membrane closing the *bottom* of the cylinder is not permeable to sugar molecules, water sips in from the container, and increases the level of the fluid inside the cylinder

outside the cylinder remains $c_A = 0$. Chemical equilibrium between the two parts of the open system requires the concentration of any species to be equal (recall Eq. (2.19) of Chap. 2, stating that the entropy will be maximum when the concentration $c_A = c_B$; or, equivalently, the Gibbs-Duhem equation (2.44), stating that there will be a $\Delta G \neq 0$ whenever concentrations are not at equilibrium). Since the passage of glucose into the container A is not possible, the only other possibility is that water diffuses through the membrane from A to B. The level of water in the cylinder is raised by an amount h (see Fig. 5.1b), as if there were an artificial pressure that pushed the water up inside B, working against the gravity. In fact, if the experiment were to be performed in void, such as inside the orbiting Space Station, all the water would flow from A into B. On the Earth surface, the hydrostatic (gravity) pressure $p = \rho g h$, with ρ the water density, is the only force that contrasts the apparent pressure pushing the water up in the cylinder. Such an artificial pressure π is called **osmotic pressure**, and it arises only from the chemical difference of concentration. It is therefore a clear example of a generalised thermodynamical force (or entropic force). Eventually, $\pi = p$ is exactly the conventional value, e.g. in units of atm or bar, attributed to the osmotic pressure.

In 1885, the Dutch chemist A. Van't Hoff was the first to perform systematic experiments on this quite surprising effect. The surprise originates from the fact that the substance added inside the cylinder does not have any interaction with the water molecules, it does not give rise to special new compounds, it does not carry any additional type of force (e.g., electrical, magnetic...), therefore the fact that the water in the sealed cylinder would rise was somewhat mysterious. By carefully looking at his own experimental data, Van't Hoff discovered that the osmotic pressure, when expressed as a function of the concentration c and of the fluid temperature T, follows a sort of 'perfect-gas' law:

$$\pi = \Delta c(k_B T) \tag{5.6}$$

Here $\Delta c = \Delta N/V = (N_B - N_A)/V$ is the difference in concentration of the solute, i.e. the molarity of the glucose solution with N molecules in the volume V, if $c_A = N_A = 0$ at the beginning of the experiment. Such a finding was even more surprising, since a dense solution of water and glucose is anything but so far removed from the idea of a perfect gas! In fact, Van't Hoff equation is valid at relatively small concentrations of solute, up to some 5–10%, whereas π increasingly deviates from the above simple equation, as the concentration is increased. This suggests that maybe the perfect-gas idea should be taken as the equivalent of glucose molecules being enough distant from each other, at the smaller concentrations, such that the fluid water exerts a sort of screening effect, and the glucose molecules do not 'feel' each other, thus effectively behaving as a kind of perfect-gas molecules separated by a vast empty space.

In the following, we provide two simple justifications of the osmotic effect, a first one more bookkeeping and intuitive, and a second one that points at the entropic origin of the osmotic pressure.

5.2.1   Microscopic Model

The water and sugar molecules in the A and B subsystems in Fig. 5.1 are at the same temperature, therefore their average kinetic energy is the same. Just like in the closed-system ideal situation (see Chap. 2), they can exchange energy at equilibrium by means of collisions with the membrane wall, under the condition that the total energy in the sum system A+B is conserved, and that the energy of each molecule follows at any instant the equipartition law.

For the sake of simplicity, we will suppose that the pores in the membrane are simply straight channels of diameter 1 nm. We also assume that collision of the molecules with the membrane occur elastically. Let us consider water and glucose molecules flowing in the perpendicular direction toward the membrane, from the two sides, with average velocity v dictated by the law of equipartition, i.e. $\frac{1}{2}mv^2 = \frac{3}{2}k_BT$, m or M being the mass of a water, or a glucose molecule, respectively. When a water molecule in A hits a pore of the membrane, it can cross the membrane and get into B without changing its speed, and vice-versa if a molecule from B passes into A. However, when a glucose molecule in B aimed in the $-x$ direction hits a pore no passage occurs, and the molecule rebounds by changing its velocity from $-v$ to $+v$. If, for the moment, we only consider molecules traveling perpendicularly along the x direction, against the membrane lying in the yz plane, at every glucose hit the membrane receives a mechanical impulse:

$$\Delta p = M(v_{fin} - v_{in}) = 2Mv_x \tag{5.7}$$

This is also the total impulse per glucose molecule, if we consider that on average there will be a same number of water molecules hitting the membrane from either side, and their contribution summing up therefore to zero. Since there are no glucose molecules in A, such an impulse Δp results in an average pressure imbalance on one side of the membrane equal to:

$$\frac{1}{S}\frac{\Delta p}{\Delta t} = \frac{2Mv_x}{S\Delta t} \tag{5.8}$$

with S the membrane surface, and Δt the time duration of a collision. What we define as the osmotic pressure π results from the sum of all such collisions, averaged in a time Δt, since the impulse variation divided by time has the dimension of a force, $[M][L][T^{-2}]$, which divided again by a surface, $[L^2]$, gives the dimensions of pressure.

The average collision time Δt can be taken proportional to the inverse of the collision frequency, $\nu = (\Delta t)^{-1}$. By following a simple reasoning common to such physical situations, the latter can be obtained as the number of molecules crossing a given volume in a unitary time, i.e. $\nu = Sv_xc$, if c is the concentration of glucose molecules in the subsystem B. Hence:

$$\pi = \frac{2Mv_x}{S\Delta t} = 2Mcv_x^2 \tag{5.9}$$

Up to now we considered only molecules impinging perpendicularly on the membrane. In fact, the average quadratic speed appearing in the equipartition law can be written more properly as:

$$v^2 = v_x^2 + v_y^2 + v_z^2 = 3v_x^2 \tag{5.10}$$

where the last equality comes from the consideration that the three directions are all equivalent. Then, our osmotic pressure estimate should be reduced to 1/3 of the previous Eq. (5.9). Moreover, we considered all molecules aiming at the membrane along the perpendicular direction, when only half on average would travel along the $-x$ direction, the other half traveling away from the membrane along the $+x$ direction, which reduces our estimate by another factor 1/2. Therefore $v_x^2 = \frac{1}{6}v^2$ and, by using the equipartition, we obtain Van't Hoff's 'perfect gas' law for the osmotic pressure:

$$\pi = \frac{2}{3}c\left(\frac{1}{2}Mv^2\right) = ck_BT \tag{5.11}$$

5.2.2   Thermodynamic Model

As we have seen in Sect. 2.5, the mixture of a solute in a solvent with a molar fraction x brings a change in entropy of:

$$\Delta S \simeq -Nk_B \ln(1-x) \tag{5.12}$$

At equilibrium under conditions of constant internal energy, this entropy variation multiplied by the temperature equals the mechanical work of the system in the form of the product of a 'pressure' times the volume:

$$T\Delta S = -Nk_BT \ln(1-x) = \pi V \tag{5.13}$$

Since for small concentrations, $x \ll 1$, we can replace $\ln(1-x) \simeq -x$, it is:

$$\pi V = Nk_BTx \tag{5.14}$$

from which we reobtain $\pi = ck_BT$, since the molar fraction is $x = n/N$ (number of molecules of solvent n, over the total N), and the corresponding volume concentration is $c = n/V$.

This latter derivation shows that the osmotic pressure (a force per unit surface) is in fact of entropic origin. It is related to the entropy difference of the solute mixing with the solvent, and the control variable is the solute concentration. The osmotic pressure can be obtained as the derivative of the free energy (in fact, the chemical potential at constant pressure and temperature) with respect to the concentration:

$$\pi = -\frac{dG}{dx} \qquad\qquad (5.15)$$

5.2.3   Osmolarity and the Healthy Cell

Again reasoning in terms of the idealised perfect-gas law for the diffusion of solute molecules, we should consider that all the different species act independently on the cell wall (as well as on any other semi-permeable membrane, such as the epithelial tissue or the blood vases). The total osmotic pressure results from the algebraic sum of the partial pressures relative to each one of the solutes (ions, sugars, proteins,etc.). The cell reaches the **isotonic equilibrium** when the outside and inside concentrations of all species result in a global mechanical equilibrium. A *hypotonic* cell has an internal pressure lower than the outer pressure, and appears flabby; a *hypertonic* cell has an internal pressure that exceed the outer pressure, and appears therefore inflated.

From the point of view of the cell, what counts is the net osmotic gradient across the membrane. The state of tonicity is important to predict the result of a change of concentrations of some species, since it takes into account the sum of all the inflows and outflows of all species across the membrane. In general, all cells in the human body are at osmotic equilibrium under normal conditions (with some exception, for example the cells in the kidney medulla, which are always hypertonic). In fact, the displacement of water molecules occurs quite rapidly on the typical cell-scale times, and continues until intracellular and extracellular pressure, and concentrations, are equal (see Fig. 5.2). A solution that maintains the isotonic equilibrium is called **osmolar** (a solution that maintains a good osmosis).

If an hyper-osmolar solution is administered to a patient, the extra supply of solutes would drive water to flow out of the cells. On the other hand, if the solute responsible for the hyper-osmolarity can cross the cell membrane, such as urea, it could enter the cell and arrest the water loss. Hyperglycemia in non-treated diabetic patients accumulates an excess of glucose in the plasma, with the direct result of forcing blood cells to expel water to bring intracellular glucose concentration at the same level as the extracellular. The opposite behaviour would be observed with an hypo-osmolar solution, leading to cell swelling and ultimately to the rupture, or **lysis**, of the membrane.

Osmotic pressure and surface tension in the cell

To get a more precise idea of the importance of the osmotic pressure in biology, let us consider a cell as a spherical vesicle with radius $R \sim 10\,\mu$m, filled with water. Some proteins are diluted in this idealised cell, in the form of spheroidal objects of radius $r \sim 10$ nm, in the proportion of, e.g., 30 % vol. (such an example is not far from the case of an erythrocyte with dissolved haemoglobin molecules). The number concentration is given as:

$$0.3 = c\tfrac{4}{3}(10\,\text{nm})^3 \quad \rightarrow \quad c \simeq 7 \times 10^{22}\,\text{m}^{-3} \tag{5.16}$$

from which the corresponding osmotic pressure exerted by the molecules on the vesicle cell (from the inside) is $\pi = k_B T c = 1.38 \times 10^{-23} \cdot 300 \cdot 7 \times 10^{22} = 290$ Pa (or J/m^3). This does not seem like a very large value when compared to the atmospheric pressure (remember that 1 bar = 10^5 Pa). However, could it be important for a cell?

A spherical vesicle filled with liquid, subject to the pressure exerted from the interior, tends to expand. As we will see in the following Chap. 9, any material can only resist the mechanical pressure up to the point permitted by its elastic resistance, after which it will start to deform irreversibly and eventually break. In Chap. 9 it will be made clear that, for a two-dimensional object like a cell membrane, such a mechanical resistance is equivalent to the *surface tension*, Σ (in units of [Energy]/[surface]). To obtain here a rough estimate of this quantity, let us imagine that the radial forces generated by the internal pressure tend to increase the spherical surface of the vesicle. The mechanical work done by the pressure force to expand the volume by an infinitesimal radius ΔR is (to first order in ΔR):

$$\Delta W = p\Delta V = p\tfrac{4}{3}\pi[(R + \Delta R)^3 - R^3] \simeq 4p\pi R^2 \Delta R \tag{5.17}$$

Such a mechanical work must be equilibrated by the internal work done by the surface tension, equal to $\Sigma \Delta S$, with ΔS the increase in the surface element upon the infinitesimal expansion of the sphere radius ΔR. If we integrate such an increase over all the spherical surface, it is $\Delta S = 4\pi(R + \Delta R)^2 - 4\pi R^2$, and the internal work to first order is:

$$\Delta W' = \Sigma \Delta S \simeq 8\pi \Sigma R \Delta R \tag{5.18}$$

By equating the two expressions, $\Delta W = \Delta W'$, we obtain an estimate for the surface tension:

$$\Sigma = \frac{pR}{2} \tag{5.19}$$

This is nothing else but the well-known Laplace's law. For our idealised cell: $\Sigma = pR/2 = 290 \cdot 10^{-5}/2 = 1.5 \times 10^{-3}$ J/m^2, or 1.5 mN/m. This value is the minimum of surface tension a cell membrane should display, in order to support an inner pressure of about 300 Pa. It is known that the value of surface tension for most cell membranes is smaller than this, being of the order of a few 10^{-4} J/m^2. A pressure of the order of a few hundreds Pa is enough to break the cell membrane, thus leading to cell death. Pressures of about 400 Pa are measured at the event of mitosis, when a cell splits into two, indicating that a value of surface tension of about 2 mN/m is a limit for the membrane resistance. It should be noted that the model of a cell as a spherical sack full of water does not hold, whenever substantial rearrangements of the cell shape (by the cytoskeleton) are involved, such as during mitosis. Note also that the total osmotic pressure acting on the membrane is the sum of the concentration differences between all the species, inside and outside the cell. These limiting values of Σ are realistic, however even larger osmotic pressures can be generated by the smaller solutes, such as Na and Cl ions: experiments show that erythrocytes in pure (i.e., unsalted) water may readily explode because of the excess internal pressure.

(a) **(b)** **(c)**

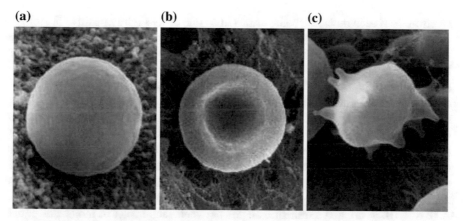

Fig. 5.2 Tonicity of red blood cells (RBCs) shown by electron scanning microphotography. **a** In a hypotonic medium such as distilled water, RBCs absorb water, swell, and may burst. **b** In an isotonic medium such as 0.9 % NaCl, RBCs gain and lose water at equal rates and maintain their normal, concave disc shape. **c** In a hypertonic medium such as 2 % NaCl, RBCs lose more water than they gain and become shrunken and spiky (crenated). [Image © RR Nursing School www. rrnursingschool.biz, repr. under CC-BY 3.0 licence, see (*) for terms.]

A behaviour similar to that described in Fig. 5.2 would be observed also for plant cells. The main difference with respect to animal cells would be that water in this case is mostly contained in the **vacuole**, a large reservoir inside the cell, which must be constantly filled to maintain the cells and plant's condition of turgidity.

5.3 Hydrophobicity, Depletion and Other Entropic Forces

An often cited example of entropic force is the **hydrophobic** force. Despite the fact that the hydrophobic effect includes also a substantial enthalpy contribution, the largest part of the force in water at standard pressure and temperature comes from the entropy of the rearrangement of the tridimensional network of hydrogen bonds between water molecules.

A hydrogen bond between two molecules involves a donor species, whose charge distribution is slightly unbalanced so as to appear slightly positive, and an acceptor species, whose charge distribution appears in turn slightly negative. However, because of charge conservation, if some part of a molecule becomes negatively charged, some other part must get positively charged (thus generating a **dipole moment** in the polar molecule). This means that every molecule is at the same time a donor and an acceptor, depending on the geometry of bonding. Water molecules are quite unique in their capability of forming hydrogen bonds with similar molecules. In fact, their chemical structure with two 'lone' electrons makes each water molecule capable of accepting two such bonds from two neighbouring molecules (one lone electron from each molecule being attracted to one of its slightly positive H atoms), while at the same time donating two more bonds to two other molecules (its two lone electrons being attracted toward two H atoms from two different molecules). Other H-bond-forming molecular species have a reduced capability in this respect: for example hydrofluoric acid, HF, could accept three bonds but can donate only one; ammonia, NH_3, could donate three but accepts only one. Under such conditions, species like HF and NH_3 in a dense, liquid-like environment can only form chains of molecules. On the other hand, water molecules can form a symmetric tetrahedral structure, in which each molecule is at the center of a tetrahedron, with four other molecules (the two donors and the two acceptors) situated at the four corners of the tetrahedron. In such a tridimensional structure, water in the liquid phase maintains a rather regular geometrical structure, with a constant O-O bond length of about 2.8 Å (2.8×10^{-10} m), and tetrahedral angles between each triplet of molecules equal to about 109.5°. Since water in these conditions is a liquid its molecules are highly mobile, and constantly exchange their location at the tetrahedral sites, however keeping the tetrahedral geometry with a remarkable regularity.[1]

Note that mixing two polar molecules, such as acetone and water, gives similar results since, being both polar, the two types of molecules can mix and maintain a more or less compact network of hydrogen bonds. We say in this case that acetone is readily dissolved in water. On the other hand, if we mix an assembly of polar and non-polar molecules, for example water and gasoline (which is a combination of many different hydrocarbons), such molecules try to avoid each other. Droplets of pure gasoline form in water (or vice versa, depending on which one is the majority component) to minimise the contact surface. Ideally, all the minority molecules would like to form a single spherical bubble, whose shape has the minimum surface to volume ratio.

Introducing a non-polar (for example, a plastic) object in water (Fig. 5.3a) partially destroys the ordered tetrahedral structure, since non-polar surfaces, with their molecules being insensitive to charge displacement, do not allow the formation of hydrogen bonds. As a consequence, the water molecules in direct contact with the plastic surface loose part of their hydrogen bonds, and try to adjust their configuration in order to minimise the number of broken H-bonds. The result is an interface structure in which some water molecules are constrained in a sort of cage, their mobility is reduced and, consequently, their entropy is also decreased in comparison with the free liquid. Seeking to maximise their entropy, such constrained water molecules try to reduce their contact with the non-polar surface by escaping this region, thus creating a layer of reduced density.

Now, let us think of two non-polar surfaces immersed in water (Fig. 5.3b). If these surfaces are approached, the simultaneous density reduction in the space comprised between the two surfaces will result in an effective attraction between the two objects.

---

[1] Note that the number of hydrogen bonds in water is a rapidly decreasing function of the temperature. It is equal to 4 at $T = 250$ K, about 3.85 at $T = 300$ K, and 3 at $T = 400$ K. The fact that ice has on average more hydrogen bonds than liquid water is also responsible for the increased density of water upon freezing.

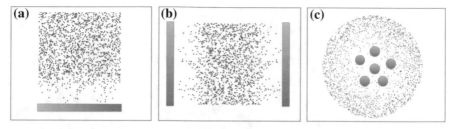

Fig. 5.3 The hydrophobic effect between water molecules (*green*) and a non-polar (*blue*) material. **a** A non-polar surface (for example, a plastic sheet) is introduced in water; as a result, the water molecules tend to maximise their entropy by retreating from the plastic surface. **b** Two non-polar surfaces in water have the effect of creating two such regions of reduced density: as soon as the two surfaces are approached, water molecules tend to escape from the region between the two. **c** The effect of introducing in water several non-polar surfaces, for example a bunch of plastic microspheres, leads to a kind of voiding effect in the space comprised between the plastic surfaces, thus promoting an effective *hydrophobic attraction* between the non-polar objects

By following this same pattern, if several non-polar particles are immersed in water (Fig. 5.3c) they tend to aggregate because of this effect of reduced density: it is like water "escapes" from the region comprised between the particles, which now seem to attract each other as if in the presence of some force. In fact, there is no force at all: the thermodynamic drive coming just from the requirement of maximising the solvent entropy. This is the basis of the **hydrophobic attraction** (from the ancient Greek words ύδωρ, *water*, and φόβος, *fear*) between non-polar objects in water.

The hydrophobic effect is at the basis of the spontaneous formation of biological membranes starting from **amphiphilic** molecules, as described in the Appendix D at the end of the chapter.

5.3.1 The Depletion Force Between Large Objects in Solution

Like in a shopping bag, inside the cell membrane we find a great variety of different objects with largely variable sizes and concentrations: from the ribosomes (300 Å), necessary to 'read' the RNA and produce proteins, to various globular proteins (50–100 Å), to the smaller glucose and other simple molecules (10 Å), to large amounts of ions ($\sim$1 Å). Such a hierarchy of sizes results in another surprising entropic effect: the **depletion force**.

The most definitive experimental observation of such a force dates from no longer than sixty years ago, by the two Japanese physicists Sho Asakura and Fumio Oosawa, of the University of Nagoya [1]. By looking at a colloidal mix[2] of large and small

---

[2]A **colloid** is a mixture of at least two different kinds of particles, which are not as small as in a solution, and not as large as in a suspension, but are intermediate in size. In a suspension, large particles would settle at the bottom of the container after some time. In a solution, the dispersed and host particles (at the scale of atoms or molecules) would form a single continuous phase. In

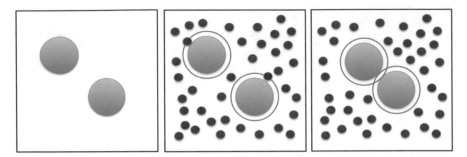

Fig. 5.4 The depletion effect between large and small solutes. *Left* Non-polar, non interacting, large colloidal particles with size R are put in a non-polar solvent. Under such conditions, no forces can possibly act on the particles, which fluctuate freely in the fluid. *Middle* A concentration of small particles with size $d << R$ is introduced in the container. Such particles can approach the large-particle surfaces, and their presence creates an 'exclusion zone' all around the large particles (*red ring*) against the other approaching particles. *Right* The small solutes try to maximise their available free volume (i.e., their entropy) by 'pushing' the large particles to share part of their exclusion zones, thus inducing an effective attraction between the large particles

particles, with radius R and d, respectively, they found that the largest objects are surrounded by a 'inaccessible' layer of thickness d. Just like in the case of the hydrophobic effect, the presence of such a layer makes the larger objects to experience a kind of effective attraction, as if the simple addition of small particles to the mixture would induce an extra force, pushing together the big colloids.

To get a qualitative explanation of this, again purely entropic effect, let us look at the sketch in Fig. 5.4. When two large particles with radius R and volume $v = \frac{4}{3}\pi R^3$ are introduced in a container of volume V, the available (free) volume for them to diffuse, is given by the difference between the total available volume and their own occupied volume:

$$V' = V - 2v \tag{5.20}$$

If n small particles with radius $d << R$ and volume $w = \frac{4}{3}\pi d^3$ are now added to the container, the available volume for both the large and small particles should be $V' = V - 2v - nw$. However, if the large particles are taken as rigid, the small particles cannot approach the surface of the large ones to a distance smaller than $R + d$, thereby creating a sort of exclusion zone around the volume of the large particles. In practice, it works as if the large particles have now an effective volume $v' = \frac{4}{3}\pi(R + d)^3$. The available volume for free diffusion becomes $V'' = V - 2v' - nw$, and evidently $V'' < V'$.

(Footnote 2 continued)
a colloid the dispersed particles and the solvent remain two distinct phases. In practice, colloidal particles can have dimensions ranging from about 1 to 1,000 nm.

As we remember from Chap. 2, the entropy of an ensemble of non interacting particles enclosed in a box is proportional to the logarithm of the box volume, $S \propto k_b \ln V$. Therefore, entropy can be increased if the volume V'' can be increased. One solution possible is that the large particles tend to approach each other, as shown in the right panel of Fig. 5.1. In this doing, a part of their respective exclusion zones overlap, thereby liberating some amount of volume for the smaller particles, while not affecting their own. In this way the entropy of the solutes can increase. The overall effect is that the simple adding of some concentration of small solutes induces an attraction between the large solutes, which would not exist when the large solutes are the only ones present in the solution. Again, this is a force of purely entropic origin. Differently from the hydrophobic effect, however, which originates from the entropy of the solvent molecules, the depletion force originates from the competition between solutes of largely different sizes.

Clearly, inside a solution of various components all such effects (as well as the others we are to describe in the foregoing) are present at the same time: the thermodynamic goal being that of maximising the *total* system entropy, in some instances it will be the solvent to lead, while in some other it could be the solutes, depending on the relative constraints of pressure, volume, temperature, concentration, ionicity, and so forth.

A more quantitative model of the depletion effect by the same Asakura and Oosawa shows that for small- to medium-diluted solutions (of the order of 40 % vol. max.) the average effective potential $U(r)$ experienced at short distances r between the large solutes should be given by the following expression:

$$U(r) = \Pi \left(\frac{\pi d^3}{6} \right) \left(\frac{\lambda}{1-\lambda} \right)^3 \left[1 - \frac{3r}{2R\lambda} + \frac{1}{2} \left(\frac{3}{R\lambda} \right)^3 \right] \qquad (5.21)$$

with $\lambda = 1 + d/R$, and Π a concentration-dependent constant of the order of $k_B T$. By integrating for $d < r < \infty$, the gain in free energy upon approaching the two large particles can be approximated as:

$$\Delta G_{depl} = \Pi \left(1 + \frac{3R}{2d} \right) \qquad (5.22)$$

We used the symbol Π for the force constant to indicate that the depletion force can also be interpreted as another form of osmotic force. In fact, the small particles perceive the network of large particles as a kind of semi-permeable membrane, allowing the passage of the solvent molecules while blocking their own flow across, since the space between the large particles after approaching each other (see left panel of Fig. 5.4) is smaller than their size d.

For ratios R/d of the order of 10, the above expression gives a free-energy gain of several $k_B T$s, to be compared with the energy associated to Van der Waals attraction

($\sim$0.1 $k_B T$), hydrogen bond (1–10 $k_B T$), or a covalent bond (50–100 $k_B T$). Clearly, the depletion force can become important at the length- and energy-scale of cells. For example, it is today well recognised that the capability of large DNA fragments to fold, or the decrease of hemoglobin solubility in sickle-cell anemia, improve in the presence of small solutes like the bovine serum albumin (BSA) or polyethylene glycol (PEG). Such small solutes are sometimes called **crowding agents**, for their capability of inducing the densification of large molecules. For example, the self-assembly rate of actin filaments in cell motors can be increased by up to 2 orders of magnitude, by adding a small concentration of PEG.

It might seem strange that entropic interactions, such as the hydrophobic and depletion force, could promote the self-assembly of some molecules, and even of the so-complex cell membrane: being entropic in origin, they should favour the disordering rather than leading to increasingly ordered structures. In fact, already by simple calculations it can be estimated that the disorder of the total system, solvent + small solutes + large solutes, will correctly increase overall, despite the superficial impression of a local increase of order.

5.3.2 Steric Forces and Excluded Volume

Steric effects (from the ancient Greek στερός, *solid*) are connected with the finite volume occupied by each molecule, inside or outside the cell. Although at the atomic scale electric charges are highly mobile, with the result of a considerable flexibility of the molecular shape, there is a limit below which the molecular volume becomes impenetrable, giving the concrete appearance of solid matter to our daily objects. If some of the atoms from two adjacent molecules approach each other at too short a distance, the energetic cost of the superposition of the respective electronic orbitals (Pauli repulsion between the electrons) becomes the leading energy term, thereby implying a repulsion which goes beyond the pure electrostatic repulsion between negative charges. Such a superposition would imply both a change of the shape of the molecule, as well as of its reactivity. The Pauli repulsion, ultimately of quantum origin because electrons obey Fermi-Dirac statistics, defines an exclusion volume around the atoms, and therefore all around the molecule.

The steric volume, or steric resistance, is apparent when the size of some molecular substructure does not allow some chemical reaction, which is readily observed in another similar molecule. For example, if the substructure blocks an otherwise accessible reaction site of the molecule. While the steric volume can be a source of problems, however it is often exploited by the chemists, to modify the behaviour of a given molecule in the course of a chemical reaction, for example to stop another parasitic reaction, or to avoid aggregation (*steric protection*, Fig. 5.5a, b).

Steric repulsion occurs whenever some electrically charged group at the surface of a molecule is spatially protected by another group of lesser or opposite charge. This also includes the screening coming from ions in the solution (Debye repulsion, Fig. 5.5c, and greybox on p. 175). In some cases, for an atom to interact with another atom that is sterically protected, it will have to follow a different kinetic path along a less protected direction. By attaching the proper protecting functions, chemists can use such effect to control both the timing and the direction of a molecular interaction. The opposite effect is steric attraction, which occurs when two molecules have a geometry respectively optimised for the mutual interaction, in what is called a *lock-and-key* configuration, thus increasing the exclusive specificity of the interaction.

Steric interactions are to a large extent of entropic origin. For the case of polymer-coated surfaces (Fig. 5.5b), it can be shown that the main force driving the repulsion originates from the entropy restriction to the free fluctuation of the polymer chain, when it approaches another surface covered by a similar polymer. Just like in the case of the hydrophobic effect, polymer molecules try to increase their entropy by avoiding each other, which leads to the steric repulsion effect. Biological realisations of such an entropic stabilisation may include, e.g., the role of flexible proteins inserted in the surface of interacting cell membranes, collagen polymers that reduce friction in synovial joints, or DNA adsorbed on histone proteins in the chromatin structure. For polymerically stabilised systems, the repulsive energy per unit area of the interacting surfaces at a distance R has an approximately exponential decay:

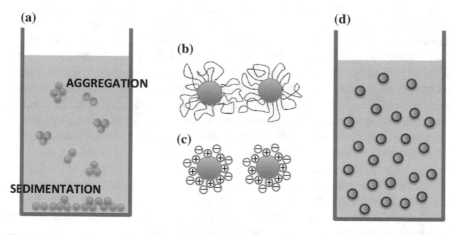

(a)

(d)

AGGREGATION

(b)

(c)

SEDIMENTATION

Fig. 5.5 The steric repulsion effect can be exploited to avoid unwanted reactions. **a** colloidal particles dispersed in a solvent can aggregate and precipitate from the solution, thus making the suspension unstable. Particles can be sterically protected, for example **b** by coating their surface by a layer of long-chain polymers, or **c** by adding ions to the solution, so as to form the (Debye) electrical double layer of oppositely charged ions at the surface of the particles. **d** With such a protection, the particles can avoid aggregation and form a stable suspension

$$U(r) = 36k_B T e^{-R/R_g} \tag{5.23}$$

where R_g is the **gyration radius** of the free polymer.[3]

For the case of ion-coated surfaces, it might be thought that the repulsion effect should be essentially electrostatic in origin, since it comes from the repulsion between ions of the same charge attached to the surface of the molecule or colloidal particle. However, it must be remembered that these ions come from a solution, and they are not firmly stuck to the surface, but rather have a dynamical distribution both in space and time. In other words, the surface of the colloidal particle, or of the protein immersed in the solvent, is only *on average* charged by some amount of positive and negative ions, organised in the double layer; however these ions are constantly moving, and can be exchanged with other identical ions from the solution. The equilibrium between association and dissociation of the ions to/from the surface is at the origin of the entropic term in the electrostatically stabilised systems. Calculating the relative entropic and enthalpic contributions in this case can be a very complicate task, however various authors (see e.g., [2, 3]) reported for many systems a majority contribution of the entropy to the overall free energy of the solution.

5.4   Diffusion Across a Membrane

Membranes are ubiquitous in cells, from the outer plasmic membrane enclosing the entire cell volume, to the internal membranes separating the nucleus and the double membrane of the mitochondria, to the endoplasmic reticulum found in eukaryote cell, entirely composed by a labyrinth of multiply folded membranes, and so on. The fundamental function of the membrane is to divide a volume into separate compartments, so that different concentrations of chemicals can be maintained on either side. We have seen in the case of osmotic pressure how concentration gradients are a powerful source of chemical force, which can drive the entire cell metabolism. Moreover, we will see later on concentration gradients permitting chemical messages to be exchanged between neurons in the nervous systems, as well as driving the polymerisation and depolymerisation of cytoskeleton filamentary protein, to modify the cell shape and to regulate cell motility. And the list could continue. Every cell function demands the displacement and transport of chemical species across one or more membranes, to the point that (with bold exaggeration) one could define life as being a sequence of chemical reactions separated by membranes!

---

[3]The gyration radius R_g is a measure of the average spatial extension of the polymer chain (see p. 321). For a polymer made up of N monomers, each of length b, the *contour length* is $L = Nb$, corresponding to the length of the fully extended, linear polymer. However, when the polymer fluctuates (either in void or in a solvent), it occupies a much smaller volume, of variable shape and extension, which can be described as being proportional to R_g^3. To give a practical example, the (contour) length of the DNA of the bacterium *E. coli*, with its about 4.6 millions of nucleotide pairs, is $L \sim 1.5$ mm, whereas its $R_g \sim 9\,\mu$m.

Electrostatics of colloidal particles in ionic solutions

A colloidal particle dispersed in a ionic fluid may attract point charges on its charged surface, providing an effective screened Coulomb potential $\psi(z)$. A first layer of counter-ions (*Stern layer*) is solidly bound on the surface. A second, thicker layer (*diffuse layer*) contains more mobile counter-ions. The *shear plane* defines the depth at which mobile counter-ions can be still exchanged with those in the solution. The red curve in the Figure indicates the shape of the electrostatic potential ψ. The **Zeta potential** is the value of the ψ at the shear plane. Basically, the higher the Z value, the larger is the repulsion between the screened colloidal particles, and the more stable is the solution (little tendency to aggregation and precipitation, see e.g. acid milk curdling into cheese).

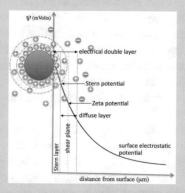

The theory of the **electrical double layer** is the empirical approximation to the (much complex) problem of describing the electrostatic interactions of charged molecules dispersed in ionic solutions. For a distribution of counter-ions $n_i(z)$, with charges q_i, the electrostatic potential $\psi(z)$ must satisfy the *Poisson-Boltzmann equation*:

$$\frac{\partial^2 \psi}{\partial z^2} = \frac{\partial \Theta[\psi]}{\partial \psi} \tag{5.24}$$

in which $\Theta = \frac{k_B T}{\varepsilon} \sum_i n_i^0 \exp(-Q_i \psi / k_B T)$, and $n_i^0 = n(z_0)$ at $\psi(z_0) = 0$.

The P-B equation can be used to model implicit solvation, an approximation of the effects of solvent on the structures and interactions of proteins, DNA, RNA, and other molecules in solutions of different ionic strength. In practical cases, it is difficult to solve the Poisson-Boltzmann equation for complex systems, but can be solved numerically on a computer. For small potential values ($\psi \simeq 40-50$ mV), the Debye-Huckel approximation of the P-B equation gives an exponential solution, $\psi(z) = \psi_0 \exp(-z/\lambda_D)$. The decay of $\psi(z)$ with the distance z from the surface, is measured by the **Debye length**:

$$\lambda_D = \left(\frac{\varepsilon \varepsilon_0 k_B T}{\sum_i n_i q_i^2} \right)^{1/2} \tag{5.25}$$

with ε the relative permittivity of the solution, in units of $\varepsilon_0 = 8.85 \times 10^{-12}$ Farad/m. For example, 1 mM NaCl has a $\lambda_D = 9.6$ nm, and 0.3 nm at 1M; pure water at pH 7 has $\lambda_D = 960$ nm ($\simeq 1\,\mu$m), a very slowly decaying potential.

The Zeta-potential is an important quantity for technological applications of colloids, for example to predict the aggregation of proteins, and can be easily measured. When an electric field E is applied to a charged particle of size R, in a fluid with viscous drag F proportional to the viscosity (ex. $F = 6\pi R\eta$), the particle drifts with a steady velocity $v = \mu E$, with μ the mobility (cm^2/V/s). The mobility can be measured by tracking the change in particle velocity as a function of E. The *Smoluchowski equation* establishes a relation between the mobility and the Zeta potential:

$$\mu = \frac{Z\varepsilon}{\eta} \tag{5.26}$$

The above linear relation is applicable for $Z \lesssim 120$ mV and for $R/\lambda_D > 100$.

The key property of a membrane is not just that of being an impenetrable wall separating two worlds. If that were the case, chemical species would maintain their concentrations indefinitely, and no chemical reaction would ever occur.[4] Therefore, the important property of the membrane is that of being indeed *permeable*, i.e. having the capability of letting some particles flow across its thickness.

Previously in this chapter, we considered the membrane as a completely deterministic gatekeeper: either it was entirely transparent to the passage of some molecules, or entirely impenetrable to other species (like water and glucose, respectively, in the discussion of the osmosis experiment). In reality, a membrane is a more complex object, with a typically porous structure, a distribution of sizes for the pores centred around some value, and more or less spread around that value. The pores of different size create a maze of contorted channels inside the membrane, such that even for the species that are smaller than the average pore, the passage is not automatic, and the membrane represents a barrier that slows down the flux of particles between two compartments.

The passage of a particle across a membrane is a **diffusion process**, controlled by the jump probability of the particle into/out each pore. If we maintain a deterministic description, we would assign only integer values of 0 or 1 to such probability. If, on the other hand, we want to describe the physical process of diffusion by which the particle jumps in and out the network of connected pores, possibly reaching the other side of the membrane after a long series of bounces back and forth, we would better adopt a stochastic description in which the probability of passage is a continuous variable ranging between 0 and 1.

By looking at Fig. 5.6, let us imagine to have a membrane of thickness L, separating over a surface S two containers A and B filled with water, for simplicity taken of equal volume $V_A = V_B$. Suppose that some particles are diluted in the water with different concentrations, $N_A/V > N_B/V$, or $c_A > c_B$. In their stochastic

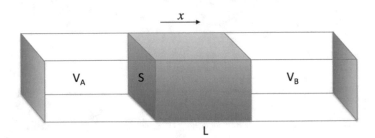

Fig. 5.6 Schematic of a membrane of thickness L and surface S, perpendicular to the x direction. A flow of particles across the membrane in the x direction is determined by a difference of concentrations $c_A \neq c_B$ between the two volumes V_A and V_B

---

[4]Remember that a concentration is a number of particles divided by a volume, i.e. with dimensions of $[L^{-3}]$. Concentrations are usually expressed as "molarity", indicating the number of moles of a substance per liter of volume. In most cases of biological relevance, the volume is filled with water, therefore the concentration is interpreted as the number of moles dissolved in one liter of water.

movements, particles from the container A will try to enter the porous membrane, and start jumping inside the pores; some will flip back into A, but some will make it across the membrane, and enter into B. The same and opposite will happen for some particles from the container B. However, due to the larger number of particles in A, the probability of finding some particles from A in B, will be larger than the probability of finding particles from B in A. In other words, despite the random, uncorrelated movement of all the particles, we will see over a long enough time, a net flow of particles from A to B.

Now, let us look more closely at what is happening inside the membrane. Since we are interested in the unidirectional flow of particles, from A to B and vice versa, we can for simplicity assume a strictly one-dimensional description of the process. Let us take the direction of the flow perpendicular to the separating surface S, as our x axis, and let us imagine to slice the membrane volume into thin layers of thickness dx. The probability $P(x)$ of jumping into/out of a layer is independent on the position x, but is only proportional to the difference between the number of particles present in the thin layer at x and those in the thin layer at $x + dx$. Therefore we can write:

$$P(x)dx = N(x + dx) - N(x) \qquad (5.27)$$

where the signs $+/-$ are adjusted to reflect the larger concentration at the left of the membrane (i.e., we conventionally assume more particles to be moving towards the positive x, than to the opposite direction). If we normalise the probability to the unit volume, by dividing P by the membrane volume $V = SL$, it is:

$$\frac{P(x)}{V} = \frac{1}{V}\frac{dN}{dx} = \frac{dc}{dx} \qquad (5.28)$$

The **particle flux**, with dimensions of $[L^{-2}][T^{-1}]$, is the number of particles traveling across the surface S in a unit time, and is therefore proportional to the probability of jumping:

$$j = -D\frac{dc}{dx} \qquad (5.29)$$

where we use the negative sign to indicate that conventionally j decreases when particles flow to the positive end of the x axis. The proportionality constant D, with dimensions of $[L^2][T^{-1}]$, is the **diffusion coefficient**.

On the other hand, the introduction of the time-dependent concept of flux suggests us the alternate possibility of considering rather the variation of concentration c at a given point x, as a function of the time. Just from dimensional arguments, it is evident that the ratio between concentration and time must give a quantity with dimensions of $[L^3][T^{-1}]$, which can then be equated to a spatial derivative of the flux itself:

$$\frac{dc}{dt} = -\frac{dj}{dx} \qquad (5.30)$$

Random walks and the Brownian motion

When considering the diffusive motion of a heavy object among a dense medium made of smaller particles (imagine for example a large protein moving among water molecules), a great many collisions between the large and small particles take place, each with a very small exchange of energy and momentum, due to the large size mismatch. The large particles would appear to move at random, as if pushed by an invisible troop. Such a phenomenon was experimentally visualised by the Scottish botanist Robert Brown in 1896, while studying the movement of the tiny particles suspended in the water around the roots of aquatic plants. (It is funny to mention that the movements he actually observed were later attributed to the vibrations of the table, however the 'brownian' naming of the phenomenon remained attached to his family name...)

For simplicity, we start by looking at particles jumping along only one direction in space, to show that the distribution of the random steps taken by a tracer moving back and forth on a straight line (the problem also know as 'drunken walk'), gives a Gaussian distribution of the traveled distance. We imagine the particle to start from $x = 0$ at $t = 0$, and to make, after a time t, N_R steps to the right, plus N_L steps of the same length to the left, with $N = N_R + N_L$; further, we take that each step has a probability r of going to the right, and a probability $l = 1 - r$ of going to the left. Then, the distribution of steps is a binomial:

$$P(N_R; r) = \binom{N}{N_R} r^{N_R} l^{N_L} = \frac{N!}{N_R!(N - N_R)!} r^{N_R}(1 - r)^{N - N_R} \tag{5.31}$$

It is easily proved that the average of the binomial distribution is equal to $\mu = Nr$, and its variance is $\sigma^2 = Nrl$. For large N, the Stirling approximation for the factorial (see Appendix A) is $N! \simeq N^N e^{-N}\sqrt{2\pi N}$. By substituting in the expression for $P(N_R; r)$, it is:

$$P(N_R; r) \simeq \frac{N^N e^{-N}\sqrt{2\pi N}}{N_R^{N_R} e^{-N_R}\sqrt{2\pi N_R}(N - N_R)^{N - N_R} e^{-(N - N_R)}\sqrt{2\pi(N - N_R)}} r^{N_R}(1 - r)^{N - N_R} =$$

$$= \left(\frac{r}{N_R}\right)^{N_R}\left(\frac{1 - r}{N - N_R}\right)^{N - N_R} N^N \sqrt{\frac{N}{2\pi N_R(N - N_R)}}$$

$$= \left(\frac{Nr}{N_R}\right)^{N_R}\left(\frac{Nl}{N_L}\right)^{N_L}\sqrt{\frac{N}{2\pi N_R N_L}} \tag{5.32}$$

If the distribution becomes increasingly narrow about the mean, i.e., $N_R \simeq Nr + \delta$ and $N_L \simeq Nl - \delta$, with $\delta \ll 1$, the logarithm of the first two terms is approximated as:

$$\left(N_R \ln \frac{Nr}{N_R}\right) + \left(N_L \ln \frac{Nl}{N_L}\right) \simeq -N_R \ln\left(1 + \frac{\delta}{Nr}\right) - N_L \ln\left(1 - \frac{\delta}{Nl}\right) \tag{5.33}$$

Then, by using $\ln(1 + x) \simeq x - \frac{x^2}{2}$, and neglecting terms higher than δ^2, we have:

$$-N_R \ln\left(1 + \frac{\delta}{Nr}\right) - N_L \ln\left(1 - \frac{\delta}{Nl}\right) \simeq$$

$$\simeq -(Nr + \delta)\left(\frac{\delta}{Nr} - \frac{\delta^2}{2(Nr)^2}\right) - (Nl - \delta)\left(-\frac{\delta}{Nr} - \frac{\delta^2}{2(Nl)^2}\right) =$$

$$= -\left(\frac{\delta^2}{2Nr} + \frac{\delta^2}{2Nl}\right) = -\frac{\delta^2}{2Nrl} \tag{5.34}$$

By the same approximations, the prefactor turns to:

$$\sqrt{\frac{N}{2\pi N_R N_L}} \simeq \sqrt{\frac{N}{2\pi(Nr+\delta)(Nl-\delta)}} \simeq \sqrt{\frac{1}{2\pi Nrl}} \tag{5.35}$$

Therefore, the original binomial distribution is eventually approximated by:

$$P(N_R;r) = \binom{N}{N_R} r^{N_R} l^{N_L} \simeq \sqrt{\frac{1}{2\pi\sigma^2}} e^{-\delta^2/2\sigma^2} \tag{5.36}$$

that is, a Gaussian with variance $\sigma^2 = Nrl$. Moreover, for the case of equal probability of jumping in either direction, $r = l = 1/2$, it is $\delta = N_R - N_L$, that is the difference between the number of right and left steps. Therefore $\delta = x$, the position of the tracer at any time t, and $\delta^2 = x^2$. This demonstrates the equivalence between the Brownian motion and the diffusion process, which end up having the same distribution, Eq. (5.41), the variance being identified with the product of the diffusion constant and the elapsed time, $\sigma^2 = Dt$.

Two important properties of the Brownian motion are that the average displacement is zero, and that the first interesting (non-zero) measure is the root mean squared displacement. This can be easily verified by calculating the respective averages over the Gaussian distribution (see Appendix A):

$$\langle x \rangle = \frac{1}{\mathcal{N}} \int_{-\infty}^{+\infty} x e^{-x^2/2\sigma^2} = 0 \tag{5.37}$$

$$\langle x^2 \rangle = \frac{1}{\mathcal{N}} \int_{-\infty}^{+\infty} x^2 e^{-x^2/2\sigma^2} = 2\sigma^2 = 2Dt \tag{5.38}$$

with the normalisation integral $\mathcal{N} = \int \exp(-x^2/2\sigma^2) = \sqrt{2\pi\sigma^2}$. The second equation tells that the random Brownian movement of the tracer will explore increasingly larger intervals over time, while constantly passing back and forth through the zero, such that its *average* displacement remains equal to zero. In terms of the variance of the gaussian, which grows linearly with t, or, equivalently, with the number of jumps N, this same concept can be seen as the distribution (whose integral must remain constant) progressively becoming broader, while its maximum value at $x = 0$ decreases. Equation (5.38) had been established by Albert Einstein in his work on the kinetic theory of gases, and can be easily extended to 2 and 3 dimensions, by the equivalence $\langle r^2 \rangle = \langle x^2 \rangle + \langle y^2 \rangle + \langle z^2 \rangle$, giving $\langle r^2 \rangle = 6Dt$, if the jumps (fluctuations) in the three directions are uncorrelated.

In that same article, which appeared in 1905 in the journal *Annalen de Physik*, Einstein obtained a second equation, linking the diffusion coefficient of molecules/particles of size R fluctuating in a medium, to the temperature T and viscosity η of the medium itself:

$$D = \frac{k_B T}{6\pi\eta R} \tag{5.39}$$

One relevant question about the Brownian movement was asked by the Polish physicist Marian Smolu-chowski. If the random motion derives from the great many collisions of small particles with the big particles, hitting with equal probability from any direction and therefore providing zero net force on average, why the big particle does move at all?

A special property of the binomial distribution contains the answer (known as the *ballot theorem*, proved in 1878 by W. A. Whitworth). For a tracer moving left or right with equal probability, the average position is zero at *very long* times. However, over some short time there could be more collisions, e.g., from the left than from the right, therefore it would be $N_R > N_L$. Over a finite time, this event has a binomial probability $(N_R - N_L)/N$, and its binomial average is of order $N^{1/2}$. Take a particle of a few μm in water, each collision contributes a velocity $mv/M \sim 10^{-8}$ cm/s. If the number of collisions in one second is of the order of 10^{23}, a net velocity of several m/s can result (although rapidly changing in direction), which keeps the particle in constant motion.

This is the **equation of continuity**, well known in the physics of continuous media to ensure that a quantity (here, the concentration) changes continuously without jumps. The equations says that the variation over a time dt of the concentration at a point x, is equal to (minus) the variation of flux between x and $x+dx$. By putting together the previous two equations, we are led to write:

$$\frac{dc}{dt} = D\frac{d^2c}{dx^2} \tag{5.40}$$

This is the celebrated **diffusion equation**, or Second Fick's Law (the first one being Eq. (5.29)), from the name of the German physiologist Adolf Fick.[5] This equation allows to determine the concentration profile $c(x, t)$ for any position x and at any time t.

It can be shown after some algebra that a particular solution of the (5.40) is:

$$c(x, t) = \frac{N}{\sqrt{2d\pi Dt}}e^{-x^2/2dDt} \tag{5.41}$$

the coefficient d being equal to 1, 2 or 3 for diffusion along 1 (a channel), 2 (a surface) or 3 (a volume) spatial dimensions. This solution applies to the simple case of a point source of N particles injected at the origin at time $t = 0$, in a 1-, 2- or 3-dimensional space of (practically) infinite extent. A general analytic solution for a real membrane is more complicated (see for example Ref. [4]).

If the membrane is permeable to the particles of the type considered, this is equivalent to say that their diffusion coefficient is $D > 0$. For a fixed concentration c_A at time $t = 0$, a flux from A to B will establish; after some time (function of D), the concentrations will became equal, $c_A = c_B$, and the flux will stop. Note that, from a microscopic point of view, this does not mean that particles will stop jumping from A to B, and vice versa: they will indeed continue to move to/from A/B, but their average numbers in A and B will remain equal. In this sense, the diffusion process is a typical phenomenon of **Brownian motion**, originating from the random molecular motion of elementary particles (see the greybox on p. 178).

If, on the other hand, we can maintain by some means a *constant* concentration of particles, e.g. in the volume A, or if particles in the volume B are constantly removed, thus maintaining $c_B = 0$, the flux will become stationary, and the concentrations across the membrane will become constant. In fact, at the stationary state it is by definition $dc/dt = 0$, and the diffusion equation reduces to:

---

[5]A. E. Fick discovered the law of diffusion in 1855, by performing an experiment exactly alike the one sketched in Fig. 5.6. He used tubes filled with water and salt, joined across a membrane. Around the same years, other laws related to the diffusion of something other than particles in a fluid were established, all with the same mathematical form of Eqs. (5.29) and (5.40), such as Ohm's law of charge transport (1827), Fourier's law of heat transport (1822), and Darcy's law of fluid flow (1856).

$$\frac{d^2c}{dx^2} = 0 \qquad\qquad (5.42)$$

whose simple, time-independent solution is: $c(x) = a + bx$.

5.4.1   Permeability and the Partition Coefficient

The flux can also be defined in terms of the **permeability**, P_M, a quantity also relative to diffusing particles, describing both the properties of the particles and those of the membrane:

$$j_m = -P_M \Delta c \qquad\qquad (5.43)$$

This definition is only valid under stationary flow conditions, since it requires as input the (constant) concentration difference Δc at the two sides of the membrane. However, note that the flux j_m is not exactly the flux as defined from Fick's equations, $[L^{-2}][T^{-1}]$, but rather a number of moles per unit surface and time. Therefore, to go back to the ordinary flux we must divide by the Avogadro's number, $j_m = j/N_{Av}$. If we now identify the two definitions, it is:

$$P_M \Delta c = \frac{D}{N_A}\frac{dc}{dx} \qquad\qquad (5.44)$$

Let us now ask what happens to the diffusing particles, if the medium is made of two **immiscible fluids** in contact, such as water and oil. For such fluids, phase separation occurs. On the other hand, the diffusion coefficient and solubility of the particles under consideration are generally different in the two fluids. Therefore, if some quantity of these molecules is dissolved in the fluid mixture, it would be found with different concentrations in each of the two separate phases. For example, when adding table salt to a mixture of water and oil, the NaCl microcrystals will readily dissolve into separate Na^+ and Cl^- ions in water (polar solvent); but the charges do not interact with the (non-polar) oil, therefore NaCl does not mix with oil. If instead we try the same home experiment with sugar, which has both $-CH_2$ hydrophobic terminations, and $-OH$ hydrophilic terminations, it will dissolve partly in the water and partly in the oil. A new quantity can then be introduced, the **partition coefficient**, given by the ratio between the concentrations of the species [A] in the two solvents:

$$K = \frac{[A]_{oil}}{[A]_{water}} \qquad\qquad (5.45)$$

To provide a standard, octanol ($CH_3(CH_2)_7OH$) is frequently used as the non-polar solvent, and the base-10 logarithm of K is tabulated: values below 1 denote more hydrophilic, and values above 1 more hydrophobic substances. The partition coefficient is a very useful quantity in pharmacokinetics, to characterise drugs that concentrate in the lipid-based cell membrane (more hydrophobic), or in water-based regions (hydrophilic), such as the cell cytoplasm or blood serum. A good correlation exists between $\log P_M$ and $\log K$ for many drugs. In fact, for a membrane of thickness d it is found:

$$P_M = \frac{KD}{d} \tag{5.46}$$

The greater the solubility of a substance, the higher its partition coefficient; the higher the partition coefficient, the higher the permeability of the membrane to that particular substance.

5.5   Forced Flow in a Channel

Nearly all living organisms (plants, insects, mammals,...) are endowed by one or more networks of vessels, whose role is to transport blood, air, sap, lymph and so on, to every part of the body. Such vessels form intricate paths, by successive branching into thinner and thinner channels. For example, in the case of arteries, the aorta starting from the heart has a diameter of some cm, then splits into smaller vessels, such as the iliac artery of size a few mm, down to the smallest capillaries whose diameter is even smaller than the size of the red and white blood cells, which are supposed to pass through them.

In the previous Sections we studied the diffusive motion of particles diluted in a fluid medium, which remained itself immobile on average. In this Section, instead, we will focus on the movement of the fluid itself when it is confined in a given volume, such as a tube or a vessel, and set in motion by an external gradient of pressure at the extremities of the tube. Clearly, inside such a moving fluid all the disordered molecular motions leading to diffusion would continue to take place. These molecular movements manifest in the form of *viscous resistance*, more or less important as a function of the internal structure and chemistry of the fluid. Nevertheless, at least in the cases that we are going to consider, the flux is mostly dominated by the imposed external force. Therefore, there will be a global, net movement of the fluid, superimposed to the disordered molecular diffusion. The net movement is oriented in the direction of the gradient of external force (or pressure), and will contrast the disordered, fundamentally isotropic, Brownian movement of the molecules.

As a function of the density ρ (with dimensions $[M][L^{-3}]$) and viscosity η ($[M][L^{-1}][T^{-1}]$) of the fluid, and of the velocity v of the flow, the fluid motion can be steady, or be more or less perturbed by phenomena of turbulence. A convenient way of characterising the flow in all such different conditions is the dimensionless quantity:

$$Re = \frac{lv\rho}{\eta} \tag{5.47}$$

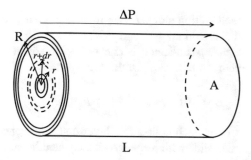

Fig. 5.7 Representation of a segment of a blood vessel, schematized as a rigid wall of a cylindrical section with length L and radius R. The fluid fills the volume with constant density. For the purposes of studying laminar blood flow in the vessel, induced by a constant difference of pressure ΔP at its extremities, the fluid cylinder can be thought of being composed by a dense series of concentric, thin cylindrical layers of thickness dr, with the radial variable r spanning the length from the center, $r = 0$, to the vessel wall, $r = R$

called the **Reynolds number**. The length l is a characteristic length of the fluid motion (for example, the diameter of the tube), or that of a large body moving in the fluid. When applied to a moving object, the Reynolds number allows to compare the motion of different objects in different fluids. For example a beetle ($l = 3$ cm) flying in air ($\eta/\rho = 0.16\,\text{cm}^2$/s at $T = 300$ K) at a speed of $v = 500$ cm/s, has approximately the same Re as a goldfish ($l = 15$ cm) swimming in water ($\eta/\rho = 0.009$ cm^2/s) at a speed of $v = 5$ cm/s. This means that the beetle and the goldfish feel the same medium resistance while moving at their cruise velocity, despite the fact that air is one thousand times less dense than water. When applied to the fluid itself, Re is, instead, a measure of the relative importance of inertial force versus internal friction. A large value $Re > 10^4$ indicates a flux strongly perturbed by the own inertia of the fluid, and therefore displaying turbulence and disordered flow. Conversely, smaller values $Re < 10^3$ indicate a dominance of internal friction, and therefore a smooth and 'laminar' flow, in which each layer of the fluid moves alike any other.

Let us make the simplifying hypothesis that blood vases are rigid, cylindrical structures of constant cross section $\pi R^2 = A$ (al least over some given length L), and that the flow of blood remains in the laminar flow regime (a reasonable approximation, since typical values for the aorta artery are about $Re \sim 10^3$). In this case, we can study the laminar flow by imagining that the blood in the cylinder moves within thin concentric layers of thickness dr, with the radial distance $0 < r < R$ (see Fig. 5.7), such that the velocity inside each discrete layer is constant, and there are no turbulence perturbations at the interfaces between r and $r + dr$, where the fluid slightly changes speed. To maintain a steady flux of blood, i.e., the same flow of blood, constant along the whole length L, there must be a constant pressure difference (provided by the heart) between the two extremes of the length of cylinder considered. Under such conditions, since the vessel walls remain immobile, the thin layer of fluid at $r = R$ directly in contact with the vessel wall must have $v(R) = 0$ (a condition technically called *stick-slip*). Since the velocity must be different from zero somewhere, and the flow is the same in each section along L, we then ask: what would be the velocity profile $v(r)$ in the radial direction?

The pressure difference p (indicated as ΔP in Fig. 5.7) creates a driving force on each cylindrical element $dA = 2\pi r dr$, of the cross section:

$$df^P = pdA = 2\pi prdr \tag{5.48}$$

Since the fluid speed in each concentric layer could be generally different, we admit that each layer can slide with respect to its adjacent layers, with a shear resistance measured by the fluid viscosity, η. For the concentric cylindrical layers, the lateral contact surface between the thin sections of radius r and $r + dr$, is $S = 2\pi r L$. The force originated by the sliding resistance on the unit lateral surface is, in fact, a **shear stress** $\tau = df/dS$ (see Appendix H). How the shear stress is related to the fluid velocity is a very complicate question, and here the assumption of laminar flow splits from the possibly turbulent flow. While writing a general equation for $\tau(v)$ is impossible, without making several simplifying assumptions and mathematical models, for a laminar flow the relation can be taken to be of simple proportionality to the velocity gradient, as also suggested by a dimensional analysis:

$$\frac{df^S}{dS} = -\eta\frac{dv}{dr} \tag{5.49}$$

This is the very definition of a **Newtonian fluid**, which, together with the previous equation, can be combined into Newton's Second Law, to express the instantaneous equilibrium of the sum of all the forces acting on the i-th layer of radius r and lateral surface $2\pi r L$:

$$df_i^S + df_{i+1}^p + df_{i-1}^p = 2\pi prdr - 2\pi r\eta L\left(\frac{dv}{dr}\right) + 2\pi(r+dr)\eta L\left(\frac{dv'}{dr'}\right) = 0 \tag{5.50}$$

the prime indicating that the value of v must be derived at the radial coordinate $r' = r + dr$. Note that the contributions from the two adjacent layers have opposite signs, since they have different relative velocities and therefore the viscous force changes sign.

As shown in the greybox on p. 185, the solution to the above force balance equation gives a parabolic profile for the velocity, with the maximum at the center of the cylindrical vessel, $v(0) = pR^2/4\eta L$. If we multiply and divide by a length squared $[L^2]$ its natural dimension of $[L][T^{-1}]$, this fluid velocity can be interpreted as a **volumetric flux**, $j_V = jV$ (j being the ordinary flux), displacing a volume of fluid V across a surface A in a time t. It is worth noting that this could be also taken as the true interpretation of the fluid velocity.[6]

---

[6]The conventional definition of the velocity as the ratio dx/dt between the distance dx traveled in an infinitesimal time dt is appropriate for a particle, whose trajectory can be traced according to classical or relativistic (electro)dynamics, but it is hardly useful in the context of a continuum medium.

The forced viscous flow equation

Starting from Eq. (5.50), the derivative of the (unknown) velocity profile $v(r)$ at a radius dr slightly different from r can be expressed as a Taylor expansion:

$$\frac{dv'}{dr'} = \frac{dv}{dr} + dr\left(\frac{d^2v}{dr^2}\right) + O(dr^2) \tag{5.51}$$

Hence (apart from the common $2\pi\eta L$ factor):

$$-r\left(\frac{dv}{dr}\right) + (r+dr)\left(\frac{dv'}{dr'}\right) \simeq -r\left(\frac{dv}{dr}\right) + (r+dr)\left[\frac{dv}{dr} + dr\left(\frac{d^2v}{dr^2}\right)\right] \tag{5.52}$$

which, after neglecting terms of higher order in dr, becomes:

$$-r\left(\frac{dv}{dr}\right) + (r+dr)\left(\frac{dv'}{dr'}\right) \simeq \left(\frac{dv}{dr}\right)dr + \left(\frac{d^2v}{dr^2}\right)r\,dr \tag{5.53}$$

Inserted in the above force balance equation, this approximate expression gives finally:

$$prdr + \eta L\left(\frac{dv}{dr}\right)dr + \eta L\left(\frac{d^2v}{dr^2}\right)r\,dr =$$
$$= \frac{pr}{\eta L} + \left(\frac{dv}{dr}\right) + \left(\frac{d^2v}{dr^2}\right)r = 0 \tag{5.54}$$

This is a second-order, non-homogeneous differential equation, which can be solved by finding firstly the general solution of the corresponding homogeneous equation, and then summing one particular solution of the non-homogeneous one.

For the <u>homogeneous</u> equation:

$$\frac{1}{r}\left(\frac{dv}{dr}\right) + \left(\frac{d^2v}{dr^2}\right) = 0 \tag{5.55}$$

we can substitute $y = dv/dr$ and obtain $dy/dr = -y/r$, or also $dy/y = -dr/r$, which is easily integrated to obtain $\ln y = -\ln r$. Then, by taking the exponential of both members, it is $dv/dr = 1/r$, which finally gives:

$$v(r) = A + B\ln r \tag{5.56}$$

For the <u>non homogeneous</u> equation (5.54), it is easily seen that one possible solution is $v(r) = -(p/4\eta L)r^2$. In fact, the first and second derivatives being simply $v'(r) = -(p/2\eta L)r$, and $v''(r) = -(p/2\eta L)$, we have:

$$\frac{1}{r}v' + v'' + \frac{p}{\eta L} = -\frac{1}{r}\frac{p}{2\eta L}r - \frac{p}{2\eta L} + \frac{p}{\eta L} = 0 \tag{5.57}$$

Hence, the complete solution is given by:

$$v(r) = A + B\ln r - \frac{pr^2}{4\eta L} \tag{5.58}$$

Note that, in order to have a finite velocity at any r, it must be $B = 0$. On the other hand, the 'stick-slip' condition, $v(R) = 0$, implies $A = pR^2/4\eta L$. Therefore, the radial velocity in a cylindrical tube under laminar flow has a parabolic profile:

$$v(r) = \frac{p(R^2 - r^2)}{4\eta L} \tag{5.59}$$

The volumetric flux of fluid can then be integrated all over the surface A, to obtain the volumetric flow-rate, with dimensions of $[L^3][T^{-1}]$:

$$Q = \int_0^R 2\pi r v(r) dr = \frac{\pi R^4}{8\eta L} p \tag{5.60}$$

This expression is known as the *Hagen-Poiseuille equation*, from the world of fluid dynamics, and tells us how much fluid is pushed by the pressure p, across a given circular section, in a unit time. Note that it is strictly valid only for a Newtonian fluid under laminar flow conditions, in practice at values of $Re \lesssim 2{,}000$. Such values include almost all blood vessels in the human body, possibly with the exception of the largest arteries next to the heart.

The ratio R_h between the pressure and the flow-rate:

$$R_h = \frac{p}{Q} = \frac{8\eta L}{\pi R^4} \tag{5.61}$$

is the **hydraulic resistance**, with dimensions of $[\text{Pressure}][T][L^{-3}]$, a more general quantity than simply the resistance between two fluids moving at different velocities. The resistance R_h varies very rapidly with the blood size, as the fourth power of R (for a section different from circular, a geometric factor different from $8/\pi$ must be included in Eq. (5.61)). This allows the easy adjustment to variable blood pressure by a small change in the diameter.

At the branching points of the vascular network, where a vein or artery splits into vases of smaller diameter, the above equations are no longer valid. However, until the flow remains in the laminar regime, such junction regions can be neglected with a minor error. If this is the case, the hydraulic resistances of the different sections of the network, split into one, two, four, etc. vases, can be treated like ohmic resistances in an electrical circuit, and can be summed in series or in parallel. For two sections of different diameter joined together (Fig. 5.8a), the pressure drops at the extremes of each section are summed, $\Delta p = \Delta p_1 + \Delta p_2$, and the hydraulic resistance is the series of the two respective resistances, $\Delta p = Q(R_1 + R_2)$. On the other hand, for a section splitting into, e.g., two vases of smaller diameter (Fig. 5.8b), the flux is constant in the two sections, $Q = Q_1 + Q_2$, and the pressure drop is given by the parallel combination of the hydraulic resistances, $Q = \Delta p(R_1^{-1} + R_2^{-1})$.

For a fluid traversing a membrane of cross section A, the following expression can also be introduced:

$$R_h = \frac{1}{A K_f} \tag{5.62}$$

The constant K_f is the **filtration coefficient** of the membrane, with dimensions of $[L][\text{Pressure}^{-1}][T^{-1}]$, i.e. a velocity divided by a pressure, or a volumetric flux per unit pressure. Therefore, it is a measure of the fluid flow induced by a given pressure across a given membrane. In terms of the volumetric flux, it is as well $j_V = p K_f$. Note that the flux j_V is the amount of fluid crossing the membrane, i.e. the flow of

(a)   **(b)**

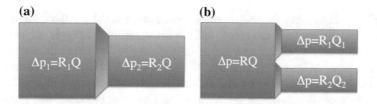

Fig. 5.8 Series and parallel sum of hydraulic resistance of blood vessels. **a** For two vessels of different diameter joined, the pressure must be continuous, equal to the sum of the two pressure drops. The two resistances R_1 and R_2 are summed as two electrical resistors in series. **b** For a vessel of radius R splitting into two vessels of radius R_1 and R_2, the flux Q must be constant in each section. The two resistances R_1 and R_2 are summed as two electrical resistors in parallel, and the result is summed in series to the resistor R

solvent. If the solvent contains solute particles in suspension these may or may not cross the barrier, depending instead on their permeability P_M, Eq. (5.43), which is related to the diffusivity of each particular species in the medium composing the membrane. Particles can be carried by the physical flow of the solvent, or they can diffuse following the chemical concentration gradient: in general, the movement of solutes will be a combination of both driving forces.

It should be noted that the filtration coefficient is also function both of the type of fluid and of the membrane. A typical value for water and the epithelial membrane of the human capillaries is $K_f = 6.9 \times 10^{-6}$ cm/(atm s), and for the membrane of the red blood cells $K_f = 9.1 \times 10^{-6}$ cm/(atm s). A relatively large value of filtration coefficient means that water flows out of the capillary system, and accumulates in the neighbouring tissues, likely causing an *edema*. This outgoing flux is partly equilibrated by an opposing pressure, originating from the concentration gradient of those large proteins that cannot easily cross the capillary membrane, i.e. the osmotic pressure of the large solutes. In medical physiology, the particular osmotic pressure associated with such large blood solutes, e.g. albumin, globulin, fibrinogen, is called **oncotic pressure**. Such large solutes inside the capillary create a negative pressure, tending to pump water into the capillary itself, thereby opposing the water loss due to the internal pressure. The balance at the capillary membrane between these pressures differences (of different origin) is summarised in physiology by the so-called *Starling equation*:

$$j_V = K(\Delta p - \Delta \pi) \tag{5.63}$$

If the overall flux is positive, the capillary is said to be in the *filtration* state, if negative it is instead in the *absorption* phase. All along the circulatory system, capillaries must adapt to largely different physiological conditions, and must adapt their characteristic K and p_M accordingly. The way K can be adjusted is more likely a mechanical one, by varying the density of endothelial cells making up the capillary tissue, and their lining membranes. The permeability P_M is instead adjusted by locally changing the cell membrane composition, with the increase of concentration of some particular proteins (cytokines, prostaglandins, histamine etc.).

Capillary membranes in most tissues are permeable to the small molecular weight solutes normally found in the blood, and are impermeable to bigger solutes. Therefore, big proteins like albumin are the main responsible for setting the osmotic pressure difference. On the other hand, the situation is quite different with the capillary in the brain. Thanks to very tight cell junctions, these capillaries are impermeable also to small solutes (the so-called *brain-blood barrier*, a defence against brain infections), such as the Na^+ and Cl^- ions, which are thus even more important than the big proteins in establishing the osmotic equilibrium, since π is proportional to the concentration and not to the mass, and ion concentrations are 100–1,000 times higher for ions than for proteins. Because of this, neuronal cells are extremely sensitive to small variations of osmotic equilibrium and can easily get dehydrated. This would lead potentially to hypertonicity of neurons in the brain and loss of intracranial fluid; however, this effect is balanced by an increased water permeability of the neuron membrane, which has a much lower filtration coefficient K, compared to other cell membranes.

5.6 Moving Around in a Fluid World

The preceding two Sections introduced two diverse modes of moving particles through a medium, the **diffusive** and the **advective** transport. The former is related to the difference (gradient) in concentration, the latter to the existence of a driving force (pressure) inducing a directed flow. (If the driving force is heat, the transport is called *convective*.) Such processes are of great importance in many biological situations, such as oxygen transport between air and tissues, the structure and function of cell membranes, the nature of the intermediary metabolism, and so on.

The overall expression for the unidimensional flux of particles would be:

$$j_x = -D\frac{dc}{dx} + v_x c \tag{5.64}$$

combining the diffusive and advective contributions. The typically small values of the diffusion coefficient, in the range of 10^{-10}–10^{-12} m²/s, set limits to biological flows to occur either over long times (slow velocities), or highly surface-convoluted structures, and generally over small sizes. Organisms relying exclusively on diffusion for the transport of nutrients and oxygen are typically micron-sized and thin-walled. On the other hand, maintaining a pressure gradient over relatively long lengths requires a lot of energy (ATP) consumption and a substantial waste of body parts to make up pipes and transport fluids.

The equilibrium between these two competing aspects of matter transport can be judged by another dimensionless quantity, the **Péclet number**: :

$$Pe = \frac{vL}{D} \tag{5.65}$$

(a)   **(b)**   **(c)**

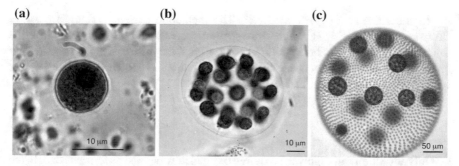

Fig. 5.9 Species of volvocalean green algae spanning a large range in size. **a** Unicellular *Chlamy-domonas*. **b** A colony of about 30 cells of *Eudorina elegans*. **c** Colony of *Volvox aureus*, about 2,000 cells (larger spots are daughter colonies). [*Photos* (**a, b**) public domain © of U.S. Environment Protection Agency; (**c**) courtesy of Aurora Nedelcu, University of New Brunswick www2.unb.ca/]

Values of Pe of the order or less than 1 indicate a predominance of diffusion, while larger values indicate a system dominated by advection [5]. A typical case is provided by Volvocalean algae, which exist in a range of social structures (Fig. 5.9): from the isolated unicellular *Chlamydomonas*, to the pluricellular *Gonium*, which form small colonies of ~50 cells, to the multicellular *Volvox*, whose dense, spherical societies can count up to many thousands of citizens. Correspondingly, the size of individual cells declines from about 30 μm to less than 5 μm. As the size of the colony grows, and that of each single cell shrinks, a constraint on diffusive transport of oxygen from the environment appears, because of the decrease of the surface to volume ratio.

The steady-state diffusion equation (5.42) can be solved for the radial coordinate r, measured from the centre of the colony of radius R, with the boundary conditions $c(r = R) = 0$ and $c(r = \infty) = c_0$ (see Problems 5.3, 5.4). These specify, respectively, that at the colony surface the oxygen concentration is zero, by assuming that all the oxygen entering is immediately distributed to the cells and absorbed; and that at a sufficient distance from the colony, the concentration tends to the constant value c_0. The solution is:

$$c(r \geq R) = c_0 \left(1 - \frac{R}{r}\right) \tag{5.66}$$

The diffusive flux, obtained from Eq. (5.29), is:

$$j(r) = -Dc_0 \left(\frac{R}{r^2}\right) \tag{5.67}$$

that is, a diffusive oxygen current (flux at $R \times$ surface) $I_R = -4\pi R D c_0$ at the surface of the colony. On the other hand, the colony requires an apport of oxygen to survive, expressed as a *metabolic* current crossing the surface of the cells. By applying a requirement of constant volume to a colony of N cells each of size r_0, the total surface is $N(\frac{4}{3}\pi r_0^3) = \frac{4}{3}\pi R^3$. The metabolic current is therefore $I_M = \beta N(4\pi r_0^2) = \beta N^{1/3}4\pi R^2$, with β a metabolic coefficient. By the two different functional dependencies on R and R^2, it is seen that the diffusive flux will no longer be sufficient to feed oxygen into the colony, when it has grown beyond the radius corresponding to $I_R = I_M$, that is:

$$R_c = \frac{Dc_0}{N^{1/3}\beta} \tag{5.68}$$

Only advection can supply the required oxygen flow beyond the critical size. In the case of *Volvox*, the cells situated at the surface of the colony provide the forced fluid flow by the agitation of their flagella, and the resulting flow velocity is proportional to the size R of the colony [6]. As a consequence, the advection current can be written as $I_A \sim Dc_0R^2/R_a$, and it will take over diffusion when the colony size reaches about $R_a > R_c$. With such a R^2 dependence, advection parallels the increase of I_M, and can in principle supply enough metabolic oxygen for any colony size.

The competition between diffusion and advection is crucial to many biological problems, and the Péclet number is a practical way to judge whether an organism exploits one or the other, for a particular function. For example, consider the blood flow in capillaries, with their typical diameter is of $\sim 2-3\,\mu$m: the diffusion coefficient of oxygen across the endothelial vessel membrane is $D = 18 \times 10^{-10}$ m^2/s, and the typical flow velocity is 0.7 mm/s. This sums up to a $Pe = 1.15$, which looks a value "evolutionarily chosen" just to properly equilibrate diffusion and flow. According to the reasoning of the previous Section and the Hagen-Poiseuille equation (5.61), which impose conditions of continuity of the pressure drop Δp and of the blood flux Q at the junctions and branchings of the vascular network, the diameter of the final sections (the capillaries) should determine backwards the sizes of all the veins and arteries in the network. Such sizes, in turn, determine the total volume of blood circulating, and the average pressure the heart must provide to reach the farmost blood vessels. It seems therefore that the condition of correct balance between diffusion and advection ($Pe \sim 1$) represents a strong constraint on the overall design of the circulatory network.

Oxygen is crucial to respiration of animals both living on the surface and in the water. Fishes and other sea creatures must extract the oxygen dissolved in water, by using gills as specialised filtering apparatus. Some fishes use gills also for feeding, by capturing small food particles that float in water. Because of their multilayered structure offering a huge surface to volume ratio, gills seem ideal organs to exploit diffusion in order to intake oxygen gas. However, for the purpose of capturing the sparse microscopic food particles, a lot of water must be filtered, therefore a high-rate flux seems rather necessary. The Péclet number in this case may help in indicating which is the predominant function of the gills in each animal. Vogel [7] cites two opposite situations, the gastropod *Diodora aspera* and the bivalve mollusk *Mytilus edulis* (the common black mussel). The former has gills filaments spaced by 10 μm and an apparent flow rate of 0.3 mm/s, giving $Pe \simeq 1.5$, adequate for oxygen breathing by diffusion; the latter has gills spaced at 200 μm and a flow rate of 2 mm/s, resulting in $Pe \simeq 100$. Such a value is clearly excessive for a purely respiratory function, likely an indicator that the mussel uses its organ mostly for feeding.

5.6.1 Brownian Swimmers

Consider a solution of particles of mass m and size(radius) R, diluted in a container of finite size L. As a function of the fluid density, particle mass and temperature, the particles will progressively settle within the fluid, according to the Archimedes' principle. This phenomenon is called **sedimentation**. The vertical profile of particle density is an exponential, $n(z) = n_0 \exp(-z/l_g)$, with l_g a characteristic sedimentation length, given by the ratio between the thermal (Brownian motion) energy $k_B T$ and the total forces acting on the particle. These are the gravitational pull $-mg$, and the Archimedes buoyant force $F_b = \rho V g$, with ρ the fluid density and V the particle volume. If we define $\Delta \rho$ the difference of density between the fluid and the particle, it is:

$$l_g = \frac{k_B T}{\Delta \rho V g} \tag{5.69}$$

The dimensionless ratio R/l_g is a measure of the relative importance of Brownian fluctuation versus sedimentation. For $R/l_g \ll 1$, density variations are apparent only over lengths much larger than the size R of the particles. On the other hand, when $R/l_g \gg 1$, gravity is stronger than thermal fluctuations, and local density fluctuations can arise. In the first case, the diffusion time is much shorter than the settling time, since the ordering action of gravity is a small perturbation compared to the disordered thermal agitation. Because the diffusion time is $t_D \sim R^2/D$, and the settling time is rather $t_s \sim R/v_s$, with v_s the sedimentation velocity, the ratio of the two times is:

$$\frac{t_D}{t_s} = \frac{v_s R}{D} \tag{5.70}$$

that is the Péclet number, Pe. However, remember that the diffusion coefficient is linked to the temperature by the particle mobility, as $D = \mu k_B T$ (in the linear flow regime), and the mobility is the ratio of the drift velocity to the force, $\mu = v/F$, or $\mu = v_s/F_b$ in this case of particle sedimentation. Therefore, $D = \mu k_B T = v_s k_B T/F_b$, or $D = v_s l_g$, and consequently:

$$Pe = \frac{v_s R}{D} = \frac{R}{l_g} \tag{5.71}$$

In other words, the above defined dimensionless number distinguishing Brownian versus settling particles, is nothing else but another definition of the Péclet number (which we used to characterise diffusive vs. advective flow).

Fig. 5.10 True-color image of the dynamic growth of a springtime phytoplankton bloom in the Bay of Biscay, off the coast of France. Image captured on April 2013, by the Moderate Resolution Imaging Spectroradiometer (MODIS) aboard NASA's Aqua satellite. The swirling colours indicate the presence of vast amounts of phytoplankton, tiny plant-like microorganisms that live in both fresh and salt water. [Public-domain image, © of NASA GSFC Archive]

Plankton is a collective name for a variety of micro-organisms, crucial to the oceanic food chain (see Chap. 12), which spend their life by drifting in the water column[7] of vast sea areas, being incapable of swimming against the currents. A large part of the plankton is represented by the *phytoplankton*, photosynthesising microscopic plants often unicellular, which are thought to provide from 50 up to 80 % of the total oxygen in the Earth's biosphere (Fig. 5.10).

Most of the time such micro-organisms are negatively buoyant, i.e. they sink, although at very low sedimentation velocities of about 4 μm/s, or 35 cm/day. It is believed that sinking could improve access to CO_2 (their "food" that is transformed into oxygen by photosynthesis), since the slowly moving cell consumes all its neighbouring source of carbon dioxide. However, sinking too much takes the cell down to depths at which photosynthesis becomes less and less efficient. Therefore, some evolutionary reason must have pushed these widespread micro-organisms to select such a strategy. For a diatom sinking over a length of 100 μm (about 10 times its own size) at a rate of 4 μm/s, with the diffusion coefficient of CO_2 equal to 14×10^{-10} m^2/s, the Péclet number is $Pe \simeq 0.3$. This means that carbon dioxide diffusion is the main factor, while advection (by sinking to different depths) does not significantly improve the capitation of CO_2. In his already cited study [7], S. Vogel suggested that phytoplankton could be sinking for a different reason, maybe to escape trapping at the sea surface by surface tension. However, it appears that phytoplankton cells could

---

[7]A **water column** is a fictitious column of water defined over a given area, from the surface of a sea, river or lake, to the bottom. It is an important concept in environmental studies, since many aquatic phenomena are explained by the vertical stratification and mixing of chemical, physical or biological parameters, measured at different depths.

not have hydrophobic surfaces, therefore the reason for their sinking still remains another one of Nature's mysteries.

5.7 Squeezing Blood Cells in a Capillary

Blood is composed by a transport fluid, the **plasma**, which flows in the vascular network transporting a variety of cells. Among these, the most important are undoubtedly the **erythrocytes**, or red blood cells (RBC), which transport oxygen to each and every cell in the body, and the **leucocytes**, or white cells (WBC), which intervene in many functions of immune response, open wounds, foreign infections, inflammations. The former have a typically flat, biconcave shape, while the latter have a roughly spherical shape, which can however be adapted and modified very easily following the specific function required. In any case, both RBC and WBC must be capable of considerable deformations, to be able to fit in the terminal capillaries of the vascular network. The diameter of the thinner capillaries is of the order of, or even smaller than, a single cell. The great deforming ability of the blood cells comes form their cytoskeleton inner structure (see Appendix E to the next Chapter), which is however very different in the two types of cell.

The deformation of WBC can be described in a simplistic model as the transformation of a spherical into an elongated, cigar-like membrane structure. The condition of constant volume must also be ensured, since the cytoplasm is a practically incompressible fluid. Since the sphere is the solid with the smallest surface to volume ratio, the amount of surface at given volume of a WBC must considerably increase, when squeezing within smaller and smaller capillaries.

Mechanically, the cell membrane cannot tolerate a deformation larger than about ~4 %, before being broken (**lysis**, see greybox on p. 166). WBCs dispose of a sort of "reserve" of membrane material, in the form of pockets and folds, which allow expanding the surface up to 2.5 times the isotonic size (the size of a sphere in osmotic equilibrium). The cell can maintain its spherical shape even with this excess membrane thanks to the surface tension Σ, ensured by the layer of actin filaments supporting the membrane from the inner side. This cortical **actin layer** forms part of the cytoskeleton (see Appendix E), integrating a network of flexible actin and a main frame of more rigid microtubules, emanating from a central structure placed next to the cell nucleus, the **centromere**. The membrane *cortical tension* can be constantly adjusted by remodelling the surface-bound actin layer.

Although these days more precise measurements can be carried out by the atomic-force microscope, the membrane surface tension Σ has been classically studied by means of a conceptually very simple experiment [8], schematically depicted in Fig. 5.11. A micropipette aspires the membrane surface with a controlled value of (negative) pressure, and the value of Σ is deduced from Laplace's law:

$$p = \frac{2\Sigma}{R} \tag{5.72}$$

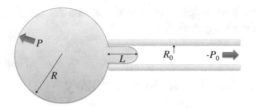

Fig. 5.11 The membrane of a spherical cell or lipid vesicle is deformed by a micropipette. Pp is the pressure in the pipette, and Po is the pressure in the reservoir. Ro and Rp are the radii of the cell or vesicle and the pipette. The resulting isotropic stress in the membrane is the surface tension T and is determined by Eq. (5.75)

valid for the cell at isotonic equilibrium, with a roughly spherical shape with average radius R.

Let us consider firstly the equilibrium of the forces acting on a diametral circumference of the sphere. The equilibrium condition between the pressure acting on the half-surface and the tension holding the perimeter, is written as:

$$F_{tot} = 0 = p(\pi R^2) - \Sigma(2\pi R) \tag{5.73}$$

In the micropipette experiment (Fig. 5.11), the cell of isotonic size R, and unknown Σ, is attached by the end of a glass tube (the micropipette) with internal diameter $R_0 \ll R$. A negative (sucking) pressure p_0 is applied from the other end of the tube. As a consequence, a portion of the cell membrane is aspired, and forms a protrusion of length L inside the tube. In practice, if a too large pressure is applied, the cell is entirely squeezed within the pipette (in fact, it is shattered); if the pressure is too small, the cell detaches from the glass surface. Only one value of p_0 can keep the system in equilibrium, by satisfying a modified Laplace equation. It can be shown that this value corresponds to a length of membrane $L = R_0$. The new equilibrium condition is obtained by imposing that the forces acting on the border separating the protrusion of size R_0 from the cell, sum to zero:

$$F_{tot} = 0 = p(\pi R_0^2) + p_0(\pi R_0^2) - \Sigma(2\pi R_0) \tag{5.74}$$

the $+$ and $-$ sign corresponding to tensile or compressive forces, respectively. Combining this latter with Laplace's equation (5.73), the value of surface tension can be extracted:

$$2\Sigma = \frac{p - p_0}{\left(\frac{1}{R_0} - \frac{1}{R}\right)} \tag{5.75}$$

Note that all the quantities at the right side are known, since p_0 and R_0 are parameters set by the experiment, and R is directly measurable, for example by an optical microscope. This equation is an equilibrium condition. In fact, if the pressure p_0 is increased, the term $1/R_0$ gets smaller, with $L > R_0$. As a result no equilibrium can be attained, with the numerator increasing while the denominator decreases. The opposite occurs if p_0 is reduced. Therefore, by manipulating the pressure until reaching the unique equilibrium condition ensured by Eq. (5.75), the experimenter can determine the value of Σ. The precision of the estimate can be increased by repeating the experiment with pipettes of different R_0, giving different equilibrium values of p_0. However, compared to an ideally homogeneous lipid bilayer, the value of Σ for the cell is not constant but depends on the adherent cytoskeletal structures, and detailed distribution of proteins and cholesterol in the membrane. Therefore, average values can be obtained for the same type of cell, which can be largely scattered about the mean. Moreover, this kind of experiment is difficult to perform on cell with highly irregular shapes, such as neurons.

Red cells do not have the reserve membrane as the WBC, therefore they adopt a different strategy to fit inside the thin capillaries. Their biconcave shape suggests that at isotonic equilibrium the integral of the forces over the entire surface of the membrane (actually the *stress*, see Appendix H) is zero. This contrasts with the situation of spheroidal cells and vesicles, which have a non-zero cortical tension, ideally defined by Laplace's equation. For the RBC to have a partly concave, and partly convex surface, it must be under zero net pressure, from which also the surface tension must be zero. (This does not mean, however, that the membrane has zero mechanical resistance, because $\Sigma \neq 0$.) While for a sphere it is not possible to find a *compatible* deformation that can reduce the pressure (compatible here means "continuous, without breaking or tearing the membrane"), a biconcave shape has an infinite variety of geometrical transformations allowed, at constant volume. This also means that, even if both volume and surface area cannot change, such a membrane can sustain very large deformations without breaking, and by this way a RBC can squeeze into a capillary of width even half its normal diameter. The fact that the RBC is stress free at isotonic equilibrium indicates that such a biconcave shape is indeed the minimum of total membrane energy. Such peculiar mechanical properties originate from the internal cytoskeleton (Appendix E) that, in the case of RBC, is entirely made from a network of very thin and flexible filamentary proteins, the **spectrines**, connected by very short actin fragments.

Appendix D: Membranes, Micelles and Liposomes

All the membranes that are found in the cell are constituted by phospholipids, for not less than 50% in mass. Artificial membranes composed by only phospholipids are also stable, and can occur in a variety of shapes. The main characteristic of **phospholipid** molecules, besides the large number of variants, is that they are *amphiphilic*, being constituted by a polar head, therefore with hydrophilic character, and two non-polar, thereby hydrophobic, tails. Each tail is composed by **two** long aliphatic chains (each one a fatty acid), containing about 13–23 CH_2 groups, plus one terminal CH_3 (Fig. 5.12a, b). If all bonds between the carbon atoms in the chain are simple bonds (called *trans*), the fatty acid is said to be **saturated**; on the contrary, if one or more bonds are double (or *cis*, with the removal of one H from each group) the fatty acid is **unsaturated**. Each double bond implies formation of a "knee", or bend in the chain, which affects the capacity of the phospholipid to assemble into a densely packed structure. Often in a phospholipid, one of the lipids is saturated and the other is unsaturated. Typical chain lengths of the lipids found in biological membranes are 16–18. While the most energetically favoured state is a straight, *all-trans* carbon chain, deviations of 120° (called *gauche* bonds) cost a bending energy of only about 0.8 $k_B T$, and so can be be thermally excited. These thermally excited kinks of a normal carbon chain are not to be confused with the permanent kinks provided by the double (*cis*) bonds.

As shown in Fig. 5.12c, the phospholipid molecule is built by attaching the terminal -OH oxygens from the two lipid chains (after liberating the H) to a central phosphate, RPO_4 (hence the prefix *phospho*), to which a side group R is attached. The nature of the latter is variable, the two most commonly found in biological membranes being ethanolamine, $CH_2CH_2N^+H_3$, and choline, $CH_2CH_2N^+(CH_3)_3$. The head groups, with the negatively charged oxygen and the positively charged nitrogen, have a dipole which interacts with the dipoles of water, thereby making the phospholipid head strongly polar. Also, the choline is significantly bigger than the ethanolamine, as the H attached to the nitrogen is replaced by the much larger methyl group, CH_3.

When phospholipids are mixed with water, their hydrophobic tails try to minimise the contact with water molecules, leaving only the hydrophilic heads exposed. Above some threshold concentration, $c_{mc} \simeq 10^{-10}$ M, phospholipids start to assemble spontaneously into a patch, which further gets curved into a globular aggregate,

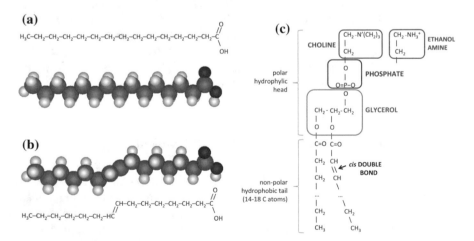

Fig. 5.12 Atomistic model and formula of: **a** saturated, and **b** unsaturated palmitic acid. The saturated molecule has all single C–C bonds, and has a linear shape. The unsaturated molecule has (at least) one double C=C bond, which forms a kink in the carbon chain. **c** The structure of typical phospholipids: a polar head (choline or ethanolamine) linked via a glycerol to two fatty acids of variable length and saturation

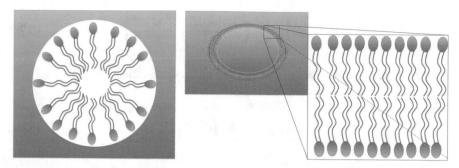

Fig. 5.13 Self-assembly of phospholipids. *Left* Above some critical micellar concentration, phospholipids minimise the interaction energy with the solvent (water) by clustering into spherical micelles. The non-polar tails point toward the interior of the sphere, to avoid contact with water, while the polar heads make up the spherical surface. *Right* If the concentration is further increased, the spheroidal micelles grow and the layer of phospholipids splits, forming a double-layer membrane. Within the two layers the tails are facing each other, providing a water-exclusion zone; heads form two outer surfaces, with which water can make contact on both sides of the membrane. The membrane bends and folds into a closed shape, thus forming a closed vesicle that resembles the structure of a cell

or **micelle**, with the hydrophobic tails grouped together so as to exclude contact with water (Fig. 5.13, left). Upon increasing the concentration, many spherical micelles can come together and pack into a dense lattice. Further concentration increase leads to elongated micelles, which eventually fusion together into a **double layer** (Fig. 5.13, right). This structure is the beginning of the cell membrane, a sandwiched structure with all the hydrophobic tails facing each other, while the hydrophilic heads are in contact with water on both sides. This is an outstanding result of the hydrophobic attraction effect, as described in Sect. 5.4, despite the fact that assembling the molecules together reduces their entropy.

Amphiphilic species can be obtained by attaching a polar headgroup to one, two, even three lipid tails. All such molecules have the capability of self-assembling, when the respective critical concentration is attained. The value of c_{mc} can be inferred, at least qualitatively, from considerations about the free energy of assembly of one individual molecule to an existing cluster: the single molecule gains binding enthalpy in attaching to the cluster, but loses entropy of its free motion in the solvent. By schematising a long-tail molecule as a cylinder of radius r and length nl_C, with n the number of CH_2 groups and $l_C \simeq 1.25$ Å the length of a C-C carbon bond (it is 1.54 Å projected along the vertical direction), its lateral surface coming into contact with the other molecules in the cluster can be estimated as $S = 2\pi r n l_C$; the gain in binding enthalpy is $E_b = \Sigma A$, for a generic lipid-lipid interfacial tension Σ (see p. 338, for the membrane surface tension). The entropy gain can be estimated by considering the single amphiphile in the solvent as an "ideal gas" molecule. Then, from Eq. (2.11) in Chap. 2, by setting $N = 1$, $c = N/V$, and $E = \frac{3}{2}k_B T$, the free molecule entropy is:

$$S_f = k_B \left\{ \ln\left[\frac{1}{c} \left(\frac{2\pi m k_B T}{h^2} \right)^{3/2} \right] + \frac{5}{2} \right\}$$ (5.76)

The critical concentration c_{mc} corresponds to the equilibrium, $G = E_b - TS_f = 0$, from which we obtain the condition:

$$\frac{E_b}{k_B T} = \frac{5}{2} + \ln\left[\frac{1}{c_{cm}} \left(\frac{2\pi m k_B T}{h^2} \right)^{3/2} \right]$$ (5.77)

or else:

$$c_{cm} = \left(\frac{2\pi m k_B T}{h^2} \right)^{3/2} \exp\left(\frac{5}{2} - \frac{E_b}{k_B T} \right)$$ (5.78)

This shows that the critical concentration decreases rapidly with the binding enthalpy from the surface tension. This latter depends on the lateral contact surface of the amphiphile in the cluster, whose radius r to a first approximation can be taken to vary as $r = 0.2$ nm, $\sqrt{2}r$, $\sqrt{3}r$, for 1, 2, or 3 lipid tails in the molecule, respectively. With $\Sigma \simeq 30$ mJ/m^2 (see below), the binding enthalpy of amphiphilic molecules with $n = 15$ is, respectively, $E_b = 17, 24, 30$ $k_B T$, for 1, 2, or 3 tails. As a result, the critical concentration of 2-chain phospholipids, from Eq. (5.78) above, is about 10^{-3} smaller than for single-chain amphiphiles (and about 500 times larger than for 3-chain ones). It takes a much higher molar concentration for single-chain amphiphiles to assemble into a double-layer, compared to 2-chain phospholipids. This is the main reason why cell membranes should be composed by 2-chain, and not by single-chain amphiphilic molecules. The latter prefer to form spherical micelles, unless the molar concentration becomes very high.

The bilayer is a highly stable configuration, in which the phospholipid molecules maintain a considerable in-plane mobility thus giving the membrane a fluid-like consistency. However, also in this configuration the free edges of the bilayer expose some of the fatty acid tails to the contact with water molecules. This is the driving force for folding the flat bilayer into a curved structure, which will eventually close to form a vesicle, superficially similar to a closed cell wall. In this configuration, the water confined inside the double layer is in contact with the hydrophilic heads, as well as the water outside, while the hydrophobic tails are completely screened by any contact with water.

The thickness of a bilayer is usually about 5 nm (50 Å). A number of factors can affect the membrane free energy, by changing either E_b or S_f. Notably, the total membrane surface can change by cutting away or adding portions of surface. Since no free patches can be found in solution because of the too high energy cost of the free borders, the patch will suddenly fold into a spherical shape (see Chap. 8). Such double-layer lipid spheroids, called **vesicles** or **liposomes,** can be found at several stages during the cell life, for example in the processes of exocytosis and endocytosis, by which a protein or a neurotransmitter is expelled from, or incorporated in the cell membrane. Also micelles can be observed in cells, when a hydrophobic species must be transported within the cell plasma: the single-lipid layer of the micelle provides a hydrophobic inner cavity, which can host and transport hydrophobic species in water. The membrane tension Σ is affected by the mechanics of the underlying cytoskeleton, with the actin cortical layer exerting more or less pressure on the cell membrane during various stages of cell life (see Chap. 6). Also, transmembrane and cytoskeletal proteins can affect the contractility or elasticity of the membrane itself, and of the cortical layer immediately in contact with it. Figure 5.14 summarises some such processes.

In a real cell, the membrane is composed not only by lipids of various types, but contains in notable proportion **cholesterol** molecules, as well as a variety of proteins, which perform numerous specialised functions at the interface of the cell with the external world. Since the bilayer is in a fluid state, the diffusion of the proteins is sufficiently rapid. At low temperature, a phase of pure lipids undergoes a structural transition to a **gel phase**, in which the tails are more ordered, (i.e., have fewer gauche bonds) as are the headgroups. Presumably the heads possess long-range orientational order, but no long range positional order (this is called a *hexatic* phase, and is usually indicated as S_o). The increased order of the tails permits the lipids to pack more efficiently, with the consequence that the diffusion constant of the proteins decreases. At higher temperatures, this phase starts forming surface undulations, or "ripples"; the bilayer in this phase, indicated as P_β, is still very ordered despite the wavy appearance. Because the proteins cannot do their job in a timely fashion, these gel-like phases are biologically useless. Notably, lipids with two fully saturated tails of length about 16 are in this useless state already at body temperature. Replacement of one of the saturated chains by an unsaturated one causes a permanent kink in that

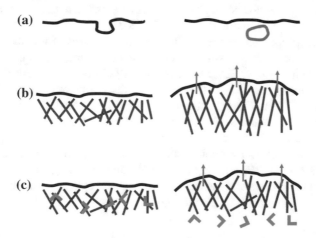

Fig. 5.14 Interaction between the tension and area of the cell membrane, and various cellular processes. When the term ΣA in the free energy is too high, the events on the *right side* occur; when it is too low, the events on the *left* take place. **a** In exocytosis, a patch of the membrane under low tension is detached and forms a closed liposome, or vesicle (*blue*), the remaining membrane (*black*) increases its tension. **b** The actin cortical layer (*red*) can increase its pressure on the membrane by fast polymerisation, leading to elongation of the actin filaments. **c** The expulsion of myosin (*purple*) from the actin layer decreases the contractility, leading to an increased actin tension on the membrane

chain and makes the system difficult to pack. With a large enough fraction of such "defects", the transition temperature is lowered to well below 300 K, likely providing a reason for the widespread presence of unsaturated fatty acids in the cell membrane of all living organisms.

At yet increasing temperatures, the bilayer goes into a state with the tails very disordered. This is a "tail-melting", and the corresponding phase, indicated as L_d, retains the properties of a fluid, with increased in-plane disorder.

Cholesterol, like any sterols, is a lipid with a very small polar head (just the OH hydroxyl termination), next to a planar structure of 3 hexagonal + 1 pentagonal aromatic rings, and terminated by short a hydrophobic tail, $(CH_2)_4CH_3$. Adding cholesterol to a lipid bilayer in the fluid phase decreases the membrane permeability to water, since cholesterol tends to occupy part of the free volume within the long lipid chains, thus decreasing their flexibility. A partial phase diagram of a synthetic membrane of DPMC phospholipids with increasing molar concentrations of cholesterol is shown in Fig. 5.15. The pure lipid phases are found by looking along the vertical line at zero concentration. Adding cholesterol to the gel phase disrupts the local order, increasing the in-plane diffusivity and reducing the membrane elastic modulus. A liquid-ordered phase is formed, on the right side of the phase diagram. This would be the "normal" phase also for a biological membrane, however considering that in a cell membrane islands of different lipids can exist, with locally variable concentrations of cholesterol. The average cholesterol concentration in the cell membrane can be about 40–50 % on a molar basis (about 15–20 % when expressed in weight fraction, since cholesterol is a smaller molecule compared to typical phospholipids). In the simple phase diagram in the figure, liquid-disordered and liquid-ordered phases, as well as liquid-ordered and solid-ordered phases are seen to coexist in the homogeneous lipid bilayer, at the normal temperatures $T \sim 36-38\,°C$.

The elastic properties of the membrane chiefly derive from the competition between attractive (Van der Waals + electrostatic) interactions between the side chains, and their entropy. Since each phospholipid occupies about 0.4 nm^2, and the interaction energy is $\sim 3k_BT$, the energy required to stretch a patch of membrane is of the order of 7.5 $k_BT/$nm^2, or 30 mJ/m^2. The elastic moduli of membranes will be better defined in Chap. 8. Typical cell membranes have a low shear modulus, $4-10 \times 10^{-3}$ N/m; a high elastic modulus, due to the small stretching allowed in lipid bilayers, 10^3 N/m^2; a variable viscosity, which depends on membrane composition, $0.36-2.1 \times 10^{-3}$ Pa-s for red blood cells; and a bending stiffness κ_b, strongly influenced by the presence of membrane proteins and cytoskeleton elements, of the order of 10^{-19} N-m, or ~ 100 pN-nm.

Problems

5.1 Stationary flux
Show that the time-independent solution of the diffusion equation (5.42) corresponds to a constant flux across the membrane.

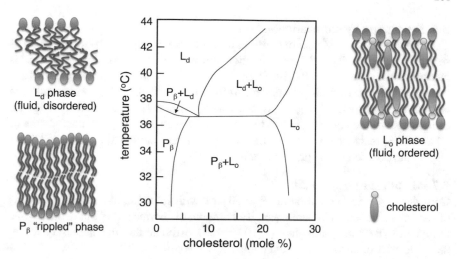

Fig. 5.15 Partial phase diagram of DPMC phospholipids with cholesterol, in excess water solvent (experimental data taken from [9]). The curved lines indicate coexistence points (temperature, concentration) between the various phases. Within each area of the diagram one phase is formed or, between two coexistence lines, a mixture of two phases (A+B) appears. The S_o phase is not visible, since it appears at lower temperatures

5.2 Artificial blood

In your laboratory, someone is trying to make artificial blood. Therefore, they start preparing spherical vesicles from a phospholipid suspension, with average size R. A concentration of about 30 % vol. of haemoglobin is introduced in the vesicles. When such artificial "red blood cells" are placed in pure water, the membrane is ripped open, and the protein diffuses in the water. After some test, you discover that if the vesicles containing haemoglobin are placed in a 1mM solution of NaCl, they do not explode and remain quite spherical. Explain the result. Moreover, if 1mM is good, do you think that 2 mM should be better?

5.3 A cell spewing glucose

Take a spherical cell of radius $R = 10$ μm, whose membrane has a permeability for glucose of $P_M = 20$ μm/s. Calculate the time variation of the glucose concentration inside the cell, after it is immersed in a large tank of pure water at time $t = 0$.

5.4 A breathing bacterium

Consider a bacterium as a sphere of radius R_0. Our bacterium lives in a pond, from where it takes the oxygen to breathe at a concentration c_0. Take that as soon as the oxygen molecules pass the outer bacterial membrane they are instantaneously turned into CO_2, and compute the oxygen concentration profile around the bacterium.

5.5 Haute cuisine

You are preparing a strawberry pie in the kitchen. So, you cut your berries in half and sprinkle them with powdered sugar. After just a few minutes, your fruits are softened and float in juice. What happened? Where the water comes from?

5.6 Separation by sedimentation

The *sedimentation coefficient* of a species in solution, $s = v/a$, is the ratio between its sedimentation velocity and the acceleration applied (causing the sedimentation); it is measured in a special unit, the Svedberg, $1\ S = 10^{-13}$ seconds. Consider a centrifuge turning at 10^3 rpm (rounds per minute). At time $t = 0$ a beaker containing a solution of two mixed proteins A and B is placed at a position $r_0 = 5$ cm away from the central axis of the centrifuge. The two proteins have sedimentation coefficients of 10 and 30 S, respectively. At what time the protein A is found at $r = 10$ cm? What will be the position of the protein B at that time?

5.7 Membrane permeability

Consider a spherical cell of radius $R = 10\,\mu m$, with some initial concentration $[c_{in}]$ of a species, immersed in pure water. By knowing the permeability of the membrane to glycerol (10^{-8} m/s) and glucose (10^{-12} m/s), estimate the time necessary for all the molecules of each type to void completely the cell.

5.8 Blood flow in the arteries

Compute the increase in cardiac pressure necessary to transport the same amount of blood, from a single artery of radius R, into two branched arteries of equal radius $R/2$.

5.9 The osmose on Mars

You and your friend who lives on the planet Mars are repeating the osmotic pressure experiment of Van t'Hoff. You both build the water container, a glass cylinder sealed by the same type of semi-permeable membrane, and do the experience of adding variable concentrations of glucose inside the cylinder. However, after a Skype call to Mars, you discover that the level variations of the water in the cylinder are very different between the Earth and Mars. Can you explain to your martian friend why?

References

1. S. Asakura, F. Oosawa, On the interaction between two bodies immersed in a solution of macromolecules. J. Chem. Phys. **22**, 1255 (1954)
2. D. Stigter, K.A. Dill, Free energy of electrical double layers: entropy of adsorbed ions and the binding polynomial. J. Phys. Chem. **93**, 6737–6743 (1989)
3. J.T.G. Overbeek, The role of energy and entropy in the electrical double layer. Colloids & Surf. **51**, 61–75 (1990)
4. J. Crank, *The Mathematics of Diffusion*, 2nd ed. (Clarendon Press, Oxford, 1975)

5. C.A. Solari, S. Ganguly, J.O. Kessler, R.E. Michod, R.E. Goldstein, Multicellularity and the functional interdependence of motility and molecular transport. Proc. Natl. Acad. Sci. USA, **103**, 1353 (2001)
6. M.B. Short et al., Flows driven by flagella of multicellular organisms enhance long-range molecular transport. Proc. Natl. Acad. Sci. USA **103**, 8315–8319 (2006)
7. S. Vogel, Living in a physical world I. Two ways to move material. J. Biosci. **29**, 391–397 (2004) [First of a series of review articles on biophysics by S. Vogel, appeared in this journal between 2004 and 2007]
8. A. Diz-Munoz, D.A. Fletcher, O.D. Weiner, Use the force: membrane tension as an organizer of cell shape and motility. Trends Cell Biol. **23**, 47–53 (2013)
9. M.E. Vist, J.H. Davis, Phase equilibria of cholesterol/dipalmitoylphosphatidylcholine mixtures: 2H nuclear magnetic resonance and differential scanning calorimetry. Biochem. **29**, 451–464 (1990)

Further Reading

10. N.J. Shirtcliffe, G. McHale, S. Atherton, M.I. Newton, An introduction to superhydrophobicity. Adv. Colloid & Interface Sci. **161**, 124–138 (2010)
11. R. Phillips, J. Kondeev, J. Theriot, H. Garcia, *Physical Biology of the Cell*, 2nd edn., Chap. 14 (Garland Science, New York, 2012)
12. W. Bialek, *Biophysics: Searching for Principles*, Chap. 4 (Princeton Univ. Press, New Jersey, 2012)
13. P. Nelson, *Biological Physics: Energy, Information, Life*, Revised 1st edn., Chap. 4–5 (Freeman, New York, 2013)
14. R. Piazza, Settled and unsettled issues in particle settling. Rep. Progr. Phys. **77**, 056602 (2014)

Chapter 6
Molecular Motors in the Cell

Abstract Many molecules found in living organisms can bind ATP, and use its energy to perform mechanical actions such as bending, twisting, rotating. In some special proteins such an action can be performed cyclically, as the same molecule can use ATP units at regular intervals, to repeat continuously its mechanical action. If this may not appear at all surprising from a purely chemical perspective, being just one more case of enzymatic chain reaction, it becomes a fascinating subject when seen under an engineering perspective. In fact, such molecules are nothing less than true molecular-scale motors. Dozens of different motor proteins exist in every eukaryotic cell to perform the most diverse functions, and prokaryotic cells also have their share, by employing sophisticated rotating or flapping molecular structures, in their swimming movements.

6.1 Molecular Motors

In the previous Chapter, we saw that how the Brownian motion of molecules provides a microscopic basis for diffusive phenomena. Seen under another point of view, diffusion also connects with the concept of irreversibility, since any process that spreads energy, heat, concentrations, over wider and wider distances increases the entropy of the total system. If the process leading to diffusion were to be reversed, free energy should be supplied to the system to invert each collision and reduce the entropy. Therefore, free energy is being dissipated by the system during diffusive spreading, in the form of entropy production released to the surroundings (the thermal bath).

The important connection between random Brownian motion and diffusion was based on the Einstein's equation, $\langle x^2 \rangle = 2dDt$, stating that the root-mean squared displacement of a particle increases linearly with time, through the diffusion coefficient D and a constant $d = 1, 2, 3$ for diffusion in 1 (channel), 2 (surface) or 3 (volume) dimensions. What this equation means is that the random motion of the particle spreads over larger and larger regions of space with time, and the corresponding probability of finding the particle at a given point, for example the origin from where it started, becomes exponentially small as time goes by.

© Springer International Publishing Switzerland 2016
F. Cleri, *The Physics of Living Systems*, Undergraduate Lecture
Notes in Physics, DOI 10.1007/978-3-319-30647-6_6

As discussed in the greybox on p. 178, a second fundamental equation was derived by Einstein, expressing the numerical value of the diffusion coefficient as a function of the local temperature T, the viscosity of the medium η, and some geometrical parameter of the diffusing object, which for a spherical particle coincides with its radius R, $D = k_B T / 6\pi\eta R$. Note that in deriving this expression, use was made of the Stokes' relationship for the **particle mobility** μ, as the inverse of the linear drag coefficient for a spherical particle, $\zeta = 1/\mu = 6\pi\eta R$.

The viscosity of water is $\eta = 0.001$ kg/(m s) at $T = 300$ K. For a big protein with radius $R = 10$ nm, the diffusion coefficient is $D \simeq 10^{-10}$ m^2/s, a value that drops to 10^{-12} m^2/s for a lipid medium with viscosity about 100 times larger than water. With such values of diffusivity, a protein in water can move over a distance of 1 μm in about 1 s, a reasonable time for a protein to, e.g., move from the nucleus to some organelle inside the cytoplasm; but it would take about 300 years to cover a distance of 1 m, making it impossible to move a neurotransmitter or ATP molecules along a long nerve axon, from the brain to a limb.

Because the mechanism is dissipative, it is not enough to provide information about the free energy difference between the initial and final states in order to estimate the work performed in the diffusive spreading process. As we will see later, dissipative forces depend, among other parameters, also on the velocity: accelerating the rate of diffusion by any mechanisms, such as the molecular-scale equivalent of a "motor", has the effect of increasing the dissipation of free energy, to a first approximation at a rate proportional to the velocity.

Cells need several microscopic mechanisms capable of assisting and enhancing their movements, in a number of different instances, such as: cell division, growth and expansion of tissues, search for food and escaping predators (for unicellular organisms), transport of materials, ions, proteins, inside and outside the cell. Cellular movement is also connected with the change of the cell shape, by reshaping the membrane and the internal structure, via rearranging and remodelling the **cytoskeleton** (see Appendix E in this chapter). In either case, the evolution is accompanied by a forced transport or displacement of matter inside and, to a variable extent, outside the cell.

These additional mechanisms cannot be based on the simple Brownian motion, which are at the basis of simple diffusive processes, since as we have already seen the Brownian motion produces on average a zero net displacement. Such additional mechanisms are just real motors but at the molecular scale, which function by transforming the chemical energy stored in ATP into mechanical energy for movement and actuation functions, in every instance in which the cell needs to exploit a mechanical force. On the other hand, such molecular motors are, as their very name says, nothing more than highly specialized molecules; therefore at the smallest scale their elementary movements are subject to temperature fluctuations, and could not be anything else than Brownian. The question is then: how is it possible that from a Brownian movement, a net displacement and transport of matter could result?

In Chap. 2 it was shown that the efficiency of transformation of free energy into work is increased, whenever the process is broken into smaller and smaller elementary steps. In that case the demonstration was done for a thermal engine, but it is obviously valid also for a chemical engine. Such an observation is central to the idea that natural mechanisms are invariably based on multi-level hierarchical structures, ranging from the molecular scale up to the macroscopic. Muscles, to be described in more detail in Chap. 10, represent a good example of this multi-step architecture:

- a muscle is constituted of a parallel bundle of long fibers, on the cm scale and 0.1 mm thickness, each fiber being a multinuclear cell wrapped in its membrane;
- each fiber is assembled from parallel individual myofibrils, with the same length of the fiber but about 2 μm thick;
- each myofibril is highly structured bundle of myosin and actin long molecules, subdivided into elements extensible between 1.25 and 2.5 μm length (the sarcomeres);
- each myosin and actin unit results from the assembly of individual molecular filaments, made from many molecular subunits, with sizes in the range of 1–2 μm length and tens of nm in diameter.

At the very bottom of the mechanical actuation process, the origin of the force exerted by the muscle lies in the microscopic displacement of the millions of individual actin and myosin molecules. This is likely the best studied example of a molecular motor, for which a simple mathematical model will also be discussed in the next Section. The microscopic mechanism by which each individual myosin molecules can actuate a force on the actin filament has been elucidated by J. Finer and coworkers in 1994 [2], and since then their experiment, which goes under the general name of **motility assay**, has been repeated a number of times by many other laboratories (Fig. 6.1).

Myosin molecules are braided in bundles, so that the entire bundle can exert and amplified force on the actin, pretty much like a gymnast pulling himself along a hanging rope. On the other hand, there are several examples of molecular motor proteins which work in isolation, instead of grouped. Important examples in this class are the dyneins, and the kinesins. Such molecules are capable of translating along a fixed support structure, such as a rigid microtubule in the interior of a cell (Fig. 6.2), by alternating between two metastable conformations of the same molecule. During their movement, they can transport some large cargo, such as a vesicle of 50–100 nm size, containing e.g. molecules of a neurotransmitter, along quite far distances, in a very well directed walk that defies the idea of random, Brownian diffusion. Examples of such processes can be seen in the micro-photographies shown in the figure, showing the interior of neuronal cells from the mouse spinal chord (photo taken with a cryo-electron microscope, a particular electron scanning microscope capable of working at very low temperatures, on frozen biological samples).

6.2 The Mechanics of Cyclic Motor Proteins

Myosins and kinesins are two example of *cyclic* engines: they can repeat indefinitely the same sequence of elementary steps, by changing between two metastable

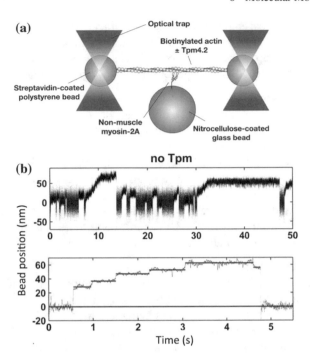

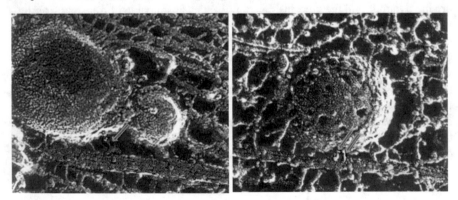

Fig. 6.1 Schematic of the myosin motility assay. **a** A glass or plastic bead is sparsely covered with myosin molecules. A single F-actin filament is held with its extremities fixed at two suspended microbeads. The experimental apparatus, not shown in the figure, is a double laser trap, which maintains the beads, and can measure the recall force exerted by the actin filament. **b** The effect of adding 2 mM concentration of ATP (fuel for the molecular motor): myosin generates a force that displaces the actin filament. The time-trace above measures the displacement of one of the plastic beads; the trace below this a zoom on a single event, with displacement steps highlighted in *red*. [Adapted from Ref. [1], under CC-BY 3.0 licence, see (**) for terms.]

Fig. 6.2 Cryo-electron micrographs of membrane organelles (highlighted in *blue*, size of the vesicles $\simeq$100 nm), transported along microtubules (highlighted in *green*) in a nerve axon, obtained by quick-freezing of a cell section and deep-etching. Short crossbridges, which are supposed to correspond to different molecular motors (highlighted in *red*, and indicated by a *yellow arrow*), can be noted between membrane organelles and microtubules. [Adapted from Ref. [3], w. permission.]

conformations, and without altering their basic molecular structure, provided that ATP energy is constantly available to power their sequential conformational trans-formations.

Let us look a bit more closely at one molecule from these very large families: myosin-II, exactly the one implicated powering the contractile stroke in muscle fibrils. Graphical descriptions of the molecular structures and processes in which this motor protein is implicated are shown in Fig. 6.3.

Myosin-II results from the braiding of two filaments, each one about 150 nm long and carrying a globular end, of about ∼20 nm size; the myosin head is capable of binding to an active site of the actin monomer, carried by the actin filament (Fig. 6.3a). It may be observed that kinesin has a very similar structure to myosin-II, however its processes are shorter (∼60–70 nm) and with smaller globular heads (∼10–15 nm). The globular head, a structure which is very much conserved across the more than 17 different types of myosins, each with many variants, displays an ATP-binding pocket, and an actin-binding site; the head is linked to the tail by a neck, or 'lever' region, flexible and typically hosting various thinner filamentary proteins, participating in the stroke regulation; the tails are quite variable among the different myosins, those of myosin-II being among the longest. The basic functioning of the myosin-II sliding and pulling the actin fiber has been reconstructed based on X-ray diffraction and microscopy observations, although many important details are still missing or unclear.

The model in five steps described in Fig. 6.3b is today the accepted version of the chemo-mechanical actioning of the force stroke, as proposed by Lymn and Taylor in 1971. (1) Starting from the *rigor* position, (2) the head attaches one ATP molecule: this 'cocks' the lever mechanisms, and releases the head from the starting actin binding site. (3) The subsequent hydrolysis of ATP into ADP plus Pi lets loose the lever arm. (4) Release of the phosphate allows the head to bind to a different actin site, some 5 nm distant from the previous one. (5) The final release of the ADP makes the lever to fold back into position (1), thus transferring the power stroke to the fixed actin filament, and making the myosin to advance by the same length of about 5 nm. The frequency of this cyclic movement is measured in ∼0.5–1 Hz.

The overall mechanism of actin-myosin combined power stroke is supposed to involve more than 300 different molecular species, in which ATP plays a major regulating role. The ATP attachment is a random process, ensured by the ATP con-centration inside the myofibril. As it will be shown in Chap. 10, the muscle fibre is composed by a tight interconnection of sarcomeres, and sarcosomes (the equivalent of mitochondria for muscle cells); therefore, ATP in sufficient amounts is normally available for actin-myosin contraction. However, the way this mechanism is regulated is still under active debate. The primary neural input comes in the form of Ca^{2+} ions invading the sarcoplasmic reticulum (equivalent of endoplasmic reticulum in other cells), a network that entirely wraps around the fibre (see Fig. 7.4 in Chap. 7). One possible regulatory mechanism has been identified in the protection/deprotection of

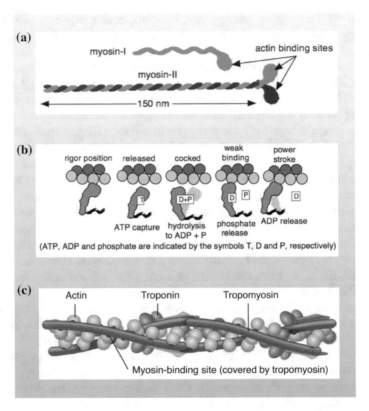

Fig. 6.3 a Schematic structure of the myosin-I and myosin-II proteins. Both have a globular head at one end, very similar in all types of myosins. The bending neck is the site of regulation, where smaller filamentary proteins can loop. In myosin-II, the tails of the two monomers (*light and dark grey*) are twisted about each other. **b** The five-steps power-stroke mechanisms that is suggested as the basis of the movement of myosin-II (lower shape with black tail) along the actin filaments (represented as two series of linked spheres). **c** Schematic of the protection/deprotection mechanism, by which tropomyosin filaments, actuated by a conformational change of troponin, can cover or expose the active sites of actin (*black spots*), to which myosin-II can subsequently bind

the accessible binding sites on the actin filaments (Fig. 6.3c): the actin filaments are not exposed naked to the myosin heads, but are wrapped themselves in a double filament of a long molecule, the tropomyosin, which in resting condition 'covers' the actin active site; the calcium ions activate the change of shape of a secondary protein, troponin; this, in turn, makes the tropomyosin filament to move with respect to the actin filament, thus making actin binding sites accessible to the myosin heads. However, the way troponin conformation switching and myosin attachment could be synchronised is yet poorly known.

6.2.1 Two-State Model of a Machine

Here the molecular motor is simply described as an object with two possible internal states, a and b, as schematically represented in Fig. 6.4. Movement can occur in finite steps of unit length along the discrete coordinate N, with all positions N independently reachable, i.e. at each position the motor can step forward or backward, with given probabilities identical for all positions. To fix ideas, a may be thought of a bound state in which the motor is (non-covalently) attached to the rail, like a myosin in the *rigor* position (Fig. 6.3b); and b could be the released state in the same Fig. 6.3b, in which the motor is loosely bound. The two states are separated by an energy difference $\Delta E = E_b - E_a$, likely corresponding to a different molecular conformation; and by a chemical potential difference $\Delta \mu = \mu_b - \mu_a$, for example corresponding to the binding of ATP in one state and ADP in the other.

Let us further assume, for the sake of simplicity, that the a and b states correspond to different discrete positions N along the rail, so that the motor can be in only one state a or b at each position. Once it is in a position N corresponding to the internal state a, the motor can jump 'forward' from state a to b with a rate (probability per unit time) $r_{\vec{a}}$, and 'backward' with a rate $r_{\overleftarrow{b}}$; on the other hand, if it is at a position corresponding to the internal state b, it can jump forward from b to the next a with a rate $r_{\vec{b}}$, and backward to the previous a with a rate $r_{\overleftarrow{a}}$.

Transitions between the two states can therefore occur by two channels: either by a chemical transformation, or by a thermal fluctuation. The corresponding rates can be schematically written as:

$$r_{\vec{a}} = \left(u e^{\Delta \mu / k_B T} + w\right) e^{-\Delta E / k_B T}$$
$$r_{\overleftarrow{b}} = (u + w)$$
$$r_{\overleftarrow{a}} = \left(u' e^{\Delta \mu / k_B T} + w'\right) e^{-\Delta E / k_B T}$$
$$r_{\vec{b}} = \left(u' + w'\right) \tag{6.1}$$

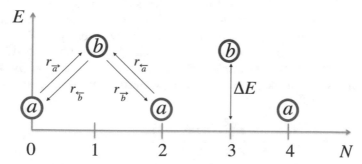

Fig. 6.4 Schematic representation of the discrete two-state model for a molecular motor. The two internal states a and b have an energy difference ΔE, and a different chemical potential $\Delta \mu$. The locations N of the motor along the rail alternate between the two states, so that the motor can be in only one of the two states at each location. Forward and backward transition rates between the two internal states are indicated by a subscripted r

with coefficients u, u', w, w' depending on the particular molecule and its chemical reactivity. Note that the coefficients are in general different for the $a \rightarrow b$ transition and the reverse $b \rightarrow a$ transition, to represent the fact that the two reaction paths may not be symmetrical (for example, by looking at Fig. 6.3b, it is seen that the $a \rightarrow b$ transition corresponds to the capture of ATP, while the $a \rightarrow b$ transition corresponds to the release of ADP+P_i).

Now, the probability of moving from one a position to another one in the forward direction is equal to the product of the rates $(r_{\vec{a}} \cdot r_{\vec{b}})$:

$$W_{N \rightarrow N+2} = \left(u \ e^{\Delta\mu/k_B T} + w\right) e^{-\Delta\varepsilon/k_B T} \left(u' + w'\right) \tag{6.2}$$

and the probability of jumping backwards from the same a is:

$$W_{N \rightarrow N-2} = \left(u' e^{\Delta\mu/k_B T} + w'\right) e^{-\Delta\varepsilon/k_B T} \left(u + w\right) \tag{6.3}$$

The difference in energy ΔE between the movement in the forward versus backward direction is proportional to the ratio between these two probabilities (see Eq. (3.4) in Sect. 2):

$$\Delta E = k_B T \ln \frac{W_{N \rightarrow N-2}}{W_{N \rightarrow N+2}} = k_B T \ln \frac{\left(u' e^{\Delta\mu/k_B T} + w'\right) (u + w)}{\left(u \ e^{\Delta\mu/k_B T} + w\right) (u' + w')} \tag{6.4}$$

Three things can immediately be noticed. Firstly, the dependence on the thermal fluctuations has disappeared. Secondly, if there is no chemical affinity difference, $\Delta\mu = 0$, then also the energy difference is $\Delta E = 0$, meaning that the system has no preference for the movement in the forward or backward direction: it will make as many jumps to the right as to the left, and there will be no net displacement from its average position. Third, this same indifference holds if, whatever the value of $\Delta\mu$, the forward and backward coefficients are equal, $u = u'$ and $w = w'$.

For the molecular motor to produce a net displacement, thus expressing a sort of 'preferential diffusion' in one direction, the conditions $\Delta\mu \neq 0$, and at least one of the $u \neq u'$, $w \neq w'$, are strictly necessary. If this is the case, the positions at $N, N + 2, N + 4, \ldots$ are all separated by an energy difference ΔE; the energy landscape for the molecular motor would be a straight line with slope $\Delta E/2$, and the motor could perform a net displacement in the direction of the negative slope.

For such a simple model, the drift velocity, v, and diffusion coefficient, D, can be explicitly calculated:

$$v = 2 \frac{(uw' - u'w) \left(e^{\Delta\mu/k_B T} - 1\right)}{u + u' + w + w'} \tag{6.5}$$

$$D = \frac{(2uu' + uw' + u'w) e^{\Delta\mu/k_B T} + 2ww' + uw' + u'w}{2(u + u' + w + w')} \tag{6.6}$$

Coherently with the previous findings, the diffusion coefficient D is always different from zero, even if the conditions for biased diffusion are not met, while the drift velocity is zero if $\Delta\mu = 0$, or if both $u = u'$ and $w = w'$.

6.2.2   Continuous Energy Surfaces

Any machine, macroscopic or microscopic, thermal, electrical or chemical, performs its work on a well defined **energy surface**, or energy landscape. In the two-state model it was assumed that the machine could switch periodically between two discrete states. Now, by introducing the concept of energy surface we rather see the machine evolution as operating a continuous transformation of its states, along one or more coordinates (control parameters). Let us take as an example a very simple machine, a bucket lifting a mass of water m from an underground pit, pulled by a long rope. This machine has an energy surface governed by one single control parameter: the height h of the bucket inside the pit, giving the gravitational energy $U = mgh$ that the tension on the rope must balance. Let us take that the pit has a depth $-L$, and we fix a time T to pull the bucket up to the ground level at constant velocity. If we wish to represent the energy U of our water-lifting machine in a time diagram, this would be simply a zigzag line changing between $-L$ and 0 in height, with a period T (as in Fig. 6.4). In this case the energy surface is unidimensional, with the only dimension being the height variable h.

The more general energy surface $E(\xi_1, ..., \xi_p)$ has a number of dimensions equal to the number p of variables that can control its value and evolution, or control parameters $\xi_i, i = 1, ..., p$ (we always speak of a "surface", even if such a mathematical entity has $p = 3$ or more dimensions). For the machine to be **cyclic**, at least one fixed point $E = E_0$ must exist on this surface, to which the machine returns at more or less regular intervals of time. To move between the fixed point(s), the machine must necessarily to go through points of higher energy, i.e. the energy surface has one or more *barriers* to cross. Our simple water-lifting machine in the one-dimensional parameter space, can be described by a diagram in which the energy appears in ordinates, and h is on the abscissa: the energy barrier would have a triangular shape, with base $2T$ (the time necessary to perform one up-and-down cycle) and height L. Some examples of simple machines are depicted in Fig. 6.5.

Each of the mechanical machines of the types shown in the figure can be defined by an input function and an output function. The output is a result of the input via a mathematical relationship, the *transfer function*. If we think of the simple action of kicking a ball with our foot, the output is a linear function of the input: the stronger the kick, the farther the distance we send the ball, and no chance to see it coming back. A cyclic machine, however, is characterised by a non-linear transfer function between the input and output. It would be the case of a ball linked with a long rubber

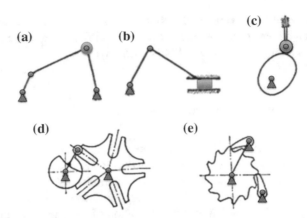

Fig. 6.5 Examples of cyclic machines; *small green triangles* indicate fixed elements, *small red circles* indicated mobile elements. **a** Two-point lever, to raise the blue point from ground to a maximum height; **b** constrained lever, to slide the blue object back and forth along the straight guide; **c** cam, where the circular movement of the eccentric (*egg-shaped part*) translates into the lifting of the blue element; **d** maltese gear, in which the circular movement of the moon-shaped element on the *left* turns, by equal discrete steps (5 in this case), the central shaft on the *right* to which the cross is attached; **e**: ratchet, to turn by discrete angular steps the central shaft without possibility of stepping back. The last two machines are called "steppers"

rope to our foot: if the kick is not enough to send the ball to a distance longer than the rope length, the ball does not come back; but if we kick the ball with enough energy to cross the "barrier", the rubber stretches and springs back, and we see the ball coming back to our foot.

A machine with an energy surface characterised by one or more barriers, has one or more minimum energy positions between the barriers. These lowest energy positions coincide with the fixed point(s). The highest points on each barrier, instead, are metastable states. The derivatives of the energy function taken with respect to all the control parameters, $\partial E/\partial \xi_i$, are zero at both the minima and maxima. Notably, once the machine is at a metastable point, it can take any direction in the parameter space $\xi_1, ..., \xi_p$, since the energy has a negative slope in every direction ξ_i (because any other point around a metastable point has a slightly lower energy). The amount of energy necessary to set in motion the machine must be at least equal to the nearest barrier height, and if the machine is returning to the same point, this amount of energy must be supplied periodically, in order to maintain the cyclic movement.

The myosin-II moving along the actin filament in discrete steps, and consuming one ATP molecule at each step, is just an example of such a cyclic machine going between a periodic sequence of energy minima and maxima. A simple representation of its unidimensional energy surface could be written as:

$$E(x) = E_0 \sin\left(\frac{2\pi x}{L}\right) \tag{6.7}$$

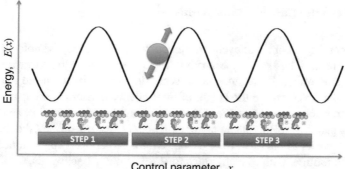

Fig. 6.6 Energy surface for the myosin-II climbing along the actin filament. The position of the myosin is symbolically represented by the *blue ball* sliding above the unidimensional energy surface, defined by Eq. (6.7). The 5-step model of Fig. 6.3b is reported under each period of motion, with the first rigor position coinciding with the resting position of minimum energy, and the position at the *top* of the hill being the metastable state

The only control parameter is the position x along the actin filament. The period L is equal to about 5 nm for this case, and the periodicity of the sine function ensures that $E(x+L) = E(x)$, as represented in Fig. 6.6. At each step, the myosin head returns to its minimum (the *rigor* position in Fig. 6.3b). Each time an ATP molecule is captured, the molecule climbs the energy barrier, up to the metastable point represented by the "cocked" position in which the ATP is hydrolysed into ADP+Pi. From this point, the myosin can attach to a novel position along the actin filament, thus falling into a new minimum, a stable position.

Such an idealised description of the myosin-actin energy surface reveals two important shortcomings. Firstly, the process of barrier climbing, powered by the attachment of ATP, is perfectly symmetric here: the blue ball can climb indifferently the barrier to its left or to its right with equal probability. On average, it will make as many steps to the left as many to the right, therefore its average displacement will be $\langle x \rangle = 0$, as for any diffusion process. If we want to push the myosin toward a definite direction, it is necessary either to "tilt" the energy landscape, by adding some energy penalty $-\beta$ for the movement in one direction, e.g. $E(x) \propto \sin(2\pi x/L) - \beta x$; or, alternatively, we must find a way to break the left-right symmetry of the unidimensional Brownian motion, by some *rectification* mechanism which would make the random jumps more probable in one direction. Secondly, as shown by the Fig. 6.3b, the 'cocked' position does not coincide exactly with the maximum of the energy surface. In fact, it is experimentally observed that the mechanical movement of the lever-arm of the motor protein is much shorter than the ∼5 nm covered by each myosin step. Therefore, the mechanical stepping mechanism powered by the ATP cannot be the only explanation of the longitudinal displacement.

6.3   The Thermal Ratchet Model

If we look at the examples of cyclic machines in Fig. 6.5, it may be noted that all but
the (e) are reversible, i.e., they can run in a 'forward' and a 'backwards' direction
with the same effect. If we imagine to scale down the size of these machines to
the molecular level, where the height of their energy barriers is comparable to the
energy of thermal fluctuations, the probability of moving in either direction would be
identically given by a Boltzmann factor $\exp(-\Delta G/k_B T)$, with ΔG the free-energy
height of a barrier in the forward or the backwards direction. The machine performs
a Brownian motion, with as many steps forward as many backwards.

The ratchet, instead is an example of irreversible machine: once the cogwheel has
turned by enough an angle, so that the pawl clicks on the next tooth, the backwards
movement becomes impossible. In energy terms, it is like the forward and backwards
barriers are no longer symmetrical, in fact the ΔE for the backwards motion has
become practically infinite. In this case, the Brownian motion is said to be *rectified*,
in that one direction of motion has a larger probability than any other. The currently
accepted models of molecular motors, such as a myosin-II traveling along the actin
filament, or a dynein moving along a microtubule, are all based on the rectified
Brownian motion.

Similar 'ratcheting' mechanisms begun to be identified in many types of molecular
motors, and they are all practical realisations of the same idea of **thermal ratchet**,
with 'thermal' having the same meaning above, of a machine whose energy scales
are comparable with the energy of thermal fluctuations (Fig. 6.7).

If we consider a particle with its diffusion coefficient D, diffusing in a medium
limited by boundaries $[0, L]$, the Einstein formula allows to estimate the average
time required to traverse this domain as:

$$\langle t \rangle = \frac{L}{2D} \tag{6.8}$$

Fig. 6.7 Schematic of the
thermal ratchet model of
rectified diffusion. The
diffusion path is divided into
many boxes of equal length
δ, separated by a one-way
door, or 'ratchet'. Once the
red particle diffuses in the
adjacent box, after an
average time τ, it cannot get
back to the previous one

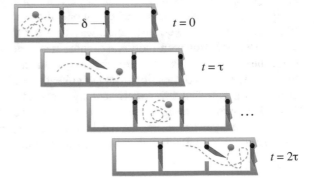

As already noted, in the absence of any perturbing action, the diffusion length at times $t' >> \{t\}$, is $\langle x \rangle = 0$. Now, let us imagine that the domain $[0, L]$ is split into N sub-intervals of length δ, such that $L = N\delta$, and that each barrier between any two adjacent intervals is a ratchet, i.e. a device which allows the passage of the particle in n to the interval $n + 1$, but not to the $n - 1$. After an average time:

$$\tau = \frac{\delta^2}{2D} \tag{6.9}$$

the particle crosses the barrier, and changes it position from $n\delta$ to $(n+1)\delta$ irreversibly.

After a time $N\tau$, the particle has crossed the entire domain $[0, L]$, therefore its diffusion mechanism has been rectified to some arbitrary 'forward' direction. Its average velocity along the path would be:

$$\langle v \rangle = \frac{L}{N\tau} = L\frac{2D}{N\delta^2} = \frac{2D}{\delta} \tag{6.10}$$

The average apparent speed of this rectified Brownian motion increases, for smaller and smaller δ intervals into which the path length is split, because the frequency of smaller and smaller random steps increases faster than the steps shrink. Note that the width of δ is limited, for the myosin-actin case for example by the minimum distance between two actin monomers, and in any instance by the thermal mean free path length.

It should immediately come to the physicist's mind that such a device violates the venerable Second Principle of thermodynamics. If the disordered thermal motion of the molecule can be converted into a directed flow, this would amount to a spontaneous reduction of the entropy. In the long history of failed physics miracles, this goes under the name of "Perpetual motion of the second kind", a motion that is in fact physically impossible. (The ratchet device, however, does not violate the First Principle of the conservation of energy, since it does not make up energy from nothing; this would have been called a perpetual motion of first kind.) In fact, in the macroscopic ratchet of Fig. 6.5e, the pawl is kept in place by a spring; if the spring were not there, the pawl once disengaged could jump in the opposite direction for a time sufficient for the cogwheel to turn backwards. The energy and entropy of the Hookean spring, in that case, compound to ensure that the two fundamental principles of thermodynamics are not violated. But what happens at the microscopic scale?

The answer was provided by Richard Feynman. Whatever the "pawl" mechanism is intended to be, at the molecular scale (it could be, for example, another protein acting; or a part of the same molecule changing shape; or an electric charge displacement inducing a local dipole, etc.), it would be at the same temperature of the "cogwheel", or the motor molecule; therefore, the pawl would be subject to the same thermal fluctuations as the wheel, and it could move in either direction, allowing the wheel to do the same. In other words, no rectification of the Brownian motion would be allowed, since all parts of the device are subject to the same destructive random fluctuations.

The rectified Brownian motion

The simplest potential that satisfies the first two requirements is again the sinusoidal: $E(x) = E_0 \sin(2\pi x/L)$. In the absence of other external forces than thermal fluctuations, the equation of motion that gives the molecule position x as a function of time, is a *dissipative* equation:

$$\eta \left(\frac{dx}{dt} \right) t = \eta v t = -E\left[x(t)\right] + \xi(t) \tag{6.11}$$

The dissipative character comes from the presence of a term depending on the velocity $v = dx/dt$, multiplied by an effective viscosity coefficient η, which models the resistance to the motion from the substrate. The term $\xi(t)$ describes implicitly the particle position as a thermal noise, in the form of a Gaussian probability distribution of random jumps of the position x in time, with average $\langle x \rangle = 0$ and variance $\sigma^2 \sim k_B T$.

An equation of this kind is called a *stochastic dynamics* equation, and requires special tools to be solved. However, even without solving explicitly this equation, we already know that its solution at $t \to \infty$, is zero average velocity, $\langle v \rangle = 0$, since the average displacement is zero for a purely thermal noise. Moreover, this remains true also if we make the potential asymmetric by adding a constant shift, $(-E[x(t)] + G)$. The amplitude of the potential remains symmetrical over an interval $[-L/2, +L/2]$, therefore the integral of the equation is zero, with a positive and a negative half equal and opposite. We anticipated such results in our discussions on the Brownian motion in Chap. 5, where the root-mean squared displacement was identified as good the quantity, and not the absolute displacement.

To add a mathematical description capable of 'rectifying' the Brownian motion, a symmetry-breaking element must be introduced in the potential. An idealized example can be the double-sine potential:

$$E'(x) = E_0 \left[\sin \left(\frac{2\pi x}{L} \right) + \tfrac{1}{4} \sin \left(\frac{4\pi x}{L} \right) \right] + G \tag{6.12}$$

with the constant G required to shift the integral to a non-zero value over the symmetric interval $[-L/2, +L/2]$.

The action of the simple-sine and double-sine potentials can be compared by a graphical representation, as given in the two following plots. The plots show the superposition of the Gaussian distribution of particle probability density onto the shapes of the two potentials, centered in $x = \tfrac{3}{2}L$ for the sake of example.

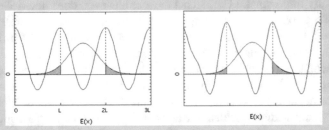

For the sinusoidal potential on the left, the particle probability density is the same on the right and on the left after a time t. The grey tails on the diagram in the figure represent the fraction of the initial Gaussian distribution, which smeared beyond the two maxima $x = L$ and $x = 2L$ of the barriers adjacent to the central minimum. By taking the difference between the two grey tails, the net probability of motion is zero.

If, on the other hand, the potential $E(x)$ is left-right asymmetrical, as shown on the right, the superposition of the Gaussian distribution to this potential gives a probability density that spills over the right direction, i.e. the probability of finding the particle is higher on the right than on the left.

Of special interest is the case of two-headed motors, such as the kinesin moving on a microtubule. In general, the corresponding ratchet model will be characterised by two locations (on the microtubule) with enzymatic activity. Each head of the motor is an enzymatic domain which can be activated only at one of these locations, the situation is similar to that described in the two-state model. However, the situation is more complex, because the motor can have more than just two internal states. For example, in the model of Lymn and Taylor discussed in Sect. 6.2, five different internal states are enumerated, each with a different molecular conformation and chemical state. Here we present another simplified model with $M = 3$ internal states and transitions occurring at $K = 2$ adjacent sites, described in the figure below.

The ground state of the model, $m = 0$, correspond to a resting state with both heads bound to the microtubule; the first excited state, $m = 1$, has one of the heads, h_1, detached, and the other h_2 attached to its location; and the second excited state, $m = 2$, has h_1 attached and h_2 detached. Since both heads are taken to be identical (as for dimeric kinesin), the two potentials $E_1(x)$ and $E_2(x)$ for the two excited states have the same period l and are shifted along the microtubule as $E_2(x) = E_1(x - l/2)$.

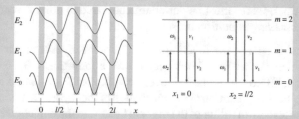

By using the same language of the "rate transition theory" already introduced for the two-state model, four transition rates are defined (see green-shaded bands in the figure), from the ground state to $m = 1, 2$ ("unbinding") and back ("rebinding"), which share the obvious relations $W_{0\to2}(x) = W_{0\to1}(x)(x-l/2)$, and $W_{2\to0}(x) = W_{1\to0}(x)(x - l/2)$. The binding reactions depend on the two rate constants w_1, w_2, and the rebinding on the v_1, v_2, so that the transition rates can be formally written, by using appropriate Dirac's delta functions, as:

$$W_{0\to1}(x) = w_1\delta(x - l/2) + w_2\delta(x)$$

$$W_{1\to0}(x) = v_1\delta(x - l/2) + v_2\delta(x)$$

(6.13)

Therefore, a three-state model with two distinct binding sites has 8 independent transition rates, as shown in the figure above. The unbinding rate constants w_1, w_2 depend on ATP concentration c_A. A reasonable hypothesis is that such constants follow a Michaelis-Menten (see greybox on p. 234), with two constants α, β:

$$\frac{1}{w} = \frac{1}{\alpha c_A} + \frac{1}{\beta}$$

(6.14)

Under the further hypothesis that $w_1 \sim w_2$, it is possible to show that the drift velocity has a ATP-concentration dependence, as:

$$v = v_0 + (v_\infty - v_0)\frac{c_A}{c_A^* + c_A}$$

(6.15)

where v_0 and v_∞ are the limits of the molecular velocity at zero or infinite ATP concentration, respectively, and c_A^* is a fitting parameter to match experimental data.

However, the device can function if the parts corresponding to the pawl and the ratchet are kept at different temperatures. In that case, the difference in negative entropy from the rectified motion would be compensated by the positive entropy from the thermostat, needed to maintain the temperature difference. Of course, it is very difficult in practice to hold a temperature difference of even a few degrees over such a small distance as a few nm, which would correspond to a temperature gradient of 10^9 K/m! At the molecular scale, however, some conditions have been identified for such devices to work. The basic, minimal requirements for a thermal ratchet to work are:

- the energy surface must be periodic;
- all forces must average to zero, in time, space and temperature;
- random (thermal) forces must be dominant compared to other forces;
- a symmetry breaking condition must be present.

The chemical details of the functioning of the actin-myosin motor based on the rectified Brownian ratchet have been debated for a long time (Fig. 6.8). A widely accepted description was the so-called **lever-arm model** (Fig. 6.8a), which relied on a tight coupling between a conformational change of the myosin head induced by the ATP, such as the lever-arm tilting of the myosin head, and the detachment/attachment between two adjacent actin binding sites. In this model, the displacement per ATP cycle is expected to be proportional to the length of a domain, about 5 nm.

Although the basic concept was laid out already in the late 50's [4], starting from the beginning of the 2000's the rectified Brownian motion model has been gaining credibility, since it allows to better explain the dynamics and characteristic

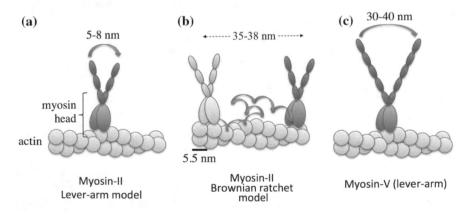

Fig. 6.8 Two models of actin-myosin motors. **a** The "lever-arm" model is based on a tight coupling: a tilting of the lever-arm (conformational change) is coupled to each ATPase cycle in a one-to-one fashion, and each step is equal to the minimum 5 nm distance between two actin binding sites. **b** The "loose coupling" model is based on the rectified Brownian ratchet model: the myosin head, activated by the ATP, thermally diffuses along the actin filament by making random jumps (sometimes backwards), summing several small steps. **c** Myosin-V operates similarly to myosin-II, however its longer lever arms allow for a larger unit step

time-energy scales of muscle contraction (**loose-coupling model**, Fig. 6.8b). Here, the action of ATP is restricted to favouring the detachment of the myosin head and promoting the 'clicking' of the lever mechanism. However, at the moment when the head is in a weak binding configuration, the thermal fluctuations play a major role, by allowing the molecule to jump by several units of 5 nm forward (and sometimes backwards), until a new actin binding site is found [5].

Experiments of the type described in Fig. 6.1 established that the relaxation time is inversely proportional to the ATP concentration, and that the relaxation time of force and displacement are different: this lends support to the model, in that the time duration of force release is not correlated with the time of random jumping around before finding the binding site. The average jump is around 35–38 nm, performed at a rate (depending on the ATP concentration) between 100 and 200 s^{-1}, which turns into an average speed of myosin-II relative to actin of 350–750 nm/s. The force exerted by a single myosin head is about 7 pN. Since the maximum force deduced from macroscopic muscle contraction measurements is about 530 pN per myosin filament (see Chap. 10), and the average number of cross bridges per filament is $\sim$100, this implies that about 40–60 % of the heads work simultaneously in each contraction step.

6.4 Symmetry-Breaking Transformations

From the point of view of our physical analysis, we look for modes of breaking the translational symmetry of the energy landscape in Fig. 6.6. In more mathematical terms, symmetry-preserving transformations of the energy landscape $E[x(t)]$ leave the equation of motion Eq. (6.11) unchanged, while reversing the sign of the velocity $v \rightarrow -v$. Trajectories with opposite velocity vectors are equivalent, with a net null contribution to the motion, i.e. the directed current turns out to be zero. Symmetry-breaking transformations, by contrast, transform E such that the trajectories with v and $-v$ are different, thus giving a net current.

The "deformation" potential, naively represented in Eq. (6.12) in the greybox on p. 218 may originate from some irreversible modification of the molecular structure upon energy transformation. It may happen that once the ATP molecule is hydrolized, some modification of the motor molecule occurs, such that the stepping is forced in one particular direction. In other words, there is not an explicit "pawl" in the ratchet, but it is the moving part itself that is cyclically modified, so as to favour one direction with respect to the other. In another version of the symmetry breaking, it can be the pathway along which the molecule moves (e.g., the microtubule such as in Fig. 6.2), which has a different molecular structure in the two opposite directions. We will see some examples of both cases.

Like other molecular motors, myosin works directionally, with forward steps being favoured over backward steps. The efficiency of such a forward-directed process can be measured directly, by applying a force load that tends to oppose the molecule motion. The myosin-V motor can step continuously on cytoskeletal actin filaments,

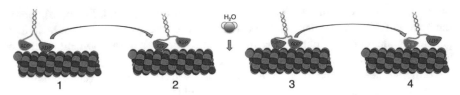

Fig. 6.9 *Above* Kinesin moving along a microtubule. (*1*) A two-headed kinesin molecule, initially with both heads in the ADP form, binds to a microtubule; (*2*) release of ADP and binding of ATP results in a conformational change that locks the head to the microtubule and pulls the neck linker (*orange*) to the head domain, throwing the second domain toward the plus end of the microtubule; (*3*) ATP hydrolysis occurs while the second head interacts with the microtubule. (*4*) The exchange of ATP for ADP in the second head pulls the first head off the microtubule, releasing Pi and moving the first domain along the microtubule. The cycle can repeat, the kinesin moving farther down along the microtubule

against a load of up to $\sim$2–3 pN, consistent with using one ATP molecule for each $\sim$36-nm step (see also Fig. 6.8c). Above a counter force of $\sim$3 pN, the molecular motor "stalls" by pausing with both heads attached to the actin filament. In such a configuration the myosin-V is unable to step forward because the work involved exceeds the energy available from ATP hydrolysis. Importantly, experiments suggest that in this stalled state ATP turnover is halted, so that the myosin motor only consumes ATP when it is actively stepping. In recent experiments [6], it was found that pulling backward on a walking myosin-V molecule causes the motor to reverse its mechanical action, always in $\sim$36-nm steps, but without a requirement for ATP binding. The maximum speed observed was in the range of 1000 nm/s.

The basic functioning of such families of motor proteins as myosins, kinesins, dynein, observed in tens or hundreds of variants in eukaryote cells, are often very similar to the one described for the myosin/actin motor complex. The displacement of a kinesin along a microtubule follows a very similar qualitative path, although the chemical details may differ substantially (Fig. 6.9). The kinesin heads take a different conformation when they bind an ADP or an ATP molecule. The cyclic binding of ATP, hydrolysis to ADP+Pi, release of ADP, and binding of a new ATP, performed alternately by the two heads, permits the displacement of the motor protein along the microtubule at a quite constant rate. Since kinesin hydrolyses ATP at a rate of approximately 80 molecules per second, given the step size of 8 nm, kinesin moves along a microtubule at a speed of 640 nm/s (considerably slower than the maximum rate quoted above for the cooperative movement of myosin-V: clearly, a steady velocity is the best quality of kinesin, while maximum velocity is the best quality of myosin, on an evolutionary selective basis).

An outstanding example of how this mechanism is utilised in the cell, and of its regulation, is provided by mitochondrial transport in neuron cells ([7], see also next Chap. 7), between the soma (the main part of the cell, including the nucleus) and distal processes as the axon or neural synapses. The long neural axon consumes a high amount of ATP molecules for transmitting the electrical impulse along its length, which can be of many centimetres. Therefore, mitochondria have to be distributed along the whole length of the axon protruding from the central soma, to

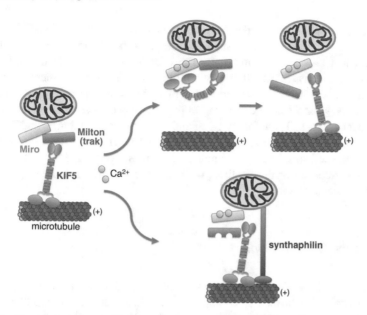

Fig. 6.10 Local variations in Ca^{2+} concentration (*pink spheres*) regulate the mitochondrial transport, in possibly alternative ways. The Miro-Milton (or Miro-Trak) adaptor complex mediate attachment of the mitochondrion to KIF5 (*blue*, a motor protein of the kinesin family). *Above* Ca^{2+} binds to the Miro promoting disconnection of the kinesin from the microtubule, followed by decoupling of Miro-Milton, and detachment of the mitochondrion. *Below* increased Ca^{2+} recruits syntaphilin, which provides a "stop" to the mitochondrial transport

make rapidly available the necessary ATP on site, which could never be realisable if ATP were to diffuse from the centre of the cell to the distant periphery. Mitochondrial transport depends upon microtubule-based motors, which drive their cargos via mechanisms requiring ATP hydrolysis. Microtubules are uniformly arranged in axons. As we remember from this chapter, they have a definite +/– polarity originating from their basic heterodimer unit: their plus ends are oriented distally, and the minus ends are directed toward the neuron body (soma). Such a uniform polarity has made axons particularly useful for exploring mechanisms regulating bi-directional transport: dynein molecular motors drive the retrograde movement, whereas kinesin motors mediate anterograde transport. Of the 45 kinesin motor genes identified, the kinesin-1 family (KIF5) is the key motor driving mitochondrial transport along the neuron axon. KIF5 kinesin motors attach to mitochondria through adaptor proteins, such as the Miro-Milton protein complex (also called a "trak" complex, from the acronym of "trafficking-kinesin"), and the local Ca^{2+} concentration is thought to regulate the transport in different ways (Fig. 6.10).

Note that when we say "attach", be it the motor to the filament, or the cargo to the motor protein, we are speaking of weak chemical forces: typically, a few hydrogen bonds are formed, plus some electrostatic and Van der Waals forces. In any case, no covalent bonds are formed because these would be too strong, and neither the motor protein could move nor the cargo could be delivered, with such bond energies. The flip side of the coin is that such non-covalent bonds are of limited lifetime: they must just be resistant enough for the process to be completed. However, in the Boltzmann-statistical world of single molecules, surviving the thermal and fluid Brownian fluctuations is always a matter of probability, for these molecular acrobats on a rope. The typical distances that motor proteins can travel before dissociation are of the order of 800–1200 nm for kinesin-I and dynein on microtubules, and 700–2100 nm for myosin-V on actin filaments.[1] Such lengths are enough to cover a substantial fraction of the cell size. Because of their longer steps, dynein and myosin-V process about 30–60 steps during this lifetime, while kinesin-I with its smaller step can make 100–120 consecutive moves.

6.4.1 The Tubulin Code

As it was seen in the Appendix E to this chapter, microtubules are non-covalent cylindrical polymers formed by α- and β-tubulin heterodimer building blocks, with apparently contrasting properties: they are highly dynamic, exhibiting rapid growth and shrinkage of their ends, but are also very rigid, with persistence length λ_p much larger than the cell size. In their functioning as a kind of "highways", along which motor proteins can move at much faster rates than by pure diffusion, microtubules may appear as passive structures. However, rather than changing cyclically the shape and bonding of the moving motor protein, another way of breaking the symmetry of the Brownian motion can be to modify the chemical structure of the tubulin blocks. In a way, it would be like painting molecular-scale traffic signs over the intricate road network of the cytoskeleton, and the traffic routing becomes directly inscribed on the microtubules.

Such modifications of the tubulins are called *post-translational*, since they are not expressed in the DNA coding for these specific proteins, but rather are the result of the enzymatic processes completing the mRNA translation of the genetic code (see also *epigenetics*, Appendix B to Chap. 3). Tubulin post-translational modifications are chemically diverse (phosphorylation, acetylation, polyamination, and so on), and are generally reversible, evolutionarily conserved and abundantly represented in cellular microtubules. Most importantly, their distribution is very much stereotyped in cells. For example, microtubules observed during stable cell life (interphase) are enriched in tyrosination, whereas the microtubules observed during the various steps of the cell splitting (mitosis) are enriched in detyrosination and glutamylation; microtubules in neural axon are enriched in detyrosination, acetylation, and glutamylation;

---

[1]Myosin-II moving along actin in muscle sarcomeres is less concerned by such problem, because of the much larger number and high density, and their strictly fixed arrangement.

microtubules in cilia and flagella are especially heavily glutamylated; in some cases, adjacent microtubules have completely different post-translational modifications, such as in axonemes (see below, central pair in Fig. 6.17d), where the B-tubule is highly glutamylated, whereas the adjoining A-tubule is enriched in tyrosination (for a review, see [8]).

Such a microtubule chemical diversity was proposed to form the basis of a **tubulin code** that is read by cellular agents. Despite the widespread appreciation for the ubiquity and functional importance of these modifications, and their stereotyped distribution in organisms and cells, we do not currently understand how complex microtubule modification patterns are written and interpreted by cells.

Single-molecule tracking experiments in cells revealed a special reactivity of some kinesins for such modified microtubules, and the differential regulation of several kinesin and dynein variants by modified tubulin isoforms. Decreased glutamylation on axonal microtubules lowers the affinity of kinesin-3 and reduces synaptic vesicle trafficking. Tubulin missing C-terminal tails was shown to self-assemble, in addition to microtubules, other structures such as sheets, rings, and aggregates. These and many other experiments indicate that tubulin tails and their modifications can tune both the basic properties of the microtubule and its interaction with cellular motor proteins. A striking example is provided by the experiment of Fig. 6.11, in which the movement of kinesin-1 proteins is individually tracked inside a fibroblast cell, by means of a special microscopy technique. The figure shows the path followed by the kinesin over several seconds, a static image showing a portion of the cell microtubules, and a graphic superposition of the two images. In this latter, it can be clearly seen that the kinesin (red) follows quite closely the pattern of the acety-lated microtubules (green), and mostly avoids other microtubules treated by different enzymes.

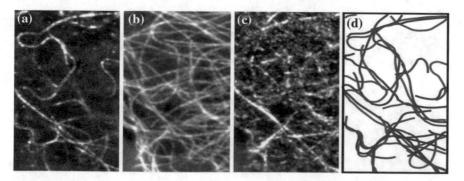

Fig. 6.11 **a** TIRF-microscopy image of the path followed by fluorescently-labelled kinesin-1 motor proteins. **b** Microscopy image revealing the ensemble of the microtubules. **c** Microscopy image highlighting the acetylated subset of microtubules. **d** Graphic reconstruction of the superposition of the kinesin path in (**a**) (with *green* segments) and the acetylated microtubules in (**c**) (with *red* segments). [Images from Ref. [9], repr. under CC-BY 3.0 licence, see (**) for terms.]

6.5 Cell Shape and Cytoskeleton Polymerisation

Compared to the case of myosin or dynein, which are reusable motor proteins, other cellular motors work on disposable mechanisms. In this case, the force is actuated by the assembly and disassembly of the protein itself, at the site where the mechanical action is needed. A characteristic example is provided by the regulated polymerisation of actin filaments in various instances of the cell life (Fig. 6.12). In this case, that same molecule actin that we will see in Chap. 9 as being a fundamental *static* component of the muscle sarcomere machinery, here works as a modifiable force actuator, and can, e.g., push the cell membrane to extrude a pseudopod, or shrink and split portions of the membrane during the process of cell separation. This mechanism can be considered as disposable, or dynamic, since the actin polymers are continuously elongated and shortened, clustered, bundled and disassembled, at various sites around the cell and notably at the membrane. The source material is the pool of available actin monomers, present with variable concentrations in the cytoplasm solution. Similar dynamic mechanisms operate also for other components of the cytoskeleton, such as the microtubules.

Even if quite different from the more direct force action of myosin or kinesins, this is yet another example of molecular motor, since also in this case a certain quantity of chemical energy ΔQ (supplied as usual by ATP molecules) is converted into mechanical work ΔW. As it will be shown in the following, both for the case of actin and microtubules the symmetry breaking comes from the very structure of the monomers, in that they have a peculiar orientation. For actin, with two opposite "pointed" and "bearded" ends of the molecule; for the microtubules, with a heterodimer, i.e., a basic unit formed by two different monomers. In either case, the two ends of each filament have different kinetic rates, therefore the direction of growth of the microfilament has a natural orientation.

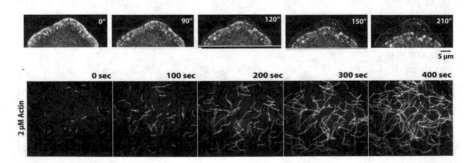

Fig. 6.12 Polymerisation of actin filaments is responsible for many cell membrane deformations. In the sequence above, actin assembles at the border of the outer membrane of a *Xenopus* fibroblast (XTC cells) pushing against cell membrane to follow cell movement. In the sequence below, purified actin filaments placed in a solution of 2 μM actin monomers grow at a rapid rate of about 10^{-2} μm/s. [Adapted from Ref. [10], under CC-BY 3.0 licence, see (**) for terms.]

6.5.1 Polymerisation Dynamics and the Treadmill Effect

To describe the length variation of an actin filament, let us represent it as a homopolymer made of a chain on n monomers, placed in a solution in which a concentration $[M]$ of the same monomers is diluted. Such monomers are normally available in the cytoplasm, to be recruited to modify the cytoskeleton structure. The monomer capture rate by the filaments may be taken as proportional to the concentration, times a rate constant k_+, leading to a filament growth rate:

$$\frac{\Delta n_+}{\Delta t} = k_+[M] \tag{6.16}$$

At the same time, the filament may loose some monomers, with a constant rate k_-, independent on the concentration, thereby leading to a filament shortening at a rate:

$$\frac{\Delta n_-}{\Delta t} = -k_- \tag{6.17}$$

The overall length change $n = n_+ + n_-$ is:

$$\frac{\Delta n}{\Delta t} = k_+[M] - k_- \tag{6.18}$$

This simple linear equation expresses the fact that the filament will grow, $\Delta n/\Delta t > 0$, when the monomer concentration exceeds a critical value $[M]_c$:

$$[M]_c = \frac{k_+}{k_-} \tag{6.19}$$

At steady state the polymerisation speed is given by:

$$v_p = \delta \frac{\Delta n}{\Delta t} \tag{6.20}$$

with δ the size of one monomer.

Typical values of the rate constants k_+ et k_- can be measured in the laboratory, e.g., for monomers of actin (G-actin) or microtubules (tubulin). For typical values of $[M] \simeq 0.12 - 0.6\,\mu M$, it is found $k_+ \simeq 1-10\,\mu M^{-1}\,s^{-1}$, $k_- \simeq 1-2\,s^{-1}$. The theoretical polymerisation speed of actin with such values is of the order of $0.7-1.3\,\mu m/s$.

The energetic aspects of such a simple model can be studied by the following considerations. The ratio between the fraction of attached and detached monomers at any time t is proportional to the respective probabilities (k_+M) and (k_-):

$$\frac{\Delta n_+}{\Delta n_-} = \frac{(\Delta n_+/\Delta t)}{(\Delta n_-/\Delta t)} = \frac{k_+M}{k_-} \tag{6.21}$$

which, by considering E_+ and E_- the enthalpies ($\sim$ energies) of attachment and detachment of a monomer, can as well be expressed by the ratio of the respective Boltzmann factors:

$$\frac{k_+ M}{k_-} = \frac{\exp(-E_+/k_B T)}{\exp(-E_-/k_B T)} = \exp(-\Delta E/k_B T) \qquad (6.22)$$

The energy difference $\Delta E = E_- - E_+$ is a chemical energy term, which gives a mechanical work, equal to a mechanical force times the elementary displacement δ, as $\Delta W = F\delta$. Therefore, the effective force generated by an elongation of the filament by an elementary unit is found as:

$$F = \frac{k_B T}{\delta} \ln \left(\frac{k_+ M}{k_-} \right) \simeq 2 \text{ to } 7 \text{ pN} \qquad (6.23)$$

Upon a more careful observation, molecular biologists noticed that the two ends of the actin monomer are not identical: one extremity has a 'pointed' shape (P), while the other rather looks like a 'barbed' shape (B). Mechanistically, monomers could attach and detach to/from either the pointed end of the filament, or the barbed end. The molecular details of the actin polymerisation are not yet completely elucidated, it is known that both ATP and ADP participate in the process. Given the different conformations, it can be also supposed that the attachment/detachment rates should generally take on different values, k_+^P, k_+^B, et k_-^P, k_-^B. It is found that the k_+ and k_- values are always larger at the B than at the P extremity.

Two separate equations like the (6.18) above can be written for the B and P ends of the filament, which can grow or shrink at the same time, with different rates:

$$\frac{\Delta n^{P/B}}{\Delta t} = k_+^{P/B} M^{P/B} - k_-^{P/B} \qquad (6.24)$$

The different possibilities are shown in Fig. 6.13, as a function of the different values of the four constants. If the respective values of critical concentrations are nearly identical, $[M^B]_c \simeq [M^P]_c$, the situation shown on the left of the figure is realized, with both ends growing or shrinking in parallel; this occurs rather typically with tubulin proteins in microtubules. If on the other hand, as it happens more usually with actin filaments, the two critical values are sensibly different from each other, the situation shown on the right of the figure is realized: for an intermediate interval of concentrations, one end grows while the other shrinks. For a particular value of the concentration $[M]$, at which the shrinking rate of one end is equal to the growth rate of the other, $\Delta n^P/\Delta t \sim \Delta n^B/\Delta t$, a peculiar condition of "treadmilling" can be realised:

$$M_c' = \frac{k_-^B + k_-^P}{k_+^B + k_+^P} \qquad (6.25)$$

The filament is this case is in a steady state, as shown in the middle of Fig. 6.13: its average length does not change, however there is a net displacement towards one direction (in the case written above, towards the B end).

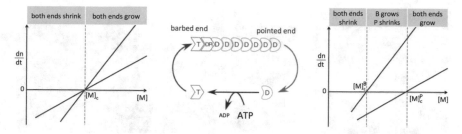

Fig. 6.13 *Middle* Actin treadmilling effect when growth and shrinking rates are equal. *Left* Growth rates of the pointed end (*blue*) and barbed end (*red*) of filament, when the critical concentrations of the two types of monomers are nearly equivalent (e.g., tubulin in microtubules). *Right* Growth rates of the two ends, for largely different critical concentrations of the two monomers, such as in actin (ADP-bound or ATP-bound actin). The central region of the plot corresponds to treadmilling, with one end growing while the opposite end is shrinking

In Chap. 9 the cytoskeleton filaments polymerisation will be coupled to membrane mechanics to build models of cell deformation, for example by looking at the mechanics of the protrusion of pseudopods by unicellular organisms.

6.6 Variations on a Theme of Polymers

The mechanism by which a chain-like molecule grows by the addition of chemically distinct units (monomers) described in the preceding Section was limited to the kinetics of cytoskeletal filaments. However, the same conceptual framework applies to the replication of nucleic acids, DNA and RNA, as well as to the building of proteins (see Chap. 3 and the Appendix B). What all such process have in common is the fact that the monomer units are added or removed from the growing filamentary structure with a rate (probability) proportional to the monomer concentration in the surrounding environment.

According to a definition introduced by the American chemist and Nobel laureate Paul Flory in 1953, the kinetics can be alternatively described as **stepwise** or **chain** polymerization. The two mechanisms are schematically described in Fig. 6.14.

In the *stepwise polymerisation*, any two monomers present in the reaction mixture can link together at any time, therefore the growth of the polymer is not limited to chains that are already formed. Monomer addition typically proceeds through a condensation reaction, in which a small molecule (e.g. water) is eliminated in each step. The reaction kinetics between two monomers can be written as the decrease in the monomer concentration $[M]$ by an amount proportional to the probability of random encounter:

$$\frac{d[M]}{dt} = -k[M]^2 \tag{6.26}$$

that is, a constant k times the square of the concentration, $[M] \cdot [M] = [M]^2$. By integrating the above equation, one gets:

stepwise polymerisation

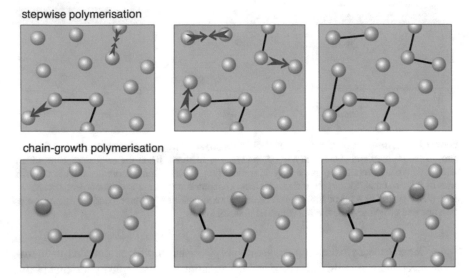

chain-growth polymerisation

Fig. 6.14 The two mechanisms of polymerisation according to Flory (1953). *Above* In the stepwise addition, monomers undergo random collisions and condense, usually by liberating a small molecule byproduct. Chains of any length can grow at steady state. *Below* In the chain growth of the polymer, free monomers (*grey*) must be firstly activated (*red*). Polymerisation reaction can only occur between the (*blue*) free ends of the growing chain, and activated monomers

$$[M] = \frac{[M]_0}{1 + kt[M]_0} \qquad (6.27)$$

with $[M]_0$ the initial concentration at time $t = 0$.

Hence, the fractional variation in concentration is proportional to the actual concentration at time t, as:

$$p = \frac{[M] - [M]_0}{[M]_0} = \frac{kt[M]_0}{1 + kt[M]_0} = kt[M] \qquad (6.28)$$

The degree of polymerisation, also know as the *Carothers equation*, is the average number of monomers per chain at time t:

$$\langle n \rangle = \frac{[M]_0}{[M]} = \frac{1}{1 - p} \qquad (6.29)$$

By reworking the previous equation, it is seen that $\langle n \rangle = 1 + [M]_0 kt$, i.e. the degree of polymerisation grows linearly with t, apparently without limits (the obvious limit is the maximum amount of monomers available). Because the condensation reaction can occur between molecules containing any number of monomer units, chains of many different lengths can grow in the reaction mixture. Moreover, the reaction mechanisms is assumed to be constant and independent on the concentration of monomers, leading to a steady-state growth.

The probability of finding a chain of length n is the product of the probability that $2,3,...n-1$ monomers have sequentially collided, p^{n-1}, times the probability $(1-p)$ to meet an n-th monomer still free:

$$P_n = (1-p)p^{n-1} \tag{6.30}$$

that is also equal to the fraction $[M]_n/[M]$ of the polymers of length n from the total. If we want the weight distribution of the growing polymers, this is obtained as:

$$\frac{W_n}{W_0} = \frac{n[M]_n \cdot m_0}{[M]_0 \cdot m_0} = n\frac{[M]_n}{[M]_0} \tag{6.31}$$

where m_0 is the mass of a monomer. By multiplying by the unit ratio $[M]/[M]$, Eqs. (6.29) and (6.30) can be combined, giving:

$$\frac{W_n}{W_0} = n\frac{[M]_n}{[M]}\frac{[M]}{[M]_0} = n(1-p)^2 p^{n-1} \tag{6.32}$$

It may be noted that such a distribution has always a maximum in correspondence of a given chain size, for any value of p, despite the above observation that polymers of ever increasing length can be found as time goes by.

In the case of *chain polymerisation*, activated monomers are linked to the growing chain one after another, at one or both ends of the chain. Differently from the previous case, however, in which the reaction is identical in all steps (allowing to describe the growth as a random collision process), chain polymerisation requires different steps, namely:

- chain nucleation, by means of an initiator which starts the chemical process;
- chain propagation, in which reactive end-groups of the chain react in each step with a new monomer, transferring the reactive group to that last unit to regenerate the active site;
- chain termination, which can stop the elongation, or transfer the reactive group to a new chain, thus leading to a branched polymer.

Actin polymerisation can be schematically broken into the above steps, as shown in Fig. 6.15 below. G-actin in solution bound to ADP must firstly be *activated* to ATP-actin; this occurs at least partly with the help of the protein cofilin. The actin *nucleus* is a complex of three ATP-actin monomers, from which an actin filament may start to elongate; since the trimer is highly unstable, actin nucleation requires additional proteins, such as Arp2/3, to promote the formation of a stable nucleus. During the *propagation*, or elongation, ATP-activated G-actin monomers are rapidly added to the "+" end of the actin filament; this process is also mediated by proteins that translocate along the growing filament, and simultaneously catalyse the addition of monomers. Once actin monomers are incorporated in the growing filament, the bound ATP is slowly hydrolysed to ADP. Filament growth can be *terminated* and protected by a specialised CP capping protein. Elongation proceeds until the rate of elongation is

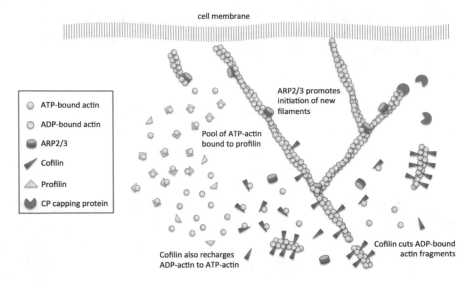

Fig. 6.15 Schematic of the three steps of actin polymerisation. Nucleation occurs with the help of Arp2/3 protein, which also promotes branching. When G-actin is activated by ATP, it can attach at the pointed end of the growing F-actin filament, which elongates with the help of profilin. Stable ends are protected by the CP capping protein. Cofilin severes fragments of ADP-actin along the filament, but it also contributes to recharging free ADP-actin monomers into ATP-actin

greater than the loss of ADP-actin from the pointed end. The protein profilin binds to ATP-actin, inhibits nucleation and accelerates elongation. When the dissociation rate of ADP-actin exceeds the rate of ATP-actin association, the filament shrinks, aided by the protein cofilin, which can severe filaments into short fragments and promote monomer loss from the pointed ends. Clearly, all such competing reactions depend on the relative concentrations of the various actors, for nucleation, growth and termination of the chains. For example, in the prokaryote *Acanthamoeba* for a concentration of [F-actin] = 100 (all concentrations in μM), it is found [G-actin] = 100, [Cofilin] = 20, [Profilin] = 100, [Arp2/3] = 2–4, [CP] = 1.

In this case, we must introduce the nucleation step, with its own rate k_n, concentration of "initiator" nuclei [I]; the termination step, with rate k_t, and the elongation, by which an activated monomer $[M']$ is added to an existing chain of length $[M]_n$ at a rate k_p. Model equations for these events can be written as:

$$\text{nucleation:} \qquad \frac{d[M']}{dt} = uk_i[I]$$

$$\text{elongation:} \qquad \frac{d[M]}{dt} = -k_p[M][M'] \qquad (6.33)$$

$$\text{termination:} \qquad \frac{d[M']}{dt} = -k_t[M']^w$$

where u is the number of monomer implicated in the nucleation (e.g., 3 for G-actin), and w is the number of activated chain ends implicated in the termination (it can be $w = 1$ for a capping protein, $w = 2$ for two chains colliding, etc.).

At the steady state, the speed of nucleation equals the rate of termination, $v_n = v_t$:

$$[M'] = \left(\frac{u\,k_i}{k_t}[I]\right)^{1/w} \tag{6.34}$$

giving the well-known dependence of the free radical concentration on the square-root of the nucleation centres, $[M'] \propto [I]^{1/2}$, as is the case of many artificial polymers synthesised by radical addition, e.g. polystyrene. Therefore, the steady-state polymerisation rate (from the second of Eqs. (5.33)) is:

$$v_p = k_p[M]\left(\frac{u\,k_i}{k_t}[I]\right)^{1/w} \tag{6.35}$$

The average chain length and chain mass distribution are given by the same expressions as Eqs. (6.30) and (6.32), respectively, by replacing p for the probability of elongation, and $(1 - p)$ for the probability of termination. The *kinetic* chain length is the ratio between the velocity of propagation divided by the velocity of nucleation, i.e., a measure of how fast the polymer grows compared to the rate of creating new chains:

$$l_K = \frac{k_p[M][M']}{uk_i[I]} = k_0\frac{[M]}{[I]^{1/w}} \tag{6.36}$$

with $k_0 = \frac{1}{2}k_p(uk_ik_t)^{-1/w}$.

6.6.1   Enzymatic Reactions and Kinetics

In Chap. 4 a greybox (see p. 126) illustrated the role of **enzymes** in modifying the free-energy landscape of chemical reactions. Enzymes in biological reactions act as catalysts, meaning substances that can accelerate a reaction between a reagent and a reactant (or "substrate"), but undergo no net chemical change between the initial and final states. If a spontaneous reaction turning a substrate into a product is $S \rightarrow P$, a non-spontaneous reaction occurring with the help of the enzyme would be $S + E \rightarrow P + E$. Many of the molecular reactions occurring inside the cell can be described as occurring with the help of enzymes. Although RNA variants are capable of catalysing some reactions, most biological reactions are catalysed by proteins, such as the DNA polymerase linking nucleotides to DNA during the replication. Sometimes, further help for a reaction is required from smaller molecules, called **coenzymes**, such as in the Krebs cycle described in Chap. 4, with coenzymes NADH and $FADH_2$ among others.

Michaelis-Menten kinetics

In 1913, the German chemist Leonor Michaelis and the Canadian physician Maud Menten, both working on the mechanism of hydrolysis of sugars, proposed a mathematical model that is still valuable to describe the basic behaviour of **enzymatic catalysis**, p. 235. The rate at which the product is formed is:

$$\frac{d[P]}{dt} = k_P[ES] \tag{6.37}$$

in which the concentration of the bound ES species is the unknown, obtained from the second rate equation:

$$\frac{d[ES]}{dt} = k_f[E][S] - k_r[ES] - k_P[ES] \tag{6.38}$$

At steady state, it is $k_f[E][S] - k_r[ES] - k_P[ES] = 0$, hence:

$$[ES] = \frac{k_f}{k_r + k_P}[E][S] = \frac{1}{K_M}[E][S] \tag{6.39}$$

Now, observe that, for $[E]_0$ the concentration of enzyme at $t = 0$, it is $[E]_0 = [E]+[ES]$ at any $t > 0$. Moreover, it can be considered that the substrate is present always in large concentrations, so that $[S] \simeq [S]_0$. Under such conditions, we get:

$$[ES] = \frac{[E]_0}{1 + \frac{K_M}{[S]}} \tag{6.40}$$

By putting this result in the rate equation for $[P]$, the *Michaelis-Menten kinetics equation* is obtained:

$$\frac{d[P]}{dt} = v_P = \frac{k_P[E]_0}{1 + \frac{K_M}{[S]}} \tag{6.41}$$

demonstrating a saturation behaviour of v_P to the value $v_{max} = k_P[E]_0$ for the product-formation rate. The constant $K_M = \frac{k_r + k_P}{k_f}$, also equal to $[E][S]/[ES]$, is the Michaelis constant, and represents a measure of the 'affinity' of the substrate toward the enzyme: a relatively small value of K_M indicating that v_{max} is attained more quickly. In particular, it is seen that for $[S] \ll K_M$ the product-formation rate is proportional to $[S]$:

$$v_P = \frac{k_P}{K_M}[E]_0[S] \tag{6.42}$$

while if $[S] \gg K_M$, v_P goes rapidly to v_{max} and becomes independent of $[S]$.

The ratio $e_P = k_P/K_M$ is called the *catalytic efficiency* of the enzyme. The M-M equation can be rearranged as follows, to construct the Lineweaver-Burk plot of the catalytic efficiency:

$$\frac{1}{v_P} = \frac{1}{v_{max}} + \left(\frac{K_M}{v_{max}}\right)\frac{1}{[S]} \tag{6.43}$$

Three principal features of enzyme-catalyzed reactions are the following:

(i) for a given initial concentration of substrate, $[S]_0$, the initial rate of product formation is proportional to the total concentration of enzyme, $[E]$;

(ii) for a given $[E]$ and *low* values of $[S]$, the rate of product formation is proportional to $[S]$;

(iii) for a given $[E]$ and *high* values of $[S]$, the rate of product formation becomes independent of $[S]$, reaching a maximum value known as the maximum velocity, v_{max}.

The simplest enzymatic reaction above can be written in more detail as:

$$S + E \underset{k_r}{\overset{k_f}{\rightleftharpoons}} ES \overset{k_P}{\rightarrow} P + E$$

to underscore the fact that the combination of enzyme with substrate $S + E$ is a reversible equilibrium reaction, governed by the respective concentrations, with two generally different 'forward' k_f, and 'reverse' k_r rate constants; the final step of the enzyme-catalysed reaction is irreversible, definitely downhill in free energy (single arrow), to give the product P with a rate constant k_P.

The Michaelis-Menten model, described in the greybox on p. 234, has been applied to studies of enzymatic kinetics for over a century. The two basic features of a reaction according to this model are: (1) the reaction velocity v_P increases with the substrate concentration $[S]$ up to a maximum saturation value, and (2) v_P decreases if k_r (unbinding of the $[ES]$ complex) increases. With the notation of the greybox, the M-M equation can be simplified to:

$$v_P = \frac{v_{max}[S]}{K_M + [S]} \tag{6.44}$$

very clearly showing both features, since K_M is proportional to the $[ES]$ unbinding rate k_r.

The overall time evolution of the concentrations $[S], [E], [ES], [P]$ is qualitatively shown in Fig. 6.16a. For an initial value of $[S]_0 = 1$ (in arbitrary units), the product $[P]$ starts from 0 at $t = 0$ and goes to 1 at long times, while $[S]$ goes to 0; the $[ES]$ complex grows at the beginning, and then goes to 0 following the depletion of $[S]$; the enzyme $[E]$ starts (in this example) from a concentration of 0.5, and returns to the same value when the reaction is completed.

The catalytic efficiency $e_P = k_P / K_M$ can be obtained as shown in Fig. 6.16b, by constructing the Lineweaver-Burk plot. The values of $1/v_P$ are plotted as a function of $1/[S]_0$, for varying initial concentrations of the substrate, and a constant concentration of the enzyme $[E]_0$. The resulting linear plot, Eq. (6.43), has a slope equal to K_M/v_{max} and intercept equal to $1/v_{max}$, from which e_P can be obtained.

The M-M model has proven useful also at the single-molecule level, albeit with a slight change in interpretation [11, 12]. Binding, unbinding, and catalysis are now considered to be stochastic processes, whose rates are defined to be the reciprocals of

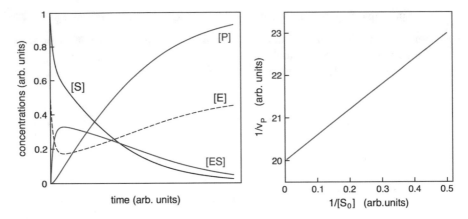

Fig. 6.16 *Left* Time evolution of the concentrations of substrate $[S]$, enzyme $[E]$, complex $[ES]$, and product $[P]$. The value of the rate constants are $k_f = 1$, $k_r = 0.2$, $k_P = 0.1$, in arbitrary units. Initial concentrations $[S]_0 = 1$ and $[E]_0 = 0.5$. *Right* Example of Lineweaver-Burk plot for the catalytic efficiency. With the same values of the constants, the intercept is $v_P^{-1} = 20$ and the slope $K_M/v_{max} = 6$, resulting in a value of $e_P = 1/3$

the respective mean molecular lifetime τ. If the molecular-scale probability follows Poisson's statistics, i.e., when the times between consecutive events are sampled from an exponential distribution, a single-molecule analog of the Michaelis-Menten equation can be written:

$$k_{turn} = \frac{k_P[S]}{K_M + [S]} \tag{6.45}$$

where k_{turn} is the reciprocal of the "turnover" time, the average time it takes the enzyme to make one molecule of product.

6.7   The Movement of Unicellular Organisms

Cell movement and migration of unicellular organisms occurs, evidently, in the absence of proper muscles. Nevertheless, movements are possible at speeds which obviously exceed simple diffusional translation, as shown in the following Table 6.1.

Cells may use several different mechanisms to generate an inner mechanical force, capable of imparting a global movement to the cell, and to overcome the viscous resistance, of water or of the extracellular matrix.

The swimming movement of most unicellular organisms can be accomplished via specialised structures placed across the cell membrane. These can be **cilia**, microscopic whips that surround in large number the outer surface of the membrane; or **flagella**, a thicker and longer whip attached at one end of the cell singularly or in small number, performing undulating and/or rotating movement. In eukaryotes, both cilia and flagella have a similar structure and composition, despite their different length, and are used by cells for various functions, for example to expel mucus from lungs, to propel spermatozoids, to transduce air waves into sound in the cochlear cells, and so on. In the case of eukaryotic cilia, the inner microtubule structure is connected to the transmembrane filament by specialised proteins, the dyneins, which transmit the flexion of the microtubules to the cilium, and make it swing in the extracellular fluid like a paddle. Cilia are much shorter than flagella (5–10 vs. 30–40 nm), and perform a simpler power stroke.

Prokaryotes have a rapidly rotating flagellum, while that of eukaryotes undulates by a sliding mechanism. Figure 6.17 summarises the structure of prokaryote (a, b) and eukaryote (c–e) flagella. In the case of the prokaryote flagellum, a complex structure is found inserted within the cell membrane, the **axoneme**. This is a system made by several proteins, fulfilling at the same time the role of supporting structure, and of transducer of the rotatory movement. The axoneme is made by a static part, fixed to the membrane, and a rotating (transmembrane part), the two acting somewhat like the stator and rotor in an electric motor. The rotating part is powered by a proton pump, in a way analogous to the mitochondrial ATPase proton pump (see Fig. 4.8 on p. 133), and can turn at speeds of up to 10^5 rpm, although the flagellum would rotate at slower rates, between 500 and 1000 rpm. During the bacterial cell evolution, the fixed parts are assembled first, then the tubular structure is "extruded" through the axoneme. The tubular structure of the flagellum is formed by special proteins, the flagellines, arranged in a hollow cylinder of about 20 nm in diameter.[2] The eukaryote flagellum, which lacks the rotating axoneme and is directly inserted in the cell membrane, has in turn a more complex structure. Figure 6.17c shows the

Table 6.1 Velocity (μm/s) of some typical celllular movements.

Type of movement	Velocity (μm/s)	Example
Growing actin filament	0.01–1	
Projection of pseudopodes	0.01–1	Fibroblast
Myosin-actin relative displacement	0.1–1	Sarcomer
Growing microtubule	~0.3	
Retreating microtubule	0.4–0.6	
Fast axon transport	1–4	Kinesin on MT
Slow axon transport	0.001–0.1	
Flagellate bacterium swimming	1–5	

---

[2]It may be interesting to note that the base structure of the bacterial axoneme is strictly related to, and might have evolved from, the so-called Type-III secretory system found in many bacteria, a sort of proteic "syringe" by which a bacterium can inject a protein or enzyme across and into the membrane of another cell (this is the way in which, e.g., *Yersinia pestis* infects human cells with the bubonic plague).

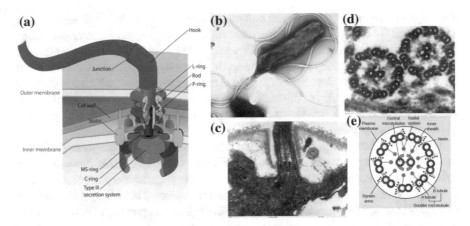

Fig. 6.17 a Schematic of the bacterial flagellum insertion in the membrane via the axoneme. **b** Microphotography of *Helicobacter pylori* swimming, with its few flagella in motion. **c** Scanning-electron microscope image of a vertical cross section of the flagellum in the eukaryote alga *Chlamydomonas*, showing the membrane insertion and parallel bundle structure of the flagellum. **d** Transverse cross section of two flagella next to each other (**d**) and schematic representation (**e**), showing the arrangement of microtubules in pairs (9 peripheral plus one central pair) and of the network of actuating dynein motor proteins. [Public-domain images © of: **a** M.R. Villareal, **b** Y. Tsutsumi, **c**, **d** Dartmouth College Electron Microscope Facility; **e** unknown (Wikipedia). All repr. under CC-BY-SA-3.0 licence, see (*) for terms.]

vertical cross-section arrangement of the microtubule bundle in a flagellate eukaryotic cell (*Chlamydomonas r.*). A horizontal cross-section (Fig. 6.17d, e) shows that the microtubule bundle is organised into pairs, arranged in a circle about a central pair. The mutual sliding of one tubule with respect to the other, in each pair, induces the undulating flexion of the flagellum, which turns into cell movement.

6.7.1   Linear Translation with Drag

For a nearly-spherical object of radius R, moving at a low speed v, in a fluid of macroscopic viscosity η, the Stokes' law gives an estimate of the viscous resistance:

$$F = 6\pi R \eta v \qquad (6.46)$$

Since this force is also $F = m(\Delta v / \Delta t)$, such an object starting with an initial velocity v_0 will be arrested over a distance:

$$x_0 = \frac{m v_0}{6\pi R \eta} \qquad (6.47)$$

The viscosity of fluids of biological relevance ranges over very different values, from $\eta = 10^{-3}$ kg m^{-1}s^{-1} of water, to $\eta = 1.34$ for glycerine, up to $\eta = 10^{13}$ kg m^{-1}s^{-1} for glucose. Cytoplasm can be considered to have a viscosity some 100 times that of water, due to the fraction of proteins and other molecular species diluted. For example, a vesicle with $R = 50$ nm in the cytoplasm transport by a kinesin motor protein along a microtubule, at a velocity of 0.5 μm/s, must overcome a viscous force of:

$$F = 6\pi \ (5 \times 10^{-8}) \cdot 0.1 \cdot (0.5 \times 10^{-7}) = 5 \times 10^{-14} \ \text{N} = 0.05 \ \text{pN}$$

indeed a value within the range of protein motors. As it was discussed in Chap. 5, the Reynolds number $Re = R\rho v/\eta$ gives a proportion of the relative importance of inertial forces to viscous forces. By putting the values for water, our vesicle has a $Re \sim 10^{-8}$, i.e. a very low Reynolds, meaning a null role of its inertia mg.

As another example, let us consider a unicellular organism swimming in water, for simplicity taken again to be nearly spherical with $R = 1$ μm. By assuming the cell interior to have the same density as water, the cell mass is $M = \rho\frac{4}{3}\pi R^3 = 4.2 \times 10^{-12}$ g. Considering an average swimming speed of about 10 μm/s, the drag force would be:

$$F = 6\pi R\eta v = 6\pi \cdot 10^{-6} \cdot 10^{-3} \cdot 10^{-5} = 0.19 \ \text{pN}$$

and the distance of arrest:

$$x = \frac{Mv^2}{F} = \frac{(4.2 \times 10^{-15}) \cdot (10^{-5})^2}{0.19 \times 10^{-12}} = 0.002 \ \text{Å}$$

This distance is zero compared to the size of the cell. The propulsion does not result from free swimming, but it requires to be constantly powered at a high rate to overcome the viscous resistance. For the bacterium with its $Re \sim 10^{-6}$, water looks like molasses for a human swimmer. Again, the inertia does not play a role, meaning its motion does not have memory of forces acting in the past, it only cares about the force that is applied instantaneously to propel its swimming. Amazingly, this is a world in which Aristotle could be right![3]

In a famous 1976 lecture delivered at the American Institute of Physics, E.M. Purcell pronounced the (unproven) "theorem of the scallop". The theorem states that a movement performed by one single degree of freedom cannot produce a net translation. What the theorem means is that to swim any animal has to move in the medium in some way that breaks the symmetry: if the animal makes one movement and then makes the reverse movement, there will be no net motion. An animal with only one degree of freedom has no other choice than to perform always the same

---

[3]The Aristotelic view of motion was based on the idea that the "natural" state of a body is rest, therefore he built a whole theory around the concept that all that moves is moved by something else, and that a body in movement slows down to stop if is not continuously pushed by some force. Only many centuries later Galileo laid the basis for Newton's Principle of Inertia, according to which a body set in motion will continue to move indefinitely, until some other force stops it.

movement, back and forth, and as such it could not move. This is why the theorem is linked to the scallop, since the bivalve animal can only open and close the valves about one hinge, and as such it has only one degree of freedom (the angle between the valves). In order to produce motion the scallop has learned to spit out water when the valves are open, and this gives it the net movement. By analogy, a human swimmer moving just one arm forward and then backwards cyclically would not go anywhere, she has to alternate the movement of the limbs with a swinging stroke of the body, to break the symmetry.

The movement of the cilia and flagella is probably the simplest way that animals have devised to propel their bodies, however with one notable difference. The cilia, like the arm of the human swimmer, has to return at its original position after each stroke, so its forward and backwards paths must be different, otherwise there would be no motion; the flagellum on the other hand may work just as one big cilium, or rather as a propeller; in the latter case it would be turning always in the same sense like a worm gear, the symmetry being broken by the asymmetric winding of the screw.

The power expenditure by the linear motion with drag can be estimated as:

$$P = Fv \tag{6.48}$$

resulting equal to about 2×10^{-18} W. Since a mole of ATP gives off 30.5 kJ of energy, or 5.07×10^{-20} J per molecule, the bacterium must use about 100–200 ATP molecules per second, to maintain its swimming speed.

6.7.2 Rotatory Translation with Drag

Let us now consider the same unicellular organism propelled by a rotating flagellum. The mechanical torque acting on an object that rotates at an angular velocity ω is:

$$T = F \cdot r = f\omega \tag{6.49}$$

where r is the radial distance between the rotation axis and the point where the force is applied, and the drag coefficient is in this case proportional to the third power of the nearly-spherical object size R:

$$f = 8\pi R^3 \eta \tag{6.50}$$

Some observed values of rotational velocity of monoflagellate bacteria give values in the range of $\omega \sim 10$ s^{-1}, or $\omega \sim 20\pi$ rad/s. By using the same values of the constants and taking $r \simeq R$, the rotational drag force results $F = 8\pi \times 10^{-21}$ N; the resulting torque is then: $T = 160\pi^2 \times 10^{-21}$ Nm, and the power consumption:

$$P = T\omega \tag{6.51}$$

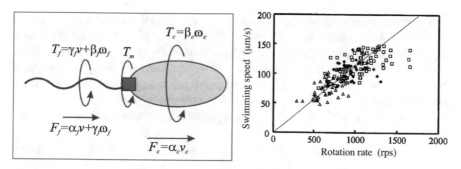

Fig. 6.18 *Left* Schematic of the model for flagellar bacterium rotational motion, with definition of the geometrical kinematical parameters. *Right* Experimental results for the translational swimming speed versus rotation rate. [From Ref. [13], repr. w. permission.]

that is about 10^{-16} W, apparently 100 times larger than for the translational movement. However, such a value does not tell us nothing about the translational velocity that could be attained by converting the rotational motion into rectilinear. In fact, it turns out that cells using rotatory actuation may typically reach much higher velocities, compared to cells adopting simple rectilinear swimming.

By using a slightly more detailed model, the characteristics of the rotatory motion cam be better appreciated [13]. In this model, schematically illustrated in Fig. 6.18, the difference between the rotational velocity of the flagellum, ω_f and that of the cell ω_c is explicitly taken into account. Therefore, separate equations are written for the force and torque on the cell:

$$F_c = \alpha_c v$$
$$T_c = \beta_c \omega_c \tag{6.52}$$

and for the flagellum:

$$F_f = \alpha_f v + \gamma \omega_f$$
$$T_f = \beta_f \omega_f + \gamma v \tag{6.53}$$

The translational velocity v must be clearly the same for both the cell and the flagellum; however, a cross-term coupling the rotational and translational velocity appears in the equations for the force and torque of the flagellum, via the coefficient γ. It may be worth noticing that the sign of the two independent rotational velocities is arbitrary, and not necessarily correlated.

Two separate balance equations can be written at equilibrium, for the total force and total torque:

$$F_f + F_c = 0$$
$$T_f + T_c = 0 \tag{6.54}$$

The geometric and kinematic coefficients appearing in the above model equations are defined for the bacterium cell as: $\alpha_c = 6\pi R\eta$, $\beta_c = 8\pi R^3\eta$ (the same as in Eq. (6.50)). The coefficients for the flagellum contain the length L, the screw thread p and screw flight r of the element, taken as a cylindrical screw; by taking typical values of $L = 5\,\mu m$, $r = 0.15\,\mu m$, and $p = 1.5\,\mu m$, the geometric-kinematic parameters are: $\alpha_f \simeq \frac{4}{7}\pi\eta L$, $\beta_f \simeq \frac{2}{5}\pi\eta r^2 L$, and $\gamma \simeq -\frac{4}{21}p\eta L$.

From the solution of the force balance equation, it is easily obtained:

$$v = -\frac{\gamma}{\alpha_c + \alpha_f}\omega_f \tag{6.55}$$

Notably, the ratio v/ω_f between translational velocity and flagellum rotational velocity is independent on the medium viscosity. Moreover, because of the negative sign of γ, a positive translation velocity is associated to the clockwise rotational velocity of the flagellum. From the torque balance equation is is furthermore obtained:

$$\frac{\omega_f}{\omega_c} = \frac{\beta_c(\alpha_c + \alpha_f)}{\gamma^2 - \beta_f(\alpha_c + \alpha_f)} \tag{6.56}$$

As it appears from the good correlation observed between v and ω_f in the right side of Fig. 6.18, this model gives a quite correct explanation of the kinematics of the coupled rotational motion of the flagellum and the cell. It may be noticed that, in the second balance Eq. (6.56), the denominator of the right side can take either positive or negative values, according to the structure and size of the flagellum; therefore, it may happen that for some cells the flagellum could rotate in one sense, while the whole cell rotates in the opposite sense. At the higher temperatures the ratio v/ω_f appears to saturate to a constant value, a phenomenon whose elucidation would require to take into account the detailed molecular structure of the mechanism.

6.7.3   Swimming Without Paddling

Euglenids are unicellular eukaryotes that situate somewhere between animals and plants. They are believed to have originated from a unicellular organism that may have endocytosed a green unicellular alga. Such cells seemingly lack a real cytoskeleton, but rather have a homogeneously distributed network of diffuse peripheral elements (called *pellicula*) capable of deforming the plasma membrane. These organisms have developed a unique low-Reynolds number swimming technique (Fig. 6.19a), sometimes called metaboly. Although it also may have a flagellum, the power stroke of an euglenid consists of a large distortion of the membrane that propagates back along its slender body (Fig. 6.19b). This large and nearly axisymmetric shape deformation moves opposite to the euglenid's direction of motion. Once the large distortion reaches the end, the body of the euglenid is rearranged, so that the next stroke is initiated. During this recovery process in which the euglenoid retreats, symmetry is

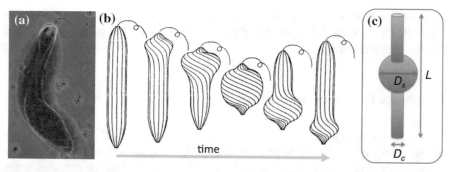

Fig. 6.19 a An euglenid (*Euglena proxima*), real size is length $L \sim 40\,\mu$m, diameter $D_c \sim 10\,\mu$m. **b** A computer simulation model of the deformation of the euglenid body, during its transnational movement. **c** Scheme of the simple model described in the text: the euglenid body is represented as a sphere sliding up and down along a cylinder. [Image **a** from Microscopic World www.youtube.com/user/TheMicrobiology09/ repr. under Youtube Standard lic.; **b** from Ref. [14], repr. w. permission.]

broken, hence forward motion is ensured. The flagellum seems to be used as a sort of tail, to keep a balance and oriented motion.

As shown in Fig. 6.19c, the motion of the euglenoid may be very schematically represented like that of a sphere sliding along a coaxial cylindrical rod, which is moving in the opposite direction, such as a spherical bead sliding on a necklace, as the necklace is being pulled through a fluid. The sphere, whose diameter D_s is equal to the amplitude of the distortion, travels backwards at a speed $v - u$; on the other hand, the cylindrical body, which is moving forward at a speed u, has a diameter D_c and its exposed length is $L - D_s$. In the limit $D_c \ll D_s$, the viscous drag force exerted on the sphere and on the cylinder can be calculated by the same Stokes' law:

$$F_s = 3\pi\eta D_s(v - u)$$
$$F_c = \frac{2\pi\eta(L - D_s)u}{\ln\left[0.6(L - D_s)/D_c\right]} \tag{6.57}$$

Since the drag forces on the sphere and on the cylinder are acting in opposite directions, we must have $F_s = F_c$. Therefore, the ratio of the velocities is:

$$\frac{u}{v} = \left(1 + \frac{2}{3}\frac{L/D_s - 1}{\ln\left[0.6(L - D_s)/D_c\right]}\right)^{-1} \tag{6.58}$$

The real dimensions of the euglenid like the one in Fig. 6.19a are $L = 40\,\mu$m, $D_s = 10.2\,\mu$m, and $D_c = 3.4\,\mu$m. With such values, the previous equation gives an estimate $u/v = 0.46$ for the forward stroke, to be compared to the experimental value of 0.43. The shift from I to II in Fig. 6.19b takes about 1.6 s, therefore by taking that the difference between the two positions of the "sphere" is about $L/4 \sim 10\,\mu$m, the euglenid's swimming velocity can be estimated from this simple model at $u = 5\,\mu$m/s. The experimentally measured values vary quite a lot with the degree of illumination (euglenids use mostly, while not only, photosynthesis to get

their energy), however they were found to be around 2–3 μm/s in the work cited in
Fig. 6.19. Given such a naive model of this complex, still unexplained cell motion,
the agreement is not bad.

Appendix E: The Cytoskeleton

From a structural point of view, an eukaryote cell is substantially different from the
simple picture of a lipid bag filled with water and proteins (and other stuff), because
of the presence of a scaffolding structure, the **cytoskeleton**. The cell cytoskeleton is
the dynamically organised ensemble of biological polymers that impart the system
the essential component of its mechanical properties. (In Chap. 8, the mechanics of
filaments and membrane structures will be separately treated.) The different components of the cytoskeleton are capable of actuating internal forces and responding
to the application of external forces, not just opposing a 'passive' resistance to the
cell deformation. Because of its name, it is tempting to attribute to the cytoskeleton
a similar role to that of the skeleton in superior animals; however, the cytoskeleton
provides the cell, at the same time, both its internal structure and its force actuating
system, thereby resembling at an organ that sums in one all the functions of the bones
and the muscles (Fig. 6.20).

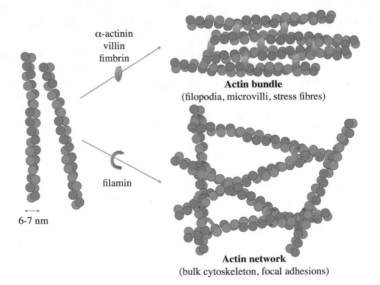

Fig. 6.20 Scheme of actin filament structures. G-actin (globular actin) monomers with bound ATP
can polymerize, to form F-actin (filamentous actin). They are shown in *different colors* to highlight
the helical winding of the two strings. F-actin may hydrolyze its bound ATP to ADP + Pi and release
Pi. However, ADP release from the filament does not occur because the cleft opening is blocked.
With the help of many species of actin-binding proteins, F filaments can quickly assemble and
disassemble into superstructures, such as actin parallel bundles (above, by binding e.g. α-actinin,
villin, fimbrin), or actin networks (below, by binding filamin dimers)

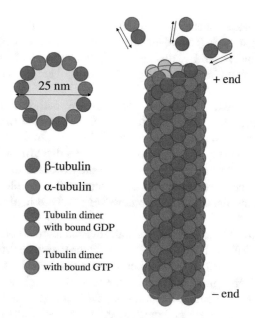

Fig. 6.21 Scheme of microtubule structure and growth. α- and β-tubulin monomers combine to form heterodimers; many such heterodimers assemble to form a single microtubule. Thirteen linear *protofilaments*, made of vertically-stacked tubulin heterodimers, are arranged side-by-side to form a hollow, cylindrical microtubule. The plus end (*top*) is the faster growing end of the microtubule, and heterodimers are arranged with the β-tubulin monomers facing the plus end. To the *left*, a cross-section through the microtubule, with diameter of about 25 nm. At the plus end of the microtubule, some tubulin heterodimers are shown while attaching to the microtubule (polymerisation). These, and the dimers already present at the + end, are bound to guanosine tri-phosphate (GTP, *red-green spheres*). Over time, GTP loses a phosphate and becomes G-diphosphate (GDP, *orange-green spheres*)

The cytoskeleton structure and components are quite similar in all eukaryote cells (from uni- to pluricellular organisms), although important differences exists between animal and plant cells. It is constituted of different types of long-chain molecules, or *polymers*, sometimes called fibres because of their sizeable length compared to the cell scale. These are usually classified into three categories (Fig. 6.21):

- **Actin filaments**, or F-actin, is formed by monomers of the actin protein (G-actin, of which several variants are known). The long F-actin polymers are twisted in pairs, which then assemble in bundles of variable thickness. Actin is ubiquitous in cells, and is mostly important in muscle cells (*myocytes*), where it couples with myosin in the sarcomeres. F-actin have a diameter of 6.5–7 nm, a contour length ranging from very small up to a fraction of the cell diameter, and a persistence length (see Chap. 7) $\lambda_p \sim 17\,\mu$m: therefore they are semi-flexible, since their average length is comparable to their persistence length. They have an orientation, due to the asymmetry of the actin monomer (which moreover changes shapes according to whether ATP or ADP is bound), and to their helical self-assembly in which monomers are attached head-to-tail. In particular, this leads to the fact that one of the filament ends (termed +) can polymerise faster than the other end (−). If the length of individual F-actin is generally between 2 and 3 μm, however they are usually assembled in tight bundles, whose overall length is rather in the 10–20 μm range.

- **Intermediate filaments**. These are the less dynamic components of the cytoskeleton, little known yet but undergoing intense research. In the variant of *lamin*, they are most important in the structure of the cell nucleus. Intermediate filaments are assembled from a family of related proteins, which share many common features. The definition of 'intermediate' comes from their average diameter of 8–10 nm, which is in-between that of F-actin and microtubules. While most abundant in epithelial and neuronal cells, they are however observed in almost all animal cell types.
- **Microtubules**, the stiffest constituents of the cytoskeleton. Their persistence length is of several mm, largely beyond the cell size, which therefore limits their flexibility. Their diameter is typically ∼25 nm, but in some cases can be smaller. The largest contribution to their rigidity comes from the special arrangement of parallel protofilaments in a hollow tubular structure, since the bending modulus κ_b (see greybox on p. 322) grows as the fourth power of the diameter. Compared to the densely twisted F-actin, the gain in stiffness is by a factor of about 80–100. Microtubules are polarised, like actin filaments, but their biochemistry is different. In particular, it is known that a dynamic instability can lead to a very sudden shortening of microtubules, thus originating a large impulsive force (Fig. 6.22).

Polymers (see again Chap. 7) can be organised into fibres, bundles or networks, according to the function they are performing. Sometimes, such as for the case of

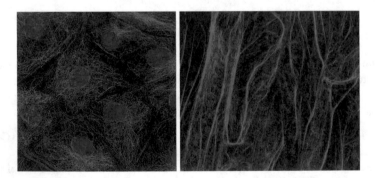

Fig. 6.22 Optical microscopy images of animal cells (*left*) and plant cells (*right*). Different cell elements are stained with fluorescent labels, to visualise the cytoskeleton. In the *left image*, microtubules are shown in *green*, actin filaments in *red*, and the nucleus in *blue*. Note that microtubules constitute a main scaffold architecture, while actins are in this case concentrated in dense networks just below the cell membrane. In the *right image*, actin networks are in *green*, and plastids in *red*. [Image © *left* M. Shipman, J. Blyth and L. Cramer, University College London; *right* E. Blancaflor, The S.R. Noble Foundation, [15]. Repr. w. kind permission from the authors.]

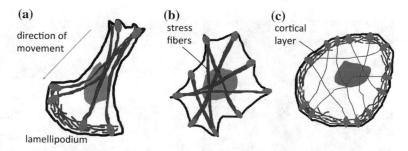

Fig. 6.23 Some possible organisation of actin filaments in the cell. **a** A migrating cell with a *lamellipodium* extending in the direction of cell migration, terminal focal adhesions (*yellow spots*), a meshwork of actin filaments (*red stripes* next to the lamellipodium membrane), and actin stress fibres (*thick, long red stripes*). **b** Isolated, adherent cell with stable focal adhesions and thick stress fibres. **c** Adherent cell from a dense tissue, with stable focal adhesions, a few thin stress fibres, and a dense meshwork of cortical actin filaments. The *grey* mass indicates the cell nucleus

actin, the same polymer fibres can auto-organise into any of the different structures, by dynamically rearranging their configuration, or by breaking and reforming their bonds (Fig. 6.23). Such an outstanding level of organisation is also made possible with the help of a large number of **auxiliary proteins**:

- **Crosslinking proteins**. The denomination comes from the physics of polymers (see Chap. 7), where some molecular components can be added to induce cross bridges between long fibres, thereby completely modifying the physical properties of the material (such as adding sulphur to rubber mixtures in the process of vulcanization). This is what happens, in an even more spectacular way, in the cytoskeleton. Most crosslinkers are controlled cyclically by a regulatory network of other proteins, thus allowing a rapid reorganisation of the cytoskeleton.
- **Branching proteins**, sometimes considered as a special case of the crosslinking proteins, as the name explains these permit the branching out of lateral chains from a central one. They are important for actin filaments.
- **Capping and severing proteins**, necessary to regulate and eventually arrest the polymerisation rate of the polymer filaments at their extremities.
- **Anchor proteins**, serve the purpose of transmitting the forces at the interface between cytoskeleton filaments and cell membrane, and participate in cell adhesion. They can attach to integral membrane proteins, or penetrate in the lipid bilayer. Their attachment is reversible, i.e. they can be detached and reconnected at a different place in the cell.

The operative keyword of the cytoskeleton structure (as well as for other cell structures) is the concept of **remodelling**: all the components of the cytoskeleton systems are active structures, which change and evolve during the biological activity of the cell, with a continuous functional reorganisation of the elementary proteins, available as raw material in the cytoplasm.

The cytoskeleton is linked to the cell membrane by specialised proteins, which cluster into **focal adhesions**, to some extent analog to "feet" allowing the cell to

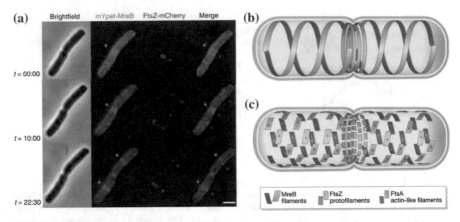

Fig. 6.24 Microscopy imaging of the localisation of MreB (actin-like) and FtsZ (tubulin-like) proteins in *E. coli* bacteria. **a** The *left column* is a bright-field electron microscopy image of *E. coli* bacteria. In the *right columns*, fluorescence microscopy images. *Green* indicates MreB fluorescence; *red* indicates anti-FtsZ immunofluorescence; the *rightmost image* shows the superposition of the two separate images (the scale bar on the *left* is 1 μm). The equipartition of the MreB cytoskeleton into splitting cells is triggered by the membrane association of the FtsZ protein in the cell equatorial plane, and is eventually accomplished by division and segregation of the MreB array. This process ensures that each daughter cell, after splitting at the region indicated by the red FtsZ, inherits one copy of the MreB cytoskeleton. **b** MreB (*purple*) has long been thought of as a spiral filament twisting along the cell length, to control cell shape. Likewise, FtsZ proto-filaments (*blue*) were once thought to wrap around the cell midpoint to organise the division. **c** Recent work using high-resolution microscopy has revealed that long cytoskeletal filaments are more likely to be short patches of polymers. [Image **a** from Ref. [16], **b, c** from Ref. [17], repr. w. permission.]

sense and measure the relative stiffness of the surrounding environment (other cell walls, the extracellular matrix, a foreign surface). In response to the measured force, the cytoskeleton can react and rearrange, to adapt the cell shape, to make the cell adherent to (or detach from) the substrate, or follow the commands to activate cell displacement. The cytoskeleton elements also actively participate to the complex process of cell duplication (mitosis), by providing the necessary mechanical forces for chromosome separation and membrane splitting (see Sect. 7.5).

Prokaryote cells seem to lack a properly organised cytoskeleton. This is one of the major differences between nuclear and non-nuclear cells, aside of the different organisation of their respective DNA, the general absence of internal membranes, and the very different sequence of cell reproduction. However, analogues for all major eukaryote cytoskeletal proteins have been found in prokaryotes, and scaffolding structures with similar cytoskeletal function, despite a largely different shape organisation, are being discovered (Fig. 6.24). The functional equivalent of actin contractile rings is thought to be the protein FtsZ, participating in cell division, however from a structural point of view this protein resembles tubulin; by contrast, a protein like ParM resembles actin in its structure, but performs functions resembling those of tubulin in eukaryotes; the structural analog of actin, albeit with a completely different geometric distribution, should be the MreB protein, a polymerisable long filament, running in a coil-like shape about the bacterial membrane; even some analog of the intermediate filaments have been identified, such as the crescentin (CreS) family, which share similarities with both the eukaryotic cytokeratin and lamin-A. Moreover, 'structural' proteins with no known eukaryotic homologues have also been discovered. Such polymerisable molecules play essential roles in prokaryote cell division, protection, shape determination, and polarity determination.

Problems

6.1 Swimming bacterium

Consider a bacterium with spherical shape, and radius $R = 15\,\mu m$, swimming in pure water (viscosity $\eta = 10^{-3}$ kg/(m s)), with a speed $v_0 = 0.3$ cm/s at time $t = 0$. Calculate the stopping distance $t > 0$, by taking a mass $m = \rho V$ for the bacterium.

6.2 Actin polymerisation velocity

Calculate the velocity of polymerisation of a filament of F-actin, formed by monomers of G-actin of average size $\delta = 5$ nm, with reaction constants $k_+ = 7.50$ $(\mu M\ s)^{-1}$, $k_- = 1.25\ s^{-1}$, in two different solutions of G-actin, with concentration 0.1 and 0.5 μM. Comment on the difference between the two cases.

6.3 Chain polymerisation

Let us examine a free-radical addition polymerisation with $k_i = 5.0 \times 10^{-5}\ s^{-1}$, $u = 0.5$, $k_t = 2 \times 10^7$ dm^3 mol^{-1} s^{-1}, and $k_p = 2640$ dm^3 mol^{-1} s^{-1}, and with initial concentrations $[M] = 2.0$ M and $[I] = 8 \times 10^{-3}$ M. Assume the chain termination occurs by combination. Calculate: (a) The steady-state concentration of free radicals. (b) The average kinetic length of the chain. (c) The production rate of the polymer.

6.4 Microtubules association/dissociation constants

In your lab you have only a centrifuge and an UV-absorption spectrometer, and you can play with a solution of tubulin in physiologic medium (pure water with 0.15 M NaCl). With these two instruments, you should design an experiment to measure the association constant of microtubules. (You can do an internet search for additional parameters needed in this case, check for example the websites www. rcsb.org/pdb/ and www.web.expasy.org/protparam/, to obtain some quantity relevant to your problem such as the sequence and the **extinction coefficient** of the protein tubulin.)

6.5 DNA replication

The mechanisms of DNA replication are very similar in both prokaryotes and eukaryotes, proceeding at the rate of up to 1,000 nucleotides per second in the former, while being slower (50–100 nucl/s) in the latter. Consider the replication of DNA in the *E. coli* bacterium, which occurs about every 30 min. The replication fork is a structure created by enzyme helicase, which breaks the hydrogen bonds holding the two

DNA strands together. The two strands of the DNA open up, and are used as templates by the enzyme polymerase for making the two identical copies. In this action, helicase makes the DNA to turn, with the help of other specialised proteins (the topoisomerases). (a) How much time is needed to make two complete copies of the bacterial DNA? What this tells us about the replication mechanism in prokaryotes? And what about eukaryotes? (b) How fast does the template DNA spins? (c) What is the velocity of a DNA-polymerase-III relative to the template?

6.6 Active and passive diffusion

A membrane with thickness L separates two volumes of fluid V_A and V_B, in each of which a constant concentration c_A and c_B of some protein X is maintained. Describe the profile of concentration $c(x)$ inside the membrane. What this has to do with the phenomenon of Brownian motion?

Subsequently, a pressure difference is applied between A and B, which establishes a flux across the membrane at constant velocity. What is the physical coefficient characterising the passage of the protein across the membrane in this condition? What are its physical dimensions? Is Brownian motion playing the same role as before?

6.7 Michaelis-Menten kinetics

The enzyme carbonic anhydrase catalyses the hydration of CO_2 in red blood cells to give bicarbonate ion:

$$CO_2 + H_2O \rightarrow HCO_3^- + H^+$$

CO_2 is converted to bicarbonate ion, which is transported in the bloodstream and converted back to CO_2 in the lungs, a reaction that can be catalyzed by carbonic anhydrase. In an experiment, the following data were obtained for the reaction at pH $= 7.1$, $T = 273.5$ K, and anhydrase enzyme concentration of 2.3 nmol L^{-1}:

$[CO_2]/(mmol\,L^{-1})$ 1.25   2.5   5.0   20.0

$v_P/(mol\,L^{-1}s^{-1})$ 2.78×10^{-5} 5.02×10^{-5} 8.33×10^{-5} 1.67×10^{-4}

Determine the catalytic efficiency e_P of the enzyme carbonic anhydrase at 273.5 K.

References

1. N. Hundt, W. Steffen, S. Pathan-Chhatbar, M.H. Taft, D.J. Manstein, Load-dependent modulation of non-muscle myosin-2A function by tropomyosin 4.2. Scientific Reports **6**, 20554 (2016)
2. T. Finer, R.M. Simmons, J.A. Spudich, Single myosin molecule mechanics: piconewton forces and nanometre steps. Nature **368**, 113 (1994)
3. N. Hirokawa, Organelle transport along microtubules: the role of KIFs. Trends Cell Biol. **6**, 135–141 (1996)
4. A.F. Huxley, Muscle structure and theories of contraction. Prog. Biophys. Biophys. Chem. **7**, 255–318 (1957)
5. T. Yanagida, A.H. Iwane, A large step for myosin. Proc. Natl. Acad. Sci. USA **97**, 9357–9359 (2001)
6. J.C.M. Gebhardt, A.E.-M. Clemen, J. Jaud, M. Rief, Myosin-V is a mechanical ratchet. Proc. Natl. Acad. Sci. USA **103**, 8680–8685 (2006)
7. Z.-H. Sheng, Mitochondrial trafficking and anchoring in neurons: new insights and implications. J. Cell Biol. **204**, 1087–1098 (2014)
8. I. Yu, C.P. Garnham, A. Roll-Mecak, Writing and reading the tubulin code. J. Biol. Chem. **290**, 17163–17172 (2015)
9. D. Cai, Single-molecule imaging reveals differences in microtubule track selection between kinesin motors. PLoS Biol. **7**, e1000216 (2009)
10. S.D. Hansen, R.D. Mullins, Lamellipodin promotes actin assembly by clustering Ena/VASP proteins and tethering them to actin filaments. eLife **4**, e06585 (2015)
11. S.C. Kou et al., Single-molecule Michaelis-Menten equations. J. Phys. Chem. B **109**, 19068 (2005)
12. V.I. Claessen et al., Single-biomolecule kinetics: the art of studying a single enzyme. Annu. Rev. Anal. Chem. **3**, 319 (2010)
13. Y. Magariyama et al., Simultaneous measurement of bacterial flagellar rotation rate and swimming speed. Biophys. J. **69**, 2154 (1995)
14. T. Suzaki, E. Richard, R.E. Williamson, Cell surface displacement during euglenoid movement and its computer simulation. Cytoskeleton **6**, 186–192 (1986)
15. E.B. Blancaflor, Regulation of plant gravity sensing and signaling by the actin cytoskeleton. Am. J. Botany **100**, 143–152 (2013)
16. A.K. Fenton, K. Gerdes, Direct interaction of FtsZ and MreB is required for septum synthesis and cell division in Escherichia coli. EMBO J. **32**, 1953–1965 (2013)
17. J.R. Juarez, W. Margolin, A bacterial actin unites to divide bacterial cells. EMBO J. **12**, 2235–2236 (2012)

Further Reading

18. R. Dean Astumian, Making molecules into motors. Sci. Am. **285**, 57–64 (2001)
19. M. Schliwa, G. Woehlke, Molecular motors. Nature **422**, 759–765 (2003)

20. R.D. Vale, The molecular motor toolbox for intracellular transport. Cell **112**, 467–480 (2003)
21. F. Ritort, Single-molecule experiments in biological physics: methods and applications. J. Phys.: Cond. Matter **18**, R531–583 (2006)
22. C. Bustamante, Unfolding single RNA molecules: bridging the gap between equilibrium and non-equilibrium statistical thermodynamics. Quarterly Rev. Biophys. **38**, 291–301 (2006)

Chapter 7
Bioelectricity, Hearts and Brains

Abstract It may be surprising to think that animal bodies are just full of strong electric fields and, to a lesser extent, magnetic fields, which surround every cell. Such fields are created by electrochemical gradients, very localised, and—luckily for us— they don't extend beyond some fractions of a micrometer around each cell. Otherwise, we would go around attracting or being pushed away from metallic objects. The yet little known functioning of brain, and the much better known functioning of heart, would each require an entire library to be described; of course, these modest pages could not cover the physiology of such complex organs. Nevertheless, some of the biggest advances in the understanding of brain physiology came from physics, with electrical studies of single animal neurons, between the 1950s and the 1960s. Here we will use simple electrical circuit models, to learn some of the most important effects induced by the fast movement of electric charges inside cells.

7.1 Cells Processing Electromagnetic Information

The lack of appropriate sources of electrical current hampered for a long time the development of electrophysiology, compared to other fields such as biomechanics, more easily accessible to experimentation. The first known record of a bioelectric phenomenon is an ancient Egyptian hieroglyph of 3000 B.C., the stone palette of King Narmer (Fig. 7.1), depicting an electric catfish with "rays" emanating from its head. In the year 46 A.D. Scribonius Largus, court physician to the Roman emperor Claudius, in his *Compositiones Medicae* recommended the use of torpedo fish for curing headaches and gouty arthritis. In practice the electric fishes, capable of peak voltage spikes reaching more than 400 V, remained the only means of producing electricity for bioelectric and therapeutic experiments until the 17th century. Still by the end of the 18th century, the Italian scientist Luigi Galvani, already cited in the Introduction as the "first biophysicist", was forced to perform his studies on frogs by using atmospheric electricity. He connected an electric conductor between the side of the house and the nerve of the frog leg; then he grounded the frog muscle with another conductor in an adjacent well. Contractions were obtained when, by chance, a lightning flashed. In later experiments, the ingenious Luigi contacted nerve and

© Springer International Publishing Switzerland 2016
F. Cleri, *The Physics of Living Systems*, Undergraduate Lecture
Notes in Physics, DOI 10.1007/978-3-319-30647-6_7

Fig. 7.1 Detail from the palette of King Narmer, from Hierakonpolis, Egypt, Predynastic, c. 3000-2920 B.C.E., slate, 2′ 1″high (Egyptian Museum, Cairo). In the two *yellow frames*, the first known representation of an electric catfish with rays emanating from the head

muscle with a conductor made of two different metals and observed electrical stimulation, however without understanding the origin of the current. Galvani supposed, in fact, the electricity to come from the frog tissues. It was Alessandro Volta, another Italian professor of "Fisica particolare" (molecular physics, in modern terms) in the University of Pavia, who made clear that the frog was not the generator, but the detector of electricity, generated instead by the contact between dissimilar metals. This finding led Volta in 1799 to the discovery of the voltaic pile, made by stacking discs of copper and zinc, which eventually gave rest to fishes for producing electricity, until in the 1870s the electric generator was invented.

Since then, the field of bioelectricity and biomagnetism has known a rapid development, and in the 20th century the medical applications have bloomed. Today it is impossible to imagine a hospital without electrocardiography and electroencephalography, and magnetic resonance is among the most widely used diagnostic techniques. Implantable cardiac pacemakers have allowed millions of people to live a normal life. The latest advances in the measurement of electric currents flowing through a single ion channel of the cell membrane (the *patch clamp* technique) have opened up completely new applications in molecular biology.

Bioelectromagnetism is a complex discipline, dealing with the electric, electromagnetic, and magnetic phenomena which arise in biological tissues. These include: (i) the behaviour of excitable tissues, i.e. the electric and magnetic sources; (ii) the electric currents and potentials induced in the conducting medium; (iii) the response of excitable cells to electromagnetic fields; (iv) the intrinsic electric and magnetic properties of cells and biological tissues.

The space limits of this (already too extensive!) book will not permit to deal in details with biological sensing, that is the way by which external stimuli are conveyed into the brain. Vision, hearing, olfaction, touch, taste, are the human senses to which we are accustomed, although a more complete list should include at least perception of temperature, pain, equilibrium. But the physical distinction between these different stimuli turns rather into a problem of the type and distribution of sensors, specialised proteins usually located in the cell membrane. These proteins act as "channels" for charged ions (see p. 263 below), whose flux in and out the membrane induces electric currents, to be further amplified, and treated as sensorial information by neurons in the brain. However, the differences between senses can be blurred for organisms living in very different environments. The following two examples of unusual detectors for light or magnetic fields are typical, but not unique cases.

7.1.1 The Eyes of a Plant

It us usually assumed that a **visual system** made of light-activated specialised sensors should be peculiar to insects and other animals "higher" in the evolutionary scale. However, many "lower" eukaryotes have a crucial need to use light. In fact, photosynthetic organisms are observed to migrate to regions of optimal light intensity, on the basis of the input coming from some kind of light-sensing device. Surprising as it may be, some plants do have eyes.

Unicellular monoflagellate algae such as *Chlamydomonas*, *Euglena*, *Trachelomonas*, *Eutreptiella*, the colony-forming *Volvox*, all have an identifiable "eyespot". These primitive organs constitute the simplest and most common visual system found in nature [2]. They contain optics, photoreceptors, and the elementary components of a signal-transduction chain (Fig. 7.2). The above unicellular algae have two different

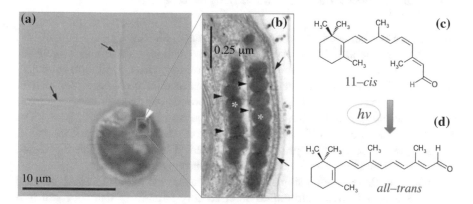

Fig. 7.2 **a** Differential contrast microscopy image of *Chlamydomonas reinhardtii*. The eyespot is indicated by a *white arrow*, flagella by *black arrows*. **b** Thin-section scanning electron micrography of the eyespot, zoomed by a factor ×40. *White asterisks* indicate the lipid globules. *Long arrows* indicate the outer membrane, *short arrows* the thylakoid membrane. **c** Chemical structure of the 11-*cis* retinal, the light-activated chromophore, which turns into *all-trans* structure upon photon capture (*hv*). [Photos (a, b) adapted from Ref. [1], w. permission]

responses to light: at low light they exhibit positive **phototaxis**, i.e., they swim toward the light source, with typical speeds of 120 μm/s; at higher light intensity, the cell shifts to backward swimming, that is negative phototaxis, at a speed reduced to less than 20 μm/s. Compared to the cell body, of about 20–100 μm size, the size of the eyespot is very small, about 1 μm. It comprises a sort of "lens", made of lipid globules rich in light-sensitive carotene proteins, usually arranged into highly ordered parallel layers. The fact that cell mutants deprived of the eyespot still perceive light but lose the correct orientation, seems to prove that the eyespot is an optical device to sense and concentrate light, but is distinct from the photoreceptor. The highly regular structure of the bent parallel layers (see figure) suggests that the eyespot may act as a mirror, converging the light intensity to a focus where the photoreceptors are hosted. These are clusters of retinal-binding proteins, as it has been verified by alternately suppressing and restoring the synthesis of retinal molecules in genetically mutated cells. Upon capturing a light photon, the retinal chromophore makes a *cis* to *trans* isomerisation (Fig. 7.2, right), which starts a protein modification response at the origin of the "vision". The mechanotransduction mechanism, which translates light detection into movement, is not completely clear. However, the eyespot is often found close to the microtubules attachment of the cell flagella. It is supposed that the eyespot may be part of an antenna, which scans the incident light intensity as soon as the cell rotates while moving in water (see Chap. 6, on the rotational movement of ciliates and flagellates). Someway, the higher light intensity produces a signal controlling the flagellar beat, such that the cell progressively shifts its motion towards the light source.

7.1.2   Birds and Flies Can See a Magnetic Field

Humans are not believed to have a **magnetic sense**. However, many animals use the Earth's magnetic field for orientation and navigation, although this was once dismissed as a physical impossibility. Two types of magnetic information are potentially available to the animal. The simplest is directional information, which enables to maintain a consistent heading, for example towards the north or south. Animals with this ability are said to have a *magnetic compass*. By contrast, at least a few animals can also obtain a kind of positional information from the magnetic field, for example to assess their approximate geographical location. Animals that derive positional information from the field are said to have a *magnetic map*.

Up to now, there is not a single "magnetic sensor" identified, in any biological organism. Magnetic fields are unlike other sensory stimuli, in that they can pass freely through biological tissue. Whereas receptors for senses such as olfaction and vision must make contact with the external environment, a "magnetoreceptor" could be located almost anywhere inside an animal's body. In addition, the large accessory structures needed for focusing and manipulating the field (magnetic lenses) are unlikely to exist, because few materials of biological origin can affect magnetic fields. Tiny magnetoreceptors could be dispersed in a large volume of tissue, or the transduction process might occur as a set of chemical reactions, which means that no obvious organ or structure devoted to this sensory system should necessarily exist.

Currently, three mechanisms are suggested to explain magnetic sensing by animals: one invokes electromagnetic induction, another involves magnetite, and the third chemical magnetoreception. The first two require either the animal's body to be conductive, or the animal to have magnetic nanoparticles dispersed in the tissues, the displacement of such particles due to the magnetic field possibly initiating a signalisation cascade. Nanoparticles of this kind (with size $\sim$50 nm) have been actually identified in some animals and bacteria, whereas the induction mechanism might apply to some fishes like sharks or rays, whose skin is sensitive to electricity.

The third mechanism is much more involved, but might be indeed the most promising. The idea is that magnetoreception could occur through unusual biochemical reactions that are influenced by a magnetic field: for example, reactions involving the formation of free radicals, in which the lone-electron spin can interact with the magnetic field, like in a MRI. Many free-radical reactions are initiated by the absorption of light, and this led to the suggestion that chemical magnetoreceptors, if they exist, might also be photoreceptors. A special class of proteins, the cryptochromes, have the required chemical properties (they contain an element that forms radical pairs after photoexcitation by blue light), and are concentrated in the cells of the eye retina of many different animals. For example, experiments on migratory birds show that such retina cells increase their activity when the bird orients magnetically. Strong evidence for cryptochrome involvement in magnetic sensing came from behavioural experiments with the fruitfly *Drosophila melanogaster* [3].

Such a coupling of magnetic field sensing with vision-related receptors is indeed fascinating. When magnetically-sensitive animals look out at the world, they should see superimposed on the normal visual field an additional signal, consisting of a pattern of lights, or colours, which changes depending on the direction the animal faces. If so, the animal might learn to associate a particular visual signal with a particular magnetic direction. Moreover, in some transgenic experiments, a cryptochrome protein commonly found in the human retina was reintroduced in the *Drosophila* [4], demonstrating that this human protein can nicely function either as a light-sensitive magnetosensor or as part of a magnetosensing pathway, in the insect. Although it is not known whether such properties might translate into a similar biological response also in the human retina, this finding could reopen an area of sensory biology for further exploration also in humans.

7.1.3 The Neuron

And again the Egyptians! The "Edwin Smith" papyrus, from the XVI-XVII Dynasty (about 1600 BC), contains the earliest descriptions of the cranial structures, the meninges, the external surface of the brain, the cerebrospinal fluid, and the intracranial pulsations. Here, the word *brain* appears for the first time in any language. According to the historians of medicine, the procedures described in this papyrus demonstrate a level of knowledge surpassing that of the Greek Hippocrates, who lived 1000 years later. Galenus, in the 2nd century, proposed that the brain is the centre of the mind, at a time when this role was generally attributed to the heart, and proved experimentally that the brain commands the muscles. However for nearly two millennia, while the anatomical studies slowly progressed, the functions of the brain remained metaphysically obscure. When in 1838 Theodore Schwann and Matthias Schleiden proposed that cells are the basic functional units of all living things, this theory was not believed to apply to the nervous system, and it was not until towards the end of the 19th century that it became generally accepted that the brain, too, consisted of cells. The discovery of the **neuron** was a milestone in brain research, and paved the way for modern neuroscience.

The Italian anatomist Camillo Golgi invented about 1870 a staining technique for neurons (*la reazione nera*, or black reaction). It consisted in fixing silver chromate particles to the neuron membrane, resulting in a stark black deposit on the soma as well as on the axon and all dendrites. This provided an exceedingly clear and well contrasted picture of single neurons against a yellow background (without such a stain, brain tissue under a microscope appears as an impenetrable tangle of protoplasmic fibres, in which it is impossible to detect any structure). Golgi viewed the nervous system as being a seamless, continuous network of interconnected cells, with nerve signals firing along in all directions. The Spanish physiologist Santiago Ramón y Cajal, on the other hand, proposed that the brain is composed of billions of individual cells, or neurons (a term coined some years earlier by the German anatomist Heinrich Wilhelm Gottfried von Waldeyer-Hart), receiving information at one end, and transmitting it unidirectionally to the next cell. Cajal was almost clairvoyant in proposing that the increase in the number of synapses could be one of the mechanisms of learning and memory, a fact that was ascertained only much later. Only recently, the discovery of gap junctions and a prevalence of electrically coupled neurons in some regions of the brain (see below), seemed to revive the argument for a "reticular" network, in parallel to this predominant "neuron doctrine".

By looking at images showing the structure of typical neurons (Fig. 7.3), we see something that looks quite distant from the characteristic cell we have been taking as a model example until now. Instead of a roundish, compact mass enclosed by a rather smooth membrane, containing a large nucleus plus a bunch of other organelles, the typical neuron appears as a small central core from which many filaments emanate in all directions, with a thicker one and a seemingly infinite branching into thinner and thinner stems. Although our model-spheroidal cell is certainly a highly idealised picture, and most real cells may have quite weird shapes, neurons look more like microscopic plant roots, rather than what we thought as cells.

The neuron may be divided on the basis of its structure and function into three main parts: (1) the cell body, also called the **soma**; (2) numerous short processes of the soma, called the **dendrites**; and (3) the single long nerve fibre, the **axon**. The body of a neuron is, in fact, similar to that of most other cells, including the nucleus, mitochondria, endoplasmic reticulum, ribosomes, and other organelles. The cell size is extremely variable, its volume ranging between 500 and 70,000 μm^3. The short processes of the cell, the dendrites, receive impulses from other cells and transfer

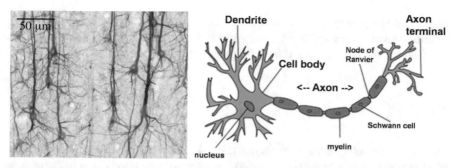

Fig. 7.3 *Left* SMI32-stained pyramidal neurons in cerebral cortex (recoloured and enhanced). *Right* Sketch of a typical neuron. The soma (*left end*) includes the nucleus and most of the usual cell organelles, and branches out with dendrites, to receive input currents from other neurons. The axon, extremely variable in length, is partly covered by the myelin sheath, carried by Schwann's cells and interrupted by the Ranvier nodes. At the right end, the axon splits into many terminals, which make contact to other neurons' dendrites, via individual synapses. [Images © *left* UC Regents Davis campus, www.brainmaps.org, *right* Q. Jarosz. Adapted under CC-BY-SA 3.0 licence, see (*) for terms.]

them to the cell body. The effect of these impulses may be excitatory or inhibitory. A cortical neuron, located in the brain cortex, may receive impulses from tens, or even hundreds of thousands of neurons. The long nerve fiber, the axon, transfers the signal from the cell body to another neuron, or to a muscle cell. Mammalian axons are usually a few micrometer in diameter, while their length in larger animals may reach up to several meters. The axon may be covered with an insulating layer, called the **myelin sheath**, which is formed by Schwann cells (named for the German physiologist Theodor Schwann, who first observed the myelin sheath in 1838). When present, the myelin sheath is not continuous but is split into sections separated at regular intervals by the nodes of Ranvier (named for the French anatomist Louis Antoine Ranvier, who observed them in 1878).

Similar to any cell, the neuron is enclosed by a membrane whose thickness is about 7.5–10 nm. However, the membrane of the neuron includes a large concentration of proteins that regulate the inflow and outflow of charged ions, mainly Na^+, K^+, Cl^- and Ca^{2+}, the already mentioned ion channels. These make the neuron membrane an excitable medium, notably all along the axon, and constitute the most important feature of the neuronal cell, being at the basis of the bioelectric phenomena such as generation, amplification and transmission of electric potential pulses.

The junction between an axon and the next cell with which it communicates is the **synapse**. In most synapses, information coded into electrical impulses proceeds from the cell body unidirectionally, first along the axon and then across the synapse, to the next nerve or muscle cell. The part of the synapse that is on the side of the axon is called the *pre-synaptic terminal*; that part on the side of the adjacent cell is called the *post-synaptic terminal*. Between these terminals, there exists a thin gap, the **synaptic cleft**, with a thickness of 10–50 nm. The unidirectionality of impulse transmission in *chemical* synapses is due to the release of a chemical transmitter by the pre-synaptic cell. This transmitter, when released, activates the post-synaptic terminal. An exception to this rule, *electrical* synapses have a much simpler activation mechanism, and can transmit electric impulses in both directions.

7.1.4 The Neuromuscular Junction

The electric potential of the neuronal cell takes the form of a propagating voltage peak of a few tens of mV, which carries the information from the brain down to the muscle cells. At the microscopic scale, the mechanical action of the muscle contraction (i.e., the beginning of the sarcomere contraction, apparent in the relative displacements of the actin and myosin filaments which make up each sarcomere, see Chap. 10) is activated by a complex chain of events, taking place in the **neuromuscular junction**. This is a special type of synapse densely connecting the axon of a motor neuron to a muscle fibre, as shown in Fig. 7.4. For example, the flight muscles of the *Drosophila* fruit fly are innervated by a giant axon, $\sim$5 μm in diameter, carrying about one such junction every 1–2 μm.

The neuromuscular junction has some important differences compared to synapses formed between neurons. The contact region is here very large, with multiple **postjunctional folds**, which increase the surface area of the membrane exposed to the synaptic cleft. The neuron terminal (A in the right figure) contains a quantity of synaptic vesicles. Each vesicle is a small spheroid made of lipids, with diameter in

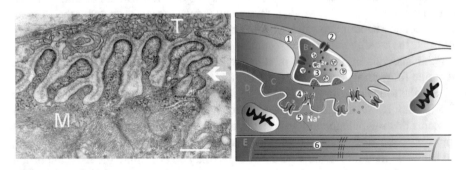

Fig. 7.4 *Left* Electron micrograph showing a cross-section through the neuromuscular junction. T is the axon terminal, M is the muscle fibre, the *white arrow* points at junctional folds. The scale bar is 0.3 μm. *Right* Schematic of the neuromuscular junction, a special type of synapse formed not between two neurons, but between a neuron and a muscle cell (myofibre). The neural cell terminal (A) contains a quantity of synaptic vesicles (B). Upon arrival of the action potential (*1*), Ca^{2+} flows in (*2*), and induces fusion of the vesicles to the multiply folded membrane. This liberates acetylcholine (*3*) in the $\sim$10–20 nm narrow space (C), between T and M, which in turn opens ion channels (*4*) on the M side. The flux of ions starts a new action potential (*5*) on the M side, which spreads in the muscle cell (E). Finally, calcium ions diffusing (*6*) through the muscle cell initiate the muscular contraction. [Public-domain photo *left* by National Institute of Mental Health; image *right* repr. under CC BY-SA 4.0 lic., see (*) for terms.]

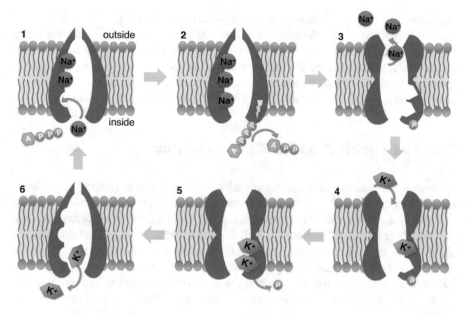

Fig. 7.5 Schematic of the cyclic working mechanism of the Na-K pump in the cell membrane. The cyclic sequence proceeds clockwise, 1–6, starting from the *upper-left*. Up to three Na⁺ ions accumulate inside the inner opening of the protein, coming from the cytoplasm. One ATP molecule can attach to a side of the pump, and be hydrolysed to ADP. The phosphate remains attached to the rim of the protein, inducing a structural change that pushes the three sodium ions to get out in the extracellular fluid. Once this process is completed, two K⁺ ions enter the pump from the rim exposed to the extracellular space. The inorganic phosphate (P) is now detached from the inner side of the pump, which changes again its structure back to the original shape, and lets the potassium ions to enter the cytoplasm

the few nanometers, filled by a constant number of acetylcholine (ACh) molecules, a neurotransmitter. When an electric impulse (or action potential, see below) arrives through the nerve axon to the synapse, the membrane channels on the T side open up, and let flow the Ca^{2+} ions from the extracellular fluid in the synaptic cleft (greenish zone in the figure on the right), into the pre-synaptic T side. This flux of calcium ions allows the vesicles to adhere to the membrane and break open (exocytosis); the ACh neurotransmitter molecules are liberated in the synaptic cleft, and diffuse to reach the receptors on the membrane of the post-synaptic terminal M. These receptors, with a density of about 10^4 per μm^2, are ligand-gated ion channels (see below, Fig. 7.5) whose ligand is the ACh. Once the ACh is captured at the receptors, the ion channels open up, and let Na^+ flow into the muscle cell membrane, while K^+ ions flow out. Since the amount of sodium entering is larger than the potassium exiting, the membrane becomes more positively charged (in fact, less negative) and this further induces the formation of a second electric pulse in the muscle.

When this electric potential spike reaches the muscle cell membrane (*sarcolemma*), the adjacent sarcoplasmic reticulum releases Ca^{2+} ions. Now the calcium ions diffusing through the entire muscle cell volume (the *sarcomeres*) can initiate the muscular contraction. As described in more details in Chap. 6, Ca^{2+} is chemically bound to some intermediate proteins standing between the actin and myosin (the troponin-C), thus providing the initial 'loading' of the actin-myosin motor. Subsequently, by consuming ATP molecules, the activated molecular motor displaces the myosin and actin relative to each other, and the muscle contraction takes place.

7.2   The Electric Potential of the Membrane

Although any analogy between brain and computer could be merely superficial, neurons appear as the fundamental elements for the treatment of information in the nervous system. Since each neuron is connected to thousands of other neurons, the neuronal "computation" is strongly parallel, as opposed to the sequential treatment of the information in our computer processors. Although the nervous system is capable of many diverse functions, the repertory of elementary operations which encode all such functions is rather limited, namely: (1) generation and propagation of an action potential, (2) synaptic transmission, and (3) production of graded synaptic or sensory potentials.

The electrical activity of nerve cells can be analysed in terms of several analogies with conventional electric circuits. The important elements to be considered in this respect include at least the following:

- a battery, which represents the source of electric power,
- a resistance, which represents the potential drop between two points, ensuring the continuity of the flux of electric charge,
- a capacitor, which represents the potential drop but with spatial separation of the positive and negative charge carriers.

The voltage drop ΔV_m between the two sides of the cell membrane (unfortunately, this quantity is often called membrane or transmembrane "tension", leading to a confusion with the concept of mechanical tension), is defined as the difference of electric potential between the internal surface, Φ_i, and the external surface of the membrane, Φ_e:

$$\Delta V_m = \Phi_i - \Phi_e \qquad (7.1)$$

Such a definition is independent on which is the origin of the potential difference, and whether its value is constant, periodic, or even non periodic at all.

7.2.1   Passive and Active Diffusion

In practical terms, the cell membrane separates two electricity-conducting fluids, the cytoplasm and the extracellular fluid, in fact two electrolytic solutions with different concentrations of various ions. As a consequence, both concentration and charge gradients are established between the inside and the outside of the cell. Because of the thick inner layer formed by the lipid tails of the amphiphilic molecules, any cell membrane is very little permeable to ions of any species. Values of intrinsic permeability of typical cell membranes are 10^{-14}–10^{-12} m/s for Na^+, K^+, or Cl^-, to be compared to 10^{-8}–10^{-7} m/s for glycerol, urea, indole, and other bigger molecules, while water may have a permeability of 10^{-6}–10^{-4} m/s depending on the type of cell. It may be noted that glucose is a polar molecule, with mass of 180 Da, but its permeability is small and close to that of the tiny ions. Tryptophan, on the other hand, has a mass of 204 Da, but being non-polar it has a permeability about 1,000 times than that of glucose. Size is clearly a much less important feature than polarity and charge, in the permeability game.

As we saw in Chap. 5, the permeability P_k for an ion of type k is proportional to the diffusion coefficient D_k, to the partition coefficient K, and inversely proportional to the membrane thickness d :

$$P_k = \frac{K D_k}{d} \tag{7.2}$$

Due to their strongly hydrophilic character, all ions have very low values of K, therefore the cell membrane (actually, the inner double layer of lipid tails) is practically impermeable to ions. In the absence of other facilitating mechanisms, any differences of ionic concentration on the two sides of the membrane would remain constant, and as a consequence a diffusion potential is established. This potential acts like a battery, with a drop equal to the voltage difference, and a resistance equal to the permeability of the membrane for that particular ion. Interestingly, the application of an equal and opposite voltage across the membrane allows to stop the ionic flux, thereby providing a way to measure the equilibrium membrane potential.

The **passive** diffusion of a particle, charged or neutral, is described by the Fick's equation (see again Chap. 5):

$$J = -D\nabla c \tag{7.3}$$

with J (in mol cm^{-2} s^{-1}) is the particle flux, D (in cm^2 s^{-1}) is the diffusion coefficient, and ∇c is the concentration gradient (in mol cm^3) between two points. For k different types of ions, it must be considered k different diffusion coefficients D_k, k gradients ∇c_k, and k independent fluxes J_k.

In the case of nerve cells, there are three main types of ions involved in determining the electrochemical potential between the two sides of the membrane: Na^+, K^+, and Cl^-. Other ions can be implied in different processes in other types of cells, such as hydrogen carbonate HCO_3^-, or divalent cations such as Mg^{2+}, Ca^{2+}. In all cases, however, the diffusion coefficients of charged species in the (hydrophobic) lipid membrane is exceedingly small.

Since the passive, or spontaneous, diffusion is so low for all ions, concentration gradients must be created by specialised proteins in the cell membrane, and especially in neurons, to facilitate the diffusion. These proteins go under the generic name of **ion channels**, and can operate according to different mechanisms, by opening and closing the passage to ions ("gating"), by using different means:

- **ligand**, if the attachment of a small ligand molecule at a particular site of the ion channel operates the switching, as it occurs for example when acetylcholine binds at a neuromuscular synapse;
- **mechanical**, when the deformation of the protein structure opens and closes the channel, for example by a sound wave as it occurs in the hair cells of the inner ear;
- **voltage**, when a protein channel is opened by the passage of an electric potential pulse along the membrane, as is the case of sodium channels along the neuronal axon that relay the electric pulse over long distances;
- **light**, occurring for example in photosynthetic bacteria, in which the light pulse opens a special channel allowing the flow of protons, resulting in a movement of the microorganism towards the light source (phototaxis).

A similar 'facilitating' mechanism allows the passage also of particular molecules through the cell membrane, with the help of other specialised channels that go under the generic name of **porins**. For example, maltoporin is a protein located on the outer surface of the membrane of some bacteria; it is made by 18 loops of amino acids shaped as a funnel, providing a sort of "sleigh" for maltodextrin to enter the cell.

Another important example of facilitating channels, which earned the Nobel prize in Chemistry in 2003 to the American physician Peter Agre, regards the diffusion of water molecules. Despite the already high permeability of water to the plasma membrane, almost all animal and plant cells also have a variety of **aquaporin** channels, which accelerate the passage of water in and out the cell. A single aquaporin-1 channel can transfer up to 3×10^9 water molecules per second, in both directions. To compare this value against passive diffusion, let us take a cell of radius $R = 10\,\mu m$, and consider the gradient of $\Delta c = 3 \times 10^9$ across the membrane thickness $\Delta x = 10\,nm$. The diffusion coefficient for water through the plasma membrane ranges from 8×10^{-10} to 2×10^{-8} cm^2/s. Hence, from Eq. (7.3) the maximum flux is $J = 6 \times 10^7$ molecules/(cm^{2}s). If all the 3 billion molecules must diffuse through the cell surface, $S = 4\pi R^2$, we get a flow rate of $J \cdot S = 750$ molecules per second, i.e., ordinary passive diffusion is a factor of 4×10^6 less efficient than one single aquaporin channel in transporting water.

A different class of proteins can perform the **active transport** of ions, sugars or salts in and out from the cytoplasm, by imposing a flow *against* the concentration gradient. Clearly, this cannot come without a corresponding expenditure of free energy, typically obtained from the consumption of ATP. The most important of these integral proteins is the so-called **sodium-potassium pump**, or Na$^+$/K$^+$-ATPase (Fig. 7.5), which uses the energy of one ATP molecule to export 3 Na$^+$ ions outside the cell, for every 2 K$^+$ ions that enter the cytoplasm. The process is clearly not electrically neutral, and makes for a disequilibrium also of the negative charges, such as the Cl$^-$ ions. Besides neurons, the sodium-potassium pump is found in almost any other types of cells, where it serves to maintain a gradient of Na$^+$, necessary for various cellular functions.

Another similar ATP-powered pump is the H$^+$/K$^+$-ATPase, needed to increase the pH of the gastric juice to about 1. This pump belongs to the larger group of

proton pumps, it transports 1 or 2 protons outside the cell, for 1 or 2 K^+ getting inside (therefore it is electro-neutral). In this way, it increases the extracellular H^+ concentration from the value of 4×10^{-8} M, found inside the epithelial cells of the stomach, to a humongous concentration of about 0.15 M in the gastric juice. (To this purpose millions of ATP units must be consumed, and the epithelial cells of the stomach must contain an important proportion of mitochondria, compared to other cells.)

A final category of ion channels which is of interest for neuronal functioning is that of **indirect active transporters**. Such pumps exploit the flow of one active ion (usually Na^+, whose gradient is separately established by the working of Na/K-ATPase) to transport some other molecule or ion against its own gradient. If the indirectly transported species flows in the same direction of the active transporter, the pump is called a *symporter*; if, conversely, the species flows in the opposite direction as the active transporter, it is called an *antiporter*. An example of the former is the Na^+/glucose transporter, which transports excess glucose out of the intestine or of the kidney tubules, back into the blood flow: the Na/K-ATPase pump firstly establishes a Na^+ gradient outside the cell, then the Na^+ and glucose flow back together into the cell through the Na^+/glucose pump. An example of antiporter is the Na^+/Ca^{2+} exchanger, which pumps 2 Ca^{2+} outside the muscle cells for 3 Na^+ entering; also in this case both the sodium and calcium ions flow in the direction of their respective concentration gradients (the Na^+ gradient being established again by the Na/K-ATPase, while the Ca^{2+} excess would result, in the case of a muscle cell, from the release of neurotransmitters at a neuro-muscular synapse).

7.2.2 The Nernst Equation

Given that the neuronal cell membrane is so specialized in creating and maintaining ionic concentration gradients, let us see what are the implications for its electrochemical properties. The entropy of an ion k in solution is proportional to the number of available microstates Ω that, in turn, is proportional to the volume V:

$$S_k = k_B \ln \Omega = k_B \ln V = -k_B \ln(const \cdot [c_k]) \tag{7.4}$$

the last equivalence coming from the observation that the concentration is inversely proportional to the volume, $[c_k] \propto const/V$.

If we consider the difference in concentrations of the ion k between the outside and the inside of the membrane, the corresponding entropy difference is:

$$\Delta S_k = -k_B \ln \frac{[c_k]_{out}}{[c_k]_{in}} \tag{7.5}$$

Now, charged ions separated by an impermeable membrane are analogous to the electric charges in a battery, separated by a dielectric. In the battery, the difference of electric potential V (voltage) coincides with the product of the chemical potential $\mu = \Delta G = \Delta H - T \Delta S$, times the difference of accumulated charges on the two sides of the dielectric, $-Ne$, or:

$$\Delta G = \Delta H - T \Delta S = -NeV =$$
$$= G_0 + k_B T \ln \frac{[c_k]_{out}}{[c_k]_{in}} \tag{7.6}$$

From this expression, we can write for the membrane potential V_m:

$$V_m = V_0 - \frac{k_B T}{Ne} \ln \frac{[c_k]_{out}}{[c_k]_{in}} \tag{7.7}$$

This is the *Nerst equation*, a typical tool of electrochemistry. Note that the prefactor of the logarithm can also be written as (RT/zF), where $R = k_B N_{Av} = 8.3145$ J/(K mol) is the universal gas constant, and $F = eN_{Av} = 9.6485 \times 10^4$ C/mol is the Faraday constant, or the total charge of one mole of electrons. In the routine calculation of electrophysiology, the potential is expressed in mV (millivolts), the logarithm is taken in base-10, the sign of the potential is taken with respect to *inside* negative ions (i.e., $z = +1$ for monovalent positives and $z = -1$ for the negatives), and all the constants are adjusted to the body temperature of $T = 310$ K (Fig. 7.6). Therefore, the Nernst equation looks like:

$$V_m = \frac{61.5}{z} \log_{10} \frac{[c_k]_{out}}{[c_k]_{in}} \tag{7.8}$$

Differently from most other types of cells, which mainly use a flow of Na^+ for their functions, neuronal cells need also to be (actively) permeable to K^+ and Cl^-, thanks to the various types of passive and active pumps described above. For a neuronal cell of squid, definitely the most studied animal in human electrophysiology,[1] the concentrations of Na^+ are : $[Na^+]_{in} = 50$ mM, $[Na^+]_{out} = 440$ mM, so that the corresponding electric potential of the "sodium battery" is:

$$V_{Na} = 61.5 \log_{10}\left(\tfrac{440}{50}\right) = +58 \text{ mV (positive inside)}$$

Similarly, for potassium ions $[K^+]_{in} = 400$ mM, $[K^+]_{out} = 20$ mM, and for chlorine ions $[Cl^-]_{in} = 52$ mM, $[Cl^-]_{out} = 560$ mM, with electric potentials:

---

[1] The squid giant synapse was first discovered by the English zoologist John Zachary Young in 1939, as the main controller of the muscular contraction, at the basis of the jet propulsion of the animal in water. It is linked to the giant axon, which was used by Hodgkin and Huxley in their Nobel-winning experiments on the action potential (see p. 278 below). Because of its large size (up to 1 mm) the squid axon was a relatively easy subject for early neurophysiology experiments.

Fig. 7.6 Measurement of the voltage inside a neuronal cell. The pipette on the *left* contains an electrode that can penetrate the cell membrane, while the electrode on the *right* immersed in the extracellular fluid is at ground potential. Under standard resting conditions, a human neuron is at about $V_{rest} = -70$ mV

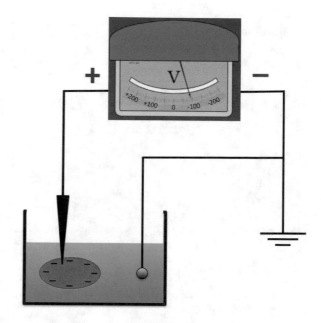

$$V_K = 61.5 \log_{10} \left(\tfrac{20}{400}\right) = -80 \ \text{mV (positive outside)}$$

$$V_{Cl} = -61.5 \log_{10} \left(\tfrac{560}{52}\right) = -63.5 \ \text{mV (negative inside)}$$

The three battery components can be arranged in a parallel setup, with voltages imposed by the concentration differences, and with their right charge signs, as shown in Fig. 7.7. Such concentration differences are not at equilibrium, but are constantly maintained or adjusted by energy consumption in the form of ATP molecules. The capacity C_m in the Figure takes into account the fact that any voltage, i.e. concentration, variations cannot be instantaneous, but will set in a characteristic time depending on a number of structural and physiological factors.

7.2.3 Polarisation of the Membrane

In a simple model, a neuronal membrane can be represented as a RC circuit (see Appendix G), Typical values of the specific membrane resistivity and capacity, due to the hydrophobic (i.e., dielectric) lipid layer, are $\rho_m \simeq 10^4 \ \Omega \ \text{cm}^2$ and $c_m \simeq 1 \ \mu\text{F/cm}^2$. For a typical neuronal cell with diameter $\delta \sim 50 \ \mu\text{m}$ (excluding the long axon) the lipid membrane resistance can be estimated as $R_m = \rho_m/(4\pi\delta^2) \simeq 200 \ \text{M}\Omega$, and its capacitance $C_m = (4\pi\delta^2)c_m \simeq 80$ pF.

Ohm's law and the diffusion of ionic charges

Before establishing this connection, it will be practical to introduce two auxiliary quantities:

1. the current density, $j = i/A$, that is the current per unit surface of the conductor;
2. the electric field, $E = V/L$, that is the difference of electric potential per unit length of the conductor.

With these new quantities, the familiar Ohm's law that relates current and potential via the resistance of the conductor, can be rewritten as:

$$jA = i = V/R = \frac{EL}{R} \tag{7.9}$$

or, equivalently:

$$j = \frac{L}{AR} E \tag{7.10}$$

It should be obvious that the specific conductivity, inverse of the resistance per unit length:

$$\sigma = \frac{L}{AR} \tag{7.11}$$

cannot change for any elementary volume of a given conductor material. Therefore, the "updated" version of the Ohm's law reads:

$$j = \sigma E \tag{7.12}$$

In the context of the diffusion theory, a material current of particles crossing a surface A in a time t, is described by the flux:

$$J = \frac{N}{At} \tag{7.13}$$

If the particles are also charged, all with the same charge q, the corresponding charge current is:

$$J = \frac{Nq}{At} \tag{7.14}$$

Let us consider the distance $L = vDt$ covered by the diffusing particles in the time t, and take t such that the volume $V = AL$ swept in this time be unitary, with $\rho = N/V$ the particle density in this 'tube' of volume V. Then:

$$J = \frac{Nq}{At} = \frac{qV\rho}{At} = q\frac{AL\rho}{At} = q\rho vD \tag{7.15}$$

This is a fundamental law of conductivity in the linear regime (i.e., when a linear relation between current and electric field holds). If we compare this flow of charge with the current resulting from Ohm's law, it is:

$$J = q\rho vD = \sigma E \tag{7.16}$$

Since the ratio $\sigma = (q\rho vD)/E$ is constant, and both the charge q and density ρ are constants as well, it turns out that the ratio vD/E must be a constant. This defines a new important quantity:

$$\mu = \frac{vD}{E} \tag{7.17}$$

that is the **mobility** of the charged particle in a given electric field, in units of $m^2V^{-1}s^{-1}$.

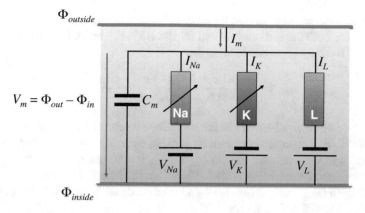

Fig. 7.7 The arrangement of the membrane ion "batteries", in the equivalent circuit of the Hodgkin-Huxley model. With the normal extracellular medium, V_{Na} is positive while V_K and V_L have negative values. During an action impulse, G_{Na} and G_K vary as a function of transmembrane voltage and time. *Red arrows* indicate the current flow direction

However, the very high resistance of the membrane to ions makes little sense here, because ions flow across the transmembrane channels. The equivalent electric resistance R_{eq} of this circuit, or its inverse the *conductance* $G_{eq} = 1/R_{eq}$, is a measure of the relative facility of ionic diffusion across the membrane (see the greybox on p. 268), via the specific ion channels and ATP-powered pumps, which were briefly described above. By analogy, the resistance of the membrane for each ion species is inversely proportional to its permeability, $R_k \propto 1/P_k$: the higher the permeability for a ionic species, the lower the equivalent resistance.

The net electric potential drop across the membrane, or **resting potential** is determined by the celebrated Goldman-Hodgkin-Katz equation (the (7.70) derived in the Appendix F), that is obtained from the current flux equation for each of the three monovalent ion species:

$$V_{rest} = 65.1 \log_{10} \left(\frac{P_{Na}[\text{Na}^+]_{out} + P_K[\text{K}^+]_{out} + P_{Cl}[\text{Cl}^-]_{in}}{P_{Na}[\text{Na}^+]_{in} + P_K[\text{K}^+]_{in} + P_{Cl}[\text{Cl}^-]_{out}} \right) \qquad (7.18)$$

The G-H-K equation requires the values of ion permeability to be known, a rather difficult quantity to obtain, since it requires to measure mass displacement in a fluid environment. It may also be noted that for one single ion species, the G-H-K equation turns into the Nernst equation (7.8).

After all transients have equilibrated, the membrane potential settles at its resting value, which can be expressed by the equivalent conductance. By assuming that each ion acts independently in the "parallel" scheme of Fig. (7.7), this is given by the inverse of the equivalent resistance (see Appendix G) as:

$$\frac{1}{R_{eq}} = \frac{1}{R_{Na^+}} + \frac{1}{R_{K^+}} + \frac{1}{R_{Cl^-}} = G_{Na^+} + G_{K^+} + G_{Cl^-} \qquad (7.19)$$

The total current is the sum of the three independent ion currents. Then, by using $I = V/R$ for both the total current and for the Na^+, K^+ and Cl^- currents, we have in terms of the respective conductances:

$$V_{rest} = \frac{G_{Na^+} V_{Na^+} + G_{K^+} V_{K^+} + G_{Cl^-} V_{Cl^-}}{G_{Na^+} + G_{K^+} + G_{Cl^-}} \qquad (7.20)$$

In the resting state, when the cell is not excited by neuronal impulses, K^+ channels are open and the potassium permeability is elevated (and so is its conductance G_{K^+}, while Na^+ and Cl^- channels are closed, thus their resistance is high and both their conductance and permeability is low). Therefore, the linearised equation above tells us that, if both G_{Na^+}, $G_{Cl^-} \to 0$, the membrane potential gets close to the potassium value, $V_{rest} = -75$ mV $\simeq V_{K^+}$.

The charge on the capacitor-membrane at this stage is equal to $Q = VC = (75 \times 10^{-3}) \cdot (80 \times 10^{-12}) = 6 \times 10^{-12}$ coulomb. By counting only the K^+'s, this amounts to about 3.7×10^7 ions. It can be noted that this amount of charge is negligible compared to the ordinary cell concentration, which in the case of potassium is about 400 mM. For the typical cell with diameter $\delta = 50$ μm, the volume is 2.6×10^{-12} l, the number of ions corresponding to this molarity is therefore $(400 \times 10^{-3}) \cdot (2.6 \times 10^{-12}) \cdot N_{Av} = 6.2 \times 10^{11}$. Therefore, the normal ionic concentrations are not significantly changed by the influx/outflux of ions corresponding to charging and discharging of the membrane. In particular, this also reassures that the electrical neutrality of the cell (and of our bodies!) is not significantly violated, except in a very thin layer of cytoplasm immediately adjacent to the cell membrane.

Now, starting from a condition in which the membrane conductance is dominated by only the K^+ ions, let us consider the effect of opening a few Na^+ channels: the Na^+ ions will start migrating inside the cell, under the combined effect of their own concentration gradient ([Na^+] being higher outside), and of the dominating potential imposed by the potassium, which is positive at the outside.

At this stage, the membrane starts to **depolarise**, meaning that its voltage V_m raises above V_{rest} (although the sodium potential, at $V_{Na^+} = +55$ mV, remains still far above). The fact that $V_m > V_{K^+}$ has also the effect of arresting the influx of K^+ ions, which instead start getting out of the cell, trying to compensate the extra charge of the entering Na^+ ions. In the transitory regime, the G-H-K equation tells us how the voltage changes in time. At the final stationary state, the currents from the two ionic species must equilibrate, $I_{Na^+} = I_{K^+}$, or:

$$G_{Na^+}(V_m - V_{Na^+}) + G_{K^+}(V_m - V_{K^+}) = 0 \qquad (7.21)$$

that is nothing but Ohm's law again. However, in this case the equation tells us that the ratio of the driving force for ionic transport inside/outside the cell is equal to the inverse ratio of the conductances. Until the sodium conductance remains substantially lower than that of K^+, the corresponding driving force for Na^+ to enter the cell is higher. The new value of the equilibrium potential is usually found at $V_{rest} = -60$ mV, meaning that in resting conditions there are just a few sodium channels opened.

This is actually the meaning of the G-H-K equation: being derived from the sum of the diffusive plus electromigration flow of all ionic species present, it states that the equilibrium among all the combined processes is reached when the concentration gradients of all species correspond to the respective Nernst potentials. In the absence of external stimulations, ions will flow in and out the membrane until the equilibrium corresponding to the combined conductances is reached. As we will see in the following, when a nerve stimulus reaches the membrane, a quantity of ion channels are opened or closed, thereby changing the conductances, and transitorily setting the concentration gradients to quite different values from those corresponding to V_{rest}.

7.3 The Membrane as a Cable

Already in 1905, Ludimar Hermann had shown that the propagation of a signal along the long axon of the neuronal cell could be described in analogy with the propagation of the current along an electric cable. The model of the equivalent circuit is a series connection of a number of identical elements, each composed of a battery, a resistance and a capacitor. These elements are connected by another resistance as described in Fig. 7.8. There, the R_i and R_o are the resistance per unit length of the intracellular fluid, or **axoplasm**, and of the extracellular fluid, respectively ($k\Omega$/cm of the axon length); R_m is the equivalent membrane resistance, and C_m is its capacity; the i_k are the different components of the current; V_m is the membrane potential (Volts) and $V' = V_m - V_{rest}$ is its deviation with respect to the resting state.

The schematic on the right panel of the same figure describes how a voltage pulse (square signal) propagates at each discrete element of the membrane. If we imagine to insert a measuring electrode at the position of the capacitor (note that in reality the membrane is continuous, i.e. the "capacitor" and the "resistance" are spread all over its length), we would measure a growing ramp of the voltage (see Eq. (7.81) in the Appendix G), followed by the exponential decrease, as it is the case for an ordinary RC circuit. The amplitude of the pulse is as well decreasing, as soon as the pulse propagate over longer distances.

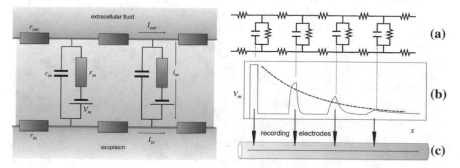

Fig. 7.8 *Left* Schematic of the equivalent circuit of the element of neuronal membrane, idealised as an electric cable. *Right* Arrangement of several elements in series (**a**), receiving a depolarising input in the form of, e.g., a square signal, which goes attenuated through successive RC membrane elements (**b**). The recording with electrodes placed at regular positions along the axon (**c**) would measure an exponential decay of the current (average *red-dashed curve*)

The cable equation

We start with considering the neuronal axon as an infinite cable, which in practice means very long compared to the region we are studying. We are looking for current-voltage equations that describe the **passive** response of the membrane. Since the resting potential is constant, we can make the substitution:

$$\frac{dV'}{dx} = \frac{dV_m}{dx} \quad \text{and} \quad \frac{dV'}{dt} = \frac{dV_m}{dt}$$

In the following, uppercase letters R, I, C indicate the integrated values of resistance (in ohms, Ω), current (in amperes, A), and capacity (in farads, F); lowercase letters r, i, c indicate values per unit length. The membrane resistance is a special case, since it represents the resistance of a transverse section of the cable per unit length, therefore r_m has dimensions of $[\Omega][L^2][L^{-1}] = [\Omega][L]$.

To derive the cable equation, let us begin by supposing that the membrane resistance is infinite, $R_m = \infty$ and its capacity is $C_m = 0$. This means that the current flows inside the axon without any loss, like water in a garden hose. Moreover, let us also consider $R_o \simeq 0$.

The variation of potential along the direction x for an amount of charge injected a a point is:

$$dV_m = -I_i r_i dx \tag{7.22}$$

that is:

$$\frac{1}{r_i}\frac{dV_m}{dx} = -I_i \tag{7.23}$$

This expression describes the decreasing amplitude of the current along the cable, with the only loss (difference of potential between x and $x + dx$) coming from the resistance r_i. By pushing the analogy with the garden hose, is like injecting a small amount of water at one end, and observing how the friction on the hose walls slows down the flow. Now, if we remove the hypothesis that $r_m = \infty$ but consider that the resistance has a finite value, it is the analog of opening holes in the hose, from which the water can be lost. The more holes, the less water will arrive at the other end. In other words, the membrane resistance represent a dissipative contribution to the overall current, which can be written as $dI_i = -i_m dx$, or:

$$\frac{dI_i}{dx} = -i_m \tag{7.24}$$

The membrane current can be decomposed into a resistive and a capacitive contribution, $i_m = i_r + i_c$. The first component is just obtained from the Ohm's law, $i_r = V_m/r_m$. The capacitive component, also called displacement current, exists only until the "capacitor" is fully charged, and is:

$$i_c = c_m \frac{dV_m}{dt} \tag{7.25}$$

Then we obtain the following expression for the axoplasmic current, including also the losses to the membrane:

$$\frac{dI_i}{dx} = -(i_r + i_c) = -\left(\frac{V_m}{r_m} + c_m \frac{dV_m}{dt}\right) \tag{7.26}$$

Now we can use the (7.23) for I_i, and obtain:

$$\frac{dI_i}{dx} = -\frac{1}{r_i}\frac{d}{dx}\left(\frac{dV_m}{dx}\right) \tag{7.27}$$

By equating the last two expressions (partial derivatives are in order, since we consider the simultaneous variation of current with time and position) the complete **cable equation** reads:

$$\frac{1}{r_i}\frac{\partial^2 V_m}{\partial x^2} = \frac{V_m}{r_m} + c_m\frac{dV_m}{dt} \tag{7.28}$$

The **steady-state** version of the equation lends itself to a simple study, which gives additional insight:

$$V_m = \frac{r_m}{r_i}\frac{d^2 V_m}{dx^2} \tag{7.29}$$

whose general solution is:

$$V_m(x) = A\exp(-x/\lambda) + B\exp(x/\lambda) \tag{7.30}$$

with $\lambda^2 = r_i/r_m$. Note that $r_m = R_m/\pi d$ and $r_i = (R_i/\pi(d/2)^2)$, which makes λ proportional to the diameter of the axon. For practical solutions, it is easier to introduce the scaled length $X = x/\lambda$, and the *electrotonic length* of the cable $L = l/\lambda$.

For the **infinite cable**, the solution must be finite at $X = \pm\infty$, therefore the solution becomes $V_m(X) = V_0\exp(-|X|)$. Alternatively, the current at $X = 0$ can be specified, $-I_0 =$ from which $V_m(X) = \lambda r_i I_0\exp(-|X|)$. Then, the axon **input resistance** R_n is the ratio of the input voltage to the input current:

$$R_\infty = \lambda r_i = \sqrt{\frac{dR_m}{4R_l}\frac{4R_i}{\pi d^2}} = \frac{2\sqrt{R_i R_m}}{\pi d^{3/2}} \tag{7.31}$$

and the input conductance is $G_\infty = 1/R_\infty = (r_i r_m)^{-1/2} = (r_i\lambda)^{-1}$.

For a **semi-infinite cable**, the potential must go to zero at $X = \infty$, therefore $B = 0$ and $V_m(X) = V_0\exp(-X)$, with . With a little algebra, it can be also shown that the input resistance for the semi-infinite cable is twice that of the infinite cable (that is intuitively obvious, since the same amount of current is split in the two directions, therefore each half of the infinite cable sees half of the current).

For a **finite length cable** $X \in [0, L]$, additional boundary conditions must be imposed at the L-end. In this case a special solution is:

$$V_m(x) = V_0\frac{\cosh(L - X) + B\sinh(L - X)}{\cosh(L) + B\sinh(L)} \tag{7.32}$$

Three cases must be distinguished:
(a) the *sealed-end* axon, e.g. terminating at a soma (cell body). In this case $B = 0$, $V_0 = -r_i I_0\coth(L)$, and the input resistance is $R_L = R_\infty\coth(L)$.
(b) the *open-ended* axon (or "killed end"), terminating at a point of zero resistance (experimentally, this means grounding the potential at L: $V_m(L) = 0$). In the equation, this corresponds to $B = \infty$, $V_0 = -r_i I_0\tanh(L)$, and the input resistance becomes $R_L = R_\infty\tanh(L)$.
(c) the *clamped-end* axon, terminating at a point polarised at some finite voltage $V_m(L) = V_L$. In this case the solution contains both terms, and the behaviour of the voltage (e.g., if $V_L > V_0$) can also be non-monotonic, despite the fact that the membrane/cable is supposed passive.

The normalised input resistance is always greater for the sealed condition than for the open-ended, because in the former case the current is prevented from escaping the cable. Problems 7.3–7.5 provide examples of solutions of the cable equation in special cases.

The equation describing quantitatively this type of response is the *cable equation*, derived in the greybox on p. 272. (It may be noticed that the mathematical structure of this equation is just the same as the heat equation, described at the end of Chap. 4, or the diffusion equation, described in Chap. 5.) The cable equation had been derived at the end of 19th century by Lord Kelvin, in a study of submarine cables, and between 1905–1920 it was applied with success by physiologists like L. Hermann, M. Cremer, K.S. Cole, A. Hodgkin and others, to study the propagation of ionic currents in the nerve axon.

The meaning of the equation can be better appreciated by introducing two scales or time and length, respectively. If we multiply both sides of Eq. (7.28) by r_m, we get:

$$\lambda^2 \frac{\partial^2 V_m}{\partial x^2} = V_m + \tau \frac{d V_m}{dt} \tag{7.33}$$

The factor $\lambda = (r_m/r_i)^{1/2}$ is the length constant, and despite being the ratio between two resistances it has dimensions of length, since r_m is $[\Omega][L]$, and r_i is $[\Omega][L^{-1}]$. The meaning of λ is that a membrane resistance much larger than the axoplasmic (internal fluid) resistance allows the impulse to travel to a longer distance or, conversely, to a shorter one if the membrane resistance is decreased (a leaky membrane). With the typical values of r_m and r_i, for a myelin-sheated axon of diameter 50 μm, it is $\lambda \sim 2 - 3$ mm, whereas for non-myelinated axon it is of the order of 30–50 μm. It is also important to note that, according to its special definition, the value of r_m is proportional to the area of the axon cross section (circular, for a simple cylindrical shape approximation), therefore λ^2 is proportional to the axon diameter.

The factor $\tau = r_m c_m$ is just the time constant of the equivalent circuit element of membrane, expressing the time during which the current charging is not negligible, i.e. the duration of the transitory current regime. For times much longer than the charging time τ, a steady state condition is reached and the time-independent cable equation (7.29) can be used.

The cable equation is, in fact, too simple to be able to describe all the complex transport phenomena in neuronal cells, because of a number of incorrect assumptions (besides considering the membrane as purely passive):

i. The resistance R_m or r_m is taken to be independent on the voltage V_m, when experiments instead show that many ion channels found along the axon membrane are gated (opened/closed) by the variations of electric potential (voltage-gated channels).
ii. The diameter of the axon is considered constant along the whole length, while it tapers as far as it gets away from the soma, the central body of the neuron.
iii. The membrane currents are considered to be in the linear regime, while in reality synaptic currents are not just algebraically summed to one another.

Nevertheless, this equation qualitatively gives a realistic impression of the phenomena accompanying the passage of a voltage or current pulse along the axon.

7.4 Excitation of the Neurons

Fluctuations in the membrane potential can be classified according to their character. One scheme, proposed in 1959 by Th. H. Bullock and still in use today (Fig. 7.9), distinguishes between the resting potential value, and the variations about this value due to different types of neuronal activity. The most important are:

- **pacemaker potential**, induced by the intrinsic activity of the cell, they are voltage fluctuations not generated by an external stimulation;
- **transducer** or **receptor potential**, a transmembrane voltage difference caused by external factors, such as chemical neurotransmitters released in the neuron, or a change of the synaptic potential, which can be either excitatory or inhibitory (see below);
- **action potential** (the actual nerve impulse), which follows an "all-or-nothing" response: following a sequence of receptor potential pulses the neuron response level can increase, but until the summed potential remains below a threshold the response does not propagate (local, or electrotonic potential); however, if the threshold is exceeded, a voltage spike is launched.

The most noticeable electrical response of the neuron is certainly the action potential, whose form results from a complex sequence of molecular-scale events. In a rapid sequence, the membrane is *depolarised* (the potential V_m increases to values way above V_{rest}), and subsequently *repolarised* after the passage of the potential spike, going back to V_{rest} and even exceeding it on the negative side (*hyperpolarisation*), until it comes back to the resting condition. The time duration of the passage of an

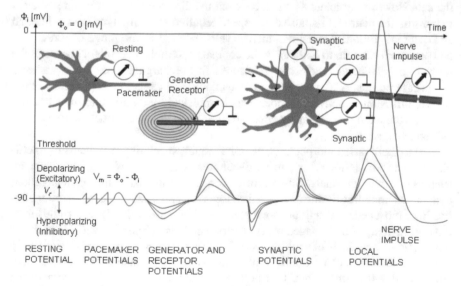

Fig. 7.9 The Bullock's classification of the different neuron potentials. [From Ref. [5], repr. w. permission]

action potential is in the order of a few milliseconds, and it propagates all along the neuron with a constant amplitude.

It should also be noted that there exist some neurons which are instead activated by a **graded potential**, i.e. a voltage ramp that gradually brings the membrane V_m to a depolarised state; such neurons are usually capable of responding themselves with a graded impulse to the graded stimulation.

As it was seen in the previous section, a voltage pulse launched at one extremity of the neuron axon decays very shortly, over a distance of a few λ's (at most a few mm). How it is then possible that a nervous impulse can travel over meters of distance from the brain to the limbs?

The answer, which will be explained in more details in the following, is that the propagation of the electric impulse is constantly relayed as it travels along the axon, by the "battery" provided by the ion concentration gradients accumulated during the resting state. This also implies that after the passage of the impulse, the gradients must be restored, i.e. the battery has to be constantly recharged. The ionic pumps work to reestablish the ionic equilibrium, after the re-permeabilisation of the membrane (opening and closing of the different channels), by consuming the energy supplied by ATP molecules. A cortical neuron in human brain utilises up to ~4.7 billion ATP molecules per second, to power also various other biological functions, such as synapse assembly and synaptic transmission (for a review see [6]).

This observation raises the correlated problem of how efficiently ATP can be transported along the neuronal axon, which can be up to many tenths of cm in length. As it was shown in Chap. 6, diffusion alone cannot be the way to overcome long distances, and specialised molecular motors much improve the situation, by efficiently transporting vesicles, liposomes etc., along the microtubule structure of the cell. However, efficient as it can be, even the directed transport is not sufficient to transmit the neuronal impulse at the speed required for animal reaction to internal and external stimuli. Axonal conduction velocities are in the range of a few m/s, and can reach even up to 150 m/s, to be compared to molecular motor speeds in the range of 0.5–10 μm/s. Therefore, to ensure a fast recharging of the ionic gradients by ATPase pumps, ATP must be readily available and cannot be supplied by either passive diffusion or active transport. This is achieved by distributing mitochondria all along the axon, as well as by concentrating them at some particular spots, such as near synapses.

Due to their varied morphology, neurons require specialised mechanisms to efficiently distribute mitochondria to far distal areas where energy is in high demand, such as synaptic terminals, active growth cones, and axonal branches, undergoing dynamic remodelling during neuronal development, and in response to synaptic activity. Since most neuronal cells are not replenished but will survive for the whole life of the organism, aged or damaged mitochondria need also to be removed. Upon direct observation, about 30 % of neuronal mitochondria are found to move bi-directionally over long distances, pause briefly, move again, frequently change their direction, at speeds of 0.3–0.9 μm/s. Their transport is actuated mainly by KIF5 motors, of the kinesin-1 family, along the microtubule structure following the axon length (see Sect. 6.3 for a discussion of the biophysics behind this process).

Fig. 7.10 Types of inhibitory (*1*) and excitatory (*2, 3, 4*) stimuli: in the plot above the idealized (step) stimulus current, I_s, is reported; in the plot below, the corresponding potential (mV) elicited in the neuron. The current stimulus *2* is excitatory and subthreshold, therefore only a passive neuron response is produced. For the excitatory stimulus *3*, threshold is barely reached: the membrane is marginally activated (potential 3b), or just a local response is produced (potential 3a). For any above-threshold stimulus *4*, a nerve impulse, or action potential, is invariably initiated. [Adapted from Ref. [5], w. permission]

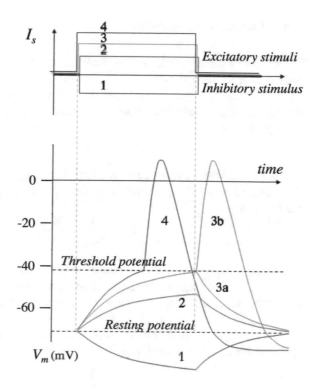

The neuronal potential can be **excitatory** or **inhibitory** (Fig. 7.10). The former depolarises the membrane, since it is formed by a variation of V_m in the positive direction with respect to the outside, thereby increasing the potential towards positive values (or less negative, starting from about -75 mV). The latter, by contrast, is hyperpolarising the membrane, as it is characterised by an opposite variation of V_m towards the (even more) negative values.

According to Bullock's classification, if a transmembrane stimulus is insufficient to bring V_m to the threshold, the response to this stimulus will be passive and remain localised. If on the other hand the height of the pulse is sufficient to reach, or even pass the threshold, the whole neuron launches the action potential in response. Once activated, the pulse shape of this potential is constant, independently on the actual intensity of the stimulus, be it just at, or largely above the threshold. This is called a "all-or-nothing" response. Moreover, if an inhibitory stimulus arrives at about the same time as an excitatory stimulus, the result is that the threshold to launch the action potential is raised to a higher level.

7.5   The Action Potential

In the passive-cable model, each local element of the neuron membrane is approximated as a RC circuit element (see Appendix F). The current I_s, represented in Fig. 7.10 as a simple square step, meets the membrane resistance R_m and must charge the capacity C_m; thereby the resulting membrane potential grows from its (negative) resting value as:

$$V_m = V_{rest} + I_s R_m \left(1 - e^{-t/\tau}\right) \tag{7.34}$$

that is the exponential voltage ramps shown in the figure for the responses 2 or 3, if the threshold value V_{tr} is not met. The corresponding amount of current needed to get to V_m is:

$$I_s = \frac{V_m - V_{rest}}{R_m \left(1 - e^{-t/\tau}\right)} \tag{7.35}$$

From this expression, the minimum current needed to get to the threshold V_{tr} and generate an action potential, called the **rheobasic current**, is obtained by considering that at this value the stimulus must last an amount of time $t \to \infty$, therefore $I_{rh} = (V_{tr} - V_{rest})/R_m$.

The time t_c to get to threshold with a stimulus twice higher than the rheobasic is called **chronaxy**. Since the time to V_m is:

$$t = -\tau \ln \left(1 - \frac{V_m - V_{rest}}{2 I_s R_m}\right) \tag{7.36}$$

by substituting I_{tr} for I_s, and V_{tr} for V_m, the chronaxy time is $t_c = \tau \ln 2$, a simple function of the time constant of the equivalent RC circuit. It should be noted that the RC-analogy is a reasonable approximation for sub-threshold response, but it becomes quickly inappropriate when approaching and surpassing V_{tr}: in that region the membrane response becomes strongly non-linear, and moreover the phenomena of *accommodation* makes the threshold voltage to change with time. However, the experimentally measured response curves allow to deduce values of the rheobase and chronaxy, which characterise the tissue response quite precisely. Typical values of chronaxy are in the range 0.1–0.7 ms for most animal muscles, around 1–3 ms for receptors located in the tongue or the retina, around 3–10 ms for involuntary muscles such as heart or stomach.

7.5.1   The Hodgkin-Huxley Model of the Membrane

Let us now look at the microscopic mechanisms elicited by the transitory currents in the neuronal membrane. Once the action potential is fired, it travels along the axon. As it was hinted above, at the passage of the voltage pulse a number of molecular-level actions follow, to relay the pulse in the forward direction, as well as to recharge the battery immediately after the passage, to prepare for a subsequent pulse. Following the arrival of the action potential at a given spot on the axon membrane, some voltage-gated sodium channels get open, while some potassium channels close. This increases the sodium conductance G_{Na^+} above the potassium conductance G_{K^+}, thus changing the effective resistance and capacity values of the local membrane element.

By considering the different types of ionic channels to act independently as current generators (see the Fig. 7.7 on p. 269 for reference), the time-dependent current equation for the local membrane element can be approximated as:

$$I_m = c_m \frac{dV_m}{dt} + G_{Na^+}(V_m - V_{Na^+}) + G_{K^+}(V_m - V_{K^+}) + G_l(V_m - V_l) \quad (7.37)$$

with each of the ionic reference potentials given by the respective Nernst equation (7.8). This equation was derived and used by Hodgkin and Huxley in their 1952 theoretical-experimental studies on the giant squid axon, which brought them the Nobel prize for medicine in 1963. The last term, labelled with an 'l', represents membrane leakage currents from ions other than Na^+ and K^+, in practice mostly Cl^- but generically any other charged species (we can take the Cl^- potential for V_l; however, the cell concentration of chloride is usually very low, therefore small Cl^- flux are enough to ensure equilibrium, and its role is negligible in establishing the membrane potential).

Whereas the cable equation (7.28) describes the ideal behaviour of an infinite or semi-infinite axon, in which the membrane response changes both in time *and* in space, this equation rather describes the local behaviour of a portion of membrane swept by the passage of a voltage spike. Correspondingly, the different terms of the time-but-not-space-dependent equation (7.37) can contribute to the membrane current I_m: the capacitive charging, the sodium or potassium current, or the leakage term, as a function of the respective (and time-dependent) local values of capacitance and conductances. The interest of this local description can be fully understood by using a technique to study locally the time-dependent behaviour of the current at constant voltage, like the **voltage-clamp** experiment.[2] This method was used also by Hodgkin and Huxley, in the footsteps of K. Cole, who had laid the foundations of the technique.

When the transmembrane current is measured by the voltage-clamp method, schematically shown in the left panel of Fig. 7.11, the typical response curves shown on the right of the figure are obtained. At the very beginning, a small spike of capac-

---

[2]In 1947 George Marmont, at the Marine Biological Laboratory in Woods Hole, Massachusetts, had invented a technique to inject current in the entire length of the squid axon at a constant (but uncontrolled) potential, by inserting a long electrode along its axis; it seems that he did a vast number of experiments with this apparatus but, being unable to sort out any theoretical explanations for his results, he never published them and abandoned the field. Kenneth Cole shared Marmont's lab, and modified the technique to obtain control of the voltage instead of the current. He used a second wire electrode to set the voltage to a chosen value, and used the current-injecting electrode as a feedback to keep voltage constant. In this way, the amount of injected electron current gave a measure of the opposite ionic currents produced by the neuron in response to the constant potential.

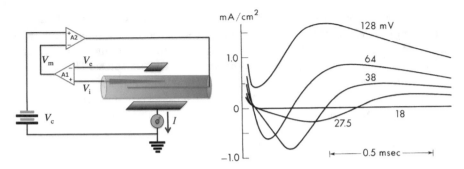

Fig. 7.11 *Left* Schematic of the voltage-clamp apparatus. V_i is the potential measured in the squid axon; V_e is the extracellular potential; the operational element A1 computes the transmembrane potential $V_m = V_i - V_e$, and sends the input to A2; there, V_m is compared to the desired "clamping" potential V_c and, if different, a current is sent to the wire electrode in the axon, to correct until $V_m = V_c$. The required current correction I is detected (*red*), and gives a measure of the (opposite) ionic current generated by the axon in response to V_c. *Right* K. Cole's original voltage clamp records in 1947, showing current density measurements in the squid axone after changing the membrane potential above V_{rest} by +18 to +128 mV. Cole identified the way to obtain a continuous change of potential, getting around the threshold V_{tr} (see footnote); however, he supposed potassium ions to be the only responsible for the shape of the curves, showing a inward current at the beginning (negative branch of the curves, disappearing at the highest potential values) and an outward current at the end (positive branch). It was up to the work of Hodgkin and Huxley to discover the competition between outflowing K^+ and inflowing Na^+ ions, leading to the correct explanation of the action potential. [Image *right* from Ref. [7], repr. w. permission]

itive current is generated by the initial voltage step from V_{rest} to $V_{rest} + V_c$, but it is immediately eliminated since the capacitive current is given by the time derivative dV/dt, which is zero when the potential is kept constant. Besides that, the current plot shows a first branch descending to negative values, meaning that there is an inflow of current inside the cell; then, the current starts to smoothly increase to more positive values, signalling an outflow of current (since only positive ions, K^+ and Na^+ are implicated), before starting to die off back to zero.

The correct interpretation of this behaviour, given by Hodgkin and Huxley, is that at the beginning Na^+ ions start flowing inside the cell, following the opening of a number of voltage-gated sodium channels (i.e., the sodium conductance is increased); the subsequent outflow is instead due to the retarded response of the potassium voltage-gated channels, whose opening at a slightly later time coupled to the closing back of many sodium channels, brings up the current; when the potential V_c is switched off, the ion concentrations regain their equilibrium values. The K^+, ion channels open with a slow time constant varying between about 2–1 ms, when V_c spans 50–100 mV. For Na^+, instead, two types of ion channels exist: voltage-gated, with a much faster time constant of about 0.15 ms, nearly independent on the value of applied voltage; and ligand-gated, with a time constant comparable to that of potassium opening. These latter are found only in the neuromuscular junction, and display some permeability also to K^+.

Maxwell's equations and extracellular potentials

Recording the membrane potential and currents generated by cells in the brain and in the heart is very difficult, especially in clinical and in vivo conditions. Therefore, in most situations it is an **extracellular potential** that is recorded. Maxwell's equations (summarised in the Appendix G) can be used to derive a relationship between the extracellular potentials measured, and membrane currents circulating in the neural or heart cells.

The mechanism by which an electric field may arise in biological tissue is that of a current source embedded in a large conductive bath of fluid, called a *volume conductor*. If a small current I is assumed to emanate from a point in a uniform conducting medium, of infinite extent and average conductivity σ, the current will flow radially in all directions. The vector flux $\mathbf{J}$ through a small sphere of radius r concentric with the source is:

$$\mathbf{J} = \frac{I}{4\pi r^2}\mathbf{n} \tag{7.38}$$

with $\mathbf{n}$ the unit vector normal to the spherical surface.

The propagation velocity of electromagnetic signals in the biological tissue is very much slower than the speed of light in vacuum, therefore it may be safely assumed that the currents are quasi-static, i.e., what is measured at a distance even quite far from the source represents the actual current at that instant. The electric field is therefore the gradient of the electric potential, $\mathbf{E} = -\nabla\Phi(r)$, and the more general form of Ohm's law (7.12) becomes:

$$\mathbf{J} = -\sigma\nabla\Phi \tag{7.39}$$

By equating the two previous expressions for $\mathbf{J}$, and after observing that for a point source only the radial components of the gradient are non-zero, it is:

$$\frac{I}{4\pi r^2} = -\sigma\frac{d\Phi}{dr} \tag{7.40}$$

The potential $\Phi(R)$ as seen by an electrode placed outside the cell, at a distance R from the source, is obtained by integrating the last equation

$$\Phi(R) = \frac{I}{4\pi R\sigma} \tag{7.41}$$

If more than one source is present around the electrode, the respective currents can be summed. It is assumed that each current is generated within some small volume dw', centred at the point R', and has an incremental effect $d\Phi$ on the total potential at R, as:

$$d\Phi(R) = \frac{I'dw'}{4\pi\overline{\overline{R}}\sigma} \tag{7.42}$$

where $\overline{R} = |\mathbf{R'}-\mathbf{R}| = \sqrt{(R'_x - R_x)^2 + (R'_y - R_y)^2 + (R'_y - R_y)^2}$ is the distance between the electrode at R, and each source at R'.

Then the total potential recorded at R is obtained by integrating over the entire volume:

$$\Phi(R) = \int_w \frac{I'(\overline{R})}{4\pi\overline{R}\sigma}dw' \tag{7.43}$$

From the Eq. (7.30), we know that the amplitude of quasi-stationary currents decays exponentially over the distance, $I(\overline{R}) = I_0 \exp(-|R' - R|/\lambda)$, with $\lambda \sim 2$–3 mm.

7.6   Transmission of the Nerve Impulse

The German physiologist Ludimar Hermann had correctly identified, already at the beginning of the 20th century [8], that the impulse should propagate in the nerve fibre without attenuation. He even went on to propose that the difference of potential between the excited and resting regions should generate small local currents, today called *local circuit currents*, to propagate the excitation to nearby regions. Although the excitatory input can be collected both in the dendrites and in the soma, the action potential starts always from the latter, from where it then starts propagating along the axon. Notably, if the excitatory impulse is artificially launched midway in the axon, the propagation goes in both directions.

One important physical property of the neuronal membrane is the change of sodium conductivity at the passage of the voltage pulse. The early beginning of the action potential increases the local circuit currents, and increases sodium conductance. This makes for a higher sodium current, and a steeper potential rise. The excitation is faster, and so is the propagation speed of the impulse.

The axon can be wrapped in a myelin sheath (constituted by the so-called Schwann cells), which makes a dielectric lipid layer acting like the insulator around a coaxial cable (Fig. 7.12). This insulating structure is interrupted at regular intervals, at the Ranvier nodes. These are gaps in the myelin cover, of width about 1–2 μm and spaced by a shielded length much longer, up to a few mm. Being uninsulated, the Ranvier nodes are the only places where the electric current can develop (also because of a high concentration of ion channels), and pass directly from one node to the next in the process called **saltatory conduction**. The effective capacitance per unit length of the myelinated axon is much smaller than that of the naked axon, thereby largely increasing the propagation speed of the action potential. Myelin, discovered in 1854 by the German physiologist Rudolf Virchow, is typical of the vertebrates, but is found also in some invertebrates, albeit with different characteristics. Neurons in the brain are generally not myelinated, since the impulse travels a short distance, whereas both motor and sensory neurons in the central and peripheral nervous system are myelinated (although some exceptions exist).

An empirical expression obtained by Muller and Markin [9] for the propagation speed in the myelinated axon is the following:

$$v = \sqrt{\frac{i_{Na^+}}{r_i c_m^2 V_{tr}}} \qquad (7.44)$$

The resistance per unit length of the axoplasm is inversely proportional to the cross-section area, therefore to the squared diameter, $r_i \propto \delta^{-2}$; the membrane capacitance per unit length, in turn is proportional to the diameter, $c_m \propto \delta$. At the threshold the membrane current is practically due to the sodium only, plus the capacitive component; therefore, the velocity is inversely proportional to $r_m c_m$, or $v \propto \tau^{-1}$, which is not unreasonable. This also implies that the velocity in the myelinated axon is propor-

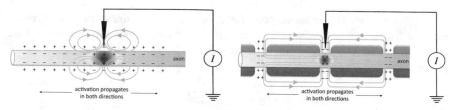

Fig. 7.12 Schematic of the unmyelinated (*left*), and myelinated axon (*right*), with its myelin sheath (*orange*) interrupted at the Ranvier nodes. A current artificially injected by an electrode midway along the axon will propagate in both directions. However, for the unmyelinated axon the active region (*red*) propagates continuously along the length, whereas for the myelinated axon the current jumps from one node to the next ("saltatory" conduction)

tional to the axon diameter, as actually observed experimentally with the approximate empirical law $v \simeq 7 \times 10^6 \delta$ m/s.

The propagation speed depends primarily on the shape and size of the axon. In principle, the membrane capacitance per unit length determines the amount of charge needed to build up a given potential pulse, and therefore it affects the time to threshold. Increasing the capacity increases the time constant, and reduces the speed. Moreover, also the changing conductance of the electrolytic solution inside and outside the membrane has a similar effect. Being also inversely proportional to the time constant, the lower the resistance the shorter the τ, and therefore the faster the propagation speed. Also temperature plays a secondary role, a lowered temperature adjusting the propagation speed to proportionally lower values because of a reduced sodium conductance.

7.6.1 Wave-Like Propagation of the Impulse

The original form of the cable equation (7.28) describes the coupled space and time variation of the transmembrane current, at any given point along the axon:

$$\frac{\lambda^2}{r_m} \frac{\partial^2 V_m}{\partial x^2} = \frac{V_m}{r_m} + c_m \frac{\partial V}{\partial t} = I_m \qquad (7.45)$$

This is a complicated equation that can only be solved numerically on a computer, by specifying the appropriate values of the parameters and boundary conditions. If instead of the formal middle term of the equation, we take for the local membrane current the Hodgkin-Huxley model, Eq. (7.37), and equate the two expressions for I_m, we get:

$$\frac{\lambda^2}{r_m} \frac{\partial^2 V_m}{\partial x^2} = c_m \frac{\partial V_m}{\partial t} + G_{Na^+}(V_m - V_{Na^+}) + G_{K^+}(V_m - V_{K^+}) + G_l(V_m - V_l) \quad (7.46)$$

Let us imagine that the action potential propagates along the axon in a wave-like manner. Under steady-state condition, the amplitude and shape of the pulse wave are constant, and so are its phase and group velocity. Therefore, a wave equation for the potential should hold:

$$\frac{\partial^2 V_m}{\partial x^2} = \frac{1}{v^2} \frac{\partial^2 V_m}{\partial t^2} \tag{7.47}$$

If the isolated voltage pulse is considered as a wavepacket of width given by the time duration of the pulse, v is its propagation velocity. Now, by substituting the second-order spatial derivative with the second-order time derivative in the previous equation, it is:

$$\frac{\lambda^2}{r_m v^2} \frac{\partial^2 V_m}{\partial t^2} = c_m \frac{\partial V_m}{\partial t} +$$
$$+ G_{Na^+}(V_m - V_{Na^+}) + G_{K^+}(V_m - V_{K^+}) + G_l(V_m - V_l) \tag{7.48}$$

This second-order equation in time is easier to solve, and can give some insight in the features of the propagating voltage pulse. Firstly, we note that the equation is not changed if the velocity changes, while keeping constant the coefficient of the left member. Therefore, the propagation velocity can be expressed as:

$$v = \frac{\lambda}{\sqrt{r_m}} \tag{7.49}$$

Since both λ and r_m are proportional to the axon diameter δ, it turns out that the propagation velocity is proportional to the square-root of the diameter, $v \propto \sqrt{\delta}$. This is in agreement with several experimental results for the unmyelinated axon, at variance with the linear proportionality found for the myelinated axon.

Moreover, the Hodgkin-Huxley model describes correctly the shape of the voltage pulse. The results of a numerical solution of Eq. (7.48) are displayed in Fig. 7.13, left. At the very beginning, the current is mostly capacitive, since the conductances of both K^+ and Na^+ are very low. Then, G_{Na^+} starts increasing, faster than G_{K^+}; note that the threshold voltage coincides with the two conductances being equal (white arrow). The sodium conductance increases in correspondence of the rise of the voltage, and starts declining when $V_m + V_{rest} \sim V_{Na^+}$, about $t \sim 0.5$ ms. In the meantime, the potassium conductance has attained a higher value: potassium channels are opening, and K^+ ions are flowing out of the cell membrane, to balance the high fraction of Na^+ that entered. The maximum of the action potential coincides with the moment at which the declining sodium conductance and the growing potassium one are again equal (black arrow); at this moment, the capacitive current is zero, and the current is totally ionic. The potassium conductance has its maximum around the time sodium conductance is back to its resting state value, $t \sim 1$ ms, then it starts to decrease; right after, the action potential goes down to zero, and the stimulus of that portion of membrane is completed.

7.6.2 The Refractory Period and Orthodromic Conduction

The initial peak phase, covering the first ~ 1 ms of the $V(t)$ plot, is called the **absolute refractory period** of the neuron, since during this time the cell does not accept any further propagating stimulation, no matter how strong. However, it appears that after the stimulus has passed, the local voltage goes further negative, before going slowly back to a true zero. The Hodgkin-Huxley model explains also this peculiar feature. Due to the slower time constant of potassium channels, the excess of K^+ ions accumulated outside the membrane renders the cell interior even more negative, with V_m approaching now the more negative V_{K+}, i.e. hyperpolarising the membrane. After the potassium channels (slowly) close, G_{K+} goes back to its resting state; the Na/K-ATPase, and other pumps, can now restore the normal ionic gradients, by pushing out the Na^+ and taking back in the K^+ ions (see Fig. 7.5 on p. 261).

During the absolute refractory period, the value of V_{tr} is so high it appears virtually infinite. On the other hand, in the second part of the activation (from ~ 1 up to 3–4 ms) the cell could be activated, but with a stimulus much higher than threshold. In this phase, called **relative refractory period** and occurring after the maximum of conductance of K^+, the membrane sees an additional barrier to the current. The action potential thus generated will be shallower than the ordinary one.

Whether an excitable cell is activated or not largely depends on the intensity and duration of the stimulus. The membrane potential can attain the threshold V_{tr} by either a short and intense stimulus, or by a long but weaker one. The diagram illustrating such a phenomenon is the "force-duration" plot, for which we already defined the smallest current, the rheobase, and its characteristic time, the chronaxy. Moreover, such near- or sub-threshold stimulation can also be very long, or quickly repeated; the cell reacts to such peculiar stimulation by raising the value of V_{tr}, in what is called **accommodation**. In other conditions, the threshold can be lowered, if a hyperpolarising current is applied, the potential drops quickly and the threshold follows; once the hyperpolarisation terminates, there is a time during which the potential is back to resting value, but the threshold is still lower, therefore a new action potential can be fired, in a process called anode-break excitation.

The mechanism that allows the action potential to proceed in a well defined direction, towards the end of the axon and not to go back to the central soma of the cell, is the unidirectional, or **orthodromic conduction**. This is derived from the alternance on each patch of axon membrane of the polarisation-hyperpolarisation-depolarisation cycle. The normal depolarisation is sufficient to keep a patch of membrane at its resting value and accept an incoming stimulus, whereas the hyperpolarisation holds the membrane quite below the V_{rest}, and makes the membrane refractory to any incoming stimulus. This asymmetry permits the unidirectional propagation. By looking at the explicatory diagram in Fig. 7.13, right, it is seen that a moving action potential encounters on its track patches of membrane at V_{rest}, and leaves behind patches at $V_{hyp} < V_{rest}$; therefore, even if in principle the membrane is conducting in both directions, the equivalent resistance on the back of the propagation direction is quite

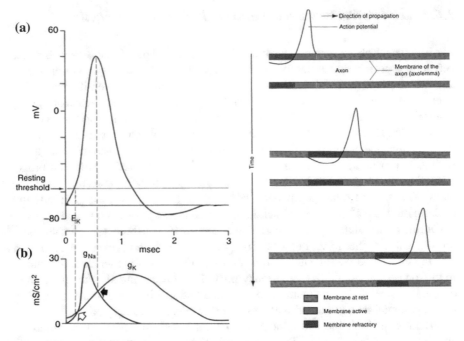

Fig. 7.13 *Left* The shape of the action-potential voltage pulse (**a**, *above*), and the time evolution of the Na$^+$ and K$^+$ membrane conductance (**b**, *below*) in voltage clamp. The *green curve*, calculated from the Hodgkin-Huxley model, follows very closely the experimentally measured shape of the action potential propagating along the squid axon. In the *bottom plot*, the *red* and *blue curves* are the result of the calculation, however no direct experimental counterpart is available for comparison; the *white* and *black arrows* indicate the two times at which sodium and potassium conductances are estimated to be identical. *Right* The mechanism of unidirectional, or 'orthodromic' propagation. The action potential, prepared by the slight rise of the local circuit currents, moves to the regions of resting potential ($V_m \sim V_{rest}$, *green*), always leaving behind a region of hyperpolarisation ($V_m \sim V_{K+}$, *red*) that is unfavourable to propagation. Note that the temporal shape of the pulse in the propagating direction is specular to the shape of the steady-state voltage-clamp. [Adapted from Ref. [5], w. permission]

higher than in the facing direction, for a time long enough for the impulse to advance always in the same direction.

The intensity of the current becomes lower as far as the end of the axon is reached, since the propagation speed is seen to decrease with the axon diameter, linearly with δ for the myelinated axon, and as $\delta^{1/2}$ for the non-myelinated one. To be noted that the thinner portions of the axon, below about 1 μm, are deprived of the myelin sheath (Fig. 7.14).

Fig. 7.14 Human brain anatomy from different sections (from above: axial, sagittal and coronal), detected by magnetic resonance imaging (MRI). [Image courtesy of www.MRImaster.com, London. Repr. w. permission.]

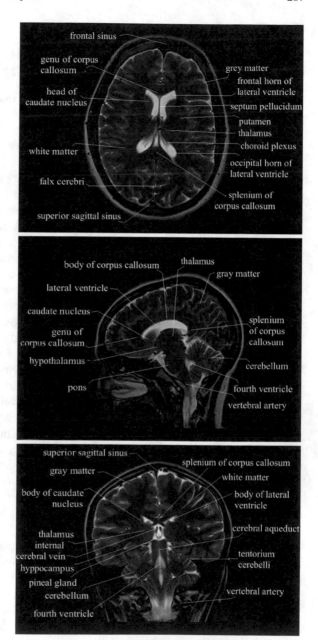

7.7   Brain, Synapses, Information

The human brain contains around 100 billion neurons and, by most estimates, somewhere between 10 and 50 times as many glial cells. Neurons receive, transform, transmit, and exchange information, to and from the brain, in the form of electrochemical impulses. Brain regions consisting of neuron bodies together with their associated dendrites and axon terminals, are termed *grey matter*.[3] The part of grey matter located on the surface of the brain is the cerebral cortex, while grey matter located deeper in the brain constitutes the so-called nuclei. Grey matter contains both the beginnings and endings of axons. On the other hand, regions consisting of the main body of neuronal axons gathered into bundles constitute the brain's *white matter*.[4] White matter is constituted by axons going relatively long distances, which make up the nervous pathways of the central and peripheral system. If white matter is cut, the cell body at one end of each axon is disconnected from its distal axon terminals at the other end. In some white matter regions, axons travel in parallel bundles, therefore all action potentials propagate in the same direction. For example, most axons in the dorsal columns are ascending, while those in the cortico-spinal tract are descending. However, in other white matter regions adjacent axons may carry signals in opposite directions, or be interwoven. For example, axons in the *corpus callosum* crisscross back and forth, interconnecting the frontal, parietal, temporal and occipital lobes of the two cerebral hemispheres.

If you ask a neurophysiologist how many different types of neurons there are in the brain, the answer will probably be hundreds, or thousands. While it is certainly true that neurons show vastly different morphological as well as functional differences, from the reductionist point of view of information processing they all should fall in just three major classes: neurons that collect and bring information to the brain, neurons that export information ("orders") from the brain to the organs, and neurons that just connect to other neurons (**interneurons**). The first ones are sensory neurons, like those found in the eye retina, ear, skin, etc.; the second ones are the motor neurons, transmitting actions to the muscles; however, the third class simply represents the vast majority of all the brain's neurons. In some sense they constitute the "computing network" and "memory storage" of the brain.

Collecting the information is the first step in this chain of events. Since interneurons are majority, it can be said that most neurons receive their input from other neurons. In general, the flow of information goes from the dendrites of one neuron, to the soma of the other, where it is processed, and the result of the information (the action potential) goes through the axon of this second to its dendrites, and so forth. Besides this general scheme, there are neurons which receive their input directly at the soma, and some even close to the axon terminal. No matter what is the site of

---

[3]In fact, 'grey' matter is such coloured only in dead tissue sections, normally grey matter is pale pink because of the numerous blood capillaries.

[4]The term 'white' comes from the fact that most axons, except the thinnest, are covered in myelin, and myelin (from Schwann's cells) is essentially fat, which appears white-coloured both in living and dead samples.

connection, however, all the relay of neuron-neuron information goes through the **synapse**, the real point of exchange of transmitted and received information.

Although the synapse is just a physiological entity (in fact, it is the terminal part of any of the dendrites, branching out of the main axon body), it is often treated as a separate cell component, with its own proper morphological, and electrochemical characteristics. Note that it is common to speak of 'synaptic current', while the action of the synapse (see below) is essentially to produce a variation in the membrane conductance. Up to date, we know two types of synapses: **electrical**, and **chemical**, these latter being the majority in superior animals.

In a simplified way, the arrival of the action potential from the axon in the chemical synapse region triggers the following series of events (Fig. 7.15):

  i. the action potential opens voltage-gated calcium channels in the terminal part of the axon membrane (the pre-synaptic terminal), where small vesicles (about 35 nm in diameter) containing a fixed amount of neurotransmitter molecules are found, bind to the terminal membrane via calcium-sensitive proteins;

 ii. upon opening of the ion channels, Ca^{2+} ions, present in substantial amount in the extracellular liquid, can flow inside the pre-synaptic terminal;

iii. Ca^{2+} ions bind to a protein present on the surface of the vesicles, the synaptotgmin, triggering a proteic interaction chain that ultimately (time delay of about 100–200 μs) leads the vesicle membrane to merge with the cell membrane;

 iv. the vesicles are opened towards the extracellular space (the synaptic cleft), and there liberate by exocytosis their content of neurotransmitter molecules;

  v. some of the neurotransmitter molecules can traverse the narrow synaptic cleft (about 20 nm, in ~0.6 μs), and reach the membrane of the adjacent neuron (the post-synaptic terminal), where they can bind to specialised receptors;

 vi. the receptors modify their conformation, thus inducing a direct, or indirect opening of nearby ion channels, from which other ions (Na^+, K^+, Cl^-) can enter the post-synaptic terminal, by following their respective concentration gradient.

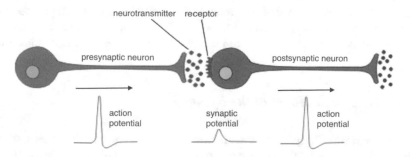

Fig. 7.15 Each neuron transmits and receives informations to/from other neurons via one or more synapses. The action potential from the pre-synaptic neuron allows a neurotransmitter to be released from the synapse, into the synaptic cleft. The neurotransmitter is captured by the receptors on the post-synaptic neuron, which may fire a second action potential in response

At this point, the synapse action can be considered completed. The new ionic concentrations, also coming from other synapses, will sum up, and determine an electrochemical response, possibly ending with the post-synaptic neuron firing an excitatory, or inhibitory action potential. The lifetime of the ligand-receptor bound complex is rather short, but lasts enough to allow the opening of the ion channels, within a time of 1–2 ms, the membrane impedance adding another 2–3 ms of delay.

The fact that the neurotransmitter vesicles are found only in the pre-synaptic terminal (at the end of the dendrite), and the receptors are found only at the post-synaptic terminal, makes the information flow in the chemical synapse unidirectional. Moreover, the intensity of the synaptic response is highly variable, and depends only in part on the amount of neurotransmitter released across the synaptic cleft. It was proved already in 1954 [10] that each vesicle contains a constant, fixed amount of neurotransmitter molecules, therefore the amount of neurotransmitter is proportional to the number of vesicles.

Among the typical neurotransmitters, we find the biogenic amines (dopamine, histamine, acetylcholine or ACh, mostly affecting Na^+ and K^+ channels [11]), some amino acids (dynorphin, glutamate, gamma-aminobutyric acid or GABA), nucleotides like the adenosine, neuropeptides like somatostatin, and even some gases, like NO or CO. It should be noted that ACh and GABA are synthesized on site, directly in the terminal part of the axon, but the enzymes necessary to their synthesis are produced in the cell soma, and are transported down to the axon terminals.

Compared to chemical synapses, electrical synapses conduct nerve impulses faster, with a typical delay time of the order of 0.2 ms compared to about 2 ms in chemical synapses. But, unlike chemical synapses, they do not amplify the signal: the signal in the post-synaptic neuron is the same or smaller than that of the originating neuron. Electrical synapses are often found in neural systems that require the fastest possible response, such as defensive reflexes. An important characteristic of electrical synapses is that they are bidirectional, allowing impulse transmission in either direction. The relative speed of electrical synapses also allows for many neurons to fire synchronously, although with less plastic and more simple behaviour. This can be seen as a revival of the "reticular" model of Golgi, which proposed that the nervous systems is organised as a network of interconnected cells, as opposed to the (today widely accepted) model of individual neuron-to-neuron communication.

7.7.1   Electrical Model of the Synapse

What happens when the post-synaptic membrane becomes permeable to ions, such as sodium and potassium? The change in the relative concentrations is going to modify the membrane potential of the relaxed neuron, from its V_{rest} value, close to the Nernst potential of potassium, to some new value. The influx of ions makes a potential burst, positive if excitatory, negative if inhibitory, called the **synaptic potential** (see again Fig. 7.15). This potential pulse approaches some weighted average of the equilibrium potentials, and may be sufficient to fire an action potential in the post-synaptic neuron.

By applying an equal and opposite polarisation to the membrane, by means of an external electrode, it is possible to measure the value of such threshold synaptic potential, the **inversion potential**, and to a good approximation it is found that:

$$V_{inv} = \frac{V_{Na^+} + V_{K^+}}{2} \qquad (7.50)$$

For example, in the neuromuscular junction of skeletal muscles, $V_{inv} \simeq -15$ mV, that is quite close to the average of sodium and potassium. Several synaptic potentials coming from different synapses to a same neuron can be summed (see below), and in this case their summed potential may reach the threshold V_{tr} to fire an action potential in the post-synaptic neuron.

The electrical behaviour of a synapse, considered as a circuit element put in series between two neurons, can be described as a modified membrane equivalent circuit, in which ACh, GABA, or some other neurotransmitter acts as a switch, increasing by some amount ΔG the conductance of some ionic species, as schematized in Fig. 7.16.[5] Of course, this model must be considered only as a highly simplified representation with discrete, localised elements, while in the neuronal and synapse membrane these variations are instead continuously distributed. In more specific terms, if the neurotransmitter acts mostly or only on Na^+ and/or K^+ ion channels, the response of the post-synaptic neuron potential will follow a positive, i.e. excitatory action potential; this is the case, for example, of ACh in neuromuscular junctions. If on the other hand the neurotransmitter acts on negative ion channels, typically Cl^-, the response will be a hyperpolarising, i.e. an inhibitory action potential. Correspondingly, excitatory or inhibitory synapses can be defined.

By looking for example at a neuromuscular synapse, after the shutting of the switch in the Figure, corresponding to the binding of the neurotransmitter at the post-synaptic terminal membrane, the currents are changed to:

$$\Delta I_{Na^+} = \Delta G_{Na^+}(V_m - V_{Na^+}) \qquad (7.51)$$

$$\Delta I_{K^+} = \Delta G_{K^+}(V_m - V_{K^+}) \qquad (7.52)$$

When V_m attains to the inversion voltage, V_{inv}, the two currents are equal in modulus and opposite in sign, $\Delta I_{Na^+} = -\Delta I_{K^+}$, since they are both from positively charged ions entering (Na^+) and exiting (K^+) the membrane. By equating the two expressions with $V_m = V_{inv}$, and rearranging the terms, the value of V_{inv} is obtained:

$$V_{inv} = \frac{V_{K^+} + q V_{Na^+}}{1 + q} \qquad (7.53)$$

with $q = \Delta G_{Na^+}/\Delta G_{K^+}$. If we assume that the binding of a given amount of ACh to the membrane receptors may lead to about the same amount of conductance increase

---

[5]Be careful not to confuse the electrical model of a synapse with an electrical synapse; the former is a mathematical model, the latter is a special type of neuron-to-neuron connection.

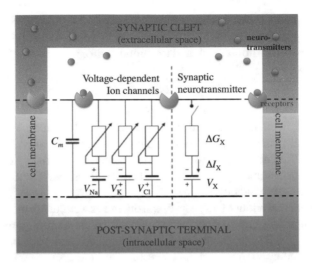

Fig. 7.16 Equivalent circuit element representing a synapse as an individual electrical component. Neurotransmitter molecules (*red spheres*) issued from the pre-synaptic terminal reach by passive diffusion the receptors (*purple*), on the post-synaptic terminal. The electric result of the ligand-receptor binding, is the opening of some ion channels (*left side*) on the post-synaptic terminal membrane. This can be represented as an additional membrane conductance for the type of ion X (*right side*), switched on/off by the arrival or the transmitter

for the two ions, it is $q = 1$, and V_{inv} is exactly given by the average of the respective Nernst potentials, Eq. (7.50).

The interplay between excitatory and inhibitory transmission has a critical role in the life-long process of creating, changing, and removing synaptic connections between neurons, called **brain plasticity**. Generally, each neuron ends with excitatory-only, or inhibitory-only synapses. Therefore, the combination of excitatory and inhibitory stimulation originates from the arrival at a same neuron of many different signals from different types of neurons. Excitatory synapses on excitatory neurons are localised to small protrusions, called **dendritic spines** (discovered by Cajal). Earlier studies have used dendritic spine dynamics to monitor excitatory synapses; however, until recently the lack of a morphological surrogate for inhibitory synapses had precluded their observation. Recent studies have found that inhibitory synapses are rather uniformly distributed, while excitatory synapses are more commonly found at the end of the dendrites. Moreover, inhibitory synapses are more dynamic, in that they are formed or removed in about half the time needed to form or remove excitatory synapses. However, once formed, both excitatory and inhibitory synapses may last very long. The way the brain combines the various inputs to form a stored information, or memory, is summarised in the experience-dependent plasticity: a highly orchestrated process, integrating the changes in excitatory connectivity, and the active elimination and formation of inhibitory synapses.

7.7.2 Treatment of the Neuronal Information

The action potential in the post-synaptic neuron is the result of a non-linear summation of all the synaptic potentials arriving at each single neuron, which can get
up to tens of thousands of synaptic contacts. The conductance variation ΔG_X of
the X-th ionic species makes ions to enter the cell soma; this active current must
induce a passive current to 'close the circuit' along the neuron membrane, including
the trigger zone at the beginning of the axon, from where the action potential could
be eventually fired. Excitatory synapses increase Na^+ concentration, thus displacing
V_m closer to V_{tr}. Inhibitory synapses, on the other hand, change the concentrations
of other ions, chiefly Cl^-, with the net result of a positive ionic current exiting the
membrane to keep charge equilibrium; if the neuron membrane potential is already
at $V_m \sim V_{rest}$, this current produces a hyperpolarisation of the membrane, in fact
a low-resistance path (called a "shunt" in electronics) where current escapes before
reaching the trigger zone.

The current summation can be **spatial** or **temporal** (Fig. 7.17). In the first case,
synaptic currents are summed in an almost linear way on the conductance, before
the action potential can be fired; in the second case, the synaptic potentials arrive in
a time shorter than their typical duration, i.e. with a high frequency, therefore their
crowded signals get summed on the membrane capacity. If the shape of the curves
$V_m(t)$ in the Figure is simply written as:

$$V_m = at \exp(-bt) \tag{7.54}$$

the spatial summation amounts to adding potentials which have roughly the same b
and different a values, i.e. they simply add different contributions to the conductance;
on the other hand, the temporal summation amounts to adding potentials having the
same a but different b, i.e. they add non-linearly on the RC constant of the membrane
capacity.

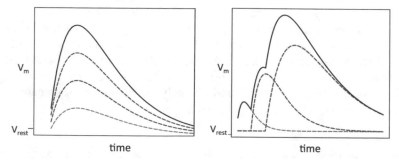

Fig. 7.17 Representation of the spatial summation and temporal summation of the synaptic potentials in the post-synaptic neuron membrane

After the arrival of the right amount of synaptic stimulation to code a given information (visual, auditive, neuromuscular, memory), the action potential is fired, with an amplitude comprised in the dynamic range of the neuron (see below). As a function of the signal amplitude, the response can increase either the signal frequency, or the number of synapses implicated in the downstream transmission, performing a more 'time-like' or 'space-like' encoding. On the next post-synaptic neuron, this signal would then be integrated time-wise or space-wise, respectively. By this cascade, neural signals are encored toward the central nervous system, up to the brain. There, deeper neurons receive the ensemble of excitatory and inhibitory signal from the various sources, for example producing a cerebral, or a neuromotor response, the detection of an image, or a sound coming from a remote source.

Information in the central nervous system is treated by **position-dependent codes**. For example, in the case of a sensorial stimulation, the code represents the position of the stimulus on a layer of sensory receptors: nerve axons under the skin are branched, with the branches from a same axon terminating on a group of neighbouring receptors; in the retina, position and other physical qualities of a light stimulus are recorded at well-defined positions of the network of receptors; in the auditive systems, receptors are distributed along a membrane sensitive to frequency, whose positional information translates into a sound information. In all cases, the positional code representing the position of the stimulus is conserved along the path to the brain, via a network of selective paths. Connexions coming from different axons are encoded with a spatial hierarchy in each deep nucleus.[6]

As a result of this hierarchical architecture, the spatial arrangement of the cells in a nucleus resembles the spatial arrangement of the receptors in the corresponding organ. In simple terms, each cerebral nucleus can be considered as sort of 'topographic map' of the organ. In the example shown in Fig. 7.18, right, the image of a disk expanding and rotating is mapped in the cerebral cortex, with a geometrical correspondence between the detection of the visual stimulus and the excitation of different areas in the cortex. Moreover, a 'shadow' effect is often observed: if an artificial stimulation is applied to a nucleus, by an electrode placed at a given site in the somato-sensory brain cortex, it can elicit exactly the same 'real' sensation of the organ corresponding to that nucleus, such as a touch on the skin.

The frequency of the response encodes the amplitude of the recorded stimulus, even if the relationship between amplitude and frequency is not linear (i.e., not simply directly proportional). In fact, during the pathway to the brain, at least two transformations must be accounted, a first one from the sensory stimulus to the amplitude of the receptor potential, and secondly from the receptor potential to the discharge frequency of the action potential.

---

[6]**Nuclei** are substructures of the central nervous system composed by grey matter (neuron cell soma, glial cells, short dendritic processes, capillaries), acting as a transit gate for the electric signals coming from a given neural subsystem. For example, the lateral geniculate nucleus is the mediator of the visual signals in vertebrates; the vestibular nucleus records the movements of the head and rives the movements of the eyes; the so-called Raphe nuclei are implicated in the command of the sleep/wake cycle; the suprachiasmatic nucleus controls the circadian rhythms; and so on.

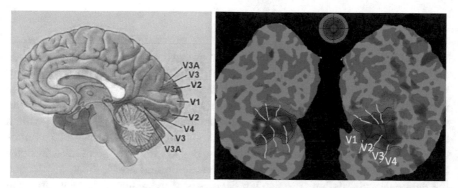

Fig. 7.18 Spatial topographic representation of a visual stimulus in the visual cortex, or **retinotopy**. *Left* Location of the visual cortex in the occipital lobe of both human and primate brain. V1 is the primary area, receiving the collected input signal from the thalamus; V2 to V5 are different parts of the visual cortex, connected in sequence, whose extent and detailed functions in signal encoding are still under active investigation. *Right* Median cross section of the visual cortex, split open in a functional-MRI image. A visual signal (a marked disk expanding in radius and rotating in angle) is shown to a subject, and the MRI images allow to reconstruct the cortical map of the different locations of the recorded information. The colour code helps in retrieving the topographic placement of the different signals. It can be seen that there is a meaningful correspondence between the spatial radial and angular variation of the visual stimulus, and its spatial recording in the different areas of the visual cortex. [Image courtesy of: *left* "The Brain", at McGill University, www.thebrain.mcgill. ca; *right* K.E. Mathewson, at Beckman Institute, Univ. of Illinois Urbana-Champaign. Repr. w. permission.]

Receptor potentials have as well a threshold, which however can be extremely low: in the case of human retina, even one single light photon (i.e., an amount of energy of the order of 3–5×10^{-19} Joules, or 2–3 eV) can start a visual stimulus. The threshold of the receptor combines with the threshold of the action potential in the immediately adjacent nerve axon (for example, of the optic nerve from the retina receptors). Evidently, a stimulus cannot be perceived unless it is higher than the most sensitive axon in the neuro-transmission chain. Another limit to the sensory stimulation comes from the upper value of the frequency of the impulse, whose exceedingly high frequency can lead to the **saturation** of the receptors. The frequency/amplitude interval comprised between the threshold and the saturation is the **dynamic range** of the sensory system. Each subsystem in the central nervous system has its own dynamic range, defined by the combination of the respective thresholds and saturation levels of all the sensors and axons recruited for the detection and response to a given stimulus. For example, the human ear has a frequency range comprised roughly between 20 and 20,000 Hz, outside which no frequency produces an audible stimulus; the retina has a wavelength range comprised between roughly 390 and 700 nm, outside which no wavelength can produce a visible signal.

7.8 Cells in the Heart

The life of all animals, from early embryogenesis and throughout adulthood, depends on the uninterrupted functioning of the heart. The heart of a normal functioning human beats at nearly constant rhythm for more than 150×10^9 times, during the whole life. The human heart cycle consists of a contraction phase (*systole*), during which blood is pushed to the body, and a relaxation phase (*diastole*), during which the blood is returned to the heart. From a structural point of view, birds and mammals have their heart subdivided into four chambers, two atriums above and two ventricles below, which ensure at each heartbeat two parallel blood flows: the systemic, and the pulmonary circulation. The two left chambers of the mammalian heart drive the oxygen-rich blood into the entire circulatory system, via the arteries; the oxygen-poor blood comes back to the right chambers of the heart, to be pushed through the lungs, where CO_2 is released from the blood and fresh oxygen is loaded, and then shipped back to the left heart. Because of the much higher pressure necessary for the general circulation, tissues on the left side of the heart are much thicker than those on the right side, despite they have to manage the same amount of blood.

By contrast, the heart of amphibians and reptiles has three chambers, and that of fishes only two. The big difference between fishes and other superior animals is due to the fact that they have their oxygen-exchange organs, the gills (corresponding to the lungs in all other vertebrates), placed in series between the heart and the blood circulatory system. In reptiles and amphibians, the membrane separating the two ventricles is either incomplete or absent. Their pulmonary artery is equipped with a sphincter muscle, a sort of valve that allows a second possible route of blood flow. Instead of blood flowing to the lungs directly from the heart, the artery sphincter may be contracted to alternately divert the blood flow to the lungs. This process is useful to ectothermic (cold-blooded) animals for the regulation of their body temperature.

The heart is essentially a muscle, an involuntary one both when normally functioning and when we fall in love. Its materials consist of different cell types (Fig. 7.19), which contribute to structural, biochemical, mechanical and electrical properties of the functional heart. The muscular walls of the heart, called **myocardium**, are formed by cardiomyocytes, elastic cells with a structure similar to muscle fibres, organised into bands that permit an easy contraction of the whole heart body (Fig. 7.19a). More than half of the cells of the heart are cardiac **fibroblasts**. Endothelial cells form the **endocardium**, the interior lining of blood vessels, and cardiac valves; slightly different "smooth" muscle cells contribute to the coronary arteries and inflow and outflow vasculature; the **epicardium** makes up the precursors of the coronary vases and cardiac fibroblasts. The most important cells in the heart, from an electromagnetic point of view, are the **pacemaker cells** and **Purkinje fibres**: these are specialised cardiomyocytes that generate and conduct electric impulses. These "electro-mechanical" cells (see Fig. 7.19b) make up the sinoatrial node (SAN), generating impulses to initiate heart contraction in the atria, and the atrioventricular node (AVN), which conducts the electric impulse to the ventricles.

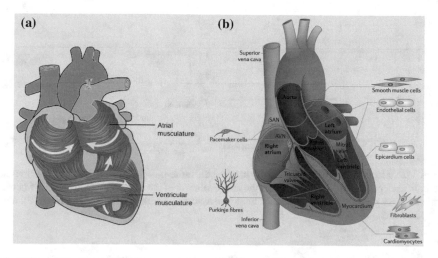

Fig. 7.19 a The pattern of muscular fibres (myocites) of the heart's walls (myocardium) has a swirling structure, which facilitates the contraction of the entire organ. **b** The different types of cells making up the heart. The electromechanical pacemaker cells are located in the AVN, above the right atrium, and initiate the electric stimulation; the Purkinje fibres surround the ventricles and transmit the impulse to the myocites. [Image (**a**) from: Anatomy & Physiology Connexions http:// cnx.org/, repr. under CC-BY-3.0 licence, see (**) for terms; **b** repr. from Ref. [12] w. permission]

In contractile heart cells the electric activation takes place by means of the same mechanism already seen in the neurons, that is the inflow of Na^+ ions across the cell membrane. The amplitude of the action potential is also similar, being about 100 mV base-to-peak. However, the duration of the cardiac muscle impulse is ∼300 ms, two orders of magnitude longer than in neurons and skeletal muscles. A *plateau* phase follows depolarisation, and thereafter repolarisation takes place. As in the neuron, repolarisation is a consequence of the outflow of K^+ ions. The mechanical contraction of the muscle cell, activated by the action potential, occurs with a little time delay. Figure 7.20 compares the electrical activity (transmembrane voltage) and mechanical contraction of a frog's skeletal muscle (the *Sartorius*) and cardiac muscle; note the very different time scales of the two figures (a few milliseconds vs. seconds), the long plateau in (b) as opposed to the sharp peak in (a), and the substantial delay between electrical stimulus (blue) and mechanical response (red) in (b).

Differently from skeletal muscles, in the cardiac muscle the electric activation can propagate from one cell to another in any direction. The only exception is the boundary between the atria and ventricles, where the activation wave is arrested by a non-conducting barrier of fibrous tissue. As a result of this lack of directionality, the activation wavefronts in the heart have more complex shapes, compared to the relatively simpler orthodromic conduction along the nerve axon.

Normal pacemaking in the heart depends on the coordinated discharge frequency of the of pacemaker cells of the SAN. Indeed, coordinated behaviour is essential to generate rhythmic activity and produce a single impulse with each cardiac cycle,

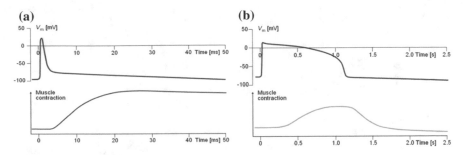

Fig. 7.20 Electric and mechanical activity in **a** frog Sartorius muscle cell, and **b** frog cardiac muscle cell. In each panel, the *blue curve* shows the time evolution of the transmembrane voltage, whereas the *red curve* describes the mechanical contraction associated with it. [Adapted from Ref. [5], repr. w. permission]

but the mechanisms leading to this behaviour are poorly understood. Two modes of synchronisation have been proposed. Under one scheme, a cell, or small group of cells, would serve as the "dominant" pacemaker, and it would drive all other pacemakers in the SAN to fire at its own intrinsic frequency. A second hypothesis, more credited in the recent years, suggests that the synchronisation may be a more "democratic" process whereby the individual cells, each beating at slightly different intrinsic frequencies, mutually interact via some form of coupling, to achieve a consensus frequency as to when to fire [13].

The spontaneous synchronisation of a large ensemble of oscillators to a common frequency is a widespread phenomenon, in both natural and artificial systems. Besides heart cells, it can be observed in the brain, with the generation of alpha-rhythm, or synchronised hormone release from hypothalamic neurons; in yeast cell suspensions achieving metabolic synchronism, in congregations of synchronously flashing fire-flies, of crickets that chirp in unison, or even the rhythmic clapping of hands by a theatre audience; in physics, in the coupling of arrays of lasers, microwave oscillators, and superconducting Josephson junctions, not to mention the popular experiments with arrays of musical metronomes on a table [14].

In the Kuramoto model, described synthetically in the greybox on p. 299, an array of idealised oscillators running at arbitrary intrinsic frequencies, are coupled through their phase differences. This model is simple enough to be mathematically tractable, yet sufficiently complex to be non-trivial. It is rich enough to display a large variety of synchronisation patterns, including a well-defined transition from disordered oscillations, to a partially or fully synchronised population. The beating of heart cells could be mathematically described by such a model of frequency synchronisation, despite the fact that the precise origin of the coupling (mechanical, electrical, chemical) is still debated.

Synchronisation of random oscillators

The synchronisation of two pendulum clocks hanging from a wall was first observed by Huygens during the 17th century. The behaviour of a large ensemble of coupled oscillators can display striking phenomena of synchronisation, even starting from random and noisy conditions. The most successful explanation is due to Yoshiki Kuramoto in 1975, who analysed a model of N phase oscillators at arbitrary intrinsic frequencies ω_i, and coupled through the sine of their phase differences θ_i. The basic equation of the model reads:

$$\frac{d\theta_i}{dt} = \omega_i + \sum_{j=1}^{N} K_{ij} \sin(\theta_j - \theta_i) \tag{7.55}$$

for each $i = 1, ..., N$. Oscillators tend to run independently, each at its own ω_i, until the mutual coupling K_{ij} is small; however, when the coupling increases beyond a certain threshold, synchronisation spontaneously emerges. This can be seen by transforming the equation via the "order parameter" $0 < r(t) < 1$, which measures the time-dependent amount of coherence in the population, and the average phase $\psi(t)$, defined as:

$$r e^{i\psi} = N^{-1} \sum_{j=1}^{N} e^{i\theta_j} \tag{7.56}$$

This last equation can be multiplied by $e^{-i\theta_i}$, then retain the imaginary part of it (since $e^{i\beta} = \sin\beta + i\cos\beta$), and take its time derivative, thus obtaining:

$$r \sin(\psi - \theta_i) = N^{-1} \sum_{j=1}^{N} \sin(\theta_j - \theta_i) \tag{7.57}$$

which, substituted in the (7.55), and by assuming for the sake of simplicity that all $K_{ij} = K$, gives:

$$\frac{d\theta_i}{dt} = \omega_i + Kr \sin(\psi - \theta_i) \tag{7.58}$$

Now each oscillator is independent, and coupled to the common average phase ψ with coupling strength Kr. In the limit of a very large N, the frequencies can be thought of being distributed according to some $g(\omega)$, and the oscillators take up values of ω and θ at each time t with probability $\rho(\theta, \omega, t)$. Sums are now replaced by integrals over $d\omega$ and $d\theta$, as:

$$r e^{i\psi} = \int_{-\pi}^{\pi} e^{i\theta} \int_{-\infty}^{\infty} \rho(\theta, \omega, t) g(\omega) d\omega d\theta \tag{7.59}$$

The meaning of the order parameter r becomes evident when considering a vanishing coupling, $K \to 0$, in Eq. (7.58). In this case $\theta_i = \omega_i t$, and by substituting $\theta = \omega t$ in the previous equation, it is $r = 0$. On the other hand, in the limit of very strong coupling, $K \to \infty$, it is $\theta \to \psi$ and $r \to 1$, meaning that all the oscillators have the same phase and are therefore synchronised. For intermediate values of K, r is between 0 and 1, meaning that on average some oscillators are synchronised in phase, $d\theta_i/dt = 0$, while some are rotating out of sync.

7.8.1 The Rhythm and the Beat

The heart beat is initiated at the **sinoatrial node** (SAN). The SAN receives several different inputs that regulate the instantaneous heart rate and its variation. Respiration gives rise to waves in heart rate, mediated by the sympathetic and the parasympathetic nervous system; other factors affecting the input nervous signal are the baroreflex (automatic regulation of the blood pressure), thermoregulation, hormones, sleep-wake cycle, meals, physical activity, stress.

The SAN in human heart is a small mass of cells in the shape of a crescent, about 15 mm long and 5 mm wide, receiving a high density of nerve terminations. The SAN pacemaker cells generate an action potential at the intrinsic rate of about 70 pulses/min. As in the neuron, this is actually a *depolarising* potential, since it starts from a negative resting level of about −70 mV. From the SAN the impulse propagates throughout the upper half of the heart, but cannot propagate directly across the boundary between atria and ventricles, as noted above. The impulse is then relayed by the atrioventricular node (AVN), a sort of button located at the boundary between the atria and ventricles. The AVN has an intrinsic frequency of about 50 pulses/min, however it can adjust to higher frequency if required. Propagation from the AVN to the lower half of the heart (the ventricles) is provided by a specialised conduction system. The **bundle of His** (named after the German physician Wilhelm His) starts from the AVN, and then splits into two branches of fibres running around the two ventricles. These further ramify into Purkinje fibres (named after the Czech physician Jan Evangelista Purkinje[7]). From the Purkinje fibres the impulse can diffuse to all myocytes, via cell-to-cell multi-directional activation. Contrary to this complex (de)polarisation phase, the following repolarisation, which brings each cell's potential back to its resting value, does not require propagation since each cell does it independently, after the impulse wave has passed. The time-shift plots in Fig. 7.21 summarise the electric events taking place during a single heartbeat and their electric waveforms, together with a table providing the correspondence between events, timings, and propagation velocities in the tissue. It may be worth to note that propagation along the conduction system takes place at a relatively high speed, once the impulse has reached the lower ventricular region, whereas it was much slower before. Also, in case of failure of the transmission at some place, the region cut off will beat at its intrinsic rhythm, which may be considerably lower (about 20–40 per min, in the lower heart).

The heart *beat* above described occurs at the level of single cells. The *rhythm* of the heart, instead, results from the sum of all these electric waves, rebounding from the different tissues. The lower diagram in Fig. 7.21 represents schematically this sum. The **electrocardiogram** (ECG) is a recording of the overall electric potential generated by the electric activity of the heart, taken on the surface of the thorax. The ECG thus represents the integrated electric behaviour of the cardiac muscle tissue, in which some of the single-cell events can be identified (see uppercase letters in the figure).

---

[7]When Purkinje died in 1869, Wilhelm His was just 5 years old.

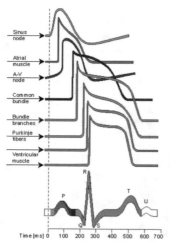

Location	Event	Time (ms)	ECG	Velocity (m/s)
SAN	start impulse	0		0.05
Right atrium	depolarisation	5	P	0.8-1.0
Left atrium	depolarisation	85	P	0.8-1.0
AVN	impulse arrival	50	P-Q	0.03
AVN	impulse restart	125	P-Q	0.03
bundle of His	activation	130		1.0-1.5
Purkinje fibres	activation	150		3.0-3.5
Left ventricle	depolarisation	225	QRS	0.3-0.8
Right ventricle	depolarisation	250	QRS	0.3-0.8
Right ventricle	repolarisation	400	T	0.5
Left ventricle	repolarisation	600	T	0.5

Fig. 7.21 Electrophysiology of the human heart. The different waveforms for each of the specialised cells found in the heart are shown on the *left*, and the corresponding events are listed in the table on the *right*. The time delays between the different impulses (latency) are similar to those normally found in the healthy heart. The lower diagram regroups the sum of the different impulses as could be registered in a ECG trace. The letters correspond to the labelling given in the adjacent table. [Adapted from Ref. [5], repr. w. permission]

The genesis of the ECG signal can be represented with a highly idealised **single dipole model** [15]. The electric source of the ECG, as measured at the surface of the torso, is the intracellular current that is generated as the action potential propagates through the heart tissues; charge conservation, on the other hand, imposes that there is also an equal and opposite extracellular current, flowing against the direction of propagation. All the current loops in the conductive tissues close upon themselves forming a dipole field, whose amplitude and direction changes in time following the moving boundary between depolarised and polarised tissue. The net equivalent dipole moment is the time-dependent **heart vector M**(t), with its origin at the centre of the chest. As a further, reasonable approximation, the dipole model ignores the anisotropy and inhomogeneity, and treats the chest as an ideal spherical conductor of radius R and conductivity σ. The Laplace's problem for the potential, $\nabla^2 \Phi = 0$, may then be solved in this simple geometry, to give the potential distribution at the surface of the body as:

$$\Phi(t) = \frac{3 \cos \theta(t)}{4 \pi \sigma R^2} |\mathbf{M}(t)| \tag{7.60}$$

where $\theta(t)$ is the angle between the direction of the heart vector $\mathbf{M}(t)$, and the fixed vector from the centre of the sphere to the point R of observation (where each one of the ECG electrodes is placed).

A typical pulse of amplitude $\sim$100 mV, traveling at average speed 0.5 m/s, crosses a cell of size $\delta = 50 \, \mu\text{m}$ in about 0.1 ms. Therefore, by assuming a linear resistivity of the 5-nm thick membrane of $2 \times 10^{10} \Omega$-cm, or $R =$

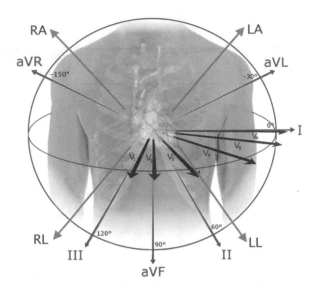

Fig. 7.22 The positioning of the six standard chest leads in ECG, which report the heart's electrical activity in the horizontal plane. Each *blue vector* is a component of time-varying heart dipole vector **M**. Each electrode defines a fixed (*red*) vector, at an angle θ with **M**. The *light-red arrows* indicate the four electrodes to the *left/right* arm, and *left/right* leg. [Adapted from: N. Patchett, Boston Medical Center, repr. under CC BY-SA 4.0 lic., see (*) for terms.]

$2 \times 10^{10}/((\pi\delta^2) = 2.5 \times 10^{14}\,\Omega/\text{cm}$, the typical current at the cellular level is $I = (0.1/10^{-4})(1/2.5 \times 10^{14})(1/0.005) \sim 0.8$ nA (10^{-9} amperes). If such a transient current is distributed over a moving dipole length $d = 0.5$ (m/s) $\times$ 0.1 ms = 5 cm, the typical dipole moment is of the order of $|\mathbf{M}| \simeq 4 - 5 \times 10^{-9}$ A-cm.

The recording is made by placing a series of six electrodes around the chest (Fig. 7.22), which record the activity in the horizontal plane (cross section through the chest), and four more electrodes at each limb, which record the activity in a frontal plane (cross section between front and back of the body). The measured voltage V_{IJ} gives the difference of electric potential (7.60) between pairs of electrodes I and J. Take $\overline{OI}$ and $\overline{OJ}$ as the two vectors from the origin O to each electrode's position (two different "red" vectors in Fig. 7.22); each vector has a modulus equal to $3/(4\pi\sigma L^2)$, if L is the distance of the electrode from the centre. The Eq. (7.60) can also be written as a scalar product $\Phi_K = \mathbf{M} \cdot \overline{OK}$ for any electrode K. Therefore the voltage is given by the scalar product of the two vectors:

$$V_{IJ}(t) = \Phi_I - \Phi_J = \mathbf{M}(t) \cdot \mathbf{L}_{IJ} \tag{7.61}$$

with $\mathbf{L}_{IJ}$ the difference vector between electrodes I and J, called the **lead vector**. A variety of lead vectors may be formed by attaching electrodes to the body in various positions. The standard clinical practice is based on $6+6$ lead vectors, defined by cutting the circle around the heart into angular sectors of 30° (Fig. 7.22). In this way, impulses traveling along different directions can be monitored, if the electrodes are correctly placed, and give spatial information about the heart's electrical activity in the (approximately) orthogonal directions: right to left, superior to inferior, anterior to posterior.

Impulses traveling toward a lead give a negative contribution to $V(t)$, while impulses traveling in the opposite direction give a positive contribution to $V(t)$. The $V(t)$ traces from the blue arrows in the figure indicate: I, difference between left arm and right arm; II, difference between the left leg and the right arm; III, difference between the left leg and the left arm; the aVF, aVL, aVR are respectively perpendicular to these. The typical $V(t)$ pulse shape obtained in Fig. 7.21 (bottom left) can thus be interpreted. For example, the P wave describes the sequential activation of the right and left atria; the QRS portion is the right and left ventricular depolarisation (normally the ventricles are activated simultaneously); the ST-T wave is the ventricular repolarisation, and so on.

An order of magnitude estimate of the amplitude of $V(t)$ can be obtained, by taking all the heart cells in a patch of ~ 1 cm^2 to have the same size δ, with a density is about 5×10^4 cells/cm^2. The total dipole is of the order of $(0.8 \times 10^{-9})(5 \times 10^4) \sim 4 - 5 \times 10^{-4}$ A/cm. For an average distance $L = 20$ cm, and $\sigma = 3 \times 10^{-3} \Omega^{-1}$ cm^{-1}, the typical voltage between a couple of electrodes, estimated with the simple dipole model, comes down to $V = 3(5 \times 10^{-4})/(4\pi(3 \times 10^{-3})20^2) \sim 0.1$ mV. Such values, of the order of fractions of millivolts, correspond well to experimentally measured values early in the sequence of ventricular depolarisation.

7.9 Electricity in Plants?

As far as we know, plants and trees do not have organs with structure and functions resembling hearts or brains, therefore this Chapter has been concerned mainly with animal cells. However, electric forces and fields exists and pervade plants as well. Already in 1873, E. Burdon-Sanderson described to the British Royal Society how the rapid clenching response of the leaves of *Dionea muscipula* to an insect touching its surface is induced by an electric action potential propagating through the two lobes of the leaf, causing them to snap close to trap the insect. (Fig. 7.23; such plants often live in nitrogen and mineral-depleted areas and may get their required nitrogen and minerals by capturing and digesting insects.)

The cells of most, perhaps all, plants are excitable. Stimuli such as chilling, heating, cutting, touching, electric stimulus or changes in external osmolarity, result in action potentials, a transient depolarisation of cell membrane which is electrotonically transmitted at rates of 10–40 mm/s (much slower than axonal currents), and which resemble primitive nerve action potentials. Until recently, plant biologists were quite reluctant to view action potentials as of primary significance in plant responses. The principal reason for this was the discovery of the ubiquitous chemical signal auxin, which seemed to rule out of relevance any other electrically transmitted signals. However, since at least the late 1980s, some prominent plant electrophysiol-

Fig. 7.23 The carnivorous plant *Dionea muscipula* uses its attractive wide-lobed leaves to attract insects, worms, even small frogs. Once the presence of the extraneous body is detected, an electric action potential spreads over the lobe surface, and causes the two halves to snap close in fractions of a second. The inside compartment at the junction of the lobes then forms a closed cavity, a sort of stomach filled with digestive enzymes, which turns the prey into a meal for the *Dionea*

ogists have argued that multifunctional electric signals (action potentials) could be primarily responsible for coordinating plant responses to the environment. Barbara Pickard [16] and E. Davies [17] proposed that a 'protease inhibitor inducing factor', a wound signal, could be electrical rather than chemical. This view has been confirmed by further research, [18] which showed that initiation of some inhibitor genes in response to wounding in tomato leaves is not brought about by a chemical signal, but electrically, by transmitted action potentials.

Today, the same intracellular microelectrode experiments performed on neuronal cells, allow to measure the electric potentials across the membrane of plant cells (Fig. 7.24, [20]). In the early studies giant characaean algae cells were particularly

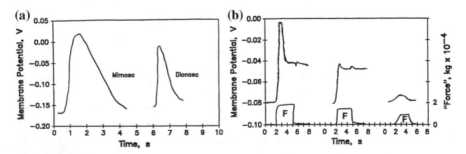

Fig. 7.24 *Left* Recording of action potentials in *Mimosa pudica* and *Dionea muscipula*. Note the vertical voltage scale, quite larger than for animal stimulation, and the horizontal time scale, quite slower. *Right* Mechanical stimulation of a single *Dionea* leaf's hair. Three levels of stimulation are shown to elicit an action potential (first), a near-threshold local response (second), and a simple electrical replica of the sub-threshold mechanical stimulus (third). [From Ref. [19], repr. w. permission]

suitable, but subsequently this kind of measurement was extended to many plant species. In the experiment described in the Figure, the action potential recorded in the *Mimosa pudica*, whose delicate branches retract and fold at the slightest pressure, and in the already cited *Dionea*, are seen to be quite higher ($\Delta V \sim 250 \, \text{mV}$) and on a quite longer time-scale (a few seconds) compared to animal cells. Also, the experiment carried out on a single leaf's hair of *Dionea* (right in the Figure) demonstrates a graded reaction to the foreign stimulation, which is able to elicit an action potential, local currents, or simple replica of the stimulus according to its relative intensity. (Note that in the original paper, the force was strangely measured in kg and not newtons.)

Compared to animal cells, in which we saw the close interplay of Na^+ and K^+, in plants the most important ions in the process of generating the action potential are Ca^{2+} and Cl^- [21]. After the arrival of the initial stimulus, free Ca^{2+} concentration in the cytoplasm increases. This Ca^{2+} originates from extracellular and intracellular spaces, through voltage-dependent ion channels and from vacuoles via secondary transduction pathways. Depolarisation occurs due to Ca^{2+} activation of Ca^{2+}-dependent anion channels, and massive efflux of Cl^-. Depolarisation leads to opening of K^+-channels, and the K^+ efflux repolarises the plasma membrane. The characteristics of action potential can be modified by changing of Cl^- or K^+ concentration. In plants, most cells have cell-to-cell conduction through the plasmodesmata (microscopic channels which traverse the cell walls) and this connection has high solute permeability and electric conductivity. Plasmodesmata are nearly identical to gap junctions of animal tissues, and could be considered as a kind of "plant synapses", in which auxin might be taken as the mediator (equivalent of the neurotransmitter in brain synapses).

It is today clear that electrophysiological properties of plants can change seasonally and with cell age. For example, in the giant alga *Chara*, the cell membrane is significantly less hyperpolarised (less negative) in winter, when plants are vegetative, and hyperpolarises again immediately before fructification in the spring, a phenomenon associated with changes in patterns of cell-to-cell communication. Sucrose concentration and ion content, particularly of K^+, varies seasonally in *Chara vulgaris*, again with a precise timing associated with the reproductive cycle. Also the cells of *Dionea* show a more negative membrane resting potential over winter. The action potential shows a definite temperature dependency in the alga *Nitella*; in another alga, the unicellular *Eremosphaera*, darkening after illumination causes a transient hyperpolarisation of the cell membrane due to divalent cation, and anion currents. The turgor pressure of plant cells and their electrophysiology are indeed linked. Hypotonic shock in the cells of *Lamprothamnium*, a salt-tolerant charophyte, results in membrane depolarisation, opening of Ca^{2+} channels, efflux of Cl^- ions followed by K^+, resulting in turgor pressure regulation, this process differing in plants of different age and from different environments.

Appendix F: The G-H-K equations

The ionic flow across the membrane can originate from a purely diffusive (passive) mechanism, given by Eq. (7.3) above:

$$J = -D\nabla[c] \tag{7.62}$$

or by an active diffusion pushed by the electric field across the membrane, $E = V/d$, V being the total voltage drop across a membrane of thickness d. This second contribution is $J = [c]v$, (the square parenthesis [...] indicates molar concentration), and by using the definition of ionic mobility (see greybox *Ohm's law and diffusion*), it is $J = \mu E[c] = (\mu V/d)[c]$.

By recalling the Stokes-Einstein equation (see Chap. 5, Eq. (5.39), and p. 179) that, in the linear response regime, relates the mobility to the diffusion coefficient as $D = \mu k_B T$, the flow under the electric field (or *electrophoretic* current) is also written as:

$$J = \frac{D}{k_B T} \frac{V}{d}[c] \tag{7.63}$$

The sum of the two components, passive and active diffusion, is therefore for each ion species k (for simplicity, we consider only monovalent ions with $z = \pm 1$):

$$J = -D_k \left(\nabla[c_k] - \frac{zF}{RT} \frac{V}{d}[c_k] \right) \tag{7.64}$$

By taking unidimensional diffusion, e.g. along the x thickness of the membrane, this differential equation of the type $A + Bc = dc/dx$ can be integrated by separation of the variables, from $x = 0$ (inside the membrane) to $x = d$ (outside the membrane). This yields the solution:

$$J = wP_k \frac{([c_k]_{out} - e^{zw}[c_k]_{in})}{1 - e^{zw}} \tag{7.65}$$

with the constant $w = FV/RT$, and $P_k = D_k/d$ the permeability of the k-th ion species. This first equation is known as the **Goldman equation for the ionic flux**, and is an expression of the balance between the diffusional and electromigration flow of the ions of each given species k, for a given permeability (i.e., a given fraction of membrane ion channels opened/closed).

On the other hand, the overall electric potential is established by the combined ionic "batteries" of the three principal ionic species (plus any other eventual ions). The total is obtained by considering the parallel superposition of the three elements, at a reference voltage V_m for which the charge is in equilibrium, i.e. the net flow of positive plus negative charges across the membrane is zero. By summing all the k positive and the n negative (monovalent) ions, and equating their respective flux, we obtain:

$$\frac{\sum_k P_k([c_k]_{out} - e^w[c_k]_{in})}{1 - e^w} = \frac{\sum_n P_n([c_n]_{out} - e^{-w}[c_n]_{in})}{1 - e^{-w}} \tag{7.66}$$

The terms on the right side, corresponding to negatively charged ions, can be multiplied and divided by e^w. In this way the denominators cancel out, and:

$$\sum_k P_k([c_k]_{out} - e^w[c_k]_{in}) = -\sum_n P_n([c_n]_{in} - e^w[c_n]_{out}) \tag{7.67}$$

(note the switch between the *in* and *out* for the negative species) or:

$$\sum_k P_k[c_k]_{out} + \sum_n P_n[c_n]_{in} = e^w\left(\sum_k P_k[c_k]_{in} + \sum_n P_n[c_n]_{out}\right) \tag{7.68}$$

from which it is obtained:

$$w = \ln \frac{\sum_k P_k[c_k]_{out} + \sum_n P_n[c_n]_{in}}{\sum_k P_k[c_k]_{in} + \sum_n P_n[c_n]_{out}} \tag{7.69}$$

This is the **Goldman-Hodgkin-Katz equation for the membrane potential**, which in a more usual form reads:

$$V = \frac{RT}{F} \ln \frac{\sum_k P_k[c_k]_{out} + \sum_n P_n[c_n]_{in}}{\sum_k P_k[c_k]_{in} + \sum_n P_n[c_n]_{out}} \tag{7.70}$$

Appendix G: Electric Currents for Dummies

Current and Resistance

When a difference of electric potential V (or voltage, in units of volts) is applied at the ends of a material with resistance R, an electron current I develops in the closed circuit, from the negative to the positive pole:

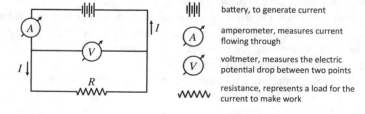

┤╟	battery, to generate current
(A)	amperometer, measures current flowing through
(V)	voltmeter, measures the electric potential drop between two points
WWW	resistance, represents a load for the current to make work

The relationship between current and potential in the linear regime is the Ohm's law:

$$V = RI \qquad (7.71)$$

The current is measured in units of amperes, 1 A being equal to 1 coulomb ($6.24 \times 10^{18}e$) per second. The difference of potential V must be maintained by a continuous supply of energy, which turns the potential energy stored in the battery into kinetic energy of electrons and heat. The power, delivered by the battery and dissipated by the resistance, is the work done by the electromotive force in a unit time:

$$P = IV \qquad (7.72)$$

measured in (volts $\times$ coulomb/second).

The resistance is the tendency to slow or arrest the flow of the current. Each material has a specific value of the electric resistance defined by its microscopic electric conductivity σ, as $R = L/\sigma S$, for a length L and a cross section area S of the conductor (here assumed cylindrical, as in a wire or a cable). Often R is expressed as resistance per unit length, and is indicated with lowercase r. A conductor is a medium (material, solution) with a quite small resistance, an insulator conversely is a medium with a high resistance to current flow. The conductance is defined as the inverse of the resistance, $G = 1/R$.

Several parts of a system of conductors may exhibit different resistance. The overall response of the system (a "circuit", if the ends are connected) to an applied voltage drop can be described by connecting the resistances "in series":

the total resistance being $R = R_1 + ...R_n$; or "in parallel":

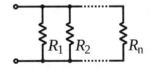

the total resistance being instead $1/R = 1/R_1 + ...1/R_n$.

A **direct current** (DC) is the unidirectional flow of electric charges. It can produced, e.g., by a battery, a thermocouple, a solar cell. A DC current may flow in a conductor such as a wire, but also through semiconductors, insulators, or even through a vacuum as in electron or ion beams.

In an **alternating current** (AC), the direction of charge flow alternates forward and backward over the time, between the two points at different potential (for example a wire). An AC current is necessarily generated by an alternating voltage, in most applications taken as a sinusoidal wave, $V(t) = V_0 \sin \omega t$, with period $\tau = 2\pi/\omega$. Therefore, by Ohm's law, the current oscillation is in phase with the voltage wave:

$$I(t) = \frac{V_0}{R} \sin \omega t \qquad (7.73)$$

Despite the constantly changing direction of the charge flow, an AC current delivers power, because of the electromotive power definition (7.72):

$$P(t) = I(t)V(t) = \frac{V_0^2}{R} \sin^2 \omega t \qquad (7.74)$$

Capacitor

A capacitor, or condenser, is obtained when two elements of a conductor, e.g. two parallel plates, or two concentric cylinders, are separated by an insulator, called the dielectric. When the positive and negative of a battery are connected to the two ends of the capacitor, the positive charges accumulate on one side and the negative charges accumulate on the other, thus producing a spatial charge separation. The capacity of a condenser is defined by the amount of charge Q accumulated under a given voltage V:

$$C = Q/V \qquad (7.75)$$

Capacity is measured in units of farad, 1 F being equal to a charge of 1 coulomb under a voltage drop of 1 volt. (Because capacity is charge/voltage, 1 F is dimensionally [Charge^2]/[Energy], or $(A\,s)^2/(kg\,m^2 s^{-2})$). Note that a capacitor, differently from a resistance, does not dissipate energy but rather stores it (of course, in a real capacitor there will be some non-ideal losses, so that some energy is necessarily lost according to the Second Principle of thermodynamics). The work done to accumulate this charge, that is the energy stored, is:

$$W = \int_0^Q V(q)dq = \int_0^Q (q/C)dq = \frac{1}{2}\frac{Q^2}{C} = \frac{1}{2}CV^2 \qquad (7.76)$$

If the voltage V is applied to a capacitor charged with any value Q' smaller than Q, a transitory current will develop between the two ends. The relationship between this time-dependent current and the voltage is:

$$I(t) = \frac{dQ}{dt} = C\frac{dV}{dt} \qquad (7.77)$$

Clearly, the current exists until $Q' = Q$, the maximum capacity of the condenser for the voltage V. In chemical terms, the separation between positive and negative charges is maintained until the dielectric separating the two ends is polarised by the electric field $-eE = V$. If the battery is disconnected, the capacitor will discharge from Q to 0, with the same time constant τ, therefore a transitory current will again be observed for a time τ. In fact, a battery could be considered as a capacitor with an extremely long time constant.

Since a conductor will necessarily display at least some finite resistance to the passage of the current, the minimum circuit that can be realised with a capacitor is in fact the sum of the resistance of the conductor plus the capacitance of the condenser, schematised as:

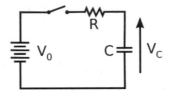

The voltage V_C measured between the two ends of the capacitor varies between V_0 and zero, as a function of the charge accumulated and, correspondingly, of the charge flow. By looking at the preceding equation, the total voltage can be written as the sum of the drop across R plus the drop across C. Assume that at time $t = 0$ the capacitor is uncharged, and the switch is closed. The voltage drop at any time $t > 0$ changes both in R and C, but its sum must be always equal to V_0:

$$V_0 = I(t)R + \frac{1}{C} \int_0^t I(t')dt' \tag{7.78}$$

which says that also the current will be distributed between R and C. By taking the derivative with respect to time of this equation:

$$0 = \frac{dI(t)}{dt}R + \frac{I(t)}{C} \tag{7.79}$$

The solution of this equation is found by separating the variables:

$$\frac{dI(t)}{I(t)} = -\frac{dt}{RC} \tag{7.80}$$

and is an exponential $I(t) = I_0 \exp(-t/\tau)$. The time $\tau = RC$, is the time needed to fully charge or discharge the capacitor by the amount of charge Q, and is called the **time constant** of the capacitor. Its value depends on the geometry and materials making up the element.

The constant I_0 is found by considering that at the time $t = 0$ the charge on the capacitor is zero; therefore the voltage across C is also zero, and the voltage across R is V_0. Then, $I_0 = V_0/R$, and the same exponential solution for the current gives an exponential variation of the voltage drop:

$$V(t) = V_0(1 - e^{-t/\tau}) \tag{7.81}$$

Kirchhoff's Laws

Although useful to reduce series and parallel resistors in a circuit whenever they occur, circuits in general are not composed exclusively of such combinations. For the more general cases there are a powerful set of relations called *Kirchhoff's laws*, which enable to analyse arbitrarily complex circuits.

The German physicist Gustav Kirchhoff in 1854 established the following two laws:

- "junction rule": for a given junction or node in a circuit, the sum of the currents entering equals the sum of the currents leaving;
- "loop rule": around any closed loop in a circuit, the sum of the potential differences across all elements is zero.

The first law is a statement of charge conservation, while the second is a statement of energy conservation, in that any charge that starts and ends up at the same point with the same velocity, must have gained as much energy as it did lose. Both laws can be derived from the Maxwell equations (see below), and are strictly valid for DC electric currents, or AC currents in the low-frequency limit. The two laws can be schematically described according to the two following diagrams:

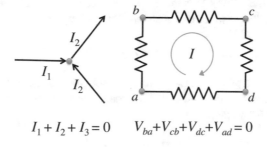

$$I_1 + I_2 + I_3 = 0 \qquad V_{ba} + V_{cb} + V_{dc} + V_{ad} = 0$$

Kirchhoff's laws are typically used by removing blocks of a circuit and replacing them with a simpler element, whose I and V correspond to the block removed. The resulting circuit is made simpler and simpler, by repeatedly applying the two laws, until it cannot be simplified anymore.

Maxwell's Equations

The celebrated equations describing the behaviour of electric and magnetic fields, which the genius of the American physicist James Clerk Maxwell donated to the humanity between 1861 and 1862 when he was barely aged 30, are the subject of entire books and ponderous treatises on electromagnetism. Here we recall them only for the reader's convenience, in the form most useful for the applications to the propagation of electric and magnetic fields in dielectric media (i.e., biological tissues). Detailed treatment can be found, e.g., in J.D. Jackson's book, listed under "Further reading" at the end of this chapter.

The ingredients of electromagnetic field equations are the electric field vector $\mathbf{E}$, the magnetic field vector $\mathbf{H}$, the electric flux density $\mathbf{D} = \varepsilon\mathbf{E}$, and the magnetic flux density $\mathbf{B} = \mu\mathbf{H}$.

The two field vectors $\mathbf{E}$ and $\mathbf{H}$ are generated by the force fields existing around, respectively, charges and currents.

The two flux density vectors $\mathbf{D}$ and $\mathbf{B}$ are related to their respective fields by the two coefficients ε, the **relative electric permittivity** of the medium, and μ the **relative magnetic permeability**. These quantities are experimentally determined for each medium, and can be expressed relative to their counterparts for the vacuum: $\varepsilon_0 = 8.854 \times 10^{-12}$ F/m (the Farad unit is equal to $[Ampere]^2[T]^2/[Energy]$), and $\mu_0 = 4\pi \times 10^{-7}$ Henry/m (the Henry unit is equal to $[Energy]/[Ampere]^2$).

Each of the four Maxwell's equations can be interpreted in terms of the elementary laws and facts of electromagnetism, established by experiments often performed well before his theoretical treatment. However, before Maxwell, the very concept of *electromagnetism* did not exist: electric and magnetic phenomena were considered as manifestations of entirely distinct properties of the nature of matter.

(1) Gauss' law (circa 1835). Dictates how the electric flux density $\mathbf{D}$ behaves at any point in space, in the presence of a distribution of charge density $\rho(\mathbf{r})$:

$$\nabla \cdot \mathbf{D(r)} = \rho(\mathbf{r}) \tag{7.82}$$

(2) Sometimes called "Gauss' law for magnetism", was originally written down by Maxwell. It summarises the evidence that magnetic monopoles do not exist (or at least, we haven't found them yet):

$$\nabla \cdot \mathbf{B(r)} = 0 \tag{7.83}$$

(3) Faraday's law (1837). States that a time-varying magnetic field creates an electric current of intensity equal to the rotor (or "curl") of the vector $\mathbf{E}$. (Remember the geometric meaning of the rotor, from Appendix A: it is the integral of the vector about a circle drawn around the central line, in this case the line of flow of $\mathbf{J}$. If $\mathbf{B}$ varies around a circle the resulting current goes perpendicular to the circle, while if $\mathbf{B}$ varies along a line the resulting current will move in closed loops about the magnetic field lines.)

$$\nabla \times \mathbf{E(r)} = -\frac{\partial \mathbf{B}}{\partial t} \tag{7.84}$$

(4) Ampere's law (1826). States that a current $\mathbf{J}$ creates a magnetic field of intensity equal to the rotor (or "curl") of the vector $\mathbf{H}$. (Remember the geometric meaning of the rotor, from Appendix A: it is the integral of the vector about a circle drawn around the central line, in this case the line of flow of $\mathbf{J}$. Clearly, since $\mathbf{J}$ is locally constant, the intensity of $\mathbf{H}$ is inversely proportional to the radius of the circle.)

$$\nabla \times \mathbf{H}(\mathbf{r}) = \mathbf{J} + \frac{\partial \mathbf{D}}{\partial t} \qquad (7.85)$$

The second term on the right is the "displacement current", added by Maxwell to include the magnetic intensity generated by a time-varying electric field.

For biological tissues, $\mu \simeq \mu_0$. However, ε varies widely with the frequency of the electromagnetic field. Most soft tissues have $\varepsilon \sim 10^3 - 10^4 \varepsilon_0$ at low frequency (< 100 Hz); bones, heart and liver tissue have $\varepsilon \sim 10^7 - 10^8 \varepsilon_0$ and above. At higher frequencies (kHz to GHz) the value of ε decreases down to 100–$10 \varepsilon_0$. Overall, this makes the propagation velocity of currents in the tissue of the order of $10^{-4} c \sim 10^4$ m/s at low frequencies, typical of signals from the brain or the heart. (*Note that this is different from the propagation velocity of the electric stimulation in the cells, which depends chiefly on the opening and closing of ion channels.*)

Problems

7.1 Absolute and relative refractory period
In a typical vertebrate axon, the absolute refractory period is 1.0 ms and the relative refractory period is 4.0 ms. Thus, the neuron is insensitive to stimulation for a total of 5.0 ms. If the cell is continuously stimulated with a train of impulses of amplitude $V \simeq V_{rest}$, what is the highest frequency of action potentials that can be generated? And what would be the maximum frequency in the case $V > V_{rest}$?

7.2 The GHK equation
Consider a cell with only Na^+ and K^+ ions, at rest with concentrations: $[c_K]_{out} = 4$ mM, $[c_K]_{in} = 140$ mM, $[c_{Na}]_{out} = 142$ mM, $[c_{Na}]_{in} = 14$ mM, at ambient temperature $T = 23°C$.
(a) Given the *relative* permeabilities $P_K = 1$ and $P_{Na} = 0.002\%$, calculate the membrane resting potential. Compare it with the values of V_K and V_{Na}.
(b) Note that the ions may still be able to cross the membrane as long as the total current sums to zero. Which direction would Na^+ and K^+ go?

7.3 The cable equation
Show that the functions $V_m(x) = A \exp(X) + B \exp(-X)$, $V_m(x) = A \cosh(X) + B \sinh(X)$, $V_m(x) = A \cosh(L - X) + B \sinh(L - X)$, are all solutions of the steady-state cable equation $V_m = \frac{\partial^2 V_m}{\partial x^2}$.

7.4 Axon resistance
Consider three axons: (1) infinite (=very long), (2) semi-infinite (=starting from a neuron and proceeding far away), (3) finite, 3 cm long (=a small section between two neurons). In each case, take the diameter as constant and equal to $d = 40 \,\mu m$; $R_m = 5000 \,\Omega \, cm^2$; $R_i = 100 \,\Omega cm$. (a) Calculate the input resistance at $x = 0$ in each case. (b) If a potential of 200 mV is imposed at $x = 0$, calculate the voltage at $x = 1.5$ cm.

7.5 Triple junction

One of the most evident properties of neurons is that they are extensively branched. Consider the simple branching depicted in the following figure: an axon starting at $X = 0$, with a branching at $X = L$ into two segments, extending to L_a and L_b. In principle, each branch could have a different λ and resistances R_i or R_m, however for simplicity they are taken to be equal here. Find the expression for the potentials and currents at L, L_A, L_B, for a condition of "open ends".

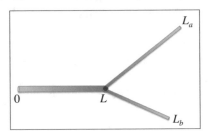

7.6 Electric frog

Consider the propagation of an action potential along a muscle fibre bundle, in a frog's leg. The action potential travels at a velocity $v = 10$ m/s. Model the current *entering* the muscle from the extracellular medium, at the onset of action potential, as a current pulse of amplitude $-I_0 = 10$ nA, traveling at velocity v; and a second current *exiting* the muscle at repolarisation, as a current pulse $+I_0 = 10$ nA, trailing at a distance $d = 1$ cm from the first pulse. An electrode placed on the leg surface at a distance $D = 3$ cm from the muscle, measures the electromyographic (EMG) signal relative to ground. Assume an average tissue conductivity $\sigma = 1.9\,\Omega^{-1}\mathrm{m}^{-1}$, and fix $t = 0$ when the midpoint of the two traveling sources (represented by the green and yellow spot) face the electrode. Derive and plot this EMG signal as a function of time, as the action potential passes by the detector.

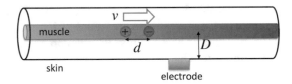

7.7 A mouse's ear

The gross size of a human ear is of the order of 70 mm in major dimension, while that of the mouse is of the order of 7 mm. The figure below represents the sensitivity and frequency range of three condenser microphones of different sizes. We can model the ear receptor (tympanus) pretty much as a microphone.

(a) Assuming that the physical dimensions of the ear of the mouse are scaled relative to those of humans in the same proportion, what does that suggest about the frequency range of hearing of the mouse?

(b) As the basic transduction mechanism is likely the same in both mouse and human, what this implies about the sound sensitivity of the mouse?

(c) Contrary to the conclusion of (b), the mouse ear is as sensitive in its frequency range as that of a human. What this implies?

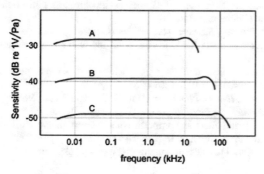

7.8 A bird's ear

In birds, a special type of neuron is responsible for computing the time difference between sounds arriving to the two ears. These neurons respond only if the inputs arriving from both ears coincide within well-defined time intervals, typically 10–100 μs, while avoid a response when the double input comes from only one ear. (a) Draw a scheme of the synapses fetching input to a dendrite of these auditory neurons. (b) What type of summation could be employed, to distinguish inputs from one, or both ears? Draw a scheme of the algorithm. (c) Draw a schematic plot of the output potential in the two cases.

References

1. G. Kreimer, The green algal eyespot apparatus: a primordial visual system and more? Curr. Genet. **55**, 19–43 (2009)
2. K.W. Forster, R.D. Smyth, Light antennas in phototactic algae. Microb. Rev. **44**, 572–630 (1980)

3. R.J. Gegear, A. Casselman, S. Waddell, S.M. Reppert, Cryptochrome mediates light-dependent magnetosensitivity in Drosophila. Nature **454**, 1014–1018 (2008)
4. L.E. Foley, R.J. Gegear, S.M. Reppert, Human cryptochrome exhibits light-dependent magnetosensitivity. Nat. Commun. **2**, 356 (2011)
5. J. Malmivuo, R. Plonsey, *Bioelectromagnetism* (Oxford University Press, New York (1995) [This great resource is made freely available on the web by the two authors: www.bem.fi/book/.]
6. Z.-H. Sheng, Mitochondrial trafficking and anchoring in neurons: new insights and implications. J. Cell Biol. **204**, 1087–1098 (2014)
7. K. Cole, *Membranes, Ions and Impulses* (Univ, California Press, 1968)
8. L. Hermann, *Lehrbuch der Physiologie* (A. Hirschwald, Berlin, 1905)
9. A.L. Muller, V.S. Markin, Electrical properties of anisotropic nerve-muscle syncytia - II. Spread of flat front of excitation. Biophys. J. **22**, 536–41 (1978)
10. J. Del Castillo, B. Katz, Quantal components of the end-plate potential. J. Physiol. **124**, 560–573 (1954)
11. P. Fatt, B. Katz, An analysis of the end-plate potential recorded with an intra-cellular electrode. J. Physiol. **115**, 320–370 (1951)
12. M. Xin, E.N. Olson, R. Bassel-Duby, Mending broken hearts: cardiac development as a basis for adult heart regeneration and repair. Nat. Rev. **14**, 529–541 (2013)
13. D.C. Michaels, E.P. Matyas, J. Jalife, Mechanisms of sinoatrial pacemaker synchronization: a new hypothesis. Circulation Res. **61**, 704–714 (1987)
14. J. Pantaleone, Synchronization of metronomes. Am. J. Phys. **70**, 992–1000 (2002)
15. G.D. Clifford, F. Azuaje, P.E. McSharry, *Advanced methods and tools for ECG data analysis* (Artech House, Norwood (USA), 2006)
16. B.G. Pickard, Action potentials in higher plants. Botan. Rev. **39**, 172–201 (1973)
17. E. Davies, Action potentials as multifunctional signals in plants: a unifying hypothesis to explain apparently disparate wound responses. Plant Cell Environ. **10**, 623–631 (1987)
18. D.C. Wildon et al., Electrical signalling and systemic proteinase inhibitor induction in the wounded plant. Nature **360**, 62–65 (1992)
19. R. Wayne, The excitability of plant cells: with a special emphasis on characean internodal cells. Botanical Rev. **60**, 265–367 (1970)
20. R.M. Benolken, S.L. Jacobson, Response properties of a sensory hair excised from Venus' fly-trap. J. Gen. Physiol. **56**, 64–82 (1973)

Further Reading

21. T. Visnovitz, I. Világi, P. Varró, Z. Kristof, Mechanoreceptor cells on the tertiary pulvini of Mimosa pudica L. Plant Signal. Behav. **2**, 462–466 (2007)
22. T.C. Ruch, H.D. Patton (eds.), *Physiological Biophysics*, 20th edn. (W. B. Saunders, Philadelphia, 1982)
23. R. Wayne, The excitability of plant cells. Botan. Rev. **60**, 266–363 (1994)
24. P. Hegelmann, Vision in microalgae. Planta **203**, 265–274 (1997)
25. J.D. Jackson, *Classical Electrodynamics*, 3rd edn. (John Wiley, New York, 1998)
26. J. Stiles, T.L. Jernigan, The basics of brain development. Neuropsychol. Rev. **20**, 327–348 (2010)
27. A.G. Volkov (ed.), *Plant Electrophysiology* (Springer, New York, 2013)

Chapter 8
Molecular Mechanics of the Cell

Abstract The mathematical models of polymers and membranes are introduced, to describe the microscopic deformation of biological materials. The mechanical properties at the molecular scale are inherently statistical, and remain hidden below macroscopic averages. But we can now see and manipulate a single molecule, and follow its movements in real time. This is made possible thanks to the advent of revolutionary tools, such as the atomic-force microscope and the optical or magnetic tweezers, which allow to test the mechanics of biological materials with unprecedented quality and accuracy, down to the molecular scale. Today it is possible to measure the elasticity of a single molecule, of a piece of DNA, of a fragment of cell membrane, and from such strictly physical comparisons, a totally new wealth of information has started to invade the already flooded desk of the biologist.

8.1 Elastic Models of Polymers

As it was described in the previous Chap. 6 and Appendix E, a large fraction of the components of a cell are in the form of long filaments: DNA, RNA, F-actin, spectrin, microtubules, long sugar chains such as proteoglycans, are all constituted by a large number of distinguishable units connected one after another, often linear but in some important cases branching out from a main chain: they are **polymers**.

A polymer is generically defined as a long and repeated supramolecular structure, made up of many identical copies of one or more elementary molecular fragments, the **monomers**. Such monomers are individual entities, jointed to one another by a strong covalent bond, about which two monomers can turn and bend. Polymers can be shaped according to different structures, by means of chemical design, as well as by different processing routes, e.g. by changing temperature, pressure, density, during the natural or artificial assembly process. A main classification, shown in the Fig. 8.1, is between single-chain, branched-chain, cross-linked, or networked polymers. In a cell, several of such structures can be found: actin filaments can form cross-linked structures, as in the stress fibres, as well as interconnected networks; proteins are single-chain polymers, which can however get cross linked at several sites; DNA

© Springer International Publishing Switzerland 2016
F. Cleri, *The Physics of Living Systems*, Undergraduate Lecture
Notes in Physics, DOI 10.1007/978-3-319-30647-6_8

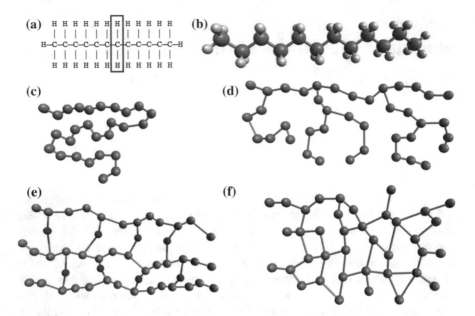

Fig. 8.1 **a** Schematic representations of polyethylene showing the monomer unit, an ethylene molecule CH$_2$ (*in red box*). **b** The zigzag backbone structure of a polyethylene macromolecule with 12 monomers, as obtained from a molecular dynamics computer simulation: carbon atoms in *grey*, hydrogens in *white*, sticks represent covalent bonds. **c** Linear polymer: a sequence of monomers linked to each other by covalent bonds (e.g., polyethylene, nylon, fluorocarbons). **d** Branched polymer: side-branch chains connect to the main one. **e** Crosslinked polymer. Adjacent linear chains are covalently connected by a foreign species' monomer (*blue spheres*); many of the rubber materials consist of e.g. polybutadiene chains crosslinked with Sulphur atoms, in the process called vulcanisation of rubber. **f** Polymer network: several monomer units with more than one active covalent bonds form a three-dimensional networks, such as in epoxies

and RNA are linear polymers, however RNA can form knots and superstructures that make it behave as a branched superstructure.

Monomers in a **homopolymer** are all identical, such as in carbohydrates and sugars, or varied from a class of similar elements, thus forming a *heteropolymer*. In the Appendix B to Chap. 3, we saw that all the proteins are in fact long, filamentary heteropolymers constituted by immensely varied combinations of the 20 basic amino acids. A globular protein like haemoglobin is made up of four nested amino acid chains, with 2×141 and 2×146 amino acids: if elongated into a single chain, it would measure about 0.5 μm in length; however, when folded it measures just about 2.5 nm in average diameter. In the same way, all nucleic acids like DNA and RNA are very long heteropolymers constituted by a combination of the four bases A, C, G, T (or U for RNA). All the components of the cytoskeleton are filamentary, or fibrous proteins, constituted by smaller monomers, G-actin for the actin, tubulin for the microtubules, and so on.

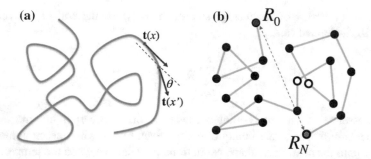

Fig. 8.2 **a** A chain polymer can be seen as a flexible, continuous wire, with thickness negligible compared to its length. In the *worm-like chain* model, a continuous variable x spans the length of the filament. The tangent vector $\mathbf{t}(x)$ can be defined at each point, and the angle θ between the tangent at two different points x and x' defines the local curvature radius R and curvature γ (see text). **b** Representation of the polymer as a *freely-jointed chain*. The *black* monomers are connected by rigid, non-extensible bonds; bonds can rotate in any direction at the joints; the *red* and *yellow* monomers indicate the start (R_0) and end (R_N) of the polymer. The *red dashed line* connecting R_0 and R_N is the geometrical definition of the "end-to-end" distance, R_{ee}

A polymer by its definition is a very flexible entity, in which for example the curvature may change continuously all along its structure (Fig. 8.2a). Many biopolymers found inside a cell can be described as filamentary, non-branched chains, which can give rise to extremely elongated supramolecular structures, by assembling either into nearly parallel bundles, as cytoskeletal proteins do; or into very compact structures, by folding onto itself, as most globular proteins will do.

For the happiness of the physicist, some of the fundamental properties of these macromolecular objects can already be captured and compared to experimental results, by constructing very simple mathematical models of their structure, and largely forgetting about their complex chemistry. As shown in Fig. 8.2b, sometimes even describing the polymer as a sequence of linked balls and sticks may be enough to infer some of its general physical features.

8.1.1 The Freely-Jointed Chain

This is possibly the simplest model of a molecule ever invented (Fig. 8.2b). In this case, the polymer is described as a sequence of $N+1$ monomers located at positions $\mathbf{R}_i$, $i = 0, ..., N$, and connected in adjacent pairs by a rigid rod of length $b = |\mathbf{R}_i - \mathbf{R}_{i-1}|$. The rod length is the modulus of the difference vector $\mathbf{r}_i = \mathbf{R}_i - \mathbf{R}_{i-1}$, and the total polymer length when extended, or *contour length*, is $L = Nb$.

To give a measure of the average size of the polymer, the **end-to-end distance**
vector $\mathbf{R}_{ee}$ is defined as:

$$\mathbf{R}_{ee} = \sum_{i=1}^{N} \mathbf{r}_i = \sum_{i=1}^{N} \mathbf{R}_i - \mathbf{R}_{i-1} = \mathbf{R}_N - \mathbf{R}_0 \tag{8.1}$$

The essence of the freely-jointed model is that at every joint, the adjacent
monomers are free to rotate in any direction, even by 2π with one monomer folding over onto the next. Such shape excitations may happen when the temperature is
$T > 0$ K, and the monomers acquire an average kinetic energy proportional to $k_B T$.
However, it is assumed that monomers cannot bend or stretch their bond lengths
$b = |\mathbf{r}|$. Such a model therefore has always zero potential energy, and its whole thermodynamic behaviour is only described by its entropy contents, pretty much like a
perfect gas of monomers with only the constraint of remaining at fixed relative pair
distances. (In fact, by looking just at black dots in the Fig. 8.2b we get the impression
of a gas.) Because of this choice, the thermally averaged value of the end-to-end
distance is also zero, $\langle \mathbf{R}_{ee} \rangle$. Just like in a random diffusion problem, the first nonzero
moment of the probability distribution of $\mathbf{R}_{ee}$ is its square:

$$\langle \mathbf{R}_{ee}^2 \rangle = \sum_{i,j=1}^{N} \langle \mathbf{r}_i \cdot \mathbf{r}_j \rangle = \sum_{i,j=1}^{N} \langle \mathbf{r}_i \rangle \langle \mathbf{r}_j \rangle = \sum_{i=1}^{N} \langle \mathbf{r}_i^2 \rangle = N b^2 \tag{8.2}$$

This is an interesting result. Since the average RMS end-to-end distance, $R_{ee} = \langle \mathbf{R}_{ee}^2 \rangle^{1/2}$, can be considered as a measure of the folded size of the fluctuating polymer,
it says that this size is proportional to the square root of the length, $N^{1/2}$, and the
volume occupied is proportional to $N^{3/2}$.

By means of simple arguments, it can be proved that the freely-jointed chain
is indeed akin to a diffusion problem in which monomers perform a random walk
in space, and the distribution function $P(R)$ of the monomers within the average
volume is, as usual, a Gaussian: $P(R) \propto \exp(-3R^2/2Nb^2)$. Moreover, the above
findings remain unchanged even if the model is modified, by adding the restriction
that two monomers cannot occupy the same space (which is in principle possible for
the totally free chain).

In order to find an improvement to this simple model, one must add the further
restriction that distant branches cannot cross any other branches of the polymer.
Look for example at the two white monomers in Fig. 8.2b: these are non adjacent,
but could come in close contact, since the freely-jointed model puts conditions only
on adjacent monomers. A further condition must be added, that segments around $\mathbf{R}_k$
must avoid segments around $\mathbf{R}_l$, with $k \neq l \pm 1$. This is called a *self-avoiding chain*,
and can be practically constructed by dividing the entire volume disponible for the
system, $V = R^3$, into small subvolumes of size v such that only one monomer b

could fit in; therefore, the probability for each fluctuating monomer, of finding a free space in V is proportional to $(1 - v/R^3)$, and the Gaussian distribution function is modified as:

$$P(R) \propto \exp\left(-\frac{3R^2}{2Nb^2} - \frac{N^2 v}{2R^3}\right) \tag{8.3}$$

With a little additional effort, it can be shown that the average size of the polymer in this case grows as the power $N^{3/5}$ of the length, and the occupied volume is therefore proportional to $N^{9/5}$, i.e. slightly below N^2. This is a result in reasonable agreement with the experimental results, in which exponents of about 0.588 are typically measured by x-ray scattering, indeed not far from 3/5.

The end-to-end distance is practically impossible to measure experimentally, unless the two ends of each polymer are marked, for example by some fluorescent label (and even in that case, it would require a very fine spatial resolution of the measurement). Fortunately, R_{ee} is not the only parameter providing a measure of the average size of the polymer. Other quantities have been introduced, such as the **gyration radius**:

$$R_g = \frac{1}{2N^2} \sum_{n,m} \langle (\mathbf{R}_n - \mathbf{R}_m)^2 \rangle = \frac{1}{N} \sum_n \langle (\mathbf{R}_n - \mathbf{R}_{CM})^2 \rangle \tag{8.4}$$

requiring the definition of the polymer center-of-mass $\mathbf{R}_{CM} = N^{-1} \sum_n \mathbf{R}_n$. Interestingly, the gyration radius can be directly measured by x-ray diffraction, being related to the small-angle contribution of the distribution of x-rays scattered by the polymer in a solution.

Another quantity that is directly related to an experimentally measurable quantity is the **hydrodynamic radius**:

$$R_H = \frac{k_B T}{6\pi \eta D} \tag{8.5}$$

While being a useful and easily measurable quantity, routinely used to estimate the size of globular proteins over a large range of sizes by the technique of Dynamic Light Scattering (DLS), the determination of R_H however relies on measuring the mobility of the molecule in solution, rather than its physical size; moreover, knowledge of the molecular diffusion coefficient D is required. To correlate R_H to the actual size, as measured by R_g and to a lesser extent by R_{ee}, further information about the structure of the polymer, its branching, local chemistry, is needed.

Bending of rods and membranes[1]
The bending deformation, as well as the torsion (or twisting), is an example of a *non-homogeneous* deformation, in which every point in the volume undergoes a different displacement. For the study of **bending of a long thin rod**, let us consider the sketch in the following figure. The rod is aligned with its length L along the x axis and held fixed at the extremes, and we assume that its cross section has a constant area A; let us consider the two points m and n along the middle axis, and a perpendicular force f tending to bend the rod at the midpoint.

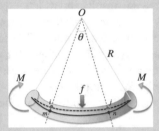

The effect of the force, in the regime of small deformation, is to impose a nearly circular curvature to the rod by an angle $\Delta\theta$, with curvature radius R. We can calculate the resulting deformation by looking at what happens to any other pair of points i and j, originally parallel to m and n in the undeformed rod:

$$\overline{mn} = R\Delta\theta \qquad \overline{ij} = (R-h)\Delta\theta$$

where h is the radial distance between the middle axis, and the (originally parallel) axis joining i and j. The angular deformation, or strain, is:

$$\varepsilon_\theta = \frac{(R-h)\Delta\theta - R\Delta\theta}{R\Delta\theta} = -\frac{h}{R} = -\gamma h \tag{8.6}$$

where we indicate as $\gamma = 1/R$ the local curvature.

It should be noted that with respect to the "neutral" $\overline{mn}$ axis (whose length does not change after bending), the lines having $h > 0$ (i.e., closer to the center of curvature O) are contracted in length, while lines with $h < 0$ (more distant from the center O) are stretched. Evidently, the strain ε_θ varies with $h(x)$ all along the length of the rod, being zero at the fixed points, and maximum at the middle of the rod; the stress σ (roughly the force per unit area; see Appendix H for the precise definition of stress) must follow the same profile. The *bending moment* M is defined as the integral of the bending force $f = \sigma dA$, taken over the values of r spanning the cross section, from $r = 0$ at the middle axis to $r = (A/\pi)^{1/2}$ at the surface:

$$M(x) = \int_A \sigma(x) r \, dA \tag{8.7}$$

In the most general case, $M(x)$ varies as well along the whole length $0 < x < L$. Note that the previous definition contains σ as the unknown, therefore it is not particularly useful, unless some solution for the stress field inside the rod is known (or can be guessed).

The Euler-Bernoulli equation provides the general relationship between the applied force and the bending displacement $h(x)$:

$$\frac{d^2}{dx^2}\left(E\mathscr{I}\frac{d^2h}{dx^2}\right) = \sigma \tag{8.8}$$

with E the Young's modulus of the material. By double integration along x, the equation linking the variation of $h(x)$ and $M(x)$ is obtained as:

$$M(x) = -E\mathscr{I}\frac{d^2h}{dx^2} \tag{8.9}$$

The quantity $\mathscr{I}$ in the previous equations is the **area moment of inertia** of the cross section perpendicular to x (not to be confused with the moment of inertia of the rod), defined as:

$$\mathscr{I} = \int dy\,dz(y^2 + z^2) \tag{8.10}$$

and contains the dependence of the bending moduli on the geometry of the rod. It varies very rapidly, with the fourth power of the transverse size, e.g., it is equal to $\mathscr{I} = \pi D^4/32$ for a thick cylinder of diameter D, or $\mathscr{I} = \pi(D^4 - d^4)/32$ for a hollow cylinder with internal cavity of diameter d; again, $\mathscr{I} = ab(a^2 + b^2)/12$ for a rectangular section with sides a and b, which becomes $\mathscr{I} = a^4/6$ if the section is square. The rapidly varying value of $\mathscr{I}$ explains, for example, why an H-shaped beam is more resistant to bending than a square or cylindrical beam (for the same mass of material).

From the geometrical definition of the problem, we obtain the following identities:

$$\theta(x) = \frac{dh}{dx} \quad ; \quad \gamma(x) = \frac{1}{R(x)} = \frac{d\theta}{dx} = \frac{d^2h}{dx^2} \tag{8.11}$$

from which it is also $R(x) = E\mathscr{I}/M(x)$. In the classical "three-point-bending" experiment, described in the figure, the radius of curvature changes at every point along L; the solution of the Euler-Bernoulli equation gives a third-order polynomial for $h(x)$, therefore both R and M are linear functions of x, with their maximum at the centre of the rod.

The elastic bending energy (stored in the rod by the applied force) is the integral of the bending moment over the half-length of the rod, from zero to the maximum, since the moment of the force acts in a symmetrical manner with respect to $0 < x < L/2$ and $L/2 < x < L$:

$$E_{el} = \int_0^{\theta/2} M d\theta = \int_0^{L/2} M(x) \frac{dx}{R(x)} \tag{8.12}$$

For small deformations, it can be considered that in each small segment of length the curvature radius is constant, and locally only axial stretching is acting, so that $\sigma \simeq E\varepsilon$. With such approximations, is is obtained:

$$E_{el} = \frac{1}{2}(E\mathscr{I})\frac{L}{R^2} = \frac{1}{2}\kappa_b L\gamma^2 \tag{8.13}$$

The constant $\kappa_b = E\mathscr{I}$ is the **bending modulus** of the rod, with dimensions of [energy] $\times$ [length] (differently from all the other elastic moduli, which are a volumetric energy density). Notably, κ_b includes both the information about the material property, embedded in the Young's modulus, and about the geometry of the rod, embedded in the moment of inertia.

The **bending of a flat membrane** with finite thickness w, shown in the next figure (left) can be treated by similar arguments by making reference to the definitions of local curvatures γ_1, γ_2 at a point P on the surface, and the $\{\mathbf{t}_1, \mathbf{t}_2, \mathbf{t}_3\}$ local reference frame at that point (see Fig. 11.6 on p. 487).

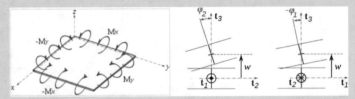

The drawings on the right of the previous figure describe the local rotation of the plate with thickness w by effect of the bending force, applied perpendicular to P (i.e., parallel to $\mathbf{t}_3$). The equivalent of the angle θ for the rod are now two distinct angles ϕ_1 and ϕ_2 about the $\mathbf{t}_2$ and $\mathbf{t}_1$, respectively, and the bending amplitude h is now function of the two directions $h(x, y)$. Every point $P'(x, y)$ of the surface about $P(0, 0)$ is lifted with respect to the initial quota $z = 0$ by an amount $h(x, y)$, with the resulting strain tensor components:

$$\begin{cases} \varepsilon_{11} = h\dfrac{\partial\phi_2}{\partial x} \\[2mm] \varepsilon_{22} = -h\dfrac{\partial\phi_1}{\partial y} \\[2mm] \varepsilon_{33} = h \end{cases} \tag{8.14}$$

The two curvatures along the directions $\mathbf{t}_1$ and $\mathbf{t}_2$ are defined as:

$$\gamma_{11} = -\frac{\partial \phi_2}{\partial x} = \frac{\partial^2 h}{\partial x^2} \quad ; \quad \gamma_{22} = \frac{\partial \phi_2}{\partial y} = \frac{\partial^2 h}{\partial y^2} \tag{8.15}$$

The bending moments M_{11} and M_{22} act symmetrically along the x and y borders of the membrane, while $M_{12} = M_{21}$ acts to twist the borders; by analogy with the rod, the moments are defined as:

$$M_{ij} dy = \int_{-w/2}^{w/2} (\sigma_{ij}) z \, dz \tag{8.16}$$

As we already did for the long thin rod, we can simplify the problem by assuming local biaxial stretching/compression, by using the corresponding linear stress-strain relations (see Appendix H in the next Chapter):

$$\begin{cases} \sigma_{11} = \dfrac{E}{1 - v^2}(\varepsilon_{11} + v\varepsilon_{22}) = -\dfrac{E}{1 - v^2}(\gamma_{11} + v\gamma_{22})h(x, y) \\[2mm] \sigma_{22} = \dfrac{E}{1 - v^2}(\varepsilon_{22} + v\varepsilon_{11}) = -\dfrac{E}{1 - v^2}(\gamma_{22} + v\gamma_{11})h(x, y) \\[2mm] \sigma_{12} = \dfrac{E}{1 + v}\varepsilon_{12} = \end{cases} \tag{8.17}$$

from which the explicit formulae for the (linearized) bending moments are:

$$M_{11} = K_b(\gamma_{11} + v\gamma_{22}) \quad ; \quad M_{22} = K_b(\gamma_{22} + v\gamma_{11}) \quad ; \quad M_{12} = K_b(1 - v)\gamma_{12} \tag{8.18}$$

or in compact form $M_{ij} = K_b[(1 - v)\gamma_i\gamma_j + v\gamma_k^2\delta_{ij}]$, with δ_{ij} the Kronecker symbol.

The elastic bending energy of the plate membrane is then:

$$E_{el} = \frac{1}{2}\sum_{ij} M_{ij}\gamma_{ij} = \frac{K_b}{2}[(1 - v)\gamma_{ij} + v\gamma_{kk}\delta_{ij}]\gamma_{ij} =$$

$$= \frac{K_b}{2}\left\{(\gamma_{11} + \gamma_{22})^2 - 2(1 - v)\left[\gamma_{11}\gamma_{22} - (\gamma_{12})^2\right]\right\} \tag{8.19}$$

The constant $K_b = Eh^3/12(1 - v^2)$ is the **bending modulus** of the membrane, with dimensions of [energy]. Note that for a locally spherical or cylindrical curvature γ_{12} and M_{12} are both zero. The average curvature is often indicated as $H = \frac{1}{2}(\gamma_{11} + \gamma_{22})$, while the product $G = \gamma_{11}\gamma_{22}$ is called Gaussian curvature.

---

[1]For the definitions of the stress/strain tensors (σ, ε) and elastic moduli (E, v), see Appendix H on p. 422.

8.1.2   The Worm-Like Chain

If the freely-jointed chain model is successful in describing easily bendable polymer, such as RNA or single-strand DNA, proteins like titin, or long sugars like polyglycans, it starts to fail when the monomer chain becomes increasingly rigid, and its bending occurs over lengths several times longer than the typical monomer size b. Such kind of polymers require a model in which the chain can be described as a smooth arrangement, with a curvature distributed and correlated over many segments. In the extreme limit, the monomer size becomes so small compared to the typical bending length that the chain can be described as a continuous filament (such as that in Fig. 8.2a), abandoning the individual tracking of each monomer. By analogy with the snaking movement that it evokes, this description of the polymer is called the *worm-like* chain model.

In this model, firstly introduced in 1949 by the Austrian physicists Otto Kratky and Günther Porod in the context of quantum spin chains, the chain is described by a continuous variable x spanning the polymer length L. A point on the chain is identified by its continuous position vector $\mathbf{r}(x)$, and at each point two unitary *tangent* and *normal* vectors can be defined, as:

$$\mathbf{t}(x) = \frac{d\mathbf{r}}{dx} \quad ; \quad \mathbf{n}(x) = \frac{d\mathbf{t}}{dx} \tag{8.20}$$

The end-to-end distance is given in this model by the integral:

$$R_{ee} = \int_0^L \mathbf{t}(x)dx \tag{8.21}$$

Also a *binormal* vector perpendicular to the plane identified by $\mathbf{t}$ and $\mathbf{n}$ can be defined, as the vector product of the two, $\mathbf{b}(x) = \mathbf{r}(x) \times \mathbf{t}(x)$. The normal vector measures the local curvature of the polymer in the plane containing $\mathbf{t}$ and $\mathbf{n}$, as the variation of the tangent orientation θ between any to close points x and x' (see again Fig. 8.2a).

The change of the tangent vector along the curved path also gives a measure of the amount of elastic energy needed to bend the filament, with respect to the straight configuration representing the zero of the energy, as:

$$E_{el} = \kappa_b \int_0^L \left(\frac{d\mathbf{t}}{dx}\right)^2 dx \tag{8.22}$$

Here κ_b is a constant characterising the resistance to bending of the material composing the filament, called the **bending modulus** (see greybox on p. 322). If we imagine to discretise back the continuous polymer into M small segments of length u (not necessarily coinciding with b) within which the curvature is nearly constant, the elastic energy can be also written as:

$$E_{el} = \frac{\kappa_b}{u} \sum_{i=0}^{M-1} (\mathbf{t}_i \cdot \mathbf{t}_{i+1}) = \frac{\kappa_b}{u} \sum_{i=0}^{M-1} \cos \theta_i \tag{8.23}$$

of which the Eq. (8.22) represents the continuous limit for $u \to dx$.

By statistical mechanics arguments, it can be shown that the average value of the correlation of the tangent vector at distant points along the chain decays exponentially:

$$\langle \mathbf{t}(x) \cdot \mathbf{t}(x') \rangle = \langle \cos \theta \rangle \propto e^{-|x-x'|/\lambda_p} \tag{8.24}$$

The dot product between the two tangent vectors, averaged over all points and all distances $x - x'$, is a measure of the probability of observing a given orientation of the tangent along the length L: if the curvature is slowly varying, the dot product is ~ 1 for long portions of the length, while if the curvature changes wildly over short distances, the dot product fluctuates randomly between 1 and -1, averaging to ~ 0 over short lengths. The typical length, λ_p in the previous equation, over which the tangent retains a good correlation is called for this reason the **persistence length**.

A further meaning of this new quantity λ_p can be found, by looking at its relationship with R_{ee}:

$$\langle R_{ee}^2 \rangle = \langle (\mathbf{r}(L) - \mathbf{r}(0))^2 \rangle = \int_0^L dx \int_0^L dx' \langle \frac{d\mathbf{r}}{dx} \cdot \frac{d\mathbf{r}}{dx'} \rangle =$$

$$= \int_0^L dx \int_0^L dx' \langle \mathbf{t}(x) \cdot \mathbf{t}(x') \rangle = \int_0^L dx \int_0^L dx' e^{-|x-x'|/\lambda_p} =$$

$$= \int_0^L dx \left[\int_0^x dx' e^{-(x-x')/\lambda_p} + \int_x^L dx' e^{-(x'-x)/\lambda_p} \right] \tag{8.25}$$

Note that the modulus $|x - x'|$ was split over two integrals to keep it positive valued. The result of the double integration is:

$$\langle R_{ee}^2 \rangle = \lambda_p \int_0^L dx \left[2 - e^{-x/\lambda_p} - e^{-(L-x)/\lambda_p} \right] =$$

$$= 2\lambda_p L \left[1 - \frac{\lambda_p}{L} \left(1 - e^{-L/\lambda_p} \right) \right] \tag{8.26}$$

In the limit of a highly flexible polymer, $\lambda_p \ll L$, it is $\langle R_{ee}^2 \rangle \sim 2\lambda_p L$. By comparing this expression with the corresponding one for the freely-jointed chain, $\langle R_{ee}^2 \rangle = Lb$, we can identify a sort of "equivalent" monomer length, $\bar{b} = 2\lambda_p$, over which the curvature remains constant, and which allows to describe the worm-like chain as an equivalent freely-jointed chain but with different monomer size $\bar{b}$, the so-called *Kuhn length* of the polymer.

Notably, semi-flexible biopolymers such as collagen, double-stranded DNA, F-actin, are well described by the worm-like chain model. However, it can also be seen that the freely-fluctuating polymer as described by either the freely-jointed chain or the worm-like chain models, occupy much larger volumes than their "containers", as it is easily verified by using the above Eq. (8.26) to estimate the average size of some biopolymers: for example, the T2 bacteriophage virus DNA has a contour length $L = 50\,\mu m$ and Kuhn length $\bar{b} \sim 110$ nm, therefore its average size should be about $\sqrt{L\bar{b}} \sim 2.3\,\mu m$, however the T2 viral capsid measures just 100 nm; again, the E. coli DNA with its 4.6 million base pairs is $L = 15$ mm long, for an average polymer size of 40 μm, but the whole bacterium cell is a much smaller cylinder of about $2 \times 1\,\mu m$ (length $\times$ width). Clearly, some smart packaging mechanisms are required to fit such lengths within those restricted volumes.

8.2 Biological Polymers

The polymers that are found in the cell have a bending elasticity modulus (with units of [Energy][L]) that can vary over more than six orders of magnitude: from the very flexible alkyl chains $(CH_2)_n$ with $\kappa_b \sim 10^{-2}$ eV nm, to semi-flexible proteins like the F-actin ($\kappa_b \sim 200-400$ eV nm), to very rigid objects like the microtubules ($\kappa_b \sim 10^5$ eV nm). On the micrometer scale, such polymers chan change shape as a function of several external parameters, such as pH or temperature, passing from bundles of parallel strings (like uncooked spaghetti) to tightly folded globules (like boiled spaghetti). As we are going to see, the elastic properties and the great conformational freedom of such objects are mostly of entropic origin.

8.2.1 Bending Fluctuations and the Persistence Length

If we consider an elastic string at T $= 0$ K, it has zero kinetic energy; moreover, if we take it to have no interactions with the surroundings, also its potential energy is zero. Such a thing should adopt the shape that allows it to minimise the internal elastic energy: the deformation tensor (see Appendix H) should be identically zero, $\varepsilon_{ij} = 0$, therefore the shape should be a straight line. However, at any temperature T > 0 K, on the other hand, the filament can start exchanging some elastic free energy against the energy of thermal fluctuations from the environment. With the above quoted values of typical linear bending modulus of a biopolymer, a fragment of a few nm length can be more or less severely perturbed already by thermal energies of a few 10^{-2} eV (corresponding to $T \sim 300$ K). As a result, thermal fluctuations will make the polymer explore many deformed configurations, by trading internal energy and entropy.

At constant temperature, the equilibrium is $\Delta F = \Delta E - T \Delta S = 0$. The probability of finding the polymer in a given deformed state, which costs an elastic energy ΔE_{el}, is proportional to its entropy, or to the number of equivalent microscopic states Ω by which this same deformed configuration can be obtained:

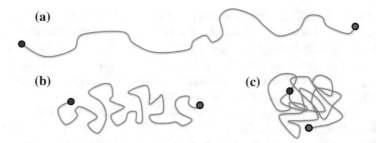

Fig. 8.3 A polymer can explore a number of configurations corresponding to the same value of energy at constant temperature. **a** At low temperature, the polymer has little excitation energy to spend, therefore it remains nearly straight; correspondingly, the number of equivalent conformations Ω is small, and so is its entropy. **b** As the temperature is increased, the polymer explores an increasing number of equivalent conformations: its elastic energy is higher, and compensated by an increased entropy because of the increased Ω. **c** At the highest temperatures, the polymer can assume a highly folded structure, allowing it to store a high elastic energy, and to explore a very large number of equivalent conformations, with a correspondingly high entropy. The dynamical transformation of the polymer from **a** to **c** is called the *coil-to-globule* transition: the R_{ee} and R_g parameters change drastically from large to small

$$P(\Delta E_{el}) \propto \Omega(\Delta E_{el}) \propto \exp(-\Delta E_{el}'/k_B T) \qquad (8.27)$$

The larger the number of equivalent conformations the polymer can adopt, the higher its entropy. But also, since the more variable conformations are also the more deformed ones, the higher its elastic energy (Fig. 8.3). The function $\exp(-\Delta E/k_B T)$ is as usual the Maxwell-Boltzmann distribution at the temperature T. Such a functional form of the probability distribution also tells that the higher the difference in elastic energy for a given conformation, the smaller its probability. Therefore, our polymer will tend to adopt less curved conformations at lower temperature, and increasingly more bent and twisted conformations as the temperature is increased.

By using the above defined variable $0 < x < L$ spanning the length L of the polymer, the elastic bending energy for a filament locally bent by some angle $\theta(x)$ (or, with a local radius of curvature $R(x)$) is found by integrating the expression:

$$E_{el} = \int_L \kappa_b \theta(x)dx = \frac{\kappa_b}{L} \int_{\theta(0)}^{\theta(L)} \theta d\theta = \frac{1}{2}\frac{\kappa_b}{L}\langle\theta\rangle^2 = \frac{\kappa_B L}{2\langle R\rangle^2} \qquad (8.28)$$

This latter equation is just the Eq. (8.22) above, in which we introduced formally the average curvature radius $\langle R\rangle$, or the average angle $\langle\theta\rangle$. Such (unknown) values can provide an approximate characterisation of the extent of folding of the polymer, which will ultimately correspond to an average elastic energy. (Of course, such a characterisation is very rough, if we hope to capture the complex fluctuating behavior of the polymer with just one simple number.)

It may be noted that the elastic energy in this small-deformation limit is always associated to a quadratic form, either in the curvature (Eqs. 8.13 and 8.19), or the normal vector (Eq. 8.22), or the curvature radius or angle (Eq. 8.28). This is always the case, whenever a linear approximation for a force, or a generalized force is adopted: if the force depends linearly on a parameter y, $f = -\frac{dU}{dy} \propto y$, the corresponding potential energy, that is the integral of $f\,dy$, is necessarily quadratic in the parameter y, $U \propto y^2$. This is called a *harmonic* approximation of the potential energy, since its prototype is the harmonic oscillator (see greybox on p. 446).

As it is usual with thermally-driven excitations, the first moment of the distribution of the fluctuating variable, in this case the average angle, is zero $\langle\theta\rangle = 0$. Therefore, to characterise the thermal fluctuation we must calculate the average root-mean-square fluctuation, $\langle\theta^2\rangle = 0$. This can be estimated by probability averaging of the fluctuating θ-squared values:

$$\langle\theta^2\rangle = \frac{\sum_i \theta_i^2 P(\theta_i^2)}{\sum_i P(\theta_i^2)} \tag{8.29}$$

But as Eq. (8.28) tells us, the values θ^2 are proportional to the elastic energy, and so must be their probability distribution. Then, $P(\theta)$ can be replaced by $P(\Delta E)$, and by replacing the sums with integrals, for finely spaced energy values:

$$\langle\theta^2\rangle = \frac{\int \theta_i^2 P(\Delta E)d\theta}{\int P(\Delta E)d\theta} = \frac{\int \theta_i^2 e^{-\Delta E/k_B T} d\theta}{\int e^{-\Delta E/k_B T} d\theta} = \frac{\int \theta_i^2 e^{-\kappa_b\theta^2/2Lk_B T} d\theta}{\int e^{-\kappa_b\theta^2/2Lk_B T} d\theta} \tag{8.30}$$

The result of the Gaussian integrals in the numerator and denominator is known (see Appendix A to Chap. 2), and gives:

$$\langle\theta^2\rangle = \frac{2Lk_B T}{\kappa_b} \tag{8.31}$$

Interestingly, since the bending modulus has dimensions of [Energy][L], a new quantity with dimensions of [L] can be identified:

$$\lambda_p = \frac{\kappa_b}{k_B T} \tag{8.32}$$

This λ_p is a characteristic length of the filament at any temperature, and is just the same *persistence length* that was previously defined. Because of its relationship with the average bending angle $\langle\theta\rangle$, such a quantity gave us a measure of the length over which the curvature is constant; here, by taking the non-dimensional ratio L/λ_p it is also a kind of "bending number" of the polymer segment at a given temperature. In other words, it gives a quick measure of the average number of bends that one can expect in a fragment of length L, thus being a kind of natural wavelength.

As a third, very interesting interpretation, it will be noted that λ_p also gives, for a polymer of given rigidity κ_b, the typical length over which the fragment appears stiff or flexible: a fragment with length L smaller or $\sim\lambda_p$ appears rigid, since it allows no bends; a fragment with $L \gg \lambda_p$ has many bends, and looks flexible. What this tells us, is that there is no such thing as an absolute stiffness or absolute flexibility when we discuss the elasticity of long filaments: it all depends on their length. It is common experience that if we take a 1 cm of chicken wire this will be quite rigid, while 1 m of the same wire will be very flexible. Furthermore, by looking at Eq. (8.32) for a given polymer type and fragment length, it can be seen that the persistence length decreases with increasing temperature, a measure of the effect of larger temperature fluctuations.

The persistence length of the more flexible biological polymers in the cell varies between the $\lambda_p = 0.5$ nm of titin (the biggest known protein in the human body), to 10–15 nm of spectrin or proto-collagen, to about 54 nm of DNA. Note that such a value of λ_p justifies the good applicability of the worm-like chain model to describe ds-DNA, since the b of its monomers (the nucleobases) is about 3.4 nm, 30 times smaller than the equivalent Kuhn length. For a typical cell diameter $D \sim 10\text{--}20$ μm, intermediate values of $\lambda_p \sim D$ are quite typical of many cytoskeletal polymers; however the extreme case of microtubules is worth mentioning: with a λ_p of several millimetres, microtubules inside the cell appear as rigid as a steel reinforcement in a concrete structure, despite their Young's modulus being less than 1/200th that of steel.

8.2.2 Elasticity From Entropy

It must be observed that the number of conformations Ω for a folded polymer is very large, compared to the corresponding number for a nearly straight polymer (see again Fig. 8.3 on p. 329). Strictly speaking, a perfectly straight filament has a value $\Omega = 1$, since it is allowed only one configuration (we neglect axial rotations about the length, which leave the conformation identical).

Therefore, we expect also the entropy, $S \propto \ln \Omega$ of a folded conformation to be higher than that of an elongated conformation. When a force f is applied to the two ends of a folded polymer to unfold and elongate it (such as the red and yellow ends of the polymer in Fig. 8.3c), this force works against the entropy of the system, which would naturally keep the folded structure with a larger S.

For the simpler freely-jointed chain model, the unfolding occurs without chemical changes of the internal molecular structure, and without stretching or bending of the individual monomers. It can be considered that the system enthalpy is not changing during the forced unfolding (in fact it is identically zero), therefore the entropy change practically coincides with the system free-energy variation, $\Delta F = T \Delta S$.

The resistance to the unfolding in this entropic deformation regime is therefore due to the natural trend of the system to search for states with the highest value of entropy. An amount of mechanical work ΔW has to be spent by the external unfolding force, pulling through an elongation Δx to reduce the entropy of the folded polymer to that of an elongated filament:

$$\mathbf{f} = \left(\frac{\Delta W}{\Delta x}\right)\hat{\mathbf{f}} = -T\left(\frac{\Delta S}{\Delta x}\right)\hat{\mathbf{f}} \tag{8.33}$$

so that $\mathbf{f} \cdot \Delta\mathbf{x} = \Delta W = T\Delta S$. Note that the elongation Δx is taken along the direction projected on the unit vector parallel to the applied force $\hat{\mathbf{f}} = \mathbf{f}/|\mathbf{f}|$; the force applied at the two ends acts on the whole length L, to increase the polymer from R_{ee} to $R_{ee} + \Delta x$ as:

$$\mathbf{f} \cdot \Delta\mathbf{x} = \int_R^{R+\Delta x} (\mathbf{f} \cdot \mathbf{t}(x)) \, dx = f \int_L \cos\alpha(x) dx \tag{8.34}$$

where α is the angle formed by the applied force $\mathbf{f}$ and the local tangent $\mathbf{t}(x)$, and $f = |\mathbf{f}|$.

The probability distribution of the freely-jointed monomers in the folded configuration was previously described to be akin to that of a gas, with a Gaussian function of average R_{ee} and variance $\sigma = N\bar{b}^2$:

$$P(R) \propto \exp(-R_{ee}^2/2N\bar{b}^2) \tag{8.35}$$

Let us try to push further our analogy of the entropic elasticity, by taking that the free energy of the folded chain polymer elongating by a small amount x can be characterised by a kind of average "entropic spring", K_S, ideally working just like a hookean spring, $\Delta F = \frac{1}{2}K_S x^2$. If we compare the distribution of the folding energies with that of the equivalent harmonic spring, we get:

$$P(\Delta F) \propto \exp(-K_S x^2/2k_B T) \tag{8.36}$$

The mean root squared displacement being $\langle x^2 \rangle = \frac{1}{3}\langle R_{ee}^2 \rangle$ (each direction x, y, z contributes 1/3 of the average displacement R), the arguments of the two exponentials can be equated, eventually finding the following result for the effective entropic "spring constant":

$$K_S = \frac{3k_B T}{N\bar{b}^2} = \frac{3k_B T}{2L\lambda_p} \tag{8.37}$$

To obtain the second identity, we used $\lambda_p = \bar{b}/2$ as an estimate of the persistence length, for the case of a semi-flexible polymer with Kuhn length $\bar{b}$. Such an elasticity is of a purely entropic origin. It comes from the resistance of the polymer to adopt any conformation that would decrease its entropy. The force-extension relationship would be of the Hooke's type:

$$x = \frac{f}{K_S} = \frac{N\bar{b}^2}{3k_B T} f \tag{8.38}$$

however valid only for small extension x of the polymer. This is because we assumed that only small variations around one single value of the free energy could take place during the deformation. The more general case, in which all the free energy contributions are weighted according to their probability of occurrence, is discussed in the greybox on p. 333.

The force-extension curve of a free polymer

An expression valid for a more general deformation state can be obtained by using the probability distribution $P(\Delta F)$ as a weight for all the deformed configurations corresponding to any value of entropy during the polymer chain extension. Let us call Z a normalization factor equal to the integral of all statistical weights:

$$Z = \int e^{-\beta F} dF \qquad (8.39)$$

with $\beta = (k_B T)^{-1}$. In physics, a function like Z is called a *partition function*.

Then, the average value of $\Delta F = F - F_0$ for each given deformation state with respect to the undeformed free energy F_0, should be:

$$\langle F \rangle = \frac{\int F e^{-\beta F}}{Z} dF + c = -\frac{1}{Z}\frac{\partial Z}{\partial \beta} = -\frac{\partial \ln Z}{\partial \beta} \qquad (8.40)$$

or:

$$\langle F \rangle = -k_B T \ln Z \qquad (8.41)$$

By looking at the Eq. (8.34), we note that each Kuhn segment $\bar{b}$ contributes the same length in the integral.

$$f\Delta x = f \int \cos\alpha\, dx = f\bar{b}\cos\alpha \qquad (8.42)$$

Therefore, the overall partition function for the freely-jointed chain is $Z = (Z_1)^N$, with:

$$Z_1 = \int e^{-\beta f\bar{b}\cos\alpha}\, d\Omega = 2\pi \int_0^\pi \sin\alpha\, e^{-\beta f\bar{b}\cos\alpha}\, d\alpha =$$
$$= -2\pi \int_1^{-1} u e^{-\beta f\bar{b}u}\, du = \frac{4\pi}{\beta f\bar{b}}\sinh(\beta f\bar{b}) \qquad (8.43)$$

for each Kuhn segment, and the average free energy during the polymer extension is:

$$\langle F \rangle = -k_B T \ln(Z_1)^N = -N k_B T \ln\left[\frac{4\pi}{\beta f b}\sinh(\beta f\bar{b})\right] \qquad (8.44)$$

The elongation Δx is the derivative of the free energy with respect to the applied force, thus:

$$\Delta x = -\frac{dF}{df} = N\beta^{-1}\frac{d}{df}\left[\ln\left(\sinh(\beta f\bar{b})\right) - \ln f\right] = N\bar{b}\left[\coth\left(\frac{f\bar{b}}{k_B T}\right) - \frac{k_B T}{f\bar{b}}\right] \qquad (8.45)$$

For small values of $f \ll k_B T/\bar{b}$, $\coth u - 1/u \simeq u$ and the same small-elongation expression as Eq. (8.38) is retrieved; however, for large forces $\coth u \to 1$ and the force is seen to diverge:

$$f = \frac{k_B T}{\bar{b}}\frac{1}{1 - z/L} \qquad (8.46)$$

as $\Delta x \to L$, when the elongation approaches the contour length, i.e. at a strain $\varepsilon \sim 1$.

It may be interesting to note that the entropic stiffness of the polymer increases with the temperature (in parallel to its persistence length decreasing, Eq. (8.32)). In fact, a very simple experience that can be done at home is to suspend a small weight with an elastic string, and heat the string with a hair-dryer: contrary to the intuition (the rubber becoming softer?) the weight will raise, since the stiffness of the equivalent spring has increased. In the deformation of polymeric chains, and notably for the polymerised filamentary proteins constituting the cell cytoskeleton, which we described in Chap. 6, the most important contribution to the elastic response comes from entropy. Only for the more rigid structural elements, such as the microtubules, the enthalpic (ordinary) elasticity of the chemical bonds plays a more relevant role.

An equivalent force-extension relationship can also be obtained for the worm-like chain model. In this case, the free energy contains both an elastic energy term and the already known entropy of the folded polymer:

$$\Delta E_{el} - T \, \Delta S = \frac{\kappa_b}{2} \int_0^L \left(\frac{d\mathbf{t}}{dx}\right)^2 dx - f \int_0^L \cos\alpha(x)dx \qquad (8.47)$$

No analytic solution exists for this equation, however a numerical solution accurate to better than 0.1 % has been obtained in the literature:

$$f = \frac{k_B T}{\lambda_p} \left[\frac{1}{4}\left(1 - \frac{x}{L}\right)^{-1/2} - \frac{1}{4} + \frac{x}{L} \right] \qquad (8.48)$$

Note that this force-extension relationship qualitatively resembles that of the freely-jointed chain (derived in the greybox), with different numerical coefficients: at small values of force it gives a linear-spring-like response, with an effective spring constant $k_B T/L\lambda_p$; while at large forces, it diverges as the extension approaches the contour length, $x \to L$. However, the divergence is $(x/L)^{-1}$ for the freely-jointed chain, while it goes with the power -2 for the worm-like chain model, therefore the former force curve would be more shallow than the latter.

8.2.3 Pulling Nanometers with Piconewtons

The characteristic features of the force-extension (or stress-strain) equations for single molecules, derived in the previous Section, can be appreciated in the Figs. 8.4 and 8.5. In the first one, the results of experiments of stretching a DNA molecule by the **optical tweezers** technique are shown. This experimental technique, introduced in 1986 by the Nobel laureate Steven Chu at Berkeley, [2] is based on the use of optical traps, by which small dielectric microparticles can be positioned and moved with great precision. An optical trap is formed by tightly focusing a laser beam with an objective lens of high numerical aperture. Dielectric microparticles in the vicinity of the focus experience a three-dimensional restoring force from the 'radiation pressure" of the laser light, directed toward the focus. Typically, for small displacements

(<150 nm) of the microparticle from its equilibrium position, the force gradient is linearly proportional to the displacement and the optical trap is well approximated as a linear spring. If a molecule or another object is attached to the microspheres (see Fig. 8.4a) it can be pulled by the movement of the trapped bead. By knowing the laser parameters, the corresponding force is deduced from the measurement. Since the typical Young's modulus of biomolecules is in the range of $E \sim 10$–100 s MPa, with sizes in the 1–10 nm, nanometer-scale displacements and pN forces can be detected.

In Fig. 8.4b it may be observed that the worm-like chain model provides an excellent fit of the entire first part of the curve, covering both the entropic regime, in which the tangled polymer is unfolded, and the "enthalpic" regime, in which the DNA should be already fully elongated and the force starts pulling on the single nucleotide bonds. The fit with the freely-jointed chain model, compared to the worm-like chain in Fig. 8.4c, as expected, predicts a more shallow force-extension curve.

From the panel (a) in Fig. 8.4, for the case of DNA, it can also be observed that at about $f \sim 65$ pN the force-extension curve is abruptly interrupted by what has been called an "overstretching transition". What happens to DNA in this part of the curve is not entirely clear, however it should probably constitute a structural transformation, either associated to the separation of the two helical strands, or possibly to a collective switching of the nucleotide bonds to a different form. In any case, it is seen that when the strain reaches $\varepsilon \sim 1.75$ this transformation is completed, and the curve resumes its rapid increase.

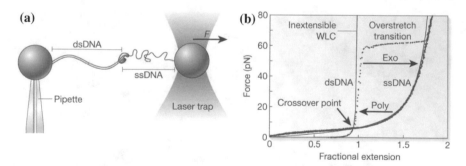

Fig. 8.4 Experiments of stretching of DNA fragments by means of the "optical tweezers" instrument. **a** The DNA molecule is fixed by its two ends on the surface of two plastic microspheres of a few μm size. One of the microspheres is held fixed by a pipette, while the other is dragged by the electric field gradient produced a tightly focused, moving laser beam. This movement can pull the DNA molecule to almost twice its original length. In this way, piconewton-scale forces can be measured, over displacements of a few nm. **b** Force-extension curve; the force is in pN, the extension is relative to the contour length $\Delta L/L$. The WLC red curve represents the fit of the first part by the "worm-like-chain" model, Eq. (8.48). The "overstretching" plateau connects the region in which DNA responds as a whole double-stranded molecule, to the region in which it appears to respond as the sum of two independent single strands. In fact, the *second red curve* (corresponding to the force release, and the DNA folding back to its non deformed state) is better fitted by a freely-jointed chain model (FJC) for single-strand DNA. [Image from Ref. [1], repr. w. permission]

The following Fig. 8.5 shows the results of stretching experiments performed on a multidomain protein, by a different experimental technique, the **atomic-force microscope probe**. The giant molecule titin, longer than 1 μm, is composed by more than 34,000 amino acids, organised into 244 subdomains belonging to two types, the immunoglobulin (Ig) and the fibronectin. In Chap. 10 we will meet again titin, the template molecule relying the extremities (Z-disk) of the muscle sarcomeres, with the fundamental role of relaxing back the muscle fibres after an effort. Titin does this exactly by stretching its many domains under stress, and by folding them back after the stress is released. In the experiments by the group of Gaub, Rief and their coworkers in Munich, [3] fragments of titin containing only eight (nearly) identical Ig domains were subject to mechanical stretching by the technique of atomic microscope force-probe (AFM), to test this molecular mechanism of folding and unfolding under stress.

In this type of experiments (Fig. 8.5a) an atomic-force microscope is used as a mechanical tester, by attaching one end of the molecule on a fixed substrate (glass, plastic), and the other end to the tip of a micron-sized mobile cantilever. The force on the cantilever can be controlled to the precision of a few pN, and its displacement is recorded by the interference pattern of a laser beam reflecting on its back surface. The main difference with respect to the optical tweezer technique, besides the obviously different technical set up, is that such an instrument provides a much higher effective

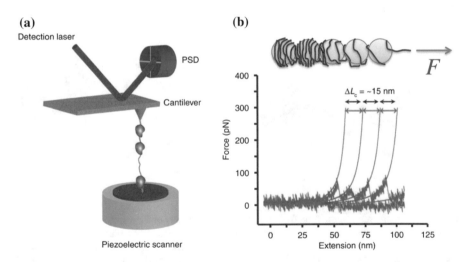

Fig. 8.5 Single molecule stretching experiment performed on a multi-domain protein by the atomic-force microscope (AFM). **a** The mobile cantilever of the AFM ends with an atomically-sharp tip, to which one end of a molecule can be attached, while the other end is fixed on a piezoelectric scanner. The displacement of the cantilever is monitored by the laser beam, reflected on a position-sensitive detector (PSD). **b** *Above* the multi-domain protein is represented as a sequence of equal domains, each folded into a globule; the applied force F pulls on one end. *Below* Each of the force peaks in the stress-strain curve corresponds to the individual unfolding of one subsequent domain, and for each domain the worm-like chain model (*dashed lines*) provides a good fit of the force-extension curve. [Image adapted from Ref. [4], repr. w. permission]

stiffness, thus allowing to perform mechanical testing of "stronger" molecules (with a larger K_S): where the optical tweezer allows a force range from fractions up to a few pN, the atomic-force microscope can apply forces in the 10–100 pN range, for similar displacements in the range of $\sim$10–1000 nm. On the other hand, the force resolution of the AFM is not so good to test "softer" molecules.

What the Fig. 8.5b shows is the typical force-extension curve for such a system: a number of peaks can be seen, after each one of which the force falls down to near zero, and then restarts with a similar pattern; each of the force rising branches are well fitted by a shifted worm-like chain force curve; note the force in the range of $\sim$2–300 pN, and the elongation of $\sim$25–28 nm, for each individual branch. The interpretation of such results is that under the applied stress, the individual subdomains of the protein unfold one after another, following a similar unfolding pattern. For each of the subdomains, the globule-to-coil unfolding scheme from Fig. 8.3 applies, and each one of them can be seen as an independent polymer, unfolding like a worm-like chain.

Experiments of the same type have been performed on various macromolecules, on fragments of chromatin (in which the DNA unfolding from the histone is observed), as well as to detect and measure the strength of receptor-ligand binding, such as in the biotin-streptavidin complex. As it will be seen in the next Section, optical tweezers and atomic-force microscope have found application also in the measurement of mechanical properties of much larger objects, such as cell organelles and even whole cells.

8.3 Mechanics of the Cell Membrane

All cells are bounded by the plasma membrane, a closed, two-dimensional sheet of 7–10 nm thickness and extending for many 1000s μm^2, which itself has a complex architecture as it was concisely illustrated in the Appendix D to Chap. 5. Its main component is a double layer of various types of amphiphilic molecules, each with a polar head and lipid tails, within which proteins such as ion channels and pumps are embedded; a loose network of sugar polymers form the outer glycocalix, a sort of protective layer of $\sim$0.5 to a few μm thickness, which is anchored to the double-lipid membrane; the cytoskeleton, on the other hand, with its tangled network of filamentary proteins such as actin, spectrin, microtubules, is attached to the same membrane from the inside surface.

On the other hand, the outer cell membrane is far from being the only example of two-dimensional material structures found in the live cell. The nucleus, mitochondria, the Golgi organ, various kinds of vesicles and endosomes, are all made of, or wrapped by a double-layer membrane; the endoplasmic reticulum is itself a huge membrane surface, 10–25 times larger than the whole plasma membrane, multiply folded and dramatically curved. Membrane-enclosed organelles may take up to half of the cell volume, with the mitochondria occupying about 20 % and the endoplasmic reticulum about 15 % of the total. Overall, various types of cells host between 10^4 to

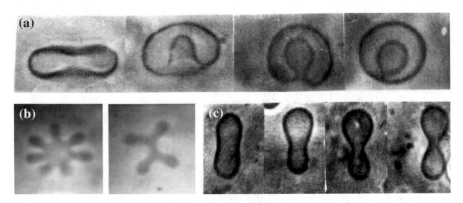

Fig. 8.6 The surprising variety of shapes observed in simple experiments of lipid bilayer self-assembly in water, as a function of temperature and osmotic pressure. **a** Discocyte-stomatocyte transition. With increasing temperature, the up/down symmetric discocyte (*left shape*) turns into a symmetry-broken stomatocyte. **b** Starfish vesicles. **c** Prolate and pear-like shapes. [Photos (**a**) and (**c**) from Ref. [5]; (**b**) from Ref. [6]. © American Physical Society, repr. w. permission.]

more than 10^5 μm^2 of membrane-like material. All such membranes have a similar composition, in which phosphatidyl-choline makes up 25–55 % of the total lipids, phosphatidyl-ethanolamine covers another 15–20 %, and other lipid variants occur in smaller concentrations. From the physical point of view, such chemical differences may be interesting to the extent that they can impart the membranes more or less different properties, such as density, rigidity, in-plane diffusivity.

However, as physicists we believe it to be our mission to try to find regularities and similarities among widely different systems, and possibly to identify the simplest model systems that can provide fundamental information about the luxuriant forest of such complex and rich structures. Cell membranes represent another challenging area in which biophysics has been working, by applying exactly the 'model-first' state of mind and the corresponding practical attitudes. In this respect, water-filled artificial vesicles formed by just a closed lipid bilayer, without all the complexity and variability of the membranes observed in the cell, have represented in the recent past an ideal playground to study the flexibility, mechanical resistance, elasticity and deformability, topological properties of membranes.

Bilayers form spontaneously in a process of self-aggregation when lipids are dissolved in aqueous solution, due to their amphiphilic nature (see the phase diagram in Fig. 5.15, Appendix D). Closed-shape vesicles with sizes in the micron range can be obtained, and can be studied by conventional video microscopy and its variants (Fig. 8.6). Such observations revealed an amazing variety of shapes, some of which may be reminiscent on the shapes of real living cells, despite the fact that such simple lipid vesicles are missing all the internal structures and chemical out-of-equilibrium conditions of a real cell.

What is the physics behind vesicle and micelle folding? From a theoretical point of view, we must search for the appropriate free energy function, whose minimisation as a function of temperature, pressure, chemical composition, yields the observed shapes.

The maths of curvature

The local curvature at any point O of a surface in 3 dimensions can be described geometrically by two parameters: the **mean curvature**, H; and the **Gaussian curvature**, G.

Their formal definition rests on the notion of a geometrical frame of reference local to the generic point of the surface, identified by the triplet of vectors $\{t_1, t_2, t_3\}$, as shown in the figure.

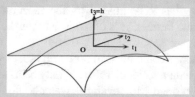

By taking the flat undeformed surface as reference (grey-shaded), the deformation at a generic point P in the neighborhood of O is defined by the height $h(x, y)$ measured from the reference flat plane. The equation for h in terms of the unit vectors t_i is:

$$h = \tfrac{1}{2} \sum_{i,j=1,2} \gamma_{ij} \mathbf{t_i t_j} \tag{8.49}$$

The 2×2 matrix of coefficients $\Gamma = \{\gamma_{ij}\}$ would be identically zero for a flat plane (zero curvature). The formal definition of the two global curvatures H and G is:

$$H = \tfrac{1}{2}\mathrm{Tr}[\Gamma] = \tfrac{1}{2}(\gamma_{11} + \gamma_{22})$$
$$\tag{8.50}$$
$$G = \mathrm{Det}[\Gamma] = \gamma_{11}\gamma_{22} - \gamma_{12}\gamma_{21}$$

Values of coefficients for simple geometrical shapes are:

$$\Gamma = \begin{pmatrix} 1/R & 0 \\ 0 & 1/R \end{pmatrix} \text{ sphere} \qquad\qquad \Gamma = \begin{pmatrix} 1/R_1 & 0 \\ 0 & 1/R_2 \end{pmatrix} \text{ ellipsoid}$$

$$\Gamma = \begin{pmatrix} 1/R & 0 \\ 0 & 0 \end{pmatrix} \text{ cylinder} \qquad\qquad \Gamma = \begin{pmatrix} 1/R & 0 \\ 0 & -1/R \end{pmatrix} \text{ saddle point}$$

The two principal curvatures at any point O on the surface define the corresponding values of the local curvature radii:

$$\gamma_{11} = \frac{1}{R_1} \qquad \gamma_{22} = \frac{1}{R_2} \tag{8.51}$$

For a surface that can locally be approximated as spherical there is only one R, and one curvature matrix element $\gamma = \gamma_{11} = \gamma_{22}$; the corresponding definitions of the curvatures are $H = \gamma$, and $G = \gamma^2$.

An alternative description of the local state of curvature of the surface at any point can be given by the triplet of vectors $\{\tau, \mathbf{b}, \mathbf{n}\}$, respectively called tangent, binormal, normal. They are typically chosen with $\tau || t_1$ and $\mathbf{b} || -t_2$, such that $\mathbf{n} || -t_3$. With this choice, it can be shown that:

$$H = \tfrac{1}{2}\nabla \cdot \mathbf{n} \tag{8.52}$$

8.3.1 The Minimal Free Energy Model

For a symmetric membrane, the chemical composition and environment of both monolayers is identical. Therefore, the flat conformation is locally the state of lowest energy. For a closed configuration, which is necessarily non-flat, the selection of the correct energy has to be guided by the essential physical properties of closed bilayer membranes. Further, it is considered that the thickness of the membrane is negligible compared to its surface, therefore the mathematical theory of two-dimensional membranes could be used (see greybox on p. 322). The membrane is much closer to a fluid than a solid, therefore it may be assumed that it cannot support shear parallel to the surface. Moreover, the water solubility of the phospholipids is very low, therefore no material is exchanged between membrane and solution. Finally, the small compressibility of the membrane implies that for a closed surface the total area A must be constant.

The free energy of a bilayer of surface A, composed of $2N$ amphiphilic molecules in water, can be symbolically written as a function of the specific area/molecule, $a = A/N$ (note that each surface element a is occupied by *two* molecules arranged tail-to-tail with their heads facing the water), equal to the sum of the hydrophobic and hydrophilic potentials per unit area (solvent-molecule interactions), and of the molecule-molecule interaction potential:

$$F = 2N\phi(a) = 2N[\phi_{phob}(a) + \phi_{phil}(a) + \phi_{int}(a)] \tag{8.53}$$

If the membrane floats in the solvent without exchanging molecules with the reservoir ($N = $ constant), and it is free of adjusting its specific area, the equilibrium state corresponds to the configuration for which:

$$\left(\frac{\partial F}{\partial a}\right) = 0 \tag{8.54}$$

In 1973 [7], Wolfgang Helfrich introduced the theoretical tool that should later became known as the Helfrich's hamiltonian.[2] The equilibrium configuration of a membrane is found in this approximate model by the free-energy balance between bending energy, E_{el}, the local curvature (induced by the pressure imbalance, P, between the inside and outside of the membrane), and the intrinsic mechanical resistance of the membrane, as expressed by its surface tension Σ:

$$F = E_{el} + \Sigma A + PV = 0 \tag{8.55}$$

where the elastic energy of bending is given by the integral over the entire, deformable membrane surface A, of the membrane bending energy from Eq. (8.19):

---

[2] As a curiosity, it may be noted that Helfrich was at that time working with Hoffmann-LaRoche Laboratories in Switzerland, on the fabrication of the first liquid-crystals displays, a subject he had started years before when working at RCA in Princeton and for which he is a credited inventor.

$$E_{el} = \int_A \left(\tfrac{1}{2} K_b H^2 + K_G G \right) dA \tag{8.56}$$

with the two curvatures H and G being defined in the greybox on p. 339.

The coefficient $K_G = 2K_b(1 - v)$, the Gaussian bending modulus K_b, and the curvatures are taken to be locally spherical, in the small deformation regime. Both Σ and the pressure P are parameters (in fact, they are Lagrange multipliers, see greybox on p. 65) to be fixed in order to impose a value of surface/volume ratio. The pressure includes the osmotic pressure difference across the membrane and, in the case of a real cell, any mechanical actions by the cytoskeleton.

Any open edge of a membrane patch exposed to water costs a large energy. This is why free membrane patches usually do not exist. Likewise, the topology of a vesicle will hardly change, since this would imply to form transient edges with high energy. A result known as the Gauss-Bonnet theorem proves that the surface integral of the second term $K_G G$ in Eq. (8.56) depends only on the topology of a vesicle but not on its shape, therefore if the topology does not change, it gives just a constant contribution to the free energy, independently on how and how much the membrane is deformed. Thanks to this result, the Gaussian curvature term is discarded when considering vesicles of fixed topology, i.e. continuous deformations not leading to cutting and/or rebinding the membrane surface.

The bending rigidity K_b is of the order of 2–4×10^{-20} J, i.e. $\sim 10\,k_B T$ meaning that thermal fluctuations may not be very important in membrane physics, at such length and energy scales (see below the persistence length μ_p of the membrane). On the other hand, we will see later that entropic effects can give rise to peculiar thermal effects.

For this **minimal model**, the solutions are found by numerically minimising the free energy equation $\delta F = 0$ with Lagrange multipliers. The resulting stationary shapes of free-standing vesicles at fixed A depend only on the reduced volume $v = V/A^{3/2}$, and the membrane bending stiffness K_b is the only energy scale in the problem. (In the experiments, the volume V would be fixed by the osmotic pressure: since the membrane is permeable to water, the volume grows until the osmotic pressure is zero). Figure 8.7, shows the free energy curves obtained for each of the shapes corresponding to a given v; The way such a free-energy diagram must be read is that, for each value of v, the vesicle would assume the shape corresponding to the curve with the smallest free energy. By looking at the plot, three types of local minima can be identified, corresponding to three different shapes: (1) the prolate ellipsoid, or *dumbbell*, (2) the oblate ellipsoid, or *discocyte*, and (3) cup-shaped forms, or *stomatocyte* (from the ancient Greek 'stoma', mouth).

While it allows to understand some of the characteristic features leading to different shapes, this minimal model is clearly too simple to cover all the deformation states available to the lipid vesicles.

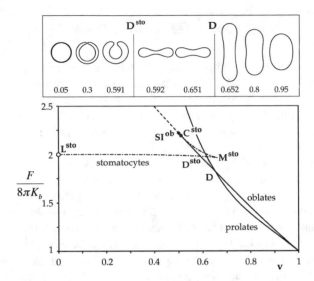

Fig. 8.7 Free-energy diagram of the minimal model of the lipid vesicle shape. The free energy on the y-axis is normalised to the value of the sphere, $F_0 = 8\pi K_b$; the x-axis is the reduced volume $v = V/A^{3/2}$. The schematic above provides a qualitative representation of the minimum free-energy shapes observed for each stability region of the $F(v)$ plot, notably stomatocyte shapes for $v < 0.591$, prolate ellipsoids for $v > 0.651$, and oblate ellipsoids for intermediate values of v. [From Ref. [8], repr. w. permission]

8.3.2 A More Refined Curvature Model

An important missing ingredient in the simple model of the membrane is the neglect of its molecular architecture. Regions of the membrane with different lipids can induce a "spontaneous" curvature, which in a real cell is also given by membrane-bound proteins and cytoskeleton actions. In some refined versions of the model, such effects are incorporated by a phenomenological parameter C_0, representing the effect of the spontaneous curvature, to be summed to the mean curvature, $dE = \frac{1}{2}K_b(H - C_0)^2 dA$.

An effect which is even more important, and which arises in artificial vesicles as well as in real membranes, is the role of lateral tension, different on the outer and inner surface of the double-lipid membrane. When bending, the lipids in the inner surface are compressed, while those on the outer surface are stretched; for both, the specific density a is not the optimal a_0 corresponding to the free energy minimum. The numbers N^+ and N^- of molecules in the outer and inner monolayer are conserved, since the exchange of lipid molecules between layers is slow compared to the experimental time-scale of these observations. The (constant) number difference $(N^+ - N^-)$ leads to a preferred area-difference: $\Delta A_0 = (N^+ - N^-)a_0$ between the two layers, which appears in the expression of the total energy.

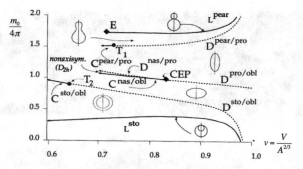

Fig. 8.8 Phase diagram of the coupled-bilayer model of the lipid vesicle shape. The plot represents a cut in the (m_0, v) plane for a particular value of $\alpha = 1.4$. Only combinations of m_0 and v comprised between the two limiting lines L_{sto} and L_{pear} lead to finite solutions of the model. First-order (continuous) transformation from one shape into another are indicated by *dashed lines*, second-order (discontinuous) transformations by *continuous lines*. The existence lines do not extend for values of v below the "critical" points T_1, T_2 and E. [From Ref. [9], repr. w. permission]

In a more complete free-energy model, we could therefore replace the simple surface tension term, ΣA, by a contribution of **area-difference elasticity** (ADE):

$$E_{ADE} = \frac{\alpha \pi \kappa}{2 A K_b^2} (\Delta A - \Delta A_0)^2 \qquad (8.57)$$

where α is a model-dependent parameter of order ~ 1. The above expression, quadratic in the preferred area difference, is the result of several attempts to include the contribution of the lipid concentration difference.

Now, we need to minimize the new free energy:

$$F = E_{el} + E_{ADE} + PV \qquad (8.58)$$

In this **coupled-bilayer model**, the minimal shapes depend not just on the reduced volume v, but also on a second parameter, $m_0 = A_0/2dR$ which is basically a constant difference of lipid molecules between the two layers, scaled by the membrane size dR. As one changes temperature or the osmotic conditions, both the parameters v and m_0 change, due to thermal expansion of the monolayer area, thus allowing to explore the (m_0, v) phase space. Since examining the three-dimensional surface of F is complicated, Fig. 8.8 gives a representation of the (m_0, v) for a particular value of $\alpha = 1.4$; similar diagrams can be traced for different values. As shown in the new phase diagram, a trajectory in the (m_0, v) plane cuts across the different regions of stability of different vesicle shapes. The new free energy function now leads to pears, shapes and a multitude of starfish shaped vesicles, comprised between the existence lines L_{sto} and L_{pear} (for values (m_0, v) outside such lines, no solution exists for the model). For example, for $v = 0.8$, values of $0.35 < m_0/4\pi < 0.7$ correspond to stomatocytes; for $0.7 < m_0/4\pi < 1.$ to oblates; for $1. < m_0/4\pi < 1.5$ to prolates; for $1.5 < m_0/4\pi < 1.75$ to pear-shaped vesicles. Transitions from one shape to

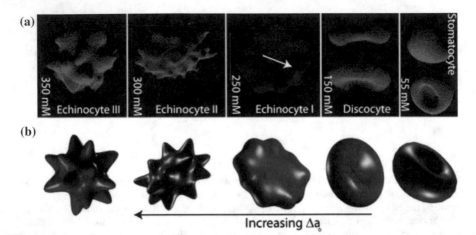

Fig. 8.9 Experimentally induced and theoretically predicted sequences of red blood cells morphology. **a** 3D confocal microscopy images of RBCs in solutions of increasing concentrations of NaCl (concentrations in mM shown on the images). **b** Theoretically predicted shapes that minimize the shape energy of the free-energy function including the intrinsic stretch-shear energy E_{SS}, under constraints of total surface area $140\,\mu m^2$, and volume $100\,\mu m^3$; the sequence is obtained by changing the value of the effective reduced area difference Δa_0. [From Ref. [10], repr. under CC-BY 3.0 licence, see (*) for terms.]

another, e.g., from prolate to pear, can be abrupt and discontinuous (transition of "first order", in the language of statistical physics), or smooth and continuous ("second order").

The role of the internal cytoskeleton in the shape of a real cell cannot be forgotten. This internal structure imparts very different properties to the membrane compared to the case of a free-standing vesicle, in which the membrane tension is only governed by the osmotic pressure. One (relatively) simple way to include some such effect, is to add yet another term in the free energy to account for intrinsic stretch and shear of the membrane material (implicitly produced by the internal forces), E_{SS}, as:

$$E_{SS} = \frac{E_{\parallel}}{2} \int_{A_0} f(\varepsilon_I)dA_0 + \mu \int_{A_0} g(\varepsilon_{II})dA_0 \qquad (8.59)$$

with f and g two non-linear functions of the in-plane strain invariant components, $\varepsilon_I = \varepsilon_{xx}\varepsilon_{yy} - 1$ and $\varepsilon_{II} = (\varepsilon_{xx} - \varepsilon_{yy})^2/2\varepsilon_{xx}\varepsilon_{yy}$, respectively, $E_{\parallel}$ the (anisotropic) membrane Young's modulus in the plane, and μ the shear modulus. The two integrals are calculated at the value of reference of the surface, A_0.

Once this new free-energy function, $F = E_{el} + E_{ADE} + PV + E_{SS}$, is minimised at fixed v and m_0, an even wider variety of shapes can be obtained (Fig. 8.9), notably the "urchin"-like shapes that red blood cells may assume by changing the salt concentration in the physiological solution.

8.3.3 Temperature and Entropy Fluctuations

The free-energy models developed until now are all taken at zero temperature. They predict the equilibrium shape of vesicles and of simple cells, in the absence of any thermally-induced fluctuations. On the other hand, video-microscopy experiments show that such membranes are never stationary at finite temperature, but exhibit visible fluctuations, likely induced by the temperature. How do we include such fluctuations?

Suppose that the membrane is confined by a harmonic potential $V(d) \sim \frac{1}{2}\varepsilon d^2$ at a distance h, for example describing the presence of another cell membrane, or some rigid obstacle [11]:

$$F = \frac{1}{2} \int_A \left[K_b H^2 + 2\Sigma(a - a_0) + \varepsilon h^2 \right] dA \simeq$$
$$\simeq \int_A \left[K_b (\nabla^2 h)^2 + \Sigma(\nabla h)^2 + \varepsilon' h^2 \right] dA \qquad (8.60)$$

To obtain the approximate expression on the second line we used the definition $H = 2\langle \gamma \rangle^2 = 2(\nabla^2 h)^2$ for the mean curvature, and the approximation $A - A' \simeq A(1 - 1/\cos\theta) \simeq \frac{1}{2}A\theta^2 = A(\nabla h)^2$ for a small portion of surface bent at a small angle θ.

Let us now introduce temperature fluctuations of the membrane shape. Following a widely diffused procedure in statistical physics, we assume that a generic shape fluctuation can be expressed as the linear combination of elementary harmonic oscillation modes. Each such "normal" mode is expressed as a combination of trigonometric functions (sine/cosine) of the wavevector $\mathbf{q} = 2\pi\mathbf{n}/L$, with $\mathbf{n} = n_x q_x + n_y q_y + n_z q_z$, and $q_{x,y,z}$ unit vectors in the momentum space. The amplitude h_{n_x,n_y,n_z} of each mode is the coefficient of the linear combination, expressing the "weight" of each mode in a particular deformed shape of the membrane. The first few normal modes in the plane (i.e., with $n_z = 0$), for a square patch of membrane, with unit length and fixed borders (i.e., $h = 0$ everywhere on the perimeter), are shown in Fig. 8.10.

By considering that all fluctuation modes of the membrane, each carrying some energy ΔE, are simultaneously excited with a probability proportional to the respective Boltzmann factor $\exp(-\Delta E/k_B T)$, it is convenient to study the power spectrum of the fluctuations, by taking the Fourier transform (FT) of the energy in the $\mathbf{q}$-space:

$$\Delta E = \frac{4\pi}{L^2} \sum_q \left(K_b q^4 + \Sigma q^2 + \varepsilon' \right) |h_q|^2 \qquad (8.61)$$

the curvature term proportional to $(d^2/dr^2)^2$ giving the q^4 term, and the surface tension term proportional to $(d/dr)^2$ giving the q^2 term, after FT (see Appendix A for the FT).

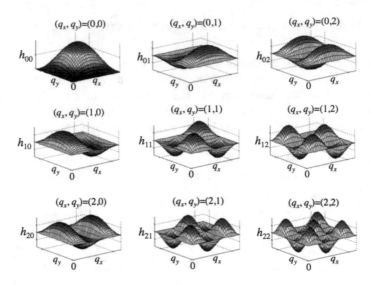

Fig. 8.10 The first 9 vibrational eigenmodes of a square membrane held fixed at the perimeter

Since we are interested in a planar system (the membrane), we can restrict the
q-space to the two x and y dimensions. The Fourier components of the membrane
vertical displacement $h(x, y)$ are:

$$h_{\mathbf{q}} = \frac{1}{L^2} \int_0^L h(x, y) e^{-i(q_x x + q_y y)} dx dy \qquad (8.62)$$

for an excitation wavevector $\mathbf{q} = (q_x, q_y)$, with modulus $q = \sqrt{q_x^2 + q_y^2}$, in the plane
of the membrane; the $h_{\mathbf{q}}$ describe the amount each fluctuation mode contributes to
the overall fluctuating shape of the membrane.

A simplified view of the role of temperature can be obtained by statistical averaging over the modes, so we replace the summing over q, by $N_q \times$ some (unknown)
average amplitude $\bar{h}$. Each of the N_q wavevectors (a normal mode) can be assigned
to carry on average $\frac{1}{2} k_B T$ of energy, according to the equipartition principle:

$$\Delta E \simeq \frac{4\pi N_q}{L^2} \left(K_b q^4 + \Sigma q^2 + \varepsilon' \right) \langle \bar{h}^2 \rangle = \frac{N_q}{2} k_B T \qquad (8.63)$$

giving the following estimate for the average vibrational amplitude:

$$\langle \bar{h}^2 \rangle \simeq \frac{L^2}{8\pi \left(K_b q^4 + \Sigma q^2 + \varepsilon' \right)} k_B T \qquad (8.64)$$

It may be noticed that for $q \to 0$ (i.e., in the long-wavelength excitation limit) the amplitude of the fluctuation grows as L^2:

$$\langle \bar{h}^2 \rangle \simeq \frac{k_B T}{8\pi \varepsilon'} L^2 \tag{8.65}$$

and is a function of only the confining potential ε'. For an order of magnitude consider $\varepsilon' = 10^{-20}$ J (that is, 2–3 $k_B T$), then the RMS amplitude of the *constrained* fluctuations is about 10% of the membrane size L at $T = 300$ K. The average free energy per unit area associated to such fluctuations is:

$$\frac{\Delta E}{L^2} \simeq \frac{k_B T}{L^2} \simeq \frac{(k_B T)^2}{8\pi \varepsilon' \langle \bar{h}^2 \rangle} \tag{8.66}$$

Actually, this is an entropic contribution to the free energy per unit area, of the order of $(k_B T)/\mu m^2$, arising from the constrained fluctuation of the membrane [11]. Thermally-excited undulations give rise to a long-range repulsion force between membrane surfaces. Just like the elastic force in long-chain polymers, such a repulsion is entropic in nature: two membranes approaching at a distance d see their "fluctuation space" restricted, therefore they tend to separate to increase the total entropy. Such a repulsive force scales as d^{-2}, with the same power law as the Van der Waals force between two parallel plates.

In analogy with the entropic fluctuation of a polymer, a persistence length of the *free* membrane can be defined, although its calculation is quite more complicate. The standard result is $\mu_p = b \exp(4\pi K_b/3k_B T)$, with b a characteristic length, for example a few nm. The exponential dependence on K_b (as opposed to the linear dependence of λ_p on κ_b) makes this parameter very large in absolute terms. With $K_b \sim 10 k_B T$ it is $\mu_p \sim 10^5$ km, meaning that for the free membrane fluctuations are unimportant.

8.4 Deformation Energy

In the previous Section we only considered spontaneous bending of the vesicle (cell) membrane seeking to optimise its shape, in order to reach the minimum free energy at $T = 0$ K. Subsequently, it was shown that thermal fluctuations can disrupt such perfect shapes, notably by inducing extra repulsive forces between approaching membranes.

However, a large body of evidence in cell mechanics focus on the rearrangement of cell shape according to internal forces, as well as the mechanical interaction with its surroundings, be it other cells, extracellular matrix (collagen, fibronectins etc.), or competing species, in the case of unicellular organisms. In this Section, we will then ask what is the energy required to bend a membrane, made of a material with given bending modulus K_b, into an arbitrary shape defined by its local curvature radii R_1 and R_2?

The answer to this question is, in principle, a simple expression, $dE = \frac{1}{2}K_b H^2 dA$, to be integrated over the whole area A of the cell membrane, to get the total *deformation energy*:

$$E_D = \frac{1}{2}\int_A K_b H^2 dA = \frac{1}{2}\int_A K_b \left(\frac{1}{R_1} + \frac{1}{R_2}\right)^2 dA \qquad (8.67)$$

where the relationship $\gamma_{kk} = R_{kk}^{-1}$ ($k = 1, 2$) between principal curvatures and curvature radius was used.

As it can be easily imagined, during the deformation of a real cell under the applied force the curvature can change, from one point to another of the surface, thus making the calculation of the previous integral a daunting task. However, for simple solid figures such as the sphere or the cylinder, the R_i's are constant on the whole surface. Therefore, it may be tempting to decompose the calculation into a sum of simpler contributions, where the deformed membrane is decomposed into a sum of pieces of sphere, cylinder etc.

For a sphere of radius R, it is $R_1 = R_2 = R$ over the whole surface of area $A = 4\pi R^2$. Therefore:

$$E_D = (4\pi R^2)\left(\frac{2K_b}{R^2}\right) = 8\pi K_b \qquad (8.68)$$

Strangely, this result is independent on the size of the sphere: the energy required to fold into a sphere a sheet of 1 or 10 m^2 of the same material, is the same! Such a bizarre property originates from the fact that the surface bending modulus K_b is already in units of energy: it already includes the length in its determination.

On the other hand, for a cylinder of diameter $2R$ and length L, we have $R_1 = R$ and $R_2 = \infty$ over the whole surface of area $A = 2\pi RL$. Therefore:

$$E_D = (2\pi RL)\left(\frac{K_b}{2R^2}\right) = \pi K_b \left(\frac{L}{R}\right) \qquad (8.69)$$

Also in this case, the energy is a function only of the aspect ratio of the cylinder, equal to L/R, but not of its absolute dimensions.

8.4.1   *Membrane Protrusions and Cell Crawling*

The stable shape of a deformable lipid bilayer results from the equilibrium between the elastic tension of the material composing the membrane, mostly short-chain fatty acids of the type -$(CH_2)_n$- with $n = 14$–18, and the sum of pressure differences acting on the two sides. For a synthetic vesicle the pressure difference is simply given by the sum of partial osmotic pressures, which are equilibrated by the surface tension with the Laplace equation (5.72) or (11.13). For the case of the cell membrane, besides the presence of cholesterol and transmembrane proteins in various proportions, the pressure difference must include also the mechanical stress exerted by the different elements of the cytoskeleton.

We saw in Chap. 6 that one very interesting mode of exerting mechanical pressure by the cytoskeleton is done via the elongation or shortening of some of its filamentary proteins, notably the F-actins, which transmit forces by pushing or pulling against the cell wall. Such mechanisms were classified under the name of polymerisation, to indicate the growth or shortening of a polymer by adding or subtracting monomers. We may now ask whether such a mechanism of polymerisation of actin fibres is actually capable of mechanically deforming the cell membrane, by forming the so-called "protrusions", and ultimately inducing the displacement of the entire cell?

By looking at the Fig. 8.11a, one of the protrusions (called *pseudopodia*, ancient Greek for "false feet") can be modelled as a long cylinder of length L, terminated by a hemispherical cap of radius R. Typical values of the surface bending modulus for, e.g., the membrane of a pseudopodium are in the range $K_b = 2 \times 10^{-12}$ dyne cm $=$

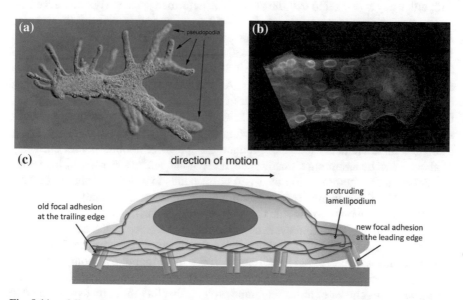

Fig. 8.11 **a** Microphotography of a unicellular *Amoeba proteus* protruding several pseudopodia. **b** Fragment of epithelial tissue with several cells; the leading cell (*extreme right*) protrudes a large lamellipodium (*green*), leading the others (*blue* cell nuclei, *green* actin, *red* myosin). **c** Schematic representation of the crawling motion of a cell. The green actin network is under active contractile stress.; the lamellipodium (*orange*) protrudes forward, and a new adhesion is formed; the actin network fixed at the focal adhesions contracts and pulls forward the cell body; the same contractile force serves to detach the old focal adhesions at the trailing edge. [Photo (**a**) courtesy of F.J. Siemensma, Microworld: world of amoeboid organisms. Kortenhoef, the Netherlands http://www.arcella.nl; **b** courtesy of O. Cochet-Escartin, Institut Curie, Paris; repr. w. permission]

2×10^{-19}J $= 1.25$eV $= 50$ $k_B T$ at $T = 300$ K (please note the units: a force multiplied by a length, giving an energy).

We may think of the corresponding deformation, as starting from an initially flat portion of the cell membrane, which is pushed by the inner cytoskeleton struts into the cylinder+hemisphere shape. By taking the energy of the flat membrane as zero, the elastic work of the bending forces stored as deformation energy in the membrane material, is:

$$W = E_D(cylinder) + \tfrac{1}{2}E_D(sphere) = \frac{\pi K_b L}{R} + \tfrac{1}{8}8\pi K_b \qquad (8.70)$$

For the membrane to resist to such a deformation, it is necessary that this amount of work be less, or at best equal to the energy of the elastic tension of the material composing the membrane:

$$W \leq \Sigma A = \Sigma \left(2\pi RL + \tfrac{1}{2}4\pi R^2\right) \qquad (8.71)$$

by taking that all the work is transformed into elastic energy (an ideal transformation, with efficiency $\eta = 1$), and that the area of the deformed surface A is conserved.

For the above typical values of $K_b = 50$ $k_B T$, and with $R = 0.05$ μm and $L = 1$ μm (such a long and thin protrusion would be called a*filopodium*), we find $W = 3770$ $k_B T$, and $\Sigma \geq W/A = 11430$ $k_B T/\mu$m^2, or 4.7×10^{-5} N/m, or again 0.047 dyne/cm at $T = 300$ K, a value fully compatible with the isotonic tension at rest of typical biological membranes. As we had already found in Chap. 5, the typical values of Σ are at least one order of magnitude larger than the above values, therefore it can be concluded that the cell membrane can withstand without problems such a deformation imposed by the growing cytoskeleton. From the data of Chap. 6, a single actin filament is capable of producing a "push" of about 5 pN for each step of about 5 nm; by assuming a steady state, and a homogeneous density of filaments, the elongation of the protrusion by 1 μm corresponds to a work of about 1000 pN nm, or 250 $k_B T$ per filament, therefore a number of F-actin filaments between 10 and 20 could be enough to provide the right amount of push.

Cells initiate the migration cycle by extending protrusions of the cell membrane towards the direction of a chemical gradient or mechanical perturbation. These protrusions comprise large, broad lamellipodia, spike-like filopodia, or both, and are all driven by the polymerisation of actin filaments. Protrusions are then stabilised by *focal adhesions*, cluster of transmembrane proteins that link the actin cytoskeleton to the underlying extra-cellular matrix. Once the link is stable, actin-myosin contraction can generate traction forces on the substrate, and the entire cell can move (Fig. 8.11c). By a reverse mechanism of contractility, the cell also promotes the disassembly of adhesions at the rear end, to allow the cell to move forward. Cell adhesion is closely coupled with the formation of protrusions at the leading edge of the cell, called *filopodia*, when these take a narrow shape, and *lamellipodia*, when they develop into wider, plate-like structure (see Fig. 6.23). Nascent adhesions initially assemble in the lamellipodium, and their rate of formation correlates with the rate of protrusion.

It has been recently shown that such a mechanism is not only capable of producing cell displacement, but also of driving and organising the group migration of cells [12]. It was observed that at the edge of a group of 30–80 epithelial cells, a sort of finger-like structure enables the pack of cells to migrate on a substrate, in a global process: cells develop a collective mechanical behaviour that overrides individual cell behaviour. One of cells takes the head of the group forward, to lead the migration (see Fig. 8.11b). This "leader" cell swells, stops dividing and leads the way, by exerting a drag force on the follower cells. These teaming cells also create a contractile pluricellular structure along the migration finger, whose role is to prevent other cells from taking the leader role within the finger, or straying away in another direction.

8.4.2   The Shape of a Bacterium

The interplay between deforming cell wall, polymerisation and depolymerisation of cytoskeletal filaments, and mechanical constraints, can be put at work to study of the growth and shape of a bacterium, a situation quite different and in many respects much simplified compared to that of an eukaryotic cell, with its rich internal structures. Already a relatively simple mathematical model can demonstrate how a pure membrane under tension could not grow to a stable state, which is instead possible to attain if the membrane deformation is equilibrated by dynamic bundles of cytoskeleton fibres [13].

As it was briefly described in the Appendix D (p. 195), one main difference between the external structure of a bacterium and an eukaryote animal cell is the presence in the former of an outer cell wall, with the functions of protecting the soft lipid membrane of the cell, and of imparting to the organism mechanical resistance, rigidity, and overall shape. The bacterial cell wall is made up by a relatively rigid network of peptidoglycan strands (a polymer consisting of sugars and amino acids), cross-linked by short peptide bonds, and constantly modified and entertained by a number of enzymes. A bacterium lacks a proper cytoskeleton, however some proteins such as FtsZ, MreB and crescentin, have been shown to be important for shaping the bacterial cell. In particular, MreB appears to play a role similar to that of actin filaments in eukaryotes, in that it grows and shrinks by analogous polymerisation mechanisms and, being attached to the cell wall, it provides about half of the overall rigidity of the bacterial cell. As it is shown in Fig. 8.12, referring to a model of the well-known *E. coli* bacterium, MreB filaments are arranged in a sort of helicoidal spring about the inner cell wall.

In the simple model schematically represented in this figure, it may be assumed that fragments dA of cell wall (proteoglycan) are added or removed randomly, while the cell is growing in size, pushed by an internal "osmotic" force P. Since each free fragment entering the cell wall must be deformed to fit the growing cell shape, the corresponding free energy is the sum of three terms, $dG = dE_D + \Sigma dA - w_0 dA$, with w_0 a chemical energy gain per unit area. Clearly, dE_D depends on the shape, size and growth direction; therefore, there could be a size and shape for which the

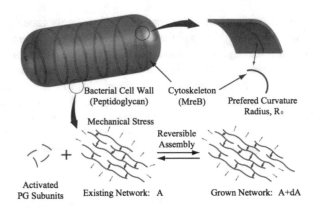

Fig. 8.12 Schematic representation of the *E. coli* cell wall. In a reversible reaction, the cell wall area increases from *A* to *A* + *dA*, by adding peptidoglycan (*green*, PG) linked by peptide bonds (*red*). A cytoskeleton substructure such as MreB is schematised as a helical bundle with preferred radius R_0 and pitch *p*, slightly smaller than the current cell radius *R*. [Adapted from Ref. [13], w. permission]

increase in elastic bending exactly balances the gain in chemical energy, while the cell grows in a particular direction. Also, it must be noted that for a "rigid" surface, Σ can no longer be considered a scalar, isotropic value, but it must be a tensor-like quantity, depending on both the direction of stretching and on the direction of the applied force (i.e., a *stress tensor*, see Appendix H to the next chapter). Moreover, this $dG = 0$ equilibrium is dynamical, in that the bacterial cell may start growing in one direction, but then change to another. This could explain why rod-like bacteria tend to grow up to a well-defined diameter, but keep growing to variable lengths in the axial direction. If we further assume that, from some time on, the shape does not change anymore, the surface can be described by a small set of parameters a_i (for example, the growing diameter and length). The driving force for surface growth can thus be formally written as:

$$F_i = -\frac{\partial G}{\partial a_i} \qquad (8.72)$$

with $dG = -\sum_i F_i da_i$, and the growth velocity of each parameter a_i:

$$\frac{\partial a_i}{\partial t} = M_i F_i \qquad (8.73)$$

By using the above equations, one can firstly ask whether some simple shapes are unconditionally stable during growth. For example, the cylindrical cell of radius $R(t)$ and length $L(t)$, can be parametrised as:

$$R(x, t) = R_s(t) + B(t) \cos(qx) \qquad (8.74)$$

where $R_s(t)$ is the steadily growing radius, and $B(t)$ is some external perturbation, periodic with wavelength $\xi = 2\pi/q$ (this is a typical way of representing small perturbations to a state of equilibrium). If the radius R_B does not diverge to infinity, nor it shrinks to zero, over long times, for any value of q and for $B << R$, then we can say that our system is dynamically stable, i.e. it can resist generic perturbations.

The average free energy for a particular perturbation q is obtained by integrating the following expression, over the length λ, and the azimuthal angle about the cylinder axis ϕ:

$$G = \frac{q}{2\pi} \int_0^{2\pi/q} dx \int_0^{2\pi} d\phi \, (E_D + \Sigma - w_0) \tag{8.75}$$

As in the previous treatment of the membrane bending, the deformation energy is just $E_D = \pi K_b(L/R)$, where both L and R are now time-varying quantities. On the other hand, deriving an expression for the "stress tensor" Σ requires a rather complex treatment, which however can be eventually reduced to:

$$\Sigma = \pi P^2 \left(\frac{LR^3}{16h}\right) \frac{(\lambda + 10\mu)}{\mu(\lambda + \mu)} \tag{8.76}$$

Starting from the above expressions, the growth velocity of the perturbation amplitude, $dB/dt = -M_B(\partial G/\partial B)$, can be obtained explicitly. The detailed calculations for a bacterium such as *E. coli* demonstrate that for any wavelengths up to about 10 μm the growth of a cylindrical shape is unstable, and should be disrupted by arbitrarily small random forces, e.g. temperature fluctuations.[3] However, real *E. coli* bacteria do not display such an instability, and can grow nicely cylindrical cells up to lengths of several μm. Something important is missing from the model.

The important role played by the peculiar cytoskeleton, provided by the helicoidal MreB scaffold (see again Fig. 8.12), contains at least part of the answer. The coil-like structure of MreB can be added to the ensemble by including its contribution in the free energy Eq. (8.75). To put it simply, this is made up of a chemical contribution, $-w_m$, and of a Hooke-like elastic contribution, $U_m = \frac{1}{2}k_m(R - R_0)$, giving the energy of a coil spring of radius R variable about the equilibrium value R_0. The effective stiffness of MreB is:

$$k_m = \frac{E_m \mathscr{I}}{p R_0^4} \tag{8.77}$$

where p and R_0 are the pitch and radius of the coil, as described in Fig. 8.12. By taking the MreB filament as a thin, dense cylinder of diameter $d = 3.9$ nm, for which $\mathscr{I} = \pi d^4/64$, and structurally homologous to F-actin, whose persistence length is $\lambda_p \sim 15$ μm, an estimate for the effective Young's modulus can be guessed from Eq. (8.32), as $E_m \sim 15k_B T/\mathscr{I} \sim 5$ GPa. Finally, by extracting a value of $R_0 \sim 0.5$ μm

---

[3]It may be worth pointing out that the perturbation amplitude is described by a parameter $\alpha = M_B(c_1 q^4 + c_2 q^2 + c_3)$, closely resembling the analysis of temperature fluctuations of the lipid membrane in Eqs. (8.61) through (8.65).

from fluorescence microscopy images of *E. coli*, the effective Hookean stiffness of the MreB scaffold can be estimated at $k_m \sim 1.5$ MPa/μm.

If now the stability against the perturbation B is carried out for the total free energy $(E_D + \Sigma + U_m - w_0 - w_m)$ in Eq. (8.75), it is found that the growth of the cylindric cell wall is stable up to very long L, and for realistic values of $R_0 \sim 0.5 - 1 \mu$m. The force applied by MreB is opposite to P, therefore the internal pressure can be balanced while the cell grows in size. The bacterial cell wall can be regarded as an elastic curved sheet with perimetral MreB reinforcement, like a sort of fibre-reinforced composite material. Of course, many other proteins can affect the observed morphology of the bacteria, some of them being even necessary to ensure a proper localisation and polymerisation of the MreB filament. Also, other types of bacteria such as *Bacillus subtilis*, may have a non-isotropic cell wall, or a less regular shape such as *Caulobacter crescentus*. Nevertheless, the simple model presented here gives a first interesting insight into the coupling of the respective stiffness of cell membranes and cytoskeletal filaments.

8.5   How a Cell Splits in Two

Cell division is the central mechanism operating in biology to grow and maintain all pluricellular organisms, and to grant reproduction to unicellular organisms. It can be estimated that a human produces many millions of new cells at every second, just to maintain the organism at steady state. The life of a cell between two successive divisions is regulated by passing through different phases, in what is called a cell cycle.[4] The details of the cell cycle can vary from one organism to the other, and between cells of a same organisms, however there are some characteristics that must remain constant. First and foremost, to produce a new cell identical to the original one, it is necessary that the DNA is duplicated and correctly split between the two new nuclei, by *segregating* the chromosomic material in two perfectly identical halves. Secondly, most cells need to double their mass, molecule by molecule, protein by protein, and duplicate all the cell organelles and subsystems. How all the necessary steps are regulated and coordinated is still poorly understood, although in the past decades scientists have been making enormous progress.

It is beyond the scope of this book to cover the biological details of the cell cycle and the biochemical cascade of complex reactions. However, as a well traveled biologist as Eric Karsenti puts it:

When you need to understand how molecules self-organise in a living system, evolution theory is not sufficient. This is a problem of physics, or chemical physics. If you pose such questions, you must look in the good direction, and ask physics for answers. [...] Cell division is an example of a process of self-organisation in which a

---

[4]For the many details of this extraordinary process, the reader can consult any good biology textbook, such as the classical *Molecular Biology of the Cell*, by Alberts, Bray, Lewis, Raff, Roberts and Watson (Garland Science Pub., New York, 4th ed. 2002).

*bunch of molecules perform a precise function, the distribution of chromosomes. The
general organising principles were already described before, such as the reaction-
diffusion equations by Turing, or the collective behaviour of molecules by statistical
physics. But we needed to apply these principles to cell biology. (Interview at CNRS,
2015, see:* http://videotheque.cnrs.fr/doc=4672.)

That important mechanical actions must be at play in the process of cell splitting
should be somewhat self-evident: how can you break apart something without applying a force? As a typical example, it has been measured that HeLa cells (an model-cell
type often used in biology experiments) increase their internal hydrostatic pressure
excess and surface tension during cell separation, from the average values of about
40 Pa and 0.2 mN m^{-1}, to 400 Pa and 1.6 mN m^{-1} during the mitosis. If we look at
the most basic features of mitosis, the so-called "M-phase" of the eukaryote cell life
cycle, schematically represented in six steps in Fig. 8.13, three fundamental *biological* events are found, all of which critically involve the action of *physical* forces:

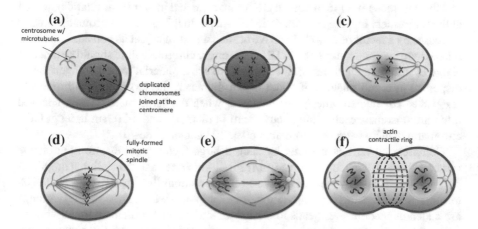

Fig. 8.13 Schematic in six steps of the division phase of the cell cycle (or "M-phase"). The cell
membrane is represented by the *green* layer, the nuclear membrane is *orange*. **a** Chromosomes have
been duplicated in the nucleus (*red*), and condensed into homologue pairs joined at the centromere
(*yellow dot*). **b** The centrosome, from which microtubules are normally sprouting, is duplicated;
the two copies take their position at the opposite sides of the nucleus. **c** The nuclear membrane
is broken, the nuclear material is dispersed in the cytoplasm; microtubules from the two opposed
centrosomes polymerise and reach the region where they can meet the chromosomes. **d** Microtubules
from opposite directions merge about the centre of the splitting cell; chromosomes are relocated
about the mid-section of the cell volume (equatorial plane), and put under tension from opposed
microtubules. **e** Chromosomes pairs are split in two, and the dispersed nuclear material starts
rebuilding two new cell nuclei at the two extremes of the mother cell. **f** Microtubules break apart
and start depolymerising; the nuclear membrane begins to be reconstituted; the physical splitting
of the membrane can be initiated, by the self-assembly of actin fibres into a contracting ring, which
squeezes the mother cell about its equatorial plane. The final result will be two cells identical to the
one in **a**, with the chromatin dispersed in the nucleus, and ready to recondense into chromosomes
for the next cell division

1. chromosome condensation [step (a) in the figure],
2. formation of the mitotic spindle [step (d)], and
3. formation of the contractile ring [step (f)].

Each of these three essential steps can be described in physical terms, by giving just the minimum biological and chemical details necessary to understand the issues.

8.5.1 Chromosome Condensation

The earliest observable manifestation of the start of the M-phase, is the progressive **condensation** of the chromatin into well-organised chromosomes. As we learned in Chap. 3 (see Appendix B), over a time covering most of the cell life the DNA is loosely arranged in a non-specific mass, wrapped around globular proteins (histones). When condensation starts the entire genetic material has already been duplicated, and the chromosomes (46 in human DNA) are arranged in pairs (chromatids), joined at the centromere into the characteristic 'X' shape. In this process, chromatin density increases by a factor of 3–400. To arrive at such an estimate, let us consider that each nucleosome wraps about 170 bp of DNA, the entire chromatin fibre should be made of about 18 million nucleosomes; if we consider such a material to be homogeneously dispersed in a cell nucleus of about 6 μm diameter, the loose chromatin density is $\sim 1.8 \times 10^5$ nucleosomes/μm^3; however, when the same material is condensed into chromosomes, each being about 35 nm in diameter and ~ 10 μm in length, the chromatin density becomes more than 60×10^6 nucleosomes/μm^3.

Compressing all that material by such a large factor requires strong attractive forces. Where do such forces come from? Firstly, histone post-translational modifications can contribute, by mediating fibre-fibre interactions that involve their N- and C-terminal tails. Histone phosphorylation, for example, in which up to six PO_3^{2-} groups are added at specific sites, leads to the expulsion of some proteins normally bound to the chromatin, thereby increasing the attraction between distant DNA strands and nucleosomes. Most importantly, however, a sort of 'scaffold' has been identified in each chromosome, made up by a number of different proteins appropriately called *condensins* and *cohesins*, and by other enzymes such as the topo-isomerase IIα. These proteins exert true mechanical actions on the DNA in the chromatin: Topo-IIα is thought to initiate chromosome condensation by clamping chromatin fibres together to form denser aggregates; condensin interacts with naked DNA and introduces coils, which (as observed at least *in vitro*) can be further transformed into DNA knots by Topo-IIα; cohesin forms a kind of molecular rings, which are thought to tether the two chromatids by either embracing them within a single ring, or by the interaction of two rings [14].

Several authors have suggested that such processing of the chromatin structure, including densification, bending, twisting, should be activated, i.e. consume an amount of free energy ΔG, to cross a barrier from the loose structure in the interphase, to the packed structure of the M-phase. A tentative estimate attributed a value of about +30 kcal/mol, to be spent in order to deform the nucleosome and pack it into the chromosome. Condensin, cohesin, and all the other motor proteins could directly consume equivalent amounts of ATP. However, an important electrostatic contribution to the packing also comes from ions, which help the stacking and densification of the chromatin fibre: a large influx of divalent Ca^{2+} and Mg^{2+} is observed in the early stages of mitosis, and likely also an energy contribution from the opening of ion channels should be included in the free energy estimate.

8.5.2 Assemby of the Mitotic Spindle

The second crucial step of cell division involving mechanical actuators, is the formation of the **mitotic spindle**, a bundle of microtubules protruding from each of the two centrosomes at the opposite ends of the nucleus (Fig. 8.14). Normally, a cell has a unique centrosome, a globule from which all the microtubules of the cytoskeleton appear to originate; however, at the start of cell division the centrosome is duplicated (Fig. 8.13a, b), so that the microtubule bundle appears to form a large spindle, attached at the two opposite centrosomes. Such a structure eventually relocates all the chromosomes at about the mid-plane of the dividing cell (see Fig. 8.13d), and tubules from opposite sides attach and start pulling on the chromosomes. The final result of this step will be the splitting of each chromosome pair into its two identical chromatids, which will relocate at the opposite ends of the cell (Fig. 8.13e), waiting for the two nuclei to recompose and the cell to split.

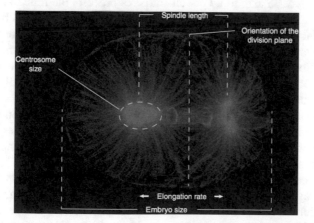

Fig. 8.14 Imaging of the spindle apparatus in the first cell division of the nematode *C. elegans*. The image shows the mother cell at the moment in which the two sets of chromosomes have been spatially separated, ready to start the actual process of splitting (see Fig. 8.13e). Microtubules are shown in *green*, chromosomes in *blue*, and the cell background (notably, actin structures) in *red*. [Image courtesy of B. Bowerman, Institute of Molecular Biology, Kansas State University]

In a simple mechanistic view, the mitotic spindle has been thought to self-assemble as a result of microtubules randomly searching for chromosomes, after which the spindle length would be maintained by a balance of outward tension, exerted by molecular motors on the tubules connecting each centrosome to the chromosomes, and compression, generated by other motors on the remaining "free" tubules, which directly connect the opposite spindle poles. However, this picture has been challenged recently, by mounting evidence indicating that spindle assembly and maintenance rely on much more complex interconnected networks of microtubules, protein motors, chromosomes and regulatory proteins. From an engineering point of view, three design principles of this molecular machine are especially important: (1) the spindle assembles quickly, (2) it assembles accurately, and (3) it is mechanically robust yet flexible. How can this design be achieved, with randomly interacting and impermanent molecular components?

The "search-and-capture" model can be already discredited on the basis of principle (1). Average velocities of microtubule polymerisation are in the range of 0.2 μm/s, in both directions. Since the typical distance between each centrosome and the cell midplane is $D \sim 10\,\mu$m, a cycle of growth and shrinking with 100% probability of getting a chromatid should take $\tau_0 = 2D/0.2 \sim 100$ s. However, most of the cycles are not successful: the probability of encountering a centromere (the middle cross of the chromosome) by a randomly thrown microtubule can be roughly estimated as $\pi d^2/4\pi D^2 \sim 0.0025$, where $d \sim 1\,\mu$m is the centromere radius; if we assume that each centrosome throws a few hundred microtubules, the probability is somewhat less than 100%, and the whole process may take even twice times the estimated 100 s. Moreover, since the search is for $\sim$100 targets in parallel, it will be over only when the last target chromatid is captured. If the process is completely stochastic, it may be akin to a Poisson-statistics process, the time to complete the search is logarithmic in the number of targets, and the 200 s should be multiplied by another factor of $\ln(100) \sim 4.5$, i.e. the total search and capture time should be in the order of $2\tau_0 \ln(100) \sim 1000$ s, and even more if we account also for the random positioning of the chromosomes at the beginning of the process. In the real conditions, however, chromatids are captured into the nascent new cell nuclei in the matter of minutes, in a clear disagreement with this picture.

One alternative view could be that microtubules start already from the chromosomes, with a local nucleation process. Already in 1985, some studies noted that a mixed origin of microtubules, both from search from the centrosome and local nucleation, was possible. Recent data [15] indeed suggest that microtubule bundles could start to be organised at the chromosomes, and then grow outward. A cooperative, hybrid pathway has been proposed for the spindle assembly, in which the centrosome-nucleated microtubules search for long chromosomal bundles that provide a larger target than centromere alone; upon capturing them, these are integrated into continuous bundles (called "K-fibers") that connect the centrosomes and centromeres. The molecular details of such a cooperative capture are still vague, but one possibility is that molecular motors such as kinesin-14 and dynein may help to crosslink, align and properly arrange the microtubules. From a mathematical point of view, a cooperative mechanism could make the spindle formation less stochastic, with a random search in the early stages and a deterministic "shooting" in the remainder of the process. The average time to capture n targets out of N is $\tau \sim \tau_0 \ln(N/N - n)$; if we take as a guess that the first 25% targets are stochastic, and 75% deterministic, the time reduces to $\tau \sim \tau_0 \ln 4/3$, a factor of 15–20 shorter, compatible with the observed experimental time.

Forces of the order of tens of piconewtons per chromosome have been measured by AFM in the mitotic spindle, summing up to a characteristic total internal force of ~ 1 nN for all the ~ 100 chromatids. Clearly, the spindle has to be firm enough to withstand internal forces of this magnitude. In other studies, the equivalent spring constant of the whole spindle was estimated at 1 nN/µm; this indicated that, if the actuating force is 1 nN, the spindle should deform by about $\Delta x \sim 1\,\mu$m, i.e. 5–10% of its size $2D$. However, spindle mechanics appear more complex than simply an elastic cage made of microtubule bundles. In the same experiments, it was also observed a *hysteresis* response (see Appendix H): the spindle resistance force is greater while compressed by the AFM cantilevers, and smaller when the cantilevers were released. Mathematically, this type of response can be modelled by dashpot elements (mechanical dampers that resist motion by way of viscous friction; see Sect. 9.2 on p. 376), arranged in parallel with the elastic elements (Fig. 8.15): both the elastic spring and the dashpot elements resist compression, but after compression is released, the spring force leads to expansion of the system while partially damped out by the dashpot element. While the elastic contribution to the stiffness should

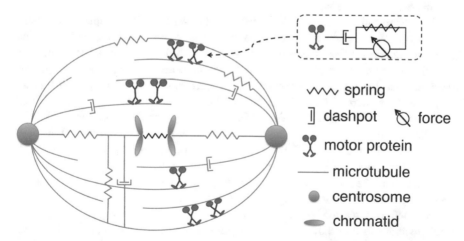

spring

dashpot   force

motor protein

microtubule

centrosome

chromatid

Fig. 8.15 Idealised view of spindle mechanics, showing a mitotic spindle built from many springs and dashpots in parallel, arranged both longitudinally and laterally. Microtubules (*green*) connect the centrosomes of the splitting nucleus, and the (still near-centered) sister chromatids (*purple*). Elastic and viscous contributions originate from the viscoelastic nature of molecular motors, which actively apply forces and feel the rectified Brownian forces from the environment, as well as from the intrinsic viscolelasticity of microtubules. Multiple motor proteins cooperate and make microtubules to slide, as well as locking parallel microtubules together

come from the microtubules, the viscous resistance could be a manifestation of the action of the molecular motor proteins acting on the structure.

This mixed response is called **viscoelastic** (see again Sect. 9.2), and it may be approximated as linearly depending on the deformation rate, rather than on the deformation itself. If the elastic contribution to the response to the applied force is characterised by Hooke-like springs k, as $F = k\Delta x$, an effective dashpot drag coefficient can also be introduced for the dashpot, relating the applied force to the deformation rate, as $F = \zeta(\Delta x/\Delta t)$. Just like the linear spring constant k has physical dimensions of $[Force][L^{-1}]$, the linear drag coefficient ζ will have dimensions of $[Force][T\, L^{-1}]$, and should be naturally related to a measure of the viscosity of the material. The experimentally observed values for the spindle are $\zeta \sim 1$ nN s/μm, which is an order of magnitude greater than the drag coefficient estimated from simply squeezing the cytoplasm.

Interestingly, the spindle exhibits a viscoelastic response both when deformed parallel or perpendicular to its axis. This suggests there might be viscoelastic elements (spring+dashpot) perpendicular to the main axis, also shown in Fig. 8.15, probably composed of microtubules laterally branching from the spindle, and the molecular motors cross-linking them. Such a network of transverse and perpendicular links resembles an architecture of crossed trusses and struts, making the spindle not just a parallel bundle, but rather like a matrix, behaving like a dense gel of interconnected elements. This gel is both anisotropic, being slightly weaker along the short axis than along the main axis, and *plastic*: if the external force exceeds a certain threshold, the spindle deforms irreversibly.

Why the mitotic spindle should have evolved towards such a complicate mechanical structure? Is the viscoelastic response actually necessary for its basic function, that is capturing and separating chromatids? One possibility is that its stiffness protects the spindle from collapse under external deformations, whereas the large viscous drag ensures that the spindle responds to perturbations very slowly, thus filtering out noise from thermal fluctuations and random "wrongly oriented" forces. On the other hand, spindle orientation also defines the mid-plane for subsequent cell splitting. During the early M-phase, the spindle moves and rotates to adjust to the "best" orientation, with the help of various protein motors; spindle movements appear to be subject to spatially distinct molecular mechanisms, which firstly mediate spindle rotation toward the long-axis of the elongating cell, and subsequently facilitate orientation maintenance along the long-axis. In this respect, the viscous response could be of great help in "steering" to maintain the correct orientation, by suppressing large-scale fluctuations. For the interconnected gel structure to be robust, the microtubule network must be homogeneous enough. It has been shown [16] that this can only be achieved if, in addition to random nucleation and dispersion within the spindle, other microtubules sprout from the sides of existing ones, with an autocatalytic nucleation mechanism; a chemical reaction is said to be autocatalytic if among its products there are some of the reactants, thus allowing the reaction to self-sustain).

8.5.3   Assembly of the Contractile Ring

The first visible sign of the splitting of the animal eukaryote cell (**cytokinesis**) is the formation of a furrow in the cell membrane, always lying in the equatorial plane perpendicular to the mitotic spindle. Experiments proved that if the spindle is moved, e.g. by inserting a microsphere in the cell, the furrow disappears and forms at the new plane identified by the new position of the spindle. In some, yet unknown way, the cytoskeleton can send the cell a clear signal as to where to start the breaking. It is also known that actin filaments, normally distributed homogeneously about the inner surface of the membrane, have some part in the formation of the spindle and interact with the microtubules. These latter could orchestrate the cleavage of the membrane, by properly redistributing actin filaments in the assembly of the **contractile ring**, a roughly circular bundle of actin filaments and myosin-II motor proteins, arranged to lie in the mid-plane of the splitting cell (see the white structure in Fig. 8.16a). This is the final step of cell division, capable of generating important mechanical forces, by

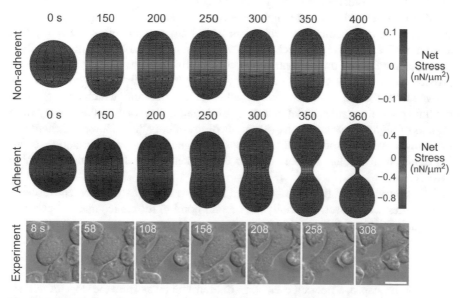

Fig. 8.16 Morphological changes in a computer simulation model of cell division (cytokinesis). The *upper row* represents a sequence of simulations for a non-adherent cell, the *middle row* a sequence for a cell adherent on a rigid surface, both starting from an initial spherical shape. *Colours* refer to the distribution of the stress, according to the scale on the right (in nN/μm^2). The *bottom row* displays experimental data for *Dictyostelium* amoeba cells (mutated to remove myosin-II) dividing on a surface. Scale bar denotes 10 μm. [Image adapted from Ref. [17], under CC-BY 3.0 licence, see (*) for terms.]

shrinking the ring and forming a deep cleavage neck in the membrane, which leads to the final physical splitting of the mother cell.[5]

The earliest experimental evidences of such a mechanical structure came already a half century ago, by observing the bending of a micro-needle inserted into the cleavage furrow of a dividing echinoderm egg. The combined contraction of myosin and actin is not new, since it had been already found to be at the origin of mechanical force in the sarcomeres, the basic units of muscle cells (see Chap. 10). Myosin motors pull the actin filament like climbers on a rope, in steps of ∼5–7 nm per 5 ms time, thus inducing the contraction of the filament bundle (see Fig. 6.3 on p. 210). Despite the similarity between the two contraction mechanisms, however, evidence of important differences has also accumulated over the years. First and foremost, in the formation of the contractile ring there is an inflow/outflow of molecules that is not observed in muscle cells: as the ring constricts, the thickness of the contractile ring remains constant, suggesting that F-actin disassembly must balance ongoing F-actin polymerisation. Secondly, the actin arrangement in the ring is quite disordered, compared to the highly ordered, periodically-repeated microstructure of the hundreds of sarcomeres composing a muscle cell.

Moreover, experiments in which the myosin-II is removed (Fig. 8.16) demonstrate that the contractile-ring mechanism works anyway, therefore indicating that alternative force-generating mechanisms must also be available. In this case, adherent junctions formed on the contact of the cell (with a surface, or with other cells) provide the necessary gradient of stress. Another major mechanical component is derived from the cell membrane tension and surface curvature, which leads to internal fluid pressure gradients that induce hydrodynamic flow of cytoplasm away from regions of high surface curvature, to regions of lower curvature. How such chemical, mechanical, and hydrodynamic forces couple together is a complex and fascinating biophysical problem in itself, besides the evident implications in understanding cytokinesis.

The central role of the contractile ring has been at the focus of recent research in cytokinesis. However, especially the second point of difference listed above may raise questions: how is it possible to generate a well-directed contractile force, starting from a random assembly of filaments and proteins? A recent review cited not less than ten different physical models suitable for force generation in the actin-myosin annular bundle, each partly meaningful but none entirely successful. As it was seen in Chap. 5, a free cell (i.e., non-adherent) far away from the stage of mitosis experiences only passive forces, coming from the Laplace pressure normal to the membrane (proportional to the effective surface tension × the surface curvature). Any perturbation leading to a non-spherical shape fluctuation would produce greatest forces at the regions of higher curvature, causing a relaxation back towards a

---

[5]Notably, plant cells operate the membrane division by an entirely different mechanism. After chromosome splitting, a portion of the rigid cell wall that surrounds most vegetable cells is nucleated in a new plaque, around the residual microtubules from the spindle that form a temporary structure called *phragmoplast*. This plaque starts with a roughly discoidal shape in the cell mid-plane, and grows in size with the help of microtubules, which dynamically reorganise around its perimeter. The process continues until the plaque entirely forms a new wall, dividing in two the old cell.

spherical morphology. Therefore, a symmetry-breaking force is required, in order to stabilise and grow the central furrow, up to membrane cleavage.

Such a force, or stress (force per unit area, see Appendix H) can indeed have different sources, given the complex cell environment. In a recently published model (see again Fig. 8.16), the splitting cell was described as a shape evolving with time from spherical, to ellipsoidal, to a dumbbell. The cell surface is described by the two coordinates z, spanning the axial length, and r, following the parallel circles in the normal planes. Stresses are directed parallel or perpendicular to the local surface element, and integrated over the entire cell surface S. The total stress is made up of several components:

$$\sigma = \sigma_{ad}D(r)D(z) + \int_S \Sigma(r)\gamma(r)dr + \sigma_p w(z,t)D(z) + \sigma_c c(z,r) \qquad (8.78)$$

The "ad" term is an adhesion force applied to the membrane by the adhesion patches, covering fractions $D(z)$, $D(r)$ of the total surface, and would be absent for a non-adherent cell. The $\Sigma\gamma$ integral is the Laplace pressure, for variable values of surface tension and curvature distributed all around the cell surface. The "p" term is associated with the protrusions of the cell membrane, originating from actin polymerisation; it is proportional to the z-component of the surface contact $D(z)$, and decays with time according to some empirical function $w(z,t)$. Finally, the "c" term describes the contraction directly induced by the myosin-II in the contractile ring, where the protein is present with a position-dependent concentration $c(z,r)$. The various terms of the model and their numerical parameters were empirically established, by fitting on a large body of experimental data for the amoeba *Dictyostelium discoideum*.

The results, partly shown in the figure, allow to establish the relative importance of the different forces in shaping the cell furrow and splitting, over realistic time and length scales. The model shows that cytokinesis may proceed through distinct phases. Actin-myosin contraction is seen to play a primary role, by providing the stress that initiates furrow ingression; such an early-stage contraction could be related to the initial densification of the ring, and therefore could work even without attaining a highly ordered microstructure. In its absence, however, this force can be supplied by a combination of adhesion and protrusion-mediated stresses. Thereafter, Laplace-like pressure takes over and, once integrated over the cell surface (several μm^2) provides large enough forces (several nN) that enable the cell to divide. In Fig. 8.16, by comparing the upper and central row, it can be seen that the role of the "ad" term is key to induce the furrowing, when the "c" term is suppressed. The *Dyctostelium* amoeba was shown to be unable to split if genetically modified to remove the myosin-II; however, its splitting ability is restored if the amoeba is contacted to a rigid surface. Both features are correctly described by the model, in the case of non-adherent (upper row) versus adherent cell (middle row). Such conclusions, although not yet definitively corroborated by supporting experiments, give support to the observations that point at cytokinesis mechanisms even more complex than the contractile ring itself.

Problems

8.1 Hollow versus filled
Calculate the persistence length and the flexural rigidity, for a hollow microtubule with inner and outer radii 10 nm and 12 nm, and Young's modulus $E = 150$ MPa. Compare the results with a solid microtubule, with the same amount of mass per unit length.

8.2 Bacterial DNA
The DNA of some bacterium contains 3.45 million base pairs. (a) Calculate the contour length. (b) Given the average DNA Young's modulus $E = 350$ MPa, calculate its persistence length. How does it compare with the contour length? (c) Given the DNA diameter of 2 nm, what is the smallest volume into which the double strand can be packed? (d) Calculate the end-to-end distance and gyration radius of this DNA. How these compare with the size of a typical bacterium?

8.3 Exocitosis
Find an explicit expression for the surface tension energy, $E_S = \Sigma A$, during the three steps of the formation of a spherical liposome, illustrated in the figure below. Starting from the membrane at rest (1), a pseudopod is extruded (2), which eventually forms a spherical vesicle of radius R, fully detached from the cell membrane.

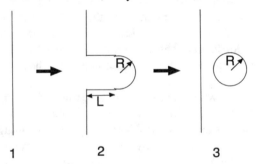

8.4 Membranes with an edge
A cell membrane is composed by two different types of phospholipids, the A with tails of length $n_A = 14$ and the B with $n_B = 18$. By assuming that bilayers form only with lipids of the same kind, and for a lipid-water interfacial tension $\Sigma = 30$ mJ/m^2, (a) what would be the energy per unit length of an A/B interface? (b) what would be the lowest-energy state for 10 domains, each with average surface 100 nm^2, dispersed in a large membrane patch? how distant in energy are the two configurations? (c) could a transition between the two configurations occur because of thermal fluctuations?

8.5 Membranes with a dimple
Consider a thermal fluctuation making a dimple of height h in a patch $S = \pi L^2$ of membrane. (a) What is the deformation energy of such a dimple? (b) For a bending constant $K_b = 15 k_B T$, at what h/L ratio the dimple could appear by thermal fluctuation of the membrane surface?

8.6 Pulling chromosomes

An average chromosome can be modelled as an elongated ellipsoid, with major diameter $b \sim 15\,\mu m$, and minor diameter $a \sim 1\,\mu m$. Observe that during mitosis, each chromosome is pulled by the microtubules by some $15\,\mu m$ in about $10\,min$ (such values can vary quite a lot according to the cell type). (a) What is an upper bound to the pulling force? (b) How much work is spent, and how many molecules of ATP are used?

8.7 Pushing cells with a laser

Consider a bacterial cell as a sphere of $2\,\mu m$ diameter, and an eukaryote cell ten times larger. A typical laser beam of power P used in optical traps can produce a force of the order of a few pN, as approximated by the expression $f = nPQ/c$, with n the refractive index of the medium, $Q \sim 1\%$ the quality factor of the laser, and c the speed of light. Estimate the power necessary to move the two cells. What does such a calculation suggest?

References

1. C. Bustamante, Z. Zev Bryant, B. Steven, S.B. Smith. Ten years of tension: single-molecule DNA mechanics. Nature **421**, 423–427 (2003)
2. A. Ashkin, J.M. Dziedzic, J.E. Bjorkholm, S. Chu, Observation of a single-beam gradient force optical trap for dielectric particles. Opt. Lett. **11**, 288 (1986)
3. M. Rief, M. Gautel, F. Oesterhelt, J.M. Fernandez, H.E. Gaub, Reversible unfolding of individual titin immunoglobulin domains by AFM. Science **276**, 1109 (1997)
4. K.C. Neuman, A. Nagy, Single-molecule force spectroscopy: optical tweezers, magnetic tweezers and atomic force microscopy. Nat. Meth. **5**, 491–505 (2008)
5. K. Berndl, J. Käs, R. Lipowsky, E. Sackmann, U. Seifert, Shape transformations of giant vesicles: extreme sensitivity to bilayer asymmetry. Europhys. Lett. **13**, 659 (1990)
6. W. Wintz, H.-G. Döbereiner, U. Seifert, Starfish vesicles. Europhys. Lett. **33**, 403 (1996)
7. W. Helfrich, Elastic properties of lipid bilayers: theory and possible experiments. Z. Naturforschung. C: Biosci. **28**, 693 (1973)
8. U. Seifert, K. Berndl, R. Lipowsky, Shape transformations of vesicles: phase diagram for spontaneous-curvature and bilayer-coupling models. Phys. Rev. A **44**, 1182–1194 (1991)
9. M. Jarić, U. Seifert, W. Wintz, M. Wortis, Vesicular instabilities: the prolate-to-oblate transition and other shape instabilities of fluid bilayer membranes. Phys. Rev. E **52**, 6623 (1995)
10. K. Khairy, J. Howard, Minimum-energy vesicle and cell shapes calculated using spherical harmonics parameterization. Soft Matter **7**, 2138 (2011)
11. N. Gov, S.N. Safran, Pinning of fluid membranes by periodic harmonic potentials. Phys. Rev. E **69**, 011101 (2004)
12. M. Reffay et al., Interplay of RhoA and mechanical forces in collective cell migration driven by leader cells. Nature Cell Biol. **16**, 217–223 (2013)

13. H. Jiang, F. Si, W. Margolin, S.X. Sun, Mechanical control of bacterial cell shape. Biophys. J. **101**, 327–335 (2001)
14. L.A. Diaz-Martinez, H. Yu, Chromosome condensation and cohesion, in "Encyclopedia of Life Sciences (ELS)" (John Wiley Ed., Chichester, 2010)
15. C.B. O'Connell, A.L. Khodjakov, Cooperative mechanisms of mitotic spindle formation. J. Cell Sci. **120**, 1717–1722 (2007)
16. T. Clausen, K. Ribbeck, Self-organization of anastral spindles by synergy of dynamic instability, autocatalytic microtubule production, and a spatial signalling gradient. PLoS ONE **2**, e244 (2007)
17. C.C. Poirier, W.P. Ng, D.N. Robinson, P.A. Iglesias, Deconvolution of the cellular force-generating subsystems that govern cytokinesis furrow ingression. PLoS Comp. Biol. **8**, e1002467 (2012)

Further Reading

18. X. Michalet, D. Bensimon, Observation of stable shapes and conformal diffusion in genus-2 vesicles. Science **269**, 666 (1995)
19. A. Mogilner, E. Craig, Towards a quantitative understanding of mitotic spindle assembly and mechanics. J. Cell Sci. **123**, 3435 (2010)
20. M. Doi, *Introduction to Polymer Physics* (Clarendon Press, Oxford, 1996)
21. Alberts, Bray, Lewis, Raff, Roberts, Watson, *Molecular Biology of the Cell*, 4th edn. (Garland Science, New York, 2002)
22. A. Mogilner, On the edge: modeling protrusions. Current Opin. Cell Biol. **18**, 32–38 (2006)
23. I. Mendes Pinto, B. Rubinstein, R. Li, *Force to divide: structural and mechanical requirements for actomyosin ring contraction*. Physical Rev. A **105**, 547–550 (2013)

Chapter 9
The Materials of the Living

Abstract This chapter delves into the mathematical toolbox of continuum mechanics, necessary to describe the mechanical deformation of biological materials. After being relegated for a long time to little more than the (useful, but quite uninspiring) role of providing healing and implants for broken body parts, the domain of biomechanics has been recently promoted to the highest level of attention. The modern view of biomaterials relies on connecting structures over multiple length-scales. Functional organisation is found at all levels, from the molecules, to fibrils and coils, to the carefully networked microstructures, up to the macroscopic scale. At odds with the variety of materials available in nature, just a small set of recurring molecule types, arranged in ever different structures, can give rise to such diverse tissues as skin, cartilage and bone, green stems, rose buds or wood. The secret is the hierarchical multi-level structure of biomaterials.

9.1 Stress and Deformation

The way the world is approached depends very much from the mental attitude of everyone: in some sense, it is the difference between brain and mind. A chemist sees molecules in every object, and reactions and transformations occurring all the time; a physicist looks for elementary constituents, universal laws and statistical principles; a philosopher would probably question the essence of reality and sometimes doubt of it; engineers consider that atoms and molecules are practically irrelevant, and have invented **continuum mechanics** to describe the world.

While the idea of continuum was around at least since the works of Leonardo da Vinci, in the late XV century, the formalism of the continuum model of matter was actually laid down by an outstanding legion of XVIII century mathematicians, chiefly Augustin-Louis Cauchy, both anticipated and followed by the like of Euler, Lagrange, Laplace, Bernoulli. The continuum theory remained a major instrument of physical analysis until the late XIX century, when physicists started to be more interested in the intimate constitution of matter, space and time at the smallest and largest scales, and left the continuum theory entirely in the hands of the engineers.

© Springer International Publishing Switzerland 2016

F. Cleri, *The Physics of Living Systems*, Undergraduate Lecture
Notes in Physics, DOI 10.1007/978-3-319-30647-6_9

In the continuum theory of material mechanics, one usually starts with rigid objects whose form and structure are not modified by the application of a force. The only admitted changes in the shape are very small deformations in the so-called *elastic* regime, such that the object retains its form and structure once the force is removed. On the other hand, it is common experience that matter can indeed be deformed to large extents, and even broken apart by an applied force. Forces can be applied on the surface of the object, like a pressure, or in the volume, like the gravity. At a microscopic level, deformation corresponds to atoms and molecules stretching and breaking their bonds, changing their positions, and possibly coming apart. Therefore the formalism of continuum mechanics must at some point include tools capable of describing large deformations, even though molecules and atoms are never explicitly described. This is even more true for biological materials, which constitute mostly deformable and flexible objects when compared to non-living matter, like rocks or steel. Moreover, biological materials have a characteristic behavior that is completely absent from engineering materials: the **remodelling**, which indicates the biological capability of "creating" entirely new material structures in response to a force that deforms or deteriorates the existing material.

Forgetting about atoms and molecules when looking at large scale deformation can be indeed a useful point of view, since the length and time scales involved are well beyond the atomic scale. A chunk of steel contains a number of atoms of the order of the Avogadro's number, $N_{Av} = 6.02 \times 10^{23}$, and even a single cell contains order of 10^{14} molecules, all which move at typical frequencies of 10^{12} Hz. Therefore attempting at a solution in terms of atoms is clearly a daunting task, which can be restricted only to special situations. However, continuum modelling comes at the price of replacing the real constituents by some mathematical objects, "continuous fields" defined everywhere in the space occupied by the object, which must try to imitate as closely as possible the behavior of the real material. If the continuum idea may seem a simpler and logical description, more adherent to our everyday experience of a world without discontinuities, it hurts against inevitable difficulties when stepping down in length scales. In fact, already at the beginning of the XIX century a rival party of French mathematicians (among which Fresnel, Navier and Poisson) had built a "molecular" theory of mechanics, which however failed to correctly predict even the simplest properties of material elasticity, thus leaving the advantage to the very successful theory of continuum fields by Cauchy, Green and the others. It would have been necessary to wait until about 1866, when the crystallography studies of Bravais opened to the idea that "molecules" are not just dimensionless points, but some kind of polyhedra that fill the space according to different shapes (symmetries) and orientations. Starting from such results, Woldemar Voigt formulated a theory capable of reconciling the continuum and molecular views, producing results in agreement with the experimental evidence.

To describe a material deformation at the continuum level it is necessary to have: (1) a formalism capable of describing the way a force applied on the surface or in the volume of a body, is distributed all over: this leads to the definition of the **stress** field; (2) a formal description of the local and global deformation of the object, after the application of the force: this leads to the definition of the **strain** field. Such quantities and their most relevant mathematical properties are synthetically described in the Appendix H to this chapter.

In general terms, a force can be applied at a point. However, in continuum mechanics a point is a mathematical singularity. The singularity must be cured by associating to it an infinitesimal volume, or an infinitesimal surface. This leads to introducing such quantities as, e.g., the mechanical pressure, actually a force divided by a surface. The latter is just the infinitesimal surface associated to the point of application, but even at the atomic scale the minimum for a "point" would be the volume or surface of one atom, a definitely small but finite quantity, observable e.g. with an atomic force microscope.

We may start by defining a very general quantity, which we call **stress**, with units of a force divided by a surface. If we think of a gas enclosed in a volume, as in the greybox on page 24, in that case the quantity with units of force/surface could be identified with the pressure of the gas itself. Pressure is one example of stress, but a very special one, in that it is perfectly isotropic in space: pressure is the same at every point of a fluid, and the same in all directions. On the other hand, we may take a solid body with an arbitrary shape, and apply a force oriented along some direction at any point P of its surface. As shown in the Appendix H, the stress is a tensor $\underline{\sigma} = \sigma_{ij}$, with $\{i, j\} = \{x, y, z\}$, a 3×3 square matrix of numbers meant to describe all the combinations of the components of the applied force vector, $\mathbf{f} = \{f_x, f_y, f_z\}$, with respect to the components of the surface normal vector $\mathbf{n} = \{n_x, n_y, n_z\}$ passing through P (some basic properties of tensors are discussed in the mathematical Appendix A).

On the other hand, **strain** characterises the continuous shape deformations of a body in response to an applied mechanical stress. Deformation is a dimensionless quantity, since it is defined by the ratio of the variation of a quantity with respect to its original value. The relative variation of the length, $\Delta L/L$, or of the volume of an object, $\Delta V/V$, are examples of strain. As shown in more detail in the Appendix H, strain is also mathematically defined as a 3×3 tensor, ε_{ij}, with indices related to the components of the deformation vector, $\mathbf{u} = \mathbf{r}' - \mathbf{r} = \{u_x, u_y, u_z\}$, with respect to those of the position vector $\mathbf{r} = \{r_x, r_y, r_z\}$, indicating any point in the body volume.

If we want to know what is the amount and distribution of the deformation for a given applied stress or, conversely, what was the stress that produced a given deformation, we need a **stress-strain relationship**. A very early example of such force-deformation relation is the Hooke's law, relating the uniaxial elastic stretching $u = (l' - l)$ of a spring of length l to the parallel applied force, as $f = ku$. In terms of stress and strain, if we take the force vector directed parallel to the x axis, Hooke's law would be written as $\sigma_{xx} = \bar{k}\varepsilon_{xx}$, with $\bar{k}$ a different constant from k, since their dimensions are different ([Energy]/[L^3] vs. [Energy]/[L]).

The deformation results from the resistance opposed by the internal forces of the body, to the external forces applied on its surface, or within its volume. 'Internal' forces are indeed the atomic and molecular forces, which keep together the material and impart its peculiar physical and chemical properties. In general, these forces are anisotropic, and depend on many conditions defining the material, such as the number and distribution of electric charges and dipoles, or its microscopic structure, from disordered to crystalline. The presence of anisotropy makes the stress-strain relation more complicated than the simple expression above, and can induce non zero off-diagonal components in the matrices of the two tensors, compared to the simple situation described by Hooke's law.

Stress-strain relations can be experimentally obtained by measuring the response of a material to the applied force along different directions. In this way, plots of the type shown in Fig. 9.24 of the Appendix H are obtained. It is worth noting that all the **elastic** theory developed in the Appendix strictly refers to the initial, linear part of the stress-strain curve, unless explicitly stated. In this regime, characterised by small deformations, the relationship between force and displacement, or stress and strain, is one of simple proportionality, just like in Hooke's law. However, experimental stress-strain plots can extend well beyond the linear-elastic limit (point 1 in Fig. 9.24), although the theory in this part of the diagram becomes much more complex: the region beyond the elastic limit is called the **plastic** part of the stress-strain curve. In this domain, the deformation begins to be irreversible, meaning that part of the mechanical work is used to induce permanent deformations, defects, broken bonds in the material, which will remain also after the stress is released. The material is still compact and can sustain some stress, but with a profoundly different behaviour, as shown in the final part of the curve in Fig. 9.24.

Even if continuum mechanics avoids the notion of an internal structure of the matter, it can however supplement a set of response coefficients depending on the material composition and phase: the linear **elastic constants** C_{ijkl}, and the **elastic compliances** S_{ijkl} (see Appendix H). These sets of material parameters allow to establish the **constitutive relation** for the material of interest, i.e. its stress-strain relationship of the most general linear form:

$$\sigma_{ij} = \sum_{kl} C_{ijkl} \varepsilon_{kl} \tag{9.1}$$

all indices $\{i, j, k, l\}$ running on the Cartesian components $\{x, y, z\}$, for a total of $3^4 = 81$ independent parameters. Because of the intrinsic symmetry of the matrices, $\sigma_{ij} = \sigma_{ji}$ and so on, these are in fact only 21.[1] Moreover, because of additional material symmetries, many of these parameters are either zero, or equal to each other, thus further reducing to a quite smaller set in most materials.

In practical cases, however, we are not so much interested in knowing the individual values of each of the constants C_{ijkl}, but rather some combinations of their values. This is because the experimental set up needed to measure such constants makes some special combinations of C's to be more easily accessible, rather than their separate values. (However, these can be retrieved once the appropriate combinations have been measured.) These combinations of the fundamental elastic constants are called the **elastic moduli** of the material, the most important ones being: the bulk

---

[1] Eventually, by using Voigt's notation the C matrix becomes a 6×6 with 36 components; because of symmetry, the matrix is diagonal symmetric, therefore with 6 (diagonal) plus $30/2 = 15$ (off-diagonal) independent components, summing to 21.

modulus, B; the shear modulus, μ or G; the Young's modulus, E, and the associated Poisson's ratio ν. All these elastic moduli are described in some detail in the Appendix H. The special cases of deformation by bending or flexion of a long rod, or a plate membrane, were already anticipated in the previous Chapter (see greybox on p. 322).

One interesting finding from the simple theory of linear elasticity, is that many materials can be described at the macroscopic scale as being essentially **isotropic**, meaning that the response to the deformation does not appear to depend on the particular direction the force is applied in the material. This is certainly the case for most fluid and amorphous substances, but even for many solid, everyday materials, such as glass, rubber, plastics. Also many metallic materials, in principle crystalline and anisotropic, display practically isotropic properties at the macroscopic scale, deriving from the fact that their intrinsically anisotropic crystalline structure is broken into many micron-sized tiny crystals, whose random assembly makes the macroscopic material to display direction-averaged, isotropic mechanical properties.

From the point of view of the defining equations, in order to describe isotropic materials we require just two elastic constants, the so-called *Lamé parameters*, which (as shown in the Appendix) are simple combinations of the fundamental elastic constants C_{ij} of the solid, namely $\lambda = C_{12}$ and $\mu = C_{44}$. Only the matrix elements $C_{11} = C_{22} = C_{33}, C_{44} = C_{55} = C_{66}$ and $C_{12} = C_{13} = C_{13}$ are non-zero for isotropic materials, satisfying the condition $C_{ii} = C_{ij} + 2C_{kk}$, with $i, j = 1, 2, 3$ and $k = 4, 5, 6$. Alternatively, the Young's modulus E and Poisson's ratio ν of the material can be used as the two independent parameters, whose relationship with the Lamé parameters is given in the Appendix. The corresponding stress-strain relationship is therefore particularly simple, and is given by the Eqs. (9.31–9.34) of the Appendix.

9.1.1   The Biologist and the Engineer

Compared to the isotropic materials with only two parameters, anisotropic materials could appear far more complex. In practice, however, their mathematical treatment is exactly the same; only the detailed calculations are more involved, because of the larger number of independent elastic coefficients. Pure crystalline materials, such as metals, semiconductors, rocks, are all anisotropic to a smaller or larger extent, yet their mathematical treatment can be carried out quite properly, to the utter delight of the engineer.

Biological materials are not only anisotropic, but they combine and mix properties in a way that goes far beyond the imagination of any engineer working with modern composite materials. They are much more complex under any respect, and it should be just amazing to think that we can make some sense at all of their mechanical behaviour, by using the simple equations of classical mechanics (also summarised in the Appendix H). Nevertheless, traditional material science, which was born to describe the behaviour of common building materials like metals and concrete, can offer some points of comparison to at least get a feeling of the staggering mechanical properties of biological materials.

By looking at inorganic materials from a mechanical perspective, the school-book classification of gas-liquid-solid translates in the observation that gases resist only compression, liquids resist tension as well as compression, while solids resist compression, tension and shear stresses. Despite being largely composed of water, nearly all biological materials appear as solids. However, they include materials ranging from the very soft (but still elastic) mucuses, to the very hard (but still deformable) dental enamels and corals: clearly, their degree of "solidity" in the commonly employed sense, seems rather irrelevant here. Also the usual engineering distinction between material, structure and system (steel is a material, a beam is a structure, and a bridge is a system) looses much of its meaning when applied to biological materials: the tightly connected levels of hierarchical organisation of the biological matter (see Figs. 9.12 and 9.13 for an example) make it quite hard, and likely pointless, to distinguish where the material starts to be a structure, and at which point exactly structures turn into systems.

A possibly interesting classification of engineering materials is based on their chemical nature. Materials are usually known as metals, semiconductors, insulators, and more precisely ceramics, silicates, polymers, and so on, based on the nature of the chemical bonds between atoms. Because of the way electrons are arranged around the constituent atoms, metals are generally ductile, while insulators tend to be more fragile; semiconductors and insulators generally conduct poorly both heat and electricity, while metals are very good in this respect. In the realm of biology, we could split materials into proteins, sugars, lipids, and nucleic acids, each with peculiar chemical bonding and reactivity characters. However, it is hard to put such a distinction into a more or less strict correspondence with their mechanical properties: silk and collagen are both proteins, but largely differ in their elasticity; silk from moths has a very different chemical composition from the silk of spiders; and even in a single spiderweb, silks of different elasticity and strength are used for the different parts.

Following Vogel ([1], Chap. 15), we could introduce another, maybe more meaningful classification, among tensile, pliant and rigid materials.

- **Tensile** materials resist to traction: basically they function like ropes in a biological context. Collagen and chitin are probably the two most common proteins found in animal bodies; collagen is the main material of tendons, which connect muscles to bones, and similarly chitin is the material connecting insect and crustacean muscles to their exoskeleton; collagen makes up also skin, cartilage, and about half of the weight of bones. Silk is another example of natural rope with exceptional properties; it is produced by all arthropods, such as spiders and silkmoths. Plants have developed their own version of molecular rope with cellulose, a sugar-based polymeric material typically used as a part of plant cell walls and composite structures, e.g. making up cotton, linen and other vegetal fibres.
- **Pliant** materials can deform to large extent, in ways also depending on their deformation rates and history (viscoelasticity, see below). Such proteins as resilin, elastin, abductin, make up powerful rubber springs that can accomodate large

deformations, thereby regulating many mechanical movements requiring large amplitudes such as the opening of shell valves or the flapping of insect wings.

- **Rigid** materials have elevated resistance to bending and torsion, and can support relatively large loads in tension, or compression, or both; compared to inorganic rigid materials (steel, concrete, ceramics) they are however strongly anisotropic, and have complex composite structures. If bone, a composite of collagen and mineral particles, is a well known example of rigid biomaterial, keratin is a widespread composite of two proteins that makes up animal hair, horns, feathers; and wood, probably the most economically viable of all natural materials, is a composite of cellulose and lignin, arranged in a carefully organised, rigid superstructure in plant cell walls.

It may be observed that in materials science it is a common practice to express the work done by mechanical forces, and the corresponding mechanical quantities of the material (elastic constants, elastic moduli), as [Energy]/[Volume]. When comparing the properties of biological materials, we deal with materials that are much lighter, less dense than most engineering solids, therefore a comparison of the energy or work spent *per unit mass* seems more appropriate. For example, if we compare the Young's modulus of steel to that of collagen in MPa units, we find an obvious difference of a factor of about 10 in favour of steel; however, if we divide the values by the respective mass density, it is found that collagen is almost as stiff as steel. Biological materials, albeit less strong on an absolute basis, can withstand very large values of deformation up to their ultimate fracture strain (the point 3 in Fig. 9.24 in the Appendix), therefore they have an amount of "area under the curve" (i.e., work stored as mechanical energy) that can equal, or largely surpass that of most engineering materials. Again by looking at collagen versus steel, we would measure values of work of $\sim$2–3 J/cm^3 versus about 1 J/cm^3 of steel, meaning that collagen is already tougher than steel on a per-volume basis; however once translated into work per unit mass, collagen outdoes steel by a factor of $\sim$15. Whereas steel can be stretched to a few per cent strain, pure collagen can take up to 30 %, and spider silk can go up to 500 % and more, thanks to their peculiar molecular hierarchical structure.

9.1.2 Brittle and Ductile

Not all the mechanical properties of a material, however, can be directly deduced from a stress-strain plot, especially concerning the properties of the material near the breaking point. The stress and strain at the point of fracture, σ_f and ε_f, can be directly read on the plot by looking at the point where the (σ, ε) curve ends, and the corresponding toughness τ_0 can be obtained by integrating the area under the curve. But the breakage itself takes an extra amount of work, distributed among the material elements: upon fracturing, the material breaks open portions of new free surface in a more or less random way, and this costs energy that is not accounted directly in

Cauchy's continuum theory of stretching a compact, homogeneous material. It is at this point that a fracture appears in the medium, and starts propagating. Generally, we define as **brittle** a material in which one single crack takes over and breaks the material in a very fast, catastrophic manner, such as the breaking of a window glass; as opposed to **ductile**, a material in which tearing and thinning starts at many points concurrently during some span of time, and the breaking is slower and progressive, such as the breaking of a piece of soft plastic.

If we look at a microscopic fracture propagating in a material put under an externally applied load, or stress σ_0, experiments show that the projected stress around the tip of the crack σ_t is generally much larger than σ_0 ("stress concentration"), and decreases proportionally to the square-root of the (growing) crack length a, in such way that the product $\sigma_t \sqrt{a}$ remains practically constant. Around 1920, the British engineer A.A. Griffith, while studying the brittle fracture of metals in large structures such as ships and airplanes, deduced that the work spent in opening the fracture must be linked to the formation of new free surface, and arrived at the conclusion that the stress at the crack tip decreases upon increasing a according to the universal law:

$$\sigma_t = \sqrt{\frac{2E\Sigma}{\pi a}} \tag{9.2}$$

where Σ is a surface energy with units of [Energy][L^{-2}] characteristic of the material surface (it could be the surface tension, for a fluid). For materials displaying ductile fracture, the value of Σ must be replaced by a different energy per unit area, actually called *work of fracture*, G; this has the same units as Σ, but takes into account the energy irreversibly dissipated in the plastic deformation of the material (see above) actually accompanying the breaking of the free surfaces. In simple conditions, the work of fracture can be obtained from the ratio:

$$G = \frac{K_I^2}{E} \tag{9.3}$$

in which the stress-intensity factor, K_I (sometimes called *fracture toughness*, but different from the toughness τ_0 defined in the Appendix) appears. The value of K_I for a material can be obtained by performing an experiment similar to that required to measure the Young's modulus, but by inserting a small micro-crack of known length in a mid-plane perpendicular to the applied force (see the diagram sketched in Fig. 9.1). The value of G in engineering materials can range (values in kJ/m^2) from the $\sim$0.005–0.040 of concrete, brick or stone, to the 10^2–10^3 of various types of steel. Typical values of G for biomaterials show as well large variations, also according to the anisotropy of the microstructure. For example, wood has $G = 0.15$ in the direction of grain, but $G = 12$ transverse to grain; mollusk shell ranges from $G = 0.15$ to 1.6, if fractured parallel or perpendicular to the surface; teeth are a singular structure, in that the outer enamel has $G = 0.2$ while the inner dentin has a higher $G = 0.55$; a stiff cow bone has a $G = 1.7$, and the very stretchable and tough skin of a rabbit measures a surprising $G = 20$ kJ/m^2.

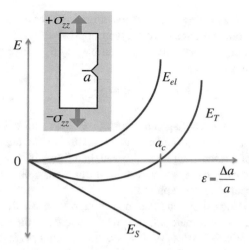

Fig. 9.1 An idealised fracture test, with a stress $\pm\sigma_{zz}$ applied on a material sample containing a microcrack of length a. Schematic plot of the total fracture energy release, E_T (*blue curve*). The sum of the (quadratic) elastic energy stored in the material, E_{el}, and of the surface creation energy, E_S, goes through zero at a critical value of the crack length a_c. Until the total energy is negative, the crack does not grow, since it costs more energy to make it advance than to make it close back. However, for a crack of length bigger than a_c, the energy is sufficient to make it grow

The dependence of the fracture progression on the actual crack length a is what makes materials resistance a complex problem. This can be appreciated by thinking of fracture growth as a competition between the different energies involved during the loading of a material by external forces: on the one hand, the external stress σ_0 continues to input energy in the system, and by increasing the strain ε this elastic energy is, as we know, proportional to the square of the deformation: $E_{el} = \frac{1}{2}E\varepsilon^2$ (note that the stress and strain can be described by linear elasticity only in the region far away from the crack tip surroundings, where instead the stress rapidly varies from σ_0 to $\sigma_t >> \sigma_0$); on the other hand, the energy needed to advance the crack surfaces is at each increment of ε proportional to the amount of surface energy times the elementary deformation, $E_S = 2\Sigma\varepsilon$ (the factor 2 comes from the fact that the crack creates two opposed facets of free surface over the length ε). By putting the total energy $E_T = E_{el} - E_S$ in a plot like the one shown in Fig. 9.1 (the + and − signs describe energy going into, or coming from the material, respectively), the crack length a_c, corresponding to the inversion of the sign of E_T, is identified as a critical value: if a crack of length $a > a_c$ were already present in the material, the applied stress can make it grow, and the crack will keep growing very fast, even if the stress is released, in the region of the diagram where the blue energy curve E_T becomes positive. Conversely, if all the microcracks in the material have lengths $a < a_c$ they can be stable and the material will not break, even if displaying localized ruptures, until a critical value of stress is reached, such that the elastic energy transferred to the longest of the crack tips wins over the surface energy.

Notably, the above reasoning allows to give an estimate of the critical crack length, which will (or will not) propagate for a given applied stress. By assuming for the sake of simplicity a linear stress-strain relation $\varepsilon = \sigma/E$, from Eq. (9.2) the critical length is of the order of:

$$a_c \sim \frac{GE}{\pi \sigma^2} = \frac{G}{\pi \sigma \varepsilon} \tag{9.4}$$

For example, by taking a collagen network (see below, animal skin) at a stress of $\sigma = 1$ MPa and strain $\varepsilon = 30\%$, and a work of fracture in the range of $G \sim 10$ kJ/m^2, the critical length of a microfracture in the collagen fibres is $a_c \sim 1$ cm. This shows that skin does not easily tear apart even with a quite large cut, unless exceedingly large deformation and stress are applied.

The above reasonings show that it is not enough to know the Young's modulus and the surface energy of a material, in order to decide if the material will break, in which way, and at which stress (unless we start from a perfectly intact and pure material, without any flaws or defects). It must be noted that the criteria for deciding whether a material is "stronger", or it is more resistant than another, are quite different, depending on the fact that the materials to be compared are brittle or ductile according to the above definitions. For example, the values of K_I for a plastic polymer are in the range of 0.5–2 MPa m$^{1/2}$, of 3–5 for silicon carbide, and between 10–100 for most metals. However, silicon carbide has a bulk modulus B (i.e., resistance to compression) more than twice bigger than steel. Steel is certainly a much stronger material than plastic, however their G values are quite close. Such striking comparisons (and the list could be very long) are due to the fact that the mode of failure is very different among the various materials. This is mostly ascribed to their internal structure, and to the presence of defects and heterogeneities of various kind.

Biological materials can tolerate large amounts of stretching and compression, thanks to their complex, heterogeneous and multi-phase internal structure, compared to the homogeneous nature of a crystalline, or even a polycrystalline bulk material. Someway, it can be said that Nature has replaced the high absolute strength, toughness, tenacity of homogeneous inorganic materials, by a carefully organised architecture at all length scales, from the molecular to the macroscopic, so as to obtain materials structures with performances equal or exceeding those of inorganic materials, but starting from much lighter and more easily available materials.

9.2   The Viscoelastic Nature of Biological Materials

Biological materials, being soft and elastic (with the exception of the likes of bone, keratin, or wood) might appear, from a macroscopic point of view, sufficiently isotropic. By touching on our skin at different places of the body we do not notice much difference in pushing, twisting or squeezing. However, nothing could be more misleading. The richness of structural organisation of biomaterials at all scales, from the molecular to the cell to the tissue, makes the problem of determining their response

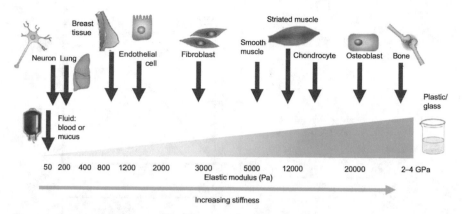

Fig. 9.2 Elastic (Young's) moduli of human biological tissues. Each type of cell finds the proper elastic environment more appropriate to its growth and functions. [Image from Ref. [2], adapted w. permission.]

even more complicated that for the most anisotropic, arbitrarily intricate crystalline material. If we look at the complex hierarchy of the molecular architecture making up, e.g., tendons, or muscle fibres, or at the multiple structures of bone material, it should be apparent that it is not possible to reduce that complexity to just two or three numerical parameters. The inside of a cell has a nucleus as one dense, tangled mass of nucleic acids and proteins plunged in some water and ions, wrapped by a tight membrane, the whole immersed in a water-based fluid containing corpuscles of size anywhere from 1 to 1000 nm, some of them with a double membrane, some with a spongy structure, entangled in a network of long filaments of the most diverse rigidity running in all directions, and the everything wrapped in another multi-layered, heterogeneous membrane. The effective modulus of elasticity of different cells spans more than four orders of magnitude (Fig. 9.2). Now, think of squeezing and squashing this tissue, and ask what is the bulk modulus, or the Young's modulus of such a thing? What quantity should be actually measured? And how?

Most importantly, however, the mechanical response of biological tissues is found to be strongly **non-linear**, and to depend on the past history of deformation, as well as to change with time for a given deformation state. These are the characteristics of **viscoelastic materials**. When such a material is stretched and held at a fixed deformation the stress declines over time, a phenomenon known as *stress relaxation*. Moreover, stress relaxation is sensitive to the velocity at which the deformation is applied, or *strain-rate sensitivity*. The complementary type of response is the phenomenon of **creep**, occurring when a material kept under a constant force continues to deform indefinitely (think of chewing gum). Additionally, the stress-strain profile obtained from a deformation experiment can be different in the loading phase, and in the unloading phase when the force is reversed. When this happens, we have a *hysteresis* phenomenon, which signifies that some of the stored elastic energy is not recovered, but has been lost in internal friction and in permanent deformation of the material.

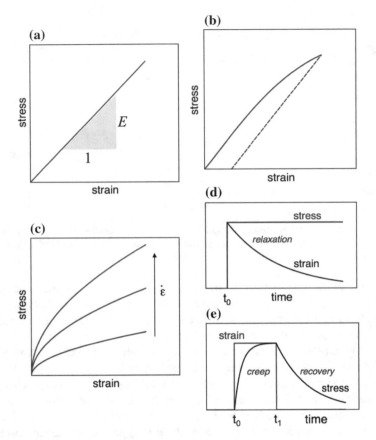

Fig. 9.3 Stress-strain plots for different materials. **a** The simple linear elastic material: the slope of the straight line gives the Young's modulus E. **b** A linear-plastic material. The response is linear up to some point, then starts to decline; the *dashed red line* indicates removal of the load, the material goes back linearly to zero stress, but some residual deformation survives (intercept of x axis at non-zero strain); the area between the *blue* and *red curve* is the hysteresis. **c** A viscoplastic material: the stress response is dependent on the strain, ε, as in (b), but also on the strain rate, $\dot{\varepsilon} = d\varepsilon/dt$. **d** A stress-relaxation material: stress is applied at time t_0 and held constant; the strain goes initially to the prescribed value, but declines in time (to a constant for a solid, or to zero for a fluid). **e** A creep-recovery material: a strain is applied at time t_0 and is released at time t_1; the stress increases from zero until to t_1, then decreases after t_1 and recovers a zero strain for a fluid (or non zero for a solid)

To understand the mechanical response of such a complex type of material, and to compare the response of different materials, we must use a plot of the stress and the corresponding strain, for a given experiment. In the case of an elastic material pulled along an axis (Fig. 9.3a) such a plot is quite simple and does not teach us very much: it is just a straight line in the stress-strain plane, with a slope equal to the Young's modulus E, Eq. (9.42). A plastic material (Fig. 9.3b) responds to the same stress by a first linear part, and then starts declining; when the load is released (dashed

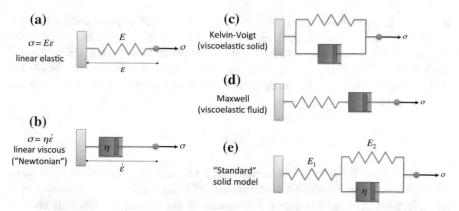

Fig. 9.4 Schematic models to describe complex materials behaviour. *Left* definition of the basic elements, the spring, **a**, describing a linear elastic material; and the dashpot, **b**, describing a linear viscous material, also called Newtonian. (*Right*) by combining the basic elements, different types of materials behaviour can be represented. **c** The Kelvin-Voigt model, describing a viscoelastic solid. **d** The Maxwell model, describing a viscoelastic fluid. **e** The "standard" solid model, appropriate to describe some complex biological materials combining solid-like and fluid-like features

red line) it goes back to a state of zero stress, but the strain is not zero anymore: some permanent deformation has occurred inside the material, which leaves behind a permanent deformation.

The response of viscoelastic materials, on the other hand, is quite more complicated. Where an elastic material has a unique stress-strain relation, the strain-rate dependence makes this a non-unique relation. Figure 9.3c shows what is meant by strain-rate sensitivity: a material can sustain increasing values of stress, if the deformation is applied at different rates $\dot{\varepsilon}$ (as usual, the dot indicates derivative with respect to time). The stress-strain relationship is symbolically written as $\sigma = \sigma(\varepsilon, t) = \sigma(\varepsilon, \dot{\varepsilon})$.

An elastic material in the linear response regime is described as a spring (see Eq. (9.43) above), eventually anisotropic, i.e. with different spring constants along different directions. By analogy, a material whose response depends on the strain rate can be approximated in a similar way, by restricting to the linear regime of proportionality between stress and strain rate:

$$\sigma = \eta \dot{\varepsilon} \tag{9.5}$$

This is a linear-viscous material, also called a *Newtonian* fluid since the coefficient η is the viscosity of the material. The analog of the spring in this case is a *dashpot*, as schematically shown in Fig. 9.4b, a kind of piston moving in a dense fluid whose *velocity* is proportional to the applied force (in the spring, the displacement is proportional to the force). By extension, a viscoelastic material, that is one displaying at the same time elastic and viscous behaviour, can described as a combination of springs and dashpots.

Two basic combinations can however be distinguished: the spring and the dashpot being connected in **parallel** or in **series**, like in an electric circuit. In the first case, depicted in Fig. 9.4c, the total stress is the sum of the stress acting on the spring and on the dashpot separately, $\sigma = \sigma_s + \sigma_d$, where σ_s and σ_d are given by Eqs. (9.42) and (9.5), respectively. The strain, on the other hand, must be the same for both elements. Therefore we have:

$$\sigma = E\varepsilon + \eta\dot{\varepsilon} = E\varepsilon + \eta\frac{d\varepsilon}{dt} \tag{9.6}$$

This is called the **Kelvin-Voigt model** of a viscoelastic material, and it is more appropriate to describe a solid, since the overall deformation is constrained by the strain in the spring. In this case the dashpot will asymptotically attain its level of stress (see also Fig. 9.3e at times $t_1 < t < t_2$).

In the second case, with spring and dashpot connected in series, the stress acting on the two elements must be the same, $\sigma = E\varepsilon = \eta\dot{\varepsilon}$, whereas their strains can be different, for a total strain $\varepsilon = \varepsilon_s + \varepsilon_d$. Let us firstly take the time derivative of the strain in the spring, that is $\dot{\varepsilon}_s = \dot{\sigma}/E$. On the other hand, it is also $\dot{\varepsilon}_d = \sigma/\eta$. Therefore, the total strain time-derivative is:

$$\dot{\varepsilon} = \frac{d\varepsilon_s}{dt} + \frac{d\varepsilon_d}{dt} = \frac{\dot{\sigma}}{E} + \frac{\sigma}{\eta}$$

or, by rearranging terms:

$$\eta E\dot{\varepsilon} = \eta\dot{\sigma} + E\sigma \tag{9.7}$$

This equation is slightly more complicated than the previous one, since it contains the time variation of both the stress and the strain. However, note that by applying a stress at one end, with the other end clamped (Fig. 9.4d), the spring will attain a fixed deformation, while the dashpot will continue to deform until the stress is maintained. This is called the **Maxwell model** of a viscoelastic material, and its behaviour is more appropriate to describe a fluid.

In reality, no materials behave exactly as in the Kelvin-Voigt, or in the Maxwell model. However, the two can be used as building blocks to construct more complicated and more realistic models of materials. In particular, many biological materials, such as cartilage, or cell membranes, display liquid-like and solid-like viscoelastic properties in different proportions. In such cases, the so-called **standard solid model** may be more appropriate.

This is described as the combination in series of one purely-elastic plus one Kelvin-Voigt block, as shown in Fig. 9.4e. Now there are three different materials parameters to play with, E_1, E_2, η. Firstly, note that the total stress must be equal in the two blocks, $\sigma = \sigma_1 = \sigma_2$. Then, the total strain is $\varepsilon = \varepsilon_1 + \varepsilon_2$, with the two contributions being respectively:

$$\varepsilon_1 = \frac{\sigma}{E_1} \qquad\qquad \varepsilon_2 = \frac{\sigma}{E_2 + \eta\frac{d}{dt}}$$

(we treated the 'd/dt' in Eq. (9.6) as an operator), hence:

$$\varepsilon = \frac{\sigma}{E_1} + \frac{\sigma}{E_2 + \eta\frac{d}{dt}}$$

or, by rearranging terms:

$$E_1 E_2 \varepsilon + \eta E_1 \dot{\varepsilon} = (E_1 + E_2)\sigma + \eta\dot{\sigma} \tag{9.8}$$

By fitting the three parameters to experimental data, the standard model can match the stress-strain relaxation curve at three points, but its ability to fit the data over the full range is usually poor. Typically, the use of a single relaxation time (see Problem 9.3) makes the transition too fast. More refined descriptions, such as Wiechert model, include many spring-dashpot elements that allow to better reproduce the competing internal relaxation mechanisms of complex materials.

9.3   Soft Tissues

With the notable exception of bone, most tissues in the animal body fall in the category of soft viscoelastic material. Four kinds of such tissues are usually acknowledged: (1) *epithelial* tissues, characterised by a dense pattern of cells tightly bound to each other, typically found on the free surface of the organs (external and internal); (2) *muscle* tissues, characterised by a high degree of contractility and to a good approximation displaying a mechanical response largely in the linear regime; (3) *nervous* tissues, composed of cells with special chemical and electrical characteristics; (4) *connective* tissues, in which a small volume fraction of cells are separated by a large amount of extracellular materials produced by the cells themselves. These latter will be treated more specifically in the next Section.

Vital organs in a animal body are usually composed by more than one tissue type, allowing to combine different functionalities. Generally speaking, it is the amount and quality of the connective tissue between cells that gives the body its mechanical strength to resist external stresses, and provides a recognisable shape that persists even in the presence of applied forces. Among all these tissues, **skin** stands out as a highly peculiar natural material. It supports all internal organs and protects the body from abrasions, blunt impact, cutting and penetration, it is moderately impermeable, and can resist the attack of various chemicals.

In humans, skin is the largest single organ taking up about 16 % of the total body mass, and makes up between 1.5–2.3 m^2 of surface, in constant contact with the external environment. In transverse section it has a total thickness going from 1.5 to 4 mm, distributed on three layers with different functions and characteristics. Mechanically, skin is heterogeneous, anisotropic and non-linear viscoelastic (Fig. 9.5).

The outer layer, or *epidermis*, is composed by several types of cells with different functions, among which the keratinocytes form the main mechanical barrier to the

(a)   (b)   (c)

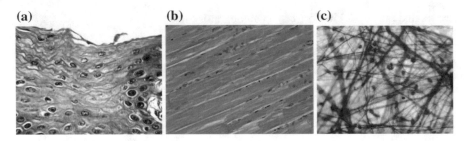

Fig. 9.5 Examples of human soft tissues from histological sections, stained to reveal cell components and nuclei. **a** Epithelial tissue from the cardiovascular system; such kind of cells cover the heart, veins and capillaries with a protective function. **b** Skeletal muscle, with extremely long fibres. Each muscle fibre has many nuclei, the majority of which are located at the periphery of the cell; transverse striations appear as a regular, cross-banding pattern. **c** Loose connective tissue, making up the mechanical structure of the elements in the cardiovascular system. [Photos courtesy of J. Oros-Montón, Universidad de Las Palmas, www.webs.ulpgc.es/vethistology/.]

exterior, mostly preventing water loss. The intermediate layer, or *dermis*, contains a substantial fraction of proteoglycans (long branched polymers made by a central protein, and numerous lateral branches of polysaccharides), collagen, and elastin protein (being similar to cartilage or tendons, but with different composition and mechanics). The inner layer, or *hypodermis*, ensures contact with the inner organs, and contains fat cells (adipocytes) and fibroblasts, aside of collagen matrix.

The mechanical properties of skin mainly depend on the nature and organisation of the dermal collagen and elastin fibres networks; secondarily on the relative fractions of water, proteins and macromolecules embedded in the extracellular matrix; and to a variable, lesser extent on the mechanical properties of the epidermis and outer keratin wall. However, in some cases these external layers can be so thick and strong, to become the true supporting structure, as it will be seen in the next Section.

Collagen is one of the most widespread proteins in the animal body. It is used in bone, cartilage, tendons, skin, and most types of soft tissues, making up between 20 and 30% of an animal's protein contents. Its elementary molecules are about 300 nm long and 1.5 nm thick, consisting of three braided helices of tropocollagen (a protein dominated by glycine), stabilised by a network of hydrogen bonds (Fig. 9.6). Although as many as 29 different types of collagen have been identified, the types I to V are the most common in the human body, with a Young's modulus $E \simeq 1$ GPa, and ultimate tensile strength of about 100 MPa; types I to III tend to form long bundles, while type V rather forms large scale networks. Together with collagen, **elastin** is a specialised protein providing skin its elasticity, most importantly the capacity to retract back to the original shape after a deformation, with a much lower Young's modulus $E = 600$ kPa.

Because of their fibrillar structure, collagens are known to have little resistance in compression, since fibrils are easy to buckle under compression. Collagen networks alone ensure most of the skin resistance under tensile stresses. However, when coupled to other components, such as elastin and proteoglycans, as well as to the fluid phase, the ensemble can withstand also compressive loads.

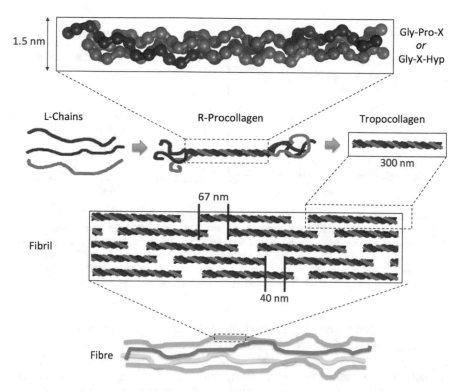

Fig. 9.6 Hierarchical structure of the collagen fibre. Three polypeptide strands, each conformed in a left-handed helix, are twisted together into a right-handed triple helix stabilised by hydrogen bonds; the repeated peptide sequences are very often Gly-X-Pro or Gly-X-Hyp, i.e. Glycine makes up about 30 % of the total. These tropocollagen units have constant size, 300 nm long by 1.5 nm in diameter. In fibrillar collagens, such as type I-III, tropocollagen units associate into a right-handed superstructure, the fibril, with a constant longitudinal spacing of 40 nm, and a shift of 67 nm; when collagen is mineralised to form bone, the interstitial space is filled by polycrystals of Hydroxyapatite, $Ca_{10}(OH)_2(PO_4)_6$. Fibrils can organise into many different structures, such as long fibres, more or less dense networks, also including combinations of various collagen types and cross-linking agents

While being all made from collagen (plus other stuff), the mechanical behaviour of skin is very different from that of denser collagenous materials, like tendons, or more hydrated collagenous materials, such as cartilage. This is due to differences in self-assembly of the microstructure, such as the tilt angle (relative orientation) of collagen fibres, as well as the average length, thickness and volume fraction of the fibres. For example, the average length of collagen fibres is about 500 μm in a human tendon, while it is 10 times smaller in the skin. The differences in Young's modulus among collagenous materials may also come from molecular-scale variations, for example the spacing shift between tropocollagen units (see again Fig. 9.6) is slightly larger in the tendon (67 nm) than in the skin (64 nm).

The mechanical properties of human skin have been measured by various scientists. Typical values derived from stress-strain curves taken at different strain rates,

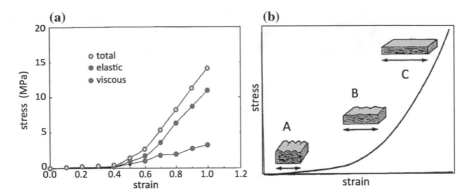

Fig. 9.7 a Stress-strain plots for human skin. Total stress was obtained from the initial stress observed at a strain rate of 10%/min while elastic stress was obtained from the equilibrium value of the stress at a fixed strain. Viscous stress was defined as the difference between the total and elastic stresses. By taking the linear parts of each component, values for E and η can be estimated. **b** Schematic of the biaxial extension of a skin sample, showing the progressive straightening of collagen fibres as far as the stress is increased. [From Ref. [3], adapted w. permission]

coupled to analysis of histological sections of skin samples (see Fig. 9.7a), give average collagen fibril length of $\sim$55 μm, Young's modulus $E \sim 18$ MPa, viscosity $\eta \sim 5$ kg m^{-1}s^{-1}. Stress-strain measurement such as those shown in Fig. 9.7 are carried out by measuring the stress corresponding to each strain increment, for various strain application rates. After the skin sample is stretched to a strain increment, the stress is allowed to relax and decay to equilibrium, before an additional strain increment is added. The elastic component of the stress is defined as the stress at equilibrium, while the viscous component is calculated from the difference between the total stress and the elastic stress component. In this way, the elastic and viscous components can be plotted separately, and from the slopes of the respective linear portions, E and η can be deduced according to Eqs. (9.5) and (9.6).

It is generally found that human skin (but also that of many furry animals) has non-linear viscoelastic properties, schematically depicted in Fig. 9.7b. Under moderate stretching, collagen fibres start to straighten and realign parallel to one another (A to B in the figure); once fibres are almost parallel, the linear part C of the curve begins, and here the moduli E and η can be measured. Beyond this region C of large deformation, a critical stress load is reached, and skin exhibits a sizeable hysteresis loop (see Appendix H). Here an amount of stored elastic energy is dissipated into reversible molecular rearrangement upon (large) deformations; collagen molecule stretching and slippage occurs; the tighter spacing of tropocollagen coupled to the quite short fibril length results in increased fibrillar slippage and energy dissipation, effectively lowering the elastic constant of collagen. The typical failure mode leading to skin tearing (indeed, a form of ductile fracture) is creep, i.e. stress accumulation under a constant level of applied strain close to the fracture strain ε_f; from a microscopic point of view, this creep phenomenon can be ascribed to the progressive dislodgement of water molecules from between the collagen fibres, leading to irreversible dehydration and distributed failure of individual collagen fibres.

9.3.1 Where Soft Turns Hard

The animals called *amniotes* (because they lay eggs on the ground or inside a mother, as opposed to the anamniotes, who use water as a egg-nurturing medium) exhibit skin appendages, such as hairs in mammals and feathers in birds, that play major roles in thermoregulation, photoprotection, camouflage, behavioral display, and defense against predators. Reptiles instead, similar to the fish anamniotes, developed various types of scales on the outer skin surface. In Chap. 12 we will see in some more detail what physics and mathematics have to say about the formation of patterns in such external material layers. However, these seem to have a common origin in the embryonal stage, by some epithelial cells that start a genetically-controlled process of differentiation into units, whose spatial organization is precisely patterned by chemical and diffusion mechanisms.

In many cases, however, the outer layer of the skin does not need to produce distinct protective appendages, but rather develops further, and thickens to an extent that makes it resemble to a sort of exoskeleton, with functions and properties similar to the tight pattern of scales covering a crocodile's body, but with an entirely different origin. The skin of a rhinoceros (see Fig. 9.8a) is popularly described as the paradigm of an impenetrable wall, endowed with exceptional hardness and resistance. In fact, the outer dermis layer in rhinoceros (as well as in elephants, and the like) is much thicker than in most animals: it is still skin, "just" water, collagen, elastin and something else, but what a fantastic material! Under the microscope (Fig. 9.8d, e) it appears as a dense and highly ordered three dimensional array of straight and highly cross-linked collagen fibres. The stress-strain curve is steep (Fig. 9.8b, c), with $E = 240$ MPa,

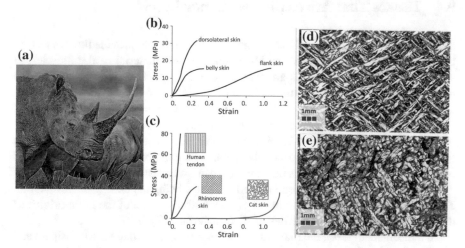

Fig. 9.8 The rhinoceros skin. **b** Stress-strain plots for different samples from the dorsolateral, belly, and flank skin. **c** Comparison of stress-strain plots for human tendon, rhinoceros skin, and cat skin; the schematic arrangement of collagen fibres is shown in the *inset*. (Data redrawn from [4]). **d, e** Histological cross sections of rhinoceros skin samples from the flank and belly, respectively, showing the cross-crossed arrangement of collagen fibres. [Image **a** courtesy of Gentside, www.maxisciences.com/; electron micrography (**d, e**) Ref. [4]. Repr. w. permission]

ultimate tensile strength of 30 MPa (1/3 of that of pure collagen), and an outstanding resistance to fracture (*toughness*) compared to other animal tissues: it takes $G \sim 80$ kJ/m^2 to break apart a piece of rhinoceros skin, compare e.g. to about 10–12 kJ/m^2 of wood, and 150–500 kJ/m^2 of hard to mild steel. The stress-strain plots in Fig. 9.8b show that rhinoceros skin is quite diversified. While being everywhere rather tough, the flank skin is however capable of stretching to twice its size under a large stress of 10–20 MPa (such a value of stress would easily bend a plate of aluminium of 3 cm thickness).

As it can be seen from Fig. 9.8c, the material properties of this thick skin layer are somewhere between those of the stiff human tendon, and those of very flexible cat skin. The diagrams in Fig. 9.8c also suggest that the main difference in the mechanical response of the different regions comes from the variation of the arrangement and size of collagen fibres. In the histological section of flank region skin (Fig. 9.8d) highly cross-linked networks of long collagen fibres are seen, averaging around 90 μm in diameter, with a good degree of long-range order. The microphotography of the harder and less deformable belly region skin, by contrast, shows thicker and shorter fibres, arranged in a more disordered, more compact network. This latter material is in fact stronger (the slope of the linear part of the $\sigma - \varepsilon$ curve indicates a bigger Young's modulus, by a factor of $\sim$6), while having a similar toughness (peak stress of $\sim$15 MPa); the dorso-lateral skin, on the other hand, has a Young's modulus similar to the belly skin, but it is much tougher, with an almost doubled peak stress and a slightly shorter fracture strain, $\varepsilon_f \sim 30\%$.

9.4   Tissues That Are Neither Solid nor Liquid

Skeletal soft tissues are specialized connective materials that provide the connection between the harder material of the bones and the soft material of the muscles, at various levels. These materials are essentially **cartilage**, covering the ends of the bones in the joints with a thickness of 1–6 mm in humans; **ligaments** and **tendons**, which connect bones to bones, or bones to muscles, respectively.

At a cellular level, such tissues are generated in large amount by sparse populations of cells: the **chondrocytes** for the cartilage, **tenocytes** and some particular stem cells for tendons and ligaments respectively. Such cells can produce a large amount of collagen-rich extracellular matrix, proteoglycan and elastin fibres; typically, only $\sim$2 % of the cartilage is composed by cells, the rest being the extracellular material. The arrangement and patterning of the collagen fibres is one of the main origins of the observed mechanical properties of such materials (Fig. 9.9).

At a mechanical level, the main functions of articular cartilage is to dissipate and distribute contact stresses during joint loading, and to provide almost frictionless articulation in joints. The primary functions of ligaments and tendons are to stabilise joints and transmit the loads, to hold the joints together, to guide the trajectory of bones, and control the joint motion. In order to accomplish these demanding tasks, articular cartilage, as well as ligament and tendons, are structured as **biphasic**

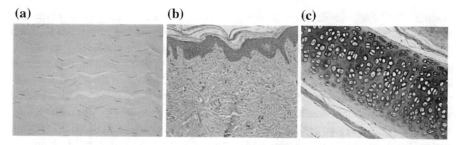

Fig. 9.9 Examples of human connective tissues from histological sections. **a** Tendon. The fibres in this dense connective tissue are more abundant than cells, with collagen fibres oriented in a regular pattern arranged in the same plane and direction. **b** Dermis, the intermediate layer of the skin between the outer and the inner layers. Here collagen fibres predominate in dense irregular connective tissue, and are generally arranged in bundles that cross each other at varying angles. **c** Cartilage (from trachea). The chondrocytes vary in size, and are dispersed in a small fraction among a much larger fraction of collagen, with a substantial fraction of fluid phase (water); deeper within the cartilage, cells tend to be larger and more polyhedral in shape. [Photos courtesy of J. Oros-Montón, Universidad de Las Palmas, http://www.webs.ulpgc.es/vethistology/.]

materials, with an anisotropic, viscoelastic and nonlinear mechanical behaviour that imparts these materials some absolutely unique mechanical properties.

The two distinct phases that make up such materials are a **fluid phase**, consisting of interstitial water and mobile ions, and a **solid phase**, consisting mainly of collagen fibrils and negatively charged proteoglycans. The main differences between these tissues are the type of molecules involved in the matrix (collagen of different kinds) and the relative proportions of the constituents. Typically, cartilage contains up to 10 % of proteoglycans and 10–20 % of collagen in wet weight, with a water contents that can reach 80 %; ligaments contain about 5 % proteoglycans and up to 30 % collagen; tendons contain somewhat more collagen and less proteoglycans than ligaments. Both the tendons and the ligaments contain less water, typically about 60 %. Moreover, the collagen molecules of cartilage are of the so-called type-II, while those of tendons and ligaments are of type-I; both types of collagens are made up of a twisted triple helix of amino acids, of which glycine, proline and alanine representing more than half, both form fibrils and are very similar in structure and functions, with differences in their respective chemical compositions. Correspondingly, the mechanical properties of these materials vary largely: the Young's modulus of cartilage is about 0.5 MPa in compression, while a ligament has a $E > 100$ MPa in tension, and a tendon can approach the GPa range.

9.4.1 Cartilage

A typical stress-strain plot for a skeletal biphasic material under uniaxial tension is shown in Fig. 9.10. It is similar in shape to those observed for skin (Figs. 9.7 and 9.8), however the scales of stress and strain are different. The first part of the curve is called the "toe", displaying a non-linear stress-strain response upon increasing load; this behaviour is due to the progressive straightening of the wavy-like collagen fibrils. After the collagen fibrils are completely straightened the elastic region begins, displaying a typical Young's modulus of the composite tissue. Strain in these first two regions is reversible, and the overall curve resembles quite closely the worm-like chain model of the previous Chapter (also in this case, the interpretation can be ascribed to the entropic elasticity of individual collagen polymers). Upon increasing further the load, the slope of the curve changes (yield point, σ_y) and the plastic region begins. In this region the tissue begins to experience irreversible, destructive changes, e.g. microfractures in the collagen fibril network. At the extreme deformation the tissue fails completely, and the stress drops rapidly to zero, i.e. the structure cannot support any further applied force.

One of the important characteristics of these complex materials is that their mechanical response is generally different in tension and compression. For the compression of cartilage some disagreement exists, in that most scientists agree that the origin the viscoelastic response should be found in the flowing of the fluid phase, whereas some believe the viscoelasticity being rather a property of the matrix structure. Part of the difficulties are due to the heterogeneity of the cartilage, whose properties change a lot across its thickness. For example, the bulk modulus in compression for bovine cartilage has been measured to vary by a factor of 20 over a few cm of depth. Compression resistance is mostly ascribed to the higher content in proteoglycans, compared to either the harder tendons, or the softer skin tissues, and in fact a good correlation exists between the increase in B and the increase in concentration of proteoglycans with depth.

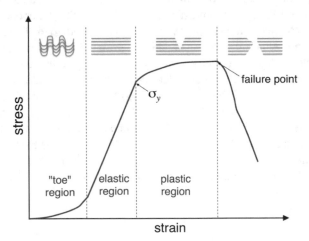

Fig. 9.10 Typical stress-strain curve for destructive tensile testing of skeletal soft tissues. In *blue* the reversible part, in *red* the irreversible part of the deformation. On the *top* of the figure, collagen fibril straightening and failure, related to different regions of the stress-strain curve, are schematically shown. The stress and strain ranges vary for each tissue, however the shape of the curve is the same. [Redrawn from Ref. [5], under CC-BY 3.0 licence, see (**) for terms.]

The biphasic model of soft tissues

The biphasic model chiefly applies to cartilaginous soft tissues. In this model, both the solid matrix and the fluid are assumed to be incompressible and non-dissipative materials; the only dissipation of energy is ascribed to viscous drag acting on the fluid flow in the tissue. The stress-strain relations for the whole tissue are written as:

$$\underline{\sigma}_S = -\phi_S p\underline{1} + \underline{\sigma}_E$$
$$\underline{\sigma}_F = -\phi_F p\underline{1}$$
$$\underline{\sigma}_T = \underline{\sigma}_S + \underline{\sigma}_F = -p\underline{1} + \underline{\sigma}_E \tag{9.9}$$

where the subscripts S, F, T stand for 'solid', 'fluid' and 'total', ϕ are volume fractions with $\phi_F = 1 - \phi_S$, p is the fluid pressure, $\underline{\sigma}_E$ is the effective stress tensor of the solid material, and $\underline{1}$ is the diagonal unit tensor (a 3×3 matrix with 1 on the diagonal, and 0 off the diagonal).

The fluid phase flow is embodied in a mass-balance equation for the fluid and solid moving with relative velocity vector fields $\mathbf{v}_F$ and $\mathbf{v}_S$, whose total divergence must be equal to zero:

$$\nabla \cdot (\phi_S \mathbf{v}_S + \phi_F \mathbf{v}_F) = 0 \tag{9.10}$$

and two momentum equations for the fluid and solid phases:

$$\nabla \cdot \underline{\sigma}_S = -\mathbf{p}_S \tag{9.11}$$
$$\nabla \cdot \underline{\sigma}_F = -\mathbf{p}_F$$

with the momentum conservation condition $\mathbf{p}_S = -\mathbf{p}_F$, which also implies $\nabla \cdot \underline{\sigma}_T = 0$.

The matrix is a porous medium, and the fluid flow across the matrix is akin to the permeation of a membrane. By defining the cartilage permeability k, the fluid viscosity η, and the effective filtration velocity as $w = \phi_F(v_F - v_S)$, the equivalent of the Darcy's law (see Sect. 5.6) can be written:

$$w = -\frac{k}{\eta}\nabla p \tag{9.12}$$

i.e., the filtration velocity is proportional to the gradient of the fluid pressure.

In the original, simple linear-elastic model (V. C. Mow et al., J. Biomech. Eng. **102** (1980) 73), together with a constant value of permeability k (leading to the linear relation between w and ∇p), also the effective stress tensor is taken to be linear and isotropic (see Eq. (9.34) of the Appendix):

$$(\sigma_{ij})_E = \lambda \varepsilon_{kk} \delta_{ij} + 2\mu \varepsilon_{ij} \tag{9.13}$$

In more refined models, the material is considered anisotropic, with a third elastic coefficient besides Lame's λ and μ, and the permeability is no longer constant, but depends on the fluid fraction $e = \phi_F / \phi_S$:

$$k = k_0 \left(\frac{1+e}{1+e_0}\right)^M \tag{9.14}$$

the '0' indicating the reference values at zero stress, and M a constant to fit on experimental data.

The viscous properties of cartilage are associated with the fluid component. The proteoglycan matrix gives cartilage a spongelike character, since it tends to hinder water movement, and the matrix can thus retain water molecules. However, water can be set in motion by the mechanical pressure applied on the joints. Since the fluid flow is proportional to the pressure difference caused by the applied stress, we can invoke an old acquaintance from Chap. 5, the matrix permeability, to describe this property. Permeability is inversely proportional to the force required for the fluid to flow at a given speed. Since the permeability of cartilage is quite low, this means that the fluid will flow at slow speed even under quite large joint pressure. With typical loading times in the few tenths of seconds, therefore, the fluid remains confined in the matrix and the cartilage can retain its stiffness up to very large loads.

In order to describe the mechanical response of such a complex material, the **biphasic model** has been developed (see the greybox on p. 389), to take also the interstitial fluid movement into account [6]. In this model, the frictional drag of the moving fluid explains the viscoelastic response of cartilage under compressive load. The solid and fluid phases are assumed to be immiscible and incompressible, the overall composite material having a Poisson's ratio $\nu = 0.5$ at equilibrium (i.e., when all fluid flow has stopped). A typical stress versus time plot for a cartilage sample subject to step loading at increasing values of stress is shown in Fig. 9.11 (actually showing the perpendicular force in N vs. time). A compressive strain is applied and held constant; the force jumps to a peak value, 1.5 N, and then relaxes exponentially to a much lower value. A second compression is applied; the force jumps again to a larger peak of 2.5 N, and relaxes to a higher value around 0.5 N. And so on, with successive compression steps. It may be noticed that the relaxation time necessary to attain the equilibrium stress (force) increases with each increasing compression step: this is one signature of the fact that the permeability (i.e. the viscoelastic coefficient)

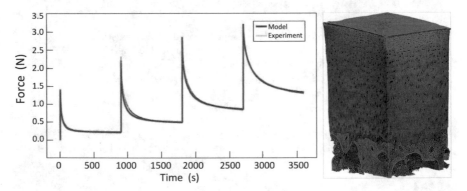

Fig. 9.11 *Left* A typical stress-relaxation measurement of articular cartilage and the corresponding theoretical fit, using a fibril reinforced poroviscoelastic model. (Compare to Fig. 9.3c). *Right* Vertical section of human articular cartilage (3D digital volumetric fluorescence microscopy). The articular surface is above, the subchondral bone is below. Note the changes in size and spatial distribution of cells (chondrocytes) through the thickness. [Image *left* from Ref. [5], repr. under CC-BY 3.0 licence, see (**) for terms; *right* from Ref. [7] p. 84, © 2008 Elsevier, repr. w. permission.]

varies upon increasing load, and the fluid velocity is progressively reduced as the composite material stiffens.

The elastic isotropic model is useful to obtain simple material parameters for the tissue. However, a more detailed description of the complex mechanical properties of skeletal soft tissues can only be obtained by using more sophisticated models.

9.4.2 Tendons

The fluid flow-dependent viscoelasticity is less important in ligaments and tendons compared to cartilage, because these elements experience mainly tensile forces under physiological loading. Moreover, it is generally understood that the fluid has but a minor role in contributing to soft tissue response in tension. Rather, the main attention is focused in this case on the molecular structure.

Tendons appear in a variety of sizes and shapes, depending on the morphological, physiological and mechanical characteristics of both the muscle and the bone to which it is attached. In every tendon, the two parts that attach respectively to the bone and to the muscle have somewhat different character from the tendon proper; this latter is constituted by 70–80 % of collagen.

The outstanding mechanical properties of tendons are due to the optimisation of their structure (Fig. 9.12) over many levels of hierarchical structures, whose respective interrelations are not yet fully elucidated. When a tensile load is applied to a tendon, the deformation is redistributed among the various components in non-obvious ways. The organisation of the molecular fibrils in the cross section plane of the tendon displays an ordered arrangement over short distances, with the fibrils approximately disposed in a hexagonal pattern, and a more loose order over longer distances, with a sort of 'polycrystalline' structure of the fibres. One of the challenges is to work out the respective influence of these different levels. Experiments done by stretching a tendon and measuring simultaneously the deformation at the level of single fibres, by synchrotron X-ray diffraction, show that the strain distribution is very inhomogeneous. In particular, the total stress is not simply retrieved by summing the stresses in the fibres and in the molecules that make up the fibres: a substantial component of the deformation is lost in the relative shearing of the fibrils with respect to each other, distributing part of the stress in the embedding proteoglycan matrix. Such molecular-level interactions are not entirely clear, but lead to a strongly non-linear and viscoelastic response of the ensemble.

The **myotendinous junction** is the region where the muscle joins with the tendon, and it is mechanically very important since it is the key to transmitting the muscle force downstream, to the tendon and the bone. The morphology of this region presents multiple folds, that largely increase the interface area, by a factor of 10–20 compared to the adjoining muscle cross section. This is one way of reducing the stress, by distributing the force over a larger surface. As a second consequence, the load transfer occurs mostly through shear, rather than by tension.

(a)

Diameter of

Collagen molecule
1.3 nm

Collagen fibril
50 - 500 nm

Fascicle
50 - 300 µm

Tendon fibre
100 - 500 µm

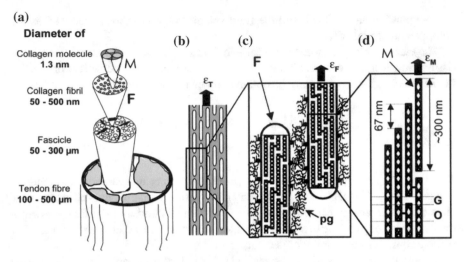

Fig. 9.12 Simplified tendon structure. **a** The tendon is made of a number of parallel fascicles containing collagen fibrils (F), which are assemblies of parallel triple-helical collagen molecules (M). **b** The tendon fascicle can be viewed as a composite of collagen fibrils (with thickness of several hundred nm and a length of $\sim$10 µm) in a proteoglycan-rich matrix, subjected to a strain ε_T. **c** Part of the total strain ε_T is accommodated by deformation of the proteoglycan matrix (pg); the remaining part, ε_F, is shared among the fibrils (F). **d** Collagen molecules are packed within the fibrils in a staggered way, with an axial spacing $D = 67$ nm, when there is no load on the tendon. Since the length of the molecules (300 nm) is not an integer multiple of the staggering period, there is a succession of gap (G) and overlap (O) zones. The lateral spacing of the molecules is approximately 1.5 nm. The full three-dimensional arrangement is not yet fully clarified, but contains both elements of crystalline order and of disorder. The strain in the molecules, ε_M, may be different from the strain in the fibril, ε_F. [From Ref. [8], repr. w. permission]

The stress-strain curve of the tendon is similar to that of other collagenous materials (see Fig. 9.10), with a "toe" region of elongation at very low stress, followed by a linear region of steep increase of stress with increasing strain, and a plastic region of permanent deformation (yield) at nearly constant stress, which terminates at the failure point. The difference is that the tendon is much stiffer and stronger than other similar materials: the toe region extends up to only about 2 % strain; the slope of the linear region is much higher, with an apparent Young's modulus in the 1–2 GPa; and the plastic yield starts quite early, around 5–6 % strain.

Tendons loaded cyclically display a **hysteresis** loop, indicating that part of the stored elastic energy in tension is lost when the load is removed. However, the area of the hysteresis loop is quite small, therefore the tendon has a high **resilience** (see Appendix H). Because of this property, tendons are capable of storing and releasing large amounts of elastic energy. This also seems to have important, albeit indirect, consequences on the evolutionary development of muscles in long-legged running animals, such as horses, deers, kangaroos, or camels. The leg muscles in these animals are relatively short, and can develop very large forces when stretched; however, because of their relatively short fibre length, the overall length variation is limited:

the mechanical work, that is the product force × elongation, or stress × strain, is reduced. Therefore, to increase the energy output part of the elongation should occur within the tendons, which are correspondingly longer for such muscles. In a number of animal studies, it was suggested that locomotor muscles keep tendons constantly taut during running; upon impact with the ground, these tendons are stretched and the energy to decelerate the mass of the animal is stored as elastic strain energy; during the subsequent propulsive phase, tendons recoil and release a large portion of the stored energy. This energy contributes to the locomotion, and may help saving metabolic power. As it will be discussed at length in the next Chap. 10, for a given cross section area, short muscles can generate the same amount of force than longer ones; however, they are also less massive and allow a slender leg shape, with a overall reduced cost of transport (see the last Chap. 12). Therefore such a structure, coupling shorter muscles and proportionally longer tendons, could be more economical for high speed running.

Tendons may be also subject to prolonged static loads, such as those imposed by postural muscles, aside of the cyclic, repetitive loads of locomotion. Application of prolonged constant forces, even below the maximum strength level, may cause failure by *creep*. By studying the constant-stress loading of wallaby tail tendons [9], it was found that tendons can break already at a stress of 20 MPa if the load is applied for several hours continuously, whereas the peak stress to break the same animal's tendon in a sudden effort would be above 150 MPa. The time to failure of the stressed tendon decreases nearly exponentially upon increasing the applied load, $T = A \exp(-\sigma/B)$, with typical values of $A = 6 \times 10^4$–8×10^5 s, and $B = 12$–16 MPa for the wallabies tested in the experiments; at a stress of 80 MPa, that is about half the peak fracture stress σ_f, the time to failure decreases to a few hundreds of seconds. The failure by creep signals the accumulation of damage in the molecular structure, with a consequent progressive degradation of the elastic moduli of the material beginning well below the final rupture point.

9.5 Rigid as Bone

By considering bone, we move from soft to hard tissues. However, note once more that "hard" in the context of a biological material is far from our daily notion of a hard material such as concrete or steel. A bone is an organ that contains aside of the proper bone tissue also other tissues, such as bone marrow, nerves and blood vessels. Bone tissue comes primarily in two forms, **trabecular** and **cortical**, similar in constitution but different in their microscopic architecture (Fig. 9.13a). As any other tissue in the animal body, also bone is continuously remodelled, replaced and repaired, by a variety of specialised cells: *osteoblasts* that form new bone, they come from the bone marrow and work in teams to make new bone ("osteoid") by building up collagen, proteins, calcium and mineral deposition; *osteocytes*, originated from osteoblasts, which get surrounded by new bone tissue, and send out long branches that connect to other osteocytes; *osteoclasts*, coming from the bone marrow and related to white

(a)   (b)   (c)   (d)

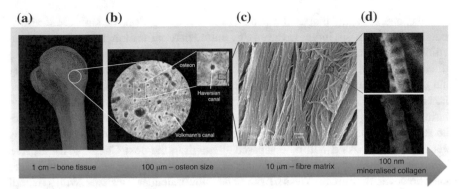

Fig. 9.13 Multiscale organisation of bone, from the macroscopic to the molecular scale. **a** Section of a tibia, showing cortical and trabecular bone tissue. **b** Zoom on a cortical bone microregion, with osteocytes and osteoblasts; one full osteon around the Haversian canal is shown in the *upper right*. **c** Scanning-electron micrograph of collagen fibres, from one layer (lamella) on the surface of human trabecular bone. **d** Zoom on a mineralised collagen fibril: above, transmission-electron micrograph, showing dark-field contrast of alternating collagen and mineral regions; below, electron energy-loss map (calcium, *red*; oxygen, *green*; carbon, *blue*). [Images adapted from: (**b**) en.wikipedia.org/wiki/Osteon/, (**c**) S. Bertazzo, repr. under CC-BY-SA 3.0 licence, see (*) for terms; **d** Ref. [10]; repr. w. permission.]

blood cells, which are directed by osteocytes to the places where high stress and breaking develops, to dissolve the damaged bone tissue.

As different as it appears from skin or cartilage, bone is made just from the same basic materials, i.e. collagen fibres, proteins and water, but with one fundamental additive: mineral crystals of hydroxyapatite, $Ca_{10}(OH)_2(PO_4)_6$, which go to fill up the interstices of collagen-I staggered fibrils (see Fig. 9.6). During bone growth, mineralization begins in the fibril gap zones and extends into other intermolecular spaces, resulting in a fully mineralised fibril. The three-dimensional arrangement of collagen molecules within a fibril is not well understood; however, collagen fibrils in bone range from 20 to 40 nm in diameter, suggesting that there are 200–800 collagen molecules in the cross section of a fibril. On a weight basis, bone is approximately 60 % inorganic, 30 % organic, and 10 % water; on a volume basis, these proportions are about 40-, 35-, and 25 %, respectively. Bone grows by accretion (Fig. 9.13b), with successive concentric layers being marked by *cement lines*. The basic unit is the *osteon*, which contains both nascent and mature material, and is vertically traversed by an empty channel, the *Haversian canal*, through which run nerves and blood vases (Fig. 9.14a). It is estimated that a human skeleton is made up by about 20 millions osteons. Within the osteon, collagen fibres are tightly organised into a dense matrix (Fig. 9.13c), with successive layers (called *lamellae*) that get progressively twisted with respect to the inner ones, to increase the resistance to torsion and bending stresses. A growing osteon is called *osteoid*. Within the osteoid, osteoblasts secrete the collagen and various proteins, among which osteocalcin that binds calcium at the right concentration to promote mineralization of the collagen fibrils (Fig. 9.13d).

The difference between trabecular and cortical bone is mostly based on the density: while trabecular bone has a spongy appearance, with a porosity (void fraction over the total volume) between 70 and 90 %, cortical bone is dense and compact, with a porosity as low as 5 % (which however increases with age). Overall, cortical bone has a density of 1.85 g/cm^3, trabecular bone varies between as low as 0.3 to about 1 g/cm^3. Both types of bone have a lamellar structure, which is the most common in adult mammals (in fast developing bone, a different mm-scale structure is found, with a 3D-woven geometry). Primary lamellar bone is new tissue, consisting of large concentric rings of lamellae, winding about each osteon canal similar to growth rings in a tree, as shown in Fig. 9.13b. Lamellae are layers of collagen fibres (Fig. 9.13c), with the typical twisted structure between adjacent layers.

As far as its mechanical properties, bone is extremely anisotropic, thus reflecting its complex hierarchical structure. Human cortical bone has Young's modulus $E = 18$–20 GPa in the longitudinal direction and $E = 10$ GPa in the transverse direction; shear modulus $\mu = 3.3$ GPa; and Poisson's ratio $\nu = 0.40$ (longitudinal) and 0.62 (transverse). The spongy trabecular bone has E between 1 and 5 GPa, depending on the porosity and age.

Figure 9.14b–d displays sample plots of the mechanical response of human cortical bone under compressive or tensile, instantaneous or steady loads. In the panel 9.14b a typical stress-strain plot under steady loading is shown, demonstrating that cortical bone is stronger in compression than in tension; this difference is indicative of its elastic anisotropy. The panel 9.14c shows bone response under a steady stress, at three increasing levels: when a low stress is applied to the bone, the strain remains constant over time and there is no permanent deformation after unloading; for stresses just below yield, σ_y, the strain starts increasing with time (creep) at a constant rate, and a small permanent deformation exists after unloading; at the highest level of stress, the rate of creep increases, and a larger permanent deformation would be observed after unloading.

Finally, in Fig. 9.14d the strain-rate sensitivity of bone under uniaxial loading in the longitudinal direction is demonstrated (compare with Fig. 9.3d). This property shows that the apparent strength is higher when the load is applied faster. Although bone is viscoelastic, like all collagen-based materials, the effect of loading rate is however quite moderate, e.g. if compared to skin. In typical experiments, both the modulus E and the yield stress σ_y (i.e., the stress at which nonlinear response sets in) increase by only a factor of about 2–3, for loading rates increasing by 6 orders of magnitude. The majority of physiological activity takes places at deformation rates up to about 0.1 s^{-1}: slow walking corresponds to a strain rate of 10^{-3}, brisk walking 10^{-2}, slow running $\sim 3 \times 10^{-2}$ s^{-1}. As shown by the receding maxima in the plots, at very high strain rates (>1000 s^{-1}) the ultimate strain decreases, and the strength increases. Cortical bone exhibits a "ductile-to-brittle" transition at such high rates, as could occur, e.g., in a car accident, or as a result of a gunshot.

Fracture properties of bone are also very anisotropic, with a fracture stress $\sigma_f = 135$ MPa (longitudinal) and 53 MPa (transverse) for tensile load, as opposed to $\sigma_f = 210$ MPa (longitudinal) and 135 MPa (transverse) for compressive load. The compressive and tensile fracture stress are therefore 1.14 % or 0.75 % of the Young's

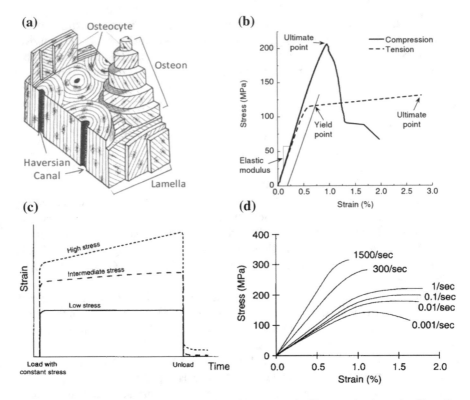

Fig. 9.14 Structure and mechanical properties of human cortical bone. **a** A schematic of lamellar micro-architecture, showing lamellae with collagen fibril bundles at different orientations. **b** Stress-strain plot in compression and tension. **c** Creep response at three different stress levels. **d** Strain-rate sensitivity for longitudinal tensile loading. [Images from: **a** Ref. [11]; **b** Ref. [12]; **c** Ref. [13]; **d** Ref. [14]. Repr. w. permissions from the publishers.]

modulus, respectively, to be compared for example to values between 0.4 and 0.7 % for high-performance aluminum or titanium alloys, thus making bone a very high-performance material as far as strength.

As we already know from the previous Chapter, bending and torsional moduli depend not only on the material characteristics, but also on the *shape and size* of the structure made from such materials. In particular, long bones (e.g. femur, tibia) satisfy the geometry of a practically empty cylinder, with inner and outer diameters d and D (taking as zero the mechanical resistance of the soft bone marrow filling the central cavity). Then, the moment of inertia of the transverse cross section may be approximated by the expression $\mathscr{I} = \pi(D - d)^4/32$. Typical values of sizes at the midshaft of human femur are $D = 2.2$–3. cm and $d = 0.8 - 2$. cm (with a variation between male and female) giving a $\mathscr{I} \simeq 3 - 9$ cm^4. Coupled to the above values of transverse E, the bending rigidity of the femur midshaft results $\kappa_b = E\mathscr{I} = 300$–900 Nm^2. Such values should be taken just as an order of magnitude, since the bone cross section is never truly cylindrical, but it takes rather an elliptical irregular section, with the top and bottom edges thickened and the wide sides thinner.

In a similar way to the bending modulus, a **twisting modulus** can be defined as $\kappa_t = G \mathcal{J}$, with G the shear modulus and $\mathcal{J}$ the "polar" moment of inertia of the transverse cross section. The latter is analogous to $\mathcal{I}$, both conceptually and practically, and for the hollow cylinder it is numerically equal to $\mathcal{J} = 2\mathcal{I}$. The twisting modulus of femur bone is therefore $\kappa_t \sim \frac{2}{3}\kappa_b$. Such a value of $\kappa_b/\kappa_t \sim 1.67$ would be more typical of a hollow cylinder made of a stiff, rigid material such as steel (for an ideally isotropic material, $E/G = 2(1+\nu)$ and $\kappa_b/\kappa_t = (1+\nu)$; most metals have values of $\nu \sim 0.4$–0.7). This indicates a material that withstands bending and torsion with about equal ability, differently from e.g. a long plant stem. However, in most of an animal's life, bones withstand essentially compressive and bending loads, while tension and torsion are relatively rare thanks to the various joints connecting them.

9.6   Strong as Wood

The cell walls of plants are made up of four basic building blocks: cellulose, hemicellulose, lignin and pectin. Depending on the arrangement and density of their hierarchical microstructure, plants give rise to a remarkably wide range of mechanical properties, with Young's moduli spanning from the few MPa of potato and apples, to tens of GPa in oak and bamboo; tensile strengths vary roughly proportionally to E, from below 1 to more than 100 MPa.

Cellulose is the main structural fibre in the plant kingdom and has remarkable mechanical properties for a polymer: its Young's modulus is about 120 GPa, and its tensile strength is $\sim$1 GPa. Cellulose fibres in wood are arranged in a peculiarly hierarchical superstructure. Figure 9.15 displays the hierarchical assembly of a slice of spruce wood, showing microfibrils with different inclination angle μ, each microfibril being the outer wall of one plant cell, called *tracheid*. The cell wall contains a large fraction of cellulose molecules (40–50 %) and lignin (25–30 %), in an orderly arrangement, plus a disordered filling of hemicellulose (10–20 %) and pectin long filaments.

At the molecular scale, cellulose is made of chains of sugar units, the basic repeat being formed by two twisted glucose rings (Fig. 9.15f). The native form of cellulose in plant cell walls are thin microfibrils containing nanometric crystals of cellulose-I. Typically very ordered (up to 90 % in crystalline form), the hexagonal axis of cellulose crystal structure follows the fibril axis. The size of the cellulose nanocrystals is species-dependent, ranging from $\sim$2.5 nm to several tens of nm. As shown in Fig. 9.15c, cellulose fibrils are wound helically with a microfibril angle μ in the dominant cell-wall layer as sketched.

In a way much similar to the role played by long collagen molecules in soft tissues, the geometric arrangement of cellulose molecules in the microfibrils and their superstructure largely dictates the mechanical properties of the resulting system (wood). Cellulose fibrils give plant cell walls most of their enormous strength, much as glass fibres embedded in an epoxy resin give strength to a fibreglass composite. At the

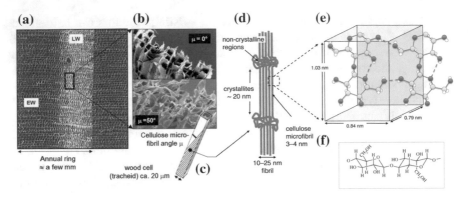

Fig. 9.15 Hierarchical structure of a wood sample. **a** Cross section parallel to the grain direction; EW is the early grown wood, LW is the late wood growth. **b** Scanning-electron microscopy images of fracture surfaces of spruce wood with two different microfibril angles. **c** A wood cell (*tracheid*) schematically drawn, to show the definition of the microfibril angle between the spiralling cellulose fibrils and the tracheid axis. **(d)** The flat, ribbon-like arrangement of cellulose chains into fibrils. **(e)** Crystalline structure of the cellulose chains, with a tetragonal unit cell of $0.79 \times 0.84 \times 1.03$ nm^3; *red spheres* oxygen, *grey* carbon, *white* hydrogen. Hydrogen bonds are indicated by *dashed lines*. **(f)** The –O– linked structure of the sugar rings of a single cellulose chain. [Scanning-electron micrographs (**a, b**) from Ref. [8], repr. w. permission.]

sub-micron scale (Fig. 9.15d) the flat, ribbon-like structure allows the chains to fit closely together, one on top of the other, over their entire lengths. At the molecular scale (Fig. 9.15e), cellulose chains are structured in a crystal lattice with a unit cell of about $0.7 \, nm^3$. Interchain associations are stabilised by hydrogen bonds with oxygen atoms linking glucose rings (Fig. 9.15f), and aggregate to form stiff crystalline rods of very considerable length and mechanical strength. The sub-nanometric molecular structure of linked glucose rings thus neatly determines the overall structure of cellulose chains, and establishes the interchain associations of the microfibrils. The stiff, crystalline rods of cellulose are clearly well suited to their biological function in the plant cell wall.

Lignin is the main biopolymer that fills the spaces in the cell wall between cellulose, together with hemicellulose, and pectin. It contributes to the mechanical strength of the cell wall, although its Young's modulus is much smaller than that of cellulose fibrils. Hemicellulose and lignin are similar to typical engineering polymers: lignin has $E \simeq 3$ GPa and $\sigma_t = 50$ MPa. This reinforcement effect comes from the fact that lignin is covalently linked (*cross-linked*) to hemicellulose, therefore it functions as a strain-accommodating filler, next to the load-carrying cellulose. Such a double-phase design is common in engineering composite materials, which have strong, stiff fibres embedded in a matrix that is weaker and less stiff; the objective is usually to make a component which is both strong and stiff, possibly with the lowest density. This very effect is obtained with the differential microstructure of the wood composite, where lignin tends to be particularly abundant in compression wood (sapwood), which make up the outer layers of the trunk under the bark, and scarce in tension wood

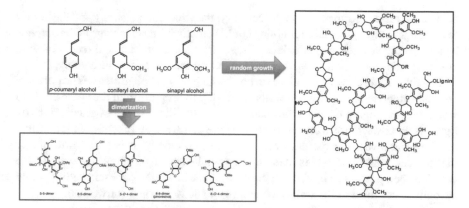

Fig. 9.16 Proposed models for the structure of lignin. The three monolignols in the *upper left box* are the most abundant in lignin. The older, random-growth models of Harkin, Freudenberg, Adler an others (1967–77) consider polymerisation by stepwise addition and merging of random branches (*right*, *orange box*). The more recent "protein directed" model by Davin (1998) introduces the formation of dimers as the first step in polymerisation, trying to explain the higher frequency of certain types of bonds observed in natural lignin (notably, the 8-O-4); however, assembly of dimers into a growing structure appears statistically unfavourable

(heartwood) found in the inner core of the trunk. The softer but covalently-linked lignin-hemicellulose matrix imparts the wood composite a high **tenacity**, i.e. the resistance to tearing and crushing forces. Because of the directional microstructural arrangement of load-carrying cellulose fibrils, the tenacity of wood is much greater in the direction of the length of its fibres than in the transverse direction.

At the molecular scale, lignin is made essentially of cross-linked phenol polymers (Fig. 9.16), formed by successive addition of alcohol monomers (*monolignols*). The chemical composition of lignin varies from species to species, with typical values of $\sim 60\%$-weight carbon, 30 % oxygen, 6 % hydrogen, 0.7 % ash, corresponding approximately to the formula $(C_{31}H_{34}O_{11})_n$. As a biopolymer, lignin is unusual because of its heterogeneity and lack of a defined primary structure. Although the enzymes and the biosynthetic ways at the origin of lignification are very well known, very little explanation exists of how lignin is built in the vegetable wall, and which its true structure is. Older structural models of lignin presented an irregular polymer, probably three-dimensional and produced by random aggregation of monomers in a stepwise polymerisation (see Chap. 6), to the disconcert of biologists and biochemists. If the proposed structures are true, lignin would be a unique case of a completely random process of biosynthesis leading to an irregular polymer. More recently, a "protein directed" model has been proposed, based on the formation of dimers to explain the preference of in vivo lignin for certain molecular bonds compared to what observed in artificially synthesised lignin, notably the high frequency of the 8-O-4 bond (see Fig. 9.16, red box). However, the statistics of aggregation from dimers are very unfavourable compared to the random growth. Moreover, to date there seem to be no experimental observations in favour of an absolute structural control over lignin formation. On the contrary, the structural plasticity and the ability to form lignin through random coupling could actually be an advantage in the defence against pathogens: for example, lack of regularity poses problems to reactions with enzymes from fungi or insects, thereby protecting the plant from invasion.

Notably, lignin plays a crucial part in conducting water in plant stems. The polysaccharide components of plant cell walls are highly hydrophilic and thus permeable to water, whereas lignin is more hydrophobic. The cross-linking of polysaccharides by lignin is an obstacle for water absorption to the cell wall, thereby makes it possible for the plant's vascular tissue to conduct water efficiently. It is also worth noting that the cell wall of woods is made up of four layers of varying composition, the inner one being richer in lignin, and the second one being the richest in cellulose and accounting for most of the wall thickness (up to 80 % in some softwoods).

9.6.1 Tension and Compression

The macroscopic structure of a tree trunk is extremely anisotropic because 90–95 % of all its cells are elongated and vertical, i.e. aligned parallel to the tree trunk; the remaining cells can be arranged in the radial direction, but no cells at all are aligned tangentially.

In the trunk there are three main sections: the inner *heartwood*, which is physiologically inactive; the intermediate *sapwood*, where all conduction and storage of nutrients (sap, water) occurs; and the outermost *bark*, which protects the interior of the trunk. Trees are usually classified as **softwoods** and **hardwoods**, because of their distinct internal structures. Coniferous trees are softwoods, with vertical cells (*tracheids*) 2–4 mm long, and roughly 30 μm wide. These cells (Fig. 9.17a) are used for support and conduction, with an open channel in the middle and a thin cell wall; storage cells are found in the radial direction. Broad-leaved trees, such as oak, are hardwoods. The vertical cells in hardwoods are mainly fibres, 1–2 mm long and 15 μm wide. These are thick-walled, with a narrow central channel (Fig. 9.17b) and are for support only; therefore other vessels are dedicated to conduction. Vessels are either *xylem*, dead cells that carry water and minerals, or *phloem*, live cells that transport energy sources made by the plant. Vessels are 0.2–1.2 mm long, open-ended and vertically stacked to form tubules <0.5 mm in diameter.

Fruits, such as apples, and root vegetables, such as potato tubers and carrots, are mostly made up of **parenchyma** tissue, to efficiently store sugars and starch. The cells of parenchyma (Fig. 9.17c) are polyhedral with thin cell walls, densely packed together like the closed cells of a foam enclosing a pressurised liquid. Their size varies anywhere between a few tens of microns to 2–300 μm, with a cell wall of ∼1 μm thickness. This latter is mostly composed of pectin and hemicellulose, does not contain lignin and just a little cellulose, therefore it is much softer than the cell wall of woods. Together with the lack of the multi-layered structure of twisted fibrils, the parenchyma tissue appears as a substantially isotropic medium, from the point of view of its mechanical properties.

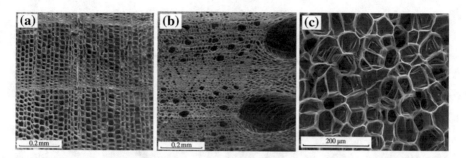

Fig. 9.17 Scanning electron micrographs of: **a** cedar (cross section), a typical softwood; **b** oak (cross section), a typical hardwood, with the large vessels for fluid conduction; **c** carrot, showing the thin foam-like cell walls of the parenchyma tissue. Note that the scale bar is 200 μm for all the three. [From Ref. [15], © 2010 Cambridge Univ. Press, repr. w. permission.]

The definition of softwood and hardwood has little relation with the material properties of wood: the softwood Scots pine has a Young's modulus 2–3 times larger than the hardwood balsa, mostly because of the much lower density of the latter. Because wood is a composite material, when stretching a wood sample it is the cellulose microfibrils that carry most of the load. The Young's modulus of cellulose fibrils is 100–120 GPa, while that of lignin and hemicellulose averages to 6 GPa. Under axial loading, an effective Young's modulus of the wood cell wall can be calculated as:

$$E_{cell-wall} = (1 - f)E_{cellulose} + f E_{lignin-hemi} \tag{9.15}$$

For a 50/50 composition, $E_{cw} \simeq 53-66$ GPa is a realistic estimate (the highest measured values reach 70 GPa). Since typical Young's moduli of various woods range between $E = 2-3$ and 20–25 GPa, it turns out that the cell wall is stiffer than the wood composite. This difference is due to the water fraction and the residual empty space, inside and between the cells. Most woods are moderately viscoelastic, their loading and unloading curves showing a relatively small hysteresis.

Data for the Young's modulus E and compressive strength σ_y of various wood samples, plotted against the relative density, are shown in Fig. 9.18a, b. The literature reference values used as scale on the abscissa and ordinates are those of the cell wall, respectively taken as $E_s = 35$ GPa, $\sigma_s = 350$ MPa, and $\rho_s = 1.5$ g cm^{-3}. As shown by the straight-line fit (labeled with numbers '1', '2', '3', the slope a of a $\ln Y = a \ln X$ plot indicating a power-law relationship $Y \propto X^a$), the E^* and σ^* parallel to the grain direction are approximately linear in the relative density, while those across the grain, in the radial or tangential direction, vary roughly with the square of relative density:

$$E_{\parallel}^* = E_s \left(\frac{\rho^*}{\rho_s}\right) \qquad E_{\perp}^* = E_s \left(\frac{\rho^*}{\rho_s}\right)^{2.5} \tag{9.16}$$

$$\sigma_{\parallel}^* = \sigma_s \left(\frac{\rho^*}{\rho_s}\right) \qquad \sigma_{\perp}^* = \sigma_s \left(\frac{\rho^*}{\rho_s}\right)^2 \tag{9.17}$$

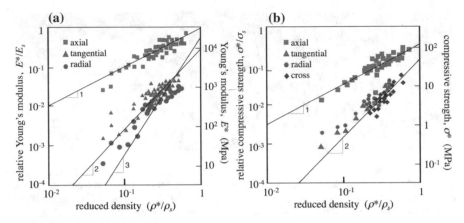

Fig. 9.18 a Young's modulus and **b** compressive strength of various wood samples plotted against the relative density (experimental points from Ref. [16]). Data labeled "across" the wood grain are for loading in the radial or tangential directions; the direction of loading is not specified. Numbers next to each *straight line* indicate the slope

   The exponents of the last two equations correlate quite well with the predictions of the mechanical theory of cellular materials with honeycomb structure. While the exponents for the parallel quantities are both predicted to be 1 and that of $\sigma_\perp$ is exactly 2, the predicted exponent for $E_\perp$ should be in fact 3; however, it can be seen that the data in Fig. 9.18 fall somewhere in-between the fit with slope 2 and 3, somewhat closer to the former. This indicates a partial agreement with the theory, however demonstrating overall that the great ranges in the elastic moduli of woods (a factor of over 1000) and the strengths (a factor of over 100) arise primarily from the honeycomb-like structure of wood cells.

   A similar theoretical description can be obtained for the foam-like structure of parenchyma. At normal or high turgor pressures the cell are tightly packed, and deformation is dominated by stretching of the cell walls. The solid mechanics model in this case predicts equations of the same type as above, but with exponents 1 for both E^* and σ^* (parenchyma being isotropic, there is no need to distinguish between the parallel and transverse directions). In this case, the model predicts values in good agreement with the measured $E^* = 3.5$–5.5 MPa and $\sigma^* = 0.27$–1.3 MPa.

9.6.2   Bending and Twisting

The composition, cell-wall structure, and cellular structure of plants and vegetables give rise to remarkable mechanical performances, when expressed on a per-mass basis. The trunks and branches of trees are loaded primarily in bending (from the wind, or their self-weight). Solid mechanics tells us that for a beam of given stiffness, span and cross-section diameter, the material that minimises the weight of a beam is

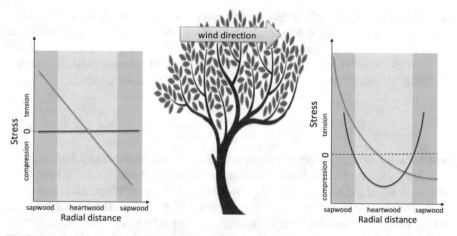

Fig. 9.19 The effect of pre-stress on the wood trunk. *Left* If the initial stress (*orange*) is zero, when the tree gets a stress from the wind (coming from the *left*) its two sides are under equal tension and compression (*blue straight line*). *Right* If at rest condition (*orange curve*) the inner hardwood is under compressive stress and the outer sapwood is under tensile stress, when the trunk is subject to the wind force (*blue curve*) the compressive stress on the *right* is much reduced compared to the tension on the *left side*

that with the maximum value of the ratio ($E^{1/2}/\rho$). By using the previous equation for E^*, this "performance ratio" can be expressed in terms of the reference value for the average cell walls material as:

$$\frac{(E^*)^{1/2}}{\rho^*} = \frac{1}{\rho^*} \left[E_s \left(\frac{\rho^*}{\rho_s} \right) \right]^{1/2} = \frac{E_s^{1/2}}{\rho_s} \left(\frac{\rho_s}{\rho^*} \right)^{1/2} \tag{9.18}$$

Therefore, the performance ratio for any wood should be equal to that of the composite cell wall, $(E_s^{1/2}/\rho_s) = 3.94 \text{ kPa}^{1/2} \text{ m}^3 \text{ kg}^{-1}$, times the square root of the ratio of the densities, $(\rho_s/\rho^*)^{1/2}$. For example, Scots pine, with $\rho^* = 0.51 \text{ g cm}^{-3}$, has a ratio of 6.8, comparable or superior to the best carbon fibre composites, like kevlar with 4, or the high-modulus, ultra-dense carbon fibre (HMCF-UD) with a ratio of 8.3 (metallic materials have too high density to be useful in this respect).

At the level of the macroscopic tree structure, there is a further point of interest: the trunk is **pre-stressed**. The centre of the trunk is in compression, and the outer layers are in tension. This condition of pre-stressing is achieved because the inner parts of the sapwood shrink as they dry and become heartwood. Since the heartwood has lower moisture content it is better able to resist compression. As shown in Fig. 9.19, such a state of pre-stress largely reduces the compressive stress on the outer layers downwind, compared to the tensile stress on the opposite side.

However, as we learned in Chap. 8, the bending stiffness of a beam is the product of the Young's modulus and the transverse moment of inertia, $\kappa_b = E\mathscr{I}$. We already noted how the mathematical expression for $\mathscr{I}$ suggests that a hollow beam resists much better to a bending load than a thick beam, for a same mass of material used. Therefore one could ask the question: why trees have not evolved a hollow-tube shape for their trunk and branches?

The Euler's theory of bending beams, which can be found in any good engineering textbook, provides a basic result for the flexion d of a beam of length L:

$$d = \alpha \frac{FL^3}{\kappa_b} \tag{9.19}$$

where α is a geometrical coefficient depending on the mode of loading. For example, $\alpha = 1/48$ if a point force is applied at the midpoint and the beam is fixed at both ends, or $\alpha = 5/384$ if the load is uniformly distributed all along the length (this is also the case of the self weight, $F = mg$ with m the mass of the beam). If one of the ends is free (a *cantilever*) and the force is concentrated at the free end, one gets $\alpha = 1/3$, while $\alpha = 1/8$ if the load is uniformly distributed. Overall, the Euler theory tells that by increasing the beam length a large price is paid in bending, proportional to L^3; but even more is gained by thickening the cross section, which appears to power 4 in the denominator from the calculation of $\mathscr{I}$.

The latter case could be interesting for the bending of trees and branches, since these can be considered as cantilevers fixed at one end and free at the other. For example, what is the longest possible branch of a pine? By taking $E = 10$ GPa, the wood density $\rho \simeq 500$ kg m^{-3}, and a diameter $D = 10$ cm for which $\mathscr{I} = \pi D^4/64$, we get $\kappa_b = 49.1$ kN m^{-2}. If we accept a maximum flexion at the free end of $d = L/10$, plug in the numbers, and we obtain a length of 10 m! By accounting the fact that the diameter of a real branch is not constant along the length, but it tapers to a very thin diameter at the free end, this branch –already unrealistically long—would almost double. And even longer branches could be predicted, if the cross section were elliptic instead of circular. In fact, for an ellipse with its longest diameter b parallel to the direction of the force (vertical, if the force is gravity), it is $\mathscr{I} = \pi ab^3/64$, and by imposing equal area of the cross section as $ab = D^2$, a factor equal to b/a is gained at the denominator of Eq. (9.19) (by the way, this is why timber beams have rectangular cross section, and are laid vertically and not flat, when building a roof). Many plant stems and branches do have elliptical sections, and the radially asymmetrical growth of wood represents a common response of trees to long-term unidirectional forces from the environment.

Another interesting result of Euler's theory concerns the bending under a vertical load: it predicts that a cylindrical stick will bend if the perpendicular force on its top end is larger than the critical value:

$$F_{crit} = \frac{\pi^2}{\beta} \frac{\kappa_b}{L^2} \tag{9.20}$$

with β equal to 4 if the upper end is free, or to 0.5 if it is held. This is called the **buckling instability**, and can be important for a stiff material that can resist fracturing under compression, at such value of force, "preferring" to bend instead of crushing.

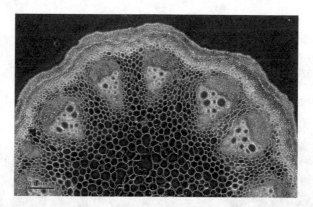

Fig. 9.20 A light-microscopy cross section of *Aristolochia* stem at ×240 magnification. This is a typical angiosperm stem with scattered vascular bundles. Scale bar on the *lower left* indicates 1 mm. [Image courtesy of Josef Reischig, repr. under CC-BY-SA 3.0 licence, see (**) for terms.]

You can try this with any thin stick, by pushing it vertically against a hard ground: it will certainly bend sideways, under a moderate force (but enough to surpass F_{crit}). If you keep pushing, the stick will continue to bend until the maximum stress at the curved side goes beyond the fracture stress σ_f, around which value it will break. Doing the same experiment on a thicker stick will not give the same result. It will resist bending and, provided enough vertical force is applied, it will rather crush under the compression (because of the rapidly increasing $\mathscr{I} \propto D^4$ at the numerator of (9.20)). For a perfect cylinder, without any defects or pre existing bends (a tree branch does not fit easily such criteria), the critical force is usually quite a high value, compared to the compressive strength, unless the diameter is very thin compared to the length.

What this brief analysis tells us, is that the self-loading received by branches and stems is not critical for bending, therefore the tree seems less interested in optimising the design of their cross section. Indeed, the resistance to bending and twisting originates mostly from the highly heterogeneous microstructure of the cell distribution in the cross section of the stem. We already saw a first example: the lignified sapwood of the outer tree structure resists tension better than compression, a good reason to use the pre-stress trick to reduce compressive loads. Similarly, if we look at the cross sections of many plant stems, such as the *Aristolochia* shown in Fig. 9.20, it can be seen that the cell size distribution is very much uneven: large and tense, liquid-filled cells are seen at the centre of the stem, providing more resistance against compression, while much smaller cells with a larger surface-to-volume ratio, and therefore a higher rigidity, are found at the outer perimeter, better suited to resist tearing and crushing.

If self-loading is not critical, however, branches and leaf stems are subject to both torsion and bending external forces, for example applied by gusts of wind coming from all directions, or by animals sitting on them. The twisting modulus κ_t, analogous to the bending modulus, and its corresponding moment of inertia $\mathscr{J}$ were introduced

(a)

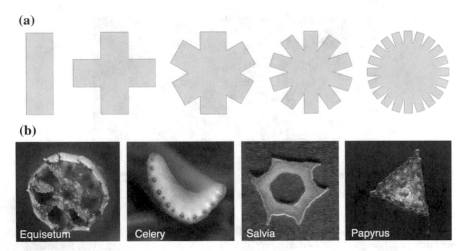

(b)

| Equisetum | Celery | Salvia | Papyrus |

Fig. 9.21 a An oblong shape has a larger value of $\mathscr{I}$ than a square. Crossing several rectangular sections may increase propensity to twisting, by inserting sharp grooves along the length. **b** Examples of plant stems with non-circular cross section

above, p. 397. However, material heterogeneity is important as well to resist torsion. We already know that for a circular cross section $\mathscr{J}$ is about twice as large as $\mathscr{I}$, this increase being compensated by a smaller shear modulus $G = E/(1 + \nu)$, compared to E. Therefore, twisting and bending resistance are generally comparable in this case, $\kappa_b/\kappa_t = E\mathscr{I}/G\mathscr{J} = 1.5$ for a Poisson's ratio of 0.5. Values of such a "bend-to-twist" ratio much larger than this ideal 1.5, should be an indication that the material tends to yield by twisting more easily than bending.

Non-circular cross sections make an interesting case in plant stems. The κ_b/κ_t ratio jumps to larger values, of $\sim$3.5 and $\sim$5.77 respectively for a square or triangular cross section. This is because of the much reduced twist resistance offered by a section with sharp angles, compared to bending resistance that, instead, is increased (for example, by about 2 for a square vs. circular cross section of equal area). Since an elliptical, or rectangular, cross section with one side much wider than the other (i.e., a large b/a) largely improves the resistance, it seems that many rectangular sections crossed together could do even better (Fig. 9.21a), with the crossed grooves adding flexibility to twisting loads. The stress-strain curve in the case of mixed load has a very unpredictable shape. It is extremely difficult to decide when and how a structure will fail under a complex load that mixes compression, traction, bending and torsion, in different parts of the body. Note that, whereas torsion is carefully avoided in engineering structures, the typically large values of the κ_b/κ_t observed for flexible plant structures seem to indicate that plant stems and tree sprouts favour torsion over bending, as their preferred mode of resisting to the random distribution of forces they may encounter. Eventually, combining one or more grooves with the hollow tube shape could be the ultimate, optimal structure to resist mixed loads. In fact, many plant stems have far from circular cross sections, often have a hollow interior, and have ridges and grooves that increase their propensity for twisting instead of bending (Fig. 9.21b). Cactus plants, which have little or none lignified trunk, often display lots of thick ridges all along the plant body. Twisting could be, indeed, the best way to spend elastic deformation energy, when one of your ends is stuck to the ground and wind comes from a given direction.

The basic lesson to be learned is that *stiffer* is not as good as *stronger*, also in the case of plants. Opposing a stiff stick may not be as convenient as gently yielding in the wind, and easy twisting is a great help when growing very slowly from fixed ends. But of course, mechanics is not the only constraint dictating plant requirements. Optimising sun exposure, water capture and retention, repelling harmful insects and undesired herbivores, dispersing spores and smell in the environment, are just a few key factors that come into play when thinking of the life of a plant, and the simple analysis restricted to the mechanical requirements may be quite far off the truth, despite providing some sound motivations.

Appendix H: Materials Elasticity Theory for Dummies

Stress

Consider a body with a generic shape, and apply a force oriented along a generic direction, at a point of its surface. If we define the perpendicular to the surface through that point by some unit vector $\mathbf{n} = (n_x, n_y, n_z)$, and the force as another vector $\mathbf{f} = (f_x, f_y, f_z)$, the stress can be defined by combining the dependence on both vectors, as shown in Fig. 9.22(left). This is a mathematical quantity with two subscript indices, $\underline{\sigma} = \sigma_{ab}$, with $a, b = (x, y, z)$, describing the possible combinations of the Cartesian components of the two vectors, called a *tensor*:

$$\sigma_{ab} = \frac{1}{A} \frac{\partial f_a}{\partial n_b} \tag{9.21}$$

The 3×3 components of the tensor can be written in the form of a matrix. If the force is acting only along one or more of the three directions, and there is no coupling between the forces in different directions, the stress tensor matrix will have non zero components only on its diagonal. If moreover the three components of the force are identical, we have $\sigma_{xx} = \sigma_{xx} = \sigma_{xx} = \sigma_0$, and it can be proved that the pressure p is the *trace* of the stress tensor:

$$p = \frac{1}{3} \text{Tr}[\underline{\sigma}] = \frac{\sigma_{xx} + \sigma_{yy} + \sigma_{zz}}{3} = \sigma_0 \tag{9.22}$$

If the stress tensor is given, the force across a surface element of the body can be defined as a mechanical *tension*:

$$t_a = \frac{1}{A} \sum_b \int_A (\underline{\sigma} \otimes \mathbf{n}) dA = \frac{1}{A} \sum_j \int_A (\sigma_{ij} \cdot n_j) dA \qquad (9.23)$$

where the symbol '$\otimes$' indicates that the product between a tensor and a vector follows the special rules of matrix multiplication.

Strain

Similar to stress, the *strain* is mathematically defined as well as a tensor, $\underline{\varepsilon}$ with indices describing the deformation $\mathbf{u} = \mathbf{r} - \mathbf{r}'$ of a vector $\mathbf{r}'$ to any point along each Cartesian direction, with respect to the original position vector $\mathbf{r}$, in that same, or another Cartesian direction, as shown in Fig. 9.22(right). The general definition of the strain tensor, symmetric in the Cartesian components, is:

$$\varepsilon_{ij} = \frac{1}{2} \left(\frac{\partial u_i}{\partial r_j} + \frac{\partial u_j}{\partial r_i} \right) \qquad (9.24)$$

The rigorous definition of the displacement vector $\mathbf{u}$ should also include local material rotation, $\mathbf{u} = \mathbf{r} - \mathbf{r}' + \omega$; however, the rotational component of deformations will not be considered in this book, except in special cases.

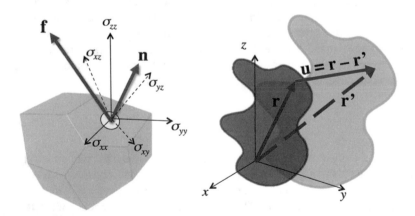

Fig. 9.22 Schematic representation of the geometrical interpretation of the stress (*left*) and strain (*right*) tensors. *Left* A force $\mathbf{f}$ is applied at a point on the surface of a body with arbitrary shape. The vector $\mathbf{n}$ indicates the perpendicular direction to the element of surface dA (*little grey circle*). The stress components $\underline{\sigma} = (\sigma_{xx}, \sigma_{xy}, ..., \sigma_{zz})$ represent the variation of each of the components f_x, f_y, f_z of the force vector, according to the components n_x, n_y or n_z of the perpendicular vector. *Right* A point located at the vector position $\mathbf{r}$ in the undeformed orange body, is displaced at $\mathbf{r}'$ in the deformed body. The strain components $\underline{\varepsilon} = (\varepsilon_{xx}, \varepsilon_{xy}, ..., \varepsilon_{zz})$ represent the variation of each of the components u_x, u_y, u_z of the displacement vector $\mathbf{u} = \mathbf{r} - \mathbf{r}'$, with respect to the components of the undeformed position vector r_x, r_y, r_z

It can be noted that the symmetry of the Euclidean space implies that the off-diagonal components of the stress and strain tensors are symmetrical, i.e. $a_{ij} = a_{ji}$ for $i \neq j$. Therefore the nine components of each tensor are reduced to six.

Any tensor $\underline{a}$ admits a unique and additive decomposition into a diagonal (or trace) component, a_{kk}, and a deviatoric (or traceless) component $\overline{a}_{ij}$:

$$a_{ij} = \tfrac{1}{3} a_{kk} \delta_{ij} + \overline{a}_{ij} \qquad (9.25)$$

(the Kronecker symbol δ_{ij} is always zero except for $i = j$).

To make a bit less cumbersome the notation in the following Sections, the so-called **Voigt convention** for numbering the tensor components can be followed, namely: $xx = 1$, $yy = 2$, $zz = 3$, $xy = yx = 4$, $yz = zy = 5$, $xz = zx = 6$. In this way, tensors have only one index running from 1 to 6.

Elastic Constants and Compliances

In the regime of small deformations for which the linear approximation can be applied, stress and strain are proportional to each other via the matrix of *elastic constants*, C, and *elastic compliances*, S:

$$\sigma_i = \sum_j C_{ij} \varepsilon_j \qquad i, j = 1, ..., 6 \qquad (9.26)$$

$$\varepsilon_i = \sum_j S_{ij} \sigma_j \qquad i, j = 1, ..., 6 \qquad (9.27)$$

where the Voigt notation of the indices was used. Note that the C have dimension of an energy density (energy/volume), and the S are just their inverse (also in the more mathematical sense of inverse matrix).

In principle, the matrices C or S have 6×6 independent components, relating each of the stress components to a different strain component, and vice versa. The same symmetry concept applies, however, and the 36 components are reduced to the diagonal plus one full triangle of the matrix, $6(6 + 1)/2 = 21$ components:

$$
\begin{pmatrix} \sigma_1 \\ \sigma_2 \\ \sigma_3 \\ \sigma_4 \\ \sigma_5 \\ \sigma_6 \end{pmatrix}
=
\begin{pmatrix}
C_{11} & C_{12} & C_{13} & C_{14} & C_{15} & C_{16} \\
 & C_{22} & C_{23} & C_{24} & C_{25} & C_{26} \\
 & & C_{33} & C_{34} & C_{35} & C_{36} \\
 & & & C_{44} & C_{45} & C_{46} \\
 & & & & C_{55} & C_{56} \\
 & & & & & C_{66}
\end{pmatrix}
\begin{pmatrix} \varepsilon_1 \\ \varepsilon_2 \\ \varepsilon_3 \\ \varepsilon_4 \\ \varepsilon_5 \\ \varepsilon_6 \end{pmatrix}
\qquad (9.28)
$$

Moreover, further symmetry considerations relative to the particular internal arrangement of the atoms and molecules of the material can further reduce the number of independent "Hooke-like" relations between the different components of stress and strain.

For a material with perfectly isotropic response to the applied forces, the three Cartesian directions are equivalent and only two coefficients are independent: C_{11} (also equal to C_{22} and C_{33}), and C_{12} (also equal to all the combinations C_{ab} with $a, b = 1, 2, 3$); the coefficients C_{aa} for $a = 4, 5, 6$ are all equal to $(C_{11} - C_{12})/2$, and all other coefficients are zero. Examples of isotropic materials are any liquid, or amorphous solid, like a glass; most mixtures of polymers and plastic materials are practically isotropic.

Elastic Moduli for Solid Materials

For an isotropic material only two elastic constants are needed to specify its response to an applied stress, in the linear regime of deformation. As we saw above, these are C_{11} and C_{12}. Just by looking at their indices '11' and '12', it is not immediate to understand what kind of deformation these coefficients relate to. On the other hand, their value is not directly accessible to a simple experiment, and a more convenient way is to deduce them from experiments in which some combination of their values occurs. It turns out that in this way, also the interpretation of their physical meaning becomes more transparent. The combinations of elastic constants are called **elastic moduli**, and each one of them corresponds to an experimentally realizable mode of deformation.

Elastic moduli for isotropic materials were identified already in the first half of the XIX century, and called Lamé coefficients. In terms of the elastic constants, they would be written as $\lambda = C_{12}$ and $\mu = (C_{11} - C_{12})/2$. The two coefficients are still derived mathematically, by looking at the symmetries of deformation.

Experimental quantities related to the C_{ij} are the *bulk* modulus, B, the *shear* modulus, G (equal to μ), the Young's modulus, E, or the Poisson's ratio, ν. Obviously, the same isotropic material will be fully described by any pair of these, but some may be more convenient than others.

By decomposing the stress tensor according to the strain components (the overbar denoting the deviatoric components), we can write:

$$\sigma_{ii} = 3B\varepsilon_{ii} \tag{9.29}$$

$$\overline{\sigma}_{ij} = 2G\overline{\varepsilon}_{ij} \tag{9.30}$$

as well as:

$$\sigma_{ij} - \tfrac{1}{3}\sigma_{kk}\delta_{ij} = \sigma_{ij} - B\varepsilon_{kk}\delta_{ij} = 2G\left[\varepsilon_{ij} - \tfrac{1}{3}\varepsilon_{kk}\delta_{ij}\right] \tag{9.31}$$

From the last two identities, it also follows that:

$$\sigma_{ij} = 2\mu\varepsilon_{ij} + \lambda\varepsilon_{kk}\delta_{ij} \tag{9.32}$$

with $\lambda = B - 2G/3$.

Alternatively, one can write the strain tensor according to the stress components, as:

$$\varepsilon_{ij} - \tfrac{1}{3}\varepsilon_{kk}\delta_{ij} = \varepsilon_{ij} - \frac{1}{9B}\sigma_{kk}\delta_{ij} = \frac{1}{2G}\left[\sigma_{ij} - \tfrac{1}{3}\sigma_{kk}\delta_{ij}\right] \qquad (9.33)$$

from which it follows the expression:

$$\varepsilon_{ij} = \frac{1}{2G}\sigma_{ij} - \left(\frac{1}{6G} - \frac{1}{9B}\right)\sigma_{kk}\delta_{ij} = \frac{1+\nu}{E}\sigma_{ij} - \frac{\nu}{E}\sigma_{kk}\delta_{ij} \qquad (9.34)$$

Accordingly, the elastic constants and compliances matrices are written in terms of E and ν as:

$$C_{ij} = \frac{E}{(1+\nu)(1-2\nu)}\begin{pmatrix} 1-\nu & \nu & \nu & 0 & 0 & 0 \\ & 1-\nu & \nu & 0 & 0 & 0 \\ & & 1-\nu & 0 & 0 & 0 \\ & & & 1-2\nu & 0 & 0 \\ & & & & 1-2\nu & 0 \\ & & & & & 1-2\nu \end{pmatrix} \qquad (9.35)$$

$$S_{ij} = \frac{1}{E}\begin{pmatrix} 1 & -\nu & -\nu & 0 & 0 & 0 \\ & 1 & -\nu & 0 & 0 & 0 \\ & & 1 & 0 & 0 & 0 \\ & & & 4(1+\nu) & 0 & 0 \\ & & & & 4(1+\nu) & 0 \\ & & & & & 4(1+\nu) \end{pmatrix} \qquad (9.36)$$

Bulk modulus

This parameter defines the relative variation of the volume induced by an isotropic compression/dilation of the solid, typically a variation of hydrostatic pressure corresponding to a stress tensor $(\sigma_0, \sigma_0, \sigma_0, 0, 0, 0)$, under the assumption that the material is homogeneous at the scale of the applied deformation (see Fig. 9.23, top):

$$\Delta P = B\left(\frac{\Delta V}{V}\right) \qquad (9.37)$$

In terms of the Lamé parameters it is $B = \lambda + (2/3)\mu$, and in terms of the independent elastic constants, $B = (C_{11} + 2C_{12})/3$. We already encountered this expression, in Eq. (9.48) above, to define elastic energy of hydrostatic compression/dilation under a strain $(\varepsilon_0, \varepsilon_0, \varepsilon_0, 0, 0, 0)$. In fact, we can also write:

$$E_{el} = \frac{1}{2} B \varepsilon^2 \tag{9.38}$$

which suggests a harmonic character of the deformation (energy proportional to a squared variation). Also, the ε^2-dependence makes the energy symmetric for compression ($\varepsilon < 0$) and dilation ($\varepsilon > 0$). In fact, this is characteristic of any perturbation in the linear regime, in which the force is proportional to the perturbation, and the energy is correspondingly proportional to its square. We will find other similar relationships also later on.

Since the strain is dimensionless, B has the same units of energy/density, or force/surface. Some typical values of B for various materials are given in Fig. 9.23.

Shear Modulus

In this type of deformation, also homogeneous within the test body, the stress is applied parallel to the surface (Fig. 9.23, middle); if we take the xy axis in the plane of the figure, the applied stress tensor would be $(0, 0, 0, \sigma_0, 0, 0)$. If we imagine the body cut in slices parallel to the direction of the force, the corresponding deformation tends to slide all slices with respect to each other. The overall deformation is d/D, occurring symmetrically with xy and yx components. This mode of deformation is called *shearing*, and the corresponding parameter is the shear modulus. The relation between the applied stress and the resulting deformation is:

$$\sigma_0 = G\left(\frac{d}{D}\right) \tag{9.39}$$

In terms of the Lamé parameters it is $G = \mu$, and in terms of the independent elastic constants, $G = (C_{11} - C_{12})/2$. For many solids, the shear modulus is of the same order of magnitude of the bulk modulus. On the other hand, it is zero by definition for any fluid, in fact the very definition of a fluid is that of a material that does not support shear stress. It may be worth noting, however, that in the context of fluid mechanics, the Lamé parameter μ is often identified with the dynamic viscosity of the medium.

Young's Modulus

Named after the same English physician Thomas Young that we mentioned in Chap. 2 for introducing the term 'energy', this elastic modulus is the most appropriate to describe an experiment of traction or compression of a body along an axis, e.g. with a stress tensor $(\sigma_0, 0, 0, 0, 0, 0)$ if deforming along x. In this case, the deformation is again homogeneous, but in a more subtle way. It is quite common to observe

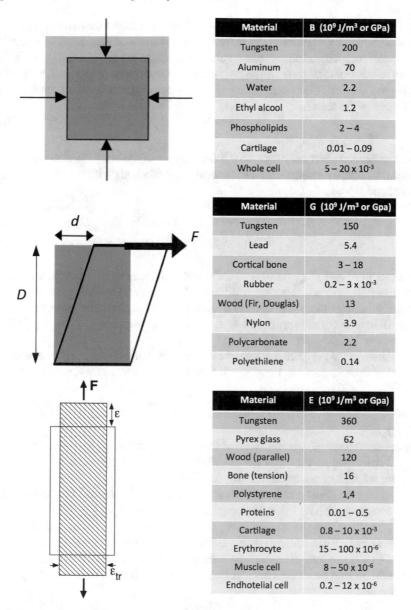

Material	B (10⁹ J/m³ or GPa)
Tungsten	200
Aluminum	70
Water	2.2
Ethyl alcool	1.2
Phospholipids	2 – 4
Cartilage	0.01 – 0.09
Whole cell	5 – 20 x 10⁻³

Material	G (10⁹ J/m³ or Gpa)
Tungsten	150
Lead	5.4
Cortical bone	3 – 18
Rubber	0.2 – 3 x 10⁻³
Wood (Fir, Douglas)	13
Nylon	3.9
Polycarbonate	2.2
Polyethilene	0.14

Material	E (10⁹ J/m³ or Gpa)
Tungsten	360
Pyrex glass	62
Wood (parallel)	120
Bone (tension)	16
Polystyrene	1,4
Proteins	0.01 – 0.5
Cartilage	0.8 – 10 x 10⁻³
Erythrocyte	15 – 100 x 10⁻⁶
Muscle cell	8 – 50 x 10⁻⁶
Endhotelial cell	0.2 – 12 x 10⁻⁶

Fig. 9.23 The simplest homogeneous deformation modes (*left column*), corresponding to: hydrostatic compression/dilation (*top*); shear (*middle*); uniaxial tension/compression (*bottom*). On the right column, some values (in units of 10^9 J/m³, or GPa) of the corresponding elastic moduli for typical materials and biological tissues: bulk modulus (*top*), shear modulus (*middle*), and Young's modulus (*bottom*)

that a material undergoing axial compression will contract along the compression direction, while it will dilate in the perpendicular plane, and vice-versa in extension. With the stress specified above, this results in a strain tensor (ε, ε_{tr}, ε_{tr}, 0, 0, 0), with ε_{tr} the relative deformation in the perpendicular directions y and z. The negative of the ratio between the two resulting values of deformation is called the Poisson's coefficient of the material:

$$\nu = -\frac{\varepsilon_{tr}}{\varepsilon} \tag{9.40}$$

It is easily seen that only for $\nu = 0.5$ the volume of the body is conserved during the deformation, for example for a traction along x of a body with lengths L_x, L_y, L_z and volume $V = L_x L_y L_z$:

$$\Delta V = L_x(1 + \varepsilon)L_y(1 - \varepsilon_{tr})L_z(1 - \varepsilon_{tr}) - L_x L_y L_z =$$
$$= V(\varepsilon - 2\varepsilon_{tr}) + O(\varepsilon^2) \simeq V\varepsilon(1 - 2\nu) \tag{9.41}$$

(the terms in ε^2 and ε^3 going to zero for small deformation). On the other hand, for most materials the Poisson's ratio is different from 0.5, and the deformation changes also the volume of the object (as it is done also by the hydrostatic compression and dilation). In terms of the elastic constants or Lamé coefficients, it is $\nu = C_{12}/(C_{11} + C_{12})$, or $\nu = \lambda/2(\lambda + \mu)$.

As shown in the Fig. 9.23, bottom, we imagine to apply a force F at the extremities of a rod with cross section A and initial length L. The stress-strain relation is:

$$\sigma = E\varepsilon \quad \rightarrow \quad \frac{F}{A} = E\frac{\Delta L}{L} \tag{9.42}$$

with the cross section A being taken at its reference, undeformed value. During the uniaxial deformation, the cross section changes as $A/(1 \pm \varepsilon)^{2\nu}$, the '+' and '−' sign being for traction or compression, respectively.

In terms of the Lamé parameters it is $E = 2\mu + \lambda$, and in terms of the independent elastic constants we have the more complicate expression $E = \frac{C_{11}-C_{12}}{C_{11}+C_{12}}(C_{11} + 2C_{12})$. It will be noticed that the traction/compression experiment is geometrically similar to Robert Hooke's experiments on his springs. Indeed, starting from the stress-strain relation (9.42) we can determine an effective "spring" constant for the material, which will depend on its geometrical shape. By multiplying both sides of Eq. (9.42) by A, we get:

$$F = \left(\frac{EA}{L}\right)\Delta L \tag{9.43}$$

from which an effective 'spring constant' of the body can be defined as $k = EA/L$.

A related elastic modulus that is of utility in biological materials is the *aggregate modulus*, H, obtained in an experiment in which only the uniaxial deformation is allowed, while no transverse strain occurs. In terms of E, this modulus is formally defined:

$$H = \frac{E(1 - v)}{(1 + v)(1 - 2v)} \tag{9.44}$$

but when looking at its expression in terms of the independent elastic constants, it is just $H = C_{11}$. To have a zero component of the transverse deformation it is necessary to apply a stress in the plane that would contrast the natural tendency of the material to contract or dilate, according to the sign of its Poisson's ratio, therefore the stress tensor in this case looks like $(\sigma, \pm\sigma_{tr}, \pm\sigma_{tr}, 0, 0, 0)$.

For the deformation in a plane upon a force applied along two directions, an equivalent expression to that of uniaxial deformation can be obtained, by using Eq. (9.35) or (9.36). For example, we may have $\{\sigma_1, \sigma_2, \sigma_4\} \neq 0$, if two forces f_x and f_y are applied in the xy plane without allowing to deform along z. The matrix product $\varepsilon_i = \sum_j S_{ij}\sigma_j$ gives:

$$\begin{cases} E\varepsilon_1 &= \sigma_1 - v\sigma_2 \\ E\varepsilon_2 &= \sigma_2 - v\sigma_1 \\ \mu\varepsilon_4 &= 2\sigma_4 \end{cases} \tag{9.45}$$

Elastic Deformation Energy

In solid mechanics, it is common practice to follow the response of a material to an applied load (in the form of an imposed stress, or deformation) by tracing its **stress-strain diagram**. Such a representation contains much information about the mechanical behaviour of our material. A typical example of such a diagram is given in Fig. 9.24, for a uniaxial force f pulling a material sample along one direction; the initial area of the cross section transverse to the direction of f is S_0, and it evolves

Fig. 9.24 Example of a stress-strain curve for a material exhibiting a mixed mechanical response to a uniaxial pulling force f. The red curve describes the "apparent" stress, $\sigma = f/S_0$, the black curve the "real" stress, $\sigma = f/S$, where S_0 and S are, respectively, the initial and instantaneous cross section area

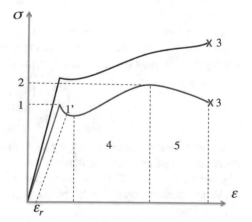

into $S(f)$ as far as the force stretches the material. The point 1 is the elastic limit: up to this point, the relation between stress and strain is linear, such as in a Hookean spring; beyond this point, the material could break if it is fragile (such as glass), or continue to deform if it is ductile (such as steel); the corresponding value of stress is called the **yield stress** σ_y of the material. The point $1'$ is a generic value of strain beyond the elastic limit: if the force is released at this point, the material will go back to zero stress but retaining a residual deformation ε_r. Upon increasing the stress, the material can continue to deform, up to the point 3 where final rupture occurs; the corresponding values of σ_f and ε_f are called **fracture stress** and **fracture strain**. The maximum value of stress supported during all the long deformation, marked as 2, is the **ultimate strength** of the material. The region 4 under the curve is called the "strain hardening" region, since the material responds in a complex way however the stress keeps increasing (the red and black curve run approximately parallel to each other). The region 5 is the "necking" region, in which the cross section rapidly decreases (note the widening difference in this region, between the red and black curve): this is also the region of maximum plastic deformation. The integral of the area under the stress-strain plot from zero to ε_f, $\tau_0 = \int_0^{\varepsilon_f} \sigma d\varepsilon$, is the **toughness**.

When a material body is deformed by an external force, this force performs a work equal to $dW = \sigma d\varepsilon$, to be integrated over the entire volume V of the deformed body. The work done by the external force is stored in the material in the form of a deformation energy, which is given back to the environment when the force is removed.

Always remaining in the limit of small deformations, so that the continuum linear elasticity theory can be applied, the elastic deformation energy in the volume V is written:

$$E_{el} = \int_V dW = \int_V \underline{\sigma} \otimes d\underline{\varepsilon} = \sum_{ij} \int_V (\varepsilon_i^T C_{ij}) d\varepsilon_j \qquad (9.46)$$

by using the formal stress-strain relation Eq. (9.26), and the explicit matrix components of stress and strain with the matrix multiplication rules (note that the writing "$\varepsilon_i^T C_{ij}$" indicates the product between the line-vector ε^T, transposed of the column-vector ε, and the matrix $\underline{C}$).

Carrying out the formal integration, in a homogeneous volume (in which the elastic constants are constant) and homogeneously deformed (i.e., every point of the volume is deformed in the same way), the elastic energy density is:

$$E_{el} = \frac{1}{2} \sum_{ij} \varepsilon_i C_{ij} \varepsilon_j \qquad (9.47)$$

For example, the compression by equal amounts ε_0 along x, y, z (or "hydrostatic" deformation) of an isotropic material (with only C_{11} and C_{12} non-zero, and $C_{44} = C_{55} = C_{66} = (C_{11} - C_{12})/2$), would result in a strain tensor $\underline{\varepsilon} = (\varepsilon_0, \varepsilon_0, \varepsilon_0, 0, 0, 0)$ (in Voigt notation). By using the matrix-vector multiplication rules, the elastic energy for this case is:

Fig. 9.25 Example of a
stress-strain curve for a
resilient material exhibiting a
hysteresis loop. The grey
area comprised between the
loading and unloading ramps
represents the amount of
work (stored elastic energy)
lost during the cyclic loading

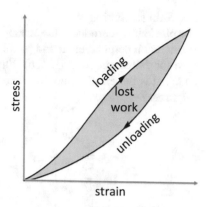

$$E_{el} = \frac{1}{2} \left(\frac{C_{11} + 2C_{12}}{3} \right) \varepsilon_0^2 \qquad (9.48)$$

Note that this is an energy density, i.e. it must be multiplied by the volume element
V, to obtain a proper energy value. Furthermore, if a body can be decomposed into
pieces of elements, each with different elastic constants and with different local
deformations, the overall elastic energy is the sum $E = \sum_i E_{el,i} V_i$, calculated for
each elementary volume V_i.

If a non-linear material is loaded cyclically between two values of strain, it
may display **hysteresis** (Fig. 9.25), i.e. the two ramps of the cyclic curve (load-
ing/unloading) are not equal. This means that a fraction of the work expended to perform the tensile deformation is not recovered when the deforming force is removed.
The lost energy (likely in the form of heat) is given by the area (grey shaded in the
figure) comprised within the hysteresis loop, and its inverse is called the **resilience**
of the material. A material that loses a large fraction of the elastic energy stored
in the cyclic loading is said to have a low resilience, and the opposite is true if the
hysteresis loop is more narrow. A resilient material is one that can efficiently cycle
back and forth in a repeated and sustained deformation, such as a tendon stretched
during the walk or run.

Problems

9.1 Average elastic modulus
Using the known volume fractions from the text, (a) calculate the volume fractions
of mineral and collagen in dry cortical bone. (b) For values of Young's modulus
$E = 54$ GPa for mineral, and 1.25 GPa for collagen, calculate the resulting Young's
modulus of wet and dry bone.

9.2 Skin stretching

The plot below reproduces the stress-strain curve of abdominal skin, stretched along the direction perpendicular and parallel to the body height. (a) What stress is developed for a stretching of 35 %? (b) What strain is developed for stretching to 5 MPa? (c) What is the elastic modulus E in the two principal directions? (d) What is the toughness?

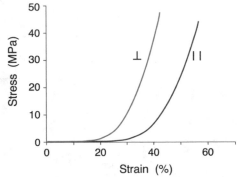

9.3 Arterial stress relaxation

In a mechanical test, a stress of 1 MPa is applied to a 2-cm aorta strip, which as a result is stretched to 2.3 cm. The strain is held constant for an hour, and the stress in the strip drops to 0.75 MPa. Assume that the mechanical properties of the tissue do not change during the experiment. (a) Use the Maxwell model of a viscoelastic material, to obtain the relaxation time of the biomaterial. (b) Calculate the stress in the tissue, if the experiment is continued up to a time of 3 h. (c) The same experiment is performed in a different way, by holding the stress constant at 1 MPa for the same time of 1 h, after which the stress is released. Use the Kelvin-Vogt model to obtain the strain relaxation time, if 1 h 25 min after the release, the strip length is back to 2.2 cm.

9.4 Stretch the leg

Compare the charge on the tendon and the bone in the calf of a man walking in the street. Take $E(\text{bone}) = 20$ GPa, $E(\text{tendon}) = 1.5$ GPa, diameter $\times$ length equal to 1.5×8 cm for the tendon, and 4×35 cm for the tibia.

9.5 Jumping cat

A cat of mass $M = 4.5$ kg jumps on the ground from a height $h = 3$ m. For simplicity, assume that the leg has two equal muscles in each of the upper and lower half, attached to the leg bones as shown in the figure. Each muscle is simulated by a cylinder of average diameter 4 cm and length 12 cm, and their attachment point (*enthesis*) is $a = 1$ cm off-axis; take the Young's modulus of striated muscle $E = 20$ kPa. Leg bones are represented as two thinner, straight cylinders of infinite rigidity, hinged at the knee. Calculate the bending angle of the legs. What does this calculation demonstrates? Compare with your answer to the previous question.

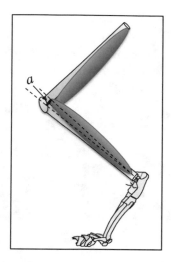

9.6 Muscles and temperature

We measure the relative elongation of an insect's muscle with the help of a *dynamome-ter* (a simple instrument that allows to impose a constant load to a structure and measure the elongation). The muscle can be represented as a homogeneous cylin-der of length L and radius R. From a measurement performed at the temperature $T = 10\,°C$, we obtain a relative elongation $\Delta L/L = +2.\%$ and a contraction in the radial direction $\Delta R/R = -0.25\,\%$; a second measurement at the same load, performed at $T = 15\,°C$, gives an elongation $\Delta L/L = +4.\%$ and a contraction $\Delta R/R = -0.5\,\%$. Again, by measuring at $T = 20\,°C$ we find a $\Delta L/L = +6.\%$ and a $\Delta R/R = -0.75\,\%$.

(a) Which elastic moduli are of interest in this kind of experiment?

(b) Find how the ratio $\Delta V/V$ varies as a function of the temperature.

(c) Predict the values of the relevant elastic moduli and the elongation at $T = 23.5\,°C$.

9.7 Implant materials

Some biocompatible materials for implants are made by a relatively soft matrix, in which a fraction h of harder fibres are dispersed for reinforcement. The overall elastic modulus is calculated on the basis of a "shear-lag" model, which considers that the fibres are too short to be in contact, and cannot share the stress on the material:

$$E = h E_f \left(1 - \frac{\tanh n_s}{n_s} \right) + (1 - h) E_m \quad ; \quad n_s \simeq \sqrt{\frac{2 E_m}{E_f \ln(1/h)}}$$

(the $\simeq$ sign in n_s comes from the assumption of a Poisson's ratio $\nu \sim 0$ for the artificial polymer matrix). Discuss the behaviour of the resulting modulus as a function of h. How does h relate to the ratio E_f/E_m?

9.8 Bend, break or twist

Compare the elastic and strength moduli of the materials in the following table, from each of which a hypothetical stick in form of a full cylinder of diameter 1 cm and length 50 cm is fabricated. Which sticks will bend, break, or twist, under the loads shown in the accompanying drawing?

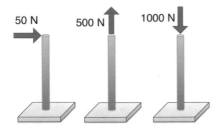

	E (MPa)	σ_r (MPa) tension	σ_r (MPa) compression	G (MPa)
Tendon collagen	2500	100	18*	350
Tooth enamel	60000	35	200	65000
Bone	20000	200	170	4000
Cartilage	20	2.5	12	1.5
Vine green stem	2000	6	20*	180
Oakwood w/grain	6500	170	45	550
Bamboo	10000	190	220	650
Dandelion stem	8	3	6*	0.3*

(*) Values deduced from indirect measurements

References

1. S. Vogel, *Comparative Biomechanics: Life's Physical World* (Princeton University Press, New Jersey, 2013)
2. D.T. Butcher, T. Alliston, V.M. Weaver, A tense situation: forcing tumour progression. Nat. Rev. Cancer **9**, 108–122 (2009)
3. F.H. Silver, J.W. Freeman, D. DeVore, Viscoelastic properties of human skin and processed dermis. Skin Res. Tech. **7**, 18 (2001)
4. R. Shadwick, The structure and mechanical design of rhinoceros dermal armour. Phil. Trans. Royal Soc. B **337**, 447 (1992)
5. R.K. Korhonen, S. Saarakkal, Biomechanics and Modeling of Skeletal Soft Tissues, in *Theoretical BiomechAnics*, ed. by V. Klika (InTech Publishing, Rijeka, 2001)
6. V.C. Mow, S.C. Kuei, W.M. Lai, C.G. Armstrong, Biphasic creep and stress relaxation of articular cartilage in compression: theory and experiments. J. Biomech. Eng. **102**, 73–84 (1980)
7. S. Standring, (ed.), *Gray's Anatomy: The Anatomical Basis of Clinical Practice*, 40th edn. (Churchill Livingstone, New York, 2008)
8. P. Fratzl, Cellulose and collagen: from fibres to tissues. Current Opin. Colloid Interf. Sci. **8**, 32–39 (2003)
9. X.T. Wang, R.F. Ker, Creep rupture of wallaby tail tendons. J. Exp. Biol. **198**, 831 (1995)

10. M. Okuda, N. Ogawa, M. Takeguchi, A. Hashimoto, M. Tagaya, S. Chen, N. Hanagata, T. Ikoma, Minerals and aligned collagen fibrils in tilapia fish scales. Microsc. Microanal. **17**, 788–798 (2011)
11. M. Fang, M.B. Holl, Variation in type I collagen fibril nanomorphology: the significance and origin. BoneKEy Rep. **2**, 394 (2013)
12. E.F. Morgan, G.L. Barnes, T.A. Einhorn, The bone organ system: form and function, in *Osteoporosis*, vol. I, ed. by R. Marcus, D. Feldman, D. Nelson, C.J. Rosen (Academic Press, New York, 2007)
13. M. Fondrk, E. Bahniuk, D.T. David, C. Michaels, Some viscoplastic characteristics of bovine and human cortical bone. J. Biomech. **21**, 623–630 (1988)
14. S. Pal, *Design of Artificial Human Joints and Organs* (Springer Science, New York, 2014)
15. L.J. Gibson, M.F. Ashby, *Cellular Solids: Structures and Properties, Series*, 2nd edn. (Cambridge University Press, Solid St. Sci, 1999)
16. L.J. Gibson, M.F. Ashby, B. Harley, *Cellular Materials in Nature and Medicine*, Solid St. Sci. Series (Cambridge University Press, 2010)

Further Reading

17. D. Klemm, B. Heublein, H.-P. Fink, A. Bohn, Cellulose: fascinating biopolymer and sustainable raw material. Angew. Chemie Int. Ed. **44**, 3358–3393 (2005)
18. M. Meyers, A.M. Lin, Y. Seki, P.-Y. Chen, B. Kad, S. Bodde, Structural biological composites: an overview. JOM **58**, 35–41 (2006)
19. L. Römer, T. Scheibel, The elaborate structure of spider silk. Structure and function of a natural high performance fiber. Prion **2**, 154–161 (2008)
20. D.R. Petersen, J.D. Bronzino (eds.), *Biomechanics: principles and applications* (CRC Press, Boca Raton, 2008)
21. C.M. Altaner, M.C. Jarvis, Polymer interactions of 'molecular Velcro' type in wood under mechanical stress. J. Theor. Biology **253**, 434–445 (2008)
22. H. Nagasawa, The crustacean cuticle: structure, composition and mineralization. Front. Bioscience **4**, 711–720 (2012)
23. J. Sun, B. Bhushan, Hierarchical structure and mechanical properties of nacre: a review. RSC Adv. **2**, 7617–7632 (2012)
24. L. Gibson, The hierarchical structure and mechanics of plant materials. J. Royal Soc. Interface **9**, 2749–2766 (2012)

Chapter 10
Of Limbs, Wings and Fins

Abstract Muscles are the engines of life, necessary for all needs of mechanical actuation of limbs, wings, fins, and any animal body parts. All animals, from invertebrates to the highest vertebrates, have developed for this function a highly specialised fibrous material, characterised by a complex molecular structure, capable of performing contraction and relaxation movements with high rapidity, under an electrical and chemical stimulation. The structure of muscle cells is remarkably conserved across the evolution, to the point that the elementary bricks of any muscle are identical, ranging from an beetle to an elephant. Insects are taken as example of extreme specialisation of muscular functions, and some secrets of their complex flight dynamics are discussed. In the second half of this chapter, dimensional analysis is introduced as a tool of paramount importance. This method allows to check the consistency of a set of variables, and to formulate interesting deductions about animal behaviour, even prior to performing any quantitative measurement.

10.1 Force and Movement Produced by a Muscle

All animals, except the very small unicellular organisms, count on their muscles for their movements, to find and manipulate food, to push blood and breathe fresh air and many other functions, essential to their survival. At the very last, animal muscles are biological engines that burn chemical energy and turn it into mechanical work.

In most books about animal physiology, the function of muscles is classified under the general chapter of **metabolism**, next to other processes such as the thermoregulation. What unifies such processes, to the eyes of a physiologist, is the fact that they consume oxygen and produce heat. As a consequence, the power supplied by a muscle is often described in terms of the limits imposed by the enzymatic chemical reactions, and their capability of making energy available for muscular work. In fact, that is a flagrant example of exchange between cause and effect.

In reality, the rate at which a muscle can deliver mechanical work is limited by three main variables, entirely mechanical in their nature: the force per unit surface the muscle can exert (or *stress*); the amount by which it can shorten (or *deformation*); and its typical contraction *frequency*. The extremal values (max or min) of such variables

© Springer International Publishing Switzerland 2016
F. Cleri, *The Physics of Living Systems*, Undergraduate Lecture
Notes in Physics, DOI 10.1007/978-3-319-30647-6_10

are constrained by purely mechanical limits, inherent to the muscle structure and geometry. Conversely, the enzymatic systems must comply and adapt to such limits: for example, it is the rate at which enzymes supply chemical energy that must adjust to the rate at which the mechanical engine can absorb and utilise it, and not the contrary.

On the one hand, it is true that any muscle can produce a force. However, some muscles are more specialised than others in this being their primary function. It should be noted that the only kind of force a muscle can develop is a traction: muscles can only pull. Differently from the engines built by engineers, muscles cannot push nor turn. Muscles contract actively and extend passively. What this means is that the phase of shortening of a muscle is active, being produced by the muscle itself by a sequence of molecular-scale mechanisms which will be outlined in the following; by contrast, the phase of elongation is passive, in that the muscle needs to be extended by either another muscle (which in this case is called the *antagonist*), or by an adjacent elastic tissue structure.

Figure 10.1 shows a schematic of the multi-level, hierarchical organisation of the striated muscle. While a human skeletal muscle is represented in the figure, this very same organisation is found in the muscles of all animals, including insects. At the **cellular** level, muscle cells (*myocytes*) appear as multi-nuclear units, much elongated and tubular in shape. They can be of various types: cardiomyocytes, found in the heart muscle; skeletal, found in the ordinary muscles; or smooth, found in blood vases and in most organs, like the intestine, stomach, oesophagus etc. While heart and skeletal muscle cells share several similarities, smooth muscle is a type of non-striated muscle, whose contraction is not under conscious control. In general, myocytes are constituted by an ensemble of fibrils with a variable number of nuclei, originating from the fusion (*syncysis*) of the membranes of adjoining pristine cells (*myoblasts*).[1]

At the **subcellular** level, each muscle fibril (*myofibril*) is organised into a linear assembly of many adjacent *sarcomeres*, attached head-to-tail. Sarcomeres are the essential mechanical actuators of the force generated by the muscle cell. This linear structure of sarcomeres is surrounded by *sarcosomes*, the equivalent of mitochondria for the muscle cell. The whole ensemble is integrated by the *sarcoplasmic reticulum*, which also ensures the connection between the muscles and the nervous system.

At the **molecular** scale, each sarcomere is built by a parallel structure of very long, filamentary proteins, combining hundreds of myosin and F-actin molecules (plus other companion proteins) in a tight geometry. The latter are solidly attached to a membrane (the Z-*line*, itself a complex aggregate of other large proteins) running perpendicular to the elongated sarcomere structure, while myosin can slide between F-actins. The sarcomere has a length spanning 1.5–2.5 μm (slightly different in the cardiac muscle), and a diameter of about 2 μm, that is the same as the myofibril. The maximum contraction is obtained when the mobile myosins, parallel sliding relative to the fixed actins, reach the Z-membrane on both sides, and the facing blocks of

---

[1] It may be interesting to note that that cell doubling (*mitosis*) in multinucleate cells can occur either in a synchronous manner, i.e. all nuclei divide simultaneously, or asynchronously, when individual nuclei divide independently, both in time and space.

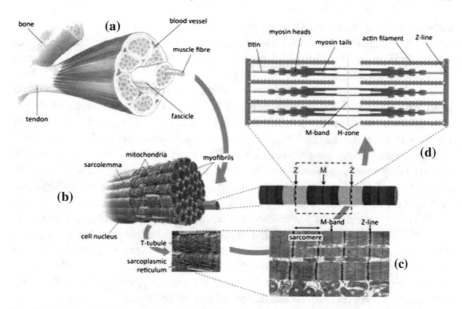

Fig. 10.1 Hierarchical structure of the skeletal muscle. **a** The bulk of the muscle is arranged in fascicles, wrapped in connective tissue. Each fascicle contains tens or hundreds of fibres. **b** Each fibre is a **myocite**, indeed a large polynuclear cell, formed by many individual **myofibrils** (primordial cells that were fused into one), and by mitochondria (**sarcosomes**), all wrapped by the **sarcoplasmic reticulum**, a meshed membrane. The T-tubules bring the neuromotor signal from the nerves into the muscle fibres. **c** Each myofibril, in turn, is constituted by a large number of elementary units, the **sarcomeres**. Each sarcomere (*below* an electron microscopy image) has a fixed size of 2–3 μm, between pairs of Z-lines (*above*), so that each myofibrils is formed by a sequence of tens of thousands such units. **d** At the molecular level, the sarcomere is formed by a parallel structure of tightly interdigitated myosin and actin filamentary proteins, carried by titin filaments held between Z-lines (actually protein membranes). [Image adapted from: a Seer, National Cancer Institute, Bethesda; b www.Blausen.com staff, "Blausen gallery 2014", Wikiversity Journal of Medicine; c Path BioResource, Perelman School of Medicine, University of Pennsylvania; (d) David Richfield, "Medical gallery of David Richfield 2014", Wikiversity Journal of Medicine. With permission, sources public domain, or licensed under CC-BY-SA 3.0/4.0, see (*) for terms.]

F-actin happen to almost touch each other. The sarcomere is maximally shortened in this condition. While sliding, myosins make labile chemical bonds (by means of Van der Waals forces and hydrogen bonding) with the actins, the **cross bridges**. Although smooth muscles do not form regular arrays of thick and thin filaments, like the sarcomeres of striated muscles, contraction is still due to the same sliding filament mechanism controlled by myosin cross bridges interacting with actin filaments.

For a linear chain made of N sarcomeres, each of length l, the total fibril length $L = Nl$ is shortened by a quantity:

$$\Delta L = \sum_{i=1}^{N} \Delta l_i = N \Delta l \qquad (10.1)$$

summed over all the sarcomeres, supposed to undergo an identical contraction (see
Fig. 10.2, right). However, the relative shortening $\Delta L/L = \Delta l/l$ (or *strain*, see
Appendix H) is the same for each sarcomere and for the overall fibril, up to the scale
of the whole muscle.

The amount of force expressed by a muscle is due to the amount of deformation,
from zero to the maximum above defined, and to the density (number/unit length)
of active cross-bridges (Fig. 10.2). According to Huxley (1985), all the myosin filaments have the same length, $l \approx 1.5\,\mu$m, and cross-section density, $5.7 \times 10^{14}\,$m^{-2},
for any muscle of any vertebrate, and even insects. It appears that this one quantity
is remarkably conserved across the evolution, thus demonstrating the optimum efficiency of the fully developed sarcomere structure as a sort of "universal" building
block for the muscle machine.

The two physical variables defined above, the total force and length variation of
a muscle, are indeed simple to define, and offer meaning easy to grasp. However,
they have the big inconvenience of being dependent on the size and the shape of the
muscle. A bigger muscle will exert a bigger total force, and it will shorten by a larger
amount, compared to a smaller muscle. This, for example, will render difficult the
comparison between the performances of animals of widely different size belonging

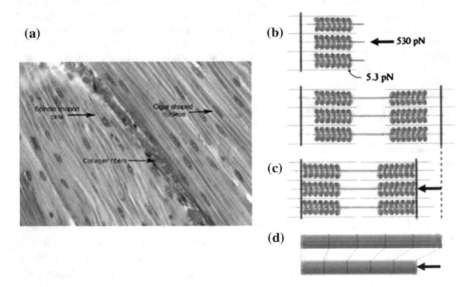

Fig. 10.2 *Left* microphotography of a cross-section of myocytes, showing their polynuclear structure. Biologically, a multinuclear cell of this kind is called a *syncytium*, the result of the fusion of
thousands of primitive mononuclear cells (myoblasts). [Photo © of J. Oros-Montón, Universidad de
Las Palmas, http://www.webs.ulpgc.es/vethistology/.] *Right* Contractile structure of a sarcomere.
b *Orange-red* strings represents myosin filaments, *blue lines* is actin, and *orange grains* are the
"cross bridges", each of them supporting a force of about 5.3 pN. **c** Each sarcomere can shorten
up to the point when myosins touch the vertical Z-lines, and actins are nearly touching each other;
d The total myofibril contraction is the sum of individual sarcomere contractions Δl; however, the
relative deformation $\Delta l/l$ is the same in each sarcomere, and in the whole myofibril

to the same species. In order to study the common properties of the tissue (or material) of the muscle, rather than the specific properties of any given muscle, it is therefore necessary to identify more appropriate variables.

Then, rather than the absolute force, it is more useful to use the concept of mechanical stress, σ, defined by the total force exerted by a muscle with (average) cross section S, and the vector $\mathbf{n}$ perpendicular to the surface (see definitions in Appendix H, and tensor notation in Appendix A) as:

$$\mathbf{F} = \int_S (\underline{\sigma} \otimes \mathbf{n}) dS \tag{10.2}$$

namely, a force per unit area of the transverse surface of the muscle (remember that the product of a rank-2 tensor times a vector gives another vector, i.e. the force).

Since each myosin filament is found to exert a force of about 530 pN (or 5.3×10^{-10} N, such a value being measured under isometric effort[2]), and the density of myosin filaments (see above) is typically of 5.7×10^{14} m^{-2}, the maximum isometric stress exerted by *any* muscle is of the order of 3×10^5 N/m^2, or 300 kPa. Because each sarcomere exerts the same stress, the stress itself is remarkably constant throughout the muscle fibre.

We already mentioned above the strain, or relative elongation, ε, as being the most appropriate variable to measure the length variation of a muscle in a way that is independent on the actual muscle size. Alike to the stress, the strain is another property of the tissue material which does not depend on the size and shape of the muscle. The strain is a 3×3 tensor as well, $\varepsilon_{\alpha\beta} = \Delta L_\alpha / L_\beta$, its components indicating the amount of deformation obtained along some direction $\alpha = x, y, z$ for an elongation/contraction imposed along some direction $\beta = x, y, z$. In practice, the locomotory muscles of most vertebrates can deform by a maximum of $\varepsilon \approx 25\,\%$, this limit being mainly set by the constraints of the supporting bone skeleton.

If we multiply stress and strain, the resulting dimensions are:

$$\sigma \times \varepsilon = [FL^{-2}][LL^{-1}] = [F][L^{-2}] \tag{10.3}$$

a force per unit surface, or else energy per unit volume.

Therefore, the mechanical work produced by a muscle can be obtained by integrating over the whole volume of the tissue, the product of stress exerted by the fibres times the strain accordingly produced (Fig. 10.3):

$$W = \int_V (\underline{\sigma} \otimes \underline{\varepsilon}) dV = \sum_{\alpha\beta} \int_V (\sigma_{\alpha\gamma} \varepsilon_{\delta\beta}) dV \tag{10.4}$$

---

[2]An effort is isometric when the muscle contracts without changing length or joint angle, as for example if holding steady to a weight on the floor.

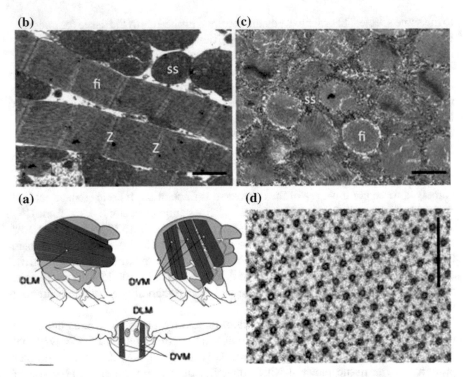

Fig. 10.3 The microscopic structure of flight muscles in the *Drosophila* fruit fly. **a** Schematic arrangement of the dorso-lateral (DLM) and dorso-ventral (DVM) muscles, in the cross-section of the thorax (*above*), and in the frontal full cross-section (*below*). **b** Longitudinal section of a wing muscle by scanning-electron microscopy, showing part of two large myofibrils (muscle cell), appearing as an ensemble of mononuclear cells merged with the neighbouring ones (syncytium). The size of sarcomeres is given by the average distance between two Z-membranes (straight segments), and is (~1.5 μm. Myofibrils (labelled "fi") have an average diameter of 2–2.5 μm. The "ss" are sarcosomes, muscle-cell equivalents of mitochondria; it is here that the pyruvate oxdation (respiration, see Chap. 3) takes place. It can be noticed that mitochondria occupy about half the volume of the myocyte. **c** Transverse section of the muscle fibre, allowing to observe the nearly cylindrical cross section of the myofibrils, as well as the structure of sarcosomes that surround the myofibrils. **d** Higher magnification of the transversal section of a sarcomere, allowing to appreciate the internal structure of myofibrils: each myofibril appears to be made up by hundreds of parallel filaments, which in turn are made of long filamentary proteins (actin and myosin) densely nested in a regular triangular lattice. Scale bars: 2 μm in **b** and **c**, and 0.25 μm in **d**. [Images (**a**) adapted from Ref. [1] under CC-BY-4.0 licence (see (*) for terms); photos (**b–c**) from Ref. [2], repr. w. permission]

(the tensor product implying the sum over the mute indices γ and δ). This quantity can also be expressed as work per unit mass, simply by dividing the result by the average muscle density, $\rho = 1060$ kg/m^3.

10.2   Dynamics of Muscle Contraction

Let us imagine to perform a simple experiment on an isolated muscle, by fixing one extremity and holding the other extremity by a dynamometer. If we load the muscle with a force F, we can measure the velocity v by which it will contract in response to the applied force. We will find that this velocity depends, among other variables, on the muscle length, L. For each muscle, we could find an empirical relationship between force and velocity, summarised in the following equation:

$$v = \frac{b(F_0 - F)}{a + F} \tag{10.5}$$

F_0 being the maximum force measure in (isometric) tension, and a, b two parameters with dimensions respectively of [F] and [L T^{-1}], to be determined for each different muscle by numerically fitting the above equation to the measured data. This equation is known as the *Hill's law*, having been established by the English physiologist Archibald V. Hill in 1938.

However, we would soon get tired of measuring the same quantity over and over for as many muscles as needed, and we would try instead to obtain a more general law expressing the behaviour of a typical muscle under an applied load. The velocity can be divided by the length, to obtain the strain rate $\dot{\varepsilon} = v/L = (dL/L)/dt = d\varepsilon/dt$, another quantity which will not be dependent by any specific muscle type or animal, since it is defined in terms of the strain ε (its first time-derivative). It is worth noticing that the strain rate has dimensions of [T^{-1}], just like a frequency.

Now the Hill's law can be recast in the form of a "universal" relationship between strain rate and stress, still having the same structure of the previous Eq. (10.5):

$$\dot{\varepsilon} = \frac{\beta(\sigma_0 - \sigma)}{\alpha + \sigma} \tag{10.6}$$

where σ_0 is the maximum isometric stress the muscle fibre can sustain up to the yielding point (300 kPa, see above), and the two "universal" constants α, β have now dimensions of [F L^{-2}] and [T^{-1}], respectively.

Hill's law is just an empirical description of the muscle dynamics, under highly idealised (and even artificial) conditions. It imagines the muscle as a homogeneous mechanical device, which shortens against a rigid external structure maintaining a constant force. Over the years, several criticisms have been raised about the universal validity of this equation, which however gives a quite good description, albeit admittedly approximate, of the fast muscle dynamics during animal locomotion.

Going to the limit at which the muscle contracts under a vanishing resistance (set $\sigma = 0$ in Eq. (10.7)), the deformation rate attains its maximum:

$$\dot{\varepsilon}_0 = \frac{\beta\sigma_0}{\alpha} \tag{10.7}$$

This number is an important characteristic value for any muscle. It was called **intrinsic rapidity** by Hill, even if its dimensions are not those of a velocity, but those of a frequency, $[T^{-1}]$. This property has a molecular-scale origin, since it is related to the microscopic rate at which the cross bridges of myosin can attach and detach from/to the actin filaments. On the basis of the relative values of $\dot{\varepsilon}_0$, a muscle can be classified as being 'rapid' or 'slow'.

10.3 Mechanical Efficiency and Cyclic Contraction

A muscle is an energy converter: it takes a given quantity of free energy (chemical energy plus entropy) at the microscopic scale, and turns it into mechanical energy (plus some heat) at the macroscopic scale. The efficiency, $0 < \eta < 1$, of this transformation is given by the ratio between the resulting work done by the muscle and the apport of free energy.

To give a proper estimate of the efficiency, it must be noted that the energy conversion in the muscle takes place at least in the two following steps:

1. Initially, a combustible from the substrate (fat, glycogen, protein) is oxidised, with the result of turning some molecules of ADP into ATP by the addition of a phosphate group (we saw in Chap. 4 that this conversion already has its own intrinsic efficiency below 1, the amount of energy being extracted as ATP being about 1/3 of the total formation enthalpy of the combustibles);
2. Secondly, the myosin/actin complexes use this ATP and retransform it into ADP, by using a fraction η of the energy available (the remaining fraction $(1 - \eta)$ being lost as heat).

When we speak of muscle efficiency we refer to the η of this second step, the losses of the first step being already accounted in the (aerobic) metabolic phase of energy conversion from the primary sources.[3]

Huxley described the microscopic aspects of the second step in terms of the rates of attachment and detachment of the myosin and actin filaments [3], and produced a diagram of the efficiency as a function of the deformation rate $\dot{\varepsilon}$ for various muscles. That study showed that the muscle efficiency is always optimal for a value $\dot{\varepsilon} \approx 0.13\dot{\varepsilon}_0$. There is no current explanation for this result, which must be therefore taken as a purely empirical result. We will come back to this point later, when discussing the optimisation of the muscle rapidity for different muscle functions.

---

[3] We limit this description to the aerobic metabolism since, as anticipated in Chap. 4, the chemical thermodynamics of the anaerobic cycles is quite more involved.

Some muscles, called **tonic**, are rather designed to maintain a constant tension, without contracting (and thus, without changing their length, $\Delta L = 0$). A muscle like this could seem a paradox, since the work, W, is the product of the force developed by the muscle times the displacement ΔL, and such a muscle would actually consume energy without doing any work. What this apparent paradox indicates, instead, is just that the efficiency η may not always be the best parameter to characterise muscle activity. As any other muscle, a tonic muscle develops a force by consuming chemical energy at a certain rate. The input power has dimensions of $[M][L^2][T^{-3}]$, the force is $[M][L][T^{-2}]$, therefore their ratio is $[L][T^{-1}]$: a velocity. This ratio of power consumed versus force developed can be taken as an alternative measure of the energy cost, χ, necessary to maintain the constant level of force expressed by such a muscle. If, as it is now usual, we wish to express this energy cost in terms which do not depend on a particular muscle length or shape, we should better take the ratio of the specific power (or power per unit volume), which has dimensions of $[M][L^{-1}][T^{-3}]$, to the stress, $[M][L^{-1}][T^{-2}]$:

$$\chi = \frac{P/V}{F/S} = \frac{p}{\sigma} \tag{10.8}$$

This ratio is now a rate, or a frequency, $[T^{-1}]$, just like the intrinsic rapidity $\dot{\varepsilon}$. In physiology it is empirically well known that slow muscles can maintain a given level of force with an energy cost lower than rapid muscles. Since the variable which determines the rapidity of a muscle has the same dimensions of the energy cost just defined, it could be tempting to identify the two variables, to describe skeletal and tonic muscles on the same level. By correlating experimental data sets, it is found that there exists a simple relationship between the two parameters:

$$\chi = \frac{\dot{\varepsilon}_0}{16} \tag{10.9}$$

Albeit empirical, such a relationship is very useful since it allows to obtain the amount energy necessary to maintain a given level of stress, by means of simple mechanical measurements, in a way which is independent on the actual size and type of muscle. For example, to find the amount of energy necessary to maintain the tension of a biceps at a given level, we have just to obtain a value for the maximum contraction speed (in m/s), divide this number by the macroscopic length of the biceps (in m), which would give us the $\dot{\varepsilon}_0$ of the biceps, and finally divide it by 16, to get the rate of energy consumption per m^3 of muscle per Pa of stress to hold. The result would be in units of $W\ Pa^{-1}\ m^{-3}$ (in fact, again s^{-1}).

10.3.1 Cyclic Contraction

Locomotory muscles are typically arranged in antagonist pairs: while one muscle of the pair contracts, it also relaxes the other, and vice versa. Therefore, when we want to estimate the rate at which a muscle consumes energy we must average between such active and passive phases of the couple. If we draw a diagram of the total deformation (elongation + retraction) of the muscle as a function of the variation of the stress, we would obtain a kind of ellipsoid, which represents a cycle of contraction and relaxation. Such a kind of closed-loop diagram is actually observed during experiments realised on living animals. In an interesting series of experiments by Boettiger [4], one extremity of the muscle of the wing of an insect was attached to a force transducer (like a minuscule spring of calibrated strength); by forcing the other extremity to contract under a load, he measured the length variations. By choosing appropriate values of load, it was possible to voce the muscle in a sequence of cyclic contractions. By sending the signals of the force-position transducers onto the X-Y entries of an oscilloscope, the ellipsoidal figure was exactly observed.

It can be simply shown that the mechanical work in one cycle is given by the integral of the area A of the ellipsoid (note that when the elongation becomes negative, i.e. contraction, also the force changes sign, therefore the contributions to the integral have always the same sign):

$$W = \oint F dl = \int_A \left(\frac{dF}{dl} \right) dA \qquad (10.10)$$

The power exploited by the muscle (energy per unit time) is the product of the work W times the contraction frequency, ν. If we go to size-independent variables, strain ($\varepsilon = \Delta L / L$) and stress σ, we get again an integral for the specific work, w, but now independent on the size and shape of the muscle:

$$w = \int_A \sigma d\varepsilon \qquad (10.11)$$

During the cycle of contraction/relaxation the length of the muscle oscillates between $L(1 - \varepsilon/2)$ and $L(1 + \varepsilon/2)$, while the stress varies between $\pm \sigma/2$. Since the work is a thermodynamic function of state, i.e. it only depends on the endpoints of a cycle and not on the way the system goes between such endpoints, we can replace the ellipsoid by a rectangle of equivalent area, comprised between $\varepsilon = (+\varepsilon/2, -\varepsilon/2)$ on the x-axis, and $\sigma = (+\sigma/2, -\sigma/2)$ on the y-axis. Therefore, we can now write the power as $P = (\sigma\varepsilon)\nu$, and the specific power (power per unit mass) as:

$$p = \frac{\sigma\varepsilon\nu}{\rho} \qquad (10.12)$$

10.4 Optimised Muscles

The Hill's equation (10.7) can be inverted to express the strain rate as the independent variable:

$$\sigma = \frac{\beta\sigma_0 - \alpha\dot{\varepsilon}}{\beta + \dot{\varepsilon}} \tag{10.13}$$

Such a writing means that the external load σ imposes a given strain rate $\dot{\varepsilon}$, as a function of which the Hill's equation allows to determine, in turn, the stress exerted by the muscle. If we take for the sake of simplicity that both the strain and stress are homogeneous during a contraction cycle, $\varepsilon\nu = \varepsilon/\tau = d\varepsilon/dt$, the power per unit volume of the muscle contraction is equal to the work per unit time, Eq. (10.12), or:

$$P = \sigma\dot{\varepsilon} \tag{10.14}$$

Since each muscle of an antagonist pair works only for one half of each cycle, only $P/2$ of the total power is due to each muscle. Therefore, by using the above Eq. (10.13) for σ, we can write:

$$P = \frac{\sigma\dot{\varepsilon}}{2} = \frac{\beta\sigma_0 - \alpha\dot{\varepsilon}}{2(\beta + \dot{\varepsilon})}\dot{\varepsilon} \tag{10.15}$$

The strain rate varies in the interval $0 < \dot{\varepsilon} < \dot{\varepsilon}_0$. If we divide the previous equation by $\dot{\varepsilon}_0$, we can introduce the new variable $q = \dot{\varepsilon}/\dot{\varepsilon}_0$, a relative rate which now varies between 0 and 1. The power is rewritten as:

$$P = \frac{(q - q^2)}{2(Kq + 1)}\sigma_0\dot{\varepsilon}_0 \tag{10.16}$$

with $K = \sigma_0/\alpha \approx 5$, valid for a majority of all the vertebrate muscles.

Now we can trace a family of curves of P versus q, for an ideal set of muscles all having the same value of σ_0 (therefore, muscles capable of supporting the same imposed charge), but having different values of $\dot{\varepsilon}_0$ (i.e., ranging from slow to rapid muscles). A plot like the one represented in Fig. 10.4a is obtained. The maximum strain rate, or intrinsic rapidity $\dot{\varepsilon}_0$, corresponds to $q = 1$. For values of $0 < q < 1$, the absolute values of $\dot{\varepsilon}$ for different curves are always in the same ratio as the corresponding $\dot{\varepsilon}_0$, and so are the corresponding values of P, since the stress σ_0 is the same for all the muscles represented in the plot. The four dots in the plot show what the result would be, if we imagine to be capable of building a bionic animal by choosing muscles of different rapidity for the same function, namely to withstand a given stress σ_0.

It may appear at first sight that the available power is simply proportional to the intrinsic rapidity $\dot{\varepsilon}_0$: if this were the case, a more powerful muscle would just be a faster one. However, if we look more attentively at this figure, we see that things are somewhat different. The four dots correspond to the same *absolute* value of $\dot{\varepsilon}$. For

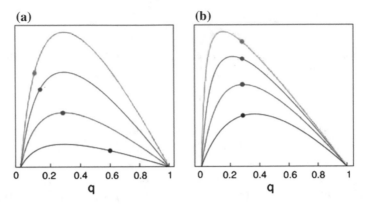

Fig. 10.4 **a** Plot of the theoretical power output (Eq. 10.16) for four muscles of increasing rapidity $\dot{\varepsilon}_0$ (small for the *black*, medium for the *red* and *blue*, maximum for the *green curve*), working on a given load σ_0. **b** Plot of the theoretical power output for a same muscle of given $\dot{\varepsilon}_0$, working on four loads of increasing σ_0 (from *black to green*). In both plots, the *four coloured dots* correspond to a same value of $\dot{\varepsilon}$

the slowest muscle, the black dot corresponds to $q \approx 0.6$, i.e. 60 % of its maximum rapidity; whereas, the red dot corresponds only to 25 % of the maximum rapidity of that muscle, and even smaller values for the other two muscles, marked by the blue and green dots. One can see that going from a muscle of rapidity 'blue' to a muscle 'green', the gain in power (intercept of the point at the y-axis) is much smaller than in going from 'black' to 'red'. This means that getting a too fast muscle becomes quickly less effective, since its *absolute* strain becomes so small that it would do a nearly isometric effort. A slower muscle, on the other hand, can output nearly the same level of power, but with a higher efficiency.

The plot in Fig. 10.4b represents again four different curves, this time for a muscle of given rapidity $\dot{\varepsilon}_0$ but working at four different levels of stress. The four dots mark a given value of strain rate $\dot{\varepsilon} = 0.3$ (the same for all four curves, since the muscle is the same). It can be seen that the optimum efficiency (maximum in the power output, red curve) corresponds to a well defined combination of $\dot{\varepsilon}_0$ and σ_0: that same muscle operated at a given contraction on different loads σ_0 (black, blue, green curves) becomes less and less efficient.

Therefore, from this very simplified, yet not too unrealistic analysis, we can conclude that the rapidity of each muscle would be adapted to its specific function: a muscle which is too rapid for a given load σ_0 would operate in a nearly-isometric condition (imagine to try lifting a 20 kg weight with one finger), and would dissipate most of the power output into heat; on the other hand, a too slow muscle for a given load would contract in a time too long compared to the necessary frequency (imagine playing a piano keyboard with your elbows), and would not give much useful work. In practice, there may be cases in which the evolution has very finely tuned the muscle efficiency, allowing the animal to adopt a quite wide range of behaviours. For example, some monkeys can use their fingers to remain suspended to the branch of a tree, therefore the $\dot{\varepsilon}_0$ of their finger flexor muscles is definitely smaller than for humans; however, not so small to hamper their hands to perform quite accurate and fast operations.

It is important to note that the value of $\dot{\varepsilon}$ increases with the environment temperature. Such a consideration might inspire the deduction that this would be one good reason to explain the fact that homeothermic animals try to keep their bodily temperature quite above the ambient value, in order to maximise the efficiency of their muscles. On the other hand, there are other evidences that appear to contradict this deduction: for example, a large bird like an eagle will use generally slower muscles compared to a colibri, but their body temperature is nearly the same; furthermore, fishes adjust their muscles for a constant water temperature, which is the main reason why cold-water fishes cannot easily adapt to tropical seas. In fact, the reason why homeothermic animals prefer a higher bodily temperature is different, and will become clear in the next Section.

The value of $\dot{\varepsilon}_0$ must be optimised also for the tonic muscles, even if these do not actuate a direct displacement. A complication in this case arises, since there is no optimal value for the intrinsic rapidity, due to the fact that a tonic muscle exists to maintain a given value of constant stress over some time, rather than during a cyclic effort. In most situations, an animal would use only the minimum necessary power consumption, therefore a slower muscle seems able to maintain a constant stress better than a rapid one. At the very limit, a tendon would be equivalent to an infinitely slow muscle, while its length can be (elastically) adjusted by only a very small amount. The optimal value of $\dot{\varepsilon}_0$ for tonic muscles should be adapted in order for the animal body to follow the adjustment between two different stable positions, but not faster than necessary.

10.4.1 Aerobic and Anaerobic Muscles

As it was already discussed in Chap. 4, aerobic muscles produce work by performing the oxidation of the combustible (fat, sugar) in the mitochondria, or sarcosomes, which occupy a large part of the muscle fibre. To complete such a task, an obvious limit resides in the physical ability for the oxygen to be transported into the fibre body. The volume taken up by sarcosomes in the fibre is a function of the rate at which energy is supplied: therefore, albeit indirectly, the volume fraction of the sarcosomes is a function of the specific power (per unit mass) of the muscle. In the muscles capables of an elevated specific power output, such as the wings of colibri and insects, which beat at very high frequency, the sarcosomes can occupy up to about half of the total cell volume. As a consequence, the volume available for the motor proteins (actin, myosin) is smaller, and the maximum stress attainable is reduced. Therefore, in aerobic muscles the power is not a simple linear function of the frequency, as Eq. (10.12) above seems to suggest.

A simple analysis rather shows that the power output of an aerobic muscle is asymptotically limited by the rate of energy supply from the mitochondria, instead of the contraction frequency. In 1984, Pennycuick and Rezende proposed the following equation for the specific power output as a function of the contraction frequency [5], for a given value of stress and strain (in other words, the power output for a given amount of work performed at a given frequency):

$$P = \frac{\sigma \varepsilon \nu}{\rho(1 + k\sigma\varepsilon\nu)} \tag{10.17}$$

where ρ is the muscle density, appearing at the denominator in order to express the power per unit mass instead than per unit volume. The interesting factor in this equation is the coefficient k, expressing the ratio between the volume fraction of the muscle cell occupied by the mitochondria, and the volume fraction of myofibrils; a value $k = 1$ means a 50/50 share of the volume.

On the other hand, anaerobic muscles perform only a very partial oxidation of the combustible, and they do not require at all oxygen transport. In this case the sarcosome volume is not a relevant variable, and the Eq. (10.12) is applicable. For a given contraction frequency, anaerobic muscles produce a power output higher than aerobic ones. Moreover, this advantage increases as the contraction frequency increases, since the aerobic power output tends to grow slower than ν, until becoming constant and independent of frequency. Anaerobic muscles are in this respect very useful, since they can produce a very quick contraction starting from a resting position, albeit with a decreased efficiency: such muscles would be useful for brisk and steep accelerations and sudden jumps, while aerobic muscles would be more useful for steady energy consumption, as required in long-range locomotion, flight or swimming. For example, a small bird can use preferably anaerobic muscle for the take-off phase, and switch to aerobic muscles for the cruise phase.

It may be interesting to note that the largest mitochondria are found in the (aerobic) cardiac muscle fibres: they are 3–9 times the size of those found in the skeletal muscle fibres. Cardiac mitochondria have the added capability of oxidising lactic acid back into pyruvic acid and pyruvate back into glucose. (The only other organ which contains such very large mitochondria is the liver.) These mitochondria are purplish in colour, and the presence of large numbers of these mitochondria gives the heart and liver tissues their dark purple-brownish colour. Aerobic muscles may be more or less rapid (called *fast-twitch* or *slow-twitch*). The aerobic fast-twitch fibre is not found in primates, and is really no longer a muscle, but is instead a bag full of tiny mitochondria with just a few contractile fibres remaining. The little mitochondria in this fibre are 1/3rd the size of those in the aerobic slow-twitch fibre, and are not able to oxidise fatty acids or ketones as can the larger mitochondria, but can oxidise only the components of glucose. These smallest mitochondria appear bright red in colour like the myoglobin which accompanies them. On the other hand, the anaerobic fast-twitch muscle fibre contains larger mitochondria that produce the enzymes needed to reduce glucose to pyruvate, and pyruvate to lactate.

In practice, many muscles contain a combination of aerobic and anaerobic fibres running next to each other, easily distinguishable by their colour. Aerobic fibres range from pink to dark red, depending on the fraction of mitochondria (whose red colour comes from the cytochrome, and from some myoglobin they contain, necessary to maintain the required partial pressure of oxygen); anaerobic fibres, on the other hand, have very small mitochondria and contain no myoglobin, their colour being thus very clear and often white.

An interesting quantity in the description of muscle operation is the respiratory quotient (RQ), defined as the ratio between eliminated CO_2 and consumed O_2 during respiration, which can be measured by a special apparatus called the respirometer. For the complete oxidation of a compound with formula $C_xH_yO_z$ (e.g., glucose with $x = 2y = z = 6$, fatty acids with $z = 2$ and $y = 2x$, etc.), the chemical equation is:

$$C_xH_yO_z + (x + y/4 - z/2)O_2 \rightarrow xCO_2 + (y/2)H_2O \qquad (10.18)$$

Hence, the complete metabolism of this compound has a respiratory quotient of:

$$RC = \frac{x}{x + y/4 - z/2} \qquad (10.19)$$

which gives $RQ = 1$ for pure carbohydrate metabolism, and $RQ = 0.67$ for pure fat metabolism (taking $x, y >> z$). Experimentally, when a runner attains a speed of about 14 km per h, $RQ \approx 1$, which indicates no fat metabolism is happening. Higher speeds in a primate (including the human runner) can be produced only by the anaerobic fast-twitch fibres, which can contract three times faster than slow-twitch fibres (25 vs. 75 ms). The fast-twitch fibres can produce a speed in excess of 25 miles per h, such as it is attained in the 100- or 200-m dashes. For the same reasoning, non-primate animals that do not dispose of a 50/50 mix of aerobic slow-twitch and anaerobic fast-twitch fibres, require aerobic fast-twitch fibres (essentially bags of mitochondria) capable of oxidizinglarge amounts of pyruvate from the anaerobic fibres.

The flight muscles of a bird are of necessity mostly all fast-twitch fibres. Photomicrographs show that out of a sample of 30 fibres, 18 are anaerobic fast-twitch, with the anaerobic fibres being five to nine times larger than the aerobic fibres. The reverse situation exists, e.g., in the calf muscles of the cat. The *soleus* muscle, smaller and closer to the bone, is made up entirely of aerobic slow-twitch fibres. This allows the cat to move with incredibly smooth slow motion when in stealth mode. To provide the quick leap in bounce mode, the other calf muscle, the larger *gastrocnemius* covering the soleus, is mostly made of fast-twitch fibres. A sample of 30 cat gastrocnemius fibres revealed seven aerobic slow-twitch fibres, 17 anaerobic fast-twitch fibres, and six aerobic fast-twitch fibres. A memory of this evolutionary step is also present in humans, whose soleus is slightly richer in slow-twitch fibres, while the gastrocnemius is slightly richer in fast-twitch ones, however remaining close to a 50/50 mix.

The rate at which energy is supplied by the mitochondria is a remarkable function of the temperature, to the point that the higher is the body temperature, the smaller becomes the volume fraction of a muscle cell dedicated to the mitochondria. This fact provides an evolutionary advantage for animals which tend to maintain a bodily temperature quite higher than the environment. Birds and mammals have body temperatures comprised between 32 and 43 °C, generally in anti-correlation with body mass (smaller animals tending to have higher temperature) [6], apart from a strange exception represented by marsupials, in which body temperature increases with body mass (and this is just another strange thing about marsupials). It is easy to observe that cold-blooded insects must firstly "warm-up" their muscles before taking off, while in flight their body temperature attains the same 37 °C as the human body's. For fishes it is more complicate to maintain a high temperature of the body, since their contact surface with water is quite extended, and the thermal conductivity of water is much larger than that of air. Therefore, they rather tend to warm up only those local parts of the body where aerobic metabolism takes place. In this way fishes can reduce the volume of mitochondria, and thereby the volume of the muscles, in their attempt to constantly optimise their overall body shape to preserve nicely hydrodynamic contours.

10.5 The Flight of an Insect

Insects are extremely fascinating animals in many respects, and most notably for their special flight abilities. In fact, insects are at the low end of the scale in terms of Reynolds number, Re, expressing the (nondimensional) ratio of inertial forces to viscous forces for an object of given shape and size L moving at a speed v in a fluid:

$$Re = \frac{\rho v L}{\eta} \tag{10.20}$$

with ρ and η the density and dynamic viscosity of the fluid, respectively ($\rho = 1.2$ kg m^{-3} and $\eta = 1 \times 10^{-3}$ kg m^{-1}s^{-1} for air at $T = 293$ K). A small fruit fly with $L = 0.3$ cm and $v \approx 1.5$ m/s has $Re = 5$; a green beetle with $L = 3$ cm and $v \approx 0.4$ m/s has $Re = 140$; a larger and fast dragonfly with $L = 7$ cm and a peak horizontal speed $v \approx 12$ m/s has $Re = 1{,}000$. For comparison, medium-sized birds have $Re \approx 10^5$, and an Airbus-380 has $Re \approx 10^7$. Under such conditions, air flow around the insect body is too viscous for true turbulence (Re should be larger than 10,000), but it is sufficiently inertial to sustain local vortices. Skin friction is a major component of the drag of a body, so that streamlining is of questionable value, in fact insects did not seem to have taken particular care in evolving particularly aerodynamic shapes. Moreover, velocity gradients during insect flight are gentle (again, with exception for dragonflies, which are capable of very abrupt changes both in speed and direction), a factor which further reduces the influence of the shape, orientation, and surface details of the flying object on its aerodynamic characteristics.

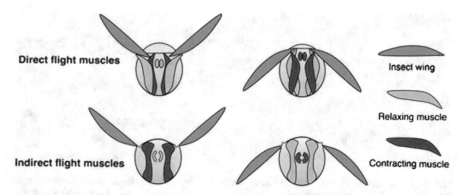

Fig. 10.5 Mechanics of direct and indirect wing muscles. *Above* In direct muscles, wings pivot up and down around a hinge, being raised by the contraction of muscles attached to the base of the wing inside the pivot point. The downstroke results by a contraction of muscles that attach to the wing outside of the pivot point. *Below* indirect flight muscles are connected to the *upper* and *lower* surfaces of the thorax. A second set of muscles attach to the front and back of the thorax. The wings are raised and lowered by the muscles attached to the top and bottom surface of the thorax. The top surface of the thorax moves *up* and *down* and, along with it, the base of the wings.

The wings of the insects usually cover a large surface compared to the size of their body, up to tens of times larger in the case of butterflies.[4] With the exception of *dyptera* (flies, mosquitoes) which have only one pair, most insects and beetles have two pairs of wings, sometimes identical as in damselflies, but often of different size and shape. Wing movement is actioned by either **direct** muscles, attaching directly to the wing base, or by **indirect** muscles, which deform the insect thorax and transmit the up- and downstroke to the wings (Fig. 10.5). Wings are attached to the thorax by a complex hinge joint that gives the wing freedom to move up and down through an arc of more than 120°. With indirect muscles, a pair running vertically provides the thorax contraction for the wings upstroke, while a pair running approximately parallel to the insect body provides thorax relaxation for the downstroke. The hinge is a **bistable oscillator**, like a pendulum having two stable positions instead of just one, which stops moving only when the wing is completely up or completely down. During flight, the wing literally "snaps" from one position to the other.

All insects can use direct muscles to control their wings during flight, however most of them use indirect muscles to perform the power stroke. Only *Palaeoptera* (dragonflies and damselflies, plus some species of *Ephemeroptera*, or mayflies),are

---

[4]The honeybee is a notable exception, its wings being quite shorter than the whole insect length. In 1934, the French entomologist August Magnan and his assistant André Sainte-Lague calculated that bee flight was aerodynamically impossible, since the too short wings could not provide enough lift according to conventional aerodynamics calculations. According to recent studies, the bee's flight ability is the result of an unconventional combination of short, choppy wing strokes, a rapid rotation of the wing as it flops over and reverses direction, and a very fast wing-beat frequency, up to 240 Hz for hovering.

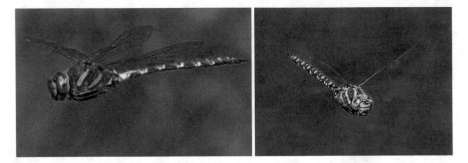

Fig. 10.6 *Left Aeshna grandis (Brown hawker dragonfly)* in fast cruise flight. *Right Aeshna juncea* (Moorland hawker) in hovering flight, with ample use of wing rotation compared to the fast flight. [Photo © **a** Roy & Marie Battell, http://www.moorhen.me.uk/; **b** Jens Buurgaard Nielsen, under CC-BY-SA-3.0 licence, see (*) for terms.]

known to use direct muscles also for the power stroke. Such a choice is apparently less efficient for the neurophysiology, since each wing has to be controlled and acted separately, with attending problems of synchronisation, while the thorax deformation by indirect muscles induces a coordinated wing movement with less neurons circuitry. However, this solution is extremely advantageous for the dragonflies, the most powerful predators in the insect world, in that they can hover, manoeuvre at low speed, and abruptly change direction and speed of flight (Fig. 10.6), by using a complex pattern of anti-phase wing flaps between the fore and hind pair, including rotation and bending [7]. Direct muscles attach to the membrane at the base of wings by resilin, an elastomeric protein known to be the most efficient energy-storage molecule: the elastic efficiency of the resilin isolated from locust tendon has been reported to be 97 %, with only 3 % of the stored elastic energy being lost as heat.

10.5.1 Synchronous and Asynchronous Muscles

During the evolution, all animals (and most notably the insects) have developed a variety of different strategies to accomplish the need of having muscles capable of working at very high contraction frequencies. We can distinguish basically between:

- *quantitative* solutions, consisting in the acceleration of each phase of the cycle of muscular work in a mostly coherent manner;
- *qualitative* solutions, in which some steps in the cycle (typically, the most time consuming) are either suppressed or grouped together with others.

The quantitative solutions are accomplished by the development of **synchronous** muscles, characterised by the fact that each excitatory nerve pulse gives rise to one contraction cycle of the muscle cells; in other words, the nerve excitation frequency is the same of the muscle contraction frequency, ν.

The quantitative solutions are, instead, implemented by **asynchronous** muscles, in which a single nerve excitatory pulse can give rise to several (from 4 to 5 up to about 10) sequential muscular contractions. It is worth noting that the qualitative solution is increasingly adopted when very high contraction frequencies are required: typically, vertebrates use synchronous muscles, while almost all the insects have developed asynchronous muscles.

As it was explained in the Fig. 7.4 of Chap. 7, the electrical nerve impulse which commands the voluntary contraction of a muscle, is transmitted from the nerve to the muscle by a network of junctions, the *neuromuscular synapses*. The electric potential of the neural cell carries the information from the brain down to the muscle cells. At the microscopic scale, the mechanical action of the contraction is activated by the influx of Ca^{2+} ions, which actually provide the initial 'loading' of the actin-myosin motor and allow the formation of actin-myosin cross-bridges (see also the previous Fig. 10.2b).

The double flux of Ca^{2+} ions, to the nerve terminal and, after liberation of acetyl-choline, to the muscle terminal, is a diffusion-driven process: the ions proceed by a random walk through the cytoplasm, and must cover by stochastic motion the few nanometers from one end to the other, with moreover all the intermediate steps described. As it turns out, this diffusion process is the slowest step (*rate-limiting*) in the whole chain of transmission.

Now, we could consider that a diffusion coefficient, D, has dimensions of $[L^2][T^{-1}]$, that is a product between a surface and a frequency. We can identify the surface with the area to be traversed by the ions, in the direction of the junction from the nerve to the muscle; and the frequency with that of the nerve impulse. Then, for a given nerve pulse frequency, the way to increase the diffusion, and therefore to speed up the process, would be to increase the contact surface available to the ions. In other words, one should let increase the amount of contacts at the level of neuromuscular junctions. This would imply that the muscle fibre is more and more invaded by the sarcoplasmic reticulum, which forms the junctions, thus leaving less and less space for the myofibrils. Such a process indeed occurs at the neuromuscular junction (see the membrane folds in Fig. 7.4), but it could not continue indefinitely. The equilibrium between the amount of contact surface, and volume available for the mechanical action, would impose an upper limit to the maximum attainable frequency, or instead a reduction in the available power. Such a hypertrophy of the contact regions, together with a reduction in size of the myofibrils, is in fact observed in some high-frequency synchronous muscles, such as the sound-producing muscles of rattlesnakes, fishes, and some insects. However, for attaining very high frequencies this would not be the best evolutionary strategy.

The solution nature has devised with the development of asynchronous muscles, is that of **decoupling** the frequency of arrival of the nerve pulses (and therefore the diffusion cycle of Ca^{2+}) from the frequency (often much higher) of muscular contraction: each nerve pulse can induce many more muscular contractions, rather than just one as it is the case in synchronous muscles. This kind of adaptation is found only in insects, to the point that about 80 % of the flying insects (beetles, flies, bees) do so with asynchronous muscle. Asynchronous muscles are likely to be more powerful, on a per-volume basis, than synchronous ones, because they do not have to invest heavily in sarcoplasmic reticulum to achieve high operating frequencies. The reduction in the volume of sarcoplasmic reticulum increases the space in a given mass of muscle for myofibrils. Moreover, asynchronous muscles are likely to be more efficient than synchronous ones because they do not have to cycle Ca^{2+} on each contraction.

10.5.2   The Power Output of an Insect's Muscle

A striking evidence of the electromechanical functioning of asynchronous muscles was provided in a series of studies by Josephson and coworkers [8, 9]. The force, mechanical work, and power output produced by the wing muscles of living insects were measured under well-controlled conditions. Measurements were realised on dissected muscles of dead animals, placed in a system of micro dynamometers allowing to impose a constant load, and to measure the force feedback (therefore, the force exerted by the muscle fibre) as a function of the mechanical stimulation (cyclic variation of ε), of the temperature (to be interpreted as the animal body temperature), and other variables. The electrical stimulus from the nerve was simulated by the introduction of thin silver electrodes, 100 μm in diameter, connected to a sinusoidal voltage pulse generator. The same studies reported also measurements realised on living insects, flapping their wings while immobilised.

Likely, the most direct proof of the decoupling between nerve impulse frequency and muscle cycling frequency is provided by the data reported in the Fig. 10.7, from Ref. [9]. Here the recordings for the beetle are compared to those for a locust, permitting to clearly appreciate the difference between the synchronous actuation in the locust, versus the asynchronous actuation in the smaller beetle. In each pair of time traces, the trace below is the electromyography (EMG) recording of a train of nerve impulses, with average height of 3–5 mV and frequency of 8 Hz in the beetle and about 16 Hz in the locust (see the scale bar in the figure indicating a period of 100 ms). The time trace above in each pair is the lift generated by the wing muscle,

Fig. 10.7 Lift Generated by wing strokes (*upper trace in each pair*) and muscle action potentials recorded from a muscle (EMG; *lower trace*) during tethered flight. *The upper pair* is from a locust (*Schistocerca americana*), the lower from a beetle (*Cotinus mutabilis*); respective wing beat frequencies are 16 and 77 Hz. [From Ref. [9], repr. w. permission]

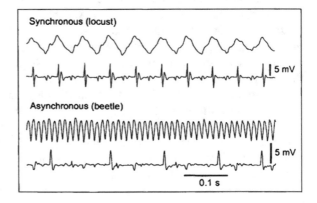

measured by the dynamometer. In the locust the wing frequency is the same as that of the nerve in the EMG; in the beetle, the wing is seen to oscillate with a much higher frequency, about 10 cycles being produced for each neural pulse. The data published in Ref. [8], in a similar set of experiments, reported a wingbeat of about 5–6 cycles in the same type of beetle. It was also measured the absolute value of lift force (from the maximum amplitude of the wingbeat trace), equal to about ±4 times the insect body weight.

The following Fig. 10.8 reports a summary of the experiments. In (a) the force output is compared with the applied strain. During this time, the muscle is stimulated by the electrodes, at a very high frequency (100 Hz), to induce tetanic contraction. When the strain is applied, in two successive steps of ±2 %, the muscle develops a

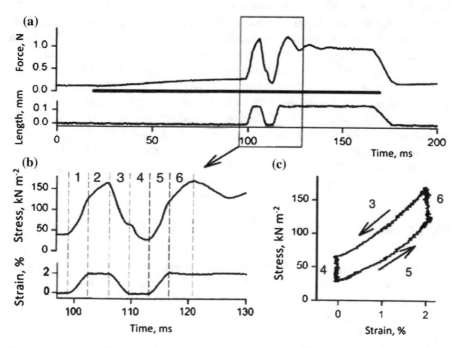

Fig. 10.8 Stretch (activation), shortening (deactivation), and work output from a beetle flight muscle. **a** The muscle was stimulated with shocks at 100 Hz and, during the tetanic contraction, subjected to two cycles of stretch-hold-release. *Upper curve* Time-trace of the force developed by the muscle. *Lower trace* Imposed elongation of the muscle. The *thick line* beneath the force trace indicates the duration of the stimulation burst. **b** The portion of (**a**) within the box is shown on an expanded time base. The values of force are translated into stress, and the values of deformation into strain. Note that the force rises during stretch (**a**), and continues to rise (stretch activation) during the interval at constant length (**b**) following the stretch. Similarly force declines during shortening (**c**), and continues to decline (shortening deactivation) for several ms following the shortening (**d**). **c** The work loop formed by plotting stress against strain for the cycle defined by c–f. The area of the loop is the work output over the cycle. The loop is traversed counterclockwise, indicating that there is net work output. [From Ref. [9], repr. w. permission]

force response. In (b) the force profile is translated into stress (in kN/m^2), by dividing the measured force by the muscle cross section of about 7 mm^2, and the absolute elongation is converted into strain (in %)m by dividing by the muscle length. The two plots can be divided into six portions. In (1) the deformation is applied and the force rises; in (2) the deformation is held constant, but the force keeps increasing; the same happens in reverse after in (3) the strain is released, but the force keeps decreasing also in (4) when the strain is held at zero. By repeating the cycle of deformation, a closed loop in the stress versus strain plot (c) is obtained. This kind of closed loop is just equivalent to the work loop described before in Eqs. (10.10) and (10.11). Then, the area within the loop represents the available power output from the muscle, according to Eq. (10.10).

In Ref. [8], this same experiment was repeated for values of total strain ranging between 2 and 8%, applied during a fixed time (therefore, a strain *rate*). It was found that the work produced by the muscle follows a curve just like our Fig. 10.8, displaying a maximum for a given value of deformation (or, equivalently, for a given value of load). Such a finding is a nice confirmation of the qualitative statement which was then made, about each muscle being best adapted for a typical load.

Overall, the maximum power expressed by the asynchronous muscle was in the range 125–140 W kg^{-1}, considered by the authors as an underestimation of the actual working power output, because of the experimental conditions. Since typical power outputs from synchronous muscles of insects range between 70 and 90 W kg^{-1}, these results suggest the hypothesis that asynchronous muscle evolved in a number of insect lines, also because it permits a greater mass-specific power output, given the high operating frequencies of small insect flight.

10.5.3 Simplified Aerodynamics of Flapping Wings

The evolutionary solution of developing asynchronous muscles rest on the central finding of adapting the muscle contraction frequency to the natural frequency of the mechanical load (the wing), rather than to the frequency of the nerve pulses, as it would be the case of synchronous muscles. Therefore, we are talking of an automatic adjustment to a *resonance* phenomenon.

The proper frequency of the load is, in fact, linked to the kinematics of wing oscillation, a mechanical structure which can be assimilated to a flat vibrating plate, or **cantilever**, clamped at one end and free at the other end. Clearly, this is a great simplification of the complex movements a real insect wing can perform, including most notably rotation and bending [10]. We will consider here only the oscillation, in order to get some ideas about the different forces acting on the wing. Still, the movements of an insect wing are quite simpler compared to those of a bird's wing. The insect wing is proportionally larger, yet much lighter than a bird's. Most importantly, a bird can break the flapping between the inner part of the wing (the "arm"), which acts more like an airplane wing providing lift with relatively small up- and downstroke, and the outer part of the wing (the "hand"), which can bend over ample spans to

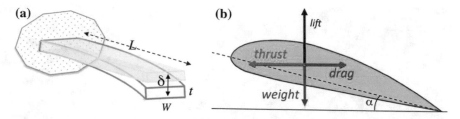

Fig. 10.9 **a** Schematic of a cantilever beam of length, width and thickness L, w, t, clamped at one end, and free to oscillate by an amplitude δ at the other end, representing a rigid insect wing doing up- and downstroke movements of the same amplitude. **b** Profile of an ideal wing seen along the cross-section, showing the four forces (lift, weight, thrust, drag) applied to the centre of gravity. The angle α formed by the line of airflow (*horizontal, full line*) and the chord (*dashed*) is the *angle of attack*. The difference between the *upper* and *lower profile* (roughly measured by the average distance perpendicular to the chord) is the *camber*

produce, aside of some lift, the necessary forward thrust. At variance with such more complex mechanics, the insect wing is very thin and rigid, and combines lift and thrust by using flapping and rotatory motion of its more flat surface.

The material composing the insect wing is a highly specialised protein cuticle with reinforcing veins, which can be characterised by an effective Young's modulus E ranging anywhere between 0.3 and 20 GPa. Figure 10.9a provides the analogy between a clamped cantilever beam and the wing profile. Figure 10.9b outlines the aerodynamic forces acting on the wing surface.

The basic equations to describe the mechanics of an oscillating cantilever, which represents the up- and downstroke movement of an insect wing, are: (i) the Stoney equation, giving the amplitude δ in the approximation of small displacements:

$$\delta = \frac{3\sigma(1-\nu)}{E}\left(\frac{L^2}{t}\right) \tag{10.21}$$

and (ii) the Hooke equation, giving the effective spring constant of the oscillating beam:

$$k_{eff} = \frac{F}{\delta} = \frac{Ewt^3}{4L^3} \tag{10.22}$$

In these equations, L, w, t are respectively the length, width and thickness of the beam-wing, resulting in a volume $V = Lwt$, E and ν are the Young and Poisson elastic moduli of the material composing the wing, σ and F the stress or force applied. Note that, in the beam scheme, the "wing" is clamped at the body end, and the force is applied somewhere between the two ends, while in the insect the force is applied at the same point of clamping (i.e., from inside the body), and the opposite end swings free. Saint-Venant's theorem on the equivalence of mechanical loads allows us to use this similarity, by exploiting the symmetry of the loading geometry.

In practice, the constant k_{eff} is linked to the natural frequency of the beam, ω_0, by the ordinary equation of the harmonic oscillator (see next greybox):

Natural, forced, damped harmonic oscillator

The harmonic oscillator is a well-known model in theoretical physics, describing the periodic, oscillatory behaviour of a wide variety of systems, e.g. from crystals, to thermomechanics, acoustics, electrical circuits, etc. In practice, any system performing small displacements about an equilibrium state can be treated as a kind of harmonic oscillator, eventually with the extra variables of including damping and forcing terms.

The **isochronous pendulum**, firstly studied by Galileo Galilei in 1602, is likely the best example of harmonic dynamical system, and has traditionally enjoyed a great importance as a model system in many areas of physics. It remained the best timekeeper device until the 1930s, and can be used also to give a proof of the equivalence of inertial and gravitational mass, although much more refined experiments are today available to prove such a fundamental property.

Figure 9.10a shows the scheme of a simple pendulum. The mass m performs a uniform movement around the circle of radius L, while the angle θ is spanned with constant angular velocity $\omega = \dot{\theta} = 2\pi\nu$, ν being the frequency of oscillation. The pendulum spans only part of the circle, $s = L\theta$, with velocity $\dot{s} = L\omega$ (maximum at $\theta = 0$, and zero at each turning point, where it changes sign). Then, the angular acceleration is $\ddot{\theta} = \dot{\omega}$. The **inertial** force induces an acceleration inversely proportional to the inertial mass m_i, directed to the fixed centre:

$$f_i = m_i \ddot{s} = m_i L^2 \ddot{\theta} \tag{10.23}$$

where $m_i L^2 = I$ is the moment of inertia of the mass about the length L.

The **gravitational** force, proportional to the gravitational mass m_g, is produced by the Earth's attraction g, and is directed toward the centre of the Earth (practically, along the vertical direction), giving a force component parallel and opposite to f_i:

$$f_g = m_g g(L \sin \theta) \tag{10.24}$$

All along the trajectory it is $f_i = f_g$. For small-amplitude oscillations, the value of $\sin\theta$ can be replaced by θ, thus obtaining the harmonic equation of motion:

$$\ddot{\theta} = \frac{m_g}{m_i} \frac{g}{L} \theta = -\omega^2 \theta \tag{10.25}$$

whose solution is:

$$\theta(t) = A \sin(\omega t + \phi) \tag{10.26}$$

a **periodic** motion (with A and ϕ fixed by the initial conditions). The experimental observation that the oscillation frequency ω is independent on the mass proves that $m_i = m_g$, i.e. the equivalence of inertial and gravitational mass.

For a generic harmonic system, moving along a generalised coordinate x, the equivalent oscillator equation is obtained by equating the actuating force to the recall force which tends to bring back the system to its equilibrium position (conventionally, $x = 0$):

$$\ddot{x} = -\omega_0^2 x \tag{10.27}$$

with $\omega_0 = \sqrt{k/m}$ the natural angular frequency of the system. The recall force can be a real mechanical spring, or any other kind of force describing the attraction basin about the potential minimum, e.g. an electric or magnetic potential, a chemical gradient, gravity from a planet, etc. (Fig. 10.10).

Similar to the treatment above, the solution to the previous equation is:

$$x(t) = A\sin(\omega_0 t + \phi) \tag{10.28}$$

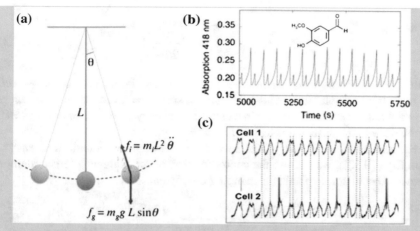

Fig. 10.10 **a** Ideal isochronous pendulum, and examples of natural oscillators. **b** Natural sustained oscillations in the oxidation of NADH by O_2, induced by the supply of vanillin; the oscillations are monitored by the characteristic peroxidase light-absorption at 418 nm (from Ref. [11], © 1998 American Chemical Society). **c** Synchronised electrical spiking from two nearby hypothalamic neurons from a rat cerebral cortex. Neurons are excited by voltage-clamp subthreshold potential (see Chap. 7; from Ref. [12], repr. w. permission)

Let us now consider the case when an external force F_0 is acting on the oscillating system, with a frequency $\omega \neq \omega_0$, the above Eq. (10.27) is modified in the **forced harmonic oscillator** as:

$$\ddot{x} + \omega_0^2 x = \frac{F_0}{m} \sin\omega t \tag{10.29}$$

with solution:

$$x(t) = \frac{F_0}{m(\omega^2 - \omega_0^2)} \sin(\omega t + \phi) \tag{10.30}$$

It can be seen that for ω approaching ω_0, the above solution becomes infinite: this is the phenomenon of *resonance*, occurring when a system is forced at its natural frequency. Even for a very small force, the amplitude accumulates and can explode the system.

In practice, in any real physical system there will be some dissipative mechanism which avoids a pure resonance amplification by introducing a damping term, typically proportional to the velocity $\dot{x}$. The above equation is again modified in the **damped harmonic oscillator**:

$$\ddot{x} + 2Z\omega_0\dot{x} + \omega_0^2 x = \frac{F_0}{m} \sin\omega t \tag{10.31}$$

whose general solution is:

$$x(t) = \frac{F_0}{m\sqrt{(2Z\omega\omega_0)^2 + (\omega^2 - \omega_0^2)^2}} \sin(\omega t + \phi) \tag{10.32}$$

The *damping coefficient* Z contains the effect of any drag force dissipating energy in the system, and removes the infinite divergent behaviour of the solution (for $Z = 0$ the resonant forced oscillator is recovered).

$$\omega_0 = \sqrt{\frac{k_{eff}}{m}} = \sqrt{\frac{Et^2}{4\rho L^4}} \qquad (10.33)$$

with $\rho_m = m/V$ the wing material density, which can be taken to be $\rho_m \approx 1000$ kg m^{-3}. For example, for a wing with typical size $L = 2$ cm, width $w = 0.5$ cm and thickness $t = 50\,\mu$m, we obtain a natural frequency $\nu_0 = \omega_0/2\pi \approx 40$–$45$ Hz, comparable to the observed values of flapping frequency.

For a given wing geometry (size Lwt, mass $m = \rho_m Lwt$, natural frequency ω_0 from Eq. (10.33)), we can use the approximation of the wing as a forced, damped harmonic oscillator, to obtain the damping coefficient Z as:

$$Z = \frac{F}{2mu\omega_0} \qquad (10.34)$$

where the force F is identified with the lift, and the displacement derivative $\dot{x}$ with the horizontal flight speed u.

If we look again at Fig. 10.7, it can be seen that the upper profile varies between ± 4 times the insect mass Mg, thus suggesting a RMS (root-mean square) value of $\sqrt{\langle f^2 \rangle} = 4Mg/\sqrt{2}$. If we consider a flight speed $u = 3$ m/s, $L = 0.02$ m, $M = 2 \times 10^{-3}$ kg, and $m = \rho_m(Lwt) = 2 \times 10^{-5}$ kg the wing mass, we obtain an estimate of $Z = 1.8$ for the nondimensional damping coefficient.

Again, by looking at the data from the previous study, we can take that the asynchronous stimulation frequency to be about 1/5 of the contraction frequency (i.e., every nerve pulse induces about five muscle contractions), $\omega = \omega_0/5$. Therefore, we can estimate the maximum amplitude of the up- and downstroke wing flapping oscillation, as:

$$\delta_{max} = \frac{F_0}{m\sqrt{(2Z\omega\omega_0)^2 + (\omega^2 - \omega_0^2)^2}} = \frac{F_0}{m\omega_0^2\sqrt{(\frac{2}{5}Z)^2 + (\frac{24}{25})^2}} \qquad (10.35)$$

Taking from Fig. 10.8a,b an average value of $F_0/S = 2 \times 10^4$ kN m^2 for the external force (from the asynchronous muscle), or $F_0 = 2$ N, we get $\delta_{max} \approx 0.5$ cm, or a maximum flapping angle of $\pm 14°$, which is indeed a reasonable value for a wing of $L = 2$ cm.

10.6 How to Choose Right Variables and Units

Physics is an attempt at solving the mysteries of Nature on a *quantitative* basis. Our science has long been described as the most exact of all sciences, since it is based on relentless experimenting and painstakingly measuring quantities by all possible means. Although nobody believes anymore than one science may be more exact than another, it remains true that precise and accurate measurements remain at the heart of physics, and a theory not leading to a comparison with at least one experimentally measurable quantity is quickly rejected as "metaphysical". If measurement is the heart of physics, observables are its blood. Observable quantities must be properly defined and translated into physical observables, in order to be unambiguously measured and quantified. Furthermore, a theory could deal with the same variables, formulate mathematical equations, and build models trying to explain and predict the results of measurements. The interplay between theory and experiment in physics is as crucial as the flow of blood through the heart. This exchange is a continued cross-checking of theoretical predictions and experimental measurements, adjustments of the theory to follow the experiment, accompanied by new predictions, then new measurements which can verify or confute the theory, thus asking for a new interpretation of the conflicting result, and so on.

In any given problem, it is necessary to identify the physical character of all the important variables. This could appear a trivial and obvious statement, but its neglect is often the possible source of the most common errors. We could cite Aristotle and his theory of motion, which deduced that to move a body at constant velocity one should apply a constant force: in this case, it was the wrong identification of the velocity instead of the acceleration, as the main variable opposing the inertia of the body, to lead him and others on the wrong track.[5]

One of the best ways to try to avoid common errors is to first take a look at the physical dimensions of variables and observables. The notion of physical dimension of a variable must be distinguished by that of the variable itself. For example Newton's Second Law stipulates that the force on a body is the product between its mass and its acceleration:

$$F = ma \tag{10.36}$$

Italic symbols F, m and a are numbers issued from some measurement process: attributing a numerical value to m corresponds to performing a measurement of the mass of the body, and replacing the symbol m by a certain number of grams or pounds. This is not a neutral statement, in that it implies that we have defined what a *mass* is, and we know how to provide a physical measure of this observable. In his experiments, Newton was thinking of the mass as measured via the acceleration

---

[5]It would be too easy today to look at Aristotle's vision as erroneous, but in the Athens of IV century BC it was very difficult to conceive the space as void, and this logical impossibility was at the basis of the Aristotelian vision of motion. In contrast with the early intuition of "atomists" as Democritus and Epicurus, who proposed the space to be occupied only by atoms of finite size, with empty space between them, the most common view at the time was that the space could not sustain any empty site (*horror vacui*, or "fear of the emptiness"). According to this vision, the local motion of a body in space after the application of a force was due to the space-filling medium (the air) being displaced from the front and filling the back of the body, thus pushing it in the space. Since the motion was identified (correctly!) as being limited by a resistance of the medium proportional to the body velocity, a medium of null resistance (like the void) would have implied an infinite velocity: hence the Aristotelian refusal of void.

provided by the gravity force, which is in turn related to the Earth's mass. Therefore, the definition seems circular: to measure a mass we must know another mass. But what if we don't know the exact mass of the Earth? (We actually don't, we only have a quite good estimate of its value).

In fact, when looking at the famous newtonic apple falling, we are *comparing* two different masses, by measuring the reciprocal force of attraction they exert on each other. What is actually done in our everyday life is to take different masses and compare their *weight*, that is the force of attraction felt by each mass while attracted by the Earth. Even if we don't know exactly the mass of our beautiful planet, we can safely assume that it will attract the two objects in the same way, provided the two are put in the same place. We will then be able to say that one object weighs X times the other object, i.e. we will have realized a weighting scale. It all boils down to properly choosing the reference mass, which, according to the convention followed by a large part of the Earth's population, is a cylindrical block of platinum-iridium alloy, defined to be the International Prototype Kilogram, or IPK. All the commercial scales of our bakeries or groceries are calibrated against this IPK standard, to allow us to weigh some pieces of bread or salami in the Earth's gravitational field, and establish to what fraction X of the IPK their gravitational attraction corresponds. In other words, what is their weight, relative to that of the IPK. Since we calibrated the scale to measure the weight at a given point on the Earth's surface, we can safely assume by proportionality that also the mass of our salami or bread corresponds to X times the mass of the IPK.

On the other hand, the physical dimension of the mass m, which we will write as [M], is not a number and does not need to be replaced by a number. [M] has nothing to do with the value of m, nor with its units (such as kilograms or tons). Rather, it is the defining character of the physical variable 'mass'. In the same way, acceleration is the variation in time of the velocity, which in turn is the variation in time of the position. In other words, a must have dimensions of a double time-derivative of the length, or $[L][T^{-2}]$. The force f, which is the product of mass and acceleration, must therefore have dimensions of $[M][L][T^{-2}]$:

$$ma = [M] \cdot \frac{[L]}{[T^2]} = f \tag{10.37}$$

a result which is valid for no matter what type of force (gravitational, electromagnetic, nuclear...). Note that we are here using a sort of "algebra of dimensions", by manipulating symbolic equations that closely resemble the corresponding equations for the numerical values, however these are void of any numerical meaning.

Once the physical dimensions of an observable are defined, choosing one unit of measure is only a matter of convenience. For example, we rarely say that a force is 1 kg m per s^2, but rather we say 1 newton (abbreviated as 1 N), which is exactly the same unit (the force that accelerates a mass of 1 kg to 1 m s^{-2}). However, constructing a coherent and non-ambiguous system of units is not a trivial task. It took many years after Newton's death, before the difference between mass and weight would be clearly stated in a system of units, and still today it is normal to hear people saying that some object "weighs" X grams (while they obviously intend that its mass is X grams, since the weight would depend on the place on Earth where it is measured). The construction of rational systems of units keeps busy the engineers since two centuries, but still some confusion exists. For example, a builder may want to actually know the weight, and not simply the mass, of a bridge to be built at a particular place, since 1 ton of iron feels a gravity force of 978 N in Congo, and of 982 N in Turkey, because of various factors affecting the gravity constant g (latitude, density, tides, altitude being the principal ones); building engineers use commonly the kilogram as unit of mass, but have introduced the "kilogram-force" (equal to about 9.807 N) to measure gravitational force, by choosing a reference value for g. Something that may appear both useless and wrong to a physicist, but very worth to them.

10.6.1 Observables, Their Dimensions, and Their Measurement

By pushing a little bit further the above ideas, we can state that the definition of the physical dimension of an observable defines the operations necessary to measure it. Force appears as a mix of concepts of mass, [M], length, [L], and time [T^{-2}], and this is a very good example of physical observable for us, since it includes all the three **fundamental quantities** of any physical variable. But, what makes a quantity 'fundamental'? Indeed, the three quantities mass, length and time define any other quantity in physics, which are hence called **derived** quantities. For those fundamental three we are unable to give definitions in terms of yet some other variable, but they are instead defined by the way they are measured. We already defined the standard of mass measurement. Similar standards are defined for the length and the time.

Length is based on the operation of translation of the object to be measured against a standard ruler, which is established to be equal to 1 m in the system of units called the International System, or SI. Time, on the other hand, is based on the operation of counting the periods of some reference oscillator, whose *frequency* has to be absolutely constant. It could be thought at this point that time is a quantity derived from frequency: in fact, from the point of view of experimental measurement, time and frequency are just one the inverse of the other, and therefore are physically one same thing. In this sense, stating that some variable has dimensions of some power of the mass, some power of the length, and some power of the time, has direct implication on the way it is measured.

For example, let us look again at the concept of force, which is mass to power 1, length to power 1, and time to power -2. The ideal way to measure gravitational force should be to take an ideal test body, such as a very small mass, move it in the field of a (large) reference mass, and measure the resulting acceleration.[6] But measuring an acceleration implies to look at how the velocity changes: so, we should take our test body moving at a given velocity, and measure how its velocity changes because of the force. To have the body moving at a given velocity we have to fix a length, and a time in which this length is traveled. Then, we would place this fixed length at different positions in the field of the reference mass, and measure the time differences for the traveled path length. In the end, we would have made one measurement of mass, one measurement of length, and (at least) two measurements of time, just as dictated by the physical dimensions of the quantity 'force'.

In many experimental situations it may be more practical to work with different, "less fundamental" quantities, which will however be always related to the fundamental mass, length and time. For example, in problems involving fluids, one may be tempted to include the pressure in the definition of the observables. Since the pressure is a force divided by a surface, force would be in this case the product [Pressure][L^2]. This may be convenient for practical purposes, since it gives us a hint about how to measure a pressure: measure the variation of the force on a given surface, with respect to a reference value. If the force is given by the gravity, build a device in which the pressure produces a movement of a mass (for example, a barometer with a column of fluid); if the force is mechanical, build a device in which the pressure produces a deformation of a material of known elasticity (for example, a curved membrane or a straightened pipe, as in a strain gauge manometer); if the force is electromagnetic, build a device in which the pressure induces a variation of electric or magnetic field (for example, a condenser with moving walls, or a magnet with moving poles, as in an aneroid barometer). However, since pressure is a derived quantity, in all cases its measurement ultimately relies on a force measurement.

In problems involving heat exchanges it would be necessary to include the temperature in the definition of the observables, since temperature (usually not considered in the realm of pure mechanics) is another quantity which can only be defined by its measurement (see the definition of absolute temperature and thermometer in Chap. 2). For example, the specific heat is defined as the amount of energy needed to increase the temperature of a given mass by one degree, so its dimensions could be conveniently defined as [Energy][M^{-1}][Θ^{-1}], or more fundamentally as [L^2][T^{-2}][Θ^{-1}], since the energy is [M][L^2][T^{-2}] (for a mnemonic, always think of the definition of kinetic energy). Electric current is another such fundamental quantity, having its own dimension (indicated by [A], from the symbol of the ampere unit). Any other electrical quantity is related to the current, combined with proper mass, For example, the electric charge is [A][T], and is measured in coulomb units (1 C = 1 A s)[7];

---

[6]The term 'ideal' is necessary here, since the test mass must be so small compared to the large source of gravitational field, to not perturb it, since both masses are reciprocally exerting attraction force.

[7]It may appear that, since 1 coulomb is 1 A/s, charge and not current is the fundamental quantity. However, the unit charge (the charge carried by one electron) is much more difficult to measure, without knowing some other quantity. Therefore the current, which relies on a simple measurement of force between two conducting wires, is taken as fundamental.

the electric potential is an energy per unit charge, that is power per unit current, $[M][L^2][T^{-3}][A^{-1}]$, and is measured in volts ($1\ V = 1\ J/C = 1\ W/A$).

At this point, it should be clear that we can manipulate algebraically the physical dimensions of observables, just as we would for their numerical values. In fact, to obtain the dimensions of the force, we simply multiplied the respective dimensions of acceleration and mass. In this way, dimensions can be obtained for any physical observable, and even for unknown observables. We may ask for example what would be the most appropriate observable to describe a phenomenon, by looking at what combinations of dimensions can be formed starting from the more fundamental observables implied in that phenomenon.

The process of deducing the dimensions from the combination of different observables is called **dimensional analysis**, and it is a very powerful technique in all the branches of science that rely on quantitative measurement.

10.7   Dimensional Analysis: Animals that Walk and Run

For a physicist, the difference between mass and weight is clear. If you let fall a mass of 1 kg near the Earth's surface, its acceleration is $9.807\ m s^{-2}$, and the mass is subject to a force of 9.807 N. Gravity may be represented in various ways, from very sophisticated ('the curvature of the space-time'), to the physiological feeling of a 'down' that we constantly perceive without much thinking about it. Likely, its most practical definition is that gravity is the ratio between the weight and the mass of an object.[8] Most people, with the exception of astronauts, will experience a very narrow variation of this ratio for their whole life. This explains why for many biologists, the idea that gravity is in fact a *variable* is often not relevant, since in their experiments they hardly have access to variations of g. As a conclusion, gravity is not considered a useful variable in biology. Nevertheless, the actual value of g represents a strong constraint on the shape and growth of most organisms (at least those who experience gravitational forces from the range of a few mN and up). Furthermore, it is acknowledged that the value of g may have suffered quite large variations during the 4.5×10^9 years of the Earth's history, with possibly interesting implications on the evolution of living organisms.

Frequency is a physical concept to which engineers, rather than biologists, seem to be mostly accustomed. Since the advent of radio waves, our world has been progressively filled up by frequencies of all kinds, from the periodic scanning of old cathode ray tubes in TV sets, to the warming frequency of our microwave ovens, to the clock frequency that keeps electronic chips alive in our computers and portable

---

[8]There is actually a subtlety in this concept, calling on the equivalence between *inertial* and *gravitational* mass. Inertial mass is the resistance to any applied force accelerating the body, for example the recall force of pendulum (see Fig. 9.10 on p. 388), proportional to m_i. Gravitational mass is a measure of the attractive force exerted on the Earth by the same body, for the same pendulum proportional to m_g. Experimentally, it is found that the oscillation frequency of the pendulum depends only on its length, and not on its mass, which is a direct proof of the identity of m_i and m_g.

phones. However, many frequencies in nature influence living beings since the earliest ages, the most obvious being the alternation of day and night, as well as the apparent motion of the Sun, the Moon phases and the eclipses, which were measured by Babylonian priests already four thousands years ago. At the individual scale, any animal has an instinctive appreciation of the variable frequency of its heartbeat and respiration, and knows well its upper and lower bounds. A whole branch of modern biology, called chronobiology, is devoted to the study of the natural rhythms, their possible origins, and their developmental and environmental consequences.

As we noted above, frequency has physical dimensions of the inverse of time, $[T^{-1}]$, and it can in all respect be considered equivalent of the concept of time, as far as measurement is concerned. Albert Einstein's definition of time ("anything that can be measured by a clock") marks a question not often challenged in basic physics courses, let alone biology ones: what is a clock? If reduced to the essential, any clock is composed by some oscillating device (a 'pendulum'), and a counting device that allows to take note of the number of cycles of oscillation. And what is an oscillator, then? It is any device that performs a cyclic process, with an intrinsic frequency that is hopefully the most constant. The pendulum, again, has been considered for long time as the epitome of frequency marker. However, cyclic processes may have the most diverse origin (see the Figure in the preceding greybox), such as an enzymatic reaction, in which one compound is transformed into another and back; protein expression in a bacterium, in which proteins self-organise in fast periodic oscillations to identify the membrane splitting region; or the synchronised firing patterns of brain cells.

Despite a certain disaffection for the concept of frequency by the biologists, cyclic phenomena abound in nature. Animal locomotion, birds in flight, fishes swimming, give us plenty of examples of periodic motion, with many interesting lessons to take home.[9]

As a start, let us examine the pendulum described in the greybox from the point of view of dimensional analysis. By taking the most candid approach, let us imagine to have just some basic scientific notions, without knowing anything about mathematical physics. By looking at the pendulum bound to the string and being constantly attracted downwards, we might suspect that gravity must implied someway in its behaviour. However, we could think it obvious that also the mass m of the ball, and the length L of the string, could play a role. Therefore, we could invent a simple equation for the frequency of the pendulum, aimed at identifying which ones among these candidates are responsible for its actual value. In the ventures of dimensional analysis, we would write something like:

$$\nu \propto m^\alpha g^\beta L^\gamma \qquad (10.38)$$

---

[9]The following parts of this Section are largely inspired by the outstanding works of McMahon & Bonner, Pennycuick, and Vogel – see "Further reading" at the end of this chapter.

The method of dimensional analysis allows us to determine the exponents α, β, γ, under certain conditions. The symbolic dimensional equation corresponding to Eq. (10.38), obtained by replacing each physical quantity by its physical dimensions, would be:

$$[T^{-1}] = [M]^{\alpha}[LT^{-2}]^{\beta}[L]^{\gamma} \tag{10.39}$$

(note that, since the equation does not refer to numerical relationships, we dropped the reference to a numerical proportionality and wrote directly '='), or more strictly:

$$[M^{0}][L^{0}][T^{-1}] = [M]^{\alpha}[LT^{-2}]^{\beta}[L]^{\gamma} \tag{10.40}$$

Now, by grouping the exponents of the similar variables on the left and on the right, we get the following three relations:

$$\text{Exponent of } [M] : 0 = \alpha$$
$$\text{Exponent of } [L] : 0 = \beta + \gamma$$
$$\text{Exponent of } [T] : -1 = -2\beta$$

The first relation shows that, contrarily to the candid expectation, mass is not at all implicated in determining the frequency since its exponent of zero removes it from Eq. (10.38). From the other two, we can deduce that $\beta = 1/2$, and that $\gamma = -\beta$. Therefore, Eq. (10.38) becomes:

$$\nu \propto \left(\frac{g}{L}\right)^{1/2} \tag{10.41}$$

This is exactly the same result already found in the greybox, i.e., the frequency is proportional to the square root of the ratio g/L. Or, in other words, the ratio $\nu^2 L/g = const$. Note that this last ratio is nondimensional, which means it does not depend on the units we use for measuring length or time.

The kinematics of a simple pendulum (greybox on p. 466) has some impact on the behaviour of walking animals. Indeed, we can imagine that the moving leg attached to a body of mass M is akin to an **inverted pendulum**, as it swings about a fixed point identified by the point of impact of the foot (see Fig. 10.11). As the animal walks, the fixed point moves. However, if the walking frequency and the stride length are both constant, we can simplify the problem by assuming that the fixed point is simply translated, and the alternate oscillations over one or the other leg are identical. Therefore, by taking all the mass concentrated in a point at the end of the leg length L, the same relationship as for the ideal pendulum can be applied. With a bit of a leap of faith, the same relationship could be applied also to quadrupeds, by assuming that animals with four legs may be the sum of two animals with two legs, and a doubled mass. This should not be entirely true in several respects, since there exist detailed phase relations between the movement of each pair of legs (extremely evident when observing the coordinated shifting motion of the many pairs of legs in

Fig. 10.11 Schematic of the walking animal as an inverted pendulum of length L, with all the animal body mass M (including the legs' mass) concentrated in the blue oscillating sphere. The fixed point, in fact, changes at every step, but it can be considered as stationary (only translated) if the stride length and walking frequency are both constant

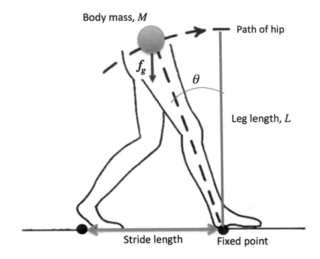

a millipede). However, as far as the walking frequency is concerned, this turns out to be a good approximation.

By following the observations of C. J. Pennycuick, it is interesting to note that the traditional approach followed by zoologists in a similar case would appear quite cumbersome. Alexander and Jaynes, in a study of 1983 [13], collected hundreds of field data for many species of animals, ranging from the size of a rat to that of a horse. By plotting their data in several different ways and looking for correlations, they consistently found, among other interesting results, that animals of largely different sized exhibit a striking similarity, in that their walking frequency is inversely proportional to the square root of the leg size. A result that does not come as a surprise to us, after the much simpler analysis above.

10.7.1 More Variables and The Buckingham π-Theorem

It should be noted that, being only an expression of dimensional relations among quantities, the method of dimensional analysis does not allow to determine the numerical values of the coefficients linking one variable to others. These must be obtained by experimental observations, or at least by informed theoretical models.

The attentive reader could further object that, quite arbitrarily, we chose a power-law form for the dependence of one variable (walking frequency) on the fundamental quantities (mass, length, time). Why we did not take, e.g., a logarithm, or a trigonometric function? There is a deep reason to this choice, with a related theorem and detailed proofs [14, 15]. The simplest way to understand it, is that a physical variable must not depend on the units in which it is measured. If we take the ratio between the values of the same physical variable for two samples, for example the masses of a horse and a mouse, this number must not change if we use kilograms or pounds. The relative theorem proves that such independence is possible, only if the defining equation in terms of the fundamental quantities has the power-law form. Any other function leads to a dependence on the particular measuring unit.

However, the chief objection to the method as we described it, is that it allows only three symbolic equations to be written down, one for each of the three fundamental quantities. This means that, at most, we can get exponents for three independent variables. But what if our problem depends on more variables than just three?

The previous harmonic development of the pendulum works only if the amplitude of oscillation is small compared to L. But for an animal walking, the inverted pendulum cannot be simply harmonic. The stride length L_s is comparable to L, therefore it must enter in the determination of the frequency. If we write $\nu \propto m^\alpha g^\beta L^\gamma L_s^\delta$, we now have four variables to determine with just three equations.

In this case, by performing the experiment, we will note that for each given ratio L_s/L, the ratio $\nu^2 L/g$ is again constant. This means that we could drop one of the variables, by replacing it with some (unknown) function of the nondimensional parameter $\Pi = L_s/L$, as:

$$\nu \propto m^\alpha g^\beta L^\gamma f(\Pi) \tag{10.42}$$

What this writing implies is that a log-log plot of the ratio $\nu^2 L/g$ versus Π should give a straight line, as it is found experimentally.

To make things yet more complicated, we may want to compare animals walking or running at different speeds. What if in our problem of walking frequency, also the velocity could u be involved? This would result in an equation like $\nu \propto m^\alpha g^\beta L^\gamma u^\delta L_s^\varepsilon$. The corresponding dimensional equation would be:

$$[T^{-1}] = [M]^\alpha [LT^{-2}]^\beta [L]^\gamma [LT^{-1}]^\delta [L]^\varepsilon \tag{10.43}$$

now with five exponents to be determined, from only three equations for mass, length and time. The two missing equations must be constructed on purpose. We should identify two equations that define two nondimensional variables, constructed by combining some of the five variables involved.

We already know that the ratio $\Pi_1 = L_s/L$ between stride length and leg size must be relevant. By looking again at Fig. 10.14, this ratio gives the angle θ of each oscillation, and it seems reasonable that two animals walking at different speed could be compared on the basis of equal swinging angles.

A second nondimensional variable may be more complicated to construct. However, let us think of the energies involved in this idealisation of a walk: there is the kinetic energy of the body translating horizontally with velocity u, and the potential energy of the gravitational field to keep the body at a height L above the ground. Their ratio is obviously dimensionless, and may be another measure of comparison between animals of different weight and correspondingly different velocities. Therefore, we may write $\Pi_2 = \frac{1}{2}mu^2/mgL$, or $\Pi_2 = u^2/gL$, by forgetting about numerical constants (which result in a relative shift of the values). This parameter is called the *Froude number*, from the English engineer J. A. Froude (1818–1894):

$$Fr = \frac{u^2}{gL} \qquad (10.44)$$

This way of reducing variables by grouping them into nondimensional numbers, goes under the name of "Buckingham's π-theorem", from the name of Edgar Buckingham, and the Greek symbol π that he firstly used to define dimensionless quantities. The theorem basically states that a physical problem described by n independent variables $q_1, \ldots q_n$, as $f(q_1, \ldots q_n) = 0$, can be restated in terms of a subset of $k < n$ variables, plus $n - k$ nondimensional parameters Π_i, constructed from the q_n as $\Pi_i = q_1^{a_1} q_2^{a_2} \ldots q_n^{a_n}$, with the exponents a_i being rational numbers. The new problem has the same solution as the old one, and is restated as $F(q_1, \ldots q_k; \Pi_1, \ldots \Pi_{n-k}) = 0$, the notation ';' indicating that the following symbols are fixed parameters, and no longer variables.

In the case at hand, Eq. (10.43) thus becomes:

$$[T^{-1}] = [M]^\alpha [LT^{-2}]^\beta [L]^\gamma \cdot \Pi_1 \cdot \Pi_2 \qquad (10.45)$$

and $\nu \propto m^\alpha g^\beta L^\gamma f(\Pi_1, \Pi_2)$, with f a homogeneous dimensionless function of the two parameters. This can be solved in the same way as above, giving the same $\nu \propto (g/L)^{1/2}$ result: the ratio $\nu^2 L/g$ is still a constant, but may change for any given combination of Π_1 and Π_2. Well, it seems that we have added a lot of complication, not to gain much in this game. But we now have the two dimensionless variables that must mean something: the unknown function $f(\Pi_1, \Pi_2) = \Pi_1^{a_1} \Pi_2^{a_2}$ means that the behaviour of frequency is unchanged if we look at animals having a similar ratio Π_1/Π_2. This is where we can learn something more than just the inverse square-root dependence of frequency versus length. Again, the same fine scientist Robert McNeill Alexander that we already quoted, was the first man to ask the momentous question: what would have been the walking (or running) speed of dinosaurs?

He observed that we actually dispose of fossil bones to determine L, and tracks of fossilised footprints to determine L_s. Firstly, he plotted a diagram of Π_1 versus Π_2 for

Fig. 10.12 Plot of Π_1 versus $(\Pi_2)^{1/2}$ for humans and various animals, recorded at different speeds. The square root of Froude number is plotted, to avoid clumping of points at the lower speeds. [From Ref. [16], repr. w. permission of The Mathematical Association.]

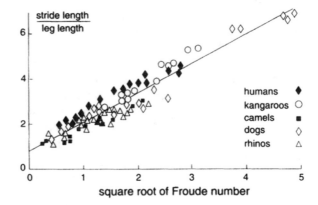

a large variety of living animals of different sizes and habits, as shown in Fig. 10.12. The fact that all the points for different animals, walking or running at different speeds, are nicely clustered along one same straight line, allows to deduce that the ratio of the two dimensionless parameters must indeed provide a representation of the similarity of their walking characteristics. Moreover, it also allows to formulate predictions for animals not being measured, by using the same straight line as a guide. For the case of dinosaurs, knowing L_s and L, Alexander deduced velocities ranging from 1 m/s for the *Brontosaurus* ($m \simeq 30$ tons), to about 2 m/s for the *Tyrannosaurus* ($m \simeq 5$ tons), up to 12 m/s for the light, two legged dinosaurs. If we try to estimate the speed of the scary *Velociraptor* from the movie "Jurassic Park", by looking at their footprints and bones found in Arizona that give $L_s = 2.5$ m and $L = 0.75$ m, we have $\Pi_1 = 3.3$; once situated on the straight line in Fig. 10.12, this suggest a Froude number $\Pi_2 \simeq 5.8$, from which a velocity of $\sim$7 m/s is obtained. This is about 25 km/h, a running speed faster than a mile-runner olympic champion, and definitely fast enough to catch the movie's actors.

The dependence of the Froude number on the inverse of g also tells that changing the conditions of gravity, the same animal may experience singular effects. When looking at astronauts walking on the Moon, we see them moving in cumbersome, 'floating' steps, to the point that after some time they start making long jumps at a funnily low speed. In fact, since the gravity acceleration on the Moon is equal to 1.62 m/s^2, the Π_2 corresponding to a same walking speed increases by a factor of 5.6: to stay on the straight line of Fig. 10.12, the Π_1 must be about 3.4, i.e. for an astronaut with a leg 1 m long, the walking stride is 3.4 m. Clearly, they need to make long jumps! The alternative to conserve a rather normal walking rhythm, would be to reduce the speed to less than 1 m/s: that would look like a slow-motion film.

10.8 Flying Animals and Wingbeat Frequency

A difficulty when approaching biological problems from a physical point of view is that a system may have apparent and deceivingly simple behaviour, however with an underlying complex ensemble of variables, and an even more complex network of mutual interrelations. For example, by looking at the wingbeat frequency of a bird, we could try to solve the problem by analogy with the previous treatment of walking animals, by taking into account just a few more variables which seem relevant in this case, such as air density and bird flight aerodynamics, among others.

As we have seen already, the method of dimensional analysis is not going to provide us a simple and unique answer, if we add too many variables to the problem. We may however hope to find correlations and a partial solution by identifying also in this case some combination of variables into dimensionless parameters. Such an approach, as a first guess, seems preferable to the empirical method of measuring speed and wingbeat frequencies for as many birds as possible, under many different conditions, and searching for statistical correlations for example by multivariate regression analysis.

The wingbeat frequency, as the walking frequency, is eventually an important information to determine the power supply from the muscles. Differently from the walk, the weight of the flying animal should now be implied, since the bird must support itself in the air during flight. The fact that air is much less dense than the body has important implications, therefore the air density ρ should enter as well. Other variables which could be considered are the wingspan, W, the wing surface, S, and wing mass m, possibly combined into the moment of inertia $\mathscr{I}$ of the wing about the shoulder joint. And maybe more.

The equation for the frequency in terms of these defining variables is therefore:

$$\nu \propto (mg)^{\alpha} W^{\beta} S^{\gamma} \mathscr{I}^{\delta} \rho^{\varepsilon} \tag{10.46}$$

and the equivalent symbolic equation for the dimensions:

$$[T^{-1}] = [MLT^{-2}]^{\alpha} [L]^{\beta} [L^2]^{\gamma} [ML^2]^{\delta} [ML^{-3}]^{\varepsilon} \tag{10.47}$$

Let us again decompose this equation into the exponents for [M], [L] and [T] :

$$\text{Exponent of [M]} : 0 = \alpha + \delta + \varepsilon$$
$$\text{Exponent of [L]} : 0 = \alpha + \beta + 2\gamma + 2\delta - 3\varepsilon$$
$$\text{Exponent of [T]} : -1 = -2\alpha$$

As above, the method gives us three equations, but for five unknown exponents. The value $\alpha = 1/2$ follows immediately from the fact that gravitational acceleration is the only variable containing time, and its role in the frequency is unambiguous. The other four are to satisfy the two equations:

$$-1/2 = \delta + \varepsilon \tag{10.48}$$
$$-1/2 = \beta + 2\gamma + 2\delta - 3\varepsilon \tag{10.49}$$

There is clearly a (doubly-)infinite number of combinations of exponents satisfying the two equations.

The search for dimensionless variables has to be based on some empirical evidence. We follow here the work of C. J. Pennycuick, who carried out an extensive series of measurements of marine birds' flight characteristics, by recording for many species their weight, wingspan, wing area, wingbeat frequency and flight speed [17]. By representing the recorded wingbeat frequencies as a function of grouped variables in linear plots, one each for frequency versus mass, frequency versus wingspan, and frequency versus wing surface, best fits are obtained, respectively, for $(\alpha + \delta)$, $(\beta + 2\delta)$ and γ (this is because in Eq. (10.47) mass, wingspan and wing surface, appear with those respective exponents). The approximate expression finally obtained by Pennycuick was:

$$\nu \propto m^{1/3} g^{1/2} W^{-1} S^{-1/4} \rho^{-1/3} \tag{10.50}$$

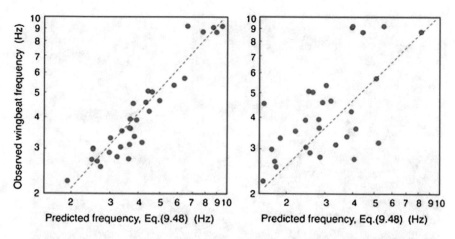

Fig. 10.13 *Left* Double-logarithmic plot of observed wingbeat frequency versus the expression on the right-hand side of Eq. (10.50) (experimental data from Ref. [17]). The slope of the straight *line is* 1.04, with a correlation coefficient of 0.947. *Right* The same data plotted according to the approximated analysis of Eq. (10.51). The slope is 0.95, with a worser correlation coefficient of 0.65

in which the moment of inertia is replaced by the simple approximation $\mathscr{I} = mW^2$. By plotting in a double-logarithmic graph the frequency versus the product $m^{1/3}g^{1/2}W^{-1}S^{-1/4}\rho^{-1/3}$, a nice clustering of the data for all birds studies was obtained, shown in Fig. 10.13 (left), the values of air density and gravity being taken as constants. The coefficient of linear regression is 0.947 for 32 experimental points (bird species), which is a very good value, also given the uncertainties inherent to the kind of measurements, done by cinematographic recording of the birds' flight in their natural environment.

But what now of all our nice arguments above, about reducing the number of variables by searching for meaningful nondimensional quantities? What do the exponents in Eq. (10.50) mean? We could push a bit further the analysis of Pennycuick, trying to simplify the Eq. (10.50) by some nondimensional auxiliary quantity. As in the case of walking frequency, we may hope that this could teach us something more, if not improve the already good numerical agreement of the solution.

The density appearing in the equation is the air density, which may experience some little variation with the altitude in the range explored by those birds. To keep itself flying (Archimede's principle) the bird should at least displace a mass of air equal to its own, therefore the ratio (m/ρ) gives this equivalent volume, V_{eq}, to the power 1/3 in the equation. This is a characteristic length, $V_{eq}^{1/3} \propto L_{eq}$, which (by analogy with the stride length of the walking) could tentatively be identified as being proportional to the 'wavelength' of the flight, i.e. the distance covered between two consecutive flapping of the wings. This length can be written as $L_{eq} = u/\nu$ for a flying velocity u. In the equation (10.50) we have now the ratio L_{eq}/W, which is the equivalent of Π_1 for a flying animal. On the other hand, the wing surface S appearing

Fig. 10.14   A few of the beautiful marine birds studied by Pennycuick in his experiments. *Above, left* Little egret (*Egretta garzetta*), average mass 300 g, wingspan 1m. *Right* a Wilson's storm petrel (*Oceanites oceanicus*), average mass 40 g, wingspan 40 cm. *Below, left* Great blue heron (*Ardea herodias*), average mass 1.9 kg, wingspan 1.75 m. *Right* Wandering albatross (*Diomedea exulans*), mass up to 10 kg, its wingspan comprised between 2.5 and 3.5 m is the largest of any living bird. [Photos respectively © by Karthik Easvur, Nanda Ramesh, Kozar Luha, J. J. Harrison; all reprinted under CC-BY-SA-3.0 licence, see (*) for terms.]

with power -1/4 is to a very good approximation linked to the square of the wing span (the log-log plot of W^α vs. S gives $\alpha = 2$ with 95 % confidence). By rearranging the equation, we find:

$$\nu \propto \frac{L_{eq}}{W}\left(\frac{g}{W}\right)^{1/2} \tag{10.51}$$

showing that the wingbeat frequency is again proportional to the square root of gravity divided by a characteristic animal size, however multiplied by a dimensionless coefficient Π_1.

If we now make a plot of the experimental frequencies versus the quantity on the right of Eq. (10.51), the data displayed in Fig. 10.13 (right panel) are obtained. Their alignment on the straight line is not bad, but clearly shows a worse correlation (coefficient 0.65) compared to the more complete representation on the left of the same Figure. This is likely due to our interpretation of the "characteristic length" L_{eq} as coinciding with the wavelength. This amounts to assuming that a flying bird (or any flying object, for that matter) relies on Archimedean buoyancy to move in the air (Fig. 10.14). In fact, the hydrostatic buoyancy force is only a minimal contribution to the force required to maintain a body floating and moving, in a medium which is about 1000 times less dense than water. As it was discussed in the previous Section, first and foremost flight requires *hydrodynamic lift*, produced by the horizontal speed and by a well-fitted shape of the wings. Evidently, the quantity with dimensions of length

(m/ρ) in Eq. (10.50) cannot be simply identified with the wavelength. Because of the large density difference between the body and the air, L_{eq} should be quite larger, also implying that, for a given size of the animal, wingbeat frequencies are proportionally faster than walking frequencies.

A final consideration can be made by looking at another nondimensional parameter, the *Strouhal number*:

$$St = \frac{\nu L}{u} \tag{10.52}$$

This is the ratio between length × frequency of the periodic motion, and the body translational velocity, and is the appropriate coefficient characterising motion in a fluid associated to an oscillatory actuator. With the few exceptions of the lightest birds (mass below $\sim$100 g), and some of those having a very large wing loading (the ratio of body mass over wing area), we find that the velocity of 85 % of the birds studied by Pennycuick varies roughly linearly with the wingbeat frequency, $u \propto \nu$, albeit with a large dispersion. Since we already found above that $\nu \propto L^{-1/2}$, it turns out that the value of the Strouhal number should vary as $St \propto \nu^{-2}$. This may be a difficult prediction to test experimentally, since the values of wingbeat frequency are spread over a small interval (between 2 and 5 Hz, see Fig. 10.13, with only a few exceptions), however the data seem to broadly confirm this deduction.

10.8.1 From Birds to Insects

Scaling down sizes and masses from birds to insects, we enter in an entirely different world. Measuring insect flying speed is more difficult than for birds, since their typical values range about 0.5–3 m/s and are comparable to normal wind speed, which therefore affects the measurement of velocity much more than for birds. However, we can begin by noting that the typical Reynolds number for insects and birds differ by about three orders of magnitude (the product velocity × length for an insect being in the range of 0.01 m²/s and $Re \simeq 1000$, while being about 5 m²/s and $Re \simeq 4 \times 10^5$ for a medium-sized bird). Secondly, the drag force expressed as $F_D = \frac{1}{2}\rho u^2 c_D A$, for a drag coefficient c_D varying between 0.5 and 1, is many thousands times smaller for insects than for birds. The Strouhal number for typical insects frequencies ($\nu = 50-100$ Hz) versus birds ($\nu = 5-10$ Hz) is $St \simeq 0.5$ versus 0.15, respectively, which, in turn, is not a major difference. Eventually, all these indications are coherent with the observation of insects flapping wings at higher frequency, and with a large inclination with respect to the body (the angle of attack is about 30–40°), and getting a considerably larger lift force from their wings, compared to birds. It is everyday's experience that insects can take off and get to cruise speed much more rapidly than birds, the latter being especially slower when having a larger wingspan, adapted to long-distance and overseas flight.

Anyway, even with ideal conditions of perfect scaling between body mass, length, wing size, area, flight speed, wingbeat frequency, the flight of insect and birds differs chiefly in the way their wings are moved in space.[10] Insects typically beat air in a kind of '∞' shape, mostly working in a plane horizontal to the direction of flight; birds move their wings in a more or less rounded–elliptical figure, working essentially in the direction perpendicular to the motion. Moreover, the keratin material of insects' wings is impenetrable to air, while the feathered bird wings are partly open to air flow. Despite being much thinner, insects' wings are considerably more rigid in proportion to their weight, being not provided with inner joints and bones. Because of the substantial rigidity of the "snapping" insect wings, we could include only the surface S as an independent variable, and leave out the wingspan and moment of inertia:

$$\nu \propto m^\alpha g^\beta S^\gamma \rho^\delta \tag{10.53}$$

Again we have four exponents to determine with three equations, therefore we need one nondimensional parameters Π. By thinking again to a kind of "buoyancy" ratio, we can construct a nondimensional parameter $\Pi = \rho S^{3/2}/m$. By subsituting in the above equation, we get the analogous of (10.42) as:

$$\nu \propto m^{\alpha+1} g^\beta S^{\gamma-\frac{3}{2}} f\left(\frac{\rho S^{3/2}}{m}\right) \tag{10.54}$$

and correspondingly:

$$[T^{-1}] = [M]^{\alpha+1}[LT^{-2}]^\beta[L^2]^{\gamma-\frac{3}{2}} \cdot \Pi \tag{10.55}$$

By solving for the three exponents, it is easily found $\alpha = -1$, $\beta = 1/2$ and $\gamma = 5/4$, from which the frequency equation now looks like:

$$\nu \propto g^{1/2} S^{-1/4} f\left(\frac{\rho S^{3/2}}{m}\right) = g^{1/2} S^{-1/4} \Pi^a \tag{10.56}$$

The last equivalence is dictated by the requirement of dimensional invariance, which forces the choice of the function $f(\Pi)$ to a power-law. Considering that g and ρ are constants, the frequency equation reduces to $\nu \propto m^a S^{-b}$, with $b = \frac{3a}{2} + \frac{1}{4}$. However, this leaves us with still the problem of determining the exponent a. Like in the previous case of bird's flight, this is as far as dimensional analysis can take us: to go beyond we need experimental information.

In a thorough study by Byrne, Buchmann and Spangler in 1988 [19], the body mass, wingbeat frequency and wingspan data for more than 150 different insects were collected. If we report the observed frequency versus the product $(m^a S^{-b})$ on a log-log plot, as in Fig. 10.15, the best fit is obtained for $a = 1/2$, which corresponds to:

---

[10] A feature already identified about 150 years ago by the elegant chrono-photographic experiments of the French polymath Etienne-Jules Marey [18].

$$\nu \propto \frac{\sqrt{m}}{S} \left(\frac{g}{\rho}\right)^{1/2} \tag{10.57}$$

The body mass of the insects studies by Byrne et al. varies over several orders of magnitude, from the $10^{-5}-10^{-4}$ g of whiteflies and aphids, to the few g of sphingid moths, some of the latter attaining 10–20 cm^2 of wing surface. Actually, by looking at the log-log plot where such data are grouped by body-mass intervals (left plot, symbols of different colours), a consistent scatter can be observed between the various groups, the very smallest and largest exhibiting big variation with respect to the central slope.

Some further observation could be made, by looking at the common (intraspecific) characters of the species studies. By grouping the insects by their family (Fig. 10.15, right), it is seen that more than 80 % of the *Apidae* and all *Saturnidae* lie on or above the average slope, while most *Syrphidae*, *Libellulidae* and *Pieridae* lie consistently below. The latter are generally characterised by relatively gracile bodies (very much elongated in the case of libellules and dragonflies) and quite larger wings, compared to *Apidae* (honey bee, bumble bee, etc.), which have much smaller wings for a compact body mass. This (quite roughly) means that our equation tends to overestimate the wingbeat frequency for lighter, large-winged insects, and to underestimate it for (relatively) heavier and smaller-winged insects. The case of *Saturnidae* is also somewhat special, since these very big moths have some of the highest wing loading (m/S) among all butterflies.

The relationship between wingbeat frequency and wing loading is of special significance. According to Byrne et al., the finding that whiteflies had a higher wingbeat frequency and a lower wing loading than aphids was unexpected, since is generally

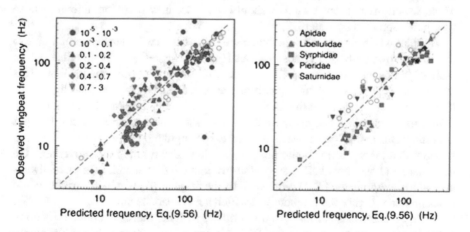

Fig. 10.15 *Left* Double-logarithmic plot of observed wingbeat frequency versus the expression on the *right-hand side* of Eq. (10.57). *Coloured symbols* correspond to different body-mass groups as indicated (in grams). The slope of the *straight line is* 1.0, with a correlation coefficient of 0.65 for 160 points, and of 0.78 if the data relative to the lighter and heavier groups are excluded from the fit. *Right* A subset of the same data, grouped according to some of the insect families

argued that, for a same body mass, insects with smaller wings should beat more rapidly than insects with larger wings. While the data collected generally confirm such empirical hypothesis, failure of this relationship for very small insects could be explained by the fact that some minute species solve the problem of staying aloft in their own unique ways. For example, whiteflies [20] and fruit flies [21], employ a "clap-and-fling" mechanism to generate extra lift: they clap their wings together at the end of each upstroke, and fling them apart at the beginning of each down-stroke. This mechanism is thought to increase the lift, and should reduce the need for exceptionally high wingbeat frequencies. Such evolutionary mechanisms cannot be simply explained on the basis of purely geometrical or dimensional arguments. By flying at extremely small Reynolds numbers, such tiny insects demonstrate an empir-ical knowledge of lift and drag that escapes our most advanced computer models of aerodynamical structures.

10.9 Dimensional Analysis: Animals Who Live in Water

The number and type of variables to be included in a dimensional analysis is chiefly guided by intuition, based on an adequate understanding of the underlying physical processes, and it represents the most controversial point of the method. For animals whose weight is supported by the Earth surface it was obvious to include gravity among the most important variables. However, already when discussing the flying animals there could be doubt, since the main force involved in keeping the animal above the ground is lift, which as we ascertained is mostly generated by forward thrust and not by buoyancy, therefore gravity could be but a secondary contribution. Then, which the relevant variables should be, if we now focus on animals whose weight is supported by water?

Animals who swim in water are immersed in a medium with density comparable to that of their own body, and viscosity about 100 times bigger, which immediately turn their Reynolds number a factor of 10 bigger than for similar movements in air. This makes for more turbulent fluid conditions around their bodies. On the other hand, they all have the advantage that their weight is entirely supported by the medium, thanks to Archimede's buoyancy force (in this case of central importance). In these conditions, the animal can move as if it had zero weight (and zero mass), and we can suppose that the mg term should not play a role in determining the movement. On the other hand, we saw in all examples above that the dimensions of g were always crucial to determine the $[\text{T}^{-1}]$ dimension of the frequency. So, which variable could determine in this case the frequency at which a fish beats its tail?

Water is much denser, more viscous, and has a higher thermal conductivity com-pared to air. Therefore, a large variability in the muscle force required for swim-ming should be expected. Differently from birds, most fishes are cold-blooded, or *ectotherm*, animals, and environment temperature is one of the most important fac-tors determining muscle performance in ectotherms. The sustained swimming speed can be reduced by a factor of 2, and the single-fibre contraction velocity by up to a factor of 6, for a change in water temperature from 25 to 10 °C [22]. Different types of muscles are involved, according to the different activity of the fish. Slow swimming involves low-rapidity muscles, running on aerobic metabolism. Recall that slow muscle fibres, with their high concentrations of myoglobin and mitochondria, and well developed capillary supply, are the main fibre type in *red* muscle. As swimming speed increases, faster contracting muscle fibres are needed. Maximum performance is achieved during fast-start, e.g. for escaping or predating, and involves the entire *white* muscle mass. White fibres are thicker, and contain a much density of myofibrils with little mitochondria, and a more extensive sarcoplasmic reticulum for fast Ca^{2+} processing. In general fast muscle is dependent on fewer physiological factors, since it is largely independent on the circulation of blood and oxygen.

Slow fibres run typically parallel to the longitudinal axis of the body, while white fibres are more intricate. This makes the contraction mechanics interesting. If we think of a fish beating his tail at varying frequency, in water of variable density (salinity) and temperature, the **stress** on its muscles is going to play an important role as well. The stress tensor σ (see Appendix H) has dimensions of a force per unit area, and its normal component σ_n is defined by the traction force exerted on the transverse cross section of the muscle. Therefore, it must be a useful variable to include in the dimensional analysis, and with dimensions of $[M\,L^{-1}T^{-2}]$ it brings a dependence on time, needed to establish the frequency. Other important variables could be the water density ρ, and at least one characteristic length, for example the fish size L. (Let's keep for the moment temperature and viscosity out of the game.) The equivalent of the Eq. (10.38) is now:

$$\nu \propto \sigma^\alpha \rho^\beta L^\gamma \tag{10.58}$$

and the equivalent of Eq. (10.39) becomes :

$$[T^{-1}] = [ML^{-1}T^{-2}]^\alpha [ML^{-3}]^\beta [L]^\gamma \tag{10.59}$$

The unknown exponents are, in this case:

$$\text{Exponent of } [M] : 0 = \alpha + \beta$$
$$\text{Exponent of } [L] : 0 = -\alpha - 3\beta + \gamma$$
$$\text{Exponent of } [T] : -1 = -2\alpha$$

from which we obtain:

$$\nu \propto \frac{1}{L} \left(\frac{\sigma}{\rho} \right)^{1/2} \tag{10.60}$$

Among the three variables on the right of Eq. (10.60), the only one readily accessible to measurement is the fish characteristic length. If we take the water density as a

(a)

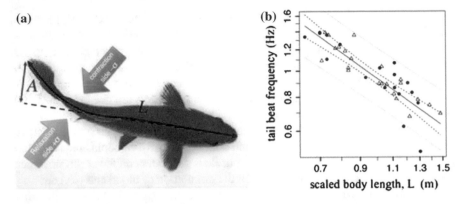

Fig. 10.16 Measuring the tail beat frequency from the lateral acceleration. **a** During swimming, the red and blue side of the body muscles alternate between compression and relaxation, i.e., contractile and tensile stress, and the acceleration oscillates between a positive and a negative value. The oscillation period is defined by the time intervals between two successive passages through zero acceleration. The amplitude A is proportional to the size L, so that a scaling relation between frequency and amplitude, or body size can be extracted. **b** Log-log plot of the normalised tail-beat frequency versus normalised body length, for saithe (*solid circles*) and sturgeon (*open triangles*). The best fit (*solid line*) is bracketed by the weighted or unweighted 95 % confidence interval (*dashed or dotted lines*). [Image (**b**) adapted from Ref. [24], repr. under CC-BY 3.0 licence, see (**) for terms.]

constant, and make the additional hypothesis that fishes of similar size/mass roughly experience the same values of muscular stress, since the microscopic structure of the muscle fibres is the same, it should be verified that $\nu \propto 1/L$ (note that both for walking and flying animals, the length dependence was instead $L^{-1/2}$). Once more, this is a very difficult prediction to check, since a fish can largely and suddenly change its tail beat frequency for various reasons (e.g., the presence of a predator, or the sight of moving food). Moreover, submarine observation in natural conditions is quite more complicate than terrestrial or airborne observation.

Recent developments in miniaturised instrumentation are starting to offer unprecedented ways to study marine life in the wild. Force-measuring devices (miniature accelerometers) can be attached to the body or tail of a fish of convenient size, so as to disturb as little as possible its natural movements, and thus follow the force developed at any instant, depth, temperature and so on [24]. As the famous French oceanographer Jacques-Yves Cousteau once put it, *The best way to observe a fish is to become a fish*. Once the instrument has recorded tail acceleration for a sufficient time, the accelerometer is recovered, the data are extracted and analysed (Fig. 10.16). By assuming that the muscles on each side of the body alternate between contraction and relaxation during each tail beat, the times at which the acceleration (stress) changes sign, by passing through zero, define the stride period. For a time-sequence of steady variation (approximately sinusoidal) of the acceleration, an average tail beat frequency can be calculated.

Which length should be taken for L can be also debated, however it has been established that the amplitude A of the tail swing is generally proportional to body length [23], so either one could be used in the dimensional analysis, with the same meaning. Broel and Taggart [24] performed a series of such measurements on two groups of fishes: pollocks (or saithe), with length ranging between 25 and 55 cm and mass 0.2–1.6 kg, swimming in an artificial pool of about 15 m diameter; and shortnose sturgeons, with length ranging between 0.56 and 1.2 m, swimming in a wild environment. The results, plotted in a log-log diagram in Fig. 10.16b, show that the tail beat frequency of both species, once scaled each by their average values, follows the law $\nu = 0.94/L$, the length exponent being equal to -1 in agreement with the prediction (10.60), with a correlation coefficient of 0.73 on the ensemble of the data.

The normalisation coefficient is different for each species, which says that the stress-dependent prefactor in Eq. (10.60) is also different. Therefore, the hypothesis that fishes of different species but of comparable same size exert the same stress must be relaxed. The non-normalised results predict that the dominant frequency for sturgeon is about twice that of saithe, when scaled at the same size, also, the estimated swimming speed for saithe, is correspondingly lower in sturgeon of the same size.

What if we wish to include the viscosity of water in the picture? Any object moving in water experiences a viscous drag, therefore the role of viscosity should be more relevant if we want to guess the swimming speed, u, rather than the stride frequency. To see this, let us firstly imagine that the velocity depends only on the drag force, D, water density, ρ, and fish size L. Since D has dimensions $[M][L][T^{-2}]$, the simplest way to combine these four variables into a nondimensional relation is:

$$D \propto \rho u^2 L^2 \qquad (10.61)$$

and since the expended power is just force $\times$ velocity, it is also $P \propto \rho u^3 L^2$. But, as it was found at the beginning of this chapter, muscle strength is proportional to the muscle cross section, therefore the power itself is proportional to the body surface, or $P \propto L^2$. Hence it would follow that $u = const$ for any fish of any size, which is obviously wrong.

Viscosity must be necessarily included in the description of swimming speed, for example as:

$$u \propto \rho^\alpha \eta^\beta \sigma^\gamma L^\delta \qquad (10.62)$$

We have four exponents for three equations, so we need to identify one nondimensional parameter Π. In this case, the obvious choice is to build a Reynolds number from our variables, $\Pi = Re$, since this parameter just represents the ratio between viscous and inertial forces.

By proceeding in the same way as above, we write:

$$u^2 \propto \eta^{\beta+1} \sigma^\gamma L^{\delta-1} f \left(\frac{\rho u L}{\eta} \right) \qquad (10.63)$$

and the equivalent of Eq. (10.55) becomes:

$$[LT^{-1}]^2 = [ML^{-1}T^{-1}]^{\beta+1}[ML^{-1}T^{-2}]^{\gamma}[L]^{\delta-1} \cdot \Pi \qquad (10.64)$$

Solving for the exponents, it is $\beta = -3$, $\gamma = 2$, $\delta = 3$, from which:

$$u \propto \left(\frac{\sigma L}{\eta}\right) \Pi^a \qquad (10.65)$$

So, we eventually found that the swimming speed is: (i) proportional to the muscle stress, i.e. to the applied force, which is an expected result in the presence of a linear drag; (ii) inversely proportional to the viscosity, which is just a restatement of Stokes' law (see Chap. 5); (iii) proportional to the fish characteristic size L, coherently with the observations [23]; and finally, (iv) it must vary as a function of some power a of the Reynolds number.

As in the case of insect flight, we have no way of deducing the exponent a without making recourse to experimental data, which as said before are much more difficult to gather for animals living underwater. In principle, we should expect a complicate dependence, because of the many factors affecting the fish hydrodynamics, such as aspect ratio, shape and position of fins, rugosity of the skin surface, body mass distribution, and so on. Moreover, Re between different fishes varies over more than 5 orders of magnitude, for a variation of speed by just a factor of 10 [25]. Therefore, we can stop here and live happily with this result, which taught us even more than we could expect.

Including also the water temperature, and the heat exchange between water and the fish skin in the picture, is an even more daunting task. However, if you are interested in temperature problems and dimensional analysis, go have a look at Problem 10.5 at the end of this chapter. And good luck!

Problems

10.1 Sarcomere stretching

The graph below [adapted from the work of D. E. Rassier, B. R. MacIntosh, W. Herzog, *Length dependence of active force production in skeletal muscle*, J. Appl. Physiol. **86**, 1445–1457 (1999)] shows the stress developed by a muscle sarcomere as a function of increasing relaxation and stretching, starting from a compression condition. By considering the sarcomere structure in Fig. 10.1, could you explain

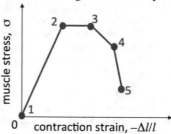

what happens at the points marked 1 to 5?

10.2 Summer training

An athlete overworks during a too hot summer day, and finds himself suddenly blocked from painful muscle cramps. Which of the following is a reasonable hypothesis to explain such cramps? 1. Muscle cells do not have enough ATP for normal muscle relaxation. 2. Excessive sweating has affected the salt balance within the muscles. 3. Prolonged contractions have temporarily interrupted blood flow to parts of the muscle.

10.3 Weightlifters

Muscle contraction involves the relative displacement of many thousands actin and myosin-II proteins. In lifting a weight of 5 kg over 25 cm, how many myosin heads have to work together? (Check out the difference between the "tight-coupling" and "loose-coupling" model.)

10.4 Cyclic muscle work

The cyclic power delivered by an insect muscle, with length $L = 1.5$ cm and cross section $S = 0.18$ cm^2, is measured with the help of a thermostatted dynamometer. It is found that the power varies periodically as a function of the stimulating frequency with the empirical law $P = -270 + 10\nu - 0.07\nu^2$ (W/kg), in the frequency interval $\nu = 50 - 100$ Hz. Find: (a) the law of variation of the work delivered by the muscle as a function of the frequency; (b) the frequencies corresponding to the maxima of P and W respectively; (c) the law for the force developed by the muscle at an average elongation $\Delta L = 10^{-3}$ m.

10.5 Dimensional analysis of sunday's oven roast

To cook a pork roast, the chunk of meat with all its condiments is put in a preheated oven at 180 °C. According to the best cooking recipes, the roast is well done when the temperature at the centre is about 70 °C. Instead of taking a meat thermometer to check every now and then the temperature, you propose a dimensional analysis of the problem, which will give you the exact time when to open the oven.

10.6 Dimensional analysis of blood pressure

By way of dimensional analysis, obtain an equation for the pressure drop, $\Delta p / \Delta L$, for blood flowing in a segment of artery of length ΔL. Consider as variables the blood viscosity η, a parameter ε (in m) measuring the rugosity of the inner arterial surface ($\varepsilon = 0$ for a flat surface), the diameter of the vase D, and the average blood velocity v.

References

1. A. Hedenström, How insect flight steering muscles work. PLoS Biol. **12**, e1001822 (2014)
2. R. Marco-Ferreres, J.J. Arredondo, B. Fraile, M. Cervera, Overexpression of troponin-T in Drosophila muscles causes a decrease in the levels of thin-filament proteins. Biochem. J. **386**, 145–152 (2005)
3. A.F. Huxley, Muscle structure and theories of contraction. Prog. Biophys. Biophys. Chem. **7**, 255 (1957)
4. E.G. Boettiger, Insect flight muscle and their basic physiology. Annual Rev. Entomol. **5**, 1–16 (1960)
5. C.J. Pennycuick, M.A. Rezende, The specific power output of aerobic muscle, related to the power density of mitochondria. J. Exp. Biol. **108**, 377–392 (1984)
6. A. Clarke, P. Rothery, Scaling of body temperature in mammals and birds. Funct. Ecol. **22**, 58–67 (2007)
7. J.R. Usherwood, F.-O. Lehmann, Phasing of dragonfly wings can improve aerodynamic efficiency by removing swirl. J R Soc. Interface **5**, 13030–13037 (2008)
8. R.K. Josephson, J.G. Malamud, D.A. Sokes, Power output by an asynchronous flight muscle from a beetle. J. Exp. Biol. **203**, 2667–2689 (2009)
9. R.K. Josephson, Comparative physiology of insect flight muscle, in *Nature's Versatile Engine: Insect Flight Muscle Inside and Out*, ed. by J.E. Vigoreaux (Landes Bioscience - Springer, New York, 2006)
10. R. Dudley, *The Biomechanics of Insect Flight* (Princeton University Press, Princeton, NJ, 2000)
11. M.J. Hauser, L.F. Olsen, Role of naturally occurring phenols in inducing oscillations in the peroxidase-oxidase reaction. Biochemistry **37**, 2458 (1988)
12. D.J. Lyons, E. Horjales-Araujo, C. Broberger, Synchronized network oscillations in rat tuberoinfundibular dopamine neurons. Neuron **21**, 7261 (2001)
13. R.McN. Alexander, A.S. Jayes, A dynamic similarity hypothesis fort he gait of quadruped mammals. J. Zool. Lond. **201** 135–152 (1983)
14. P.W. Bridgman, *Dimensional Analysis* (Yale University Press, New Haven, 1931)
15. G.I. Barenblatt, *Scaling, Self-similarity, and Intermediate Asymptotics* (Cambridge University Press, Cambridge, 1996)
16. R.McN. Alexander, Walking and running. Math Gaz. **80**, 262–266 (1996)
17. C.J. Pennycuick, Predicting wingbeat frequency and wavelength of birds. J. Expt. Biol. **150**, 171 (1990)
18. E.J. Marey, The flight of birds and insects. Am. Nat. **5**, 29–33 (1871)
19. D.M. Byrne, S.L. Buchmann, H.G. Spangler, Relationship between wing loading, wingbeat frequency and body mass in homopterous insects. J. Exp. Biol. **135**, 9–23 (1988)

20. R.J. Wootton, D.J.S. Newman, Whitefly have the highest contraction frequencies yet recorded in non-fibrillar muscles. Nature **280**, 402–403 (1979)
21. T. Weis-Fogh, Energetics of hovering flight in hummingbirds and Drosophila. J. Exp. Biol. **56**, 79–104 (1972)
22. I.A. Johnston, D. Ball, Thermal stress and muscle function in fish, p. 79–104, in *Global Warming: Implications for Freshwater and Marine Fish*, ed. by C.M. Wood, D.G. McDonald (Cambridge University Press, New York, 1997)
23. R. Bainbridge, The speed of swimming fish as related to size and to the frequency and amplitude of the tail beat. J. Exp. Biol. **35**, 109–133 (1958)
24. F. Broell, C.T. Taggart, Scaling in free-swimming fish and implications for measuring size-at-time in the wild. PLoS ONE **12**, e0144875 (2015)
25. P.W. Webb, Hydrodynamics and energetics of fish propulsion. Bull. Fish. Res. Board Can. **190**, 159 (1975)

Further Reading

26. T.A. McMahon, J.T. Bonner, *On Size and Life* (Scientific American Books, New York, 1983)
27. S. Vogel, *Life's Devices: The Physical World of Animals and Plants* (Princeton University Press, Princeton, 1988)
28. C.J. Pennycuick, *Newton Rules Biology: A Physical Approach to Biological Problems* (Oxford University Press, Oxford, 1992)
29. R. Dudley, *The Biomechanics of Insect Flight: Form, Function, Evolution* (Princeton University Press, Princeton, 2002)
30. L. Kreitzman, R. Foster, *G: Rhythms of Life: The Biological Clocks that Control the Daily Lives of Every Living Thing* (Yale University Press, New Haven, 2004)
31. Q. Boone, R. Moore, *Biology of Fishes* (Taylor & Francis, London, 2008)
32. ThY Wu, Fish swimming and bird/insect flight. Ann. Rev. Fluid Mech. **43**, 25–58 (2011)

Chapter 11
Shapes of the Living

Abstract The relationship between form and function in living systems is treated, starting from the difference between volume and surface forces, and their different scaling with size. Surface tension appears as the main player in this context, contrasting the bulk effects of gravity. Chemical gradients and synchronised oscillators are the two other protagonists. The apparent regularities of many natural patterns and forms provide the excuse to describe a range of naturally occurring shapes, also allowing to make interesting links with palaeontology and fossil remains of ancient life on Earth. This chapter owes a lot to the original works of D'Arcy Wentworth Thompson, the celebrated pioneer of mathematical biology, especially in the parts dealing with the mathematics of geometrical transformations, and their relationship with the evolution of species.

11.1 Surface Forces and Volume Forces

When observing Nature, we are often surprised by the recurrence of seemingly very regular shapes and patterns: spirals, spheres, icosahedra, hexagons... It is just obvious to ask which are the forces at play to obtain such shapes, or why, for example, an hexagonal pattern appears in such distant domains as the structure of the wings of an insect, a snowflake, or the carapace of a turtle.

Animals, trees, tissues, cells, every element of Nature has both a **shape** and a specific **function**. What is the relationship between these two terms? Is the shape depending on the function, being only determined by the external forces? For example, the shape of the muscles of animals living in the depth of the ocean or on the Earth's surface, is determined by Archimede's pull or, respectively, gravity, without any reference to the function of that muscle? Or, rather, is the function which wins over the forces, and choses an elongated or a flat shape for some muscles, irrespectively of the gravity and weight of the animal? Could a muscle take whatever shape, e.g. cylindrical or flat, once the appropriate size, fibre density, rapidity etc. are assured, without impacting on its performances?

© Springer International Publishing Switzerland 2016

475

F. Cleri, *The Physics of Living Systems*, Undergraduate Lecture
Notes in Physics, DOI 10.1007/978-3-319-30647-6_11

In many flowers, the number of petals is one of the numbers that occur in the famous Fibonacci sequence, 3, 5, 8, 13, 21, 34, 55, 89... For instance, lilies have three petals, buttercups have five, many delphiniums have eight, marigolds have thirteen, asters have twenty-one, and most daisies have thirty-four, fifty-five, or eighty-nine. (Many flowers have indeed six petals, but this would fit under a different, possible explanation!) The view of the biologist would be that the genes in the flower cells specify all such information. However, it does not follow automatically that genes determine everything, directly or indirectly. For example, genes tell plants they have to make some light-harvesting compound, which we call chlorophyll, but they do not specify what colour the chlorophyll has to be. Chlorophyll is green, most likely because the solar light spectrum maximum intensity is about the green light wavelength, and rejecting the green component is a form of "defence" from too intense light; so, it is more likely that the plant genes adapted to the green colour and not vice versa, also when noting that none of the many forms of chlorophyll (including those who give off more red or orange colours) do not absorb much in the green region. Therefore, it can be said that some features of living systems are genetic in origin, and some are a consequence of the boundaries set by physics and chemistry, and mechanical, electrical, chemical forces. Genetic driving forces have enormous flexibility, but physics, chemistry, and dynamics produce apparent mathematical regularities. After all, this is why we look for mathematical models of natural shapes and patterns.

The celebrated book by W. D'Arcy Thompson, *On Growth and Form*, already cited several times in these pages, was the first to expose with depth and clarity the idea that the shapes of natural objects, and notably the living ones, appear to be mainly decided by criteria of equilibrium between **volume forces** and **surface forces**. Volume forces act on the whole body of the object, for example gravity. On the other hand, surface forces are limited to the free surfaces of the object, for example water pressure on the skin. In this chapter, however, two other major actors of the shaping and patterning of natural objects will also be introduced: the non-homogeneous distribution of chemical species and forces, or **gradients**; and the need for spatial and temporal **synchronisation**, by which different agents ('oscillators') can operate in a coordinated fashion, either at the molecular, cellular or whole-organism level.

The origin of surface forces is entirely molecular. If we consider a liquid body, for simplicity made of only one species of molecules, the requirement of forming a homogeneous phase makes the molecules to attract each other via non-specific forces. In the liquid bulk (Fig. 11.1), a molecule is attracted on every side by its neighbours: the resulting force on each molecule at equilibrium is zero. On the other hand, in empty space a molecule does not feel any attraction. Therefore, molecules at the liquid/void border feel an attractive force on the liquid side, but nothing on the void side (considering air instead of void would not change the qualitative picture, see below). For these molecules at the surface, the resulting force is non-zero, directed towards the bulk of the liquid. This force tends to curve the free surface, to minimise the amount of surface exposed to the void.

For the case of a liquid/void interface, it is entirely the effect of internal forces inside the liquid which brings the surface to deform like an elastic membrane (purple

Fig. 11.1 Molecular origin of the surface forces. Those molecules (*purple layer*) which have unsaturated interactions with the bulk (*blue*) experience a reduced attraction, therefore a 'missing' binding energy. The tendency of the body is to reduce the number of such molecules at the smallest, therefore any surface will assume the shape that corresponds to the smallest area, compatibly with the boundary conditions

shaded area in Fig. 11.1). If there is another gas bounding the free surface (air, for example) the phenomenon is completely similar. The liquid will be submitted to the additional pressure from the gas, and the resulting force on the surface molecules will be the sum of the attraction from the bulk liquid, plus the interaction (attractive or, more likely, repulsive) with the gas molecules, plus the gravity. Since the gas is much less dense than the liquid (typically by a factor 1/1000), the interaction with the gas can be practically neglected.[1] The final shape of the surface will thus result from the equilibrium among gas pressure, bulk liquid attraction, and gravity.

At the point of contact between the three different media (solid container, liquid, surrounding gas), the interfacial energies between each couple (Σ_{SL} for solid/liquid, Σ_{SG} for solid/gas, Σ_{LG} for liquid/gas, with units of [Energy] [L^{-2}]) must satisfy a relationship, which ultimately determines the value of the contact angle θ:

$$0 = \Sigma_{SG} - \Sigma_{SL} - \Sigma_{LG} \cos\theta \qquad (11.1)$$

This is the *Young-Dupré equation*, whose practical meaning is sketched in Fig. 11.2. It was first established by T. Young (right guess, he was the same guy of the Young's modulus!), and extended by the French physicists Athanase and Paul Dupré in 1866, to take into account thermodynamic effects. Values of the contact angle smaller than 90° correspond to a good "wettability" of the surface, which is therefore termed **hydrophylic** (since in practical cases regarding living systems the liquid is always water, this term is retained even for contact with different types of liquids). For larger angles, and up to the limiting value of $\theta = 120°$, the liquid tries to lose contact with the surface, in order to maximise the overall energy. The

---

[1]A notable exception in this respect is liquid Helium, whose intermolecular interaction is weaker than any other molecule-molecule force. Because of this peculiarity He, which is a liquid at temperatures $T \lesssim 2$ K, gives rise to the famous "fountain effect": since the interaction of the He molecules with the wall is stronger than that between the molecules themselves, the fluid can climb the walls of any open container, and spill out following the direction of gravity.

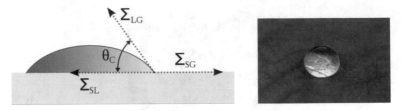

Fig. 11.2 *Left* Schematic of the interface energies acting on the triple-phase contact point, and the resulting contact angle. *Right* A droplet of water on the surface of a lotus leaf, showing the case of superhydrophobicity with a contact angle $\theta \simeq 150°$

free surface takes therefore a more and more spheroidal shape, being now termed **hydrophobic**. For even larger values of θ the surface is defined "super-hydrophobic", a phenomenon which is observed sometimes in nature (see again Fig. 11.2), and is today exploited in some technological applications.

The contrast between volume and surface forces is exemplified by their dependence on size. For an object of size L, the former is proportional to L^3 while the latter to L^2. As the English physiologist J.S.B. Haldane puts it, in his celebrated essay *On being of the right size* (1926, published in [1])), gravity is detrimental to larger animals as much as surface tension is for smaller animals. If we compare animals of different sizes, let us say L, $10L$ or $100L$ (for example, a spider, a mice and a dog) all three falling from some height, gravity will increase as L^3 while the air resistance, proportional to the surface, will increase as L^2. As a result, the spider will gently float in the air; the mice will fall straight on the ground and, after a little shock, will run away; while the dog will crash deadly. On the other hand, the same dog getting out of a river would carry a film of water, tightly adhering to its skin by surface tension, of thickness about 0.5 mm and a unconspicuous weight of about 1 kg; the mice getting out of water would instead have to carry a weight comparable with that of its own body; and the insect (if wetted) would be completely lost and never get out of water.

11.2 Capillarity, Growing Trees and Water-Walkers

The same equilibrium condition above among three different media (solid, liquid, gas) is at the basis of the phenomena of **capillarity**. This is defined as the capability of a liquid to flow within a narrow channel without any help from, and often against, external forces.

It is our common experience that, in order to make a fluid flow, it is always necessary an external force: gravity sets in motion rivers, torrents, and snow from the mountain sides; a pressure is necessary to push water through a garden sprinkler; heart pushes blood into our arteries. However, there are situations in which a fluid enclosed in a narrow space can *spontaneously* climb up against gravity. Think of what happens if you gently touch the surface of water with a piece of paper held vertically: the water invades the paper up to some height, well above its surface level, thanks to the capillary flow within the microscopic channels of the paper (which is a porous network of channels formed by the mixture of glue and cellulose).

A fluid in a narrow channel may take a positive or negative curvature at its free surface, called **meniscus** (from the ancient Greek μήνισχος, the name indicating the rising Moon). To calculate the height h reached by the meniscus, let us take the example of a narrow channel of radius R (usually called a *capillary*, the same name given to the terminal section of blood vases). We must compare the gravity acting on the mass of the fluid:

$$F_g = \rho g V = \rho g (\pi R^2 h) \tag{11.2}$$

and the adhesion forces at the triple interface, projected along the vertical:

$$F_c = 2\pi \Sigma_{LG} R \sin \theta \tag{11.3}$$

(the SL and SG tensions have zero vertical component). By equating the two forces, the equilibrium height is:

$$h = \frac{2\Sigma_{LG} \sin \theta}{\rho g R} \simeq 1.48 \times 10^{-5}/R \tag{11.4}$$

(the last approximate equality corresponding to the numerical values of the air-water interface, with R expressed in meters), giving for example $h = 0.7$ mm for a diameter $2R = 8$ cm, and $h = 70$ mm for a diameter $2R = 0.8$ mm. The attentive reader will have already noticed that Eq. (11.4) is yet another form of the Laplace's equation of Chap. 5, $\Delta P = \rho g h = 2\Sigma/R$, for the ideal case of a perfectly spherical meniscus whose tangent is perpendicular to the surface, and $\sin \theta = 1$.

This effect allows some insects to walk on the water without sinking, since for them the repulsion between solid (the insect's legs) and liquid is stronger than the gravity acting on the (very small) mass of the insect. Capillarity also explains how trees can feed themselves by pushing sap up to the highest leaves, the formation of soap bubbles, and a variety of other physical phenomena (see "Further reading" at the end of this chapter).

11.2.1 Insects Who Can Walk on the Water

Many different insects have the ability to walk on the surface of still water. By looking at the typical shape of the insect (see Fig. 11.3), it appears obvious that the very elongated and thin legs must be the key for this spectacular property, together with an overall light weight of the body. The rigid keratin structure of the legs makes a triple interface with the water surface and the surrounding air. If we look carefully at the surface of the water around the legs, it will be apparent that it is sharply curved, under

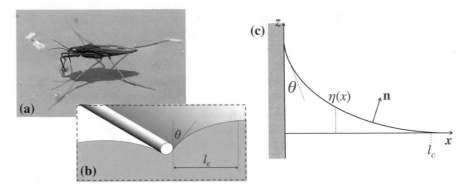

Fig. 11.3 a A water strider standing on water without sinking. **b** Schematic of a cylindrical section of the insect's leg pushing on the water surface, making a contact angle θ; the approximate extent of the capillarity length is indicated by l_c. **c** Diagram of the function $\eta(x)$ describing the curved water surface (upside down with respect to the *middle figure*) as a function of the distance x from the leg. [Photo **a** from www.en.wikipedia.org/wiki/Gerridae, repr. under CC-BY-SA 3.0 license, see (*) for terms.]

the light weight of the insect body. Compared to the previously described situations, this is a static problem in which no fluid is flowing. However, the equilibrium of the forces at the triple interface can be studied with the very same equations of the capillarity.

The shape of the meniscus formed under the weight of the insect at the air/water interface, can be found by writing the Laplace equation in geometric form:

$$P = \rho g z = \Sigma \left(\frac{2}{R} \right) = \Sigma (\nabla \cdot \mathbf{n}) \tag{11.5}$$

Here the local curvature is defined by the gradient of the unit vector $\hat{\mathbf{n}}$, perpendicular to the local tangent plane to the curved surface (see the greybox on p. 351).

By looking at the graph in Fig. 11.3c, the z-coordinate measures the depth of curvature with respect to the flat water surface, and the x-coordinate measures the parallel distance on the water surface from the insect leg. Let us introduce an (unknown) function defining the height profile of the water/air separation surface, as $z = \eta(x)$, and from this we can build a function $f(x, z) = z - \eta(x)$ which is always equal to 0 at the surface. The unit vector normal to the surface can therefore be calculated as:

$$\hat{\mathbf{n}} = \frac{\nabla f}{|\nabla f|} = \frac{-\eta_x \hat{x} + \hat{z}}{(1 + \eta_x^2)^{1/2}} \tag{11.6}$$

and its gradient:

$$\nabla \cdot \hat{\mathbf{n}} = \frac{\eta_{xx}}{(1 + \eta_x^2)^{3/2}} \simeq \eta_{xx} \tag{11.7}$$

the subscripts x and xx indicating the first and second derivative with respect to x (note that in this geometry the y component is null). The approximation in the last equation holds for a weak meniscus, $\eta_x \ll 1$.

The equation for the pressure then becomes $P = \rho g \eta = \Sigma \eta_{xx}$, to be solved with the boundary conditions $\eta_{xx}(\infty) = 0$ (meaning null curvature at infinite distance), and $\eta_x(0) = \cot \theta$ (condition of contact at the triple interface). The solution for the unknown shape of the profile is:

$$\eta(x) = l_c \cot \theta e^{-x/l_c} \tag{11.8}$$

The characteristic length $l_c = (\Sigma/\rho g)^{1/2}$ is called the **capillarity length**: it represents the distance over which the water meniscus decreases to zero, with an exponential decay.

A long and thin object like the insect leg (Fig. 11.3b) resting on the water surface, will suffer the buoyant force F_b, proportional to the displaced water volume according to Archimede's law; this will add up to the curvature force F_c produced by the curved meniscus. The ratio between the two forces is simply proportional to the ratio of the two volumes, whose characteristic lengths are respectively $V_b \propto R$ (for a cylindrical object of radius R) and $V_c \propto l_c$. As a consequence, the ratio of forces must be $F_b/F_c \propto R/l_c$. A very small object, with $R/l_c \ll 1$, will be principally supported by the curvature force of the water surface, rather than by its floating: the insect will not sink, but rather will stay on the surface. The practical criterion is embodied in the dimensionless number:

$$\frac{F_b}{F_c} \simeq \frac{Mg}{2\Sigma L \sin \theta} < 1 \tag{11.9}$$

(M is the total mass of the insect, while L is the total length of its legs contacting water), which has to be less than 1 in order for the object to float. This is a practical variant of the Bond number, $Bo = (\rho_o - \rho_m)L^2 g/\Sigma$, which more correctly underscores the mass as being in fact the difference between the actual mass of the object "o", and that of the volume of fluid medium "m" displaced.

11.2.2   The Branching of Trees

Equation (11.4), which gives the practical height of a meniscus, is also at the basis of the phenomenon of upwards capillary flow of water (in fact, a mixture of water an nutrients called *sap*) in the minuscule channels of the plant stems, or the trunk of trees, called **xylems**. Xylem vessels consist of dead cells; they have a thick, strengthened cellulose cell wall with a hollow lumen. The typical diameters of such capillaries is of the order of 20–30 μm for any plants, or $R = 10–15$ μm. If we calculate the maximum height reachable with the capillary pressure from such sizes, we obtain

Fig. 11.4 **a** Maximum-density hexagonal packing in two dimensions, for circles of radius r inside a hexagon of side $2r$; the diagonals of the hexagon are equal to $4r$. **a** Sketches 9 and 10 from Table XIX of Leonardo Da Vinci's *Trattato Della Pittura* [2], describing his intuition about trees growing by conserving the cross section at each successive branching. [Image **b** public domain from http://archive.org/.]

$h \simeq 10-15$ m. This value can be enough for a plant, but it does not explain how we can have trees growing up to tens of meters high.

A possible solution is *branching*. We can give an extremely simplified theoretical description of a tree, made of single channels of diameter $2R$ and height h_1, parallel to each other. We want to calculate what is the effect of branching the tree at its extremity, each single channel branching into N smaller channels of diameter $2r$ for a height h_2, so that the maximum height could be $H = h_1 + h_2$.

To be a bit more precise, we can also consider the best packing density η, of a parallel bundle of cylindrical channels of diameter $2r$: it is easily shown (left of Fig. 11.4) that the densest packing in plane corresponds to an hexagonal tiling, with side equal to just $2r$. Since each hexagon encloses three channels, the ratio of the useful surface to the total is $\eta = \pi/(2\sqrt{3}) = 0.907$. This is the maximum packing density in two dimensions.

To conserve the same quantity of fluid transported by capillarity in the two sections of our "tree", it is enough to impose the conservation of the transverse cross section, $N\pi r^2 = \eta \pi R^2$, an idea that was first exposed by Leonardo (right of Fig. 11.4).[2] Hence, the capillarity force acting in the lower trunk plus the N branches of the thinner xylems can be calculated as:

$$F_{c,N} = 2\pi \Sigma (R + Nr) \cos\theta \simeq 2\pi \Sigma (R + Nr) =$$

$$\tag{11.10}$$

$$= 2\pi R \Sigma \left[1 + (\eta N)^{1/2}\right] = F_{c,1} \left[1 + (\eta N)^{1/2}\right]$$

Note that this force does not depend on the height, and it increases with the square root of N. On the other hand, the total gravity force on the fluid is:

---

[2]In his *Treatise on Painting* [2], collected and published posthumously, Part VI, 813: *Ogni biforcazione di rami insieme giunta ricompone la grossezza del ramo che con essa si congiunge [...] e questo nasce perché l'umore del più grosso si divide secondo i rami.* [Each bifurcation joined together gives the same thickness of the branch from which it stems (...) and this because the fluid of the thicker branch is divided according to the thinner ones.]

$$F_{g,N} = \rho g(h_1 \pi R^2 + N h_2 \pi r^2) = F_{g,1}\left[1 + \eta\left(\tfrac{h_2}{h_1}\right)\right] \tag{11.11}$$

which evidently does not depend on N, but on h_2. It so turns out that, in order to maximise the height, such a weird shaped tree should have $N \rightarrow \infty$, but to minimise the gravity, $h_2 \rightarrow 0$. The practical considerations that we can derive from such an overly simplified model of capillary feeding, is that a tree should typically develop a large number of branches N in its upper part, and no branches in its lower part; and that the branched part should be much less important in size than the trunk, $h_2 \ll h_1$. In fact, this is just what it is observed in nature for a large variety of plants and trees. A tree structure is the outcome of a combination of hydraulics and structural constraints. This hydrological explanation seems to suggests that trees have their characteristic shape because of efficiently transporting sap, while structural explanations rather focus on the trees' ability to withstand stresses (see the discussion about stresses acting in bending of trees in the wind, in Chap. 9).

It should, however, be noted that the capillarity mechanism could operate only until mechanical equilibrium is attained, after which the transport of sap towards the top of the tree would stop. It is the evaporation of water from the leaves surface, combined with the difference in saline concentrations in the soil at the level of the roots, which provide an effective pressure (depression from evaporation, plus osmotic pressure from salt concentrations). This pressure constantly keeps the system out of equilibrium, thereby permitting the continuous flux of sap and nutrients from the bottom to the top.[3]

It is instructive to calculate the pressure difference that a tree must overcome, to push the fluid up to a certain height by such a capillarity mechanism. Just by hydraulic arguments, Stevin's law gives $P = \rho g h$. Therefore, the pressure to a height of 30 m is already $\sim 3 \times 10^5$ Pa (or about 3 atm). This pressure, in fact exerted by the surface tension of the capillary, is *negative*, as it pushes the water upwards against the gravity.[4] We encountered already in Chap. 5 the relationship among pressure drop per unit length, pipe radius, and flow speed, i.e. the Hagen-Poiseuille equation:

$$\frac{\Delta P}{L} = \frac{4\eta v_{max}}{r^2} \tag{11.12}$$

From such equation, it would seem that even higher values of pressure drop could be overcome, just by making the capillary thinner and thinner. However, another physiological limit comes from the fluid flow speed. As we remember from Chap. 4, the velocity profile of a fluid pushed through a channel is such that $v = 0$ at the

---

[3]The first detailed study of the movement of water within plants was provided by Stephen Hales in his *Vegetable Staticks* (London, 1727). Not only he described with great accuracy the transpiration stream of plants, by performing quite accurate experiments of collection of evaporated water from leaves, but also he sought to interpret his observations in light of the fluid mechanics knowledge of his time: *The sap vessels are so curiously adapted by their exceeding fineness, to raise the sap to great heights, in a reciprocal proportion to their very minute diameters.*

[4]A negative pressure must not be a surprising concept, since the fluid has a cohesive force, both internally and in contact with the walls, to spend in supporting the stress.

wall boundary, and $v = v_{max} \propto r^2$ at the centre of the stream. Evidently, making the capillary too thin implies a quickly decreasing flow speed. For a gravity pressure drop of $\rho g = 9\,810$ Pa/m, a capillary of 100 μm permits a flow speed $v_{max} \simeq 2.5$ cm/s; reducing the size to 30 μm drops the flow speed to a drastic 2 mm/s, and 5 μm to a mere 0.06 mm/s. Therefore, it can be concluded that a tree must always balance the requirement of pushing its nutrients to the required height, however in a reasonable time.

11.3 Curved Surfaces and Minimal Surfaces

Under conditions of mechanical equilibrium among all the forces acting on the volume and surface of an object, this will take a shape of its free boundary allowing to minimise the forces acting on the surface. This is because the volume forces, being related to the amount of mass, are independent on the shape of the object. On the contrary, the surface can largely change for a fixed volume, and so can the surface forces. If we take that surface forces depend only on the microscopic parameters (physical and chemical) of the material constituting the body, the equilibrium requirement therefore corresponds to minimising the extent of free surface. In other words, the object will adopt the shape which permits it to settle to the smallest free surface for a given volume. This will be called the **minimal surface**.

In void, any isolated fluid tends to a spherical shape since the sphere has the smallest possible surface for a given volume. The surface/volume ratio for some tridimensional solids is:

$$\frac{S}{V} = \frac{4\pi R^2}{\frac{4}{3}\pi R^3} = \frac{3}{R} \qquad \text{for the sphere of radius } R$$

$$\frac{S}{V} = \frac{6L^2}{L^3} = \frac{6}{L} \qquad \text{cube of side } L$$

$$\frac{S}{V} = \frac{2(\pi R^2) + 2\pi R H}{\pi R^2 H} = 2\left(\frac{1}{R} + \frac{1}{H}\right) \qquad \text{cylinder of radius } R \text{ and length } H = R/c$$

From such simple expressions, it is seen that for the same volume V, the sphere has a surface $(36\pi)^{1/3}/6 \simeq 0.81$ of that of the cube of side L ($L^3 = 4\pi R^3/3$); for the cylinder, the surface is always larger that that of the sphere for any aspect ratio c, by a factor $(4c/3)^{2/3}(1 + \frac{1}{c})$. Therefore, until the material elasticity properties allows continued deformation, the free floating object will choose the sphere as the minimal-surface solid shape.

Deviations from the spherical shape can occur when the system is not isolated in void: for example a water drop in contact with a glass surface becomes hemispherical, because the intermolecular forces between water and glass come into play (equilibrium among the interfaces glass-water, water-air, air-glass). In such cases, we realise that the boundary conditions give rise to supplementary constraints, going beyond the basic requirement of minimising the free surface.

If we take for example a free thin film of a mix of water and soap, this too will tend to make a spherical shape enclosing an air volume at ambient pressure, the familiar soap bubble. The way soap bubbles are generated, by a tool in the shape of a solid ring, made the subject of a series of outstanding, yet simple experiments by the Belgian physicist Jean-François Plateau, around 1870 (see Figs. (11.10) and (11.11)). If we consider the thin film suspended at the rim of a metal ring of radius R, as in the following Fig. 11.5, this takes initially the shape of a flat plane delimited by the ring perimeter; subsequently, the blowing of air on one side sets up a pressure difference, which starts curving the surface of the film; when the pressure difference goes beyond some limiting value (determined by the relative values of the surface tensions Σ_i and R), the film is detached from the ring, and folds back into a sphere.

With the case of the soap bubbles we moved from the simple, homogeneous fluids, studied inside an ideal container (such as the perfect gas), to a realm of more complex systems. Here, a fluid is contained by a wall, which is endowed with some material characteristic different from the material of the surrounding medium (in the example above, the metal ring, and the air). More specifically, the soap bubble gives us the example of a physical membrane that is supposed to contain a fluid medium. In order to perform this job, the membrane must have a higher mechanical strength than the fluid, which exerts a pressure from the inside. The membrane can be deformable up to a maximum failures stress, σ_f (see Appendix H), determined by the materials constituting the membrane, be it the metal body of a car, or a plastic balloon, or a soap bubble, or the membrane of a cell composed by a double layer of lipid molecules.

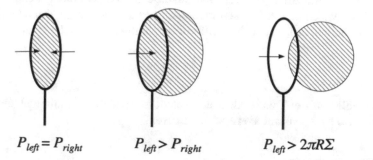

$$P_{left} = P_{right} \qquad\qquad P_{left} > P_{right} \qquad\qquad P_{left} > 2\pi R\Sigma$$

Fig. 11.5 Action of pressure on a thin film of soap-water held by a ring of radius R. *Left* No extra pressure applied, the film is in equilibrium. *Center* A slight pressure on one side, contrasted by the surface tension Σ_{SL} at the film-ring contact, bows the surface into a part of a spheroid. *Right* if the pressure is beyond the force exerted by the surface tension, the film can detach completely from the ring and fold back into its minimal surface shape, a sphere

In fact, the analogy between the soap bubble and the cell membrane is especially appropriate, since the soap bubble is itself a double layer of lipid molecules, however enclosing a thin film of water. (This is the reason for the phenomena of iridescence observed when light crosses a soap bubble: the water contained between the double thin layer of soap molecules behaves as a Newtonian prism, the solar light rebounding several times between the walls before exiting decomposed into its wavelengths.)

The surface tension Σ is a material property of the separation surface. The same symbol of the interfacial energy is adopted, to underscore that the surface tension is nothing else but the interfacial energy between two fluids.[5] The relationship between Σ, the pressure difference ΔP_i on the two sides, and the curvatures of the surface, identified by the principal curvature radii R_i, is given by the following expression, again the Laplace equation in a more general form:

$$\Delta P = P_1 - P_2 = \Sigma \left(\frac{1}{R_1} + \frac{1}{R_2} \right) \qquad (11.13)$$

We already saw a simplified form of this equation in Chap. 5, $\Sigma = PR/2$, where it was established a relationship between surface tension and osmotic pressure. In all these equations, the pressure has an algebraic sign as a function of the orientation of the force vector with respect to the vector $\mathbf{n}$. (The principal curvatures defined by the min and max curvature radii in Fig. 11.6 carry a $\pm$ sign as well, according to the convention adopted for the pressure, determined by the fact that the radius vector $\mathbf{R}$ can be parallel or antiparallel to $\mathbf{n}$.)

Equation (11.13) is at the basis of an extraordinary variety of natural shapes and patterns, ranging from the wings of some insects, to the curve of birds' eggs, networks of soap froths, bees' honeycombs, and so on (Fig. 11.7). It is the expression of an equilibrium condition: the ratio between the left-hand side, the resistance to the compression of the fluid mass enclosed (ΔP), and the right-hand side, the product of surface tension times the curvature, ($\Sigma/R_1 + \Sigma/R_2$), must be constant. The implied condition is that the surface of separation can be identified as a homogeneous pseudo-material, for which Σ is a constant parameter of the fluid, on the same level as its density or thermal capacity. This equilibrium condition leads to the result:

$$\frac{1}{R_1} + \frac{1}{R_2} = const \qquad (11.14)$$

on the whole surface. This result is equivalent to saying that a minimal surface is characterised by a constant **average curvature**.

---

[5]The interfacial energy is a more general concept, applicable to whatever kind of material composing the two sides of the interface. It can be anisotropic, if the materials on either side have different values of elastic moduli along different directions. Therefore, the interface energy is a tensor, with indices exposing the eventual anisotropy of the adjoining surfaces. A fluid is by definition isotropic, therefore its elastic moduli are independent on the direction, and the surface tension is just a scalar quantity.

Fig. 11.6 Schematic
defining the principal
curvature radii about any
point O of a generic surface
(*pink*). The vector **n** is the
oriented unit vector normal
to the plane (*grey*) locally
tangent to the point O. The
curvatures are the inverse of
these radii, $\gamma_{11} = 1/R_1$,
$\gamma_{22} = 1/R_2$. The two radii
R_1 and R_2 can each be on
either side of the surface

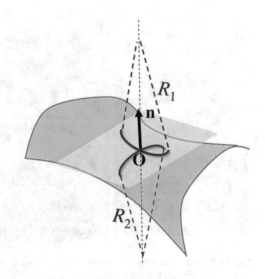

11.3.1 How the Space Can Be Filled

If we look at the examples of Fig. 11.7, and to countless many other situations of
natural patterns, one important character that should get our attention is that space
is very seldom left empty, in two as well as three dimensions. Of course, there
are cases in which there is the need for some "holes" in a natural system, but the
previous statement is very close to a general rule (even the inner body cavities of
animals and plants are always filled with some fluid). The problem of space filling
has kept mathematicians busy for centuries, often with misplaced conjectures which
could not be proven, but which couldn't be easily disproved either.

In two dimensions, a flat plane can be filled only by identical triangles or by
identical rectangles. Combinations of six adjacent triangles give a hexagon, and this
is one good reason why space-filling problems with 2-dimensional symmetry are
often solved by nearly regular hexagonal patterns, such as the honeycomb structures
shown in Fig. 11.7. The constraint of using strictly regular geometrical shapes, i.e.
having the same length of sides, perimeter, and area, is of course very often relaxed
in natural objects. However, the underlying principle is that whenever the source
material and environment are sufficiently homogeneous, nearly regular shapes are
easily obtained. If we think again of diffusion-based mechanisms as the drivers of
the growth, and assume the initial growth seeds are randomly distributed, it is very
likely that the resulting subdivisions will end up having similar lengths and areas.
If, instead, the environment presents irregularities, such as material heterogeneity
or a gradient of some parameter, then also the filling can be graded. This could be
the case of the curved (non-planar) surface of the turtle shell, in which the central
hexagons are typically larger than the surrounding ones, progressively smaller as the
surface curvature decreases away from the centre.

Fig. 11.7 In two dimensions, the plane can be divided regularly into equal-area regions by a network of segments connecting an hexagonal pattern of vertices, at each of which three segments join by forming angles of 120°. Approximately hexagonal patterns are found in a surprising variety of natural, living or non-living, systems: **a** a bees' honeycomb; **b** a green turtle from Hawaii islands; **c** columnar basalts at the Cape Stolbchatiy, Kuril islands, Russia; **d** ice crystals under scanning electron microscope (colours added). [Photos © by: **a** Wangsberg, **b** Mak Thorpe, **c** Igor Shpilenok, **d** unknown at US Dept. Agriculture. All repr. under CC-BY-SA 3.0 licence, see (*) for terms.]

If some degree of deformation of the base polygon is admitted, it turns out that also *irregular* pentagons can be used to fill the two-dimensional space. Up to now, fifteen space-filling pentagonal patterns have been found (Fig. 11.8). The first five were discovered by the German mathematician K. Reinhardt in 1918, and it was necessary to wait fifty years before R. Kershner discovered three more. A few years later R. James, a Californian computer scientist, discovered a ninth pattern, but in the same year 1975 a San Diego housewife passionate about mathematics had already discovered another four. Ten years after, the German mathematician Rolf Stein (a graduate student at the time) found a fourteenth one, and a merely thirty years later, in 2015, a fifteenth pattern was discovered. There is no way to establish whether there could be more patterns; it is however rigorously proved that no identical polygons with seven or more sides can produce a space-filling pattern, no matter how irregular they may be.

Mixed-polygon fillings are called **Archimedean tilings** in honour of the revered Greek scientist, but were actually firstly listed by Johannes Kepler in 1619: he discovered that triangles, squares, hexagons, octagons and decagons can be combined into at least eight patterns, all infinitely periodic. However, natural objects are always necessarily *finite* in size. Therefore, the constraint of periodicity in space filling need not be satisfied over the entire length of the object. This opens the way to much richer ways of building regular patterns and shapes, by using symmetric but non-periodic motifs, such as spirals (see below, Sect. 11.7), and tile-sets, or *prototiles*, made up by

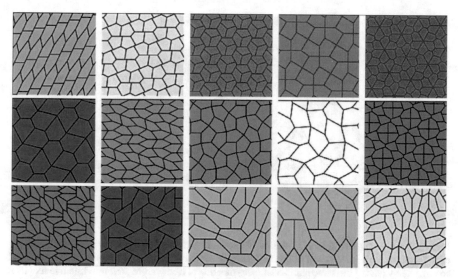

Fig. 11.8 Repeated patterns of irregular pentagons, having sides of different length and non-equal internal angles, can be used to densely fill the two-dimensional plane. The fifteen patterns discovered between 1918 and 2015 are shown. Note that vertices can be formed only by joining either three, four or six sides, and never five

more than one regular polygon. The chief example of the latter category is the *Penrose tiling*, named after the English mathematician and physicist Roger Penrose who initiated this field of study in the early 1970s. He initially found a set of six tiles that force aperiodicity, which later he was able to reduce to only two. The characteristic feature of the Penrose tiling scheme is that it allows a local pentagonal symmetry, while no infinitely-repeated pattern could be constructed with such a symmetry. Examples of both Archimedean and Penrose tiling abound in crystallography, with pentagonal quasi-crystals discovered in 1982 by Dan Shechtman, who obtained for this the Nobel prize in chemistry. However, such exotic regularities are not easily identified in the shapes and patterns of living systems.

Filling the space in three dimensions gives rise to an obviously more complicated affair. Plato, the Greek philosopher who lived between the V and IV century B.C., identified only five three-dimensional shapes as being the only **regular polyhedra**. The 'regularity' is defined by the characteristics of: (i) being made up by a certain number of faces, all from one type of polygon, and (ii) having a number of identical vertices at each of which the same number of faces meet. The only five shapes meeting such criteria are (Fig. 11.9): the tetrahedron (four triangular faces, meeting in triplets at four vertices), the cube (six square faces, meeting in triplets at eight vertices), the octahedron (eight triangular faces, meeting in quadruplets at six vertices), the dodecahedron (twelve pentagonal faces, meeting in triplets at twenty vertices), and

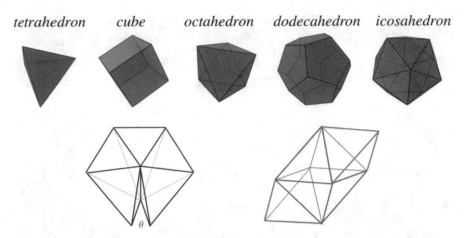

Fig. 11.9 *Above* the five "platonic" solids. These three-dimensional figures are defined by having each face from only one type of polygon (*triangles* for the tetrahedron, *squares* for the cube, and so on), and by having the same number of faces meeting at each vertex. *Below* While five identical tetrahedra cannot fill the space (*left*), a combined motif made of octahedra and tetrahedra (*right*) can indeed fill the whole 3D space

the icosahedron (twenty triangular faces, meeting in quintuplets at twelve vertices).[6] It is interesting to note that cube and octahedron are related by the fact that one can be obtained from the other by taking the centres of the faces of one as the vertices of the other; the same duality holds for the dodecahedron and the icosahedron; finally, the tetrahedron is dual of itself.

Of the five Platonic solids, only the cube, with its angles of 90° between perpendicular faces, can be used to completely fill the 3D space. The regular tetrahedron has an angle between any two faces equal to $\theta = \arctan(-1/3) = 109°47$. For more that 1800 years, it was believed (on the basis of an ancient conjecture by Aristotle) that also the tetrahedron could fill the three-dimensional space. In fact, this proves to be impossible (see again Fig. 11.9): if we join five identical tetrahedra, a small empty space is left as a thin wedge with an angle of about 7°; on the other hand, a combination of tetrahedra and octahedra can completely fill the space.

---

[6]The original Plato's description reads: *From the [equilateral] triangle [...] the three first regular solids are formed: first, the equilateral pyramid or tetrahedron; secondly, the octahedron; thirdly, the icosahedron; and from the isosceles triangle is formed the cube. And there is a fifth figure, the dodecahedron, which God used as a model for the twelvefold division of the Zodiac.* [English version by B. Jowett, The Dialogues of Plato, vol. 3, Oxford University Press, 1892].

Fig. 11.10 Joseph Plateau started experiments about equilibrium shapes with oil in a water-alcohol mixture. Later he changed to soap films, easier to manipulate, with similar or better results. When a figure made of thin wire is dipped in the liquid, elegant shapes of thin soap films form between the wires, "so light that they are not subjected to gravity" (as Plateau wrote), similar to those of oil films. Plateau used a solution he called "liquide glycérique": 3 parts of a watery, filtered solution of Marseille soap and 2 parts of pure glycerine. He mentions that it was not always simple to obtain the products in the necessary purity and concentration. [Images courtesy of the Ghent University Museum, Belgium, Collection 'History of Sciences'.]

11.3.2 Limiting Shapes, Stability and Instability

A **minimal surface** is a volume-bounding shape configuration, whose total area could be only increased by a small perturbation. In fact, a minimal surface is a two-dimensional analog of the geodesic curve (whose length also can only increase, under a small, localised perturbation).

The properties of minimal surfaces have been largely studied in the past (see "Further reading" at the end of this chapter) since being one of the most complex domains of mathematical physics. For example, one of the most important empirical discoveries in this field (again due to J. Plateau in the second half of the XIX century), namely the fact that the angles among soap bubbles can have only two possible values, of 120° in two dimensions, and 109°5 in three dimensions, could be rigorously proved only in 1976, by the American mathematician Jean Taylor [3]. Empirically, these two special values correspond well to the angles delimiting equal-area surfaces in 2D, or equal-volume regions in 3D. Therefore, they are connected with the practical realisation of minimal surfaces.

In the experiments by Plateau, the water-soap thin film was trapped within the rim of an adjustable metal ring (Fig. 11.10). The ring could be subsequently deformed, and the thin film followed its modification, by passing through a sequence of shapes and possible curvatures, always satisfying the constraint of minimising the surface for any boundary condition. Physically, such a constraint is automatically imposed by the surface tension of the water-soap solution. The stable surfaces pass from one into another via **limiting shapes**, corresponding to discontinuities imposed by the physical-chemical constraints (density, viscosity, surface tension). In fact, if the ring is deformed too quickly the soap bubble will explode, because the passage from one limiting shape to another is (topologically) discontinuous.

The properties of the limiting shapes are connected with the mathematical behaviour of the surface curvatures. There are only two cases for which the Eq. (11.13) is indefinitely stable (meaning, for any value of ΔP or T), and this is when the sum of the curvatures is zero. The first interesting case is that of the **catenoid**, a limiting shape having equally and opposite curvatures, so that the pressure (trace of the stress tensor) is zero:

$$-\frac{1}{R} + \frac{1}{R} = 0 \qquad\qquad (11.15)$$

The catenoid shares this unique property with the plane, for which both curvatures are null, $1/R_1 = 1/R_2 = 0$. No other limiting shape is characterised by having a zero average curvature.

Although not being unconditionally stable, a surface can however be locally stable up to a maximum value of an external perturbation. Plateau demonstrated empirically that only six geometrical figures could lead to shapes satisfying the conditions of local stability: (1) the sphere, (2) the plane, (3) the cylinder, (4) the catenoid, (5) the unduloid, and (6) the nodoid. In fact, the purely empirical studies by Plateau ignored that also other limiting surfaces could satisfy the mathematical conditions of local stability, such as the helicoid whose analytical shape was known already since 1776. Many other, very exotic shapes with minimal surfaces have been discovered after the 1960s, with the help of computer analysis. On the other hand, the shapes identified by Plateau remain the only ones practically observed in natural systems. For any other shape, further dimensional constraints must be added to ensure the local stability that, evidently, have no obvious physical analog to manifest themselves in naturally observed shapes.

Also in a dense fluid perturbed by effect of an external force, such as the gravity, the local stability could be lost and **instabilities** may start to develop. Plateau observed the development of instabilities in his experiments on oil and water mixtures (Fig. 11.11), in which case the external perturbing force was the centrifugal acceleration. Analogously, if we observe a column of water dripping from a faucet, and progressively reduce the speed of flow, gravity will at some point take over: the water cylinder would form thin constrictions, rather irregularly spaced, which would become thinner and thinner, up to the point when the elongated cylindrical shape will be fragmented into spherical droplets of different sizes. Similar instabilities can

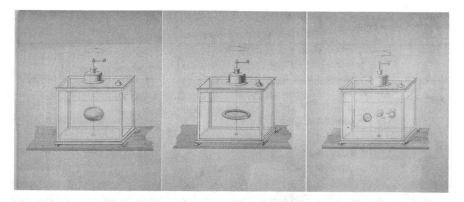

Fig. 11.11 The famous oil bubble experiment of Plateau. Oil is deposited along a vertical axis in a tank filled with a mixture of water and alcohol. The oil takes on a spherical form. If the oil and the water-alcohol mixture have the same density, the oil sphere freely floats. Then the oil sphere is "not subjected to gravity", as Plateau puts it. If one revolves the axis slowly (about one turn every 5 s) the oil sphere begins to flatten at the "poles". A maximum of this flattening occurs at 3–4 turns per second. If the angular velocity is still larger, the oil takes the form of a torus, loose from the axis. At even higher velocities, the torus breaks up into small spheres which rotate around their own axis. [Images courtesy of the Ghent University Museum, Belgium, Collection 'History of Sciences'.]

be observed in the stroboscope photos of fluids splashing on a hard surface, or in situations in which two fluid with widely different viscosity come into contact, for example dew wetting a spiderweb. Also in this case, the instabilities induced by the disequilibrium between gravity, surface tension, adhesive forces, and small random perturbations of the environment (variations in temperature, or atmospheric pressure, mechanical actions from the wind) can lead to mixed surface shapes, instead of a continuous surface.

11.4   Surfaces of Revolution, Seashells and Gastropods

The common feature shared by all the stable Plateau surfaces (sphere, cylinder, catenoid, unduloid, and nodoid; the plane being the degenerate case for which $R_1 = R_2 = \infty$), is that they are all **surfaces of revolution**.

A surface of revolution is a parametric surface in a n-dimensional space (we are here interested in the ordinary 3-D space, or $\mathbf{R}^3$), namely the surface generated by rotating a plane curve about an arbitrary axis. The general equation of the parametric surface $X(t, \theta)$, obtained by rotating about the conventional z-axis the parametric curve $c(t) = \{x(t), y(t), z(t)\}$, is given by the product of the rotation matrix about z times the curve itself:

$$X(t, \theta) = \begin{pmatrix} \cos\theta & -\sin\theta & 0 \\ \sin\theta & \cos\theta & 0 \\ 0 & 0 & 1 \end{pmatrix} \begin{pmatrix} x(t) \\ y(t) \\ z(t) \end{pmatrix} = \begin{pmatrix} x(t)\cos\theta - y(t)\sin\theta \\ x(t)\sin\theta + y(t)\cos\theta \\ z(t) \end{pmatrix} \quad (11.16)$$

The parameter t determines the span of the surface, typically ranging in the intervals $[-\infty, +\infty]$ or $[0, \infty]$. Some geometrical characteristics of these objects:

- A surface of revolution is globally invariant for any rotation about an axis, called the *axis of revolution*.
- The rotation of a curve (called *generatrix*) about the axis of revolution originates the surface of revolution.
- The sections of a surface of revolution by semi-infinite half-planes delimited by the axis of revolution, are called *meridians*; each one of them represents one instance of the generatrix.
- The sections of a surface of revolution by infinite planes perpendicular to the axis of revolution are circles of varying diameter, called *parallels*.
- A surface is identified as a surface of revolution if the normal through any of its points meets, or is parallel to, a fixed axis (in fact, the axis of revolution itself).

The following greybox contains several examples of surfaces of revolution, obtained from the rotation of trigonometric functions as the generating curve.

It may be often necessary to calculate the area of a surface of revolution, for example if we want to establish which shape gives the minimal or maximal area for a given object. For the simple surfaces shown in Fig. 11.12a, b the area is obtained by direct geometrical calculation. For the cylinder (obtained by rotating a straight segment of length h parallel to an axis), the area is just $A = 2\pi dh$, if d is the distance from the rotation axis. For a circular cone of slant height l, the area is $A = \pi rl$, if r is the radius of the limiting circle.

A surface of revolution can also be created by rotating a closed figure about an axis with which the figure has no points of contact. In this case, the axis of revolution does not intersect the surface, and the resulting volume is called a **toroid**. For example when a rectangle is rotated about an axis parallel to one of its edges, then a hollow, square-section ring is produced (Fig. 11.12c). If the revolved figure is a circle, the object is called a **torus**, with a surface equal to $A = 4\pi^2 rc$, where r is the radius of the revolving circle and c is the distance of its centre from the axis of revolution.

In general, a complex surface of revolution could be approximated by describing its generatrix as a sequence of small straight segments (Fig. 11.12d, e), each of which would give a contribution to the area as a portion of cylinder or cone (the former being a degenerate case of the latter). Each segment along the generatrix would be delimited by a pair of adjacent points, (P_{i-1}, P_i). For a segment of circular cone of slant height $l = |P_i - P_{i-1}|$, and upper and lower radii r_1 and r_2, the area of the surface segment is easily shown to be $A_i = 2\pi rl$, with $r = \frac{1}{2}(r_1 + r_2)$, and the total area is $A = \sum_i A_i$.

Surfaces of revolution

The equation of a sphere can be obtained by revolving around the z axis a circle of radius R lying in the xz plane. The parametric equation of the circle being $c(t) = \{R\cos(t), 0, R\sin(t)\}$, with $0 \le t \le \pi$, the sphere surface is defined by the equation:

$$X(t,\theta) = \begin{pmatrix} R\cos t \cos\theta \\ R\cos t \sin\theta \\ R\sin t \end{pmatrix}$$

Sphere

The torus is the surface obtained by rotating about the z axis a circle of radius R not intersecting the rotation axis, i.e. shifted by some distance d along x. For example:

$$X(t,\theta) = \begin{pmatrix} (d+R\cos t)\cos\theta \\ (d+R\cos t)\sin\theta \\ R\sin t \end{pmatrix}$$

Torus

'Gabriel's trumpet' is obtained by rotating a portion of a hyperbola, $c(t) = \{t, 0, 1/t\}, t = 0, \infty$, about its own asymptote. One possible parametrisation of the trumpet is:

$$X(t,\theta) = \begin{pmatrix} t\cos\theta \\ t\sin\theta \\ 1/t \end{pmatrix}$$

Gabriel's trumpet

The unduloid is obtained by rotating about the z axis the parametric curve $c(t) = \{d + \cos t, 0, t\}$, $t = 0, 2n\pi$, $n = 0, ...\infty$:

$$X(t,\theta) = \begin{pmatrix} (d+t\cos\theta) \\ (d+t\sin\theta) \\ 1/t \end{pmatrix}$$

The helicoid results from the parametric equation:

$$X(t,\theta) = \begin{pmatrix} t\cos\theta \\ t\sin\theta \\ \theta \end{pmatrix}$$

Unduloid

Finally, the catenoid is obtained by the parametric equation:

$$X(t,\theta) = \begin{pmatrix} t\cosh\theta \\ t\sinh\theta \\ t \end{pmatrix}$$

Helicoid

A catenoid is also described as a continuous transformation of a helicoid, from which it can be obtained without any stretching or contraction (in other terms, without changing the extent of its surface, which is called a *isometric transformation*).

Catenoid

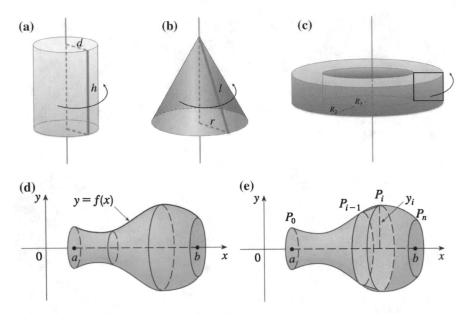

Fig. 11.12 Simple surfaces of revolution obtained by rotating a segment (**a**) about an axis, giving a cylinder, or about a point (**b**), giving a cone. If a closed figure is rotated about an axis with which it has no points of contact, **c** a toroidal ring is obtained. **d** Geometric description of a surface of revolution obtained by the rotation of a generatrix $y = f(x)$ about the x axis. **e** Break up of the surface area into elementary contributions, obtained by approximating the generatrix curve by a discrete sequence of segments $[P_{i-1}, P_i]$

Now, note that the points P_i are identified as $P_i = P(x_i, y_i) = P(x_i, f(x_i))$. In other words, the radii r_i are identified with the coordinates y_i, or $r_i = f(x_i)$. Therefore, it is also $A_i = 2\pi \frac{f(x_{i-1}) + f(x_i)}{2} |P_i - P_{i-1}|$. By taking the limit of the segment size $|P_i - P_{i-1}| \to 0$, it is:

$$\lim_{l \to 0} |P_i - P_{i-1}| = \sqrt{1 + [f'(x_i)]^2} dx \tag{11.17}$$

where f' indicates derivative with respect to x, and the total surface area is:

$$A = \int_a^b 2\pi f(x) \sqrt{1 + [f'(x)]^2} dx \tag{11.18}$$

The importance of revolution surfaces to explain many natural forms becomes evident in all those cases in which an **accretion mechanism** is important. Accretion phenomena are those phenomena of growth in which new matter is continuously added on top of existing material, during the development of the animal or vegetal organism, such as mollusks, gastropods, foraminifera, and so on.

The following, transcendental (since the variable θ appears as an exponent), parametric equation:

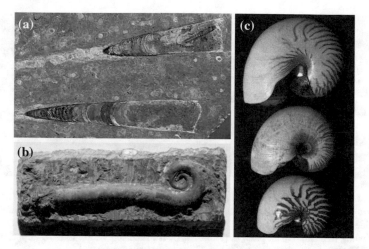

Fig. 11.13 a Fossilised cones of *Orthoceras*; **b** fossil of *Trilacinoceras*; **c** Shells of modern nautilus, from above: *N. pompilius*, *N. scrobiculatus*, *N. macromphalus*. [Photos © **a** by D. Lloyd, **b** D. W. Wade, **c** M. Gigante, repr. under CC-BY-SA 4.0 licence, see (*) for terms.]

$$
X(t,\theta) = \begin{pmatrix} k^{\theta}(R+\cos t)\cos\theta \\ k^{\theta}(R+\cos t)\sin\theta \\ k^{\theta}\sin t \end{pmatrix} \quad (11.19)
$$

for example with $R = 2.5$ and $k = 1.4$, describes the shape of a shell lying in a plane enveloped as a spiral onto itself, as shown in the adjoining figure. Such a shape is typical of the *Nautilus*, the only surviving modern species of a cephalopod with a shell. The evolution of the shell shape for such an animal is extremely interesting (Fig. 11.13). The fossil record shows how the species evolved from a straight shell with conical tip (the *Orthoceras*), to a progressively spiral-folded shell (the *Trilacinoceras*), until attaining the completely folded shell that is today visible in the nautilus. This evolution is quite difficult to trace over time, since in the fossil beds all such different species appear within a unique, large time window of about 400 million years, together with the morphologically similar, but distinct ammonoids. Ammonites had their tightly coiled shell also flat, like the nautiloids from which they are thought to descend. Straight-shelled nautiloids become extremely rare in layers after Devonian, and all nautiloid fossils seem to leave the room to the ammonoids by the end of Permian (250 millions years ago). However, a few nautiloids would

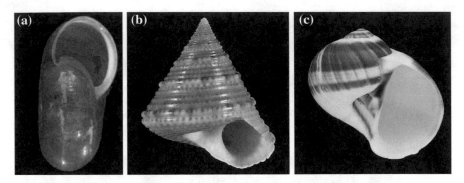

Fig. 11.14 *From the left* Quimper snail, with its almost perfectly flat, spiral shell; *Calliostoma trotini*, from the South California coast, a well-developed conical envelope corresponding to a large value of a in Eq. (11.20); *Natica vitelli*, from the Sea of Japan, a more gentle conical shape corresponding to a small helix pitch a. [Photos ⓒ by **a** Francisco W. Schultes, **b** Poppe, Tagaro & Dekker, **c** H. Zell, repr. under CC-BY-SA-3.0 licence, see (*) for terms.]

escape the Cretaceous catastrophe, about 67 million years ago, and continue up to the six modern species of *Nautilus*, while the apparently more successful ammonites disappeared.

The evolutionary advantage of such a spiral folding would be in the fact that the center of gravity lies closer to the body of the animal. This allows an optimal control of the swimming propulsion, which in all the cephalopods is typically obtained by a water-jet mechanism. For an animal carrying a heavy protective shell, it is clearly much more difficult to pilot a much elongated body compared to a compact structure. These cephalopods were among the most popular marine species at the beginning of the Paleozoic era (about 500 millions of years ago), during which they were the most effective predators. Notably, modern cephalopods such as the squid, the octopus, the cuttlefish, continued their evolution by getting rid of the external shell. The memory of the shell remains in the more or less rigid chitin gladius (vulgarly called "bone") that is now found inside the animal. Only the *Nautilus* is a kind of 'living fossil', with its beautiful spiral shell still winding around the soft animal body.

However, it must be observed that a spiral shell strictly lying within one plane is very rare, after hard-shelled cephalopods disappeared. Only rare species on land, such as the Quimper snail (*Elona quimperiana*), have a very flat spiral shell (Fig. 11.14). Instead nearly all gastropods spend their life resting on one side by a sort of foot, either on land or on the sea-shore, and consequently have their shell always folded with a climbing about the rotation axis, which yields an helicoidal spiral with a more or less conical envelope. The equation:

$$X(t,\theta) = \begin{pmatrix} k^\theta(R+\cos t)\cos\theta \\ k^\theta(R+\cos t)\sin\theta \\ k^\theta(\sin t - a) \end{pmatrix} \quad (11.20)$$

derived from the previous Eq. (11.19) by simply adding the term $-ak^\theta$ to the coordinate $z(t,\theta)$, describes a shell spirally growing about an axis, with a helix pitch different from zero. It would seem that for the less aggressive and less mobile gastropods, fast hydrodynamics did no longer represent a selective pressure. Shells with more or less pronounced conical shapes serve essentially a protective function, with a very large chamber for hosting the animal body, and a narrow tight spiral which covers a minor part of the shell volume. Depending on the helix pitch a, the conical expansion may be longer (originating from a slow growth) or almost disappear like in the abalone shell (corresponding to a fast growth). Even if with many exceptions, it is observed that high-energy sea-wave environments, such as rocky intertidal zones, are usually inhabited by mollusks whose shells have a wide aperture, a relatively low surface area, and a high growth rate per revolution, giving inconspicuous cones. Tightly-spired and highly sculptured forms are more commonly observed in quiet water environments.

11.5   Conformal Mapping and the Evolution of Species

When observing the occurrence and recurrence of natural species, one is often struck by the fact that even quite distant families of plants or animals may have very similar shapes. This, to a first approximation, may be indicative that the environmental constraints must have played a substantial role in pushing the morphology of such species into that particular direction. From a geometrical point of view, these animals or plants could appear to roughly follow common shapes, evolving with continuity across close families of curves and surfaces. Observations of such kind stimulated the reflexion about the applicability of mathematical transformations, notably geometrical mappings, to the regularity of naturally occurring shapes.

   A geometrical transformation for objects lying in a plane (x, y) can be represented by a mathematical function $F(x, y) = [p(x, y), q(x, y)]$, which transports each point of the original plane into another point of the deformed plane. As a simple example, the equation of a circle, $f(x, y) = x^2 + y^2 - r^2 = 0$, turns into an ellipse under the general transformation $F(x, y) = (x, y/2)$ (Fig. 11.15, above).

   The transformation F is a **map**, whose components p and q transform all the coordinates x and y of the plane according to a prescribed law. All the functions defined in this plane, among which our $f(x, y)$ for the circle of radius r, will be

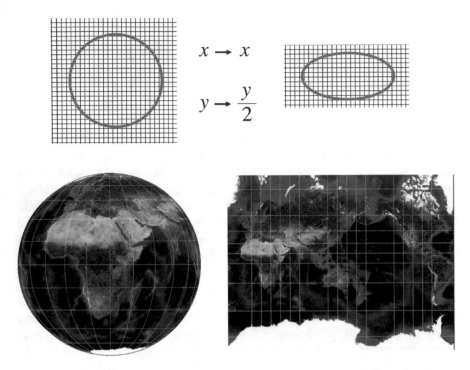

Fig. 11.15 Examples of geometrical transformations. *Above* A stretching transformation along an axis, the y axis is squeezed by a factor of 2 and the circle is transformed into an ellipse. Neither the area nor the angles are preserved in such a transformation. *Below* The Mercator projection on the right is a conformal transformation of the sphere surface into a cylinder, preserving the angles among the mesh segments (for representation purpose, the cylinder is cut open into a rectangle). [Maps © of the ICSM of Australia, repr. under CC-BY-SA-3.0 licence, see (*) for terms.]

consequently deformed in the same way. This is why F is called a *conformal map*. It will be noted, in particular that the conformal mapping preserves the angular relations among the lines defining a regular mesh on the plane (see Fig. 11.15 below: the angles of the mesh are always at $\pi/2$). In a conformal map of the Earth each parallel must cross every meridian at right angles. Also, at any point the scale distortion, either compression or dilatation, must be the same in all directions. The Mercator projection shown in the figure suffers from a huge distortion: note how Greenland appears much bigger than Australia, when in reality the surface of the latter is about 3 times bigger. Conformality is a strictly local property: angles, and consequently shapes, are not expected to be preserved much beyond the intersection point; in fact, straight lines on the sphere are usually curved along the plane, and vice versa.

D'Arcy Thompson, in his 1917 book *On Growth and Form*, applied the concept of conformal mapping to propose an explanation about how different animal species could be "obtained" from one another, by a more or less complicate distortion of their shapes. In the example shown in Fig. 11.16a he applied (entirely by means of hand-drawn geometrical sketches) the transformation $F(x, y) = (x + y/2, 3y/2)$

(namely, $p(x, y) = x + y/2$ and $q(x, y) = 3y/2$), to the bodies of two oceanic fishes, belonging to the same family (*Sternoptychidae*), to highlight their geometrical similarity[7].

It should be obvious that the concept of "transformation" between organisms or tissues must not be taken literally. A true geometric transformation in a real organism, be it an animal or a plant, implies that some matter is displaced. Therefore, a force must act for some time and length, in order for the deformation to take place. Clearly, the idea of D'Arcy Thompson was not that of opposing an ensemble of mathematical transformations to the laws of evolution. Although not a fully devout endorser of Darwin's theory, he was not such a fool to pretend to explain the origin of different species by some mysterious force that would literally mould one species into another. He rather aimed at establishing a general method to compare the evolution of superficially similar species, possibly going beyond the purely morphological aspects.

For example, by using the method of conformal transformations, Richards and Riley in 1937 proposed an interesting theory on the development of amphibians under changing environmental conditions. On the basis of the comparison permitted by the different transformations of the *Amblystoma* amphibian larva (Fig. 11.16b), they could establish that during the first few days, the development mostly occurs on the anterior part of the animal, while it increasingly involves the caudal part in the next coming days. By looking at the transformed profiles of the animal, by using conformal maps such that the body surface was constant (Fig. 11.16b, lower panel), they found that changes at the level of the larval head remain relatively modest in the following days, this being shown by the fact that the mesh lines are little modified and rather regularly spaced, with constant proportions between length and width; the same is found to hold for the size of the head compared to the total length of the developing animal. The tail becomes progressively longer in the second phase of the growth (day > 15), and the body becomes smaller compared to the length of the tail, while the larva approaches the moment of the metamorphosis. In parallel with shortening, the body also becomes wider, and the reduction of the size of the gills is observed. Such considerations would have been very difficult to expose, by looking just at the comparison of the full-scale animal shapes during the different stages of growth (Fig. 11.16, upper panel), at would be usually done.

Everywhere Nature works true to scale, and everything has a proper size accordingly—wrote Thompson.—*Cell and tissue, shell and bone, leaf and flower are so many portions of matter, and it is in obedience to the laws of physics that their particles have been moved, moulded and conformed.*

---

[7]It will be observed that the "shearing" transformation in Fig. 11.16a seems to not preserve the angles. In planar maps, the Euclidean angle is not the only one to be considered: the share and hyperbolic angles can also be calculated, and one of the three is always conserved by the map. (In more precise mathematical terms, one should look at the Jacobian of $F[p, q]$, formed by the four partial derivatives of $p(x, y)$ and $q(x, y)$ with respect to x and y: when the determinant of the Jacobian is non-zero, angles other than the Euclidean are preserved by the map.)

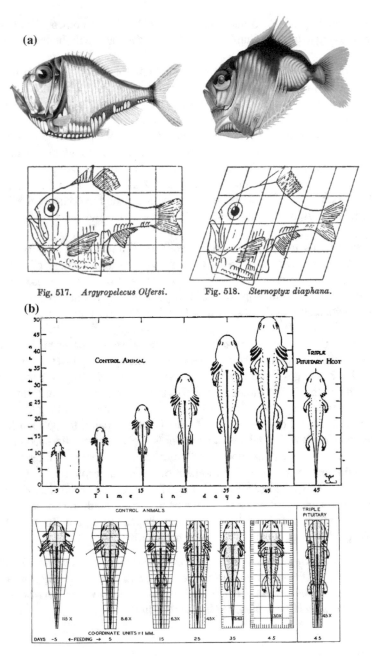

Fig. 11.16 **a** Artistic drawings of *Argyropelecus Olfersii* and *Sternoptyx diaphana*, two deep-ocean fishes belonging to the family of Sternoptychidae. In the *lower row*, the original drawings by D'Arcy Thompson describing the conformal mapping applied to the left fish, to turn it into the right side one. **b** The amphibian larva of *Amblyostoma*, represented at different stages of growth during the first 45 days before the metamorphosis. In the *lower panel*, the animal shape was drawn by empirically applying near-conformal maps, which preserve the total area projected in the plane. [Public domain images (**a**), from Ref. [4, 5]. Images (**b**) from Ref. [6], repr. w. permission]

In fact, and most importantly, the intuitions of D'Arcy Thompson, often of aesthetic as well as of mathematical nature, stimulated in the following years the development of the allometric analysis methods (notably by J. S. Huxley in the 1930s, and others to follow, see Chap. 12), by which one can compare different structures on logarithmic scales, for example to discover the growth rates of different species. In recent years, such allometric methods have found wide applications in embryology, animal and vegetal taxonomy, palaeontology and ecology.

A transformation $F = [p(x, y), q(x, y)]$ much used in various applications of geometry and architecture, is the **quadratic map**, or biquadratic application, for which both p and q are written as second-order homogeneous polynomials:

$$F(x, y) = ax^2 + by^2 + cxy + dx + ey + f \qquad (11.21)$$

with coefficients $a...f$ different for p and q. In practice, the constant term f can be dropped as inessential, therefore a quadratic map has 10 free parameters, compared to just 2 in a linear transformation. This gives the quadratic map enough latitude for adapting to complex deformation patterns. A quadratic application can be sufficient to describe a large variety of transformations, and in fact most of the maps empirically proposed by D'Arcy Thompson are often close to a quadratic.

Even if not strictly rigorous from the analytic standpoint (he used to draw his transformations by hand, and subsequently he deduced a possible mathematical form by adjusted fitting), the deformations imagined by D'Arcy Thompson to turn the *Scarus* (Parrot fish) into a *Pomachantus* (Angel fish), or to transform the skull of a hominid into that of a chimpanzee, or into that of a baboon, are often close to quadratic maps (see illustrations on pp. 1053–1083 of Ref. [5]).

11.6   The Emergence of a Body Plan

The traditional classification of animals has rested for a long time mainly on the examination of morphological differences, and is complemented today by the large bulk of data from molecular and genetic analysis. From a biophysical point of view, however, one of the most interesting features is the astounding variety of body shapes that ended up to the current differentiation, when thinking that "everything" started from a bunch of amorphous cells in the oceans of the pre-Paleozoic Earth.

Figure 11.17 reports a scheme of the phylogenetic tree of eukaryotes, starting with the ancestral organisms evolved from the unicellular protists. These should have been colonies of multicellular eukaryotes, heterotrophic (i.e., feeding on outside matter) and therefore capable of ingesting food. These primitive beings had no rigid cell walls, and a unique tissue, both nervous and muscular, allowing for minimal movements and responses to foreign stimuli. Their development led from an amorphous mass of cells, to a hollow cellular structure (the *blastula*), and eventually to the formation of an internal cavity by invagination of some outer parts of the surface (the *gastrula*). Sea sponges are modern representatives of this kind of organisms. Already at this stage, a

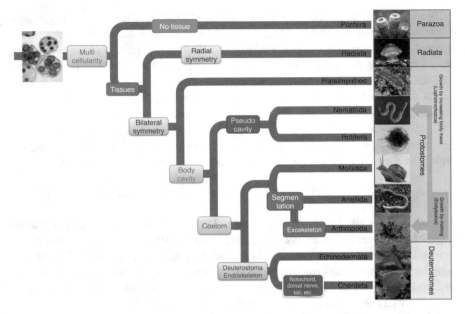

Fig. 11.17 The phylogenetic tree of eukaryotes. From the *left* to the *right*, evolution starts from ancient unicellular protists, growing into multicellular organisms. The successive addition of structural elements brings about increasingly complex body plans, from the radial to the bilateral, from the diploblastic to the triploblastic, to the development of internal cavities, segmentation, appearance of skeleton and dorso-ventral symmetry, dorsal spine and so on. On the *extreme right*, the phylogenetic classification as based on molecular sequencing

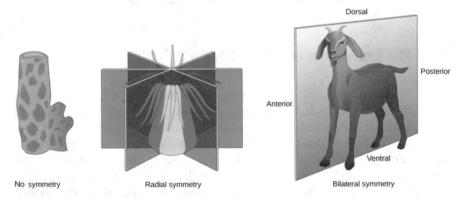

Fig. 11.18 Symmetry is a criterion to classify body plans. Starting from the more elementary organisms, lacking any symmetry, and moving to radially symmetric body, ending up with the bilateral symmetry, in which an anterior-posterior axis, and a dorsal-ventral axis can be distinguished. [Image adapted from http://cnx.org/, © Rice University, repr. under CC-BY-4.0 licence, see (**) for terms.]

first split between species should have occurred: the *Parazoa* remained with such an undifferentiated cell structure, while some of them became *Eumetazoa* and started transforming the nature of the internal layer of cells, thus forming the first germ of a true tissue. These early organisms are labelled as **diploblastic**, to underscore their two types of tissues: an external layer called *ectoderm*, and an internal one, the *endoderm*, which allowed to develop a closed body cavity and some kind of glands. Jellyfish or sea anemones are modern representatives of animals of this kind (Fig. 11.18).

From the point of view of getting to an increasingly complex body plan, an even bigger further split is the appearance of a third type of tissue, intermediate between the ectoderm and the endoderm, namely the *mesoderm*. Nearly all modern animals have such a tripartite tissue structure, and are therefore called **triploblastic**. Such a differentiation will allow to develop all the internal structures, skeleton and muscles, needed to support the cavities and organs to be developed by the endoderm layer, beginning with the earliest digestive cavity (the *coelom*).

Another important split occurred then, between animals who choose an "easier" radial symmetry, and those who go for a **bilateral symmetry**. The fact of having a body with a left and a right side will have a tremendous impact on the development of the future muscular, skeletal and nervous system. Despite their morphological diversity, most *bilaterians* are united by a handful of fundamental body plan features including bilateral symmetry, triploblasty, a coelom, a through-gut, and a central nervous system. The bilateral symmetry is a complex trait achieved by the intersection of two axes of polarity: a primary body axis (the anterior-posterior axis) and a secondary, orthogonal axis (the dorsal-ventral axis). Recognition of a front and back, and a left from a right, in a mass of embryonal cells originates from the difference in the expression of some genes, which are activated in some parts of the body and repressed in other parts (see below, p. 523). The proteins made from these genes all contain a similar 60-amino acid motif termed the *homeodomain*. Homeodomain proteins exert their function through combined mechanisms of activation and repression of multiple target genes. With many variants, such a basic mechanism is found across animals that developed over a span of more than 500 millions years, making this one of the most strictly conserved evolutionary features. It may be interesting to note that Echinodermata (sea urchin, starfishes), which are far high in the classification of Fig. 11.17 because they belong to the Deuterostomata superphylum, have a *secondary* radial symmetry: they actually start with a bilateral symmetry in the embryo, but develop a radial body plan only in the adult.

A further specialisation of the body plan comes with the **segmentation**, by which a part of the body is repeated into identical units, typically along the major (antero-posterior) axis. This characteristic is found in most animal species, from the lower anellides and arthropods, to the upper vertebrate and cephalochordate. A segmented body allows a larger flexibility in the movements, and functional structuring of the embryo. For example, in the fruit fly the segments composing the head, thorax and abdomen develop separately into the final structures of the adult. During embryonal growth, **somites** form by budding off from the anterior end of mesoderm at regular intervals, for example about 2 h in mouse embryos, 90 min in chicken, and 30 min in zebrafish. The mesoderm undergoing segmented growth loses cells in the anterior portion of the growing somite, and is replenished both by cell proliferation and ingression posteriorly. This is a dynamic structure that can for many purposes be considered to be at a steady state. Overall there is limited cell migration in segmentation, but cells do undergo a relative movement within the mesoderm, as its anterior and posterior borders move posteriorly. Segments need not to persist in the adult body, but can be joined into functional structures, or can be hidden inside the body, such as the vertebrae.

An even more profound innovation, is the development of **jointed appendages** (legs, wings, antennae, tails). Animals that developed such structures are the most successful in evolution, when success is measured by their adaptation to different environments. In this respect, insects are certainly very successful, being widespread on nearly all possible habitats, with the possible exclusion of polar environments (actually, only one single species of insect has been up to now found in the Antarctica, the *Belgica antarctica*, a flightless midge).

The upper evolutionary end of the scale (actually put at the lowest in the scheme of Fig. 11.17) is reached with the differentiation of **chordates**. Although being (for the moment) at the top of the evolution, on a purely numerical basis chordates are not numerous. The assessed number of different species is about 65,000, half of which are fishes. By comparison with non-chordates, just the single phylum of mollusks counts more than 100,000 different species, and arthropods count more than 1,000,000. However, chordates represent by far the largest part of the animal biomass on Earth. Not only they are the largest animals in existence today, but ecologically they are among the most successful. They have been able to occupy all kinds of habitats, and have adapted to more modes of existence than any other group, including the arthropods. Especially birds and mammals have been able to penetrate cold climates, because of their ability to maintain a constant body temperature, something that no other animals can do.

Four principal features characterise the chordate body plan (see Fig. 11.19): the **notochord**, the hollow **dorsal nerve chord**, the presence of **pharyngeal slits** along

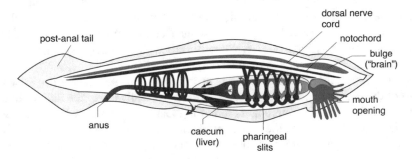

Fig. 11.19 The characteristic body plan of chordates, with the notochord, the dorsal nerve chord, the gut connecting mouth and anus, and post-anal tail. [Image © by Piotr Michal Jaworski, adapted under CC-BY-SA-3.0 licence, see (*) for terms.]

the duct connecting mouth and anus, and a musculate, post-anal **tail**. The notochord is a flexible element that forms between the primitive gut and the nerve chord, to be replaced after the embryonal development by the vertebral column. The dorsal nerve chord in vertebrates is the element that will differentiate into the brain and spinal chord. The pharyngeal slits, most often just pouches, are present in the embryo of all vertebrates, but are lost later in the development of the terrestrial ones; however, their presence provides a clue to our aquatic ancestry. Similarly, the tail is present in all embryos, but is lost for some adults like humans. Chordates have a segmented body plan, with distinct blocks of muscles (*myotomes*) already visible in the embryo. Most of them have an internal skeleton, against which the muscles make mechanical work, and which is at the basis of the extremely successful locomotion abilities of the members of this phylum.

All chordates are **tetrapods**: they have four limbs, which in some case may have turned into wings, or fins, and in some case may have entirely disappeared, such as in snakes. And all of them are **pentadactyl**: their limbs have five digits, which also signals the likely origin from a common ancestor. Even animals that today display only one or two fingers, such as horses, maintain structural vestiges of all the five fingers in their limbs. Such a structure is *homologous* in the various animals of the phylum: meaning that a same basic plan has adapted to different functions, like flying, swimming, running or walking.

11.6.1 Reaction-Diffusion and Pattern Formation

The generation of the variety of body structures from one (more or less homogeneous) single egg has long been considered to be so miraculous, that long arguments arose as to whether the laws of physics would ever be sufficient for an explanation of development. However, by looking at nature, it is clear that formation of patterns is not peculiar to living objects: galaxies, clouds, lightning, rivers, mountains, crystals, all demonstrate the ubiquitous generation of ordered structures and recurrent patterns.

Autocatalysis, a player that has already been invoked often in this book, is the mechanism by which a small perturbation in a homogeneous distribution can become amplified by a positive feedback (see also the greybox "Stability and chaos" on p. 523). However, to induce a stable pattern in a structure a second mechanism must be invoked, to stop the autocatalytic mechanism from spreading everywhere. This is **inhibition**, a negative feedback that must operate at a longer distance than the short-ranged autocatalysis and, differently from the autocatalysis, requires energy to be activated. By taking a quite radical stance, it can be said that development must be ultimately a biochemical process, consisting of interactions and movement of molecules in and across cells. In a famous paper [7], Alan Turing combined the concepts of autocatalysis and inhibition, making the very important discovery that spatial concentration patterns can be formed if two substances with different diffusion rates produced at nearby regions react with each other. This is very contrary to physical intuition, since we know that diffusion should work to smooth out any local accumulation of molecules, not to create concentration maxima.

To see these principles at work, let us assume a substance a which stimulates its own production (autocatalysis), and an antagonist h that plays the role of inhibitor. To work at longer range, the inhibitor must have a larger value of diffusion constant than the autocatalytic substance. In a population of cells, such two chemicals could not have homogeneous concentrations, since any small increase of a, e.g. from a random fluctuation, would be amplified by autocatalysis; and the inhibitor would respond to the increase in a by increasing its own concentration, and limiting the increase of a. Let us write the following two equations for the coupled evolution of a and h:

$$\frac{\partial a}{\partial t} = D_a \frac{\partial^2 a}{\partial x^2} + \alpha \frac{a^2}{h} - \beta a \tag{11.22}$$

$$\frac{\partial h}{\partial t} = D_h \frac{\partial^2 h}{\partial x^2} + \gamma a^2 - \delta h \tag{11.23}$$

These are just ordinary diffusion equations for each substance, plus the two last terms in each line to describe their 'reaction' behaviour. The constants α and γ link the time increase of a and h to the square of concentration of a itself, and the constants β and δ represent the consumption of both species. Note that the autocatalytic growth of a is divided (i.e., limited) by the concentration of h.

To study the behaviour of such a model system, let us first assume a constant $h = 1$, and a uniform distribution of a, so that its spatial derivative is zero. Then:

$$\frac{\partial a}{\partial t} = a^2 - a = a(a - 1) \tag{11.24}$$

(to further simplify things, all the constants are set equal to 1), which gives $a = 1$ at steady state when $\partial a/\partial t = 0$. The autocatalytic property of a is manifested in the fact that for a small positive perturbation, $\partial a/\partial t = \epsilon \ll 1$, the right side of the equation $a - 1$ will start increasing. This is the reason to choose (at least) a quadratic exponent in the a-production term (first on the right side).

Now let us unlock the h factor, and assume a very fast equilibration of h for any given concentration of a. At the steady state, the inhibitor concentration becomes $h = a^2$. Then, by putting this value back into the (11.22), we get:

$$\frac{\partial a}{\partial t} = \frac{a^2}{a^2} - a = 1 - a \tag{11.25}$$

Again the steady state corresponds to $a = 1$, however in this case the equilibrium is stable since for a positive perturbation, the right side of the equation stays positive only by lowering $a < 1$.

Let us consider for the sake of example an array of 100 cells, and attempt at a numerical solution of the above equations. The constants are all set to 1, and the evolution starts with all cells having the same concentration $a = 3$ and $h = 1$ (in

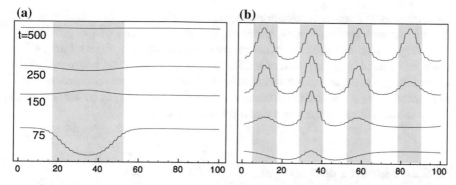

Fig. 11.20 Numerical solution of the reaction-diffusion equations for a species a and the inhibitor h, starting from homogeneous concentrations over a unidimensional array of 100 cells with periodic boundary conditions. At time $t = 0$ a perturbation in a is localised at the cell n.35. **a** Solution of the model with the two diffusion coefficients equal, $D_a = D_h = 1$, curves showing the concentration of a at successive time frames increasing from $t = 75$–500 (from bottom to top). **b** Solution of the model with the diffusion coefficients $D_a = 1$ and $D_h = 10$. The even spacing of the pattern is due to the "closed loop" boundary conditions, by which cell 1 interacts with cell 100 and vice versa

arbitrary units). Then, at time $t = 0$ the concentration of a is doubled at a single cell, for example the 35. Figure 11.20a shows the evolution of the concentration of a at subsequent times, $t = 75, 150, 250, 500$, for the diffusion coefficients of a and h equal to $D_a = D_h = 1$. It is seen that in this case the initial perturbation is reabsorbed over some time, and the concentration of a goes back to homogeneous at long times. By contrast, Fig. 11.20b shows the effect of changing the diffusion coefficient of h to a factor of 10 larger than that of a: the small perturbation over about the same time frame grows into a stable pattern, by distributing a periodic concentration profile of the autocatalytic species a over the array of cells.

What the simple Turing-like model above shows is that the combination of an "activator" and an "inhibitor" species with largely different diffusion constants is indeed capable of producing chemical patterns. The pattern originates by the local growth of the activator species, which induces the concomitant production of the inhibitor; this latter however works over a longer distance, due to its faster diffusion, and creates an "exclusion" zone, which depletes the concentration of the activator. The combination of the two diffusing species determines the actual spacing of the maxima, which can be very regular (think of the spacing of leaves on a branch) or more random (such as hairs and bristles on the skin), according to the boundary conditions imposed by the borders of the growing embryo. In the example of Fig. 11.20, cells are arranged in a closed loop, the 1 interacting with the 100 and vice versa, therefore all cells are strictly identical and a very regular spacing arises.

The combination of Eqs. (11.22 and 11.23) is just one example of possible inter-action. In a different model, the action of the inhibitor could be changed, e.g., to $\alpha(a^2/h^2)$ in the first equation, and to γa in the second one. Such a model could describe a situation in which the activator molecule is converted into a different conformation which acts as inhibitor (γa turns into h), running in competition with the pristine activator molecules. Other molecular realisations of the principle of autocatalysis-plus-inhibition are of course possible. For instance, the inhibitory effect could be realised by a depletion of a substrate, or of a precursor, consumed in the autocatalysis. Or, the autocatalysis could be realised by the mutual inhibition of two substances, or by a reaction chain consisting of many elements, or by the release of bound substances rather than by a direct production, and so on.

11.6.2   Pattern Formation and Gene Expression

In the same 1952 paper, Turing coined the term **morphogen**, to mean a substance governing the pattern of development, thus providing a mechanism by which the emission of a signal from one part of an embryo can determine the location, differentiation and fate of many surrounding cells. At the molecular level, a morphogen is a chemical whose concentration varies according to a gradient, which is set up by the diffusion. Typically, a morphogen will spread from a localised source and form a concentration gradient across a developing tissue, driving the specialisation of stem cells into the different cell types. The gradients of **transcription factors** have been extensively studied, especially in the *Drosophila melanogaster* (the common fruit fly), showing that these indeed act as morphogens. Transcription factors are proteins that bind to a specific DNA sequence, promoting or blocking the recruitment of RNA-polymerase, thereby favouring or repressing the transcription of that DNA sequence. It is a general observation that morphogen gradients are linked to the localised expression and repression of specific genes, such as in the definition of the bilateralisation. The anterior-posterior axis of all bilaterians (including humans) is known to be patterned by alternating expression domains of a family of genes called *Hox*, while the dorsal-ventral axis is patterned partially by the asymmetrical expression of a different set of genes, the *dpp/BMP4*.

Understanding morphogen gradients from a biological standpoint requires to know how cells interpret a variable morphogen concentration, and how they transduce this information to the nucleus to produce the appropriate gene or cell fate response. On the other hand, as biophysicists we might be more interested in a different type of question, that is: how a given concentration gradient is formed among the cells of a tissue?

Known morphogens seem to be effective at extremely low concentrations, 10^{-9}–10^{-11} M, and are probably not evenly distributed across their field of action. Moreover, the geometry of a morphogenetic field must be quite restricted, if the specialisation mechanisms of the cells depend critically on diffusion: since the time required for diffusion increases quadratically with the length, the spatial organisation must take place within small assemblies of cells. Embryonic fields are indeed small, of the order of 1 mm or $\sim$100 cells across, a size in which communication via diffusion can take place in a few hours. The formation of somites (somitogenesis) happens quite early in embryogenesis, and proceeds from the tip of the embryo (anterior) towards

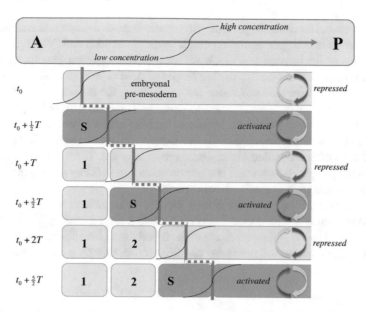

Fig. 11.21 Schematic of the clock-and-wavefront model. The concentration gradient (wave) of growth factor moves from anterior to posterior ends (A→P, shown above). The local oscillation of actuator genes switching between activated and repressed states is shown on the *right*. Cells with a low concentration of growth factor pinch off to form a somite (S) when the clock is in the correct phase (*darker colour* of the mesoderm). The wavefront is shown by the *vertical blue line*, moving from *left* to *right* at successive half-periods T of the clock. Activation occurs at $\frac{1}{2}T$, $\frac{3}{2}T$, $\frac{5}{2}T$, and so on

the tail (posterior). Each new somite forms when a block of cells splits off from a large mass of tissue called the "presomitic mesoderm". While new somites pinch off from the anterior end, the mesoderm elongates towards the posterior end.

In their 1976 study of the embryo evolution in the amphibian *Xenopus laevis*, Jonathan Cooke and Erik Christopher Zeeman proposed the **clock-and-wavefront model** of morphogenesis [8], summarised in Fig. 11.21. This is the prototype of several other models of development, relying upon the conversion of a temporal oscillation into a dynamic spatial periodic pattern.

The basic idea of the c–w model is the interaction between a wavefront which moves through the growing embryo and a ticking clock in each cell which determines its ability to respond to the signal given by the moving wavefront. The wavefront is a gradient of a chemical growth factor (morphogen) produced at the posterior end, therefore oriented with the maximum at the back and the minimum at the front. Since the source is moving to the back, there is a wave of low concentration moving in the front-to-back direction. However, such a front should move at a steady velocity; instead, somites are seen to break off at nearly regular intervals. This means that another component besides the morphogen gradient must exist.

The additional mechanism governing the spatio-temporal propagation of this wave of differentiation is a "segmentation clock" existing in each cell. When oscillating components of the internal cell clock are examined, by special genetic techniques such as *in situ* hybridisation, alternating stripes of gene expression and repression are observed. These stripes are not static, but move along the main $A \rightarrow P$ axis as the oscillation frequency changes. Numerous genes and proteins that oscillate at the rate at which somitogenesis proceeds have been identified across a number of species. For example, in the well-studied zebrafish (*Danio rerio*), all known oscillating molecules are regulators of the so-called *Notch-Delta* group of genes. The cyclic activation and repression of the oscillating genes is represented in Fig. 11.21 by the alternating light and dark colours, changing at each semi-period T.

The reaction-diffusion model above can be reformulated to display spatio-temporal oscillations. The concentrations of a and h are replaced by two variables, $u = a - a_0$ and $v = h - h_0$, representing deviation from given concentration profiles $a_0(x)$ and $h_0(x)$. Moreover, nonlinear coupling terms, mixing u and v, are added:

$$\frac{\partial u}{\partial t} = D_u \nabla^2 u + \alpha u + \beta v - c_1 u v^2 - c_2 u v \tag{11.26}$$

$$\frac{\partial v}{\partial t} = D_v \nabla^2 v - \alpha u + \beta v + c_1 u v^2 + c_2 u v \tag{11.27}$$

By numerically solving the above equations, it can be seen that when the ratio c_2^2/c_1 exceeds a critical value, stable time-oscillating patterns are formed [9]. Figure 11.22 displays four different solutions, again for the ensemble of 100 cells aligned on the x-axis with periodic boundary conditions, corresponding to different choices of the model parameters and c_2^2/c_1 above critical. The diagrams represent the course of time on the vertical axis, and the changing colour is the variation of u between -1 and $+1$ (in arbitrary units; the concentration profile of v, not shown, follows that of u).

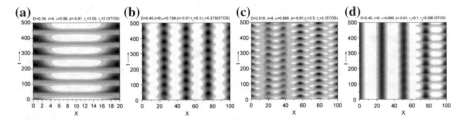

Fig. 11.22 Concentration profiles of the factor u from the modified Turing model in a space (*horizontal*)-time (*vertical*) plot, for a system of 100 cells with periodic boundary conditions, and different sets of model parameters. Four different kinds of spatial patterns are observed when the quadratic term is strong enough: **a** in-phase oscillatory patterns; **b** out-of-phase oscillatory patterns; **c** mixture of in-phase and out-of-phase oscillatory patterns; **d** combination of constant and oscillatory patterns. [From Ref. [9], repr. w. permission]

Among the different patterns, out-of-phase oscillations resembling what is expected in the c–w model are also observed (Fig. 11.22c).

However, the reaction-diffusion is a very simplified model of the complex network of interrelated chains of gene activation and repression that are taking place in the cell, in response to the chemical gradients. In particular, it is now established that the local oscillation frequency of each cell's clock is different, and depends on the relative position of the cell along the antero-posterior axis. Oscillations in the posterior end of the embryo occur at an approximately constant rate, corresponding to the frequency at which somites form, one after another. But, as the wave of differentiation moves along, the oscillation frequency of each cell decreases: somite formation is seen to occur just at the spatial position where the oscillations cease.

A possible origin for such a slowing down of the oscillation has been shown to arise from the **coupling** of the different oscillators [10]. We have already examined the coupling of oscillators in the context of heart- and brain-cell synchronisation, the Kuramoto model (p. 308). In the present case, the same Kuramoto's model is generalised by replacing the phase-coupling term of the original model, with a double (A, B) coupling:

$$\frac{d\theta_i}{dt} = \omega_i + \sum_{j=1}^{N} \{ A \sin(\theta_j - \theta_i) - B[\cos(\theta_j - \theta_i) - 1] \} \tag{11.28}$$

The attractive sinusoidal term forces synchronisation, it could e.g. represent the role of *Notch-Delta* genes; at the opposite, the cosine term pushes neighbouring oscillators out-of-phase, and could thus represent a type of coupling similar to the process of inhibition. By making the approximation that neighbouring cells, k and $k \pm 1$, have small phase differences, $|\theta_k - \theta_{k\pm 1}| \ll 1$, the sine and cosine terms can be expanded in series about each θ_k as:

$$\frac{d\theta_k}{dt} = \omega_i + A \sin(\theta_{k+1} + \theta_{k-1} - 2\theta_k) +$$

$$+ B \left[1 - \frac{(\theta_{k-1} - \theta_k)^2}{2} + 1 + \frac{(\theta_{k+1} - \theta_k)^2}{2} \right] - 2B + O(\Delta\theta^3) \simeq \tag{11.29}$$

$$\simeq \omega_i + A \sin(\theta_{k+1} + \theta_{k-1} - 2\theta_k) - \frac{B}{2} \left[(\theta_{k-1} - \theta_k)^2 + (\theta_{k+1} - \theta_k)^2 \right]$$

If one makes the further hypothesis that the coefficients scale with the (discretised) distance Δk as $A = \hat{A}/\Delta k^2$, $B = \hat{B}/\Delta k^2$, the above approximate expansion gives a second-order differential equation in the continuum limit $\Delta k \to dk$:

$$\frac{\partial\theta}{\partial t} = \omega + \hat{A} \frac{\partial^2\theta}{\partial k^2} - \hat{B} \left(\frac{\partial\theta}{\partial k} \right)^2 \tag{11.30}$$

The great advantage of such a revisited c–w model is that a travelling wave naturally emerges from the solutions, without having to make additional hypotheses on the precise nature of the syncing and dephasing terms, A and B.

It may be shown that such a travelling wave has exactly the effect of slowing down the cell clocks' oscillations along the direction of propagation. In fact, the last differential equation can be further transformed, by making the substitution $W(x, t) = \nabla\theta(x, t)$ and taking the derivative with respect to x, thus obtaining the new equation:

$$\frac{\partial W}{\partial t} + 2\hat{B}(W\nabla W) = \hat{A}\nabla^2 W \tag{11.31}$$

This is called the **advection-diffusion equation**, which has a number of applications in physics and engineering. Notably, it has a traveling-wave solution with wave velocity $v = \sqrt{\omega B}$, which implies that in this model the clock and wavefront are not two separate entities but, rather, the wavefront represents itself a gradient in clock phase. Ahead of the wave (i.e., toward the posterior end) the solution is $\theta(t) = \omega t$: all cells oscillate in phase at the same frequency. Behind the wave, the solution is space-dependent and time-independent, $\theta(x) = x\sqrt{\omega/B}$; by observing that the phase must be defined modulo 2π, the solution in this region displays periodic maxima (to be identified with the embryo somites) spaced by $2\pi\sqrt{B/\omega}$ (Fig. 11.23, right). Moreover, the local oscillator's frequency varies in space, from zero at the anterior end to the maximum ω at the posterior end (Fig. 11.23, left). With such a model, the authors were able to reproduce the variation in the number and spacing of somites across different species, from the zebrafish, to the mouse, chicken and snake. (However, the model also predicts that the somite size should be proportional to the square-root of the period T, and this seems to be in contradiction with some of the known experimental data.)

Clearly, much work remains yet to be done, for example in exploring the role of the mechanical stresses induced by the process of somite growth and separation, something that is totally neglected by purely "biochemical" models like the ones exposed in this Section. A relevant contribution to the emerging periodicity of segmented substructures should also result from some kind of mechanical self-organisation, as

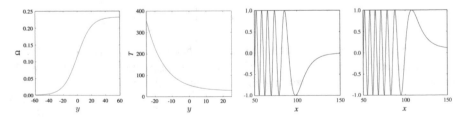

Fig. 11.23 Numerical solutions of the revisited c–w model. *Left* Variation of the oscillation frequency Ω and period T across the passing wavefront ($y = 0$ is the center of the wave). *Right* Spatial variation of the phase field at times $t = 0$ and $t = 40$ min, with parameters adjusted for the zebrafish embryo. [From Ref. [10], adapted w. permission]

observed in the examples of **strain localisation** instability. This latter phenomenon has been shown to be at the origin of periodic patterns in a variety of situations, from fracture fronts in materials, to slip and shear bands in solids, to the periodic fluctuations in the flow of granular materials, to name just a few.

Mechanical forces are chiefly present in the embryo development, beginning with the invagination of the gastrula. If the quantification of the forces involved in such delicate processes is very difficult, new experimental techniques have recently been developed to infer mechanical forces indirectly. For example, oil droplets with fluorescent coatings can be injected in the growing tissues, and the deformation that occurs to the droplets as a result of tensile or compressive forces in the surrounding can be monitored by fluorescence microscopy. Such techniques, for example, allowed to demonstrate key differences in the way gastrulation occurs, by comparing higher vertebrates and less evolved animals [11–13].

11.7 Phyllotaxis, The Spacing of Leaves

Unlike animals, which generally have to contend with a body plan which is fixed at birth, plants can continue to develop new organs and elaborate their body plan throughout their life. A beautiful example of a nearly regular and periodic patterns is the spacing of leaves, called **phyllotaxis**. Such observations had been fascinating philosophers and scientists as far back as Theophrastus, with his *Historia plantarum* in the III century BC,[8] and Plinius, with his *Naturalis historia* in the I century. Leonardo Da Vinci in the mid-XVI century described the spiral patterns of plants, and Johannes Kepler in the beginning of the XVII century was likely the first to conjecture that Fibonacci numbers were somehow involved in the structure and growth of plants. About 1836, the Franco-German botanist Wilhelm Philippe Schimper observed that after some number of complete turns around the stem of a plant, another leaf would lie almost directly above the first. He gave the name of **divergence angle** to the number of turns divided by the number of leaves in a cycle. The brothers Auguste and Louis Bravais (the first was the famous crystallographer, the second a physician passionate of botany) observed that this angle is in most plants close to $2\pi/\tau$ radians, or 137.5°, where $\tau = (1 + \sqrt{5})/2$, is the *Sectio aurea*, or "golden mean".

Such regularities, or near-regularities, are indeed commonplace in the plants' world. For example, the stalks or florets of a plant lie along intersecting spirals running clockwise and counter-clockwise. In a famous 1953 paper, the British-Canadian geometer and musician Harold Coxeter described the arrangement of scales on the surface of the pineapple fruit in terms of families of intersecting spirals [14]. Actually, what we call pineapple fruit is the result of the progressive fusion of many flowers, germinating from a common stem and growing on top of each other, over a long

---

[8]The ancient greek title is Περί φυτῶν ιστορία, the latin title is that of the first translation published in 1483 by T. Gaza, a Greek refugee in Italy, who worked on an original manuscript that has since been lost.

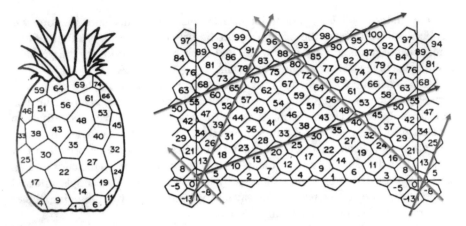

Fig. 11.24 The surface of a pineapple opened up into a periodic cylinder. The *red, blue* and *yellow arrows* indicate the three families of spirals. [From Ref. [15], adapted w. permission]

time ranging from several months to about 3 years. Figure 11.24 shows the original drawing of Coxeter, with the scales of the fruit surface numbered according to their position along the vertical axis, therefore relative to the time of successive appearance of each flower making up the final fruit. Starting from the position 0, three families of spirals can be identified: the red one starts with position 5, and five such spirals wind counter-clockwise around the fruit (they can be seen between the two red arrows); the yellow one starts with position 13, and thirteen such counter-clockwise spirals can be counted between the two yellow arrows; and the blue family, starting with position 8 and running clockwise in eight repeats. The 5, 8 and 13 positions are all tangent to the 0. Coxeter noticed that the three numbers belong to the famous Fibonacci series:

$$1, 2, 3, 5, 8, 13, 21, \ldots$$

in which each number is the sum of the preceding two. Such a series, which the medieval Italian mathematician Leonardo Fibonacci described in his *Liber abaci* of 1202, but which was already known to Indian mathematicians from centuries before, is also intimately related to the golden mean, since the ratio between any two adjacent numbers in the series represent increasingly good approximations to τ.

Similar spiral arrangements are widespread in fruits and flowers, such as the arrangement of scales in pine cones, the surface of douglas fir branches, thistle inflorescences, sunflower heads, cactus ridges, and so on. The parallel sequences of equivalent spirals are called *parastychies*, and their number is often one from the Fibonacci series. Many other such regularities, and more often pseudo-regularities, can be discovered among plants, probably owing to the more static structures of the almost immobile vegetables, when compared to the dynamical structures of moving animals. For example, by looking at the stalks of a celery in cross section, some have observed that the (approximate) projection of the (approximate) centre of mass of each successive stalk n is placed at a position that (approximately) corresponds to the spacing of the points $2\pi n/\tau$ on the unit circle. Such elegant calculation about a very vaguely defined stalk's centre of mass, leads such curious observers to claims that "stalks are optimally spaced", and "the golden-mean angles ensure that successive stalks are inserted where they have most room".

Both mathematicians and laymen are fascinated by these recurrent numerical patterns, which in practice are often only approximate, and quite less stringent than one would love to see. This could be dubbed as a very typical case of Platonism in science, in that the beauty of an idea takes over its scientific basis, in some extreme cases up to the point of forcing will against evidence. However, even if willing to accept the "regularity" and "beauty" of such patterns as an observational evidence, the really big, and even more interesting question, is: what are the physical, chemical and biological origins of such patterns?

There are currently two major approaches for the explanation of phyllotaxis, both put forward at the beginning of the XX century, and still being more or less actively pursued. A first approach supposes that leaf primordia are formed at the "first available space" [17, 18]. Such model, complemented by I. Adler's idea of a contact pressure providing a physical basis to the observation of space-filling patterns, was formalised in a "fundamental theorem of phyllotaxis" that should enable to recognise and classify patterns into mathematical objects.

However, many experiments involving surgical intervention, or treatment with plant hormones, appear to support instead a second approach, which assumes a field of inhibition around each existing primordium, such that new primordia should form where the total inhibitory influence is least [19]. This idea, outlined half a century before Turing's seminal work on morphogens, exactly summarises the behaviour of the activator-inhibitor model that was discussed in the preceding Section. Actually, in the actuator-inhibitor (a–h) model new concentration peaks can appear between existing peaks at sites where the inhibitor concentration is lowest, and several cells can start the production of the chemical activator. The emerging new peak gets sharper, since also the inhibitor starts being produced right after by all these cells, the reaction-diffusion competition starts, and only the best-located group (a or h) will win. Eventually, the resulting new maximum will have the same size and shape as the others, and will be surrounded by its own inhibitory field, as shown in the numerical example of Fig. 11.20b.

The a–h model has been used by H. Meinhardt in computer simulations of a growing shoot, approximated as a cylinder (Fig. 11.25). In the simulation, cells are doubled at fixed time intervals at the upper end of the cylinder, with a random fluctuation determining the location of the first maximum. This gives the position of a first leaf, and produces an inhibitor field that blocks the formation of other leaves in the immediate neighbourhood. After further growth, the next maximum appears, in this case on the opposite side of the cylinder, and so on, originating the *alternate* pattern. Depending on the detailed values of the model parameters, various other leaf patterns can emerge. The spiral (or *decussate*) pattern can be formed if the diameter of the stem is larger than the diffusion range of the inhibitor, especially if an inhibitory influence from the apex prevents new centres from arising near the apex.

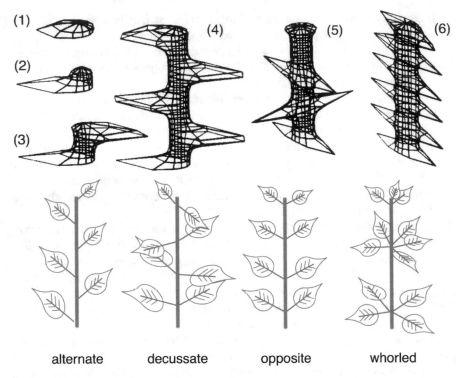

(1) (2) (3) (4) (5) (6)

alternate   decussate   opposite   whorled

Fig. 11.25 The regular spacing of the concentration peaks of the "activator" provides a model for phyllotaxis. *Above* In the steps (*1–4*) a growing shoot is computer simulated, as a cylinder at whose end cells can double at a given rate. Maxima are nucleated by random fluctuations, and start producing competing activator and inhibitor signals. Depending on the values of the parameters, (*4*) alternate, (*5*) spiral, or (*6*) opposite arrangements of activator peaks are formed, which originate different leaf distributions along the growing stem. [Public-domain [16], w. permission.] *Below* Examples of alternate, spiral, opposite, whorled leaf arrangements

On the other hand, an *opposite* arrangement of activator maxima can be formed, if the growth is fast enough so that cells have some memory that their ancestors were originally activated, or if the diffusion of the activator is facilitated in the axial direction.

11.7.1 Getting Away from Fractions

The spiral arrangement of leaves and inflorescences, and the Fibonacci series that invariably appears in such cases, can as well be explained on the basis of models of lateral inhibition with the proper geometrical constraints.

Going into a spiral pattern becomes somewhat obvious, when you are a leaf and must be born out of a central stem. To avoid crowding, leaves and flowers must

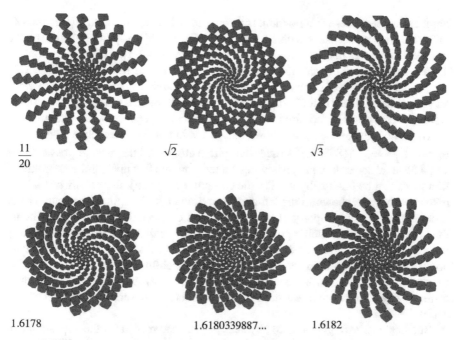

$\frac{11}{20}$ $\sqrt{2}$ $\sqrt{3}$

1.6178   1.6180339887...   1.6182

Fig. 11.26 Spirals produced by putting seeds about a central stalk, with the divergence angles indicated below each figure. Circles of increasing radius are drawn about the central axis to render visible the successive positions of the seeds. In the row above, a rational angle and two irrational values are shown. In the row below, the central panel shows the result of using the golden ratio τ as the divergence angle, while the *left and right panels* show the effect of using values just slightly below or above τ. [A Java applet to generate these figures can be freely downloaded at www. mathsisfun.com, maintained by Rod Pierce.]

necessarily sprout at some angle from the preceding ones. If this angle is nearly constant, which roughly amounts to say that the time between the birth of each leaf or flower is constant, a few spirals will result when looking from above along the stem axis. This is just the previously introduced 'divergence' angle, whose value may be obtained from the requirement of optimally filling the space, while leaving as little voids as possible. If the angle is some fraction of 2π, e.g. 3/4 or 11/20, after some turns a leaf will fall exactly on top of a preceding one: for 3/4 just the fifth leaf will lie exactly on top of the first one, and for 11/20 one has to wait 20 leaves before repeating the first one, but this will invariably happen. Therefore, some empty space will start forming between the spiralling pattern for any rational fraction of 2π, as shown by the first drawing in Fig. 11.26 for 11/20, in which case 20 spirals can be counted, and the 20 columns clearly indicate the 20-fold repetitive value of the angle. An irrational number may then seem a better choice, then. However, many irrational numbers are very close to a fraction, for example $\sqrt{2}$ is very close to 17/12, and $\sqrt{3}$ to 26/15. As a result, nests with 12 and 15 spirals, respectively, will result (see the two other drawings in the top row of the Figure), which will start displaying slightly distorted columns at increasingly larger values of the radius (the seeds in Fig. 11.26 are plotted on circles of increasing radius, to make visible the stacking of successive leaves or seeds). The irrational spiral nests are definitely denser than the first rational value, but lots of free space are nevertheless visible between the spiral pattern.

The lower row of the figure shows that if a value of the divergence angle equal to $\tau = \frac{1+\sqrt{5}}{2} = 1.6180339...$ is used, the spirals are indeed the densest possible (the slight residual separation in the drawing comes from the fact that a finite approximation to τ must be practically used in the computer program). For values just a little below or above τ, shown in the left and right drawings, the spirals are again rather widely spaced. This property of the number τ is connected to its quality of being the "worst" irrational number possible, namely the one that is farther away than any other from whatever rational fraction. What this tells us is that a structure growing about a central stalk would "choose" with high probability a divergence angle of τ, to maximise the space-filling density. However, spirals would occur also for values different from τ, although less dense and in numbers different from the Fibonacci series.

But then, can't we give a more physically-grounded reason for the appearance of Fibonacci numbers, besides the purely geometrical constraint of space filling?

In a simplified representation, let us imagine a stalk with a first leaf born at angle $0°$. If we assume some inhibitor field spreading around the just born leaf, the most probable position of the next leaf should be at $180°$. Now, if the inhibitory field from the first leaf is exhausted before that from second leaf takes on, the third leaf will just position as much as possible away from the second, therefore it will fit on top of the first. By following the same argument, the fourth will fit on top on the second, and so on, and a alternate (also called *dystichous*) pattern emerges.

However, if the inhibitor field from previously formed leaves has a somewhat longer lifetime, when leaf 3 starts, leaf 1 also repels it, although less strongly than leaf 2, as shown in Fig. 11.27a. If only one leaf at a time can be formed, the symmetry must be broken. We may take that leaf 3 will sprout at a somewhat random position, however closer to leaf 1 than to 2. Then, let us consider leaf 4: leaf 3 provides the strongest repulsive field, followed by 2 and then by 1. If a kind of exponential attenuation of the inhibitor intensity can be assumed, and the time lag between each new leaf birth is approximately constant, a stationary state will be attained after some number of leaves (Fig. 11.27b), such that the effective inhibition felt by the n-th leaf results only from a few of the preceding leaves, for example $n-1, n-2, n-3$. In a linear approximation, the relative intensities $\phi_{i,j}$ of the inhibitory fields from leaf j to leaf i should follow the proportionality law:

$$\phi_{n,n-1} : \phi_{n,n-2} = \phi_{n,n-2} : \phi_{n,n-3} \qquad (11.32)$$

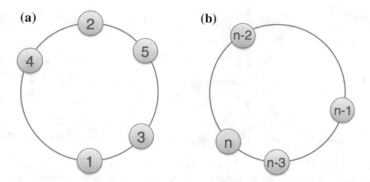

Fig. 11.27 Patterns of successive "leaves" growing about a central stalk, in a diffusion-inhibitor model. **a** The initial phase of growth: leaf 2 grows opposite to 1, then leaf 3 grows away from 2 and 1, with the inhibition from 2 being stronger than 1; leaf 4 grows away from 3 and 2, the inhibition from 1 becoming negligible. **b** The stationary phase of growth: the angles (arcs of circle) between any leaves $n, n + 1$ are constant

By looking at Fig. 11.27b, the stationary state means that the angular difference between subsequent position $n, n - 1$ is constant modulo 2π. Two relationships between pairs of successive angles can be written as $\phi_{n,n-2} = 2\pi - 2\phi_{n,n-1}$, and $\phi_{n,n-3} = -(2\pi - 3\phi_{n,n-1})$. By substituting such expressions in the above proportionality law, we get:

$$\phi_{n,n-1}^2 - 3\phi_{n,n-1}(2\pi) + (2\pi)^2 = 0 \qquad (11.33)$$

The solution of the above equation is $\phi_{n,n-1}^2 = (3 - \sqrt{5})\pi$, that is $137°5$, the "golden" angle. Clearly this is a ready-to-work analysis, in which the conditions are adjusted so as to show that within "reasonable" constraints, the golden-ratio spacing should naturally arise because of the physically motivated diffusion-inhibitor underlying dynamics. But even if the conditions are relaxed, the presence of the inhibitor field, coupled to the geometrical constraint of the growth about a central axis typical of many plant structures, would force the space filling into values approaching the golden-ratio maximal density.

This was proved in a series of experiments of beautiful simplicity carried out in 1992 by Stéphane Douady and Yves Couder, then both at the Ecole Normale in Paris [20]. In the experiment, magnetic droplets are dipped at the center of an oil-covered surface. The circular border of the surface is magnetised, such that droplets separated by a distance d repel each other, with a force proportional to d^{-4}, and stream toward the outer border at a steady velocity v limited by the viscosity of the oil. The physics of the process is governed by the single nondimensional parameter $G = vT/r$, product of the velocity times the dripping period T, and divided by the radius r of the circle. Figure 11.28 shows the results of the experiment for values of G ranging from 1 down to 0.15. A large value of G means that the system is

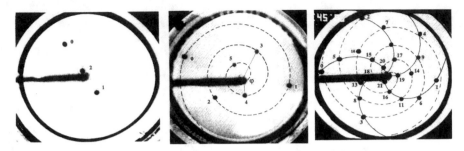

Fig. 11.28 Photographs of experimental patterns of magnetic droplets falling on a oil-covered surface. The outer border is magnetised and slowly attracts the droplets falling at the center, numbered according to the progressive order of deposition. The three photos, from *left* to *right*, correspond to values of the nondimensional parameter $G = 1., 0.7, 0.15$. Correspondingly, the "divergence angle" between successive pairs is $\simeq 180°$, $150°$, $139°$. [From Ref. [20], repr. w. permission]

dominated by the radial velocity: droplets move quickly away from each other along opposite directions, simulating the case of a very short-lived inhibitor repulsive field. For smaller values of G the mutual repulsion is increasingly long-lived; at $G \simeq 0.7$, droplets break the 180° symmetry and start to take a spiral pattern; at even smaller values $G \simeq 0.15$, more spirals are formed and the divergence angle increasingly approaches the "golden" value of 137°5. In the same experiments, it was shown that the convergence to Fibonacci-like patterns is independent of the value of the repulsive force, in that also forces different from the d^{-4} law will end up in the same qualitative results.

Such a kind of findings provide a clear demonstration that the growth process from a central stem, typical of the plant world, may approach the Fibonacci-like patterns because of spontaneous self-organisation into structures that satisfy the best space-filling constraints. It is left to the reader's imagination to consider whether this is still a magical or mysterious coincidence, or rather one of the many emergent phenomena that arise everywhere in nature, because of the cooperative action of many agents. However, the space filling is one but not the only requirement that plants try to satisfy during their growth. Other concurrent factors could be the availability of sunlight, the proximity of other plants, the presence of obstacles that direct the water flow asymmetrically about the stalk, and many others. As a result, such regularities are observed only on average. The next time you go for a walk in the garden, it will be a pleasant challenge trying to find how many flowers respect the Fibonacci numbers of their petals... but take into account that many petals and leafs could have been lost, maybe because of wind, rain, insects, or of other walkers less attentive than you!

Problems

11.1 Water walkers
Consider an aquatic hexapod insect, with a spherical body of mass M and diameter D, and six legs of negligible mass and length L. What is the criterion to keep the insect afloat on the water surface?

11.2 Climbing the tree
Knowing that the latent heat of evaporation of water is 44 kJ/mol, (a) show that water molecules climbing the interior of a tree by capillarity do not violate the second principle of thermodynamics; (b) what capillary size would be needed to this end?

11.3 Revolving parabola
Draw the section of the parabola $y = 1 + \frac{1}{2}(x - 2)^2$ in the interval $x \in [1, 4]$, and the surface of revolution obtained by rotating the section about the x axis. Calculate the area of the surface.

11.4 Morphing snails
You want to compare the growth rates of two species of snails. By graphic analysis, you determine that the two snail shells are described by the parametric equations:

$$\text{snail A} : (x = 7t \sin t, y = 7t \cos t), \text{ for } t \in [0, 3\pi + 2];$$
$$\text{snail B} : (x = \tfrac{t^2}{2} \sin t, y = \tfrac{t^2}{2} \cos t), \text{ for } t \in [0, 4\pi + 1].$$

By assuming that they grew at the same rate, which one is older?

11.5 Packing problems
(a) In this chapter it was shown that for an *infinite* arrangement of circles in the plane, the hexagonal packing gives the highest density, with the area of the circles covering about 91 % of the total. Calculate the coverage ratio for the packing of 1, 2, 3, 4 or 5 circles in a square of *finite* size. (b) Show that the best packing for spheres in infinite 3D space is by stacking 2D packed layers on top of each other, such that the spheres in each top layer coincide with the centre of three spheres in each lower layer (the so-called *face-centered close-packing*) and calculate the volume filling ratio in this case. How does it compare with the regular packing of one sphere in each unit cube?

11.6 Cell migration and gradients
Tissues are normally cultivated in a homogeneous blood-plasma gel. A small sample of spleen of a chicken embryo is placed in the position $(x = 0, y = 3)$ of a Petri dish (centre of the grey area in the figure), and left to incubation at 39 °C for 10 h, after which the culture is fixed in acetone, and the cells can be observed under a microscope. The image below shows that cells from the spleen have migrated, but not in symmetric radial directions as it could be expected. By graphical analysis, find a transformation map $(x \to x', y \to y')$ that can bring the image to a circular

symmetry. What can you deduce from such transformation, about the effects of the driving forces operating in cell migration?

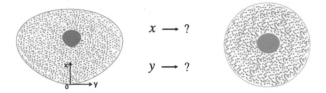

11.7 Human growth

The figure below (from Ref. [21], public domain) represents the shape of human body, from about the 5th month of foetal stage to maturity, scaled in the vertical direction so as to get unit height at any age (scale below is in years). Try to deduce growth laws for the different parts of the body as a function of age. (For this problem, you may want to take an enlarged photocopy of this page.)

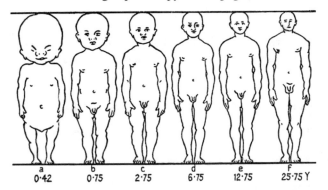

References

1. J.S.B. Haldane, *Of Possible Worlds, and Other Essays* (Chatto & Windus, London, 1932)
2. M. Tabarrini, G. Milanesi, (eds.), *Trattato della pittura di Leonardo Da Vinci, condotto sul Codex Vaticanus Urbinate 1270*. Unione Cooperativa Editrice, Roma (1890) [A modern reprint of the *Urbinate 1270*, assembled around the years 1520–25 by the Milanese painter Leonardo Melzi, former pupil of Leonardo]
3. J. Taylor, The structure of singularities in soap-bubble-like and soap-film-like minimal surfaces. Ann. Math. (2nd ser.) **103**, 489–539 (1976)
4. J. Richard, (ed.), *Résultats des campagnes scientifiques accomplies sur son yacht par Albert Ier, prince souverain de Monaco*. Fasc. X and XXXV, Imprimerie de Monaco (1896-1911) [released for public domain: archive.org/details/rsultatsdescam10albe]
5. W. D'Arcy Thompson, *On Growth and Form*, 1st edn, 1971. Cambridge University Press, New York (1945) [released for public domain: archive.org/details/ongrowthform00thom]
6. O.W. Richards, G.A. Riley, The growth of amphibian larvae illustrated by transformed coordinates. J. Exp. Zool. **77**, 159–167 (1937)
7. A.M.Turing, The chemical basis of morphogenesis. Phil. Trans. Roy. Soc. (London) Biol. Sci., B **237**, 37–72 (1952)
8. J. Cooke, E.C. Zeeman, A clock and wavefront model for control of the number of repeated structures during animal morphogenesis. J. Theor. Biol. **58**, 455–476 (1976)
9. R.-T. Liu, S.-S. Liaw, P.K. Maini, Oscillatory turing patterns in a simple reaction-diffusion system. J. Kor. Phys. Soc. **50**, 234–238 (2007)
10. P.J. Murray, P.K. Maini, R.E. Baker, The clock and wavefront model revisited. J. Theor. Biol. **283**, 227–238 (2007)
11. C.D. Stern, *Gastrulation: From Cells to Embryo* (Press, Cold Spring Harbor Lab, 2004)
12. O. Campas, T. Mammoto, S. Hasso, R.A. Sperling, D. O'Connell, A.G. Bischof, R. Maas, D.A. Weitz, L. Mahadevan, D.E. Ingber, Quantifying cell-generated mechanical forces within living embryonic tissues. Nature Meth. **11**, 183–189 (2013)
13. S. Piccolo, Developmental biology: mechanics in the embryo. Nature **504**, 223–225 (2013)
14. H.S.M. Coxeter, Golden mean phyllotaxis and Wythoff's game. Scripta Math. **XIX**(2, 3) (1953)
15. J. Kappraff, Growth in plants: a study in number. Forma **19**, 335–354 (2004)
16. H. Meinhardt, *Models of Biological Pattern Formation*, 1st edn., out of print. Academic Press, London, 1982) http://www.eb.tuebingen.mpg.de/research/emeriti/hans-meinhardt/82-book/bur82.html
17. G. van Iterson, *Mathematische und mikroskopisch-anatomische Studien über Blattstellungen* (Fischer, Jena, 1907)
18. I. Adler, A model of contact pressure in phyllotaxis. J. Theor. Biol. **45**, 1–79 (1974)
19. J.C. Schoute, Beiträge zur Blattstellung. Rec. Trav. Bot. Neerl. **10**, 153–325 (1913)
20. S. Douady, Y. Couder, Phyllotaxis as a physical self-organized growth process. Phys. Rev. Lett. **68**, 2098–2101 (1992)
21. P.B. Medawar, in *Size, Shape and Age*, eds. by W.E. Le Gros Clark, P.B. Medawar. Essays on Growth and Form (Clarendon Press, Oxford, 1945)

Further Reading

22. J. Liouville, Sur la surface de révolution dont la courbure moyenne est constante. Journal de Mathematique **6**, 309 (1841)
23. J.C. Maxwell, On the theory of rolling curves. Trans. Roy. Soc. Edinburgh **16**(519), 519 (1849)
24. J.F. Plateau, *Statique expérimentale et théorique des liquides soumis aux seules forces moléculaires* (Gauthiers-Villars, Paris, 1873)

25. J.K. Wittemore, Minimal surfaces of rotation. Ann. Math. **19**, 2 (1917)
26. G. Loria, *Curve piane speciali algebriche e trascendenti, teoria e storia* (Hoepli, Milano, 1930)
27. T.A. McMahon, J.T. Bonner, *On Size and Life* (Scientific America Library, New York, 1983)
28. K. Schmidt-Nielsen, *Scaling: Why is Animal Size so Important?* (Cambridge University Press, 1984)
29. W. D'Arcy Thompson, *On Growth and Form*, (2nd ed.,1942) abridged (Dover Publications, New York, 1992)
30. P.G. De Gennes, F. Brochard-Wyart, D. Quere, *Capillarity and Wetting Phenomena: Drops, Bubbles, Pearls, Waves* (Springer, Heidelberg, 2004)
31. P. Bourgine, A. Lesne, *Morphogenesis: Origins of Patterns and Shapes* (Springer, Heidelberg, 2011)

Chapter 12
The Hidden Mathematics of Living Systems

Abstract Albert Einstein once wondered: How can it be that mathematics, being after all a product of human thought and independent of experience, is so admirably appropriate to the objects of reality? Mathematics is a humanised way of describing patterns and regularities we see in the universe. We formulate such regularities with equations, scaling laws, invariance principles that, as long as they remain sufficiently robust on the scale of human perception, bring a sense of order to our understanding. This chapter deals with the mathematical modelling of biological species in their environment, or ecosystems. It may be surprising that the same scaling laws and similar organising principles could apply to groups of cells, single animals, and entire populations. However, this also points out the crudeness and approximation of such treatments. Nature seems to make an habit of surprising us: it always proves to be a lot stranger than we give it credit for.

12.1 Changing Size Without Changing Shape

By looking around the natural environment, it is evident that both plants and animals of similar species vary quite a lot more than us humans in their sizes, notably as a function of certain genetic characters. A Chihuahua is much a smaller dog than a Saint-Bernard, which however is as well a dog: the couple could have a baby-dog, although the result might be quite weird. Moreover, sizes can vary as a function of environmental characters, for example, the amount and quality of food during the early development phases. However, in a taxonomic context size is never considered as an important factor, since it is deemed not useful to determine the ancestry and relationships among species. What this also implies, is that the size can be adjusted according to natural selection, if some extreme conditions would favour a race of giant, or of dwarfs. In nature it is frequent to observe a same body shape with quite extreme variations in size, which is often an indication of a common ancestry. For example, whales and dolphins appear as two representatives of a same evolutionary design, although varying by factors of about 60 in size and ~2,000 in mass.

Nevertheless, extreme variations in size can have interesting consequences, as Gulliver discovered in his *Travels into the Several Remote Nations of the World*. A

© Springer International Publishing Switzerland 2016
F. Cleri, *The Physics of Living Systems*, Undergraduate Lecture
Notes in Physics, DOI 10.1007/978-3-319-30647-6_12

surgeon and captain, but definitely not a physicist, besides marvelling at the unusual size variation of the inhabitants of those fabulous countries, he could have attempted a definitely more systematic study of functional variations as a function of the variation in animal body size, by using the method of **scaling analysis**.

To illustrate the basic principles of scaling analysis, we can compare two cubes, the one with side of 2 units and the other with side of 4 units. The surface of each face is $2^2 = 4$ for the first cube, and $4^2 = 16$ for the second one; their volumes are $2^3 = 8$ and $4^3 = 64$, respectively. However, the ratio of their surfaces is 4, in any unit of length we wish to measure it, and the ratio of their volumes is 8, in any unit. Scaling analysis is not concerned with units, it looks for relationships that hold besides the particular numerical values. Moreover, we may note that the size of the surfaces of the cubes is proportional to the square of their side lengths, and the size of their volumes is proportional to the cube of their sides. In the notation of scaling analysis, we would write:

$$S \propto L^2$$
$$V \propto L^3 \tag{12.1}$$

We could actually look at these expressions as dimensional relationships, as we did in Chap. 11, meaning that the dimensions of a surface are the square of a length, and the dimensions of the volume are the third power of a length. However, the implications of the two equations are more far-reaching. They also express relations between the numerical values associated with the quantities 'surface' and 'volume'. Moreover, the relations work independently on the shape of the object. If we consider the same quantities for two spheres, with radius 2 and 4 in whatever units, we would find the same expressions.

What the relations (12.1) above tell us, is that by modifying the size of an object by a common **scaling factor**, we could deduce the length, surface and volume of all its parts. Consider for example a complex body formed by joining together n cubes of different sizes into some shape; each cube has 12 equal sides of length a_i, $i = 1, \ldots, n$, 6 equal faces of area a_i^2, and volume a_i^3. Suppose to scale one of the lengths a_j by a factor λ, such that $a_j' = \lambda a_j$, and that the same scaling is applied to the entire body. Then, the faces of any of the n cubes have the new surface $\lambda^2 a_i^2$, and each cube has the new volume $\lambda^3 a_i^3$.

The example may look somewhat banal. However, if the animals were to follow such a perfect scaling we could get, for example, the length of the intestines, or the amount of blood circulating in the veins of any dog, by scaling. To get such figures, we should firstly measure the length L of the intestines and the volume of blood V in one dog of a given size; then, by taking the ratio λ between, e.g., their nose-to-tail length, the size of the intestines of any other dog could be obtained by multiplying L by λ, or the blood volume by multiplying V by λ^3. As weird as it may seem, such a method works more often than one could think.

12.1.1 Allometry and Scaling

The hypothetical notion underlying the application of a scaling law to a group of plants or animals, geometrically similar in shape but largely varying in their sizes, may share the same similarity in their basic vital functions, their metabolic rates, their relationship with a similar environment, goes under the name of **allometry** (from the Greek words ἄλλος and μετρέιν, "to measure others"). We used already similar scaling arguments, for example in Chap. 11 when discussing the difference between surface and volume forces. Already in 1883, the German physiologist Max Rubner had used this concept, when studying animal metabolism and the heat dissipated by animals with bodies of different sizes, and D'Arcy Thompson introduced the idea in his 1905 book *On growth and form*, which was also amply cited in the preceding Chapter. The general relationship of allometry for a quantity y supposed to change as a function of the size scaling of another quantity x, is $y = \alpha x^{\beta}$, where α is called the scaling coefficient and β the scaling exponent. (Note that for $\beta = 1$ the scaling relation is linear, and the allometry becomes a simple *isometry*.)

We could take as an example a family of animals all living in very similar environment, which should have been subject to a similar kind of evolutionary pressure to adapt to such an environment. By taking again inspiration from a scientific work due to Pennicuick [1], we will look at a small fleet of oceanic birds from the order of *Procellariiformes*, defined as "pelagic" birds since they feed in the open sea instead of coming back to shore. These birds are represented by four families and more than 130 species distributed all over the Earth, with a special diversity concentrated around New Zealand. One may search for a relationship between the size of their wings, or wingspan, and the sizes or volumes of their bodies. Indeed, if measuring the volume of an animal may be quite difficult, measuring its mass is in turn a much simpler task. Then, by assuming that the average density ρ of the body materials does not change significantly upon changing the body size, we can rely on a mass-length scaling law as:

$$M = \rho V \propto L^{3} \tag{12.2}$$

In practice, it is more useful to invert this scaling relation as:

$$L \propto M^{1/3} \tag{12.3}$$

The data of mass and wingspan for 11 birds of this order, covering a wide range of mass ratios, between 1 and 400, and linear dimensions varying by more than a factor 10, show that a good linearity exists between the wingspan ratio W (measure of the end-to-end distance of their fully spread wings), and their mass ratio. It is appropriate to express these numbers as ratios with respect to the length and mass of a reference animal, since they will be plotted in a log-log scale, and it is always good to remember that a logarithm must have a non-dimensional, pure number as an argument. If the relation between wingspan and mass were of the type $W = \alpha M^{1/3}$, the linear fit on the log-log diagram should give a slope of exactly 1/3:

$$\ln W = \ln \alpha + \tfrac{1}{3} \ln M \tag{12.4}$$

Any deviation from the 1/3 exponent represent a deviation from the assumed scaling law. The wingspan plot in Fig. 12.1 has a slope of 0.370, somewhat larger than the 0.333 expected. The standard deviation of the straight line fit is 0.0142, with a confidence interval of 5%. Is this a significant deviation? If yes, this would mean that the larger birds in the order have the tendency to develop wings proportionally larger than the smaller ones. As a check, we could look at a similar relationship between the wing surface ratio and the mass ratio, $S = \alpha' M^{2/3}$ or:

$$\ln S = \ln \alpha' + \tfrac{2}{3} \ln M \tag{12.5}$$

From the plot in Fig. 12.1 a slope of 0.627 is found, slightly smaller than the 0.667 expected; the standard deviation is 0.0180. If we take the statistics by their face value, the necessary conclusion is that the geometric scaling law applies quite well in the case of these oceanic birds, with only a slight tendency for the larger birds to increase their wing size compared to the smaller members of the order.

However, as our guide Pennicuick suggests, if we look at the shape of the wings it can be easily seen that they do not look too much similar by spanning across the range of birds of increasing size (see Fig. 12.1c). By rescaling the silhouettes of the birds with their wings spread open, such that they have all the same unit width, it can be seen that the shape of the wings becomes thinner as the mass of the bird increases (from the 40 g of the Wilson's petrel, to the 9 kg of the wandering albatross).

Therefore, the two scaling laws hold quite well for the wingspan and wing surface separately, but do not seem to tell the whole story. To have a further check of this discrepancy, we could combine the two quantities into a third variable, the aspect ratio Q (in which the wingspan appears squared, to form a non-dimensional number):

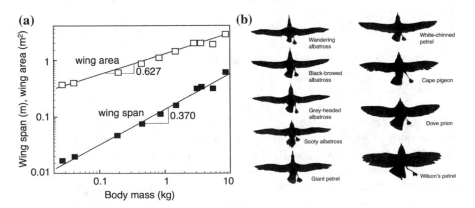

Fig. 12.1 a Log-log scaling plot of the wingspan and wing area versus body mass for 11 birds of the order of *Procellariiformes*; the slope of the *straight line* fit for the wingspan is 0.370, with a standard deviation of 0.0142; and for the wing area 0.627, with a standard deviation of 0.018. **b** Silhouettes from the same group of birds, rescaled so that they have all the same unit width, to show the importance of the aspect ratio. [Image (**b**) from Ref. [1], adapted w. permission.]

$$Q = \frac{W^2}{S} \tag{12.6}$$

This quantity is in fact the ratio between two ratios. It quantifies the "amount of perimeter" necessary to delimit a given surface, with an increased role of the numerator, in which W appears squared. A very elongated surface with a given area would have a quite large value of Q, while another more rounded surface with the same area would have a smaller value of Q; the minimum of Q is for a circular surface, which has the smallest perimeter for a given area. Given that $W \propto L^{1/3}$ and $S \propto L^{2/3}$, if the geometric scaling hypothesis is valid the aspect ratio Q should be independent on the mass. Instead, the log-log plot of Q versus M gives a slightly inclined straight line, with a slope of 0.116 and a standard deviation of 0.011, that is a significant deviation from the slope zero. This is a strong indication that the geometric scaling alone in this case fails to give a meaningful representation of the data from various birds. Wings aerodynamics must have its word.

12.2   Scaling Laws for Animal Locomotion

The English physiologist Archibald V. Hill, who had been awarded the Nobel prize in 1922 for his understanding of how heat and work are generated in animal muscles, used in the 1950s the concepts of scaling analysis in a study on the animal locomotion, which should later become a classic. He started from the observation that the mechanical work produced by a muscle while shortening, should be equal to the force developed multiplied by the muscle shortening, ΔL. As it was noticed in Chap. 10, the stress exerted by a muscle is an intrinsic property of the muscle tissue, therefore the force is anyway proportional to the muscle cross section; on the other hand, the shortening for a given deformation rate $\dot{\epsilon}$ is proportional to the muscle length, therefore the mechanical work scales with the length as:

$$W = F\Delta L \propto L^3 \tag{12.7}$$

in other words, it is $W \propto M$.

Already on the basis of such a modest result, Hill could consider the following problem. Let us imagine that a series of geometrically similar animals make a vertical jump, each one using the maximum of mechanical work it can extract from its muscles during an individual contraction. What scaling law should the jump height follow? By supposing that all the mechanical work turns into gravitational potential energy, each animal with mass M should attain a height h such that:

$$W = Mgh \tag{12.8}$$

However, since $W \propto M$, Hill came to a rather surprising result: for animals geometrically similar, the jump height is independent on the mass of the animal. This is sometimes called the "First Hill's law of animal locomotion". According to this 'law', a dog and a horse should jump to the same height, since their body shapes are enough similar: the smaller animal has a smaller muscular mass, which in turn has to lift a smaller body mass, and vice versa. This observation apparently justifies the independence on the mass. However, upon direct measurement, jump heights turn out to be quite different from one animal to another. For example, a 200 kg antelope jumps to about 2.5 m height (that is about 1.4 times its size), while a 2.5 kg cat jumps to a maximum of about 1.5 m (~4 times its size), and the small mammal galago to a surprising 2.25 m, more than 10 times its size.

In fact, it is a consequence of Newton's laws that in order to reach a same height, the animals should take off ground with the same initial speed, since the gravity acceleration is always the same (and we are neglecting the air resistance). Because the take-off speed is obtained by quickly relaxing the leg's (and other body's) muscles from their maximally compressed configuration, not just quickly but *how much quickly* is the key. A smaller animal has proportionally shorter limbs, therefore to attain the same initial velocity of a larger animal it must adopt a proportionally faster deformation rate, since the muscle frequency $v \propto \dot{\epsilon}$.

To what extent such relationships are verified in practice is not easily assessed, and conflicting data have often generated sharp controversies between scientists. If we use a typical deformation of $\epsilon \sim 0.25$, which is proper to the sarcomere structure, and therefore a quite well conserved value across the evolution of largely different animal species, knowledge of the take-off time should give the take-off speed. Remaining with the previous example, such time values have been recorded to be about 250 ms for the cat, about 100 ms for the galago, and about 430 ms for the antelope. By comparing the antelope and the cat, the ratio of their jump heights looks indeed proportional to the ratio of the respective take-off velocities, indeed growing with the mass, however quite far from a M^{-1} scaling law. But if we compare the galago and the antelope, jumping to nearly the same height, their velocity ratio is quite far from ~1. The main reason behind such incoherencies is the assumption that all muscles participating to the jump contract at the same rate, which may instead change very substantially depending on the type of animal and its metabolic requirements. Moreover, the geometric similarity may be a too vague and superficial concept: while it is true that a cat and an antelope both have four legs around an oblong body, many important details are different, such as the neck length, the distribution of mass between fore and hindlimbs, the relative proportion of internal organs (antelope, as a ruminant, has three more stomachs than a cat); frogs and toads may appear as geometrically very similar animals, however frogs can jump to longer distances than toads of similar mass, because their hind legs are much longer.

12.2.1 Scaling Law for the Characteristic Frequencies

Hill seems to have been the first to understand that the contraction frequency of muscles is slower in large sized animals than in smaller ones, and that this difference is fundamental in determining the power their muscles can supply, this latter being in turn correlated to the rate at which fuel and oxygen are consumed in the animal body (see Chap. 4).

We already demonstrated in the preceding Chap. 10 by simple analogy with an ideal pendulum, that the walking frequency is inversely proportional to the square root of the length of the walking leg. By following a different reasoning, Hill tried to find a law for the *maximum oscillation frequency* of a limb, by calculating the angular acceleration produced by a tendon actuating an angular momentum about the leg joint [2]. Take the strength of the collagen molecules of the tendon as the upper limit of the applicable stress σ_y, beyond which the tendon with cross section s would start yielding, the angular momentum $\mathcal{M}$ is written as the product of the force $F_y = \sigma_y s$ by the tendon length b (the lever arm):

$$\mathcal{M} = F_y \times b \propto L^3 \tag{12.9}$$

Newton's law, $F = ma$, is also applicable to the equation of motion for the angular movement $\theta = \theta(t)$, by replacing the force by the angular momentum, the mass by the moment of inertia $\mathcal{J}$, and the linear acceleration by the angular acceleration $\dot{\omega}$ (with $\omega = \dot{\theta}$ the angular velocity):

$$\mathcal{M} = \mathcal{J}\dot{\omega} \tag{12.10}$$

The moment of inertia has dimensions of $[M][L^2]$, therefore its scaling with the length is $\mathcal{J} \propto L^5$. If we invert Eq. (12.10) for the angular acceleration, it is:

$$\dot{\omega} = \frac{\mathcal{M}}{\mathcal{J}} \propto L^{-2} \tag{12.11}$$

After this observation, Hill moved on to consider the time necessary for the running animal leg to cover an angle θ_0, starting from $\theta = 0$ with zero angular velocity. The equation of motion is simply $\theta(t) = \frac{1}{2}\dot{\omega}t^2$, from which it is found the time as: $t \propto \dot{\omega}^{-1/2} \propto [L^{-2}]^{-1/2}$, i.e. the acceleration time is proportional to the length, $t \propto L$. In the end, the frequency ν (inverse of the time necessary to sweep the angle from zero to θ_0 at each step) should vary as the inverse of L:

$$\nu \propto L^{-1} \propto M^{-1/3} \tag{12.12}$$

This result leads to a second interesting generalisation. Consider that the length of the step, or **stride length** L_p, is the distance between two subsequent steps. The animal speed u can be defined by multiplying this characteristic length by the walking frequency, which according to Hill varies as the inverse of the length. Therefore, from a scaling point of view, it is:

$$u \propto L_p \nu \propto L^0 \tag{12.13}$$

In other words, the maximum speed should be independent on the stride length. This second, again surprising result is sometimes dubbed as the "Second Hill's law of locomotion". Alexander, in a study published in 1977, measured the maximum running speeds of ten species of african ungulates, from the gazelle to the giraffe, with mass ratios ranging from 1 to 50; since bigger animals appear to make longer strides, but with a slower frequency, it was found that—at least for the maximum velocity—the so-called second law is verified to a good approximation.

Hill's discussion of frequency scaling postulates that while running at the maximum speed, the tendon is loaded to a constant level of stress, in principle the maximum it can sustain before breaking. This is a purely theoretical statement, since we know that an overtrained athlete can strain his tendons (beyond σ_y, with permanent damage), if the muscular power developed surpasses the mechanical strength of other parts of the body, not equally reinforced. Nevertheless, Hill's argument is necessary to identify this concept of "maximum animal speed", whereas the animal may use a range of different displacement velocities in its daily life, notably its typical cruising speed being much below the as-defined maximum speed.

12.2.2 Walkin' the Dog

In terms of frequency, cruise walking identifies a natural frequency corresponding to aerobic walking with the minimum effort (and minimum energy consumption), just enough to maintain the 'body-pendulum' oscillation. High-speed running, on the other hand, defines another natural frequency, in which the animal is forcing its body parts to extreme movements, often coupled to anaerobic energy consumption. As it was described in the forced and damped harmonic oscillator (see greybox on p. 460), an oscillator can be forced to a different frequency from its natural one, but in that case the force (and mechanical work) necessary grows rapidly. As we saw in Chap. 10.2, Alexander and Jayes showed that for an animal in 'gravitational' walk (when the animal is rather well described as an inverted pendulum, see the discussion of this work on p. 471), the walking frequency varies as length of the leg to the power $-1/2$, in fact the same scaling law that was found for the pendulum. Animals at trot or fair-slow gallop increase the step frequency and stride length by walking in jumps on the terminal (distal) segment of the leg, and also in this case the frequency is found to vary as the length to power $-1/2$, with a different scaling coefficient.

The exponent $-1/2$ is definitely different from the exponent -1 of Hill's law, which therefore should hold only for the animal running at its maximum speed. However, from a kinematic point of view, it should be also noted that the leg and body movements involved in the various gaits of the animal are indeed very different. The gaits of dogs have been traditionally the best studied [3], since the advent of

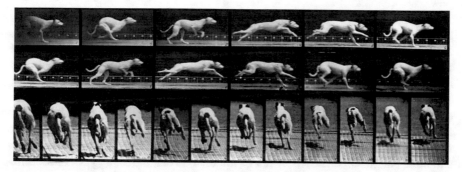

Fig. 12.2 An original photographic plate sequence by Eadweard Muybridge (born Edward James Muggeridge), showing a typical dog run cycle. Note the very long period of suspension, where all legs are off the ground, and the harmonic compression and expansion of the back spine. After 1880, Muybridge, then at the University of Pennsylvania, produced more than 100,000 images of animals and humans walking and running, thus providing a precious data base of stop-motion pictures of animal motion. [Public-domain image: E. Muybridge plate 710, Boston Public Library.]

photography in the XIX century, when Muybridge in 1888 was able to show the stride of a racing greyhound, by using stroboscopic photography (Fig. 12.2).

During the *walk*, the dog never has fewer than two feet on the ground (usually three feet), and occasionally all four feet may be on the ground. The *trot* is a symmetric gait produced when the diagonal pairs of legs move almost simultaneously, the dog usually places two feet on the ground at all times; it is noted that dogs with short body length and long legs have difficulty trotting, since their hind legs interfere with their front legs. In the *pace* the dog moves by swinging the forelimb and hindlimb on one side while bearing weight on the other side. Finally, the *gallop* is an asymmetric gait used for high-speed locomotion. Dogs can use two patterns of gallop: the transverse gallop, similar to the pattern used by the horse; and the rotary gallop, which seems to be preferred by the dog and which in the horse is referred to as a crossed-lead gallop. Notably, a dog can sustain the gallop at two different regimes of speed: a slower gallop (also known as *canter*), aerobic and easily sustained over a long period of time; and a fast, anaerobic gallop.

When the animal begins to gallop, the frequency of the stride remains almost constant while the animal increases its speed by increasing the stride length. While the gaits of the walk and trot seem for the most part to use only muscles associated with the legs, the gallop uses muscles of the trunk also, hence the arching and extension of the back. It has been postulated that, as far as the animal changes gait as its speed increases, additional elements of the body can be recruited for storage of elastic energy [4]. Very little energy is stored in elastic elements (see Chap. 9) during walking, while during trotting energy is stored in elastic elements of the limb (tendons most likely); however, the entire trunk of the body is involved in the elastic storage of energy during galloping. Large animals seem to have a maximum "whole animal" efficiency that is nearly three times greater than the maximum efficiency of their muscles.

When the animal has accelerated to its constant peak speed, the mechanics of its locomotion are similar to those of a bouncing ball: the only energy necessary to keep the animal "bouncing" is the elastic energy not recovered at each step. Therefore small amounts of energy are put into the system, thereby giving an overall efficiency that is higher than what would be obtainable by accounting for the full muscle contraction and extension at each step. As we described at some length in Chap. 9, for a muscle to use elastic energy, it must first lengthen to develop tension, then quickly shorten to release the stored elastic energy. Observations suggest that many more muscles are used as the gait speed increases, and their contraction/extension becomes more and more partial. On this basis, it can be understood that the scaling laws for cruise walking and fast running may indeed exhibit quite different scaling exponents.

12.3   Paleontology, Or When Animals Were Huge

For animals capable of different types of locomotion, the maximum frequency must be obviously higher than their natural cruise frequency. Since the former is found to scale as L^{-1} while the latter scales as $L^{-1/2}$, in a log-log plot of the frequency versus body mass (that is in turn proportional to L^3), we should find two straight lines with different slopes, respectively of $-1/3$ and $-1/6$, and decreasing upon increasing mass. In the bi-logarithmic scaling plot shown in Fig. 12.3, the two straight lines must cross at some value of the animal mass, M_{max}. This is a sort of upper limit for that animal species for which the ideal plot is being drawn: its members can grow in size, while their maximum speed decreases faster than the cruise speed, until the two practically coincide. Since it is not possible for the former to be larger than the latter, no animal in that species, or in that geometric similarity class, could grow a mass beyond M_{max}.

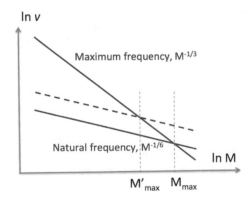

Fig. 12.3 Bi-logarithmic scaling plot of the walking/running frequency as a function of the body mass. The natural walking frequency varies as $M^{-1/6}$, the maximum running frequency as $M^{-1/3}$. However, while the first one depends explicitly on the gravity, via $g^{1/2}$, the second does not. An increase of the gravity constant, for example, should displace the natural frequency to higher values (*dashed red line*), and the upper limit animal mass M_{max} should consequently move to smaller values M'

When an animal species is subject to an evolutionary pressure that pushes them to increase their body mass, the only way to escape such a limit would be to substantially change their body shape, therefore escaping the geometric similarity constraint. For example, by increasing in a non-proportional way the size of their bone compared to the fat body mass; by enlarging the surface of muscle insertion on the bone spines; by increasing the bone joint size more than the bone thickness; by changing the aspect ratio of their wings with increasing surface; and so on. However, it should be also noticed that a substantial transformation of the body shape does not come without a price in terms of maintaining a good mechanical efficiency, compared to the original design, as it was noted already by Galileo in his writing about scaling of animal bones (Fig. 12.4).

An entirely different way to escape the evolutionary pressure would be that of changing the living environment, or **ecological niche**. The best example is probably given by the great marine mammals, or cetaceans, like whales, dolphins and orcas. These have evolved from terrestrial ancestors, and were forced to go back to the marine environment in order to support their growth to a larger and larger body mass, thanks to the Archimedes' buoyancy force. Many fossils demonstrate this evolutionary path, from the primitive artiodactyls, which had teeth with triangular shape, and ankle bones identical to those of fossil proto-whales (Fig. 12.5a, c and Ref. [5]). This transition can be considered quite rapid, on the geological time scale, since it should have occurred in the space of just about 10 millions years. The terrestrial origins of cetaceans are evident from several traits: modern whales must come to the sea surface to inhale fresh air; the bones of their pectoral fins are perfectly homologous to the upper limbs of terrestrial mammals; and their dorsal movements undulate in the vertical direction (Fig. 12.5b), more appropriate to animals adapted to walking on the rigid surface, rather than horizontally as all fishes do.

It was already noted in Chap. 10 that the tail beat frequency of fishes should follow a scaling law of the type $\nu \propto L^{-1}$, similar to Hill's law for the maximum running frequency. If our deduction has some truth to it, this should imply that the crossing point between the cruise speed and maximum speed in Fig. 12.3 should not hold for sea-bound animals, and that there should be no upper body mass limit for fishes of similar body shape. In fact, it may be observed that there seems to be no mechanical reason for limiting the size of fishes in the sea, and that if any, such a limit should rather come from physiological factors. For example, the time necessary for the blood to run from the heart down to the extremities of such a long animal, thus limiting the oxygen flow; or the time necessary for a nerve pulse to propagate to such a length. When a whale beaches accidentally, it quickly suffocates under its own weight because its supporting structure is inadequate, in the absence of buoyancy.[1]

---

[1] It is odd that Galileo, in the *Discorsi* page immediately following the one shown in Fig. 12.4, has Simplicius objecting to Salviati that whales seem indeed to have no limits to their size. The answer provided by Salviati-Galileo, although vaguely making recourse to Archimede's principle, is curiously wrong: by looking at the way fishes may float in equilibrium at any desired depth in water, he assumes that the fish body should have the same density as the water's, and since its bones are denser, its flesh should be made of a material lighter than water.

In good agreement with the scaling laws established in Chap. 10, the natural walking frequency varies as the square root of the gravity, $\nu \propto g^{1/2}$. On the other hand, the maximum Hill's frequency seems not affected by gravity. Therefore, life on a planet with a larger value of g should see the natural frequency line to be shifted towards higher values. As a consequence, the crossing with the maximum frequency line would occur for lower and lower maximum body mass: a higher gravity would make smaller the upper mass limit of any species. At the opposite, a planet with reduced gravity than the Earth's, could host larger animals. It is nothing more than a curious coincidence that during two periods of the past Earth's history, the Cretaceous (145–66 millions years ago) and the Miocene (23–5.3 millions years ago), the terrestrial animals were considerably larger in size than in any other era. The flying animals of those times were even larger than any other recorded form of life on Earth, the *Pterosaurus* in the Cretaceous having a wingspan of up to 12 m, and the *Argentavis* in the Miocene with up to 7 m. The latter bird, in particular,

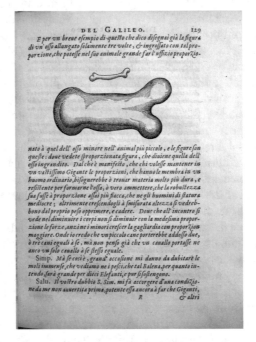

Fig. 12.4 A page from the *Discorsi e dimostrazioni matematiche intorno a due nuove scienze* (Discourses and mathematical demonstrations on two new sciences, Leiden, 1638), by Galileo Galilei. The text concerns the dialog between the two fictional characters, Simplicius and Salviati, respectively advocates of the 'ancient' and of the 'new' scientific construction of mechanics. In the brief excerpt, Salviati says to Simplicius: "I have just drawn here a bone increased by three times in length, and so much thicker as to allow this bone to perform a similar function in the larger animal, as it does in the smaller, and as you can see the bone is disproportionate. From this example, it is evident that if you were to make a giant from the proportions of a man, you should either choose a material much more resistant than human bone, or make his body lighter, […] otherwise you should see him shattered under its own body weight"

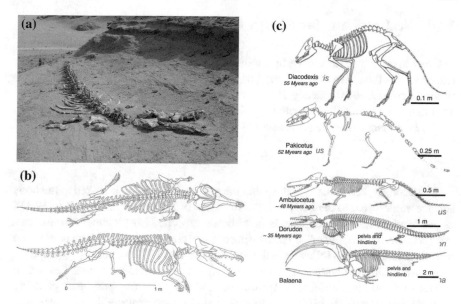

Fig. 12.5 a The discovery of a fossil of a proto-whale (*Dorudon Atrox*) in the wadi El-Hitan, Egyptian desert. **b** Reconstruction of a skeleton of proto-whale, showing how the *dorsal vertical movement* results from the adaptation to swimming of the four members (subsequently only the two fore members) originally developed for walking. **c** Genealogical tree of the whales, starting from terrestrial artiodactyl ancestors, and going through the amphibian *Ambulocetus*, to the proto-whale *Rhodocetus*. [Images © (**a**) by Christoph Rohner, and (**b**) from Ref. [6], repr. under CC BY-SA 3.0 licence; (**c**) from Ref. [5], repr. w. permission, see (*) for terms.]

was very similar to a modern condor, however its humerus was almost twice as long as the condor's. If we were to apply the scaling law $L \propto M^{1/3}$, its mass should have been 8 times that of condor, which would have posed this bird enormous mechanical problems. In fact, already the condor with its huge size is barely capable of performing a flapping flight, it slowly flaps its wings only for take off, and spends almost all of its flight time gliding along the wind currents. To support such a mass with that wingspan, much larger muscles should be necessary, attached to a much bigger sternum that is has not (yet) been found in the fossil remains.

Do such indications point at a substantial reduction of the gravity constant g on the Earth, during those geological period? This subject had been advanced in the past, unfortunately with little scientific support and much nonsense. On the other hand, the idea according to which a large variation of the gravity in the opposite (increasing) direction could have caused the much debated extinction of dinosaurs at the end of the Cretaceous, is entirely implausible, since the variation of g should have been of at least 30% of the current value. The origins of the large animals proliferation, and some reasons for their extinction, must be searched in other directions, for example (and not only) by observing that both the late Cretaceous and the Miocene appear to have been periods of considerably warmer climate.

12.4   Scaling Laws for Energy Consumption

In Chap. 9, the power developed by a muscle was calculated by multiplying the force exerted times the speed of contraction or, equivalently, the work produced in a cycle times the contraction frequency, $P = W\nu$. On the other hand, we just saw that scaling arguments can be attached to both variables, as $W \propto M$ and $\nu \propto M^{-1/3}$, from which it may be deduced that the power should scale with mass as:

$$P = W\nu \propto M^{2/3} \tag{12.14}$$

Interestingly, this leads to deduce that the power expenditure is linked to the body surface, rather than to the body volume (because of the power 2/3, see Eq. (12.1)), as one would intuitively suppose. In fact, the already cited Max Rubner formulated a "surface law of metabolism", by measuring oxygen consumption and heat production in various animals. Hill followed in the footsteps, and in 1950 (see Ref. [7]) proposed a number of different body surfaces which could be connected with the 2/3 exponent above, such as: lung surface, controlling the exchanges of oxygen; heart pumping rate and diameter of the principal veins; skin surface, controlling the dissipation of heat.

On the other hand, if Hill's treatment of the maximum running frequency is followed, the frequency should scale as $\nu \propto M^{-1/6}$, therefore the power should scale as:

$$P = W\nu \propto M \cdot M^{-1/6} = M^{5/6} \tag{12.15}$$

The practical way to measure the energy consumption in animals is to place them on a treadmill at constant velocity, and measure their consumption of oxygen while running at different forced speeds. Such measurements, taken on animals of widely different sizes [8], demonstrate that the preferred velocity scales as $M^{0.2}$ (to be compared to the value 1/6), and the stepping frequency scales as $M^{-0.15}$ (to be compared to −1/6); moreover, it is concluded that energy consumption is directly proportional to the muscle mass, just the result that Hill had used as starting point for his analysis.

Several results are reported in the literature concerning the oxygen consumption for the **basal metabolism**, measured on animals at rest (and since these are often taken in their zoo cages, there could be also an influence from a bad psychological condition…). The allometric scaling diagram of the energy consumed versus body mass generally converge to a straight line with slope 3/4. It may be noted that this value is exactly midway between the 5/6 proposed by Hill, and the 2/3 observed for the case of cruising locomotion. However, the basal metabolism is difficult to relate to the power demand during locomotion, let alone during an extreme effort. Basal consumption is related to all the mechanisms that maintain the organic functions, such as respiration, blood circulation, peripheral nervous system activity, in the absence of the voluntary muscle movements. This empirical law with exponent 3/4 is ubiquitously found in such metabolism studies, independently on the type of animals studies, however its origins remain obscure (see [9], as well as McMahon and Schmidt-Nielsen in "Further reading" to Chap. 11).

12.4.1 Choosing a Mode of Transport

Among the most common actions of an animal's life, there is the displacement from one place to another, usually to look for food sources. The prominent Norwegian-born physiologist Knut Schmidt Nielsen, defined the quantity *cost of transport*, C_m, as the amount of fuel necessary to displace a unit weight over a unit distance [10]. Since all the measurements made at that time were indirect, based on the oxygen consumption rate, this new quantity was introduced by defining a metabolic rate, R_m, the physiological measure of the total animal energy consumption, which includes the mechanical energy delivered by the muscles, as well as any other consumption of chemical energy necessary for the animal survival:

$$C_m = \frac{R_m}{uw} \tag{12.16}$$

In this definition, u is the displacement velocity of the animal, and w is its weight. The units given by Schmidt-Nielsen for this quantity were calories per gram-kilometer, as if derived by a ratio of work per unit time, divided by a mass and by a velocity. As later noted, this quantity could be more properly written as:

$$C_m = \frac{P_m}{Mgu} \tag{12.17}$$

with P_m the total metabolic power. In this way, the numerator and the denominator have both the same dimensions of $[M][L^2][T^{-3}]$, and C_m is therefore a nondimensional number.

In practical terms, the muscle power P could be obtained from physiological measurements of the total P_m, after subtracting the basal metabolic power, and multiplying by some 'conversion efficiency factor', usually 0.20–0.25. This gives a net cost of transport, that is the ratio between purely mechanical quantities. The power at the numerator can be further phrased as the product between an average translational force T, and the translation velocity, giving:

$$C_m = \frac{Tu}{Mgu} = \frac{T}{Mg} \tag{12.18}$$

The cost of transport appears therefore as the ratio between the horizontal force pushing the animal at a given speed, and the vertical force represented by its weight and, notably, it includes explicitly the gravity constant. It may be also noted that the reciprocal of this quantity, $N = 1/C$, is a coefficient well know in aeronautics, namely the ratio between **lift** and **drag**. For both flying machines and animals it is:

$$N = \frac{1}{C} = \frac{\text{Lift}}{\text{Drag}} \tag{12.19}$$

For terrestrial and sea animals, N is rather a "performance" index, but it may be useful to compare the relative efficiency of different methods of transport.

By collecting data of C_m for many animal species, either flying, swimming or running at widely different speeds, Schmidt-Nielsen deduced a series of interesting conclusions. Firstly, the ratio between power and speed is not fixed for a given mode of locomotion, but it can vary as a function of the speed. For flying animals (and machines) in particular, there exists an optimum flight speed u_m, for which N has a maximum, which is the maximum lift that can be obtained for a given drag. The value of u_m is independent of the flyer's weight, area of the wing, or wing loading, but it mainly depends on the wingspan and the total area of the body surface. A sparrow has a maximum N around 4, while an albatross nears the value 20. For comparison, the supersonic liner Concorde at high speed had $N = 7.5$, and a Boeing 747 is close to 18. A migratory bird must always fly at this maximum-efficiency speed, in order to cover the long distance by optimising its energy reserves. Swimming animals, by contrast, can always run a bit further by saving energy, since their weight is supported by water at any speed. Surface walking and running are most difficult to deal with, because of their variability, however the vast amount of experimental information accumulated allows to deduce that N could be nearly constant, except maybe at the fastest running speed.

The experimental log-log plot of C_m versus M shows that the cost of transport is decreasing linearly with the mass, for any mode of locomotion (Fig. 12.6). For the flying animals, insects are found at the extreme right of the plot, with proportionally very high cost of transport. With such a poor efficiency in using the available energy, they must often opt for different modes of transport. In fact, most insects can choose between flight and walking as a function of their environment.

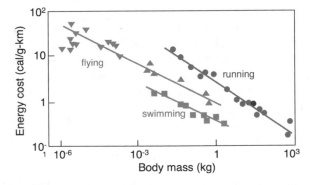

Fig. 12.6 Energy cost of locomotion for swimming (*squares*), running (*circles*), and flying animals (*triangles up for birds, down for insects*), as a function of body mass (experimental data from Ref. [10])

12.5 Energy Stocks for the Offspring

In ecology, it is often matter of discussing the relations between different organisms sharing a same environment, in terms of **energy budget**. We may consider as an example an herbivore feeding on grass for some time, and extracting free energy at a given rate. This would be its **energy accumulation rate**, i.e. a power, with dimensions of $[M][L^2][T^{-3}]$; at the same time, this animal consumes energy at a different rate, again a power, which is the sum of all its basal metabolism, plus some muscle power expenditure. The difference between the two rates represents its **power surplus**. Being negative or positive, the power surplus is reflected also in a **material surplus**, by taking the balance between the materials entering the body as food, the amount converted to energy, the amount consumed, and the materials expelled as waste. An animal living in constant power surplus is also under material surplus, therefore it will get fat. Once summed over all the organisms of an ecosystem, the mass balance is an important quantity in ecology (see the next Sect. 12.6).

A positive surplus is not necessarily a negative concept, since it may be instead useful for the reproduction, during the phases of pregnancy, and of the growth of the progeny. To ensure the survival of any species, a constant positive surplus is in fact necessary across the generations. The very same concepts of energy budget, input and output power, and surplus, can be applied to a single individual as well as to an entire herd, or to a whole animal species. A positive surplus at the level of a whole population may be due to the increase in number of individuals in the group, or to an increase in individual mass, or to a combination of both factors.

As it was seen in Hill's scaling analyses, the power expenditure is not proportional to the body mass, or volume, but rather to mass raised to a power comprised between 2/3 and 5/6. From the studies on animal metabolism, it is found that most power expenditures follow the law $M^{3/4}$, which is somewhat in-between the two extremes indicated. If also the surplus is surmised to scale following the same 3/4-exponent empirical law, a similar scaling law should be obtained for the energy and material available for reproduction. For a herd of animals of size N, the **power surplus** can be written as:

$$P_{s,pop} = N P_s \propto N M^{3/4} \tag{12.20}$$

where P_s is the average surplus for each individual in the herd.

If we neglect mass fluctuations among individuals of the herd, notably the fact that offspring have a smaller mass for some time, the population rate of change, averaged over many generations, should be proportional to the power surplus, normalised to the average mass of the individual:

$$\frac{dN}{dt} \propto \frac{P_{s,pop}}{M} \propto N M^{-1/4} \tag{12.21}$$

The solution of this equation, $N(t) \propto \exp(t/\tau)$, is a well-know law in ecology, the so-called **Malthusian exponential growth law** for a population free of any constraints (which will be discussed at some length in the next Section). In particular, no constraints are imposed on the fertility nor on the mortality, or to the presence of other species competing for food in the same environment, or representing a vital menace for the herd. Because of the time constant $\tau \propto M^{1/4}$, such a law gives a reproduction rate that is higher for animals of small size and slower for animals of larger size.

A study by David Western [11] reported an allometric scaling diagram of the reproduction rates of some african mammals, ranging from the shrew (mass of 70 g) to the elephant (2600 kg), aligned on a straight line with slope -0.325. This is somewhat larger than the -0.25 of Malthusian growth, and seemingly closer to a power variation with the exponent 2/3, which would give a mass growth rate $\tau \propto M^{1/3}$. It should however be observed that the pretension to describe such a range of animal species, possibly from a single cell to an elephant and even to an entire ecosystem, with such a reduced set of physical variables, cannot provide but very crude estimates. The actual worth of such considerations is to allow to appreciate some global trends, and to suggest possible indices to aggregate data. Nevertheless, the surprising validity of the empirical exponent $-3/4$ law allows to translate at least some basic concepts across widely varying scales, from the individual, to the population, to the entire ecosystem. In the following Sections of this chapter, we will deal with the dynamics of populations growing and declining in interaction with their environment.

12.6   Analytical Models of Population Growth

The subject of **population dynamics** is the study of the laws governing the growth and decline in size of any group of living species, and more interestingly that of sexually-reproducing species. The mass distribution, the age-group composition of the population, the environment, all natural and artificial processes influencing such evolution are as well part of this field of study. Besides predictions concerning the population size, these works aim at predicting and understanding the environmental effects on the population size. As such, these studies may reveal crucial in, e.g., the control of fishing areas, the management of protected natural zones, the control of animal species considered as harmful or invasive. By including the interaction of a population with other populations of different species, and with the constraints and resources made available by the surrounding natural and artificial environment, population dynamics merges into the wider field of **ecology**, an interdisciplinary field of research encompassing physics, chemistry, biology, climate and Earth sciences.

The population model proposed in 1798 by the British economist Robert Malthus, was simply based on the introduction of an average birth rate n, and of a mortality rate m, whose difference gives the rate of variation of a population N as a function of time:

$$\frac{dN}{dt} = n - m \tag{12.22}$$

The solution is the exponential function, $N(t) = N_0 \exp(t/\tau)$, with a coefficient $\tau = 1/(n - m)$. If $n > m$, the population increases without limits, leading to what has been dubbed as the "Malthusian catastrophe", namely the population size outgrowing the available resources, ultimately leading to starvation and regression. By referring to the original Malthus' hypotheses [12], n should be taken as independent on the environmental conditions, being mainly dictated by the basic animal physiology, while m could be changed to lower or higher, by improved or impoverished environmental conditions, food availability, climatic changes, and so on. On the other hand, a negative τ implies the disappearance of the population, therefore the malthusian model has no equilibrium, or steady-state condition.

Somewhat in response to such a 'catastrophic' vision of population evolution, the French mathematician Pierre François Verhulst around 1840, proposed a more rich and slightly more realistic model. In this case, it is imagined that both the natality and mortality rates do depend on the current size of the population at any instant t, respectively in an inversely proportional, and in a directly proportional manner. In other words, the more the population grows in size, the more its natality rate declines, while the mortality rate increases. According to the **Verhulst model**, the population size at an instant t is governed by the differential equation:

$$\frac{dN}{dt} = n(N) - m(N) \tag{12.23}$$

If m and n are taken, as said, as functions of N respectively increasing and decreasing, their difference $n - m$ is as well a decreasing function of N. If moreover we impose that for $N \to 0$ the growth rate must remain positive (the species does not become extinct), the Verhulst equation can be rewritten as:

$$\frac{dN}{dt} = N(r - kN) \tag{12.24}$$

with r and k two positive coefficients. Further, by putting $K = r/k$, we obtain:

$$\frac{dN}{dt} = rN\left(1 - \frac{N}{K}\right) \tag{12.25}$$

The general solution of Verhulst equation, with the initial condition that at time $t = 0$ the population starts from a finite size N_0, is the so-called **logistic function**:

$$N(t) = \frac{K}{1 + \left(\frac{K}{N_0} - 1\right) e^{-rt}} \tag{12.26}$$

It can be seen right away that for $t \to \infty$, and independently on the initial value of the population size, N approaches the value K, which for this reason is called the **carrying capacity** of a given ecosystem. Figure 12.7 displays some solutions of the Verhulst equation, all normalised to $K = 1$, demonstrating that starting from whatever large or small initial value of the population size, the carrying capacity is invariably the long-time limit of N.

In ecology, a species can be described as following a "r-strategy" or a "K-strategy", according to the selection process that is at the basis of its evolution. A population that relies more heavily on the r factor, tends to exploit a smaller ecological niche, with a relatively high reproduction rate accompanied by a lower individual survival probability. By contrast, species selected on the basis of their K factor are powerful competitors in a crowded environment, their offspring is reduced in number but with a higher survival probability. Such populations may also be called, respectively, "opportunists" or "equilibrists"; the former could be dominant in a rather unstable environmental setting, whereas the latter may become dominant in a more stable situation; also, their prevalence may alternate in a given ecosystem, as the external conditions change.

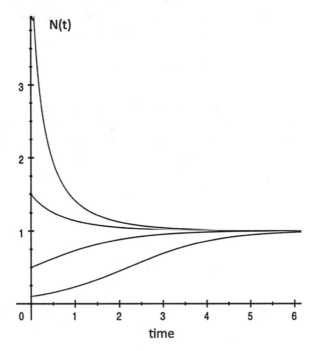

Fig. 12.7 Plot of the logistic function, solution of the Verhulst differential equation $N = N(t)$, for different values of the initial population size, and $r = K = 1$ Both N and time are plotted in arbitrary units

Stability and chaos

It may be interesting to study the mathematical stability of the solutions of the population evolution equations, in the presence of an external or internal perturbation, of the form $u(t)$. The equilibrium condition is represented by any value N^* for which $dN/dt = 0$. For example, it was seen that $N^* = K$ for the logistic function.

Let us take that $N(t) = N^* + u(t)$, with $u > 0$ a small perturbation, representing some arbitrary variations of the environmental conditions about the equilibrium value, such as an occasional availability of some more food, or a temporary shift in the environment temperature, and so on. The new perturbed equation would be:

$$\frac{du}{dt} = f\left[N^* + u(t)\right] \simeq f'(N^*)u(t) + O(u^2) \tag{12.27}$$

where the second approximate equality represents the first order of the development in series of powers of u around the stable solution (the zeroth-order term being $u = 0$).

The time derivative f' of the logistic function is:

$$\frac{df}{dt} = \frac{KWr\exp(-rt)}{(1 + W\exp(-rt))^2} \tag{12.28}$$

with $W = N/N_0 - 1$ positive or negative according to $N_0 < K$, ou $N_0 > K$, respectively.

We immediately see that in this case the outcome depends also on the initial value of the population size, N_0. In the case $N_0 < K$, du/dt is going to increase, therefore the perturbation u is amplified, and makes the system unstable towards an exploding population. Instead, in the $N_0 > K$ case du/dt will decrease together with df/dt, therefore the system can resist the action of the perturbation, which dies off in time, and brings back the system to its stable solution N^*.

The logistic equation can be reformulated also as an iterative problem, by considering finite time intervals. For example, we may look at the evolution of the population from a time n to a time $n + 1$ (such as from one year to the next), and rewrite the equation as:

$$N_{n+1} = rN_n(1 - N_n) \tag{12.29}$$

with a K normalised to 1. Starting from $n = 0$ with $N = N_0$, the equation is iteratively solved, and would give the same result as the original differential formulation. However, once written in this form (called a *map* in mathematical language), it can be easily shown that the population growth model can have pathological states, leading sometimes to **chaotic** behaviour. In fact, by varying the parameter r between 0 and 1, the population will decline to zero at long times, independently on the initial value N_0; for r between 1 and 3, the population will attain the asymptotic limit $N \to 1 - 1/r$; for $r > 3$ the stable solution is preceded by large positive and negative fluctuations; such oscillations become permanent, up to $r = 1 + \sqrt{6}$. From this value of r and on, the solution starts oscillating between four, then 8, 16, 32,... different values, until at $r \simeq 3.56995\ldots$ a completely chaotic behaviour sets in, with the population size fluctuating among random values. The "bifurcation" behaviour of the asymptotic, long-time solution for N is shown in the following plot, limited to values of $r > 2.4$.

The idea of **deterministic chaos** is associated with complex and unpredictable behaviour of some classes of phenomena over time, arising in deterministic dynamical systems. Like in the logistic map, many examples are based on mathematical models for (discrete) time series in which, starting from some initial condition, the value of the series at any time is a nonlinear function of the previous value. These processes are very intriguing, in that the solutions corresponding to different, although extremely close, initial conditions will exponentially diverge. The practical implications are that, despite the underlying determinism, one cannot predict with any reasonable precision the solutions at long times: even the slightest error in specifying the initial conditions, eventually ruins the solution.

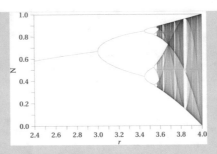

It must be noted that perfectly integrable systems (those for which the behaviour can be predicted at any time) are rare. The fact that such systems are almost exclusively treated in textbooks on classical dynamics, such as the a forced or parametric harmonic oscillator, a point mass in a three-dimensional spherically symmetric potential, N-dimensional coupled harmonic oscillators, etc., gives the impression that the well-organised behaviour of integrable cases is the rule, at least for systems with few degrees of freedom. In fact, it is rather the exception, and most natural systems are both non-integrable and chaotic. The sensitivity to the initial conditions is known as the *butterfly effect*: the state of the system at a given time can be entirely different if the initial conditions are only slightly changed, such as by a butterfly flapping its wings.

A quantitative measure of this exponential growth of the separation between initially close solutions is given by the **Lyapunov exponent**. For a map $x_{n+1} = f(x_n)$, $x \in [0, 1]$, and two solutions starting at x_0 and $x_0 + \epsilon$, $\epsilon \ll 1$, their separation after the n-th iteration is:

$$\Delta x_n = |f^n(x_0 + \epsilon) - f^n(x_0)| \tag{12.30}$$

The Russian mathematician Aleksandr Lyapunov showed that the two solutions diverge exponentially as:

$$\Delta x_n \simeq \epsilon^{n\lambda_L} \tag{12.31}$$

The Lyapunov exponent λ_L can be estimated (in practical cases only by numerical approximations) as:

$$\lambda_L = \lim_{n \to \infty} \frac{1}{n} \left(\ln \frac{\Delta x_n}{\epsilon} \right) \tag{12.32}$$

Negative values of the Lyapunov exponent indicate stability, and positive values chaotic evolution; at critical bifurcation points λ_L approaches zero. It is also instructive to note its relationship to the **loss of information** during the process of iteration. When the interval [0,1] is partitioned into N equal boxes, one needs $\ln_2 N$ bits of information to define in which box to find a point x, i.e., one has to ask on average $\ln_2 N$ 'yes/no' questions. After each iteration the box is stretched by a factor proportional to Δx, corresponding to a "loss of information" by an amount proportional to $\ln_2 \Delta x$, about the new position of the point x.

The **Lyapunov time** is the length of trajectory after which a system start showing chaotic divergence. It can be as short as a few picoseconds for the atoms of a gas, seconds to minutes for hydrodynamic or chemical mixing, up to millions of years for gravitational systems. For the entire Solar system it is of the order of 50 million years: although integrating the Newton's equation of motion for spans beyond such time would not make sense, we can still safely predict the trajectories of our planets for quite a bit longer than our lifetime!

Actually, if we want to consider that the environmental conditions may also change, a variable (but always positive) carrying capacity could be introduced as $K(t) > 0$. An important instance is that of a periodically variable K, for example linked to the coupled increasing or decreasing availability of food, and more or less favourable climatic conditions alternating during summer and winter, as $K(t + T) = K(t)$, with $T = 6$ months. It can be shown in this case that also the solution $N(t)$ is periodic, with the same period of the carrying capacity, and again independent on the initial conditions at time $t = 0$.

12.6.1 Preys and Predators

The Malthus' and Verhulst's models consider the isolated evolution of a population of individuals of the same species in a given ecosystem, this latter being represented in the most simplistic way by just one or two numerical parameters, which regulate the growth and decline of the population size. However, any species cannot survive alone in an ecosystem, and the interaction with other species, as well exploitation of the finite resources of the environment, are key factors in determining its evolutionary success or decline. Let us see what would happen if a first, external element is added to the intrinsic mortality rate of a species, such as the presence of a natural predator.

As a simple hypothesis, we can assume that the rate of killing by the predators $p(t)$ is proportional to the actual size of the population $N(t)$, except when the population is so large that the predators are saturated:

$$p(t) = \frac{bN}{a + N} \tag{12.33}$$

with a, b arbitrary constants, such that for $N \to \infty$, it is $p \to b$. The new equation for the population is:

$$\frac{dN}{dt} = rN\left(1 - \frac{N}{K}\right) - \frac{bN}{a + N} \tag{12.34}$$

Now we ask, for which values of the prey population the equilibrium $dN/dt = 0$ is reached? Excluding the banal condition $N^* = 0$, we can look for a value N^* such that:

$$rN^*\left(1 - \frac{N^*}{K}\right) - \frac{bN^*}{a + N^*} = 0 \tag{12.35}$$

This is a second order algebraic equation, which has two (or none) possible real solutions, according to the values of a, b, K, r. In particular, since b, K and r are by definition positive quantities, a positive value of a ensures the existence of real solutions.

However, it should be also taken into account the fact that predators have their own life cycle, with their own natality and mortality rates, and that their population dynamics is influenced in turn by the availability of a sufficient number of preys.

The **competition** between species was firstly formalised in the **equations of Lotka-Volterra**, which are also designed as the "prey-predator model". These are a set of two coupled non-linear differential equations of first order, describing two populations in competition. They are currently employed to describe the dynamics of ecological systems in which a species depends on the other for its food supply. However, the model provided by these equation has found use in many other domains, in which there is a competition between two entities whose size-dependence is interrelated. These equations were independently proposed by the American mathematician Alfred James Lotka in 1925 and by the Italian Vito Volterra in 1926. A classical application describes the population dynamics of the Canadian lynx and of its favoured prey, the snowshoe-hare. For these two animals, constituting a sort of closed ecosystem, a vast accumulation of experimental data were made available, by counting the number of respective animal skins collected by the Hudson Bay hunting company in late-XIX to early-XX century (Fig. 12.8), [13]).

By indicating as $N(t)$ and $P(t)$ the populations of preys and predators, respectively, the Volterra-Lotka equations are often written in the form:

$$\frac{dN}{dt} = N(\alpha - \beta P)$$
$$\frac{dP}{dt} = -P(\gamma - \delta N) \tag{12.36}$$

Here, dN/dt and dP/dt represent the (positive or negative) growth rates of the two populations, and are mutually dependent upon, via the cross terms in each equation. The four parameters characterise the interaction between the two species, as:

- α, prey reproduction rate in the absence of predators
- β, prey mortality rate, controlled by the number of alive predators
- δ, predator reproduction rate, controlled by the available number of preys
- γ, predator mortality rate in the absence of preys

In the model, preys are supposed to have access to practically unlimited sources of food, and reproduce exponentially with the rate α (according to Malthusian growth) if no predators are present. The prey killing rate by predators is represented by the cross term $-\beta PN$, a sort of prey-predator encounter probability: if either population

Fig. 12.8 Fluctuations of the populations of linx and snowshoe-hare species in Northern Canada between 1845 and 1935, as deduced by the number of hunted pelts (data redrawn from Ref. [13])

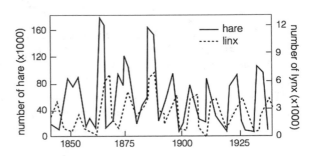

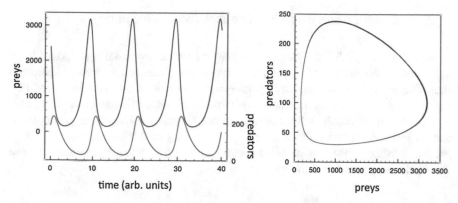

Fig. 12.9 *Left* Representation of the time evolution of the predator population $P(t)$ (*red curve*) and of the prey population $N(t)$ (*black curve*), from a linearised numerical solution of the Lotka-Volterra equations with the values $\alpha = 1.$, $\beta = 0.01$, $\gamma = 0.5$, $\delta = 5 \times 10^{-4}$, of the free parameters. *Right* Cross-plot of the two periodically-fluctuating populations from the *left graph*, showing the cyclic and coupled alternance between growth and decline of the population sizes

is zero, no predation occurs. With the above terms in the coupled equations, the model can be interpreted as: the variation of the number of preys is given by its growth rate, diminished by the probability of getting killed by a predator. It is supposed that no prey could survive until a natural death.

The term $+\delta N P$ represents the prey-dependent growth rate of the predator population, formally similar to the predation rate but governed by a different rate constant, since the rate at which the predator population grows does not coincide with the rate at which it feeds (although the two can be quantitatively linked). The predator natural death rate is again Malthusian, with a negative coefficient $-\gamma$.

In agreement with the classic experimental data from the Hudson Bay Company reported in Fig. 12.8, the coupled Volterra-Lotka equations admit periodic solutions, however not easily formalised in terms of simple trigonometric functions. A numerical solution of the linear approximation of the equations is given in Fig. 12.9 (left), demonstrating a simple periodic recurrence between the coupled variations of the two populations $N(t)$ and $P(t)$, the size growth and decline of the predators being somewhat delayed with respect to the corresponding growth and decline of preys. For a perfectly cyclic solution as the one shown, with the predators and preys numbers varying between two given maximum and minimum values, the cross-plot of P versus N is a cyclic trajectory (Fig. 12.9, right), periodically spanned by the coupled pair of values $\{P(t), N(t)\}$.

Equations of the Lotka-Volterra type have been employed with success also in other domains presenting "populations" in competition. For example, in physiology, to describe the periodic activation dynamics of families of cholinergic and aminergic neurons (the study of the "paradoxical sleep" by Hobson [14]), or in economy, to describe the dynamics of industrial competition in various sectors of economic activity (see the "growth cycle" theories by Goodwin [15]).

12.6.2   Competition and Cooperation Between Species

In the equations (12.37) the carrying capacity of the environment is supposed to be unlimited: in the absence of predators, the preys' population could grow to indefinite size. The finite carrying capacity of the ecosystem can be included in a modified version of the Lotka-Volterra model, by considering a logistic model of growth for both preys and predators, instead of the simpler Malthusian scheme. In particular, it can be written that preys follow the carrying capacity of the environment K, while predators follow a different carrying capacity $K' = hN$, linearly dependent on the size of the prey population, on which they feed:

$$\frac{dN}{dt} = r_N N \left(1 - \frac{N}{K}\right) - \frac{bN}{a+N}$$
$$\frac{dP}{dt} = r_P P \left(1 - \frac{P}{hN}\right) \tag{12.37}$$

with r_N and r_P the birth rates of N and P, respectively. Such model equations, strongly non-linear, show the typical behaviour of chaotic systems (see the greybox on p. 561), also with unstable fixed points, attractors and limiting cycles for particular combinations of the values of the free parameters.

When two species are competing for the same set of resources, available in a finite amount, it can be show that there is only one stable state, while in all other combinations of parameters the two species cannot coexist: this is the **principle of exclusive competition**. Two separate carrying capacity values can be introduced, K_N and K_P. Note that this model does not describe one species preying on the other, but two species competing for finite resources in the same ecological niche, such as two families of herbivores eating the same kind of grass. The two species are mutually harmful, in that the species P subtracts an amount b of the resources from N, and the Ns' subtract an amount b' from the Ps'. Putting this model into equations:

$$\frac{dN}{dt} = r_N N \left(1 - \frac{N - bP}{K_N}\right)$$
$$\frac{dP}{dt} = r_P P \left(1 - \frac{P - b'N}{K_P}\right) \tag{12.38}$$

The equilibrium conditions are found by setting the time derivatives of P and N to zero:

$$N - bp = K_N$$
$$P - b'N = K_P \tag{12.39}$$

These equations describe two straight lines in the plane (N, P). According to the values of K_N, K_P, b, b', the two lines can cross in four different ways, shown in Fig. 12.10(1–4), the blue line corresponding to the condition $dP/dt = 0$, and the red line to the $dN/dt = 0$. In all cases, the practical condition that will be realised is the one giving the highest value for N, or P, or both. In the possibility (1), Ps are always winning and get to their maximum by pushing the Ns to zero (*extinction*). The (2) is the opposite case, with Ns winning and Ps becoming extinct. For the possibility (3) there is just one unstable-equilibrium point, where the two red and blue conditions cross: for values slightly on the left of this crossing, P wins and N disappears, while for values slightly to the right, the opposite happens. Only for the possibility (4) the **coexistence** of the two competing species can always be obtained: P strives always toward the left of the diagram, thus letting N to grow, while N does the opposite to the right, therefore the two populations are always pushed towards the crossing point, the stable equilibrium.

On the other hand, the two species could cooperate, and help each other, for example if one is in **symbiosis** with the other. Such a situation would be described in the previous set of equations by changing the algebraic sign of the two parameters b, b', meaning that the carrying capacity of each species is increased thanks to the presence of the other. In this case, the slopes of the straight lines in Eq. (12.39) would be both positive, and their graphical representation would be that given in Fig. 12.10(5–6). In the possibility (5), the two species can grow together for any

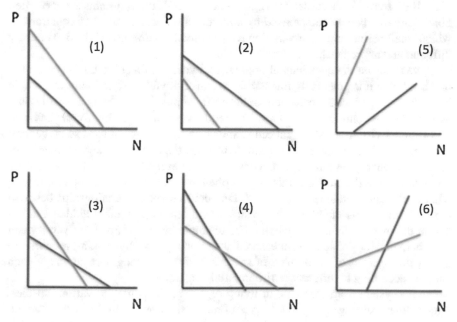

Fig. 12.10 Representation of the possible equilibrium and non-equilibrium solutions of the double-logistic model (Eq. (12.38)). The *blue lines* correspond to the condition $dP/dt = 0$, and the red lines to the condition $dN/dt = 0$

combination of the respective birth and death rates; for the possibility (6), instead, the two populations converge again to a unique stable state, represented by the sizes of N and P for which the two lines are crossing each other.

12.7   Dynamical Models in Ecology

The analytical models of the preceding Section are part of a long-standing tradition in ecology and sociological sciences, grounded on the construction of differential equations to obtain relationships between the size of a population of individuals and the many intrinsic and extrinsic perturbing factors, chiefly including reproduction and mortality rates. Such an endeavour has been undertaken by many scientists, in order to formulate predictions on the evolution of a population. For example, by dividing the individuals into age groups, and studying their evolution as a function of variables representing the physical and chemical environment, the existence of other competing populations such as preys or predators, and so on. Contrary to the hopes of the pioneers of such methods, however, deducing the complex evolutionary patterns of a population by purely analytical methods revealed an impossible task. Suffering from pretty much the same kind of difficulties as the meteorological and climate models, as well as other distant but "mathematically" similar fields, such as the somersaults of world's economy and financial markets, these are all examples of **chaotic systems**. As discussed in the greybox on p. 561, these systems are very often non-integrable, being characterised by coupled non-linear systems of equations in which small perturbations, or variations of the initial conditions, can lead to entirely different and unpredictable results.

If we focus on the questions of population dynamics this difficulty says, on the one hand, that it is practically impossible to start from the description of one single individual and deduce the behaviour of the entire population, by using this individual as the average John Doe. On the other hand, it remains still highly desirable to find strategies to formulate a reduced framework, from which at least some general principles could be deduced, for example to track the circulation and exchanges of matter and energy within a given ecosystem. The aim of such an effort would be not much that of discovering entirely new phenomena, but rather that of simplifying what is already contained in the data.[2] This could highlight relationships between the different agents of the system, which could otherwise remain hidden and get lost in the overwhelming complexity. For example, the notion of **biodiversity** can be schematically defined and measured in an ecosystem, by making recourse to aggregate variables of the number of species, and the evolving populations of each species, according to different environmental constraints.

An ecosystem is an ensemble of living organisms belonging to various species, which store, exchange and use energy and materials from the limited environment.

---

[2]However, it would be philosophically inappropriate to say "already known", since the simple availability of an enormous quantity of data cannot be sufficient to gain also the knowledge.

Organisms can be: **producers**, if they convert the energy flow into chemical energy, typically stored into carbon bonds (e.g., glucose); or **consumers**, if their main function is to break carbon bonds and retrieve the stored energy to function. Based on their way of input of carbon, consumers can be further identified as: herbivores, carnivores, omnivores, and detritivores (or decomposers). Such distinctions lead to the introduction of the concept of **food chain**, organised into **trophic** (=feeding) **levels**: producers are at the bottom, by transforming the Sun energy into chemical species; herbivores are immediately above, followed by different levels of carnivores; species feeding partly on plants and partly on meat can occupy intermediate levels; omnivores can appear at any levels above herbivores; at the top of the food chain we find the **apex predators** (which need not be hyper-carnivores); decomposers feed on detritus, and extract the last bits of energy from the food chain. Primary producers (typically, plants) thrive on the pool of nutrients provided by the environment, to work out new carbon-based chemical species with the help of energy from sunlight (see Chap. 2). It is worth underscoring that, while all the materials (carbon, oxygen, minerals, metals…) are recycled to a variable extent, at worst remaining non used in the pool, energy is never recycled but is ultimately lost as heat, and must be continuously supplied from outside the Biosphere.

A simplified view of an ecosystem is given in the flow diagram of Fig. 12.11. Each level of the system represents a trophic level, but it is also more general, by including also non-living components, such as dead matter or inorganic nutrients. We should recall here the concept of **biomass**, from Chap. 2, as the sum of all the materials circulating under different forms. It may be useful to subdivide the biomass on a elemental basis, i.e. accounting separately for all the carbon, nitrogen, phosphorus, etc., in the different molecular species in which they may be transformed, while passing from one level to another of the food chain. The biomass is the sum of all the masses of all the elements. (Note that accounting for volatile elements, such as the fraction of oxygen or carbon going into O_2 and CO_2, can complicate things.) In general, the further high is a trophic level compared to the lower ones (producers or decomposers), the less biomass it contains. This holds true both if we account for the total biomass by the mass of all the individuals, and if we account on elemental basis. The reasons for such an effect are that: (i) not all the materials in the lower levels are completely eaten at the upper level; (ii) not everything that is eaten is digested and assimilated; (iii) a fraction of energy is always lost as heat in each transformation.[3]

Each level l in the flow diagram (pool, producers, consumers, dead matter, decomposers) contains some amount of each chemical element k, let us say M_l^k. The total amount of that element is the **circulating reserve**:

$$R_k = \sum_l M_l^k \tag{12.40}$$

---

[3] An exception is represented by the sea algae: they are both outnumbered and outweighed by the organisms that feed on them. Algae can support the greater biomass of the upper trophic level thanks to their velocity of reproduction, and also because they are much more efficiently useable compared to the typical, ligneous surface plants.

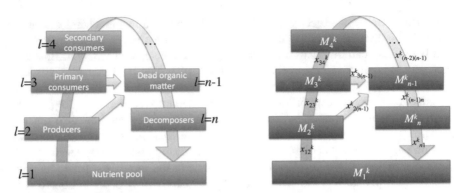

Fig. 12.11 *Left* Scheme of an ecosystem with one level of producers (plants), and two levels of consumers (herbivores, carnivores). In general, there can be more than one levels of each, with the only condition that producers do not generally feed onto other producers (except maybe parasitic plants). The level of dead matter is therefore the $(n - 1)$-th, and the single level of decomposers the n-th. However, also decomposers may represent more than just one level. *Right* Circulation of a chemical element k between the different levels. Each level l captures a fraction M_l^k / R; exchanges between levels are indicated by x_{ij}^k

Exchanges between two levels i and j are indicated as x_{ij}^k, with:

$$M_l^k = \sum_i x_{il}^k - \sum_j x_{lj}^k \tag{12.41}$$

sum of all the incoming flux minus all the outgoing flux to/from level l. The rate of variation of a chemical element k in a trophic level l is defined as:

$$r_l^k = \frac{\Delta M_l^k}{\Delta t} \tag{12.42}$$

The circulating reserve may be treated as a conserved quantity only if the ecosystem is *closed*, i.e. physically limited, for example a small lake. However, it would be very difficult to consider it constant for a large environment such as the ocean. This identification is difficult at the surface as well, with the possible exception of enclosed regions such as a valley or a basin, which can be considered to a good approximation as closed ecosystems. A special case is that of *mobile* ecosystems, such as the Serengeti in the East Africa, dominated by the great herbivores such as the wildebeest (or gnu). They migrate throughout the year, constantly seeking fresh grazing and better quality water, accompanied by large numbers of zebra, and smaller numbers of antelopes of different species. Their movement in an annual pattern is fairly predictable, with a precise timing entirely dependent upon the rainfall distribution of each year. Since the largest component of the biomass is that of those animals, it has been considered that such migrating hordes constitute a kind of closed but mobile ecosystem. On the other hand, an ecosystem can be considered as closed only for certain elements, such as calcium or phosphorus: for the volatile oxygen, no system can be truly considered as closed.

Liebig's law or the *law of the minimum*, is a principle developed in agricultural science by Carl Sprengel (1828) and later popularised by the German chemist Justus von Liebig, often considered the founding father of organic chemistry. It states that population growth is controlled not by the total amount of resources available, but by the single, most scarce resource. This acts as a limiting factor for growth, and could be either a whole resource (bread, gasoline, a particular kind of algae,…), or just one chemical element. The typical example of the latter is phosphorus, an element for which no known natural or synthetic input can stand in.[4]

A plant like the Alfalfa could thrive in a typical soil containing $\sim 0.12\%$ phosphorus, while the plant's structure contains about 0.7% phosphorus. Such a large concentration of one chemical element ($0.7/0.12 = 5.8$) is exceptional, since all other elements are about even in their need/supply ratio.

Biomagnification occurs when organisms at the bottom of the food chain concentrate some element at a value above its concentration in the surrounding soil or water. Producers take in inorganic nutrients from their surroundings. Since a lack of these nutrients (e.g., phosphorus) can limit the growth of the producer, considerable amounts of energy can be spent to pump them into their bodies and store it.

Pollutants present in the environment, such as DDT or mercury, are also brought into the producer's body and stored by the same mechanism. This first step in biomagnification is called **bioaccumulation**: the pollutant is at a higher concentration inside the producer, than it is in the environment.

The second stage of biomagnification occurs when the producer is eaten by organisms above it in the food chain. Consumers of any level consume a lot of biomass from the trophic level immediately below. If that biomass contains the pollutant, the pollutant will be taken up in large quantities by the consumer. Pollutants that biomagnify are absorbed and stored as well in the bodies of the consumers. This occurs more often with pollutants that are soluble in fat. Water-soluble pollutants usually cannot biomagnify in this way, because they would dissolve in the bodily fluids of the consumer, and be expelled.

Based on the concept of input and output rate, the concept of **bio-geochemical cycle** can be introduced. While energy does not cycle through an ecosystem, chemicals do. The inorganic nutrients cycle through the trophic levels, as well as entering into the atmosphere, oceans and rocks. Each element has its own unique cycle, but all of the cycles have some things in common. A **reservoir** is a part of the cycle where a particular element is held in large quantities for long periods of time. In an **exchange pool**, on the other hand, the element is held for only a short time (its **residence time**). For example, oceans are a reservoir for water, while a cloud is an exchange pool. In this respect, the animal and vegetal species in an ecosystem are **biovectors**, in that they serve the function of moving the elements from one level to the next in the cycle. For instance, the trees of the tropical rain forest bring water up from the forest floor

---

[4] A subject brilliantly raised by Isaac Asimov in his 1959 essay *Life's bottleneck*; see, e.g., the study by Lougheed in [16].

to be evaporated into the atmosphere. Likewise, coral endosymbionts take carbon from the water and turn it into limestone rock. Needless to say, the energy for most of the transportation of chemicals from one place to another is provided by the Sun, with a minor participation of the geothermic heat.

The **carbon cycle** is relatively simple (Fig. 12.12). From a biological perspective, the key events are the complementary chain of chemical reactions embodied in the respiration and photosynthesis. Respiration (see Chap. 4) takes carbohydrates and oxygen, and combines them to produce CO_2, H_2O, and store energy. Photosynthesis (Chap. 2) takes CO_2 and water, and produces carbohydrates (glucose) and molecular oxygen. Photosynthesis takes energy from the sun and stores it in the carbon-carbon bonds of carbohydrates; respiration releases that energy. Both plants and animals carry on respiration, but only plants can carry on photosynthesis. The main reservoirs for CO_2 are in the oceans and rocks: carbon dioxide dissolves readily in water, where it may precipitate as calcium carbonate (limestone). Corals and algae help this reaction and build up limestone reefs in the process. On land and in the water, plants take up carbon dioxide, which now has three possible fates: (i) it can be liberated to the atmosphere by the plant through respiration; (ii) it can be eaten by an herbivore, or (iii) it can remain in the plant until the plant dies. Animals obtain all of their carbon from their food, and, thus, all carbon in biological systems ultimately comes from plants. When an animal or a plant dies, the carbon can either be respired by decomposers (and released to the atmosphere), or it can be buried intact and ultimately form coal, oil, or natural gas (fossil fuels).

As shown in Fig. 12.12, the carbon cycle is tightly linked to the oxygen cycle. Oxygen is present in the CO_2, in the carbohydrates, in water, and as a O_2 molecule. Oxygen is released to the atmosphere by autotrophs during photosynthesis. In fact, all of the oxygen in the atmosphere is *biogenic*, i.e., it was at some time released from water through photosynthesis. It took about 2 billion years for cyanobacteria

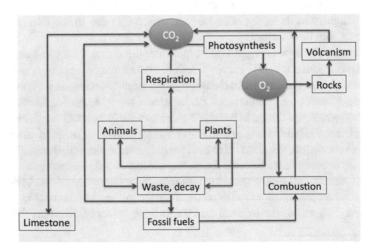

Fig. 12.12 Simplified scheme of the oxygen-carbon cycle

to raise the O_2 content of the atmosphere to the 21 % of today; this opened the way for developing complex organisms such as multicellular animals, which need a lot of oxygen.

The **nitrogen cycle** (Fig. 12.13) is more complex, because of the different chemical forms of nitrogen useful for life, and because organisms are responsible for each of the interconversions. As we know, nitrogen is critically important in forming the amino acids. The chief reservoir of nitrogen is the atmosphere, which is made up by about 78 % nitrogen. N_2 gas in the atmosphere is a non-reactive gas, which can be fixed in two basic ways. First, lightning may provide enough energy to split the molecule and fix it in the form of nitrate (NO_3). This process is replicated industrially, to produce nitrogen fertilisers, and it is the very same used by Miller and Urey in their experiments about the ancient Earth's atmosphere (see Chap. 3). The other form of nitrogen fixation is by nitrogen-fixing bacteria, who use enzymes to break apart N_2, instead of the huge amount of energy from lightning. Depending on the pH, such bacteria fix nitrogen in the form of nitrate (NH_4) or ammonia (NH_3), and can be found free-living in the soil; or in symbiotic associations with the roots of bean plants and other legumes (rhizobial bacteria); or as photosynthetic cyanobacteria (blue-green algae). Most plants can take up nitrate and convert it to amino acids. Animals acquire almost all of their amino acids by eating plants or other animals. When plants or animals die or release waste, the nitrogen is returned to the soil, most likely as ammonia. Nitrite bacteria in the soil and in the water take up ammonia and convert it to nitrite (NO_2), followed by nitrate bacteria, which convert it to NO_3. This latter is useable by the plants to continue the cycle. Nitrogen finally returns to the air with the help of denitrifying bacteria, which take the nitrate and combine the nitrogen back into N_2 gas.

The **phosphorous cycle**, shown schematically in Fig. 12.14, is the simplest. For biological purposes, the only interesting form is the phosphate, PO_4. This is a heavy,

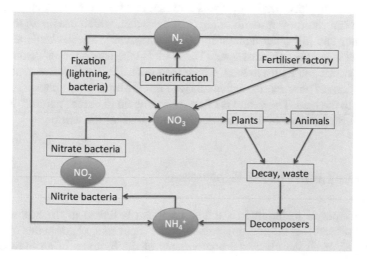

Fig. 12.13 Simplified scheme of the nitrogen cycle

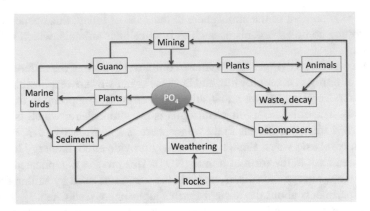

Fig. 12.14 Simplified scheme of the phosphorus cycle

non-volatile molecule, always either incorporated into an organism, dissolved in water, or in the form of rock. When a phosphate rock is exposed to water, especially if at a slightly acidic pH, the rock is weathered out and goes into solution. As we learned in Chap. 3, phosphorus is an important constituent of cell membranes, of DNA, RNA, and ATP/ADP. Animals obtain all their phosphorous from the plants they eat, and also use phosphorus as a component of bones, teeth and shells. When animals or plants die or expel waste, the phosphate can be returned to the soil or water by the decomposers. There, it can be taken up by another plant and used again. This cycle will continue until phosphorus is lost at the bottom of the ocean, where it becomes part of the sedimentary rocks. Phosphorus can be extracted from mines, for the principal purpose of producing fertilisers. As we hinted before, mining of phosphate and its use as fertiliser greatly accelerates the PO_4 cycle, and may cause local overabundance of phosphorous, particularly in coastal regions, at the mouths of rivers, and anyplace where sewage is released into the water. Local abundance of phosphate can cause overgrowth of algae in the water, which use up all the oxygen in the water and kill other aquatic life (*eutrophisation*). Marine birds also play a unique role in the phosphorous cycle. These birds take phosphorous-containing fish out of the ocean and return to land, where they defecate. Their *guano* contains high concentrations of phosphorus, and in this way marine birds return phosphorous from the ocean to the land. The guano is often mined, again to make "natural" fertilisers, and may form the basis of the economy in some areas of the world.

12.8   The Limits of the Ecosystems

Since the input of materials in a living organism is linked to the flux of energy, we could predict that the limiting values of the input rate, r_l^k, or better of the sum $r_l = \sum_k r_l^k$ of all the elemental components of the biomass of an individual (as well as of the entire population), could exhibit some form of scaling, for example by getting progressively smaller for the larger animals.

What values these rates could take is open to the most ample speculation. However, it can be imagined that for each species occupying a given trophic level, the rate could vary between a minimum and a maximum value. Since the power (energy input per unit time) follows a 3/4-scaling, Eq. (12.20), the specific power should rather scale with the exponent $-1/4$ of the mass, $P/M \propto M^{-1/4}$. Therefore, it could not be too surprising if also the ratio between r_{min} and r_{max}, should vary with the same $-1/4$ power law, or:

$$r_{l,min} \propto M^{-1/4} r_{l,max} \tag{12.43}$$

There are no experimental measurements of such energy (=food) input rates. On the other hand, the zookeepers know well that a smaller daily amount of food is necessary for one big elephant, than for a herd of small animals whose masses sum up to the same mass of the elephant. For a population of animals, an upper limit to the input of food mass can be set by the availability from the environment. If this availability, let's call it m, is known or measurable, the upper bound of the biomass of that species is given by the following relation:

$$M_{max} = \frac{m}{r_{min}} \tag{12.44}$$

This quantity is very much analogous to the concept of *carrying capacity* of the ecosystem, which was introduced in the Verhulst model above. Since m does not depend on the feeding population, it should be $M_{max} \propto M^{1/4}$. In other words, for a given availability m, a much larger biomass of big animals can be supported, compared to small animals. However, when expressed in absolute numbers, the larger mass of the individuals will prize the smaller animals. Suppose that m kg/day of hay are available in a country farm, for feeding horses and sheep. For horses of average mass 400 kg, and sheep of about 50 kg, the supported biomass should be in the ratio $(400/50)^{0.25}$, i.e. about 1.7 times more horse biomass than sheep. In absolute values, however, ten horses would correspond to $(400 \times 10)/(50 \times 1.7) = 47$ sheep (instead of $4{,}000/50 = 80$, if the scaling were even).

In the simple ecosystem diagram of Fig. (12.11), the mass of each element in any level, M_l^k, can be described by at least two variables, one input $x_{l-1,l}^k$ and one output $x_{l,l+1}^k$. If we consider the values $M_1 = \sum_{l,k} M_1^k$, for the primary producers, these could increase indefinitely in the absence of an upper level of herbivores. If, on the contrary, the environmental pressure of the herbivores is very high, M_1 risks to fall into desertification. The response of the ecosystem, in such cases, is based on the values of the rates r_l^k. For environments with high feeding pressure, plant species with very fast growth rates will develop. Because of the same relationship Eq. (12.44), plants with a very fast reproductive cycle will be typically small, such as the predominant short grass in the savannah, with just a few trees widely spaced. At the opposite, in an environment where there is a large supply of rainwater, and with a much reduced presence of herbivores, large plants and dense forests can thrive.

The same principles can be applied to the next steps up in the food chain. The total biomass of consumers of a given species, in a given level, is:

$$M_l = \sum_k M_l^k = \sum_k \left[x_{l-1,l}^k - x_{l,l+1}^k - x_{l,d}^k \right] \tag{12.45}$$

by assuming for the sake of simplicity that the consumers feed only on one single species in the lower level, $x_{l-1,l}$, are predated by one single species in the upper level, $x_{l,l+1}$, and have a natural death rate $x_{l,d}$.

If the input rates are sufficient and the output rates not too negative, such species can accumulate biomass. Progressively, the total ecosystem equilibrium will shift to a preference for species characterised by larger individual mass, since this allows to optimise the carrying capacity (see again Eq. (12.44)). If for any reasons the mortality rates increase, for example the arrival of new predators, hostile microorganisms, scarcity of food and water input, etc., the equilibrium will be displaced to species with individuals of smaller unit mass and faster reproduction rates.

Such automatic regulatory mechanisms of ecosystems, which were just presented in an extremely simplified version, may induce sometimes very drastic changes in the animal and plant populations, over times relatively short but definitely longer than the human lifespan, all by maintaining nearly constant the total circulating reserve of materials. Such changes, which we sometimes qualify of destruction of habitat, are indeed long-time adjustments of the equilibrium between different species. If the fluxes x_{ij} are not sensibly altered, biomass will be just redistributed within a perfectly closed ecosystem (even if such hypothesis is often difficult to maintain). In a well developed forest, the most part of the circulating reserve is contained in the plant bodies, leaving room for but a very moderate fauna biomass. Progressive destruction of the forest and herbivores pressure will open the outer limits to a savannah-like environment, putting back into circulation part of the reserve toward the decomposers (worms, nematodes) and the smaller plants, which in turn may favour the development of larger animals, from herbivores to carnivores.

A true, and practically irreversible, destruction of an ecological habitat comes if, instead, some part of the circulating reserve is destroyed, or displaced. This is for example the case of intensive agricultural exploitation, when for several consecutive years some edible plant species are densely cultivated over an extended area. The direct consequence is the impoverishment of the circulating reserve, in one or more of the key elements, such as magnesium, phosphorus, nitrogen, etc. The artificial fertilisation by chemicals may supply back some of the lost elements, however with many undesirable side effects, such as the eutrophisation by excess of phosphates or nitrates, soil acidification, accumulation of toxic elements (fluorides, mercury, cadmium and other heavy metals), down to more complicate feedback effects, such as the breakdown of symbiotic relationships between plant roots and mycorrhizal fungi.

12.8.1   Trophic and Non-trophic Interactions

The input/output relations displayed in Fig. 12.11 describe the direct effects of availability of food, from the lower-lying to the upper-lying levels of the food chain, and therefore are called **trophic interactions**. However, it is increasingly accepted that other kind of biological interactions, less direct but equally effective, can take place among the species occupying a same ecological niche. For example, the *mutualism* (species benefiting from the independent activity of each other) between herbivores and bacteria living inside their intestines; or the *commensalism* (feeding on the same source without mutually affecting) existing between humans and domesticated animals; or, at the opposite, the *antagonism* (reduction of a species' fitness by the presence of another), such as some Chihuahuan desert ants that interfere with the foraging of red ants, by systematically blocking their colonies access holes. Such interactions, which do not involve directly food-chain relations, are called **non-trophic**. Including them in dynamical models may prove very effective, however observing and estimating the fitness costs and benefits of such interactions among species can be very problematic, and the way interactions are interpreted can profoundly affect the ensuing conclusions.

In a one-of-a-kind experiment that was run in the years 2002–2009 at the Göttingen University in Germany [17], C. Scherber and his coworkers cultivated under extremely well-controlled conditions an ensemble of 82 identical plots, each measuring 20×20 m^2, inseminated at the beginning of the experiment with a variable number of species, each with different plantation density; numbers of species equal to 1, 2, 4, 8, 16, 60, with densities ranging from 4 to 16 units, were used, each unit being composed by a constant number of plants. Each plot was maintained under constant conditions by standard gardening techniques, and without using extra fertilisers or insecticides. During the 8 years of continued observation, the scientists registered the abundance and the richness for all vegetal and animal species, both above and below the ground.

The objective of this experiment was to precisely measure the evolution of the different populations, as a function of the availability and variability of primary food (richness) for the herbivores. In 1921, the Swedish chemist Arrhenius formulated a law to explain the variation of biodiversity among species occupying the surface A of an ecosystem:

$$S = cA^z \tag{12.46}$$

Here S is the number of different species, or *richness*, a direct measure of the biodiversity; the constant c is the richness factor, broadly comprised between about 20 and 20,000; and the exponent z is the accumulation factor, typically between 0.2 and 0.5. The **abundance** is defined as the number of individuals composing a given species, at a given site and time.

Figure 12.15 shows that the increase in richness has a positive correlation with both the abundance and biodiversity of all species, both above and below ground, at all the trophic levels, from the herbivores to the omnivores, to a variable extent (different constants c) and with different response times. After a variable-length initial transitory, nearly all species appear to settle to a power-law growth, which follows rather well the Arrhenius law $S = cA^z$. The precise role of the exponent z is demonstrated in Fig. 12.16a: a value $z \simeq 0$ corresponds to a flat abundance-richness curve, while values of $z \simeq 0.3-0.4$ show a strong dependence of the number of animal species on the number and density of different plants in each plot. Such a measure of biodiversity fits well also with the impact on the different trophic levels, where it is found that the biodiversity naturally decreases while going up in the levels, from the primary producers, up to the herbivores, carnivores, and omnivores. In particular, Fig. 12.16b, c show that the z exponent for most species is sensibly reduced, upon reducing the plant richness. This latter was in fact the main control parameter of the experiment, which could be adjusted by changing the density and diversity of plant species initially installed in each plot.

Different mathematical models were put forward, to analyse and interpret the 8-years long experimental data, all generally based on linear correlation analysis.

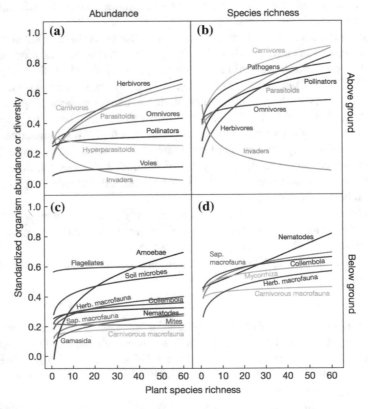

Fig. 12.15 Abundance (*left*) and biodiversity (*right*) for the organisms living *above* and *below* ground in the Göttingen experiment, as a function of the richness of species planted in each plot. For all animal and vegetal species, both abundance and biodiversity clearly increase with the richness, showing a good correlation with the Arrhenius law. [From Ref. [17], repr. w. permission]

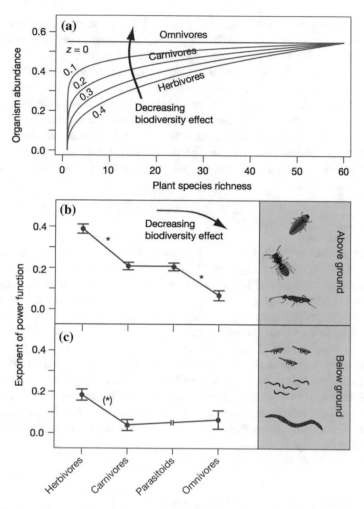

Fig. 12.16 Growth rate of the population for the animal species living *above* and *below* ground, as a function of the abundance and richness of each plot. **a** Role of the exponent z in describing the biodiversity: $z = 0$ gives a flat *curve*, values of $z \simeq 0.3$–0.4 show a strong dependence on the number of different plant species. **b, c** The exponent z of all species (or accumulation factor) decreases with reducing the plant (this was the principal parameter, controlled by changing the density and diversity of seeds). [From Ref. [17], repr. w. permission]

A first model, called *top-down*, in which the abundance of carnivores is the main controlling factor for the population of lower-lying herbivores, was found to be in clear disagreement with the data. Similarly, other models based on direct correlations were rejected; for example, one in which plant richness directly links to the carnivore biodiversity. Actually, the experiment shows that plant richness only has an influence on the primary consumers, notably herbivores and nematodes. Such variations, in turn, affect the chain in a bottom-up way, by modifying the secondary consumers and so on, going up in the food chain. Indirect effects of plant species richness were observed also on soil microbes, probably mediated either through changes in plant roots production, or through root exudates (i.e., alterations in the below-ground chemistry). The results of this unique experiment suggest that the effects of the species richness from one trophic level to others decrease progressively with trophic distance. However, probing such trophic and non-trophic correlations with a relatively limited number of control variables is a difficult endeavour.

Trophic and non-trophic interactions have been included also in generalised Lotka-Volterra models [18]. In this type of analysis, the linear equations (12.41) are replaced by a coupled set of differential equations of the type (12.37):

$$\frac{dQ_i}{dt} = \sum_j a_{ij} Q_i Q_j - b_i Q_i \tag{12.47}$$

Here the state vector Q_i indicates the interacting species, e.g., Q_1 = plants, Q_2 = herbivores, Q_3 = carnivores, Q_4 = land nutrients, Q_5 = reserve pool; the coefficients a_{ij} describe the cross interactions between the different levels, including both trophic ($j = i \pm 1$) and non-trophic ($i + 1 < j < i - 1$) interactions; and the coefficients b_i describe the mortality or degradation rates of each species. (Note that many of the coefficients a_{ij} would be nevertheless equal to zero, since not all the possible interconnections can actually take place.)

Such models have been able to theoretically predict an unexpected bottom-up control of carnivores by plants, with the carnivore biomass being indirectly controlled by plant and herbivore biomass, as opposed to a top-down control of herbivores by carnivores. These structural equation models are a powerful tool for detecting complex mutual interdependencies, greatly enhancing our understanding of biodiversity effects in multi-trophic systems.

12.8.2   Linear Models of Structured Population

In the last Sections we looked at dynamical models, as a complement and alternative to the older (both linear and non-linear) analytical models, which often result in non-integrable systems whose solution is practically impossible. Another major criticism to the analytical models is that they treat all the individuals in a population as being identical. However, natural populations display a rich differentiation, with subgroups each having diverse behaviour from the average. For instance, the death rate of the young individuals can be high at birth and decreasing with age; or the fact that young individuals not yet in their reproductive phase do not contribute to the growth of the population; and for adults, the death rate increases with age. In insect populations, we must account for the distinct life stages (egg, larva, pupa, adult), which actually have the characters of a distinct population each, but whose numbers are however strictly correlated to each other. For example, only adults can lay eggs and reproduce, and death rates can be very different in the different stages. Plants also have various life stages, such as dormant seed, seedling, non-flowering, and flowering. To describe such differences, a population model needs to be **structured**, with separate treatment for the different components (subgroups) of the population under study, each with its own characteristics.

In this conclusive Section we will focus, for the sake of simplicity, on simply linear models of structured populations. The observed behaviour could only be relatively simple, leading to typically exponential growth, as it is the case with any linear approximation (see Malthusian models above). However, with some revealing differences that may suggest roads for further development (see also "Further reading" at the end of this chapter, [21]).

In a simple model, let us imagine a population of individuals divided into a number of age groups, such as newborns $= N$, youngs $= Y$, adults $= A$, and olds $= O$. Indicate the numbers of individuals in each group at a given time with a subscript t, and their total as $T_t = N_t + Y_t + A_t + O_t$. For each group we attribute different birth, death and reproduction rates. A reasonable description of the time evolution of the entire population could be:

$$
\begin{aligned}
N_{t+1} &= b_1 Y_t + b_2 A_t + b_3 O_t + N_t(1 - d_1) \\
Y_{t+1} &= g_2 N_t + Y_t(1 - d_2) \\
A_{t+1} &= g_3 Y_t + A_t(1 - d_3) \\
O_{t+1} &= g_4 A_t + O_t(1 - d_4)
\end{aligned}
\tag{12.48}
$$

Here we stated that the number of newborns at time $t+1$ is a function of the number of youngs, adults and olds present at time t, each with their respective reproduction probability b_1, b_2, b_3, plus the number of new individuals already present at time t diminished by the natural rate d_1 of mortality at birth; with the second equation, the number of young individuals depends on the number of newborns at the time before, minus their own mortality rate; next, the number of adults depends on the number of youngsters, minus their own mortality; and finally, the number of ageing individuals at $t + 1$ is a fraction of the number of adults entering the old age, minus the old individuals dying at the same time with probability d_4. Depending on the various coefficients, the evolution of the single groups, and of the total population T, can be studied.

One way of looking at the equations above, is to turn the equations into a **linear system**. The variables are collected in a vector with four components, $\mathbf{x} = (N, Y, A, O)$, and each equation in the group above becomes one row of a matrix $\underline{\mathbf{M}}$:

$$
\underline{\mathbf{M}} =
\begin{pmatrix}
(1 - d_1) & b_1 & b_2 & b_3 \\
g_2 & (1 - d_2) & 0 & 0 \\
0 & g_3 & (1 - d_3) & 0 \\
0 & 0 & g_4 & (1 - d_4)
\end{pmatrix}
\tag{12.49}
$$

such that the populations at time $t + 1$ are obtained from the populations at time t via a matrix-vector product $\mathbf{x}_{t+1} = \underline{\mathbf{M}} \otimes \mathbf{x}_t$, or more explicitly (see Appendix A):

$$
\begin{pmatrix} N_{t+1} \\ Y_{t+1} \\ A_{t+1} \\ O_{t+1} \end{pmatrix} = \begin{pmatrix} (1-d_1) & b_1 & b_2 & b_3 \\ g_2 & (1-d_2) & 0 & 0 \\ 0 & g_3 & (1-d_3) & 0 \\ 0 & 0 & g_4 & (1-d_4) \end{pmatrix} \begin{pmatrix} N_t \\ Y_t \\ A_t \\ O_t \end{pmatrix}
\tag{12.50}
$$

A matrix like $\underline{\mathbf{M}}$ is called a *projection* or *transition matrix*, since it allows to "predict" the population at a subsequent time from the values at a given time. Variants of this simple model have known a certain success in human population modelling. A version due to Patrick Leslie [19] considered the population divided into 6 groups of 15-years span; it assumed all the coefficients $d_i = 1$, so that the projection matrix has all zeros on the diagonal; the top row coefficients, b_i, describe the age-dependent fecundity, while the sub-diagonal coefficients g_i the survival probability in passing from one age group to the subsequent. A variant proposed in 1969 by Michael Usher [20], which has found notable use in forest management, included d_i's different from zero, to examine the effect of fractions of population remaining in a given class without passing on to the next.

As an example of application of the Usher model, consider a plant that (i) needs some years before maturing into a flowering stage and (ii) that, after reaching maturity, does not always flower every year. In addition, (iii) its seeds may lie dormant for some time before germinating. The life cycle of an ensemble of these plants over several years can be described by the following life stages: U_t, number of dormant seeds at time t; Y_t, number of young plants; F_t number of mature, flowering plants; and N_t, number of mature plants not flowering in the time t. The projection matrix can be written as:

$$
\underline{\mathbf{M}} = \begin{pmatrix} (1-d_1) & 0 & b_1 & 0 \\ g_2 & (1-d_2) & b_2 & 0 \\ 0 & g_3 & (1-d_3) & b_3 \\ 0 & 0 & g_4 & (1-d_4) \end{pmatrix}
\tag{12.51}
$$

The coefficients state that: $(1 - d_1)$ is the fraction of ungerminated seeds that can still germinate in the next year, while b_1 are new seeds produced by the flowering plants; b_3 is the fraction of non-flowering plants that go back in the flowering group at the subsequent time; g_2, g_3, g_4 describe transfer of, respectively $U \rightarrow Y, Y \rightarrow F$, and $F \rightarrow N$. A numerical solution is shown in Fig. 12.17, starting from an initial population of 50 seedlings planted at time $t = 0$. Only the first few steps of the numerical solution are shown, since at long times the growth becomes merely exponential, for each of the populations (in the log-linear plot, all graphs tend to parallel straight lines for $t > 20-30$). However, at short times there are interesting oscillations, especially when looking at the interplay between mature plants cycling between the flowering and sterile states (green and red graphs, respectively).

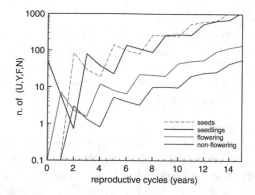

Fig. 12.17 Early stages of the simulation of the plant growth model (12.51), with the parameters $d_1 = 0.98, d_2 = 0.9, d_3 = 0.8, d_4 = 0.9, g_2 = 0.95, g_3 = 0.14, g_4 = 0.43, b_1 = 12, b_3 = 0.32$, and starting from $Y = 50$ seedlings installed at time $t = 0$

Despite their relative simplicity, examples of use of linear models in demography and ecology abound in the literature. The following Problems 12.6 and 12.7 offer a little bit of challenge to the student, to guess a linear model in a couple of realistic, albeit still idealised natural systems. Linear models of the type described can give useful insights by using the power of linear algebra analysis, by studying the behaviour of the matrix eigenvalues and eigenvectors. Important features can be extracted, like the intrinsic growth rate, stable and dominant age distribution of a population, and so on. While such a treatment goes clearly beyond the introductory level of this book, the interested reader can refer to the "Further reading" suggestions at the end of this chapter.

Problems

12.1 Savannah's beauties
By way of an allometric analysis, you are trying to determine a meaningful relationship between the body mass, and the characteristic sizes of a family of african antelopes. Find a scaling relation with the best length parameter, and comment on the results.

	Mass (Kg)	Leg length, H (m)	Nose-to-tail length, D (m)	Neck length, h (m)
Speke's gazelle	8	0.45	0.8	0.4
Royal antelope	1.5	0.3	0.40	0.3
Impala	150	1.1	2.02	0.7
Waller's gazelle	22	0.6	1.12	0.5
Antelope	48	0.8	1.45	0.55

12.2 Climbing heights

According to the metabolic scaling law, the energy consumed per day for resting body functions is $E = 75M^{0.75}$, and walking at sea level adds about 2 kcal/(kg-h). A man of 70 kg is climbing uphill a slope with 12 % inclination at a pace of 3 km/h. After 20 min of climbing, it is measured that he consumed 121 kcal. What this tells you about the efficiency of his muscles?

12.3 The life of a heart

In comparative metabolism studies across a large range of different mammals with masses ranging from a few grams to many hundreds of kg, two allometric scaling law have been established, the first one for the heart rate $h = 241M^{-0.25}$ beats/min, and the second for the average life span $T = 15M^{0.20}$ years. What do such equations suggest about the heart capacity across different mammals? Compare for example a mouse of 30 g and a man of 70 kg.

12.4 The fish hatchery

A fish hatchery is stocked with a population N_0 of juvenile fish, which is subject to a natural depletion rate r. After a time T to allow the fishes to grow, fishing commences and causes an additional depletion rate q. Obtain an expression for the population at time $t > T$, and determine the fraction of the initial N_0 that can be recovered by fishing.

12.5 Influenza spreading

Starting from the 1920s, mathematical models of epidemics begun to be formulated. In their simpler versions, they require to work with at least four parameters: (i) the size N of the population, (ii) the number Q of individuals already immune from the virus, (iii) the number of infected individuals I (usually starting with $I = 1$), and (iv) the infection spreading probability, R_0, also called the reproduction number. Therefore $N = Q + I + S$, with S the "susceptible" population, which at a given time is neither infected nor immune. In a simple population model, $R_0 = p\tau$, the product of the probability p of an I encountering an S per unit time, times the average time span of virus activity τ; if $R_0 > 1$ the infection spreads in the population. One of the basic findings of such models is that at steady state, it is $R_0 \cdot S = N$. At the start of a new epidemic, $I \ll N$, so it can be neglected, and $Q + S = N$. All this given, can you deduce a critical vaccination threshold, Q_c, that is a minimum fraction of the population that needs to be immunised at $t = 0$ to avoid spreading of the epidemic?

12.6 From egg to adult

An insect like *Tenebrio molitor*, the common flour beetle, goes through three life stages: egg, larva, and adult. We can model this evolution in discrete steps: progress from egg to larva; from larva to adult; finally, adults lay eggs and die in one more step. Let E_t = the number of eggs, L_t = the number of larvae, and A_t = the number of adults at time t. Suppose we collect data from a colony and find that only 4 % of the eggs survive to become larvae, 39 % of the larvae make it to adulthood, and adults on average produce 73 eggs each. (a) Build a linear model with recurrence equations of the type $W_{t+1} = \alpha W_t + \beta$ and show that the result can be expressed as a Malthusian growth law. (b) Next, consider the possibility that some fraction of adults, like 65 %, do not die after laying eggs, and see how the model is modified.

12.7 The constant gardener

Your are monitoring the growth of two species of plants in a plot. A_n and B_n denote the number of each species in the plot at week n. When a plant dies, a new plant grows in the empty space, but the new one can be of either species. The species A are long-lived plants, with a mortality of 1 % in any given week, while 5 % of the species B die in the same time. However, because B are rapid growers they are more likely to succeed in winning a space left by a dead plant; 75 % of all vacant spots go to species B plants, and only 25 % go to A. Build a linear model of the garden evolution and study its practical consequences.

References

1. C.J. Pennycuick, The flight of petrels and albatrosses (Procellariiformes), observed in South Georgia and its vicinity. Phil. Trans. Roy. Soc. (Lond.) B **300**, 75–106 (1982)
2. A.V. Hill, The dimensions of animals and their muscular dynamics. Sci. Progr. **38**, 209–230 (1950)
3. C.D. Newton, D.M. Nunamaker, *Textbook of Small Animal Orthopaedics* (J. B. Lippincott Co., London, 1985)
4. N.C. Heglund, C.R. Taylor, T.A. McMahon, Scaling stride frequency and gait to animal size: from mice to horses. Science **86**, 1112 (1974)
5. C. de Muizon, Walking with whales. Nature **413**, 259–260 (2001)
6. A.D. Gingerich, M. ul-Haq, W. von Koenigswald, W.J. Sanders, B. Holly Smith, I.S. Zalmoud, New protocetid whale from the Middle Eocene of Pakistan: birth on land, precocial development, and sexual dimorphism. PLoS ONE **4**, e4366 (2009)
7. A.V. Hill, Animal dimensions and muscle dynamics. Sci. Prog. **38**, 209 (1950)
8. N.C. Heglund, C.R. Taylor, Speed, stride frequency and energy cost per stride: how do they change with body size and gait? J. Expt. Biol. **138**, 301–318 (1988)
9. G.B. West, J.H. Brown, The origin of allometric scaling laws in biology from genomes to ecosystems: towards a quantitative unifying theory of biological structure and organization. J. Expt. Biol. **208**, 1575–1592 (2005)
10. K. Schmidt-Nielsen, Locomotion: energy cost of swimming, flying, and running. Science **177**, 222–228 (1972)
11. D. Wester, Linking the ecology of past and present mammal communities, p. 41, ed. by A.K. Behrensmeyer, A.P. Hill. *Fossils in the Making* (University of Chicago Press, 1980)
12. T.R. Malthus, An essay on the principle of population, as it affects the future improvement of society. J. Johnson in St. Paul, London (1798) [Public-domain 1914 reprint by J. M. Dent. http://archive.org/details/essayonpopulatio00malt]

13. D.A. MacLulich, *Fluctuations in the Numbers of the Varying Hare (Lepus americanus)* (Univ. Toronto Studies Biol. Ser. 43, University of Toronto Press, 1937)
14. J.A. Hobson, E.F. Pace-Scott, R.Robert Stickgold, Dreaming and the brain: toward a cognitive neuroscience of conscious states. Behav. Brain Sci. **23** (2000)
15. R.M. Goodwin, A growth cycle, p. 54–58, ed. by C.H. Feinstein. *Socialism, Capitalism and Economic Growth* (Cambridge University Press, 1967)
16. T. Lougheed, T.: Phosphorus paradox: scarcity and overabundance of a key nutrient. Environ. Health Perspect. **119**, A208–213 (2011)
17. C. Scherber et al., Bottom-up effects of plant diversity on multi-trophic interactions in a biodiversity experiment. Nature **468**, 553–556 (2010)
18. A. Goudard, M. Loreau, Nontrophic interactions, biodiversity, and ecosystem functioning: an interaction web model. Am. Nat. **171**, 91–106 (2007)
19. P.A. Leslie, The use of matrices in certain population mathematics. Biometrika **33**, 183–212 (1945)
20. M.B. Usher, A matrix approach to the management of renewable resources, with special reference to selection forests. J. Appl. Ecol. **3**, 355–367 (1966)

Further Reading

21. E.S. Allman, J.A. Rhodes, *Mathematical Models in Biology: An Introduction* (Cambridge University Press, Cambridge, 2004)
22. J. Maynard Smith, *Models in Ecology* (Cambridge University Press, Cambridge, 1974)
23. G. Bateson, *For an Ecology of Mind* (Kluwer, New York, 1977)
24. E.P. Odum, The emergence of ecology as a new integrative discipline. Science **195**, 1289–1293 (1977)
25. I. Stewart, *Nature's Numbers* (Basic Books (Harper), New York, 1995)
26. M. Austin, Species distribution models and ecological theory: A critical assessment and some possible new approaches. Ecol. Model. **200**, 1–19 (2007)
27. S.P. Otto, T. Day, *A Biologist's Guide to Mathematical Modelling in Ecology and Evolution* (Princeton University Press, 2011)

Chapter 13
Solutions to the Problems

The problems at the end of each Chapter are not just meant as numerical applications of the many equations displayed in the text. Rather, each one is an occasion to learn something more, to go a bit deeper on some aspects that were only hinted at in the Chapter, or to apply the concepts you just learned to some unusual, or even funny situation. This is why in this final part complete solutions to the problems are provided, with quite detailed discussions about the implications of a particular solution, and some possible developments suggested by the answer to the problem.

However, it is the duty of any good student to try to solve the problems independently, without looking at the solutions in the first place. Each problem is a little challenge, taking you a little step further. Do not take for granted that just 'reading' the solution will improve or deepen your understanding of the subject. It is the thinking around that does it. The goal is not as much to answer the exercises, as it is to figure out *how* to answer the exercises. Also, remember that in many cases a problem can have more than one solution, and that the one proposed in the text could not necessarily be the best (e.g., in terms of approximations, elegance, simplicity, formal rigour, etc.). Use the solutions only when you run out of options, but always use your own brains first!

Chapter 2

2.1 Basic nomenclature
(a) closed; (b) strictly speaking closed, but practically isolated for all we can see; (c) open; (d) closed; (e) closed; (f) open.

© Springer International Publishing Switzerland 2016
F. Cleri, *The Physics of Living Systems*, Undergraduate Lecture
Notes in Physics, DOI 10.1007/978-3-319-30647-6_13

2.2 Formal identities

(a) From the equipartition it is $E = \frac{3}{2}Nk_BT$; the equation of state of a perfect gas is $PV = Nk_BT$, from which $P = \frac{2}{3}\frac{E}{V}$.

(b) From the (Euler) equation $E = TS - pV + \mu N$, it follows immediately $F = -pV + \mu N$.

(c) Write the total differential $dG = dF + d(pV) = dF + pdV + Vdp$. On the other hand, $dF = dU - d(TS) = dU - TdS - SdT$, and the fundamental equation of thermodynamics says that $dU = TdS - pdV + \mu dN$. Then, $dF = -pdV - SdT + \mu dN$, and finally $dG = -SdT + \mu dN + Vdp$. Therefore, G depends on T, N and p, as independent variables.

2.3 Thermal engine

(a) The whole system is open, since it exchanges energy and matter (the air from the fan) with the external world. The subsystem represented by the cooling fluid is closed, since it exchanges only energy but not matter with the rest of the world.

(b) The operation can be divided into four steps, which are cyclically repeated. (1) Heat is absorbed by the circulating fluid in the warm room. It becomes a room-temperature, low-pressure gas, before entering the compressor. (2) The fluid enters the compressor and comes out as a high-pressure, hot gas. Passing in tubes outside the building, the hot gas dumps heat to outside air. (3) The warm gas from outside enters a constriction and is further pressurised to form a liquid in the condenser. (4) The liquid undergoes free expansion into a gas and cools. The cool gas then flows in pipes inside the room. Although the air conditioner pumps heat from cold to hot regions, it does not violate the Second Law of thermodynamics: the compressor adds entropy, so that the total entropy of the system actually increases.

2.4 Exchanges of entropy

One possible strategy is to assume that the two parts of water come separately at equilibrium in a slow time, and find the entropy change in each of the hot and the cold water, then add these to get the total entropy change. The entropy changes can be computed by using the 'caloric' definition $dS = \delta Q/T$. The amount of heat to bring the volume of water V from one temperature to another at ΔT degrees of difference is simply calculated from the specific heat of water, as: $c_P = 4,186$ J/(kg-K), $\delta Q = c_P\rho V \Delta T$, with $\rho = 1,000$ kg/m^3 the density. The final temperature T_f is found by imposing that the amount of heat lost from the water at higher temperature is equal to the heat gained by the water at lower temperature:

$$c_P\rho\, 50(T_f - 55) + c_P\rho\, 25(T_f - 10) = 0$$

from which $T_f = 40\,^\circ$C. Now the entropy of the two volumes of water is found by integrating the heat released/gained between the respective temperatures at start and the final temperature, i.e. their sum is zero:

$$\Delta S_{cold} = c_P \int_{283}^{313} \frac{dT}{T} = c_P \ln \frac{313}{283} = 10{,}550 \, \text{J/K}$$

$$\Delta S_{hot} = c_P \int_{328}^{313} \frac{dT}{T} = c_P \ln \frac{313}{328} = -9{,}800 \, \text{J/K}$$

The total entropy is $10{,}550 - 9{,}800 = 750$ J/K, and it is seen that the entropy of the cold water is increased, since its temperature is raised, while the opposite is true for the hot water. Note that we had to integrate the relation between heat, temperature and entropy, since the (quasi-static) process of interest is not isothermal.

2.5 Boiling, temperature and pressure

While heating the pressure cooker up to T around $100\,^\circ$C, there are three components contributing to the pressure: the expansion of water (negligible), the evaporation of water (negligible until the temperature reaches very close to the boiling point), and the expansion of the air filling the pot. For this latter, main contribution, we apply the perfect gas equation, $PV = nRT$, with n the number of moles. Since one mole is 22.4 l, $n = 0.134$ here, and $\Delta P = 0.134 \cdot 8.3145 \cdot 373/0.003 = 138.5$ kPa. By applying Clapeyron's equation, we get the new approximate value of the boiling point, $T = 100.7\,^\circ$C. Enclosing the water under pressure brings it to boil at a (slightly) later time.

2.6 Stefan-Boltzmann T^4 law

Consider that the energy E of the perfect gas depends only on the temperature. Start from:

$$T dS = dE + p dV = d(eV) + \frac{e}{3} dV = e dV + V \left(\frac{de}{dT} \right) dT + \frac{e}{3} dV$$

So:

$$T dS = V \left(\frac{de}{dT} \right) dT + \left(e + \frac{e}{3} \right) dV = V \left(\frac{de}{dT} \right) dT + \frac{4e}{3} dV$$

Now, take the derivatives of S with respect to T and V:

$$\frac{\partial S}{\partial T} = \frac{V}{T} \frac{de}{dT}$$

$$\frac{\partial S}{\partial V} = \frac{4}{3} \frac{e}{T}$$

then take the cross-derivatives, and equate them:

$$\frac{\partial}{\partial V} \frac{\partial S}{\partial T} = \frac{1}{T} \frac{de}{dT} = \frac{\partial}{\partial T} \frac{\partial S}{\partial V} = \frac{4}{3} \left(T \frac{de}{dT} - e \right) \frac{1}{T^2}$$

By rearranging the terms:

$$\frac{de}{dT} = 4\frac{e}{T}$$

which can be integrated by separating the variables to $\ln e = 4\ln T + c'$, or $e = cT^4$.

2.7 A negative temperature

For a given total energy of the system $E = Me$ (to be conserved), the particles can be variably distributed according to a number of microstates given by:

$$\Omega(E) = \frac{N!}{M!(N-M)!}$$

corresponding to the different combinations of M (distinguishable) particles occupying the excited state, while the remaining $N - M$ are in the ground state. For $N \to \infty$ we can use the Stirling's approximation for the logarithm, and obtain the system entropy as:

$$S/k_B = \ln \Omega(E) \simeq$$
$$\simeq (N \ln N - N) - (M \ln M - M) - ((N - M)\ln(N - M) - (N - M)) =$$
$$= N \ln N - M \ln M - (N - M)\ln(N - M)$$

or:

$$S = k_B \left(N \ln \frac{N}{(N-M)} - M \ln \frac{M}{(N-M)} \right) = Nk_B \left(\ln \frac{1}{1-x} - x \ln \frac{x}{1-x} \right)$$

with $x = M/N$. Now, the temperature is by definition:

$$\frac{1}{T} = \frac{\partial S}{\partial E} = \frac{\partial S}{\partial (Me)} = \frac{1}{e}\frac{\partial S}{\partial M} = \frac{1}{Ne}\frac{\partial S}{\partial x}$$

(since the particles in the ground state contribute 0 to the total energy). Therefore:

$$\frac{e}{k_B T} = \frac{\partial}{\partial x}\left(\ln \frac{1}{1-x} - x \ln \frac{x}{1-x} \right)$$
$$= \frac{\partial}{\partial x}[\ln 1 - \ln(1-x) - x \ln x + x \ln(1-x)] =$$
$$= \frac{1}{1-x} - \ln x - \frac{x}{x} + \ln(1-x) - \frac{x}{1-x} = \ln \frac{1-x}{x}$$

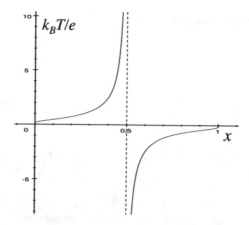

The plot of the temperature is shown in the graph above. It is seen that for $M > N/2$ the temperature is actually negative. The reason for this, is that when the energy is such that more than half of the particles have to be in the excited state, the corresponding *density of states* (that is the number of microstates per energy interval $d\Omega/dE$) becomes a negative function: in practice, this means that upon increasing the energy further, the number of available microstates actually decreases. In virtually none of known physical systems this does occur, and therefore no system would exist at a negative temperature. However, it has recently been found that some very exotic spin-systems could show a similar effect (see Ref. [16] in Chap. 2).

2.8 Greenhouse gases 1
The original Sackur-Tetrode expression for the perfect gas contains only the kinetic energy K. However, in the atmosphere a gas molecule of mass m is also subject to gravity from the Earth, and the total energy to consider is now $E = K - Nmgh$. Therefore we can write:

$$S(N, V, E) = Nk_B \ln\left[\left(\frac{V}{N}\right)\left(\frac{2m(K - Nmgh)}{N}\right)^{3/2}\right] + Nk_B T\left(\frac{5}{2} + \frac{3}{2}\ln\frac{2\pi}{3h^2}\right)$$

$$= Nk_B \ln\left[\left(\frac{V}{N}\right)\left(\frac{4\pi m(K - Nmgh)}{3h^2 N}\right)^{3/2}\right] + \frac{5}{2}Nk_B T$$

To simplify the final expression, note that $(K - Nmgh) = \frac{3}{2}k_B T$. Then, the derivative of S with respect to N is:

$$\frac{\partial S}{\partial N} = k_B \ln\left[\left(\frac{V}{N}\right)\left(\frac{2\pi mk_B T}{h^2}\right)^{3/2}\right] - \frac{mgh}{T}$$

and the chemical potential is therefore:

$$\mu(h) = -k_B T \ln \left[\left(\frac{V}{N} \right) \left(\frac{2\pi m k_B T}{h^2} \right)^{3/2} \right] + mgh$$

Now impose that $\mu(h) = \mu(0)$, for any h, however with a different number of molecules $N(h)$ at each altitude. Then:

$$-k_B T \ln \left[\left(\frac{V}{N(h)} \right) \left(\frac{2\pi m k_B T}{h^2} \right)^{3/2} \right] + mgh = -k_B T \ln \left[\left(\frac{V}{N(0)} \right) \left(\frac{2\pi m k_B T}{h^2} \right)^{3/2} \right]$$

which after simplification gives: $k_B T \ln N(h) + mgh = k_B T \ln N(0)$.
Then: $N(h) = N(0) \exp(-mgh/k_B T)$, i.e., the concentration decreases exponentially with the altitude and the mass of the molecule.

2.9 Greenhouse gases 2
By looking at Fig. 2.5, it can be seen that the important contribution to the greenhouse effect comes from the gases absorbing part of the radiation reemitted by the Earth's surface at $T = 300\,\text{K}$, i.e. the blue part of the spectrum in the figure given in the problem.

(a) Water vapour is the most effective, since it absorbs over a very large part of the spectrum, starting from a wavelength of few μm and higher.
(b) According to their concentration and absorption capability in this spectral region, the ranking is: water vapour, CO_2, methane, N_2O, ozone.
(c) Because the lifetime of water vapour in the atmosphere is very short, while N_2O and CO_2 last for about 100 years, once arrived in the upper atmospheric layers.

Chapter 3

3.1 Thermodynamic and probabilistic entropy are the same
The number of microstates is $\Omega = N!/(n_1!n_2!\ldots n_m!)$. From the Stirling approximation, $\ln x! \simeq x \ln x - x$, or $x! \simeq (x/e)^x$, so we can replace:

$$\Omega = \frac{(N/e)^N}{(n_1/e)^{n_1}(n_2/e)^{n_2}\ldots(n_m/e)^{n_m}} = \frac{N^N}{n_1^{n_1} n_2^{n_2}\ldots n_m^{n_m}}$$

If we take $p_i = n_i/N$ for the probabilities, it is:

$$\Omega = \frac{1}{(n_1/N)^{n_1}(n_2/N)^{n_2}\ldots(n_m/N)^{n_m}} = \frac{1}{p_1^{n_1} p_2^{n_2}\ldots p_m^{n_m}}$$

By taking the logarithm and multiplying by k_B/N:

$$\frac{k_B}{N} \ln \Omega = -k_B \sum_i p_i \ln p_i$$

3.2 Information entropy

Information entropy was defined as the difference between the logarithm of the number of available states for the system, $\Delta S_I = k_B(\ln \Omega_f - \ln \Omega_d) = k_B \ln(\Omega_f/\Omega_d)$, with d and f being, respectively, the fully disordered and the fully functional state of the information-carrying system considered. For a disordered polymer f is very close to d, so their difference in Ω is small. For a crystal, Ω_f is of order 1, and Ω_d is represented number of combinations of the n atoms on the p unit cell positions is the binomial coefficient $\binom{n}{p}$; since both n and p are not very large numbers, again the logarithm of the ratio of Ωs is not a big number. For a protein instead, Ω_f is of order 1, but Ω_d can be very large, since the number of amino acids n can easily be of the order of tens of thousands.

3.3 Entropy of erasure

Here we want to see the difference between losing the information, by rearranging at random the nucleotides in a DNA sequence of length N, and the physical destruction of the bonds, to return the sequence to an ensemble of disconnected nucleotides (monomers). Considering the number of bases contained in the sequence, the maximum number of combinations in a given arrangement is $2 \times N/3$, since each strand can code for $N/3$ amino acids, and each combination will correspond to one or more proteins. Each base can contain one of the four nucleotides, so the number of microstates is $4^{2N/3}$. The associated entropy is $\Delta S = k_b \ln 4^{2N/3} = \frac{2}{3} N k_B \ln 4$. Since $N \sim 6 \times 10^9$, it is $\Delta S \sim 7.6 \times 10^{-14}$ J/K.

The chemical entropy is $\Delta S = \delta Q/T$, where for δQ we can take the average enthalpy of dissociation of a single base, $\Delta H \sim 8$ kcal/mol, then $\Delta S \sim 25$ cal/mol-K ~ 100 J/mol-K (the experimentally measured value is slightly different). One human DNA is $\sim 10^{-14}$ moles of individual nucleotides, therefore $\Delta S \sim 100 \times 10^{-14}$ J/K.

3.4 Genetic mistakes

An error in transcription would lead to many erroneous protein copies, whereas an error in translation affects only one protein copy. Moreover, the transcription is based on a one-to-one correspondence, whereas the translation seeks correspondence of three nucleotides to one amino acid; this means that mRNA messages would require a higher fidelity "per letter" to achieve the same error rate as transcription.

3.5 Peptide bonds in proteins

Arrows 2 and 5 are peptide bonds; 2 and 3 are alpha-carbon bonds; 1 is an amide bond.

3.6 The Solar system has a negative heat capacity

The velocity of a planet (mass m) orbiting at a distance r around the Sun (mass M) is obtained by equating the centripetal force to the gravity, $F = GMm/r^2 = mv^2/r$, from which $v = \sqrt{GM/r}$.

If some energy is added into the system the orbits expand, planets are more loosely bound, and require less velocity to keep from collapsing. The orbiting system is in equilibrium when its kinetic energy is just enough to prevent collapse. But in this case planets have a lower "temperature" despite we *added* kinetic energy. The added energy went into the gravitational potential.

This can be verified by looking at the total energy as the sum of kinetic energy $K = \frac{1}{2}mv^2$ and potential energy from the gravitational force, $U = -GMm/r = -mv^2$, that is $E = K + U = -\frac{1}{2}mv^2$. By equating kinetic energy and temperature, it is then $E = -\frac{3}{2}k_BT$. The total energy is negative, and so it is its derivative with respect to T; so, the heat capacity is negative.

Chapter 4

4.1 The ΔG of metabolic reactions

(a) At equilibrium $\Delta G = 0 = \Delta G_0 + RT \ln K$, from which $K_{eq} = \exp(-\Delta G_0/RT)$. In this case, $K_{eq} = \exp(-7,500 \text{ J/mol})/(8.3145 \text{ J/mol-K})(298 \text{ K}) = 0.0485$.

(b) Again, $\Delta G = \Delta G_0 + RT \ln([B]/[A])$, giving:

$$\Delta G = 7,500 + (8.3145 \cdot 310) \ln[0.0001/0.0005] = 3.35 \text{ kJ/mol}$$

Being $\Delta G > 0$, the reaction is non-spontaneous.

(c) The reaction can proceed if B is the reactant for a highly exergonic, subsequently (downstream) coupled reaction, whose total $\Delta G < 0$.

4.2 Switching from ATP to ADP

It is $\Delta G = -RT \ln K$, hence the rate constant is $K = \exp(-\Delta G/RT)$. At equilibrium:

$$K_{eq} = \frac{[ATP][AMP]}{[ADP]^2} = e^{-\Delta G/RT}$$

From the tables in Ref. [2] in Chap. 4, the ΔG for the ATP+AMP/ADP interconversion is +3,700 kJ/mol, giving $K_{eq} = \exp -(-3,700 \text{ J/mol}/8.3145 \text{ J/mol-K} \cdot 298 \text{ K}) = 4.45$. Now solve for [AMP], and substitute the concentrations:

$$[AMP] = \frac{[5 \times 10^{-4}M]^2 \cdot K_{eq}}{[5 \times 10^{-3}M]} = 2.3 \times 10^{-4}M = 0.22 \text{ mM}$$

4.3 Energy harvesting

For all cases, the complete oxidation reaction is written [substrate] + $xO_2 \rightarrow yCO_2 + zH_2O$. The absolute yield in ATP is recovered by the value of about 5 ATP per mole of O_2 consumed in the respiration (oxidation phase). Then:

(a) pyruvate, $x = y = z = 3$, yield 15 ATP; (b) lactate, $x = 2.75$, $y = 3, z = 2.5$, yield 13.75 ATP; (c) glucose, $x = y = z = 6$, yield 30 ATP; (d) fructose diphosphate is just like glucose plus two HPO_3^- groups, therefore $x = y = z = 6$, yield 30 ATP.

4.4 Human blood

The equilibrium constant is in each case given by the following ratio of concentrations:

$$K_n = \frac{[H^+][I_{n-1}]}{[I_n]}$$

with I_n the potassium ion with protonation state n, for $[H_2O]=1$ M. Therefore we firstly need the concentration of $[H^+] = 10^{-7.4} = 3.98 \times 10^{-8}$ M. From this and the values of K_1, K_2, K_3 we calculate the following relative ratios:

$$\frac{[H_2PO_4^-]}{[H_3PO_4]} = \frac{7.5 \times 10^{-3}}{3.98 \times 10^{-8}} = 1.88 \times 10^5$$

$$\frac{[HPO_4^{2-}]}{[H_2PO_4^-]} = \frac{6.2 \times 10^{-8}}{3.98 \times 10^{-8}} = 1.56$$

$$\frac{[PO_4^{3-}]}{[HPO_4^{2-}]} = \frac{2.2 \times 10^{-13}}{3.98 \times 10^{-8}} = 5.53 \times 10^{-6}$$

It is seen that the most abundant ions are $H_2PO_4^-$ and HPO_4^{2-}, present with about a 2:3 ratio, while the neutral H_3PO_4 and the highly charged PO_4^{3-} have negligible concentrations. Note however that such ratios can rapidly change by small changes of the pH.

4.5 Gym doesn't slim

The net conversion of fats into glucose is forbidden because the way to get the carbon atoms from fats into oxaloacetate (the precursor to glucose) is through the citric acid cycle. However, although two carbon atoms enter the cycle as acetyl CoA, two carbon atoms are also lost as CO_2 before oxaloacetate is formed. Thus, although some carbon atoms from fats may end up as carbon atoms in glucose, we cannot obtain a net synthesis of glucose from fats.

4.6 Pigeon muscles love citrate

(a) $x = 4.5$, $y = 6, z = 4$. The number of moles of O_2 per 3 μM of is $4.5 \cdot 3 = 13.5 \mu M$.

(b) Since leads to the consumption of much more O_2 (85 − 49=36 μM) than simply by the oxidation of itself (13.5 μM), this means that facilitates O_2 consumption, by increasing the overall efficiency of the combined cycles.

4.7 Antibiotics

Because by blocking the ATP-synthase it blocks also the electron respiratory cycle, and leads to suffocation.

4.8 Transmembrane proteins

The protein structure will be arranged in such a way that the hydrophobic segments are contained within the membrane, while the hydrophilic segments are exposed to water, either outside or inside the cell, as shown in the sketch below.

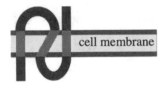

Chapter 5

5.1 Stationary flux

The solution for the stationary concentration profile is $c(x) = c_A - (x/L)\Delta c$, with $\Delta c = c_A - c_B$. The flux is constant within the membrane thickness, and equal to: $j = -D(dc/dx) = D(\Delta c/L)$, positive parallel to the direction of the concentration gradient $\Delta c > 0$.

5.2 Artificial blood

The osmotic pressure from the proteins inside is sufficient to break the membrane. By adding 1 mM of salt, this dissociates and the two ions create enough osmotic pressure from the outside, to equilibrate that of hemoglobin from the inside. Probably, doubling the NaCl concentration will make an excess of pressure from the outside, so 1 mM should be just ok.

5.3 A cell spewing glucose

The environment outside the cell is practically infinite, therefore the concentration c_{out} is constant. The time-dependent concentration inside is $c_{in} = N(t)/V$. The flux is $j(t) = P_M \Delta c(t)$. Note that the flux can be negative, if the outside concentration is higher than inside. For the cell with surface $A = 4\pi R^2$, $N(t)$ varies as $dN/dt = -Aj(t) = AP_M \Delta c(t)$. Dividing by V, it is:

$$\frac{d(\Delta c)}{dt} = \left(\frac{AP_M}{V}\right)\Delta c$$

After integration, $\Delta c(t) = \Delta c(0)\exp(-t/\tau)$, with the time constant $\tau = V/(AP_M)$.

5.4 A breathing bacterium

The surface concentration is zero at steady state, $c(R = R_0) = 0$, whereas the concentration at large distance from the bacterium is constant, $c(R = \infty) = c_0$. Imagine a family of concentric spherical surfaces with radius $R_1, R_2 \ldots R_n$, around the sphere R_0 of the bacterium. The oxygen at each surface is constant, and so is the flux $j = J/A$. Therefore:

$$j = \frac{J}{4\pi R^2}$$

On the other hand, Fick's law also gives $j = -D(dc/dR)$. By integration one finds:

$$c(R) = A + \frac{J}{4\pi R D}$$

The two constants A and J are found by imposing the above boundary conditions on c: $A = c_0$, $J = -4\pi R_0 D c_0$. Finally:

$$c(R) = c_0 \left(1 - \frac{R_0}{R}\right), \quad \text{for} \quad R < R_0$$

5.5 Haute cuisine

The glucose concentration outside the cells is much higher than inside, therefore an osmotic gradient is established. Since there is no water on the outside to try to compensate the glucose/water concentration, the water comes from inside the cells pushed by the osmotic pressure.

5.6 Separation by sedimentation

The viscous resistance for a supposedly spherical particle of size D is the Stokes' law, $f =_v 6\pi\eta D$. The centrifugal force for a spinner at angular speed ω and distance r from the centre is $f_c = m\omega^2 r$. By equating the two forces at steady state, the equation governing the process is:

$$v = \frac{dr}{dt} = s\omega^2 r$$

with the sedimentation coefficient $s = v/a = m/6\pi\eta D$. By simply separating the variables we get: $\frac{dr}{r} = s\omega^2 dt$. This can be integrated, to give: $\ln r = s\omega^2 t + c$, or $r = r_0 \exp(s\omega^2 t)$. Hence the time A gets to $r = 10$ is:

$r_A = 10 = 5\exp((30 \times 10^{-13})(1{,}000^2)t)$, $t = \ln 2/(30 \times 10^{-7}) = 2.31 \times 10^5$ s = 64 h 10 min.

At this time, the radial position of B is:

$r_B = 5\exp((10 \times 10^{-13})(1{,}000^2)(2.31 \times 10^5)) = 6.3$ cm.

5.7 Membrane permeability

Assuming that the outgoing rate is roughly given by $P_M A c_{in}$, and the number of molecules in the cell is $V c_{in}$, we can take the ratio between the two as $\tau = V c_{in}/P_M A c_{in} = (R/3)(1/P_M)$. It is found $\tau = 5.5$ min for glycerol, and $\tau = 926$ h for glucose.

5.8 Blood flow in the arteries

The flux must be equal before and after the split, i.e. $j_1 = p_1/Z_1$ and $j_1 = 2j_2 = 2(p_2/Z_2)$. Since the Z are inversely proportional to the power 4 of the respective R, all the rest being equal this gives: $p_1 R_1^4 = 2p_2 R_2^4$. Then, because $R_2 = R_1/2$, it is $p_1 R^4 = 2p_2(R/2)^4$, that is: $p_2 = 8p_1$.

5.9 The osmose on Mars

Since the osmotic pressure pushes the fluid against the hydrostatic pressure, $p = \rho g h$, the fact that Mars has a much lower gravity than the Earth ($g = 3.711$ vs. 9.807 kg m/s^2, or N) makes the fluid to go higher on the "red planet", by a factor of $9.807/3.711 = 2.64$.

Chapter 6

6.1 Swimming bacterium

The distance of arrest x_0 is given by Eq. (6.47). We take for the bacterium the same density of water, $\rho = 1$ g/cm^3, therefore $V = \frac{4}{3}\pi(15 \times 10^{-4})^3 = 1.41 \times 10^{-8}$ cm^3, and $m = 1.41 \times 10^{-8}$ g.

$$x_0 = \frac{mv_0}{6\pi R\eta} = \frac{(1.41 \times 10^{-8})(0.3)}{6\pi(15 \times 10^{-4})(10^{-2})} = 0.15\,\mu\text{m}$$

6.2 Actin polymerisation velocity

The polymerisation velocity is the product of the monomer length $\delta \times$ the number of monomers added/removed from the chain per unit time, $v = \delta(dn/dt)$. With the values given in the problem, $\frac{dn}{dt} = (7.50 \cdot 0.1) - 1.25 = -0.5$ s^{-1} in the first case, and $\frac{dn}{dt} = (7.50 \cdot 0.5) - 1.25 = +2.5$ s^{-1} in the second case. The corresponding polymerisation velocity is -2.5 nm/s, or $+12.5$ nm/s, respectively. A negative polymerisation velocity means that the F-actin filament is shortening its length by losing monomers.

6.3 Chain polymerisation

(a) The amount of free radicals available for continuing the polymerisation is:
$[M'] = \left(\frac{(0.5)(5\times10^{-5})}{2\times10^7}[8 \times 10^{-3}]\right)^{1/2} = 10^{-7}$ M, at steady state.

(b) The kinetic length of the polymer is the ratio between elongation and nucleation. We calculate firstly $k = k_p/2\sqrt{uk_ik_t}$:

$$k = \frac{2640}{2(0.5 \cdot (5 \times 10^{-5}) \cdot (2 \times 10^7))^{1/2}} = 59$$

Then, $v = 59 \cdot 2 \cdot (8 \times 10^{-3})^{1/2} = 10.55$ mol/s.

(c) $v_p = k_p[M][M'] = 5.28 \times 10^{-4}$ mol/s.

6.4 Microtubules association/dissociation constants

Making the assumption that the association is favoured over dissociation over some concentration range, for a given concentration [T] of tubulin in the solution, microtubules of average fixed length n will be in equilibrium with free tubulin $\alpha - \beta$

dimers, according to the equation: $T_{n-1} + T \leftrightarrow T_n$, with the association constant K given by:

$$K = \frac{[T_n]}{[T_{n-1}][T]}$$

The crucial observation is that at equilibrium, it is $[T_n] = [T_{n-1}]$. Therefore, the constant is inversely proportional to the concentration of free tubulin dimers, $K = [T]^{-1}$. The centrifuge can be used to separate the long microtubules from the fraction of free tubulin, at each given dilution. We will choose a rotation speed for which the microtubules will sediment, whereas the dimers will remain in suspension. From Problem 5.6 we know that the sedimentation coefficient is proportional to the mass of the microtubule and free dimer, respectively, therefore the two differ by a factor of n, and so does their sedimentation time. Subsequently, the floating suspension (so-called *supernatant*) is manually separated from the precipitate. Aromatic residues, like tyrosine and tryptophan, absorb UV light at 280 nm, so you can measure the absorbance of the suspension in the UV-spectrometer, and deduce the tubulin concentration [T].

One important requirement to use UV absorption is that your protein actually absorbs decently. In other words, there must be enough aromatic residues in its composition. A particular tubulin sequence can be obtained from the website www.rcsb.org/pdb, see for example the accession codes 5JCO or 5IJ0 (there will be many other examples). You should take the one-letter sequence of the protein, e.g. starting with MRECISI-HVGQAGV... (it could be several hundreds of amino acids long), and plug this in the webpage www.web.expasy.org/protparam/. This will give you a vast amount of informations about tubulin, among which its extinction coefficient that turns out to be slightly above 1. Therefore, the UV-absorption can be quite correctly measure the tubulin concentration.

6.5 DNA replication

(a) About 40 min (each half of 4.6×10^6 DNA nucleotides, divided by 1,000 nucleotides per second). More than one replication fork is needed (actually there are two forks moving in opposite directions, called a "replication bubble"). In eukaryotes this is even more critical, since the genome is much longer (the 150 millions of base pairs would require a month to replicate with a single fork); actually, there are many replication forks working in parallel, to complete the copy in shorter times.

(b) 96.2 revolutions per second (1,000 nucleotides per second divided by 10.4 nucleotides per turn for B-DNA gives 96.2 rps).

(c) 0.34 μm/s (1,000 nucleotides per second corresponds to 3,400 Å/s because the axial distance between nucleotides in B-DNA is 3.4 Å).

(a) About 40 min (each half of 4.6×10^6 DNA nucleotides, divided by 1,000 nucleotides per second). More than one replication fork is needed (actually there are two forks moving in opposite directions, called a "replication bubble"). In eukaryotes this is even more critical, since the genome is much longer (the 150 millions of base pairs would require a month to replicate with a single fork); actually, there are many replication forks working in parallel, to complete the copy in shorter times.

(b) 96.2 revolutions per second (1,000 nucleotides per second divided by 10.4 nucleotides per turn for B-DNA gives 96.2 rps).

(c) 0.34 μm/s (1,000 nucleotides per second corresponds to 3,400 Å/s because the axial distance between nucleotides in B-DNA is 3.4 Å).

6.6 Active and passive diffusion

(a) The profile of flux is constant in case of constant boundary concentrations, therefore the concentration profile, $\Delta c = -j \Delta x$, is linear. The Brownian motion is at the origin of the diffusion process, molecules statistically cross the membrane in both directions, however the chemical potential gradient makes the amount of molecules going in the direction of lower concentration to be on average larger. Hence a net flux is observed.

(b) It is the filtration coefficient, K_M, having dimensions of [velocity]/[pressure]. This established how efficiently a flow goes through the membrane for a given pressure applied on one side. The pressure in this case is much larger than the brownian force, therefore while the molecules continue their thermal agitation, they are steadily pushed in one direction.

6.7 Michaelis-Menten kinetics

A Lineweaver-Burk plot has to be constructured (see figure below). By putting the data on a 1/velocity-1/concentration plot, the intercept at zero gives $\frac{1}{v_{max}} = 4000\,\mathrm{M}^{-1}$, or $v_{max} = 2.5 \times 10^{-4}$ M/s. The slope is 40 s, from which $K_M = (40)(2.5 \times 10^{-4}) = 1$ mM. The formation rate of the product is:

$$k_P = \frac{v_{max}}{[E_0]} = \frac{2.5 \times 10^{-4}}{2.3 \times 10^{-9}} = 1.1 \times 10^5 \ \mathrm{s}^{-1}$$

Therefore, the catalytic efficiency is:

$$e_P = \frac{k_P}{K_M} = \frac{1.1 \times 10^5}{0.001} = 1.1 \times 10^8 \ \mathrm{M}^{-1}\mathrm{s}^{-1}$$

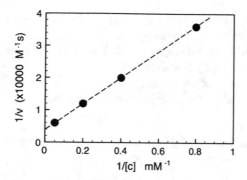

Chapter 7

7.1 Absolute and relative refractory period
Iin the first case, the response is governed by the sum of the absolute+relative refractory period, i.e. $\tau = 1 + 4 = 5$ ms. The maximum frequency is then $\nu = 1/\tau \simeq 200$ Hz. In the second case, the neuron is limited only by the absolute refractory period, so $\nu \simeq 1,000$ Hz.

7.2 The GHK equation
(a) With $R = 8.316$ J/(K mol), $T = 296.15$ K, and $F = 96,487.302$ C/mol, it is $RT/F = 0.0258$ J/C = 25.5 mV. Then:

$$V_{rest} = 25.5 \ln \left[\frac{1 \cdot 4\,\text{mM} + 0.002 \cdot 142\,\text{mM}}{1 \cdot 140\,\text{mM} + 0.002 \cdot 14\,\text{mM}} \right] = -89\,\text{mM}$$

The individual Nernst potentials for K and Na may be found from:

$$V_K = \frac{RT}{ZF} \ln \left(\frac{c_{out}}{c_{in}} \right) = 25.5 \cdot (-3.555) = -90.7$$

$$V_{Na} = \frac{RT}{ZF} \ln \left(\frac{c_{out}}{c_{in}} \right) = 25.5 \cdot (2.317) = +59.1$$

V_{rest} is much closer to V_K than to V_{Na} at rest.
(b) Each ion will flow in the direction that sends V_m closer to its own Nernst potential, therefore more K should enter the cell, while the same amount of Na ions should exit (equal charges).

7.3 The cable equation
One need to compute the second derivative of each function, and verify by substitution that each of them represents a solution. Remember that $X = x/\lambda$ and $L = l/\lambda$.

(a) $V(x) = A \exp(-X) + B \exp(X)$, $V'(x) = \lambda[-A \exp(-X) + B \exp(X)]$, and
     $V''(x) = \lambda^2[A \exp(-X) + B \exp(X)]$.
(b) $V(x) = A \cosh(X) + B \sinh(X)$, $V'(x) = \lambda[A \sinh(X) + B \cosh(X)]$, and
     $V''(x) = \lambda^2[A \cosh(X) + B \sinh(X)]$.
(c) $V(x) = A \cosh(L - X) + B \sinh(L - X)$, $V'(x) = \lambda[A \sinh(L - X) + B$
     $\cosh(L - X)]$, and $V''(x) = \lambda^2[A \cosh(L - X) + B \sinh(L - X)]$.

7.4 Axon resistance

In the foregoing, the axon radius is $a = d/2$, and the resistances are $R_m = 2\pi a r_m$,
$R_i = \pi a^2 r_i$.

(a) For the infinite axon: $R_\infty = \frac{2}{\pi}(R_m R_i)^{1/2}d^{-3/2} = \frac{2}{\pi}(707)(40 \times 10^{-4})^{-3/2} = 1.78$
     $M\Omega$.
     For the semi-infinite axon: $R_{\infty/2} = 2(1.78) = 3.56$ $M\Omega$.
     For the finite-length axon, since it is between two neurons we can impose the
     condition of "sealed-end": $R_L = (\lambda r_i) \coth(L)$. With:

$$\lambda = \left(\frac{r_m}{r_i}\right)^{1/2} = \left(\frac{R_m/(2\pi a)}{R_i/(\pi a^2)}\right)^{1/2} = \left(\frac{(40 \times 10^{-4})5,000}{2(100)}\right)^{1/2} = 0.316 \text{ cm}$$

$$r_i = \frac{R_i}{\pi a^2} = \frac{100}{\pi(40 \times 10^{-4})^2} = 2 \times 10^6 \ \Omega/\text{cm}$$

we obtain: $R_L = (0.316)(2 \times 10^6) \coth(40 \times 10^{-4}) = 157$ $M\Omega$.
(b) For both the infinite and semi-infinite axon: $V(x = 1.5) = 200 \exp(-1.5/$
     $0.316) = 1.74$ mV.
     For the finite axon: $V(x = 1.5) = 200(\cosh(3 - 1.5)/\cosh 3) = 46.7$ mV.

7.5 Triple junction

For each of the three segments we have three similar equations:

$$V_1 = A_1 \exp(-X) + B_1 \exp(X)$$
$$V_a = A_a \exp(-X) + B_a \exp(X)$$
$$V_b = A_b \exp(-X) + B_b \exp(X)$$

In the most general situation we need 6 boundary conditions to determine the six
unknown constants. For a current I_0 injected at $x = 0$, and for open ends at L_a, L_b:

$$\frac{dV_1(0)}{dx} = -\lambda r_i I_0$$
$$V_a(L_a) = V_b(L_b) = 0$$

Three more constraints come from imposing the conditions at the node L. The potential must be continuous there, $V_1(L) = V_a(L) = V_b(L)$. Moreover, according to the
first Kirchhoff's law, the current must be conserved at L:

$$\frac{1}{\lambda r_i}\frac{dV_1(L)}{dx} = \frac{1}{\lambda r_i}\frac{dV_a(L)}{dx} + \frac{1}{\lambda r_i}\frac{dV_b(L)}{dx}$$

with $\lambda r_i = \sqrt{(r_m r_i)^{-1}} = d^{3/2}\sqrt{\pi^2/(4R_m R_i)}$. If the branches have the same physical properties (which is not unrealistic) but have different diameters, the current conservation condition is:

$$d_1^{3/2}\frac{dV_1(L)}{dx} = d_a^{3/2}\frac{dV_a(L)}{dx} + d_b^{3/2}\frac{dV_b(L)}{dx}$$

By taking the derivatives and imposing the conditions, it is 'easily' obtained:

$$B_1 = A_1 - \lambda r_i I_0$$

$$A_a = -B_a e^{2L_a}$$

$$A_b = -B_b e^{2L_b}$$

$$B_a = \frac{A_1 + B_1 e^{2L}}{e^{2L} - e^{2L_a}}$$

$$B_b = \frac{A_1 + B_1 e^{2L}}{e^{2L} - e^{2L_b}}$$

The last condition to obtain A_1 from the current condition is slightly more cumbersome:

$$A_1 = \lambda r_i I_0 \left[1 + \frac{(w_1 + w_a + w_b - 1)}{(w_a + w_b - 1)e^{2L}}\right]^{-1}$$

$$w_1 = \frac{2d_1}{d_1 - d_a - d_b}$$

$$w_a = \frac{d_a w_1}{d_1}\frac{1}{e^{2(L-L_a)} - 1}$$

$$w_b = \frac{d_b w_1}{d_1}\frac{1}{e^{2(L-L_b)} - 1}$$

7.6 Electric frog

By taking the x axis parallel to the direction of v, and the y axis parallel to D, the distance to the detector of each current pulse $I_{+/-}$, varies over time as:

$$R_{+/-} = \sqrt{(vt \mp d)^2 + D^2}$$

We must use Eq. (7.41) to find the potential induced at the detector location by each moving current source, and sum the two. Note that the detector is grounded, so its potential is zero. Therefore the voltage detected is:

$$V = \frac{I_0}{4\pi\sigma}\left(\frac{1}{R_+} - \frac{1}{R_-}\right) = 4.2 \times 10^{-9}\left(\frac{1}{\sqrt{(1{,}000x - 0.5)^2 + 9}} - \frac{1}{\sqrt{(1{,}000x + 0.5)^2 + 9}}\right)$$

with the following plot:

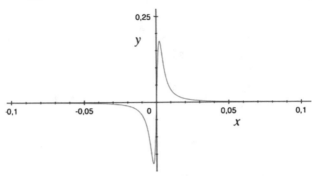

7.7 A mouse's ear

(a) The relationship between frequency and wavelength is $\nu = c/\lambda$, with c the speed of sound (340 m/s at sea level), so a larger size implies a lower frequency cut-off. If we look at the maximum frequency for each size in the plot, also the figure shows this inverse proportionality. Since the human frequency range is 20 Hz to 20 kHz, scaling of all physical dimensions by a factor of 10 would mean that the mouse's range of hearing should be about 200 Hz to 200 kHz.

(b) The figure indicates that the sensitivity increases by about 10 dB, i.e. by a factor of 10, at every doubling of the size of the "microphone". A factor of 10 in size corresponds to $\log_2 10 = 3.32$ doublings, therefore the sensitivity of the mouse's ear should be $10^{3.32} \simeq 2{,}000$ times less sensitive than the human's.

(c) Since there are no major differences in the response times and thresholds of mouse neurons compared to a man's, the mouse ear transduction mechanism must be 2,000 times more sensitive than the human mechanism.

7.8 A bird's ear

(a) The scheme could be the one depicted in (a) below, considering a dendrite receiving input from two pairs synapses coming from each ear.

(b) The summation of signals from different ears is spatial, therefore the input is linearly summed and compare to the threshold. The summation of signals from the same ear is temporal, therefore is summed non-linearly on the capacitance of the membrane, and remains sub-threshold. A possible algorithm is schematised in (b).

(c) In the first case, an action potential is fired. In the second case, the signal produces only a sub-threshold potential, plotted in (c).

(a)   **(b)**   **(c)**

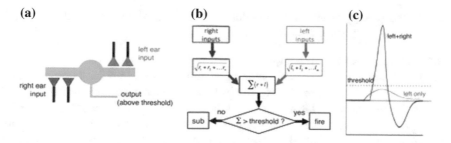

Chapter 8

8.1 Hollow versus filled

The cross-section moment of inertia for a hollow cylinder is $\mathscr{I} = \pi(D^4 - d^4)/32 = 1{,}054$ nm^4. The same area for the cross section of a filled cylinder corresponds to a diameter $D' = \sqrt{D^2 - d^2} = 6.63$ nm, therefore its moment is $\mathscr{I}' = \pi(D')^4/32 = 190$ nm^4. The bending rigidity is, respectively, $\kappa_b = (1.5 \times 10^8)(1{,}054 \times 10^{-36}) = 1.58 \times 10^{-25}$ J-m and $\kappa_b' = (1.5 \times 10^8)(190 \times 10^{-36}) = 2.85 \times 10^{-26}$ J-m. These translate into respective persistence lengths of $\lambda_p = (1.58 \times 10^{-25})/(4.14 \times 10^{-21}) = 38$ μm, and $\lambda_p' = (2.85 \times 10^{-26})/(4.14 \times 10^{-21}) = 6.9$ μm. Such a simple calculation shows why a hollow structure is more rigid than a filled one, for the same amount of material/unit length.

8.2 Bacterial DNA

(a) With $d = 0.34$ nm, it is $L = (0.34)(3.45 \times 10^6) = 1.17$ mm.
(b) The persistence length is $\lambda_p = \kappa_b/k_BT = E\mathscr{I}/k_BT$. The cross-section moment of inertia is calculated by assuming the DNA as a filled straight cylinder, $\mathscr{I} = \pi D^4/32 = 1.57$ nm^4. Therefore at $T = 300$ K, $\lambda_p = (3.5 \times 10^8)(1.57 \times 10^{-36})/(4.14 \times 10^{-21}) = 133$ nm. This is a factor of 10^{-4} shorter than the contour length, so the DNA is very flexible on this length scale.
(c) The DNA volume is $v = L\pi(D/2)^2 = (1.33 \times 10^{-3})\pi(10^{-9})^2 = 4.17 \times 10^{-21}$ m^3, which can be packed inside a sphere of radius 160 nm.
(d) For a freely fluctuating polymer, it is $R_{ee} = N^{1/2}b$, where the Kuhn's length $b \simeq d/2 = 0.17$ nm., therefore $R_{ee} = 316$ nm. The gyration radius for a flexible polymer is by definition $R_g = R_{ee}\sqrt{6} = 129$ nm. All these measures can be compared to the size of E. coli, which can be approximated as a cylinder of diameter 0.6 μm and length 2 μm.

8.3 Exocitosis

Setting to zero the energy E_1 of the flat membrane, and at $E_2 = 8\pi K_b$ that of the fully detached liposome (independent on the radius R), the energy of the intermediate configuration (cylinder + half-sphere) is $E_3 = \pi K_b(4\pi + L/R)$. The energy barrier is passed at the configuration for which $E_3 = E_2$: beyond this point it is more convenient (energy-wise) for the pseudopod to break up. This corresponds to a length of the extruded cylinder $L = R(8 - 4\pi)$.

8.4 Membranes with an edge

(a) At the interface between a region of A and B phospholipids, there are molecules of B that expose 4 extra CH_2 groups to water, each covering about $h = 0.125$ nm in height and $r = 0.2$ nm in radius (see Appendix D for a similar calculation). By assuming that half of each molecule is in contact with the B island, the other half of the lateral surface of the molecule (schematised as a half of a truncated cylinder, see green surface in the figure) is in contact with water. Therefore, each molecule along the perimeter of the interface adds an interfacial energy equal to $\Delta E = \Sigma \Delta A$, with $\Delta A = \frac{1}{2}(2\pi r(4h)) = 0.314$ nm^2, that is $\Delta E = (0.03)(0.314 \times 10^{-18}) = 9.4 \times 10^{-21}$ J. By considering that one molecule adds a length of perimeter equal to πr, the extra interface energy per unit length (of the A/B interface) is $\Delta E_l = 0.15 \times 10^{-12}$ J/m.

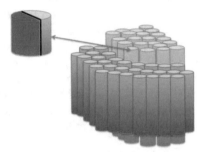

Each blue phospholipid is longer by 4 CH_2, on each end of the red ones. However, only the green half of each extra length is exposed to water, while the other remains in contact with the molecules on the back. Each green area has a perimeter equal to πr, half of the total perimeter of the cylinder of radius r.

(b) Each domain has a perimeter of about 35 nm, therefore the total energy of the A/B interfaces is $\Delta E_l = (10)(35 \times 10^{-9})(0.15 \times 10^{-12}) = 5.25 \times 10^{-21}$ J. If the whole of the B lipids were grouped into one (circular) island of 1000 nm^2, the perimeter would be 112 nm, hence $\Delta E_l = (112 \times 10^{-9})(0.15 \times 10^{-12}) = 1.68 \times 10^{-21}$ J. The coalescence of all the islands into one liberates the equivalent of 3.57×10^{-21} J.

(c) By comparing the last value to the thermal fluctuation energy at 300 K, $k_B T = 4.14 \times 10^{-21}$ J, such energies are well within the range of thermal fluctuations. We could expect to see a coexistence between B domains, dynamically merging and breaking apart.

8.5 Membranes with a dimple

(a) The zero-mode (i.e., $q_x = q_y = 0$) deformation energy is of the order of $\Delta E = (k_B T)^2 / K_b \times (L^2/h^2)$, it does not depend on the extent of fluctuating membrane but only on the L/h ratio.

(b) Given the value of $K_b = 15\,k_B T$, it is $\Delta E \sim k_B T$ for $h \sim L/\sqrt{15}$.

8.6 Pulling chromosomes

(a) To set an upper limit, let us take that one single microtubule pulls a chromosome. Given the long duration of the process, compared to the molecular time scales, it can be assumed that it occurs at steady state. The viscous force opposing the pulling is $F = 6\pi\eta vD$, with: η the viscosity of the nucleoplasm, which can be taken to be a factor of $\sim$1,000 higher than pure water; v the steady-state velocity, equal to about $v \sim (15\,\mu m)/(600\,s) = 2.5 \times 10^{-8}$ m/s; and D a geometrical factor, approximately given by $D = 3.45a$ for drag parallel to the major axis, or $D = 5.125a$ for drag perpendicular to the major axis (such geometric formulas for dragging ellipsoids can be easily found on the internet). Let us consider the worst situation, the maximum force should be about $F = 6\pi(1\,\text{kg/m-s})(2.5 \times 10^{-8}\,\text{m/s})(5.125 \times 10^{-6}\,\text{m}) \simeq 2.5$ pN.

(b) The total work is $W = (2.5\,\text{pN})(15\,\mu m) = 3.75 \times 10^{-17}$ J. Considering 30.5 kJ/mole of ATP, each ATP molecule gives 5×10^{-20} J, then about 750 ATP per chromosome are consumed, at a rate of about 1 ATP/s.

8.7 Pushing cells with a laser

To move a cell of size R, the laser trap must overcome the Stokes' drag force, $f_D = 6\pi\eta Rv$. For $\eta = 10^{-3}$ Pa-s the viscosity of water, one needs to estimate the average velocity of the Brownian motion. At the temperature of $T = 300$ K this can be obtained from the $\frac{1}{2}mv^2 = \frac{3}{2}k_BT$, as $v = \sqrt{3k_BT/m}$. The mass of the cell, by taking a density similar to water, is $m = \rho(\frac{4}{3}\pi R^3)$. Then, the velocity is:

$$v = \sqrt{\frac{3k_BT}{\rho(\frac{4}{3}\pi R^3)}}$$

which gives respectively 1.72×10^{-3} and 5.44×10^{-5} m/s for the prokaryote versus eukaryote cell. The drag force is therefore equal to $f_D = (6\pi)(10^{-3}\,\text{Pa s})(10^{-6}\,\text{m})$ $(1.72 \times 10^{-3}\,\text{m/s}) = 32.4$ pN for the bacterium, and $f_D = 10.25$ pN for the larger eukaryote cell. Compared to the equation for the laser power, to move either cell it is required respectively $P \geq (f_Dc)/(Qn) = (32.4 \times 10^{-12})(2.9979 \times 10^8)/(0.01 \cdot 1.3) = 750$ mW for the bacterium, and $P = 235$ mW for the eukaryote cell.

This calculation shows that the smaller cell is subject to a larger Brownian force, therefore more laser power is required to move it, despite the much smaller mass. Actually (since we neglected the gravity), the power varies with cell size as $R^{-1/2}$.

Chapter 9

9.1 Average elastic modulus

The average modulus is defined in terms of the volume fractions of the components, $E = f_1E_1 + f_2E_2$ ("rule of mixtures"). Water contributes zero. The average volume fractions for cortical bone are 25 % water, 40 % mineral and 35 % organic. Therefore, for the wet bone we have $E = 0.4 \cdot 54 + 0.35 \cdot 1.25 = 22$ GPa; for the dry bone $E = (0.4/0.75) \cdot 54 + (0.35/0.75) \cdot 1.25 = 29.4$ GPa.

9.2 Skin stretching

(a) 0.32 MPa in the parallel direction, and 2.16 MPa in the perpendicular.
(b) 26 and 37 % in the perpendicular and parallel direction, respectively.
(c) By extrapolating to zero the second part of each curve, it is $E_{perp} = 35$ MPa, $E_{para} = 29$ MPa.
(d) The integral of $\sigma d\varepsilon$ can be approximated by a triangle under the curve (the first small portion before the triangle contributes little). For the perpendicular direction we find $\tau_0 \simeq 37$ MPa, and for the parallel $\tau_0 \simeq 41$ MPa.

9.3 Artery relaxation

(a) The Maxwell model reads: $\eta E \dot\varepsilon = \eta \dot\sigma + E\sigma$. For the case of a constant strain applied, $\dot\varepsilon = 0$, then the stress equation is:

$$\frac{d\sigma}{dt} = -(E/\eta)\sigma$$

This is integrated to obtain: $\sigma = \sigma_0 \exp(-t/\tau)$, with the relaxation time is $\tau = \eta/E$. From the stress at 1 h, this is obtained as: $\tau = -3600/\ln(0.75/1) = 1.25 \times 10^4$ s^{-1}.

(b) By substituting the time $t = 3$ h in the same equation, $\sigma = 1\exp(-10800/12500) = 0.42$ MPa.

(c) Apply the Kelvin-Voigt model, $\sigma = E\varepsilon + \eta\dot\varepsilon$, after the stress is released $\sigma = 0$. Therefore the strain relaxation equation is:

$$\frac{d\varepsilon}{dt} = -(E/\eta)\sigma$$

As above, integration gives: $\varepsilon = \varepsilon_0 \exp(-t/\tau)$, with the relaxation time $\tau = \eta/E$. At the release, the strain is $\varepsilon_0 = (2.3 - 2)/2 = 0.15$. After 1 h 25', or 5100 s, the strain reduced to $\varepsilon = (2.2 - 2)/2 = 0.1$. Therefore, $\tau = -5100/\ln(0.1/0.15) = 1.25 \times 10^4$ s^{-1}, as expected since the relaxation times of the stress and strain are – in the *linear viscoelastic* case – the same.

9.4 Stretch the leg

The walking man of 70 kg exerts a force on the leg equal to $F = 70 \cdot 9.807 = 686.5$ N. The effective spring constant is $k = ES/L$ for a cross section area S over a length L. For the tendon loaded in compression we have $k_t = (1.5 \times 10^9)(\pi \cdot 0.0075^2)/0.08 = 3.3 \times 10^6$ N/m, and a corresponding compression $\Delta x = F/k_t = 686.5/(5.3 \times 10^6) = 0.2$ mm. For the bone it is $k_b = (20 \times 10^9)(\pi \cdot 0.02^2)/0.35 = 71.8 \times 10^6$ N/m, hence $\Delta x = F/k_b = 686.5/(71.8 \times 10^6) = 0.01$ mm. Due to the larger Young's modulus, even with a larger size the bone deforms much less than the tendon.

9.5 Jumping cat

The energy of the cat at landing is $Mgh = (4.5$ kg$)(9.807)(3$ m$)=132.4$ J, distributed on the four legs, therefore each leg receives 33.1 J, and each of the four identical muscles 8.275 J of mechanical work as input. Since the muscle volume is $V = \pi(2)^2(12) = 151$ cm³, the mechanical work corresponds to an equivalent pressure of 54.9 kPa. For a bending angle θ, the diagram below shows that the height of point a is $y_a = (L + a)\cos\theta$, while its displacement is $x_a = (L - a)\sin\theta$.

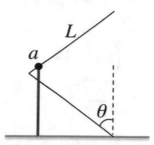

Therefore, the length of the muscle (line joining the origin O with a) is: $H^2 = (L + a)^2 \cos^2\theta + (L - a)^2 \sin^2\theta = L^2 + a^2 + 2La\cos 2\theta$. The length L is that of the bone, ideally equal to the length of the muscle minus the distance a. This expression has a maximum for $\theta = 45°$. By substituting the numerical values of the problem, we have $L = 10$ cm, $a = 2$ cm, and $H = 10.2$ cm at the maximum compression, corresponding to a strain $\varepsilon = 1.8/12 = 0.15$. The elastic energy at the maximum compression is then:

$$E_{el} = \frac{1}{2}E\varepsilon^2 = \frac{1}{2}(20 \times 10^3 \text{ Pa})(0.15)^2 = 225 \text{ J/m}^3$$

The mechanical input pressure is more than 200 times larger than this value, meaning that the muscle alone would not be able to absorb a shock of this kind. By looking at the results of Problem 9.4, it should be clear that most of the elastic energy in a shock is absorbed by the tendons (and by the bones, if too large), while the muscles take care of the slow movements of the body against gravity.

9.6 Muscles and temperature

(a) It is the Young's modulus E, and Poisson's ratio ν.
(b) By expressing the deformed volume as $V' = L(1 + \varepsilon)\pi R^2(1 - \nu\varepsilon)^2 = V_0(1 + \varepsilon)(1 - \nu\varepsilon)^2$, it is:

$$\frac{\Delta V}{V} = 1 - (1 + \varepsilon)(1 - \nu\varepsilon)^2 \simeq \varepsilon(2\nu - 1)$$

to first-order in ε. The Poisson's ratio is $\nu = 0.125$, independent on the temperature (see answer (c)). Therefore $\Delta V/V = -0.75\varepsilon$, with $\varepsilon(T) = 0.02 + 0.004(T - 10)$ (temperature T in degrees °C).
(c) The Poisson's ratio is at all temperatures $0.25/2 = 0.5/4 = 0.75/6 = 0.125$. For a fixed load $W = Y\varepsilon$, the Young's modulus decreases with temperature as $E(T) = 5E_0/(T - 5)$, with E_0 the modulus at $T = 10$ °C; then at $T = 23.5$ °C it is $E = $

$0.27 E_0$; the strain is 7.4 %, as it could be predicted also by linear extrapolation of the $\varepsilon(T)$ curve.

9.7 Implant materials

Let's rewrite the equation in terms of the ratio $\varepsilon = E_f/E_m > 1$ (a reinforcement with $E_f < E_m$ makes little sense):

$$E* = f\varepsilon \left(1 - \frac{\tanh n_s}{n_s}\right) + (1-f)$$

$$n_s \simeq \sqrt{\frac{2}{\varepsilon \ln(1/f)}}$$

and study the condition for which $E* > 1$ (the overall modulus is larger than the pure matrix's E_m). The function is quite complex and can be studied only graphically. By looking at the behaviour of the curve for $\varepsilon = 1, 5, 10$, it can be seen that the minimum fraction at which $E* > 1$, firstly decreases from 1 (blue) to about 0.75 (for $\varepsilon \simeq 5$, red), then increases again and, for large values of ε, tends to saturate to $f \sim 0.8$ (green). With such high values of filling, clearly the properties of the fibres dominate over the matrix.

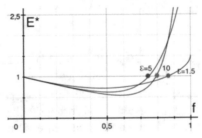

9.8 Bend, break or twist

Note that all the properties but the (d) are structure-dependent, i.e. they are not intrinsic material properties but depend on the shape, size and loading conditions of the structure (in this case, the thin cylinder fixed at its bottom end).

(a) The bending moment of inertia is $\mathcal{I} = \pi D^4/32 = 1 \times 10^{-9}\,\mathrm{m}^{-4}$. For the lateral load of $F = 50$ N, the deflections (in cm) are calculated from the equation $d = \frac{\pi}{0.5}(50\,\mathrm{N})(0.5^3\,\mathrm{m}^3)/(E \cdot (10^{-9}\,\mathrm{m}^{-4}))$. From the table, it is seen that most materials will slightly bend; the cartilage and the dandelion will likely break under a deflection larger than the length of the stem.

(b) The applied tensile stress is $\sigma = F/A$, with $A = \pi(0.005^2)\,\mathrm{m}^2$ the cross section area. From the table, the calculated stress for a pull of 500 N is larger than the tensile strength, for cartilage, vine and dandelion. These will break under tension, the other materials will stretch but resist the force.

(c) The applied compressive stress is also $\sigma = F/A$. Here we must firstly calculate the critical buckling force, from Euler's formula $F_{crit} = \frac{\pi^2}{3}(E \cdot (10^{-9}\,\mathrm{m}^{-4})/$

$(0.5^2 \ m^2)$. The critical stress for bending is $\sigma_{crit} = F_{crit}/A$. From the table, it is seen that in all cases the critical bending stress is inferior to the compressive strength, for this diameter and length of the stick. Under a compressive force of 1000 N, corresponding to a compressive stress of 12.7 MPa, all sticks will bend, and cartilage and dandelion will eventually break.

(d) For a full cylinder, the bend-to-twist ratio κ_b/κ_t can be taken equal to $E/2G$, because $\mathscr{J} \simeq 2.\mathscr{I}$. From the calculated values in the table, it is likely that cartilage, vine, oakwood, bamboo and dandelion could add a considerable twisting to the compression load. Note that this is true for practically all vegetables.

Material	Deflection at $F = 50$ N (cm)	Tensile stress at $F = 500$ N (MPa)	Critical buckling stress (MPa)	$E/2G$
Tendon collagen	0.85	100	2.5	3.6
Tooth enamel	0.03	35	60	0.45
Bone	0.1	200	20	2.5
Cartilage	$>L$	2.5	0.02	6.7
Vine green stem	1.06	6	2	5.6
Oakwood w/grain	0.33	170	6.5	5.9
Bamboo	0.21	190	10	7.7
Dandelion stem	$\gg L$	3	0.008	13.3

Chapter 10

10.1 Sarcomeres

At 1 the maximum contraction is reached, there is no longer force generation. In 3 the overlap between myosin (thick filaments) and actin (thin filaments) is maximum, therefore between 1 and 3 the force generated by the muscle increases linearly, up to the maximum. The break in linearity at the point 2 corresponds to the detachment of the thick filaments from contact with the Z-lines. Between 3 and 4 there is a 'plateau' corresponding to elongation at constant force for a width equal to the H-zone, since in the H-zone there are no overlapping myosin-actins. Extending the sarcomere beyond 4, is associated with a linear decrease in the number of actin-myosin interactions, and active force production should become zero at 5, where the sarcomere length is maximum (sum of twice the thin filaments length + the thick filament length + twice the Z-line thickness), that is between 3.6 (frog) and 4.2 μm (human). The following scheme summarises these steps:

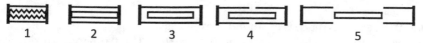

10.2 Summer training

All of the three reasons suggested may have a part in explaining cramps. ATP is needed for several processes involved in muscle relaxation. Loss of liquids implies reduced salts, which can also prevent normal muscle function. Interrupted blood flow prevents efficient delivery of oxygen, needed for ATP production during cellular respiration.

10.3 Weightlifters

The simplest way is to consider the potential energy change for lifting a 5-kg weight by 0.25 m. $E = mgh = (5 \text{ kg})(9.8 \text{ m/s}^2)(0.25 \text{ m}) = 12.25$ J. If 1 ATP is used per myosin step along, the number of ATPs needed for the task is 12.25 J/(30.5 kJ/mol)= 4.02×10^{-4} mol, or 2.42×10^{20} molecules. As it was noted (see p. 228), in the tight-coupling model all myosins work together in steps of about 5 nm; in the loose-coupling model, about 40–60 % of the myosin heads work in "loose" steps of 35–38 nm. This means 2.42×10^{20} myosin steps in the former model, and about half this value for the latter (however, with longer steps).

10.4 Cyclic muscle work

(a) The power is related to the cyclic work as $P = W\nu$ (Eq. (10.12)). Therefore:

$$W = \frac{1}{\nu}(-270 + 10\nu - 0.07\nu^2)$$

(b) Taking the derivative $dW/d\nu = 270/\nu^2 - 0.07$ and setting it to zero, the maximum of W is at $\nu_W = 62$ Hz. By contrast, the maximum of P is for $dP/d\nu = 10 - 0.14\nu = 0$ equal to $\nu_P = 71.4$ Hz.

(c) The relation between stress and cyclic power is $P = \sigma\varepsilon\nu$. For the data of the proble, it is $\varepsilon = 1/15$, and $\sigma = 15(-270 + 10\nu - 0.07\nu^2)/\nu$. The force-frequency law is: $f = \sigma S$, therefore $f = (1.8 \times 10^{-5})(-4{,}050 + 150\nu - 1.05\nu^2)/\nu$ (N).

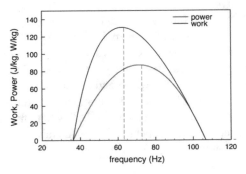

10.5 Dimensional analysis of sunday's oven roast

The physical variables that may matter in this case should be: the thermal conductivity of the meat, κ, its density, ρ, the meat's specific heat, c_P, and its characteristic dimension, like the largest diameter of the piece of meat, D. We want a variable with dimensions of time, the cooking time $\tau \propto \kappa^\alpha \rho^\beta (c_p)^\gamma d^\delta$, i.e., four exponents with four equations.

The dimensions of the variables are: $[\kappa] = [MLT^{-3}\Theta^{-1}]$, $[\rho] = [ML^{-3}]$, $[c_P] = [L^2T^{-2}\Theta^{-1}]$, and $[D] = [L]$.

The four equations for the exponents of M, L, T and Θ are:

$$0 = \alpha + \beta$$
$$0 = \alpha - 3\beta + 2\gamma + \delta$$
$$1 = -3\alpha - 2\gamma$$
$$0 = -\alpha - \gamma$$

from which $\alpha = -1, \beta = 1, \gamma = 1, \delta = 2$, giving:

$$\tau = k\left(\frac{\rho c_p D^2}{\kappa}\right)$$

After a few failed roasts to determine the constant k, you should be on business with your equation. Note that the fact that κ and c_P are the only variables containing the temperature in their denominator makes it obvious to include their ratio. Moreover, the time is proportional to D^2, that is both the volume or mass to power 2/3: it should be common barbecueing experience that the larger the roast, the shorter the relative cooking time (i.e., roasting 2 kg takes less than twice the time to roast 1 kg).

10.6 Dimensional analysis of blood pressure

The scaling equation for the pressure drop would be written as: $\Delta p/\Delta L \propto \eta^\alpha \varepsilon^\beta D^\gamma v^\delta$. The physical dimensions are the following: $[p/L] = [ML^{-2}T^{-2}]$, $[\eta] = [ML^{-1}T^{-1}]$, $[\varepsilon] = [D] = [L]$, $[v] = [LT^{-1}]$. The roughness ε and the diameter D have both units of length, and can therefore be grouped into one nondimensional parameter. Hence:

$$\frac{\Delta p}{\Delta L} \propto \eta^\alpha \varepsilon^{\beta+1} v^\delta f\left(\frac{D}{\varepsilon}\right)$$

and:

$$[ML^{-2}T^{-2}] = [ML^{-1}T^{-1}]^\alpha [L]^{\beta+1} v^\delta \cdot \Pi$$

The equations for the exponents are: $\alpha = 1, -2 = -1 + (\beta + 1) + \delta, -2 = -1 - \delta$, from which $\beta = -2, \delta = 1$. The desired equation is then:

$$\frac{\Delta p}{\Delta L} \propto \frac{\eta v}{\varepsilon^2}\left(\frac{D}{\varepsilon}\right)^a$$

This equation tells that the pressure drop is inversely proportional to the square of the roughness, and is invariant for blood vessels with a similar D/ε ratio. It is an important finding to be able to correlate with the level of cholesterol accumulation on the vessel inner surface, which determines an increase in the rugosity of the surface. With increasing thickness of the deposit on the inner wall, the pressure drop goes to zero, meaning that the flow is slowed down by cholesterol accumulation. Therefore, the heart has to increase the pump pressure to keep a steady flow. Note that this could be read as another form of the Hagen-Poiseuille equation, $\Delta p = 32\eta Lv/D^2$; however, in that derivation the length-squared in the denominator is the diameter of the vessel, while here it was emphasised the role of the arterial clotting. The generalisation of the H-P equation to a non-stick-slip surface (of which this is just the functional equivalent) would require the experimental measurement of the exponent a for the nondimensional parameter $\Pi = D/\varepsilon$.

Chapter 11

11.1   Water walkers
From the force equilibrium equation, $F_b/F_c = Mg/2\Sigma L \sin\theta < 1$, the minimum condition for the insect to float without sinking is $M/L < 12\Sigma/g$.

11.2   Climbing the tree
Water climbing a height h by capillarity must overcome a pressure difference equal to $\Delta P = \rho g h$. However, climbing keeps going until the maximum height is attained. After that, water from the top must be removed to make room for other water climbing up.

(a) In the case of a tree, with typical height of about 5–30 m, the energy to remove the water comes from evaporation. The latent heat of evaporation being 44 kJ/mol (supplied by the heat of the Sun), it is much more than the work done by the capillarity force, equal to $mgh = (0.018$ kg/mol$)(9.81$ m-s$^{-2})(30$ m$) = 5.3$ J/mol. Therefore, more energy is necessary to remove the water from the top, than it is necessary to bring it up there. Why the Second Principle? Because if this were not the case, the tree would be creating a "Perpetual Motion of the Second Kind", namely a machine that spontaneously converts thermal energy into mechanical work. Note that some people, every now and then, have been proposing hydroelectric machines based on the principle of capillarity, obviously without success.

(b) It is also interesting to compute the minimum thickness of the capillary required to equilibrate the heat of evaporation. Although this is just a first approximation, the capillarity height is $h \simeq 1.48 \times 10^{-5}/R$. Then, $h = 44$ kJ/mol$/mg = 249$ km (!), for which it should be $R = 2.49 \times 10^6/1.48 \times 10^{-5} = 0.6$ Å, that is much smaller that the size of the molecule itself.

11.3   Revolving parabola
The drawing is given in the following figure. The area of the surface of revolution is calculated by the general formula: $A = 2\pi \int_a^b f(x)\sqrt{1 + [f'(x)]^2}dx$.

In this case $f(x) = 1 + \frac{1}{2}(x - 2)^2$, and $f'(x) = x$. Therefore:

$$A = 2\pi \int_1^4 \left[1 + \tfrac{1}{2}(x-2)^2\right](1+x^2)^{1/2}dx$$

$$= 2\pi \int_1^4 \left[\tfrac{1}{2}x^2 - 2x + 3\right](1+x^2)^{1/2}dx$$

$$= 2\pi \left\{\left(\frac{x}{8} - \frac{2}{3}\right)\sqrt{R^3} + \frac{23}{16}\left[x\sqrt{R} + \ln(\sqrt{R} + x)\right]\right\}\Big|_1^4 = 82.372$$

with $R = 1 + x^2$ (the indefinite integral can be easily calculated from the tables of Gradshteyn and Ryzyk).

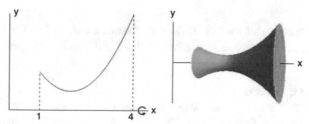

11.4 Morphing snails

The length of the spiral curve in parametric form is calculated from the general expression:

$$L = \int_a^b \sqrt{\left(\frac{dx}{dt}\right)^2 + \left(\frac{dy}{dt}\right)^2}\,dt$$

Therefore:

$$L_A = \int_0^{3\pi+2} [49(\sin t + t\cos t)^2 + 49(\cos t - t\sin t)^2]^{1/2}dt$$

$$= \int_0^{3\pi+2} [49(\sin^2 t + t^2\cos^2 t + \cos^2 t + t^2\sin^2 t)]^{1/2}dt$$

$$= \int_0^{3\pi+2} 7\sqrt{1+t^2}\,dt$$

$$L_B = \int_0^{4\pi+1} [(t\sin t + \tfrac{t^2}{2}\cos t)^2 + (t\cos t - \tfrac{t^2}{2}\sin t)^2]^{1/2}dt$$

$$= \int_0^{4\pi+1} [t^2\sin^2 t + \tfrac{t^4}{4}\cos^2 t + t^2\cos^2 t + \tfrac{t^4}{4}\sin^2 t]^{1/2}dt$$

$$= \int_0^{4\pi+1} t\sqrt{1 + \frac{t^2}{4}}\,dt$$

By using the same integration rules of the previous problem, we have:

$$L_A = \int_0^{3\pi+2} 7\sqrt{1+t^2}\,dt$$

$$= \tfrac{7}{2}\left[t\sqrt{1+t^2} + \ln\left(t + \sqrt{1+t^2}\right)\right]_0^{3\pi+2} = 469.54$$

$$L_B = \int_0^{4\pi+1} t\sqrt{1+\frac{t^2}{4}}\,dt = \tfrac{4}{3}\left[\sqrt{(1+\tfrac{t^2}{4})^3}\right]_0^{4\pi+1} = 429.78$$

Then A is slightly older than B. However, note that comparing the snails by just the shell length is not a good idea, since shells of different shape grow to different lengths over the same time.

11.5 Packing problems

(a) A circle of radius 1 has an area equal to π, that is $\pi/4 = 0.785$ that of the smallest square enclosing it, with side $L = 2$. Two unit circles can be fitted in a square with side equal to $L = (\sqrt{2}+2)$, their area is 2π, that is $2\pi/L^2 = 0.539$ times the area of the square. The solution for three circles is more tricky: turn the three touching circles such that their centres are at $(-1,0)$, $(+1,0)$, $(0,\sqrt{3})$ (see figure); the angle α is easily found to be equal to $\pi/6$; then the side if the square is $L = 2(1 + \cos(\pi/6)$, and the area of the three circles is $3\pi/L^2 = 0.609$ that of the square. Four circles is just 4 times the solution for one. Five unit circles fit in a square with side $L = (2\sqrt{2}+2)$, that is an area $5\pi/L^2 = 0.634$ times the area of the square.

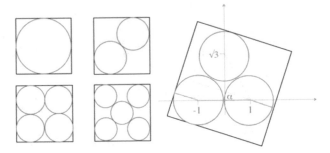

(b) A sphere of unit radius has a volume $v = \tfrac{4}{3}\pi$. The volume of the tangent cube is $2^3 = 8$, therefore the packing fraction is $\pi/6$. For the face-centred close packing, five spheres are placed in a square (see above), with three spheres exactly filling the diagonal. Therefore the side of the square is $L = 4/\sqrt{2}$ (from the centre of the first to the centre of the third sphere, see below). The second layer is made by placing four spheres above the midpoints of each side of the square. The third layer is built like the first one, and so on. With this construction, the volume fraction occupied by the spheres within the cube can be counted: six half spheres centred on each face of the cube(green, see figure), plus 1/8 of a sphere at each of eight vertices (blue).

This makes a total of 4 whole equivalent spheres contained in the cube (an infinite 3D array of such cubes and spheres fills the space). Therefore, the packing fraction is $(4\frac{4}{3}\pi)/(16\sqrt{2}) = 0.7405$.

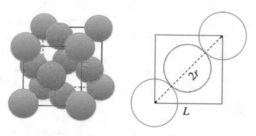

11.6 Cell migration and gradients

The cell colony has an ovoid shape. A circle can turn into an ovoid by applying a map, so the reverse map turns the ovoid back to a circle. Starting from the equation of a circle, $x^2 + y^2 = r^2$, a transformation $t(x)$ is applied to the y coordinate, as $x^2 + [t(x) \cdot y]^2 = r^2$. Examples are $t(x) = (1 - ax)^{-1}$, $t(x) = \exp(-ax)$, $t(x) = (1 + ax)$ (see example plots below, with $r = 10$ and $x_0 = 25$).

In all cases, the amplitude along one direction (e.g., the x) is modulated by a non-linear, monotonic function along the other (e.g., the y). A monotonic function along a given direction implies a directional gradient, which should be the sign of a morphogen at work in the cell culture, acting along a preferential direction. The non-linear driving force of the morphogen is here summed to the circularly-symmetric driving force from ordinary diffusion, turning its circular growth pattern into an ovoid.

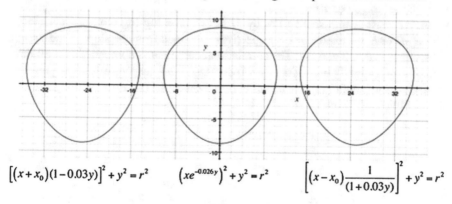

$$\left[(x+x_0)(1-0.03y)\right]^2 + y^2 = r^2 \qquad \left(xe^{-0.026y}\right)^2 + y^2 = r^2 \qquad \left[(x-x_0)\frac{1}{(1+0.03y)}\right]^2 + y^2 = r^2$$

11.7 Human growth

You can draw horizontal and vertical lines at points of interest, and measure with a ruler the ratios of such quantities, compared to the unit height. Several observations can be made (see plot below). First of all, apart from the very early stages of embryo growth, the changes over age T (in years) for all measurements seem to be rather homogeneous and smooth. By dropping the first point, they all seem to follow laws of the type $p = a \ln T$. The relative size of the head decreases steadily, both in length and width, coherently with the fact that the brain size changes relatively little after two years of age. The space of legs keeps growing in relative size, and in a nearly symmetrical way to the decrease of the head, such that the torso keeps a constant relative span over the entire 0–25-years age. The same constancy is observed for the length of the arms, whose size proportion remains practically stable relative to the growth of the whole body.

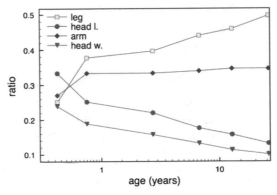

Chapter 12

12.1 Savannah's beauties

We fit the mass-length scaling law by $M \propto L^k$, on a log-log plot. By replacing L with H, D or h, the exponent k is respectively obtained from the slope of the straight lines, equal to 3.6, 3 and 5.7. The correlation coefficient is comparable for all three variables, and slightly better for h ($r = 0.85, 0.87, 0.94$, respectively). The preferred approximation is definitely that of taking $L = D$, the total (nose-to-tail) length, which gives a physically more meaningful $M \propto L^3$ dependence.

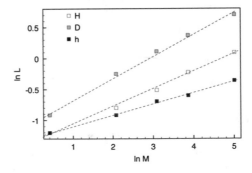

12.2   Climbing heights

By climbing for 20 min he covers a distance of 1,000 m. For a slope of 12 %, this corresponds to raising by 119 m (1,000 is the hypothenuse of the triangle). His energy consumption for normal metabolism plus horizontal walking is $E = (75 \cdot 70^{0.75})(20/1{,}440) + (70 \cdot 2/3) = 25.2 + 46.7 = 71.9$ kcal. Therefore, the amount of energy added by climbing is $121 - 71.9 = 49.1$ kcal. The mechanical work done in raising 70 kg against the gravity for 119 m is $W = (70)(9.807)(119) = 81.7$ kJ, or 19.5 kcal. By comparing this value with the climbing energy expenditure, the efficiency of the muscles is about $19.5/49.1 \simeq 40$ %, that is not bad compared to a typical engineering machine.

12.3   The life of a heart

If we take the product of the two scaling laws, [beats/minute][years·(365)(1,440)], a "law" for the number of heart beats in an animal's life is obtained, as $h \cdot T = (241 M^{-0.25})(365)(1{,}440)(15 M^{0.20}) = 1.9 \times 10^9 M^{-0.05}$ beats. Such a scaling law with a very weak (close to zero) exponent, seems to state that there should not much difference in the number of beats in a lifespan of animals even with very different body masses. For example, in a mouse of 30 g and a man of 70 kg, these numbers would be 2.26 versus 1.54 billion beats. Such a law is difficult to disprove, because the small exponent makes for large uncertainty. If for example the exponent is determined with a large uncertainty of 50 %, say $M^{-0.025}$, the variation in the predicted hear beats would be just a few per cent. This is the problem of working with small exponents. On the other hand, scaling laws are expected to give general trends, not to predict exact values. For example, the average life span of a mouse is about 3 years, and their typical heart rate is around 350 beats/min, which gives ~ 0.55 billion heart beats. While this is off by about a factor of 4 compared to our allometric estimate, the order of magnitude is correct.

12.4   The fish hatchery

For times $t < T$, the population of fishes grows according to a simple Malthusian law, $dN/dt = -rN$, $N(t) = N_0 \exp(-rt)$. For $t \geq T$, it is $dN/dt = -(r + q)t$, with solution $N(t) = N_T \exp(-(q + r)t)$. To determine the constant, set $N(T) = N_T \exp(-q + r)T = N_0 \exp(-rt)$. Hence $N_T = N_0 \exp(qT)$, and the complete solution is:

$$N(t) = N_0 e^{qT} e^{-(q+r)t}$$

The fraction of the initial N_0 recovered by fishing at a time $t > T$ is found as: $P = qN(t)/N_0$ (probability of catching $\times$ the number of fish at time t, normalised to the initial population N_0). Then:

$$P(t) = \frac{q}{N_0} \int_T^t N(t')dt' = qe^{qT} \int_T^t e^{-(q+r)t'} dt'$$

$$= \left[-\frac{qe^{qT}e^{-(q+r)t'}}{(q+r)} \right]_T^t = \frac{qe^{qT}}{q+r} \left[e^{-(q+r)T} - e^{-(q+r)t} \right]$$

$$= \frac{qe^{-rT}}{q+r} \left[1 - e^{-(q+r)(t-T)} \right]$$

12.5 Influenza spreading

It is $S = N - Q$, then $R_0(N - Q) = N$, from which $Q/N = 1 - 1/R_0$ is the fraction of immune individuals over a population of N. If we impose $R_0 > 1$ in the last equation, this gives a fraction $Q/N > 0$. This is equal to the minimum threshold Q_c, needed to keep the spreading under control.

12.6 From egg to adult

(a) The model can be described by three equations:

$$E_{t+1} = 73A_t$$
$$L_{t+1} = 0.04E_t$$
$$A_{t+1} = 0.39L_t$$

By putting the three stages (egg, larva, adult) in chain, we have $A_{t+3} = 0.39[L_{t+2} = 0.04(E_{t+1} = 73A_t)] = (0.39)(0.04)(73)A_t$. This is a finite-difference equation, that can be turned into its differential equivalent:

$$\frac{dA}{d(3t)} = 1.139A$$

whose solution is $A(t) = A_0 \exp(0.38t)$, a Malthus-like exponential growth.

(b) In this case, the third equation must be modified into: $A_{t+1} = 0.39L_t + 0.65A_t$. The model cannot be solved simply by recurrence. Instead we write the three equations in matrix form. The vector of variables is $\mathbf{x}_t = (E_t, L_t, A_t)$, and $\mathbf{x}_{t+1} = \mathbf{M}\mathbf{x}_t$, that is:

$$\begin{pmatrix} E_{t+1} \\ L_{t+1} \\ A_{t+1} \end{pmatrix} = \begin{pmatrix} 0 & 0 & 73 \\ 0.04 & 0 & 0 \\ 0 & 0.39 & 0.65 \end{pmatrix} \begin{pmatrix} E_t \\ L_t \\ A_t \end{pmatrix}$$

To obtain the equivalent of the previous finite-difference equation we must repeatedly apply the matrix, as $\mathbf{x}_{t+3} = \mathbf{M}^3 \mathbf{x}_t$. The matrix product gives:

$$\mathbf{M}^3 = \begin{pmatrix} 1.139 & 18.51 & 30.84 \\ 0 & 1.139 & 1.898 \\ 0.010 & 0.165 & 1.414 \end{pmatrix}$$

It can be seen that off-diagonal terms appear, which means that all the three stages are interacting in determining the respective population evolution. In particular, note that the M_{33} term is larger than the 1.139 of case (a) above, i.e., the growth of adult population is obviously faster if some of them survive the egg deposition stage. All the three populations (egg, larva, adult) grow exponentially.

12.7 The constant gardener

The model can be set in two recurrence equations, describing how the population of the two plant species at week $n + 1$ depends on the populations at week n. By accounting for A the fraction (99 %) surviving, plus the 25 % probability of winning a free spot from the 1 % A and the 5 % B dying, and for the B plants the same but with their own probabilities, it is:

$$A_{n+1} = 0.99 A_n + 0.25(0.01 A_n + 0.05 B_n)$$
$$B_{n+1} = 0.95 B_n + 0.75(0.01 A_n + 0.05 B_n)$$

After simplifying, the problem can be set as a linear model $\mathbf{x}_{n+1} = \mathbf{M}\mathbf{x}_n$, with $\mathbf{x}_n = (A_n, B_n)$, and:

$$\mathbf{M} = \begin{pmatrix} 0.9925 & 0.0125 \\ 0.0075 & 0.9875 \end{pmatrix}$$

This problem cannot be solved other than by numerical iteration of the matrix-vector product. The plot below shows the result of two different initial conditions, respectively $A_0 = 100$, $B_0 = 10$ (black curves) and $A_0 = 10$, $B_0 = 100$ (red curves).

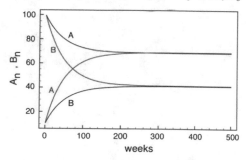

The two populations tend to a constant asymptote, irrespective of the initial values, and their ratio at long times is (in this special case) equal to 5/3. While a simple general formula for raising a matrix to a power p does not exist, by iterating the matrix-matrix product numerically for a large number of time steps, say 10^4, it is obtained:

$$\mathbf{M}^{10^4} = \begin{pmatrix} 0.62327 & 0.62327 \\ 0.37673 & 0.37673 \end{pmatrix}$$

The two columns are identical, and the ratio of the upper- and lower-row coefficients is 1.6544, very close to 5/3.

Physical Units, Constants and Conversion Factors

Here we regroup the values of the most useful physical and chemical constants used in the book. Also some practical combinations of such constants are provided. Units are expressed in the International System (SI), and conversion to the older CGS system is also given. For most quantities and units, the practical definitions currently used in the context of biophysics are also shown.

© Springer International Publishing Switzerland 2016
F. Cleri, *The Physics of Living Systems*, Undergraduate Lecture
Notes in Physics, DOI 10.1007/978-3-319-30647-6

Basic physical units (boldface indicates SI fundamental units)

	SI	cgs	Biophysics
Length (L)	meter (m) 1	centimeter (cm) 0.01 m	Angstrom (Å) 10^{-10} m
Time (T)	second (s) 1	second (s) 1	$1\,\mu s = 10^{-6}$ s $1\,ps = 10^{-12}$ s
Mass (M)	kilogram (kg) 1	gram (g) 0.001 kg	Dalton (Da) $1/N_{Av} = 1.66 \cdot 10^{-27}$ kg
Frequency (T^{-1})	s^{-1} 1	Hz 1	
Velocity (L/T)	m/s 1	cm/s 0.01 m/s	μm/s 10^{-6} m/s
Acceleration (L/T^2)	m/s^2 1	cm/s^2 0.01 m/s^2	$\mu m/s^2$ 10^{-6} m/s^2
Force (ML/T^2)	newton (N) $1\,kg \cdot m/s^2$	dyne (dy) 10^{-5} N	pN 10^{-12} N
Energy, work, heat (ML^2/T^2)	joule (J) $1\,kg \cdot m^2/s^2$ 0.239 cal $1\,C \cdot 1\,V$	erg 10^{-7} J $0.239 \cdot 10^{-7}$ cal	$k_B T$ (at $T = 300$ K) 4.114 pN · nm 25.7 meV 1 mol ATP = 30.5 kJ = 7.3 kcal/mol
Power (ML^2/T^3)	watt (W) $1\,kg \cdot m^2/s^3 = 1\,J/s$	erg/s 10^{-7} J/s	ATP/s 0.316 eV/s
Pressure (M/LT^2)	pascal (Pa) $1\,N/m^2$	atm 101325 Pa $1\,bar = 10^5$ Pa	Blood osmolarity ~ 300 mmol/kg ~300 Pa $= 3 \cdot 10^{-3}$ atm
Electric current (A)	ampere (A) 1	e.s.u. $2.998 \cdot 10^9$ A	Transmembrane current $\sim$1-2 mA/cm^2
Electric charge (A/T)	coulomb (C) 1 A·s	statCoulomb (stC) $2.998 \cdot 10^9$ C	
Electric potential ($ML^2/T^3 A$)	volt (V) 1 W/A	statCoulomb (stC) $2.998 \cdot 10^9$ C	Neuron action potential ~ 100 mV
Electrical resistance ($ML^2/T^3 A^2$)	ohm (Ω) 1 V/A	statOhm (stΩ) $1.11265 \cdot 10^{-12}\Omega$	Cell membrane (with K^+ and Na^+ channels) 1-2 k$\Omega \cdot \mu m^2$
Temperature (K)	kelvin (K) 1	celsius (°C) 1	

Useful physical constants and their combinations

Avogadro's number, $N_{Av} = 6.0221409 \times 10^{23}$ mol^{-1}

Atomic mass unit, $m_u = \frac{1}{12}m(^{12}\text{C}) = 1.66053904 \times 10^{-27}$ kg

Boltzmann's constant, $k_B = 1.3806485 \times 10^{-23}$ J/K (joule/kelvin degree) $= 8.617 \times 10^{-5}$ eV/K

Planck's constant, $h = 6.62607 \times 10^{-34}$ J s $= 4.132$ eV/fs, $\hbar = h/2\pi$

Stefan-Boltzmann's constant, $\sigma = 5.670367 \times 10^{-8}$ W/(m^{2}K^4)

Gas constant, $R = 8.31446$ J/(K· mol) $= 1.987$ cal/(K· mol) $= N_{Av} \cdot k_B$

Molar volume at STP $V_m = 22.414$ litre/mol

Speed of light, $c = 2.99792458 \times 10^8$ m/s

Elementary charge, $e = 1.60217662 \times 10^{-19}$ C (coulomb)

Vacuum dielectric constant, $\varepsilon_0 = 8.854 \cdot 10^{-12}$ F/m (faraday/metre)

Coulomb's constant, $K_c = 1/(4\pi\varepsilon_0) = 8.9875 \times 10^9$ N m^2 C^{-2}

Faraday's constant, $F = 96485.309$ C/mol

$m_u c^2 = 1.4924 \times 10^{-10}$ J $= 931.494$ MeV

$K_c e^2 = 2.3071 \times 10^{-28}$ J m $= 1.44$ eV nm

$\hbar c = 3.1615 \times 10^{26}$ J m $= 197.33$ eV nm

$RT = k_B T \times N_{Av} = 2.479$ kJ/mol (at 300 K)

$h/k_B T = 0.16$ ps (at 300 K)

Useful conversion factors (with some old-fashioned units)

To convert from:	To:	Multiply by:
radian	degree arc	57.29578
angstrom	nanometer	0.1
light-year	metre (m)	9.461×10^{15}
cm^3	litre	0.001
eV	joule (J)	1.6021×10^{-19}
eV (wavelength)	cm^{-1}	8,065.73
eV (wavelength)	nanometer	1,239.84
teraHertz (THz $= 10^{12}$Hz)	meV (wavelength)	4.13567
standard atmosphere	kilopascal (kPa)	101.325
bar	kilopascal (kPa)	100
millimeter of mercury (at 0 °C)	kilopascal (kPa)	0.1333224
millimeter of water (at 4 °C)	pascal (Pa)	9.80665
pound/sq.inch (lb/in^2, psi)	pascal (Pa)	6,900
centipoise (viscosity)	pascal-second (Pa s)	0.001
horse-power (hP)	watt (W)	746
kilowatt-hour (kWh)	megajoule (MJ)	3.6
calorie	joule (J)	4.184
eV	kcal/mole	23.0609
gauss	tesla (T)	0.0001
weber/sq.metre	tesla (T)	1
ampere-hour	coulomb (C)	3,600

Index

© Springer International Publishing Switzerland 2016
F. Cleri, *The Physics of Living Systems*, Undergraduate Lecture
Notes in Physics, DOI 10.1007/978-3-319-30647-6

Printed in the United States
By Bookmasters